EXPLAINING PHYSICS

GCSE Edition

Stephen Pople

OXFORD
UNIVERSITY PRESS

Great Clarendon Street, Oxford OX2 6DP

Oxford University Press is a department of the University of Oxford. It furthers the University's objective of excellence in research, scholarship, and education by publishing worldwide in

Oxford New York

Auckland Cape Town Dar es Salaam Hong Kong Karachi Kuala Lumpur Madrid Melbourne Mexico City Nairobi New Delhi Shanghai Taipei Toronto

With offices in

Argentina Austria Brazil Chile Czech Republic France Greece Guatemala Hungary Italy Japan Poland Portugal Singapore South Korea Switzerland Thailand Turkey Ukraine Vietnam

British Library Cataloguing in Publication

Data available

First published 1982
40 39 38 37 36 35 34 33 32

ISBN 13 978 019914272 9

line illustrations by Neil Hyslop

front cover:
CERN apparatus which provides protons for acceleration to high energies, for use in research into fundamental particles (see page 360)

back cover:
spectra for use in the unit on light (see page 182);
photograph by Paul Brierley

Typeset by Tradespools Ltd, Frome, Somerset

Printed and bound in Great Britain by
Bell & Bain Ltd., Glasgow

Introduction

This book deals with physics and its applications. You should find it useful if you are studying physics as an individual GCSE subject, as part of a GCSE co-ordinated science course, or within any science course based on Key Stage 4 of the National Curriculum. If you intend to take your studies to a more advanced level, the book will also provide you with opportunities to look at concepts which extend the normal area of study in the fourth and fifth years of secondary school.

The book contains eight major sections, each of which is introduced by a selection of short articles linking the main themes of the section with a wider view of physics. Each section is divided into 'units'. The units are usually three or four pages long, and cover one particular topic at a time. After a brief introduction, headed paragraphs guide you through the contents of the unit. Diagrams are used extensively to summarize the main principles, and hints are given on how to solve numerical problems. Suggestions for practical work are *not* included.

Each unit ends with some straightforward questions which will help you reinforce your understanding. Each section ends with a selection of questions from examination papers, which will give you an idea of the standard of questions you are likely to meet when the terrible day comes. To help you, numerical answers are given at the back of the book.

If you need to study a particular topic, and wish to find the unit which deals with it, look at the contents list given on the next two pages. If you are searching for information about a specific fact, definition, or principle, use the index at the back. There is also a detailed check list at the back to help you with your revision.

We also thank the various examining boards for their permission to use material either taken directly, or slightly adapted from their question papers. We are indebted to the various readers who have provided us with much assistance and support. Any mistakes are of course our own responsibility – if you spot any, please do not hesitate to contact us!

I hope you find the book useful – and as easy to read as my publisher told me it would be to write . . .

Stephen Pople

May 1990

Contents

Section 6
Electrical energy

Section 7
Magnets and currents

Section 8
Electrons and atoms

Acknowledgements

The author and publishers would like to thank the following people for their help with supplying photographs:

Ardea, p. 165; Stewart Bale, p. 122; Balfour Beatty, p. 315 (bottom); Barnaby's Picture Library, pp. 22 (bottom), 30, 78, 143 and 195; Barclays Bank plc, p. 202 (middle); P Barnard, pp. 94 (centre), 132, 151 and 268; BBC, p. 161; BBC Hulton Picture Library, pp. 191, 305 and 313; British Leyland, pp. 29, 63, 77, 83 and 307; British Rail, p. 126; British Telecom, pp. 124 and 206; Camera Press, pp. 35, 67 (top), 89, 115, 128, 229, 141, 166 (bottom), 279 and 333; CEGB, pp. 71 (bottom left), 127 (top right) and 327 (top); Central Press, p. 41; CERN/Science Photo Library, pp. 355 (bottom) and 360; COI, pp. 110 and 340; Bruce Coleman, p. 120; Creda, p. 145; Culham Laboratories, p. 369 (left); Tony Davis, p. 237; H E Edgerton, p. 12; Electricity Council Appliance Testing Labs, p. 264 (top); Electrolux, p. 102; Fiat Auto UK Ltd, p. 70 (bottom); Fox Photos, pp. 98, 107, 111, 166 (top), 187 (bottom), 212, 235 and 266; Government Information Service, p. 364; G R Graham, p. 201; Griffin and George, pp. 39, 125 (bottom right), 284, 324, 349, 350 and 355 (top); The Guardian, p. 340 (bottom left); Philip Harris, pp. 117, 246, 248, 299 (top), 302, 328, 349 and 352 (bottom); Chris Honeywell, p. 334 (top), p. 339; Ilford Ltd, p. 350; International Bureau of Weights and Measures, p. 9; David James, pp. 53, 56, 57, 65, 80, 94, 127, 151, 152, 169, 214, 240, 243, 247, 260, 261, 286, 295 and 312; Keystone Press, pp. 11, 21, 37, 54, 69 (bottom), 352 (top) and 358; Frank W Lane, p. 232; John Lundie, p. 230 (bottom); MK Electric, p. 270; Richard McBride, p. 71 (middle); Motor, p. 32; Mullard, pp. 192, 252, 337 and 340 (top); John Murray, p. 353 (top); NASA/Science Photo Library, pp. 33 and 99 (top); National Physical Laboratory, Teddington, p. 8; Oertling, p. 10 (centre); Olympus, p. 187 (top); Osram-GEC, p. 264 (bottom); Philips, p. 202 (bottom) and 217; Philips Ltd, p. 202 (bottom); Pifco Ltd, p. 345 (bottom); Chance Pilkington, p. 165; Poulter Compuvision Ltd, p. 341 (bottom right); Pyrometer Ltd, p. 125; R H P Bearings, p. 28 (top); Rapho/J Perrard, p. 70 (top); Reynell & Son Ltd/Edmonton Power Station, p. 71 (top); Robb & Campbell Harper Studios, pp. 178, 182, 218 and 338; Rolls Royce, p. 43; Ann Ronan, pp. 7, 89, 139, 200, 229 and 357; Royal Astronomical Society, p. 369 (right); The Science Museum (Crown Copyright), pp. 99 (bottom), 186, 210, 308 and 353 (centre and bottom); Science Photo Library, pp. 67 (bottom), 68 (top), 207 and 351 (top and bottom); Science Photo/Library/Petit Format/Nestle, p. 180; Science Photo Library/John Walsh, p. 180; Science Photo Library/Russ Kinne, p. 202 (top); Siemens–Elema, p. 311; Smiths Industries, pp. 101 and 105; A Souster, pp. 28 (bottom), 154 and 171; P Souster, pp. 90, 94 (top right), 109 and 158; Space Frontiers, p. 42; Stanley Power Tools, p. 297; Swiss National Tourist Office, p. 223; Syndication International, pp. 49, 365 and 368; J Tabberner, p. 118; TASS, p. 323; Tate and Lyle, p. 26; Timex, p. 10 (bottom); UKAEA, pp. 68 (bottom), 356, 363 and 367; Volvo, p. 49; Williams Grand Prix Engineering, p. 60

Cartoons on pgs. 7 and 229 by Ed McLachlan. Illustration on p. 115 by Norma Burgin. Additional illustrations by Quadra Graphics, Oxford. Additional photography by S Pople, J Byron and C Honeywell.

The introductory articles have been adapted from a broad variety of sources. We should specifically like to acknowledge our thanks to:

Dr. Bernard Dixon for the 'Mona Lisa';

J. Mitchell and R. Rickard's *Phenomena*, and *The Guinness Book of Records* for information in 'Lightning Never Strikes Twice';

New Scientist for 'Suckers used in American clean up' (Peter Marsh), 'Birds defend territory by changing their tune' (9.4.81), 'Magnetism puts coins on the spot' (5.7.79), and 'Death from nuclear disaster' (Lee Torrey 19.4.79);

the *Understanding Electricity* educational service for 'Life with nuclear energy'.

Every effort has been made to trace copyright holders. If there are any omissions the publisher will be happy to give full acknowledgement at the first reprint.

SECTION 1

FORCES AND MOTION

Chariots and barleycorn – Still in daily use?

Nowadays, most scientific measurements are made using units such as the metre, the kilogram, and the second.

Much older units of measurement still survive however, and some of them have a distinctly odd history. According to the Chief Information Officer of the Metrication Board:

a fathom originally meant the distance a Viking could encompass in a hug;

a foot was defined by a statute of Henry I, as the total width of thirty-six ears of barleycorn laid side by side;

an acre was the area ploughable in one day by a team of two oxen;

and the gauge of railways in Britain was chosen to be the same as the distance between the wheels of a Roman chariot!

Military Leader Raises the Standard

Napoleon authorized the introduction of a metric system in France in the nineteenth century. Discussions, sponsored by France, led to an international treaty called the Metre Conversion, in 1875. The treaty was accepted as being necessary not only because it was simpler than other measurement systems, but also because it made international trade more possible. The metre was defined as one forty millionth of the Earth's circumference, and the 'standard metre' representing that distance was marked on a metal bar.

This was good enough for most purposes – but unfortunately, metal bars undergo slight length changes over a period of time. Since 1960, the metre has been defined in terms of something which is believed never to change – the speed of light through empty space. By definition 1 metre is the distance travelled by light in a vacuum in 0.000000003335 6409 seconds. The apparatus shown on page 8 is used for making very accurate length measurements. For everyday use, the ruler is accepted by the scientific community as being more convenient . . .

One fathom was the distance that a Viking could embrace in a hug??

Victorian trickery

The strong man was a popular feature of many Victorian fairs and circuses.

A favourite trick was for him to lie with a large stone across his chest, while an assistant struck the stone with a sledge hammer. The astounded audience would see the stone split in two, and the strong man rise to his feet unharmed. A great feat of strength? Not really: the strong man has to carry the weight of the stone, certainly; but Newton's laws of motion predict that the hammer blow on the massive stone should have a negligible effect on the man underneath.

Embarrassing Problem?

Newton's laws of motion also featured in another popular Victorian parlour trick. A stick would be laid with its ends resting on two glasses full of water. A downward blow on the stick with the iron bar would break the stick but not the glasses. However, an upward blow would break the glasses but not the stick! Embarrass your physics teacher by asking them to explain why.

1.1

Units of measurement

Measure a distance and the result could be expressed in metres, fathoms, inches, feet, yards, or miles. Even then the list would be far from complete. In scientific work, life is much easier if everyone uses a common system of units.

Figure 1 Scientific apparatus used in making accurate length measurements. By definition, one metre is the distance travelled by light in a vacuum in 3.3356409×10^{-9} seconds

SI units

Most scientists make measurements using the International System of Units (Systeme International d'Unites, or SI for short). The SI system starts with three basic units for measuring length, mass, and time: the **metre**, the **kilogram**, and the **second**. From these come a whole range of units for measuring volume, speed, force, energy, and many other quantities.

Length

The SI unit of length is the **metre** [symbol **m**]. At one time, the metre was defined as the distance between two marks on a metal bar kept at the Office of Weights and Measures in Paris. A more accurate standard metre is now used based on measurements taken on the speed of orange light as shown in figure 1.

The range of lengths that scientists need to measure is enormous. The Sun for example is more than one hundred million metres across, yet the atoms of which all materials are made are less than one-thousand-millionth of a metre across. To cope with this vast range, there are several other units of length based on the metre. These are shown in figure 2.

Figure 2 Units of length and their approximate size

distance	comparison with SI unit	standard form	approximate size
1 kilometre [km]	1000 m	10^{3} m	10 football pitches
1 metre 1 m	**1 m**		
1 centimetre [cm]	$\frac{1}{100}$ m	10^{-2} m	cm 1 2 3
1 millimetre [mm]	$\frac{1}{1000}$ m	10^{-3} m	mm 10 20 30
1 micrometre [μm]	$\frac{1}{1\,000\,000}$ m	10^{-6} m	bacteria
1 nanometre [nm]	$\frac{1}{1\,000\,000\,000}$ m	10^{-9} m	atoms

Standard form

Some of the numbers in the chart on the opposite page have been written in a form using powers of 10. For example:

$1000 = 10 \times 10 \times 10$; which is written as 10^3

$\frac{1}{1000} = \frac{1}{10^3}$; which is written as 10^{-3}

Numbers expressed using powers of 10 are said to be in **standard form**. Using this notation:

2514 mm is written as 2.514×10^3 mm

0.004 mm $\left[\frac{4}{1000}\text{ mm}\right]$ is written as 4×10^{-3} mm

Standard form is particularly useful when writing down very large or very small numbers, as it avoids large numbers of noughts. It is also a convenient method of indicating how confident you are in the accuracy of a measurement.

No measurement is ever exact. But if a length measurement is expressed as 1.500×10^3 mm, it implies that the length is likely to be closer to 1500 mm than it is to 1499 mm or 1501 mm. A measurement of 1.5×10^3 mm on the other hand, only implies that the length is likely to be closer to 1500 mm than it is to 1400 mm or 1600 mm.

Figure 3 The standard kilogram is *still* based upon the mass of a lump of metal (carefully preserved!)

Mass

The SI unit of mass is the **kilogram** [**kg**]. The standard kilogram is the mass of a block of platinum alloy kept at the Office of Weights and Measures in Paris as shown in figure 3. Other units of mass based on the kilogram are shown in figure 4.

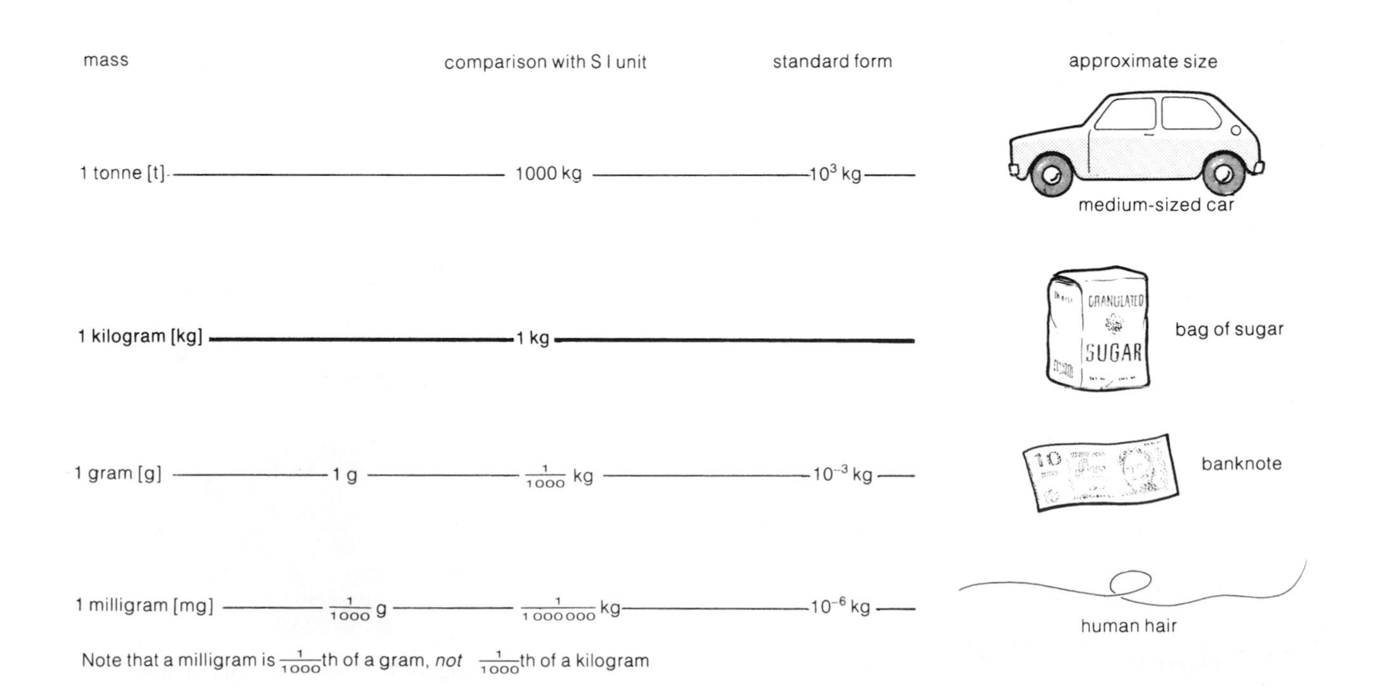

Figure 4 Units of mass and their approximate size

Mass is a mysterious property which affects the behaviour of an object in two ways. Both are examined in more detail later in the book. But you may like to remember two things:

1 All objects resist attempts to make them go faster, slower, or in a different direction. The greater the mass of an object, the greater is its resistance to any change in motion.

2 All objects are attracted to the Earth. The greater the mass of an object, the stronger is the Earth's gravitational pull upon it.

In the laboratory, the mass of an object can be found using a beam balance and a set of standard masses as shown in figure 5. The object is placed in one pan, and a selection of standard masses are added to the other pan to make the beam balance. The beam balances when the gravitational pull on the material in each pan is the same. Each pan then contains the same mass of material.

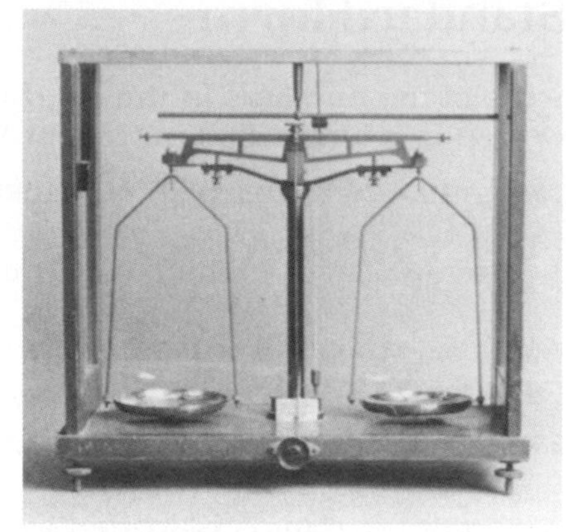

Figure 5 A beam balance

A quicker and more convenient method of measuring mass is to use a top-pan balance as shown in figure 6. Most work along similar lines to the beam balance described above, though the standard masses are not readily visible.

Figure 6 A modern top-pan balance

Time

The SI unit of time is the **second** [**s**].

All clocks and watches make use of some device that 'beats' at a steady rate. "Grandfather" clocks used the swings of a pendulum. Modern digital watches, as shown in figure 7, count the vibrations made by a tiny quartz crystal.

Smaller units of time based on the second are the **millisecond**, the **microsecond**, and the **nanosecond**:

$$1 \text{ millisecond [ms]} = \frac{1}{1000} \text{ s} = 10^{-3} \text{ s}$$

$$1 \text{ microsecond [}\mu\text{s]} = \frac{1}{1\,000\,000} \text{ s} = 10^{-6} \text{ s}$$

$$1 \text{ nanosecond [ns]} = \frac{1}{1\,000\,000\,000} \text{ s} = 10^{-9} \text{ s}$$

Such time intervals may seem very short by everyday standards, but a radio signal travels nearly $\frac{1}{3}$ metre in 1 ns.

Figure 7 A modern digital watch

Volume

The quantity of space an object takes up is called its volume, and it is measured in **metre cubed** [$\mathbf{m^3}$]. 1 m^3 is the volume of a cube with sides 1 m long, as illustrated in figure 8 – about the same volume as a wardrobe.

The metre cubed is rather a large unit of volume for everyday laboratory work, and it is often more convenient to measure volumes using the **centimetre cubed** [$\mathbf{cm^3}$]. 1 cm^3 is the volume of a cube with sides 1 cm long:

$$1 \text{ cm}^3 = \frac{1}{1\,000\,000} \text{ m}^3 = 10^{-6} \text{ m}^3$$

The volume of a modern oil tanker is about 200 000 m^3

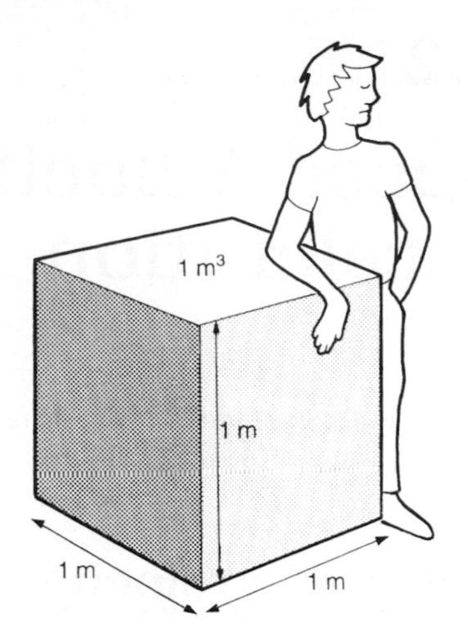

Figure 8 A cube with sides one metre long.

Volumes of regular solids If an object has a simple shape, its volume can be calculated. For example:

volume of a rectangular block = length × width × height

volume of a cylinder = π × radius^2 × height

Volumes of liquids Volumes of liquids are often measured in **litres**. A lemonade bottle has a volume of about 1 litre:

1 litre [l] = 1000 cm^3

Volumes of about a litre or so can be measured using a measuring cylinder as shown in figure 9. When the liquid is poured into the cylinder, the level on the scale gives the volume. In most cases, the liquid surface or meniscus curves upwards at the sides. The volume reading is always taken level with the flat part of the meniscus.

Most measuring cylinders have scales marked in **millilitres** [ml]. There are 1000 ml in 1 litre, so 1 ml is the same volume as 1 cm^3.

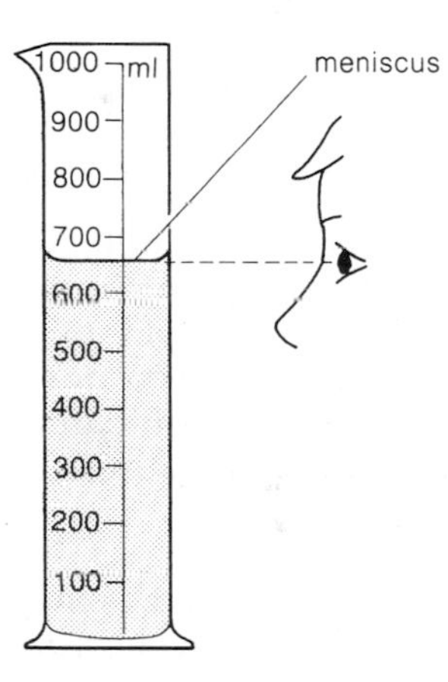

Figure 9 It is important to measure volume on a line level with the *bottom* of the meniscus

Questions

1 What are the basic SI units of:
length, mass, time?

2 What do the following represent?
μm, ns, mg, t, ms, ml, m^3

3 How many mm are there in 1 m?
How many mg are there in 1 g?
How many mg are there in 1 kg?
How many μs are there in 1 s?
How many μm are there in 1 mm?

4 A mass is measured as 6.2 kg in one experiment and as 6.20 kg in another. In what way do these two measurements differ?

5 Write down the following using standard form:
1000 kg; 1 000 000 m; 0.01 s; 0.37 kg; 0.000 25 m.

6 Express the following in mm:
2.7 m; 22.4 cm; 330 μm; 5.6×10^4 nm

7 Express the following in s:
5000 ms; 4×10^{10} μs

8 500 pages of a book have a total mass of 2.5 kg. What is the mass of each page a) in kg b) in mg?

9 How many cm^3 are there in 1 m^3?
How many cm^3 are there in 1 litre?
How many litres are there in 1 m^3?

10 A tankful of liquid has a volume of 0.2 m^3. What is the volume in a) litres b) cm^3 c) ml?

11 Calculate the volumes of the following:
a) a box 2 m × 6 m × 8 m
b) a cylinder of radius 2.0 m and height 2.5 m (assume $\pi = 3.14$)

1.2

Speed, velocity and acceleration

Some cars go faster than others. Some cars become faster more quickly than others!

Speed and velocity

If an object moves between two points in a measured time, its average speed can be calculated as follows:

$$\textbf{average speed} = \frac{\textbf{distance moved}}{\textbf{time taken}}$$

If distance is measured in metres [m], and time in seconds [s], speed is measured in metres per second [m/s]. A car which travels 1200 m in 100 s has an average speed of 12 m/s.

Unless a car is on a motorway, it is unlikely to travel 1200 m without its speed varying. This means that its actual speed at any instant is usually different from its average speed. To measure the actual speed of an object, you need to discover how far the object travels in the shortest time you can conveniently measure. The tennis ball in figure 1 travelled 0.5 m in one-twentieth of a second, so its speed was 10 m/s.

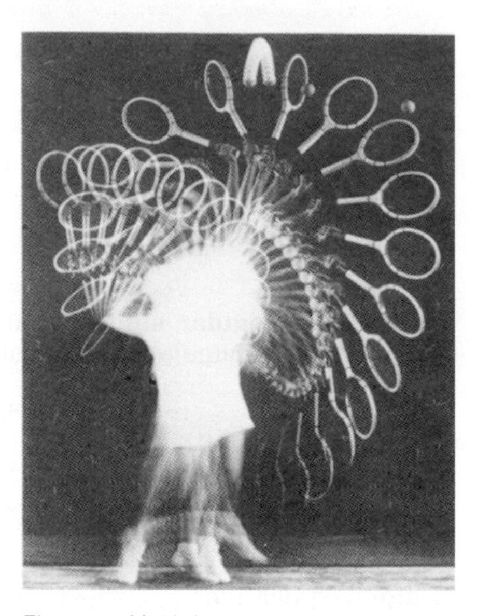

Figure 1 Multiple-exposure can show how quickly the ball moves

Velocity, like speed, is measured in m/s and indicates the speed at which an object travels. But velocity also gives the direction of travel:

$$\textbf{average velocity} = \frac{\textbf{distance moved in a particular direction}}{\textbf{time taken}}$$

On paper, a velocity is represented as in figure 2a. The velocity in this case has a magnitude of 12 m/s and a direction as indicated by the arrow. Quantities, such as velocity, which have both magnitude and direction are known as **vector** quantities. You will come across other types of vector quantity later in the book.

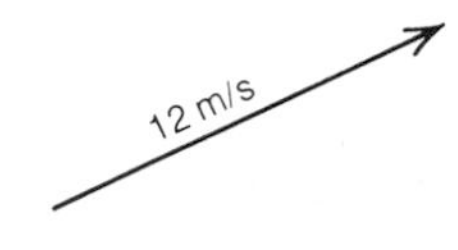

Figure 2a A *velocity vector* – the direction is given

If the velocity of an object is constant or **uniform**, both the speed and the direction remain the same. The velocity changes however, if either the speed *or* the direction changes. To keep things simple, only motion in a straight line is considered in this section, so you can assume that all velocity changes are changes in speed only.

Acceleration

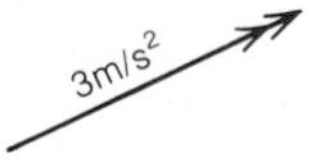

3m/s^2

Figure 2b An *acceleration vector* – the direction is given

An object accelerates whenever its velocity changes. The more rapid the change, the greater is the acceleration. Acceleration, like velocity, is a vector quantity. It is calculated as follows:

$$\textbf{average acceleration} = \frac{\textbf{gain in velocity}}{\textbf{time taken}}$$

An acceleration is represented by an arrow with two heads, as in figure 2b at the bottom of the last page.
If velocity is measured in m/s and time in s, acceleration is measured in m/s^2. For example, if a car starts from rest (0 m/s) and reaches a velocity of 12 m/s in 4 s:

$$\text{average acceleration} = \frac{12 \text{ m/s} - 0 \text{ m/s}}{4 \text{ s}} = 3 \text{ m/s}^2$$

This figure may take on more meaning if you study the progress of the car in figure 3. In this case, the velocity of the car goes up at a steady rate of 3 m/s every second; the car has a constant or **uniform** acceleration of 3 m/s^2.

If the car were to *lose* velocity at the rate of 3 m/s every second, it would have a **deceleration** or **retardation** of 3 m/s^2. Mathematically, this counts as an acceleration of -3 m/s^2.

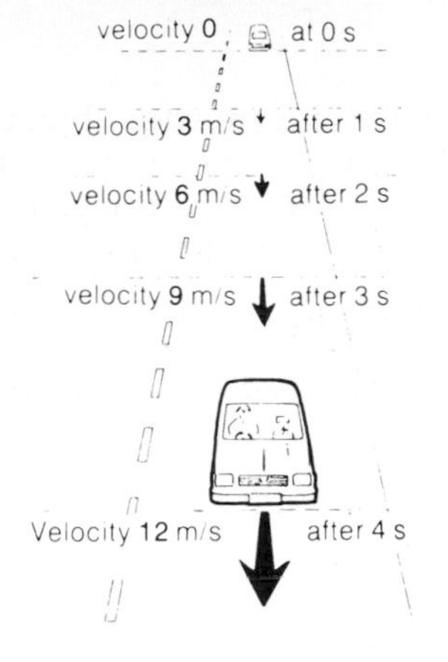

Figure 3 The velocity of the car is increasing at a steady rate. It has a *uniform acceleration* of 3 m/s^2

Distance-time graphs

In studying the progress of a moving object, it often helps to plot a graph of distance moved against time taken. Figure 4 shows a car travelling along a straight road. Distances are measured from a marker post at the side of the road, and timing starts as the car passes this post. Distance-time graphs for a car making four such journeys along a straight road are shown in figure 5.

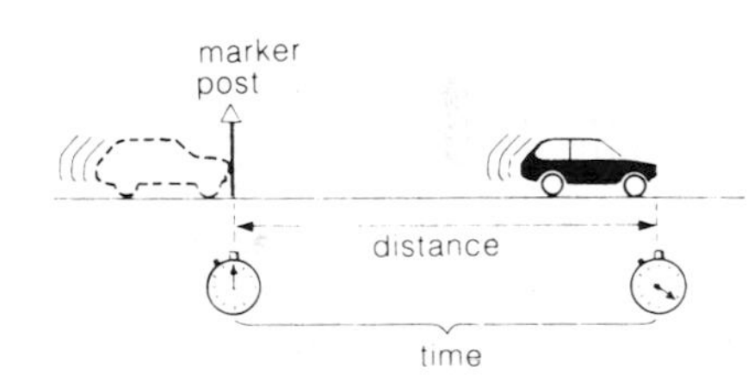

Figure 4 Car travelling along a straight road

In figure 5a, the car is travelling at a constant velocity – it moves the same extra distance further away from the marker post every second. So the graph is a line which rises the same height on the distance axis for each unit on the time axis i.e. it is a straight line.

When the car travels at a higher constant velocity as shown in figure 5b, it moves a greater distance every second, so the line of the graph rises more steeply than before. The slope of the graph is therefore an indication of the velocity of the car.

When the car accelerates as shown in figure 5c, the slope of the graph increases as time goes on. The opposite happens when the car decelerates as shown in figure 5d. When the car is stopped, the line of the graph stays at the same height because the distance of the car from the marker post does not change.

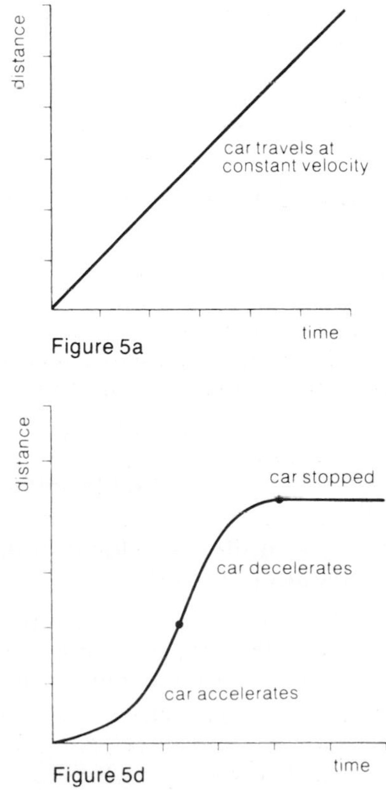

Figures 5a–5d Distance-time graphs

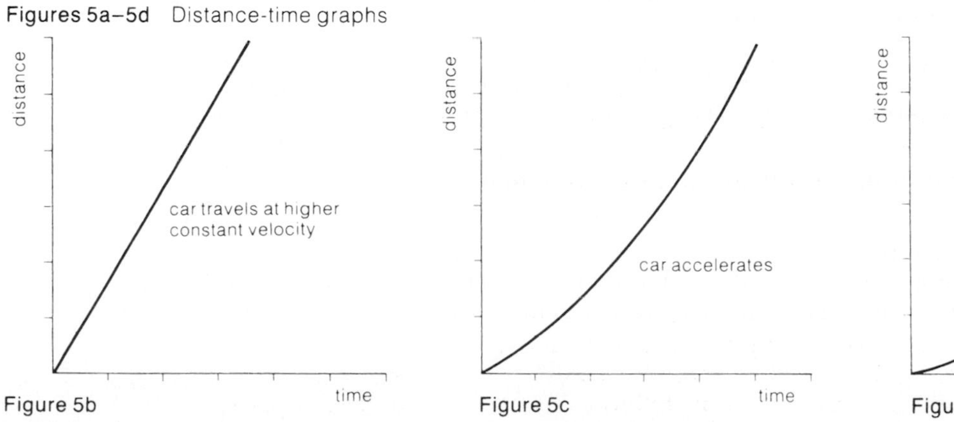

The velocity of a moving object can be found from a distance-time graph. In figure 6 for example, the line rises 12 (m) on the distance scale for every 1 (s) on the time scale, so the velocity is 12 m/s. The velocity can also be found by calculating the **gradient** of the graph.

The gradient is the ratio $\frac{y}{x}$ in the dashed triangle. Its value, 12 in this case, can also be found from any other triangle similar to the one shown.

The above example illustrates a rule which applies to any distance-time graph:

The gradient of a distance-time graph is numerically equal to the velocity.

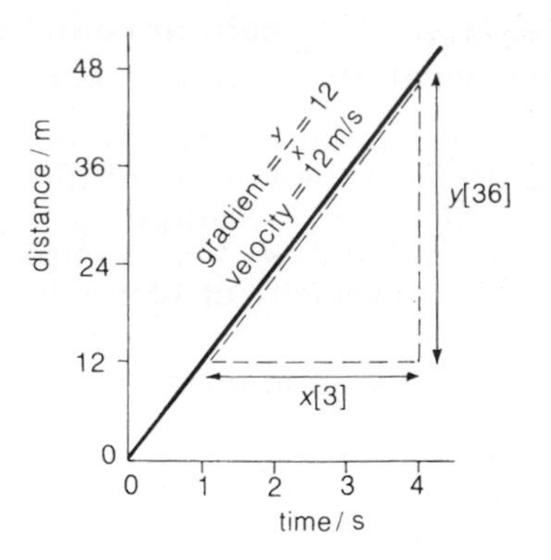

Figure 6 A detailed distance-time graph for a car travelling at a constant velocity 12 m/s

Velocity-time graphs

Graphs can also be drawn which illustrate how the *velocity* of an object changes as time passes. Velocity-time graphs for a car making different journeys along a straight road are shown in figure 7. As previously, timing starts in each case as the car passes the marker post.

When the velocity of the car is constant, as in figure 7a, the graph is a straight line of constant height.

When the car accelerates uniformly, as in figure 7b, the graph is a line which rises the same height on the velocity scale for each unit on the time scale.

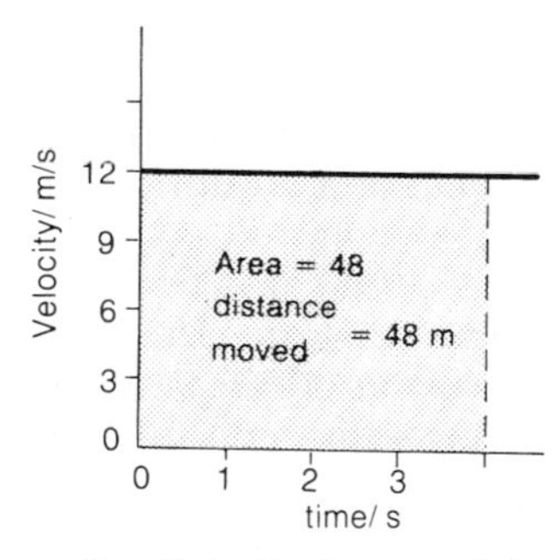

Figure 7a Velocity-time graph for a car moving with constant velocity

The acceleration of an object can be found from a velocity-time graph. In figure 7b for example, the line of the graph rises 3 [m/s] on the velocity scale for every 1 [s] on the time scale, so the acceleration is 3 m/s^2. Alternatively a triangle can be drawn under the line and the gradient calculated as described previously. The gradient in this case is 3:

The gradient of a velocity-time graph is numerically equal to the acceleration.

A velocity-time graph can also be used to find the distance moved by an object in any given time. For example, the car in figure 7a travels at a constant velocity of 12 m/s and will therefore cover a distance of 48 m in 4 s. This distance can also be found by calculating the area under the graph up to the 4 s line, provided the calculation is made using numbers on the graph scale rather than actual distances on the graph paper. This method of finding the distance moved can be applied to any velocity-time graph, whether the velocity is constant or not:

The area under a velocity-time graph is numerically equal to the distance moved.

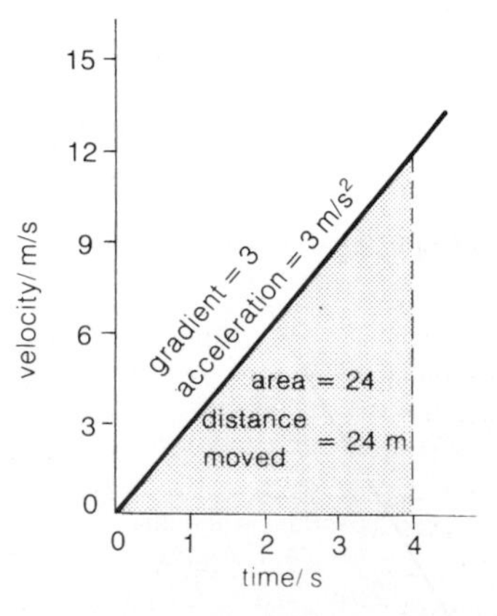

To calculate the area under the graph in figure 7b, you need to remember that the area of a triangle equals $\frac{1}{2}$ × base × height. The area up to the 4 s line works out at 24, so the car travels 24 m in 4 s.

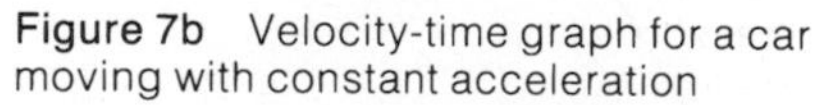

Figure 7b Velocity-time graph for a car moving with constant acceleration

Equations of motion

Many problems involving moving objects can be solved using just four equations. Taking the case of the car in figure 8, the car passes the first post at a velocity u, has a uniform acceleration a, and passes the second post at a velocity v. It takes a time t to travel between the posts, and ends up a distance s further to the right than it was previously; s is called the **displacement**.

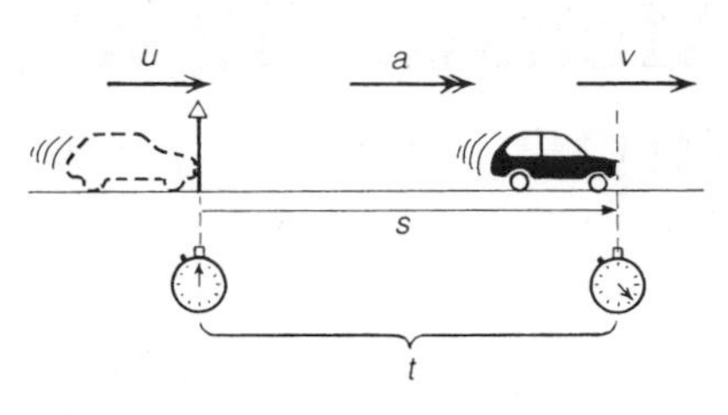

Figure 8 This car has an initial velocity u, accelerates with uniform acceleration a, and reaches a velocity v. It covers a distance s in time t

u, v, a, s, and t are related to each other by the following equations:

$v = u + at$	u = initial velocity [m/s]
$s = \frac{1}{2}(v + u)t$	v = final velocity [m/s]
$s = ut + \frac{1}{2}at^2$	a = acceleration [m/s²]
$v^2 = u^2 + 2as$	s = displacement [m]
	t = time taken [s]

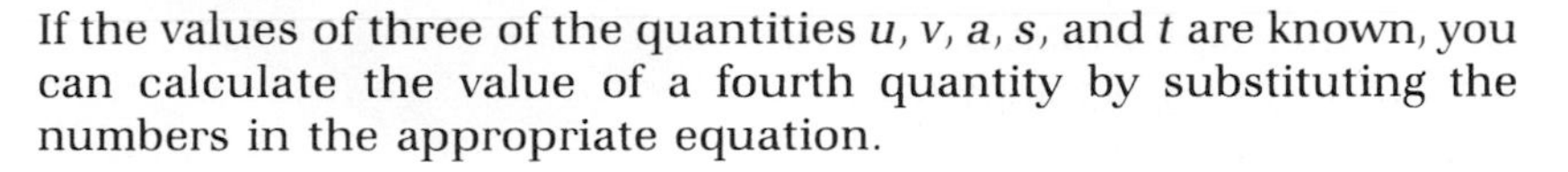

If the values of three of the quantities u, v, a, s, and t are known, you can calculate the value of a fourth quantity by substituting the numbers in the appropriate equation.

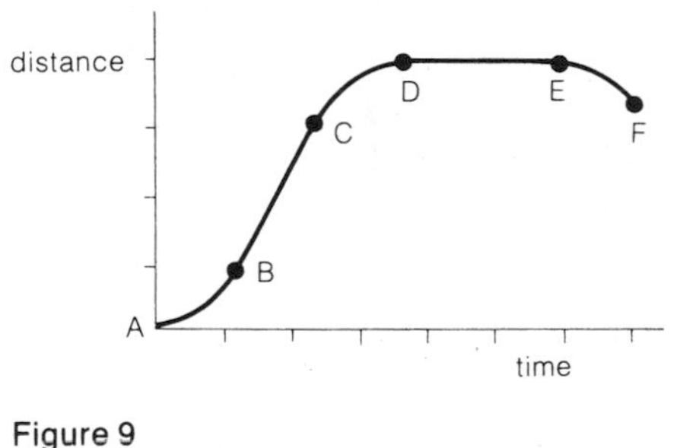

Figure 9

Example *A car starting at rest accelerates at* $3\,\text{m/s}^2$. *How far has the car travelled after* 4 s?

In this case, u is zero because the car starts at rest
v is neither asked for nor given
a is $3\,\text{m/s}^2$
s is the quantity to be found
t is 4 s

You therefore select the equation which includes u, a, s, and t, but not v:

$s = ut + \frac{1}{2}at^2$

$s = (0 \times 4 + \frac{1}{2} \times 3 \times 4^2)\text{m} = 24\,\text{m}$

The car travels a distance of 24 m, a result already worked out from the graph in figure 7b.

Negatives and positives u, v, a, and s are all vectors, so the direction of each must be considered when substituting numbers in the equations. They are normally taken as **positive** (+) when they are to the **right** and **negative** (−) when they are to the **left**. Remember also that a *de*celeration is a negative *ac*celeration. Mathematically, a deceleration in one direction is the same as an acceleration in the opposite direction.

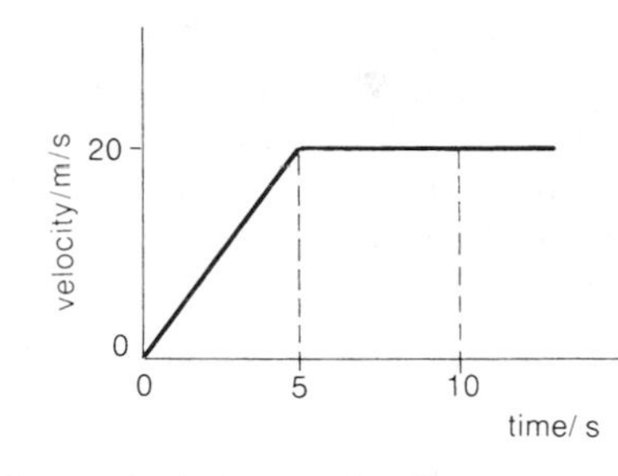

Figure 10 (see question 5)

Questions

1 A car travels 600 m in 30 s. What is its average speed? Why is its actual speed usually different from its average speed?
2 How is velocity different from speed?
3 A car has an acceleration of $2\,\text{m/s}^2$. What does this tell you about the velocity of the car? Explain what is meant by an acceleration of $-2\,\text{m/s}^2$.
4 The distance-time graph in figure 9 is for a car travelling along a straight road. Describe the motion of the car over each section of the graph.
5 The velocity-time graph in figure 10 is also for a car travelling along a straight road.
What is the acceleration of the car during the first 5 s?
How far has the car travelled after 5 s?
How far has the car travelled after 10 s?
What is the average speed of the car during the first 10 s?

Use the equations of motion to solve the following:

6 An object travelling at 10 m/s accelerates at $4\,\text{m/s}^2$ for 8 seconds. What is its final velocity? How far does it travel during the 8 seconds?
7 A motorcycle accelerates from 8 m/s to 20 m/s in 10 s. What distance does it cover in this time?
8 An object travelling at 8 m/s decelerates at $4\,\text{m/s}^2$. How far does it travel before it comes to rest?

1.3

Measuring with ticker-tape

It isn't easy to make measurements on moving cars, but trolleys trailing metres and metres of ticker-tape provide a convenient method of studying velocity and acceleration in the laboratory.

The ticker-tape timer shown in figure 1 is a small electrical vibrator which moves a metal pin up and down 50 times every second. As figure 2 shows, every time the pin moves downwards, it presses on a carbon paper disc and makes a dot on the paper tape which passes underneath. The paper tape is attached to a trolley, as shown in figure 3, and the dots on the tape form a complete record of the motion of the trolley – the further apart the dots, the faster the trolley is moving.

Figure 1 A ticker-tape timer

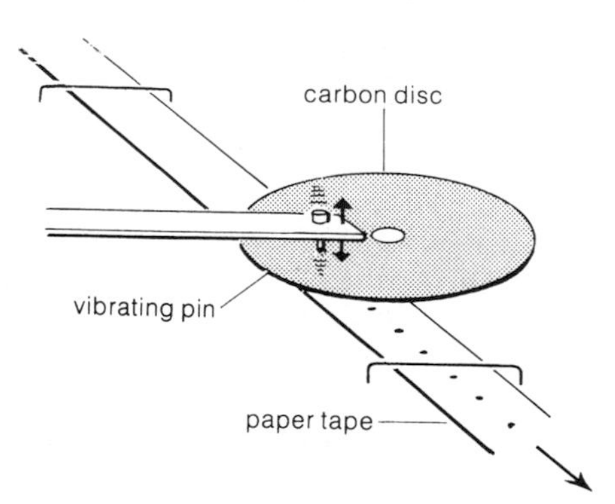

Figure 2 The ticker-tape timer makes a mark on the tape every one-fiftieth of a second. The distance between the dots is a measure of the velocity of the trolley

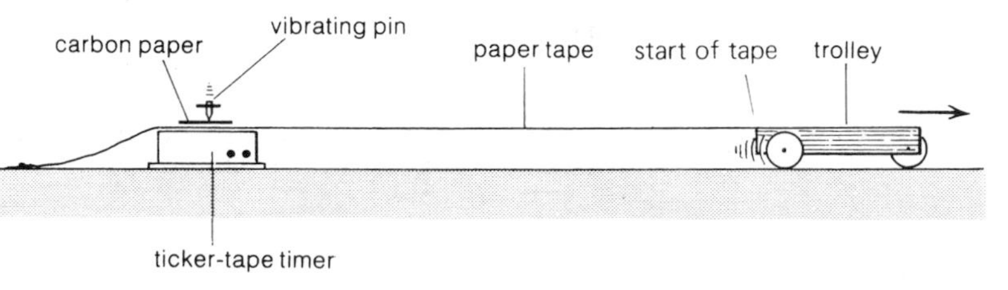

Figure 3 As the trolley moves it trails a paper tape behind it

Typical lengths of paper tape are shown in figures 4a and 4b. Figure 4a shows a tape produced by a trolley moving at constant velocity; figure 4b shows a tape from a trolley which is accelerating uniformly. The distance between dots gives the distance moved by the trolley in one-fiftieth of a second [0.02 s]. This measurement does not change when the trolley is travelling at a constant velocity, but it becomes larger and larger if the trolley is accelerating.

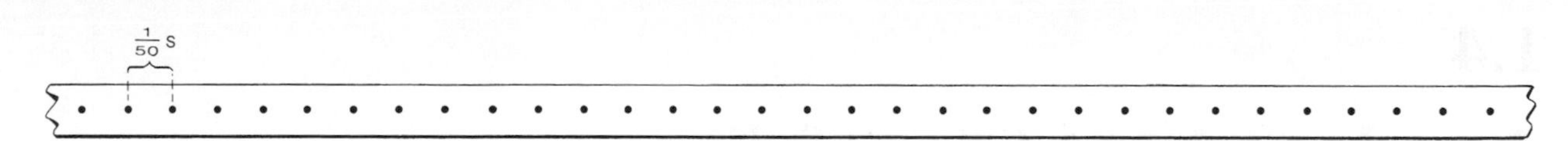

Figure 4a Tape from a trolley moving at constant velocity

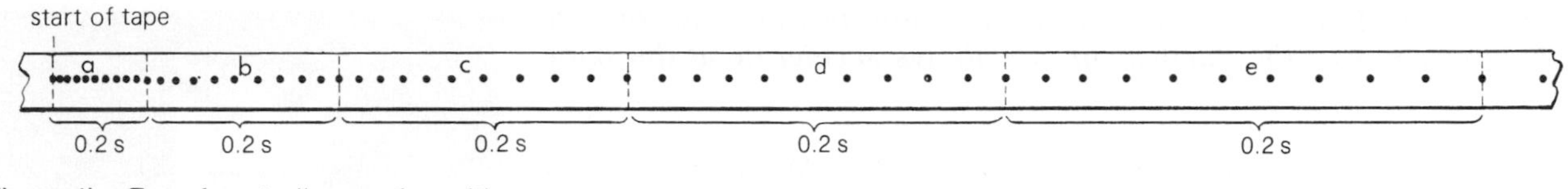

Figure 4b Tape from trolley moving with uniform acceleration

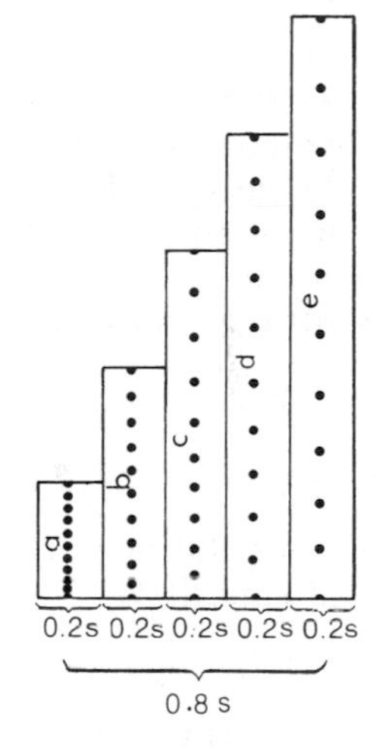

Figure 5 'Ten-dot' chart from the section of tape in figure 4b

The chart in figure 5 is made by chopping up the tape in figure 4b into sections ten dot-spaces long. The length of each section therefore gives the distance travelled by the trolley in one-fifth of a second [0.2 s]. The chart is the shape of a velocity-time graph. Side by side, the sections provide a time scale because each section is started 0.2 s after the one before. The lengths of the sections represent velocities because the trolley travels further in each 0.2 s as its velocity increases.

The acceleration of the trolley can be found from measurements made on the ticker-tape; do question 3 below to discover how.

Questions

In answering the following questions, make measurements on the diagrams above using a ruler marked in millimetres.

1 How do you calculate a) average velocity
b) acceleration?

2 In figure 4a, what distance did the trolley travel in one-fiftieth of a second? What distance did it travel in 0.2 s? How far would it have travelled in 1 s?
What was the average velocity of the trolley in mm/s?

3 In figures 4b and 5, what distance did the trolley travel in the first 0.2 s recorded on the tape?
What was its average velocity during this interval of time?
What distance did it cover during the last 0.2 s recorded on the tape?
What was its average velocity during this interval of time?
What was the change in velocity of the trolley in 0.8 s?
What was the acceleration of the trolley in mm/s^2?

1.4

Acceleration of free fall

Cars and trolleys may accelerate at a variety of rates, but objects falling freely near the Earth's surface always accelerate at the same rate.

When astronaut Alan Shepard stood on the Moon, he held a hammer and a feather side by side and dropped them to show that they would hit the ground at the same time. Both accelerated at the same rate of 1.6 m/s^2. On Earth, falling objects normally have to travel through air, and a feather is slowed much more than a hammer. Figure 1 shows that if there were no air resistance, all objects dropped on Earth would accelerate towards the ground at the same rate – in this case, 9.8 m/s^2. This is known as the **acceleration of free fall** and it is represented by the symbol ***g***.

The value of g varies slightly from one place on the Earth to another because the pull of the Earth's gravity varies. The reasons for this are examined in more detail on pages 25 and 26.

In rough calculations, the acceleration of free fall is often taken as 10 m/s^2 for simplicity.

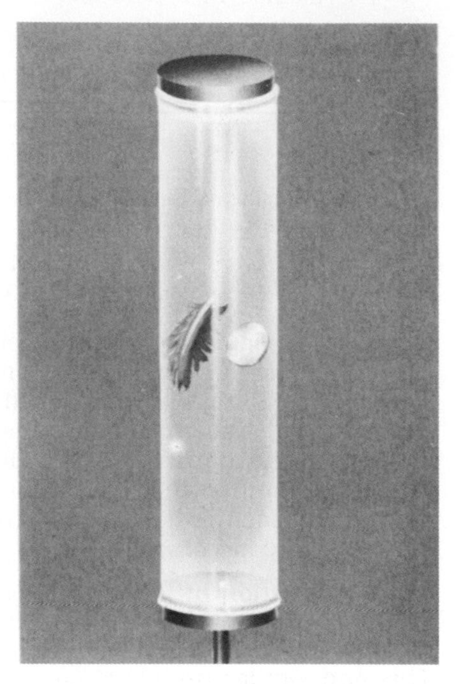

Figure 1 Demonstrating free fall in a vacuum. The coin and feather *do* fall at the same rate!

Moving up and down . . .

If the acceleration of free fall is 10 m/s^2 and air resistance is negligible, an object thrown straight upwards will change its velocity by 10 m/s every second until it reaches the ground.

The cricket ball in figure 2 is thrown upwards from the ground with a velocity of 30 m/s. The diagram shows the velocity of the ball every second as it rises to its highest point and then falls back to the ground. Remembering that an upward velocity is the same as a negative downward velocity, the motion of the ball can be described as follows.

As the ball leaves the ground, its downward velocity is −30 m/s
After 1 s, its downward velocity is −20 m/s
After 2 s, its downward velocity is −10 m/s
After 3 s, its downward velocity is 0 m/s
After 4 s, its downward velocity is 10 m/s
After 5 s, its downward velocity is 20 m/s
After 6 s, its downward velocity is 30 m/s

Note that each downward velocity given is an increase of 10 m/s on the one before. Whichever direction the ball is travelling in, it is always gaining downward velocity at the rate of 10 m/s every second, and therefore always has a downward acceleration of 10 m/s^2. Even when the ball is travelling upwards, or is stationary at its highest point, it is still accelerating downwards.

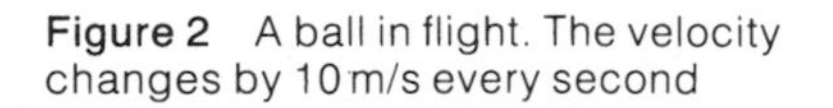

Figure 2 A ball in flight. The velocity changes by 10 m/s every second

. . . moving sideways

The photograph in figure 3 shows what happened when a ball was dropped whilst at the same instant a second ball was thrown sideways. The experiment was illuminated by a stroboscope giving out ten flashes of bright light every second, so the photograph shows the positions of the balls at regular intervals of one-tenth of a second. The position marks on two edges of the photograph show that:

1 Both balls hit the ground at the same time: the downward acceleration of both balls was exactly the same;

2 Horizontally, the second ball moves over the ground at a constant speed.

These results suggest that the horizontal and vertical movements of a falling object are quite independent of each other.

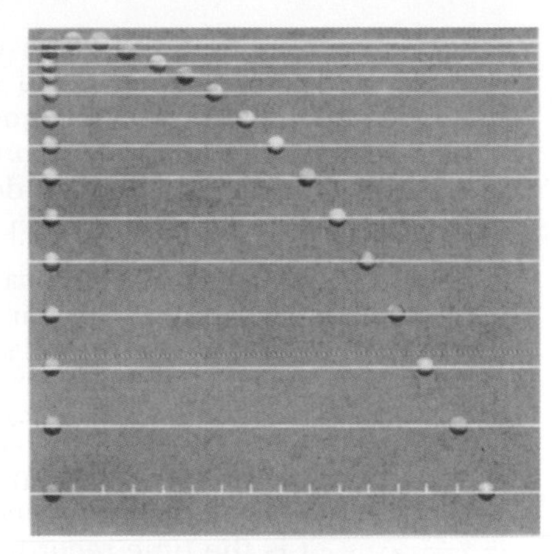

Figure 3 Both balls have the same downward acceleration. The right-hand one also moves sideways at steady speed

g and the equations of motion

The use of the equation of motion in free-fall problems is best illustrated by an example.

Example *The ball in figure 4 is thrown upwards from the ground with a velocity of* 30 m/s. *After how many seconds will it strike the ground again? Assume* $g = 10$ m/s², *and that air resistance is negligible.*

This and similar problems can be solved using the equations of motion, providing you stick to the following guidelines when substituting for u, v, a, s, or t in the equations.

1 Use, say, a positive (+) sign to indicate the downward direction. In the problem above, the initial velocity of the ball is upwards, so $u = -30$ m/s; the acceleration of the ball is downwards, so that $a = g = +10$ m/s².

2 Remember that s stands for displacement from the starting point rather than total distance travelled. The ball in the problem travels many metres. It returns to its starting point however, so its displacement is 0.

3 Look out for useful 'hidden' information. Is the object at its maximum height? If so, $v = 0$. Has the object fallen back to its starting point? If so, $s = 0$. Does it start from rest? If so, $u = 0$. In the problem given, $u = -30$ m/s; $a = 10$ m/s²; $s = 0$; t is to be found.

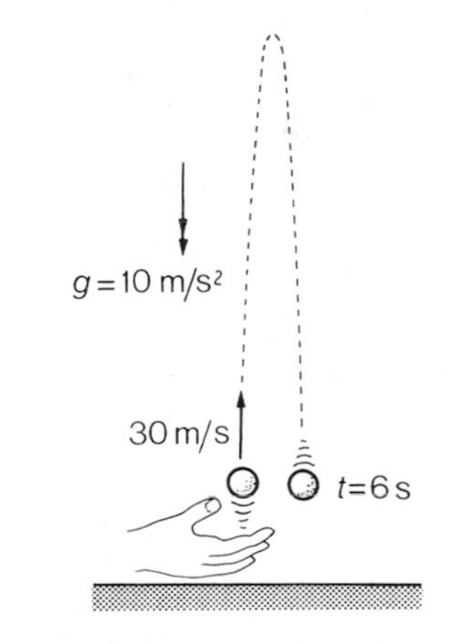

Figure 4

Using $s = ut + \frac{1}{2}at^2$ and calling the unknown time t seconds:

$$0 = (-30t + \tfrac{1}{2} \times 10 \times t^2) \text{ m}$$

Rearranged and simplified, this gives:

$t(t - 6) = 0$

so *either* $t = 6$ or 0.

The ball strikes the ground after 6 s; (the result $t = 0$ simply expresses the fact that the ball has zero displacement at the instant it is thrown upwards).

Measuring g by free fall

An experiment to find the value of g is shown in figure 5. The principle of the experiment is to measure accurately the time it takes for a ball to fall through a known height. The acceleration g of the ball is then calculated using one of the equations of motion. If the ball only falls a metre or so, it doesn't gain enough speed for air resistance to affect its acceleration significantly.

The electronic timer is automatically switched on when power to the electromagnet is cut. At this instant, the ball starts to fall. The timer is switched off when the ball strikes the gate switch and opens it.

In this case: u is 0 because the ball falls from rest
a is the acceleration of free fall
s is the height fallen h
t is the time recorded on the timer display.

$s = ut + \frac{1}{2}at^2$ can be rewritten:

$h = \frac{1}{2}gt^2$; this can be rearranged to give:

$g = 2h/t^2$, from which g can be calculated.

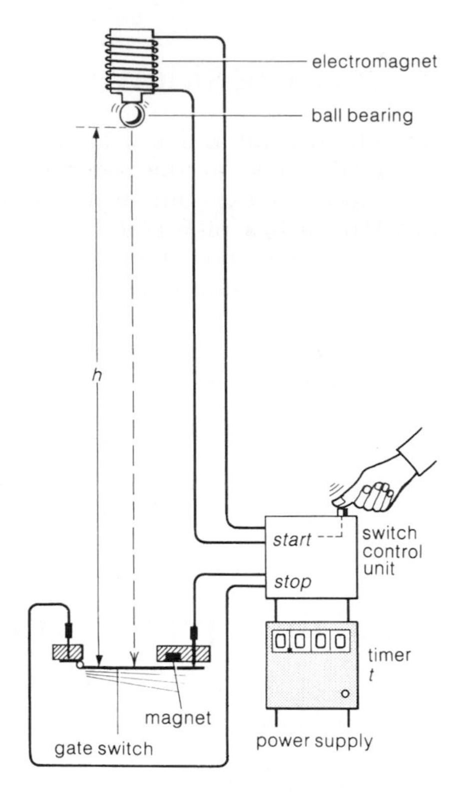

Figure 5 Experiment to measure g by free fall

Questions

Take g as 10 m/s², and assume air resistance is negligible

1. A stone takes 3 s to fall to the ground. If the stone were thrown sideways at 12 m/s, how long would it take to fall to the ground? How far would it have travelled across the ground?
2. A ball is thrown upwards from the ground with a velocity of 30 m/s. How long does it take to reach its highest point? What then is its height above the ground? What will be its velocity as it strikes the ground?
3. If an object is dropped from a height of 20 m, how long will it take to fall to the ground? What will its velocity be as it strikes the ground?
4. A stone is thrown upwards from a cliff top with a velocity of 40 m/s. What is its height above the cliff-top after a) 3 s b) 4 s c) 5 s d) 8 s?
   If the stone strikes the sea after 10 s, what is the height of the cliff-top above the sea?
5. On the Moon, an object dropped from a height of 3.2 m, takes 2 s to reach the ground. What is the acceleration of free fall on the Moon?

This waterfall is 245 m high. Assuming no air resistance, how long does it take water to fall from top to bottom?

1.5

Force, mass and acceleration

When an object accelerates, it does so because a force is acting on it. But some objects have more resistance to acceleration than others.

A force is a push or pull exerted by one object on another.

The remaining sections in this chapter deal with the ways in which forces affect motion. The basic principles were put forward by Sir Isaac Newton in a book called 'Principia Mathematica' which was published in 1687 and included his three laws of motion.

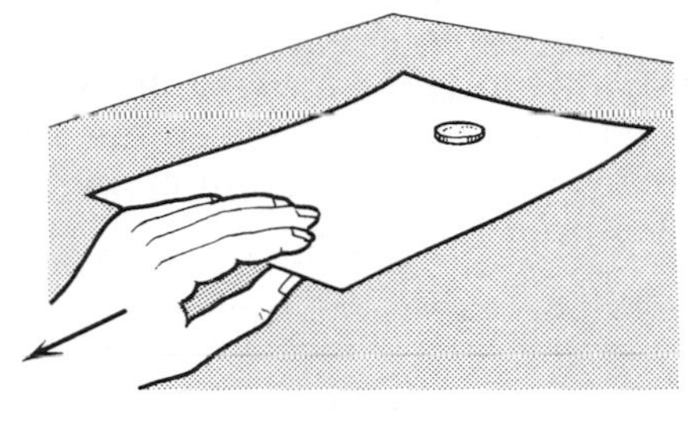

Figure 1 The *inertia* of the coin keeps it in position even though the paper is jerked sideways

Inertia and mass

All objects resist a change in velocity – even if the velocity is zero. If an object is at rest, it takes a force to make it move. If an object is moving, it takes a force to make it go faster, slower, or in a different direction. Newton called this resistance to change in velocity **inertia**. Figure 1 shows a simple demonstration of inertia; when the paper is jerked sideways, the coin stays where it is.

The more mass an object has, the greater is its inertia and the more it resists a change in velocity. The tanker in figure 2 has such a large mass that if the ship is travelling at full speed and its engines are put in reverse, it will travel nearly 10 kilometres before coming to rest.

Any change in velocity is an acceleration. Large masses have a large resistance to acceleration; small masses have a small resistance to acceleration; the mass of an object is therefore *a measure of its resistance to acceleration*.

Figure 2 This oil tanker has a very high mass and a very high inertia

Finding how force, mass and acceleration are related

An object will only accelerate if a force acts on it. The acceleration then depends on the mass of the object and the size of the force acting. Figure 3 shows an experiment to find out how force, mass and acceleration are related.

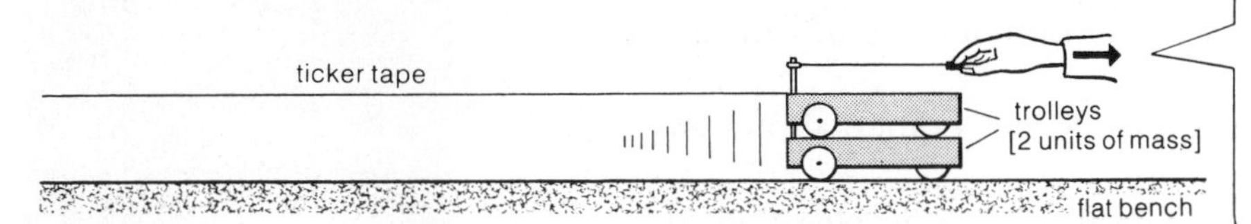

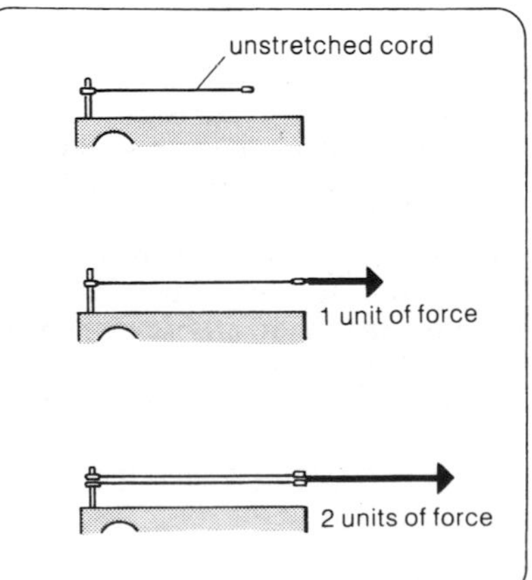

Figure 3 Experiment to find the relationship between force, mass and acceleration

The principle of the experiment is to apply different forces to different masses in the form of trolleys stacked one, two, or three high, and to measure the acceleration produced in each case using a ticker-tape timer as described on page 16. A force is applied to the trolleys by pulling on the elastic cord as shown here, in figure 4. Provided the end of the cord is always above the end of the top trolley, the stretched length of cord does not change and the force is constant. Higher forces are applied using two or more identical elastic cords arranged side by side, each stretched as before.

Figure 4 Providing a constant force to the trolleys: the 'stretch' in the elastic cord is kept the same.

Typical results are given in figure 5. The unit of force used in the experiment is the force pulled by one elastic cord stretched to the length of a trolley. To keep the arithmetic simple, the mass of one trolley is used as the unit of mass, though mass could of course be measured in kilograms.

force	mass	acceleration	mass × acceleration [numerical value]
1 unit	1 unit	0·12 m/s²	0·12
1 unit	2 units	0·06 m/s²	
1 unit	3 units	0·04 m/s²	
2 units	1 unit	0·24 m/s²	0·24
2 units	2 units	0·12 m/s²	
2 units	3 units	0·08 m/s²	
3 units	1 unit	0·36 m/s²	0·36
3 units	2 units	0·18 m/s²	
3 units	3 units	0·12 m/s²	

Figure 5 Results from the experiment shown in figure 3

Various proportions can be seen in the table.

1 Compare readings where the *mass* is the same, say 1 mass unit. Note that when the force is doubled, the acceleration is also doubled, and so on. The force and the acceleration are then said to be *in direct proportion*.

2 Compare readings where the *acceleration* is the same, say 0.12 m/s². Note that when the mass is doubled, twice the force is needed to produce the same acceleration. Force and mass are also in direct proportion.

3 Compare readings where the *force* is the same, say 1 force unit. Note that if the mass is doubled, the acceleration is halved. Acceleration and mass are said to be in *inverse proportion*.

These proportions can be seen in another form if you study the last column in the table. Note that:

1 For any given force, the quantity mass × acceleration always has the same value.

2 If the force is increased, the quantity mass × acceleration increases in the same proportion.

The quantity mass × acceleration is therefore a measure of the force acting on the trolleys in each case.

mass × acceleration: very large in this case

Defining force

In the experiment just described, the unit of force was the force from an elastic cord stretched to a certain length. In general however, the quantity 'mass × acceleration' is used as a measure of force:

force = mass × acceleration

For example:

if a mass of 1 kg has an acceleration of 1 m/s^2, there is a force of 1 kg m/s^2 acting;
if a mass of 2 kg has an acceleration of 3 m/s^2, there is a force of 6 kg m/s^2 acting.

A force of 1 kg m/s^2 is called **1 newton** [**N**], this being the SI unit of force as shown in figure 6. The newton is defined as follows:

1 newton is the force which gives a mass of 1 kg an acceleration of 1 m/s^2.

The equation: 'force = mass × acceleration' can be written using symbols: $F = ma$

F is measured in newtons if m is measured in kg, and a in m/s^2. For example:

a force of 6 N gives a mass of 2 kg an acceleration of 3 m/s^2;
a force of 12 N gives a mass of 2 kg an acceleration of 6 m/s^2.

Force has direction as well as magnitude (size) and is therefore a vector quantity. The acceleration produced by a force is always in the same direction as the force.

1 kg
force: 1N
acceleration: 1 m/s^2

Figure 6 A force of 1 N gives a mass of 1 kg an acceleration of 1 m/s^2

Measuring force

In laboratory experiments, forces are often measured using a spring balance like the one shown in figure 7. When a force pulls on the hook, the spring stretches to counterbalance the force until the spring provides an equal, balancing force. The increased length of the spring indicates the size of the force. This simple principle is also used in equipment designed to measure much larger forces.

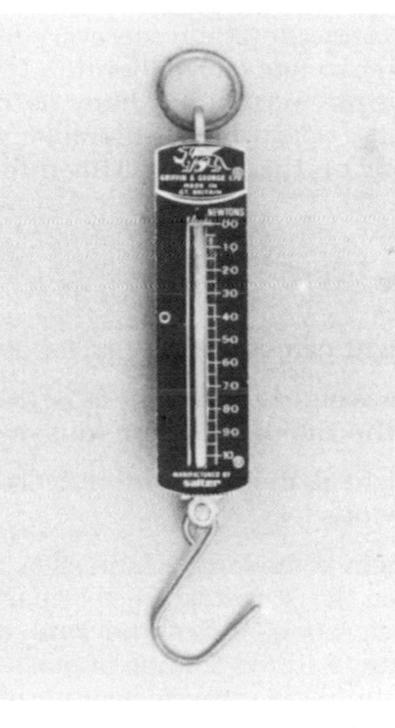

Figure 7 A spring balance measures force. As the force increases, the spring stretches

Some typical force values are given below:

Force needed to switch on a bathroom light	10 N
Force needed to pull open the top of a drink can	20 N
Force produced by a tug-of-war team	5000 N
Force produced by a jet engine	150 000 N

Questions

1 What is the SI unit of force, and how is it defined?
2 A mass of 8 kg accelerates at 5 m/s^2. What force is acting on it? If the mass is doubled, what force is needed to produce the same acceleration?
3 What is the acceleration of a mass of 24 kg when a force of 6 N acts on it? What acceleration would be produced by twice this force acting on half the mass?
4 When a force acts on an object of mass 8 kg, its velocity increases from 3 m/s to 9 m/s in 2 s. Calculate the acceleration using one of the equations of motion on page 15, then calculate the force acting.
5 A force of 4 N acts on an object of mass 0.5 kg which is initially at rest. Calculate the acceleration. Use one of the equations of motion on page 15 to find how far the object travels in 5 s.
6 Objects of mass 1 kg, 2 kg, and 10 kg are accelerating towards the ground at 10 m/s^2. Calculate the force acting in each case.

1.6

Gravitational force

All masses resist acceleration. But they have another important feature in common – they attract each other.

If an object is hung from the hook of a spring balance, the reading on the scale indicates that a force is acting. This downward pull is called a **gravitational force**, and it is exerted by the material present in the Earth.

The Earth is not the only object which causes gravitational force. Gravitational attraction exists between all objects; the greater their masses and the closer they are together, the greater is the pull between them.

The attraction between everyday objects is far too weak to detect – less than one ten-millionth N [10^{-7} N] between you and this book for example. But with a body as massive as the Earth, the pull on a nearby object is considerable, and enough in most cases to hold it firmly on the ground. This pull is called **weight**.

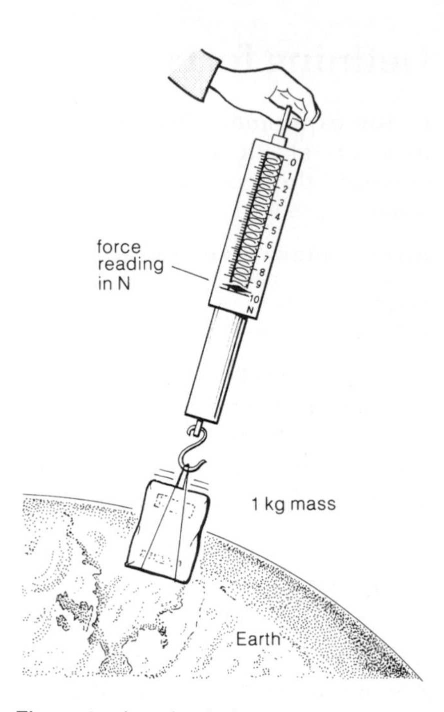

Figure 1 A spring balance can be used to measure gravitational force

Weight

Weight can be defined as follows:

The weight of an object is the gravitational force exerted on it by the Earth or other massive body.

Weight is therefore a force. Like all other forces, it is measured in newtons [N].

Weight is the force which gives an object its downward acceleration when it is falling freely near the Earth's surface. This rate of acceleration, called the **rate of free fall**, is given the symbol ***g***. Figure 2 shows a stone of mass 1 kg falling freely towards the Earth. Assuming that the acceleration of free fall g is 10 m/s², the weight of the stone can be calculated using $F = ma$:

$$\text{Weight} = \text{downward force} = 1\,\text{kg} \times 10\,\text{m/s}^2 = 10\,\text{N}$$

The stone of mass 1 kg therefore has a weight of 10 N.

Similar calculations show that a stone of mass 2 kg has a weight of 20 N near the Earth's surface. A stone of 3 kg has a weight of 30 N, and so on. These results are illustrated in figure 3. In each case:

$$\text{weight} = \text{mass} \times g \qquad \ldots\ \textit{equation 1}$$

in symbols: $\boldsymbol{W = m\,g}$

This equation can be used to calculate the weight of any mass near the Earth's surface as shown in figure 3 on the next page. In each of the examples, the weight is found from the formula $W = mg$:

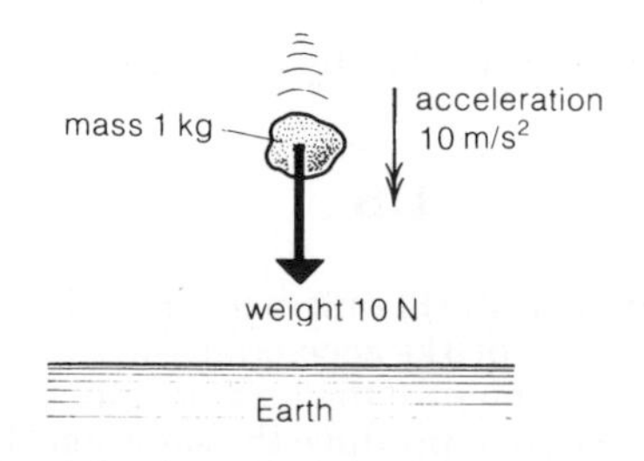

Figure 2 From $F = ma$, the weight of the stone is 10 N

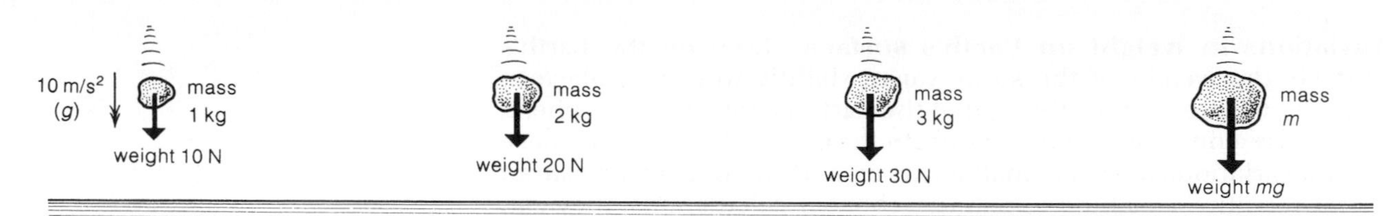

Figure 3 No matter what their mass, all objects accelerate at the same rate in the same gravitational field

Note that weight and mass are in direct proportion. It is *because* they are in direct proportion that the acceleration of free fall is the same for all objects. Objects with more mass have a greater gravitational force acting on them, but this is exactly offset by their greater resistance to acceleration.

Gravitational field strength

It is sometimes useful to think of the Earth being surrounded by a **gravitational field** which exerts a force on any mass in it.

The figures given in figure 3 show that there is a 10 N gravitational force acting on every kg of mass. The **gravitational field strength** near the Earth's surface is 10 N/kg, a value which can be found by dividing the weight of any of the objects by its mass:

$$\text{gravitational field strength} = \frac{\text{weight}}{\text{mass}}$$

Rearranged, this equation can be written:

weight = mass × gravitational field strength ... *equation 2*

Comparing this with equation 1 on the last page, it becomes clear that:

gravitational field strength = g. This is shown in figure 4.

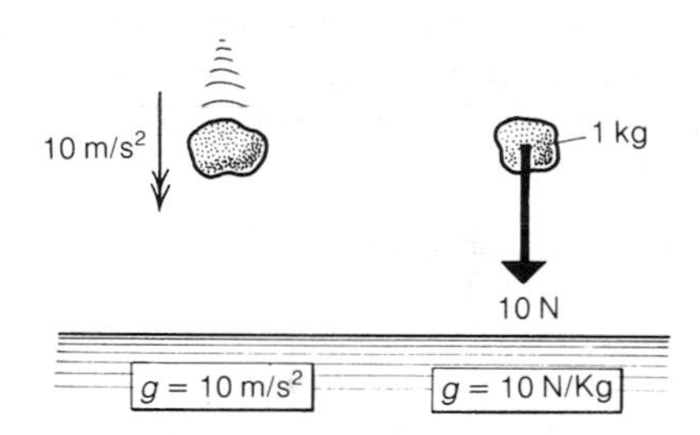

Figure 4 *g* can be thought of as an acceleration in m/s^2 or as a field strength in N/kg

The approximate value of g can therefore be written either as $10\ m/s^2$ or as 10 N/kg, and you can think of g in two ways:

1 An object falling freely near the Earth's surface will accelerate at $10\ m/s^2$.

2 Each kg of mass near the Earth's surface has a gravitational force of 10 N acting on it.

If greater accuracy is required in a calculation:

$g = 9.8\ m/s^2 = 9.8\ N/kg$

Changing weight

The weight of an object depends on its distance from the Earth's centre. If this distance is increased, the gravitational pull of the earth is reduced, so the weight becomes less. Near the Earth's surface, a stone of mass 1 kg has a weight of about 10 N. Figure 5 shows how the weight decreases as the stone is moved further and further away from the Earth.

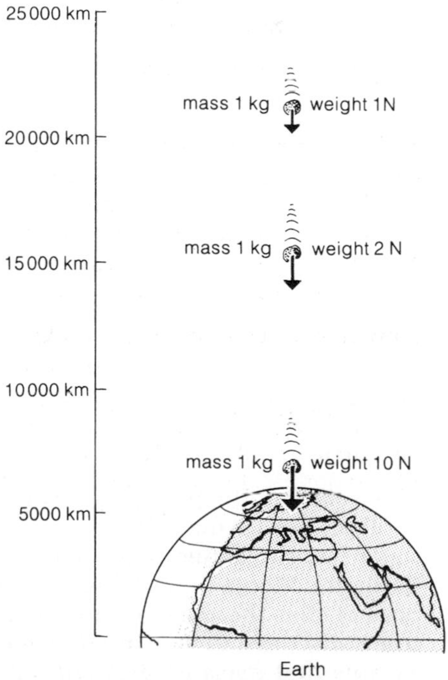

Figure 5 The weight of an object decreases as its distance from the Earth increases

Variations in weight on Earth's surface Even on the Earth's surface, the weight of the stone varies slightly from one place to another. One reason for this is that the Earth is not a perfect sphere, so some regions are further away from the centre than others. These weight variations are very small however, and can only be measured using sensitive weighing apparatus. In most calculations, they can be ignored.

A 1 kg stone would weigh most at the North or South Pole; here, its weight would be 9.83 N. It would have least weight at the equator; 9.78 N. These variations in weight cause variations in the acceleration of free fall, g (see page 18).

Weight on the Moon and in space The Moon has a smaller mass than the Earth, and a stone of mass 1 kg would only weigh about 1.6 N on the Moon's surface (as shown in figure 6). If the stone could be taken deep into space and away from all gravitational forces, it would have no weight at all.

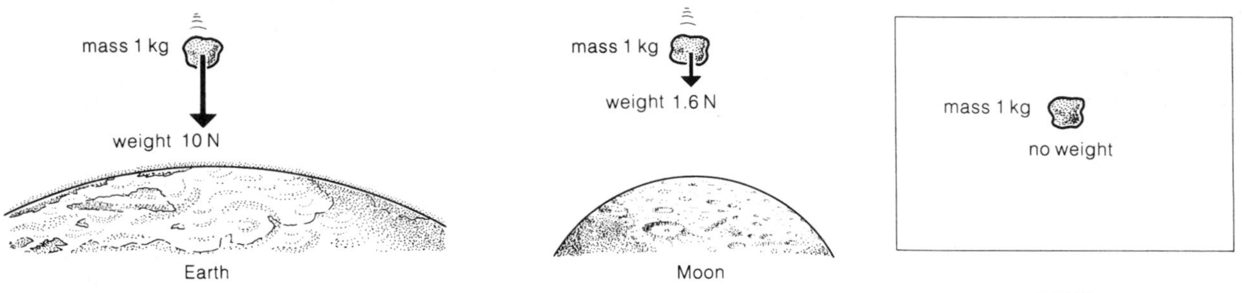

Figure 6 The mass of an object remains the same wherever it is. The weight of an object may vary

Weight and mass

As the weights of objects on Earth are in direct proportion to their masses, instruments which measure weight can also be used to measure mass. Spring balances for example often have scales marked in mass units such as kg. They give a more or less correct mass reading if used on the surface of the Earth, though mass readings taken on the Moon or in space would be entirely false.

Because of the way balances and sets of scales are marked, people tend to use the kilogram, wrongly, as a unit both of mass and of weight. Weight is a force and should always be measured in newtons in scientific work. For example, the bag of sugar in figure 7 *does not* 'weigh' 1 kg. It has a *mass* of 1 kg and a *weight* of about 10 N on Earth.

The weight of the bag of sugar may vary from one planet or moon to another, but its mass does not – at least not noticeably. The bag of sugar has the same resistance to acceleration whether it is on the Earth, the Moon, or 'floating' weightless in space. Einstein's theories predict that the mass of an object can change significantly under some circumstances, but for most practical purposes, you can assume that mass is constant.

Figure 7 This bag of sugar has a mass of 1 kg and a weight of about 10 N on Earth

Weightlessness

Moving deep into space is one way of becoming weightless. But objects can apparently become weightless even when gravitational forces are acting.

In figure 8a, the girl in the lift is standing on a set of scales. The Earth exerts a gravitational force of 500 N on her; her weight is therefore 500 N. She in turn exerts a downward force of 500 N on the scales, so the reading on the scales is 500 N. Note that the scales are measuring a force which is *equal* to the girl's weight, but they do not actually measure the weight itself.

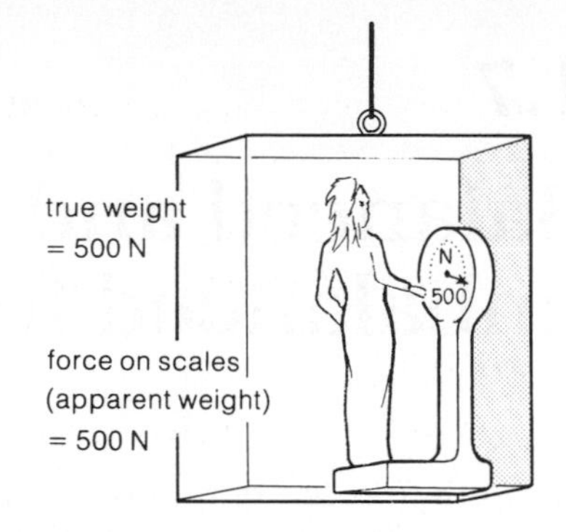

Figure 8a When the lift is not accelerating, the weight reading is normal

Figure 8b shows the situation when the lift cable snaps, and girl, lift and scales all start to fall freely towards the Earth. As fast as the girl moves towards the scales, the scales fall away from her. She therefore no longer exerts a force on the scales, and the reading drops to zero. Her true weight remains 500 N, but her *apparent* weight as indicated by the scales is zero.

If she pushes downwards slightly with her feet, she will lose contact with the scales altogether and move towards the roof of the lift – rather like an astronaut 'floating' in a space capsule. Sealed in the lift, she will be unable to tell whether she is truly weightless because no gravity is acting, or only apparently weightless because both she and the lift are falling freely towards the Earth.

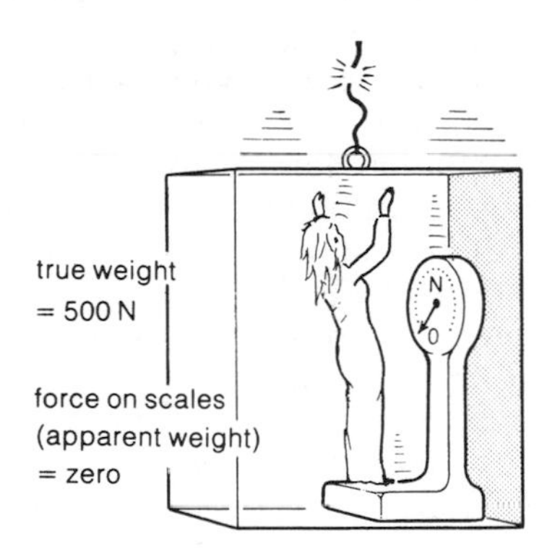

Figure 8b If the cable snaps the lift accelerates at *g*. The girl's apparent weight is zero, and she feels as if she is 'floating'. This effect will cease when the lift reaches the bottom of the shaft

Questions

(Assume $g = 10\ m/s^2 = 10\ N/kg$)

1 In what units are the following measured: force, mass, weight?

2 Write down two properties of all masses.

3 A stone resting on the ground has a gravitational force of 20 N acting on it. What is the weight of the stone? What is its mass?
The weight of the stone will vary very slightly from one part of the Earth to another. Give one reason why.

Figure 9

4 What is the weight of a 1 kilogram mass on Earth? Write down the weights of the following masses on Earth: 2 kg, 25 kg, 50 kg. What would be the mass of each on the Moon?

5 Near the surface of the Moon, the acceleration due to gravity is $1.6\ m/s^2$.
What is the gravitational field strength? What is the weight of an object of mass 20 kg?

6 Figure 9 shows two identical spacecraft falling freely towards the Earth.
   a) How does the weight of *A* compare with the weight of *B*?
   b) How does the mass of *A* compare with the mass of *B*?
   c) How does the acceleration of *A* compare with the acceleration of *B*?
   d) Why would an astronaut in either spacecraft consider him/herself to be weightless?

1.7

Balanced and unbalanced forces

Figure 1 The velocity of an object is constant if no forces are acting on it

Newton's first law of motion describes how an object behaves when no forces are acting on it. On Earth however, no object is ever free of forces.

Newton's first law of motion

The equation 'force = mass × acceleration' implies that if no force is acting on an object, its acceleration will be zero. If the object is at rest, it will stay at rest. If moving, it will continue to move at a constant velocity, i.e. at a steady speed in a straight line. An unpowered spacecraft for example, travelling deep in space and away from the influence of any force, will continue to travel in a straight line without gaining or losing speed. It does not need a force to keep it moving. It only needs a force to change its motion in some way.

These ideas are summed up in Newton's first law of motion, which states:

An object will continue in a state of rest or uniform motion in a straight line unless an external force acts upon it.

On Earth, experience suggests that moving objects soon stop moving if there is nothing to keep pushing them. But on Earth, virtually all moving objects have **frictional forces** acting against them.

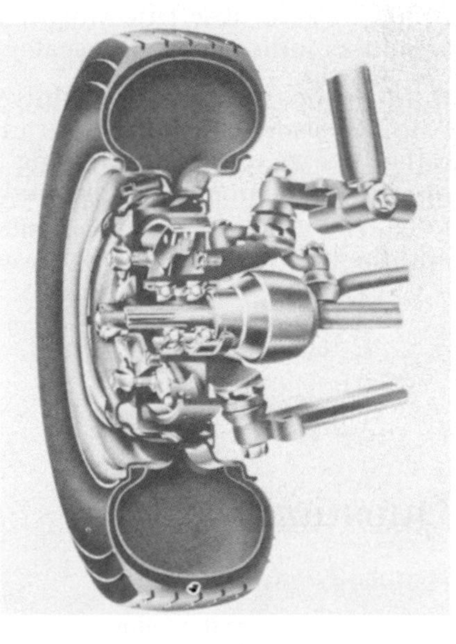

Figure 2a Reducing friction – wheels revolve around greased ball bearings

Friction – the force that slows

Friction is the name given to the force that tries to stop materials sliding across each other. There is friction between your hands when you rub them together, and there is friction between your shoes and the ground when you walk along. Friction is caused in two ways. First, rougher surfaces have ridges and bumps which catch in each other. Second, all materials are made up of tiny particles called molecules, and these have a tendency to stick to each other when materials are pressed together.

Friction prevents machinery from moving freely. It also causes any moving parts to heat up. To reduce friction, wheels are often mounted on ball or roller bearings, as shown in figure 2a, and oil is used to make moving surfaces more slippery. Friction can also be useful. It gives shoes and tyres 'grip' on the ground, and it is used in most braking systems, as shown in figure 2b.

Figure 2b Using friction – disc brakes are used to slow down moving vehicles

Static and dynamic friction A sideways force is being applied to the block in figure 3. As the force is increased the friction between the block and the bench rises, reaching its greatest value just as the block is about to slide. This is the starting or **static** friction. When the block is sliding at a steady speed, the friction is slightly less than before, since the sliding or **dynamic** friction is less than the static friction. It is easier to slide two surfaces across each other than it is to start them sliding in the first place.

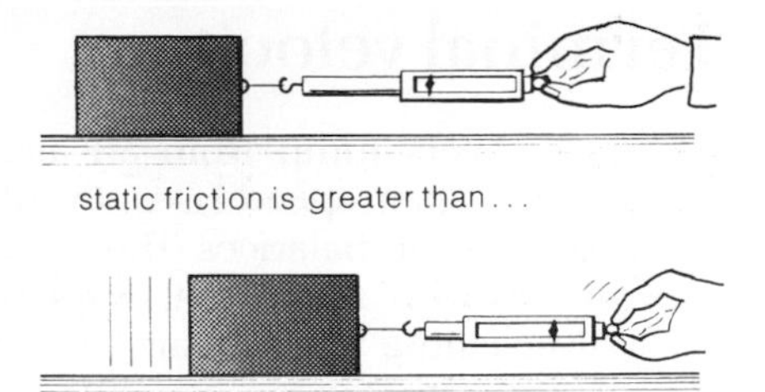

Figure 3 Static and dynamic friction

Fluid friction Gases and liquids are called **fluids**. There is friction whenever an object moves through a fluid. Air resistance is an example of fluid friction. When a car is travelling fast on a motorway, air resistance is the largest of all the frictional forces opposing its motion. Nowadays, car bodies are designed so that air resistance is reduced to a minimum, as shown in figure 4.

Figure 4 Minimizing air resistance

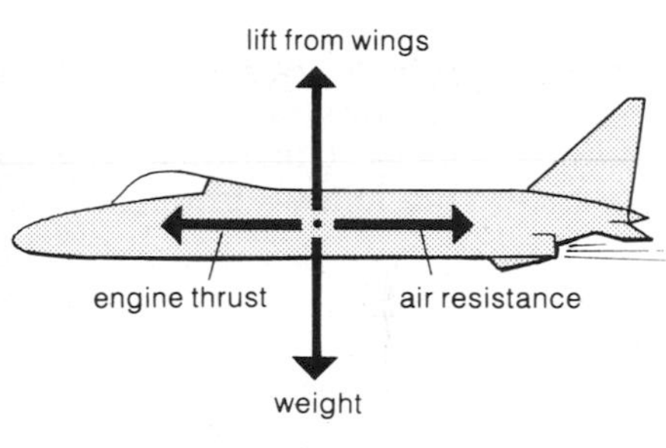

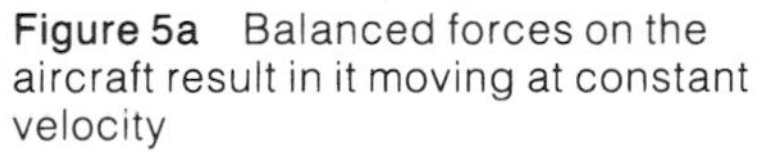

Figure 5a Balanced forces on the aircraft result in it moving at constant velocity

Balanced forces

Despite friction and gravity, objects on Earth can still be at rest or travelling at a constant velocity. In each case, this is because the forces acting are balanced; their effects cancel out and the object behaves as if no force is acting. Examples are given in figures 5a, 5b, and 5c; and also in figure 6 on the next page.

In figure 5a, the weight of the aircraft is balanced by the lift from the wings, and air resistance is balanced by the thrust of the engines. As a result, the velocity of the aircraft is constant. If one of the forces were to change, the velocity of the aircraft would change.

In figure 5b, the table sags under the weight of the rock. The sag is resisted by molecules of the table which try to return the table to its original shape, and exert an upward force on the rock in the process. This force balances the weight of the rock, so it stays at rest.

In figure 5c, the sag of the ice isn't so obvious, but the same principle applies; the weight of the skater is balanced by an upward force from the ice. No forward force is acting on the skater, but he keeps a near-constant speed because frictional forces are slight.

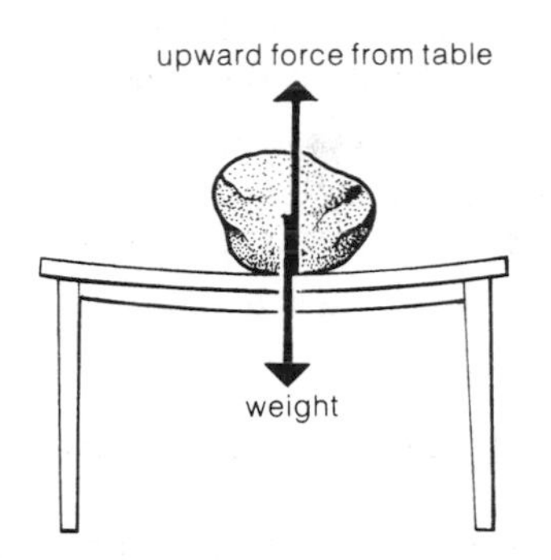

Figure 5b Balanced forces on a rock – in this case, the constant velocity is zero

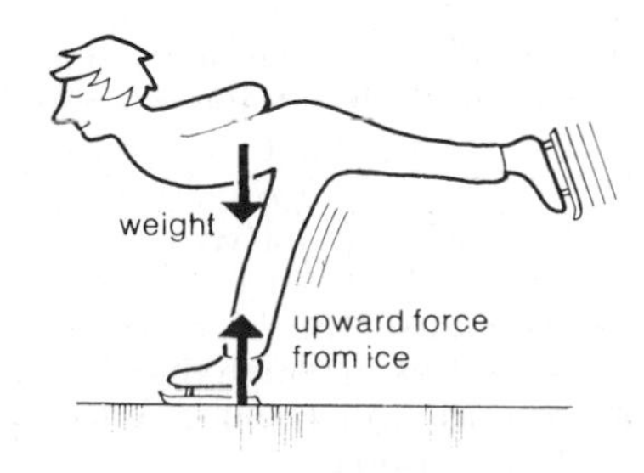

Figure 5c Balanced forces on a skater travelling at a constant velocity

Terminal velocity

When skydivers jump from an aircraft, the air resistance on them increases as their speed rises. Eventually, the air resistance becomes so great that it balances the skydivers' weight. They then travel downwards with a constant, or **terminal** velocity as shown in figure 6. A skydiver falling through air has a terminal velocity of about 60 m/s, though the actual value depends on air conditions as well as on the size, shape, and weight of the skydiver.

If air resistance balances a skydiver's weight, why doesn't he/she stay still? There would not be any air resistance if the skydiver was not moving. Only the downward force would then be acting, so they would start to gain downward velocity.

Surely the weight must be greater than the air resistance, if the skydiver is travelling downwards? No, this is only so when the skydiver is gaining velocity. Once their velocity is constant, the two forces must balance. That follows from Newton's first law of motion.

Figure 6 Balanced forces on a sky diver travelling at terminal velocity

Unbalanced forces

An object will accelerate if the forces acting on it are not balanced. Figure 7 shows a small rocket of mass 200 kg at the point of take-off. The rocket engine is producing an upward force of 3000 N; 2000 N balances the weight of the rocket, but the remaining 1000 N is unbalanced. The upward force and the weight therefore have the same effect as a single force of 1000 N upwards. There is said to be a **resultant force** on the rocket of 1000 N upwards.

The acceleration of the rocket can be calculated as follows:

resultant force = mass × acceleration

$1000\ \text{N} = 200\ \text{kg} \times a$

This gives: $a = 5\ \text{m/s}^2$

The rocket therefore has an acceleration of $5\ \text{m/s}^2$ upwards.

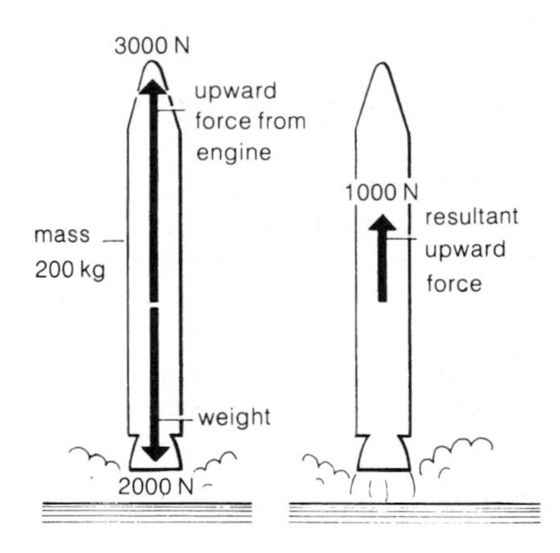

Figure 7 Two forces . . . produce one resultant force

Questions

(Assume $g = 10\ \text{m/s}^2 = 10\ \text{N/kg}$)

1 When is friction useful? Give two examples. When is friction a disadvantage? Give two examples, indicating how friction is reduced in each case.

2 What is meant by a) static friction b) dynamic friction? Which is the greater?

3 Give one example of fluid friction.

4 Explain in terms of Newton's first law of motion why you have to 'hold tight' when a bus pulls away quickly. Is it true to say that you are 'thrown forward' if the bus brakes rapidly? If not, why not?

5 An aircraft is travelling at 300 m/s. What happens to the aircraft if all the forces acting on it are balanced?

6 A skydiver falls through the air at a steady velocity of 50 m/s. If she weighs 500 N, what is the force of air resistance upon her?

7 The steel ball in figure 8 is travelling downwards through the oil at a constant velocity. What name is given to this velocity?
If the oil exerts an upward force of 5 N on the ball, what is the weight of the ball? What is the mass of the ball?

8 A rocket has a mass of 1000 kg. What is its weight on the ground? At take-off, the rocket engine exerts an upward force of 15 000 N on the rocket. What is the resultant force on the rocket? What is its acceleration? What force must the engine produce to give the rocket an upward acceleration of $30\ \text{m/s}^2$?

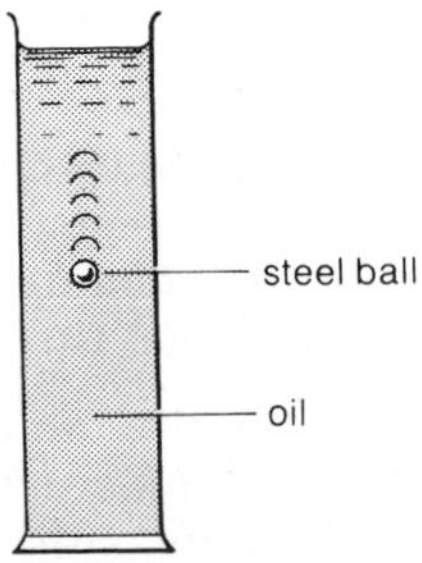

steel ball

oil

Figure 8

1.8

Moving in circles

Question: *What can have a steady speed but a changing velocity?*
Question: *What accelerates towards something, yet never gets closer to it?*
Answer: *Anything that moves in a circle.*

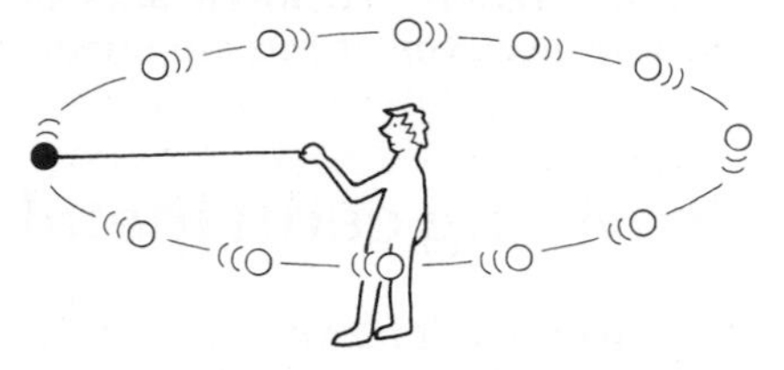

Figure 1 The *speed* of the ball is constant but the *velocity* is changing because the direction is changing

A velocity doesn't have to get bigger or smaller in order to change. Velocity is speed in a particular direction, so a change in either speed *or direction of travel* is a change in velocity. The ball in figure 1 is being whirled round in a horizontal circle at a constant speed. The direction of travel is always changing – so the ball has a changing velocity.

Any change in velocity is an acceleration. The ball has acceleration when it is moving in a circle, even though its speed is constant.

Centripetal force

The ball will not follow a circular path unless there is an inward force to make it do so, as shown in figure 2a. The tension in the string provides this inward force. If the string breaks, the force is removed and the ball does exactly what Newton's first law predicts. It travels at a steady speed in a straight line. Note that the ball is *not* 'flung outwards'; it takes the direction it had when the string broke, and moves off at a *tangent* to the circle.

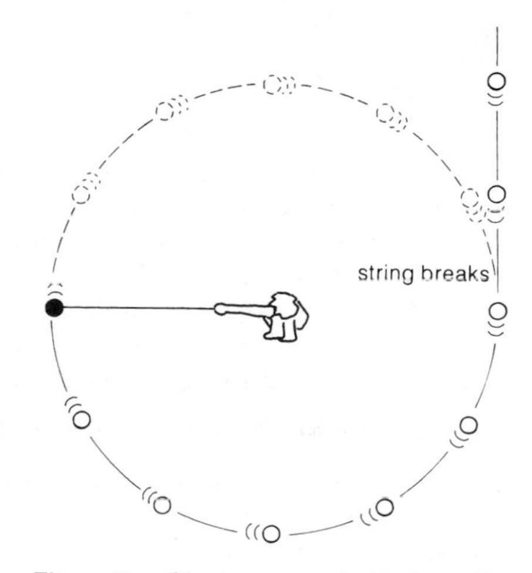

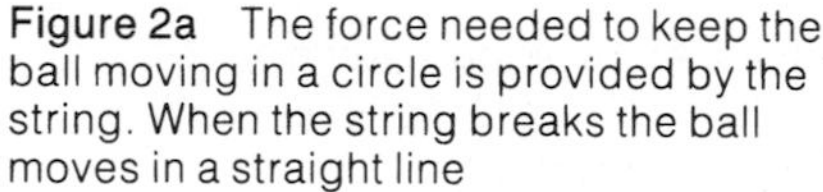

Figure 2a The force needed to keep the ball moving in a circle is provided by the string. When the string breaks the ball moves in a straight line

The inward force needed to make the ball move in a circle is called the **centripetal** force. It is this force that gives the ball its inward acceleration, as shown in figure 2b. It may be difficult to imagine the ball accelerating towards the centre of the circle without getting any closer to it, but the ball is constantly moving inwards from the position it would have had if it had moved off in a straight line.

The centripetal force needed to make the ball follow a circular path depends on several factors. A larger force is required if:

1 the mass of the ball is increased,

2 the speed of the ball is increased,

3 the radius of the circle is reduced.

If the mass of the ball is increased, a larger force is necessary because a greater mass of material is having its direction altered.

If the speed of the ball is increased, the rate at which its direction alters is also increased; the same thing applies if the radius of the circle is reduced.

It is important to realise that making something move in a circle does not *produce* a force – though people commonly talk of 'centrifugal' force. If the centripetal force in figure 2b is, say, 20 N, this simply indicates the fact that a force of 20 N is *needed* to make the ball move in the way shown.

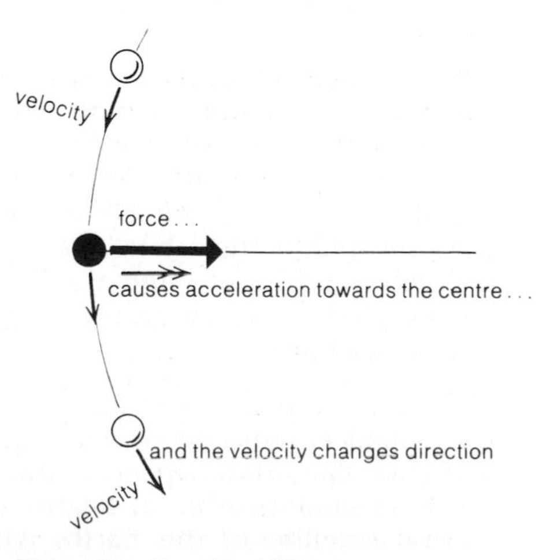

Figure 2b Centripetal force in action

Straight lines and circles Motion in a circle is just another example of force producing acceleration. Figures 3a and 3b show two possible effects of a force on a moving object. There is an acceleration in each case because the force causes a change in velocity. When the force acts at right angles to the direction of motion however, it has no effect on the speed of the object.

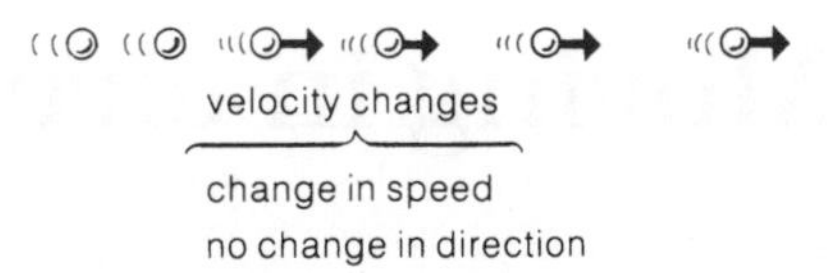

Figure 3a A force can cause a velocity change by changing the *speed* of an object . . .

Providing centripetal force

An object won't move in a circle unless centripetal force is provided. Tension in the string provides centripetal force for a whirling ball. When a car travels round a bend in the road, the centripetal force is supplied by sideways friction between the tyres and the road, as shown in figure 4.

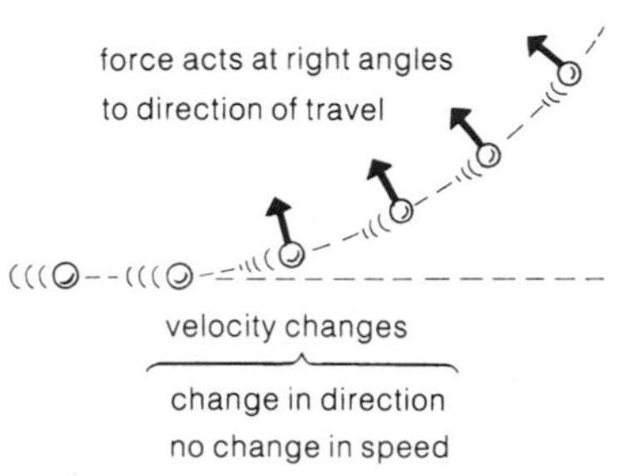

Figure 3b . . . a force can cause a velocity change by changing the *direction of travel* of an object

Figure 4 Centripetal force can be provided in other ways than by string

Satellites in orbit

Gravitational pull provides the centripetal force needed to make a satellite follow a circular path around the Earth. When a satellite is put into a circular orbit, its speed is carefully chosen so that its weight supplies exactly the right amount of centripetal force to keep it in that particular orbit. For a satellite orbiting the Earth just above the atmosphere, the orbital speed required is about 8000 m/s (29 000 kilometres per hour), and the 'burn time' of the launch vehicle's engines has to be controlled very accurately so that the correct speed is achieved.

The further out the orbit, the lower the gravitational pull, and the less speed is required to maintain the orbit; this is illustrated in figure 5a. The orbital speed of the inner satellite is 6400 m/s, while the outer satellite orbits at 4800 m/s. Much further out, the Moon is a natural satellite of the Earth, with an average orbital speed of 1000 m/s.

The mass of a satellite does not affect the speed required for a particular orbit. If the mass is doubled, twice as much centripetal force is needed, but that is supplied by the doubled gravitational pull of the Earth.

If a satellite does not have the right speed for a circular orbit, it may move in an orbit the shape of an ellipse. In some cases, its path may take it back to Earth. Given enough speed, it may escape from the Earth altogether. To escape from the Earth completely, an object near its surface needs a speed of at least 11 000 m/s. This is known as the **escape velocity**.

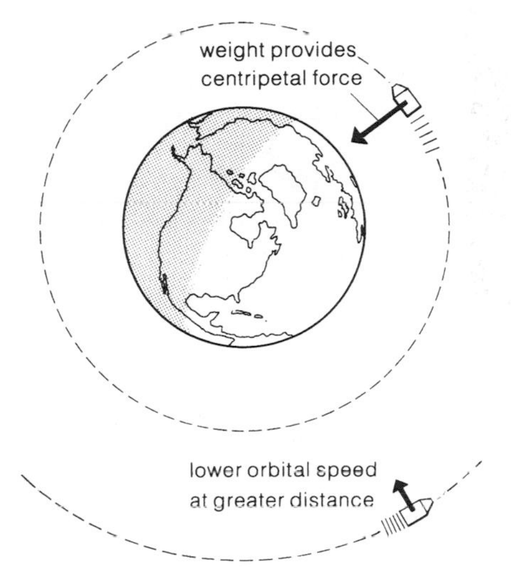

Figure 5a To stay in a circular orbit, Satellites closer to the Earth have to move faster than those further away

Figure 5b The spacecraft are about to dock. The joint mass will increase, but the orbit will stay the same

Questions

1 Give two ways in which a velocity can change.

2 A piece of Plasticine is stuck to the edge of a record turntable. Draw a diagram to show the path of the Plasticine if it comes unstuck when the turntable is rotating.

3 An object is moving freely in a straight line. Describe what happens when a force acts on the object a) in the same direction as the object is travelling b) at right angles to the direction of travel.

4 What provides the centripetal force needed to make a car travel round a bend in the road? How is the size of the necessary force affected if a) the car has less mass b) the car travels more slowly c) the car goes round a tighter curve?

5 The satellite in figure 6 is in a circular orbit around the Earth. Copy the diagram, drawing in and labelling any forces acting on the satellite. How does the mass of the satellite affect the speed required for the orbit? If the satellite is moved to a higher orbit, does its speed need to be more or less than it was before? Does it require more or less centripetal force than before?

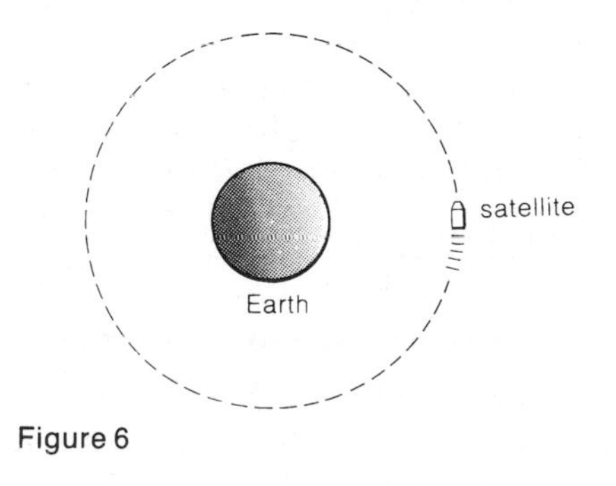

Figure 6

1.9

Changing momentum

Imagine something massive travelling fast, and you have a useful mental picture of momentum. Newton gave a more precise meaning to the idea of momentum and then went on to establish an important link between momentum and force.

Momentum

The momentum of an object is defined as follows:

momentum = mass × velocity

With mass measured in kg and velocity in m/s, momentum is measured in kg m/s. Like force and velocity, momentum is a vector quantity.

The stones in figure 1 have different masses and velocities, but they both have the same momentum, 12 kg m/s in each case.

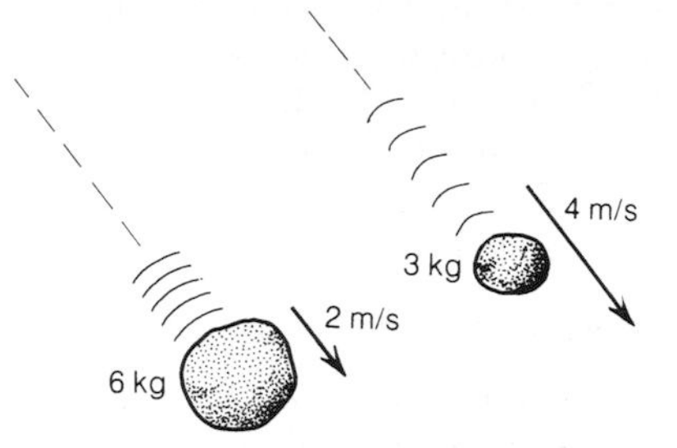

Figure 1 Each stone has a momentum of 12 kg m/s

Force and momentum

There is an important relationship between momentum and force. It emerges if you consider the equation $F = ma$ in more detail. Take the case of the spacecraft shown in figure 2.

To begin with, the spacecraft has a velocity u. Its rocket motor is fired briefly, so that a force F acts on the spacecraft for a time t. As a result, the velocity is increased to v.

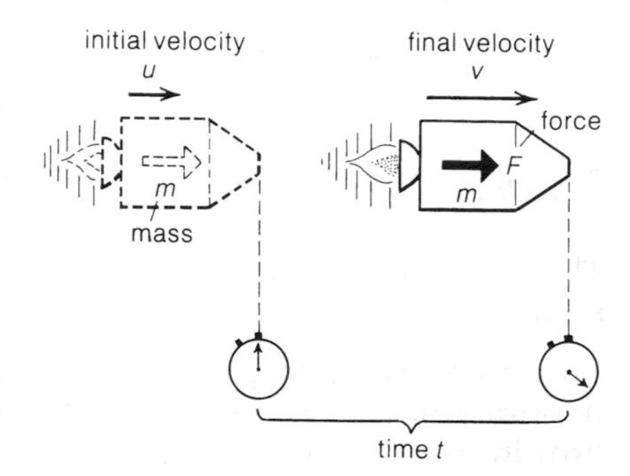

Figure 2 Under the constant force F, the spacecraft of mass m has accelerated from an initial velocity u to a final v in a time t

Then: the acceleration a of the spacecraft $= \dfrac{\text{gain in velocity}}{\text{time}} = \dfrac{v-u}{t}$

So: $F = ma$ can be written: $$F = m\left(\frac{v-u}{t}\right)$$

Multiplying out the brackets: $$F = \frac{mv - mu}{t}$$

mu is the momentum the spacecraft had originally; mv is the momentum it ends up with after the force has acted. The equation above can therefore be expressed in words:

$$\textbf{force} = \frac{\textbf{gain in momentum}}{\textbf{time}}$$

or: **force = rate of change of momentum**

It follows, for example, that a force of 1 N, applied to an object, will make its momentum increase by 1 kg m/s, each second.

A small force acting on a large mass causes very little acceleration . . .

Example *The spacecraft in figure 3 has a mass of* 1000 kg. *Its velocity increases from* 20 m/s *to* 50 m/s *in* 10 s. *What force is acting on the spacecraft?*

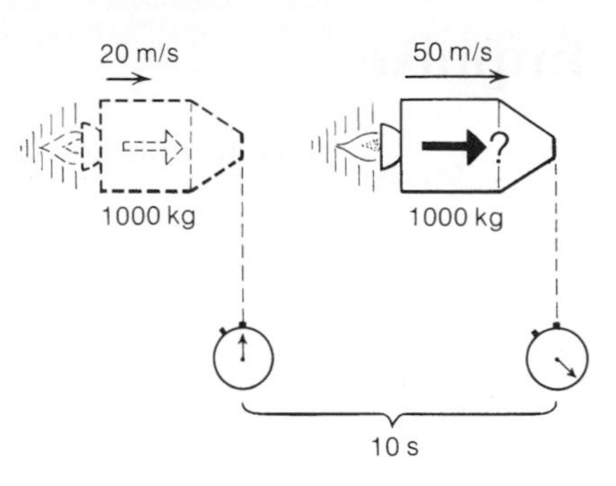

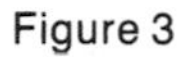

Figure 3

To begin with, the momentum *mu* of the spacecraft is 20 000 kg m/s. Ten seconds later, the momentum *mv* of the spacecraft is 50 000 kg m/s.
In those 10 seconds the gain in momentum is therefore 30 000 kg m/s:

$$\text{force} = \frac{\text{gain in momentum}}{\text{time}} = \frac{30\,000 \text{ kg m/s}}{10 \text{ s}} = 3000 \text{ N}$$

There is a force of 3000 N acting on the spacecraft.

The force in this case is giving the spacecraft an extra 3000 kg m/s of momentum every second. The spacecraft is therefore gaining momentum at the rate of 3000 kg m/s per second. In other words, the rate of change of momentum is 3000 kg m/s^2. The force acting is equal to the rate of change of momentum, so the force is 3000 kg m/s^2 or 3000 N (1 kg m/s^2 = 1 N).

Newton's second law of motion

The link between force and momentum is expressed in Newton's second law of motion. This states:

The rate of change of momentum of an object is directly proportional to the force acting, and takes place in the direction in which the force acts.

If a constant force is being considered, the law can be written in the following form:

Force is directly proportional to $\frac{\textbf{gain in momentum}}{\textbf{time}}$

The law is also sometimes expressed in the form:

Force is directly proportional to mass × acceleration

In either form, the words 'directly proportional to' can be replaced by 'equal to' provided the various quantities are measured in the appropriate SI units.

The above proportions show that there are two ways of regarding a force. There are also two ways of defining the newton:

1 newton is the force which gives a mass of 1 kg an acceleration of 1 m/s^2.

1 newton is the force which causes an object to gain momentum at the rate of 1 kg m/s per second [1 kg m/s^2].

. . . a large force acting on a similar mass, causes a very high acceleration

Impulse

As: $\text{force} = \dfrac{\text{gain in momentum}}{\text{time}}$

it follows that: force × time = gain in momentum

in symbols: $Ft = mv - mu$

When a force acts for a given time, the quantity force × time is called an **impulse**. If force is measured in N and time in s, impulse is measured in N s or kg m/s – both amount to the same thing.

The spacecraft in figure 4 has a force of 3000 N acting on it for 10 s, so the impulse delivered is 30 000 N s. As a result, the spacecraft gains an extra 30 000 kg m/s of momentum. The gain in momentum is the same whatever the mass of the spacecraft.

Newton observed that when a given force acted for a given time on any object, a larger mass gained less velocity than a smaller one, but the quantity mass × velocity increased by the same amount in every case. It was this observation that led him to define momentum (though he called it a 'quantity of motion') and to put forward his second law of motion.

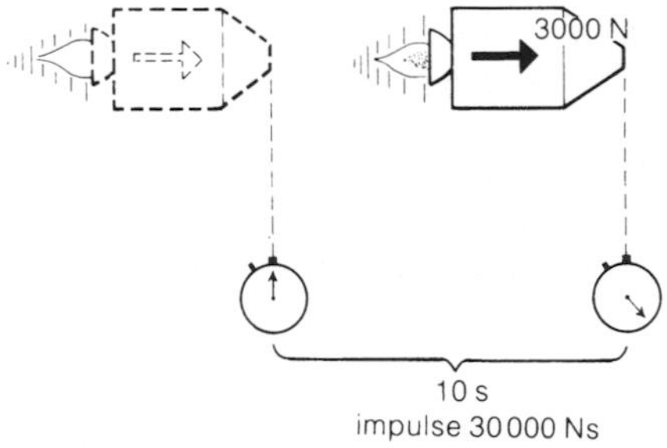

3000 N
10 s
impulse 30 000 Ns

Figure 4 The quantity *force × time* is called *impulse*

Was Newton right?

Scientists find that the equations based on Newton's second law give the right answers to most problems concerning moving objects. But there are exceptions. In particular, Einstein showed that the equation $F = ma$ was not valid for objects travelling at speeds approaching the speed of light. Not that this usually matters on or around Earth, where even the fastest rockets travel at less than one ten-thousandth of the speed of light. $F = ma$ is quite adequate for most practical purposes.

Questions

1 How do you calculate the momentum of an object? In what units is momentum measured?

2 Express Newton's second law of motion in two different ways.

3 An object of mass 12 kg is travelling at a velocity of 4 m/s with no forces acting upon it. This velocity increases to 6 m/s after a force has acted on the object for 3 s.
What is the momentum of the object before the force acts?
What is the momentum of the object after the force has acted?
How much momentum does the object gain? What is the momentum gain every second?
What is the size of the force acting?
Calculate the acceleration of the object, and use this information to calculate the size of the force by another method.

4 A force of 100 N acts for 5 s on an object of mass 20 kg.
By how much does the momentum change every second?
What is the total change in momentum? What impulse is delivered?
How much velocity does the object gain?

1.10

Action and reaction

When studying motion, it is easier to take objects one at a time and consider the forces acting on each. In reality however, forces occur when two objects push or pull on each other. A single force cannot exist by itself.

The two trolleys in figure 1 rest in contact with a flat table. One of the trolleys contains a spring-loaded piston which shoots out when a release pin is lightly tapped.

Figure 1 Trolley A contains a spring-loaded piston, held in place by a pin

When the piston is released, both trolleys shoot off in opposite directions, as in figure 2. Although a spring is released in one trolley only, two equal and opposite forces are produced, one acting on each trolley.

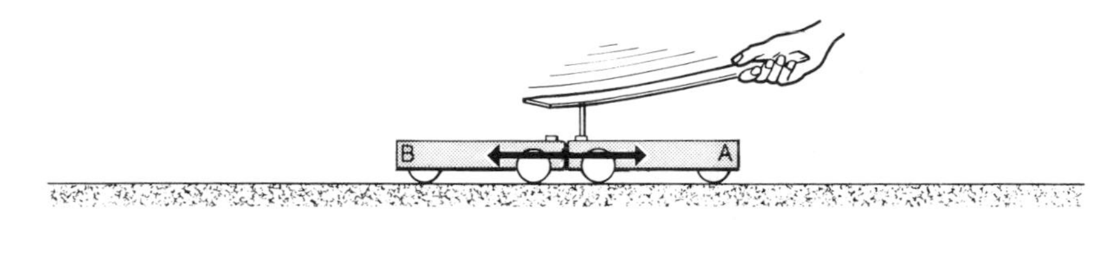

Figure 2 When the piston is released, the trolleys force one another apart, demonstrating action and reaction.

Newton's third law of motion

The effect described above is not confined to trolleys. Newton pointed out that, in every case, forces exist because *two* objects exert a push or pull on each other. A single object cannot experience a force by itself. Forces therefore always occur in pairs – an idea summed up by **Newton's third law of motion**, which states:

If object A exerts a force on object B, then object B will exert an equal but opposite force on object A.

This is sometimes expressed in the form:

To every action there is an equal but opposite reaction.

It is tempting to think that these equal but opposite forces, called an **action-reaction pair**, might cancel each other out. Remember however that the forces are acting on *different* objects, as the examples on the next page show.

The engine on the bike pushes the wheel one way; the wheel pushes the bike the other way: Barry Sheene demonstrates a 'wheelie'

Figure 3a The astronaut's force on the Earth equals the Earth's force on the astronaut

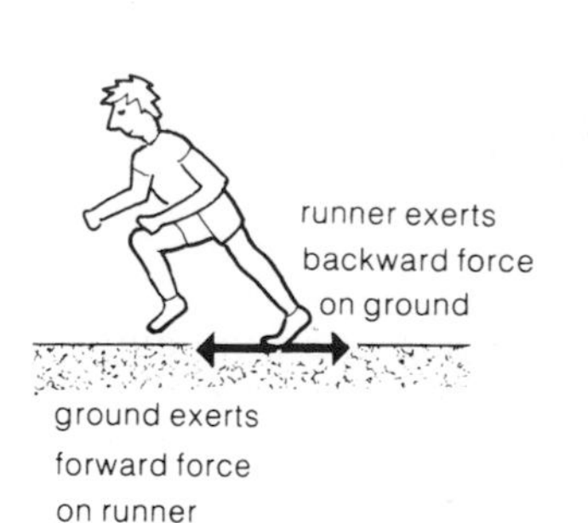

Figure 3b Here too, the forces are equal and opposite

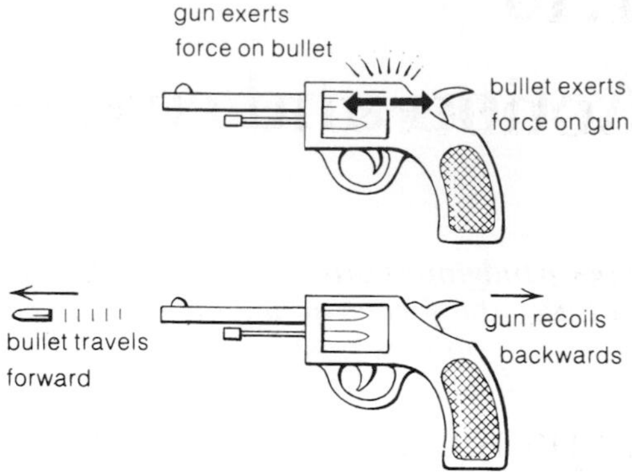

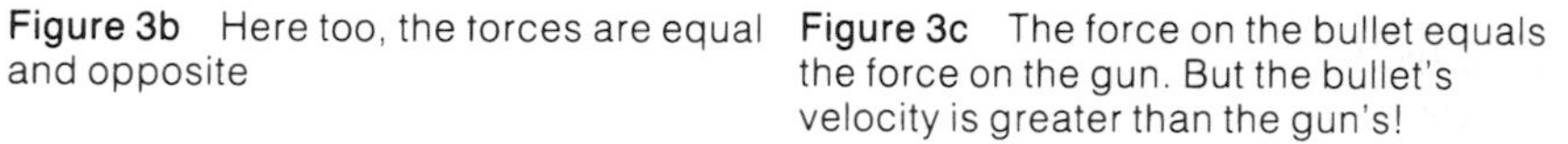

Figure 3c The force on the bullet equals the force on the gun. But the bullet's velocity is greater than the gun's!

Examples of action-reaction pairs are given in figures 3a, 3b, and 3c.

Figure 3a seems to imply that the Earth moves upwards to meet a falling object. This is exactly what happens. The Earth is so massive, however, that its upward acceleration is far too small to be measured. A similar argument applies in figure 3b – the Earth moves backwards!

In any of the diagrams, it doesn't really matter which you call the action force and which the reaction force. One cannot exist without the other.

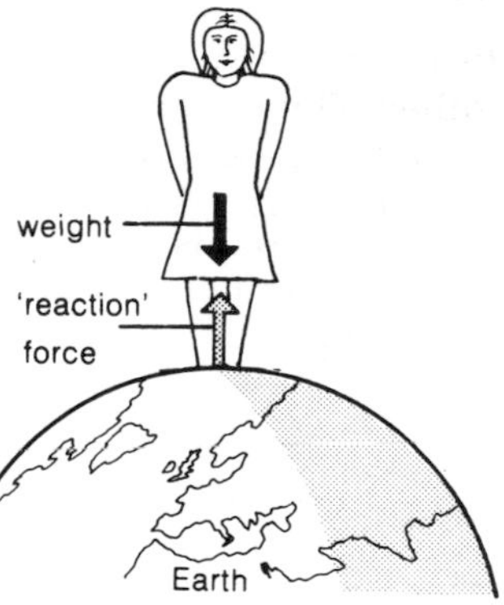

Figure 4a Two forces which are *not* part of an action-reaction pair . . .

Reactions to what? Beware of forces pretending to be what they aren't! Figure 4a shows a girl standing on the Earth's surface. The weight of the girl is balanced by an upward force from the ground. This upward force is commonly called a 'reaction' force. Ignoring complications caused by the Earth's rotation, the two forces *are* equally sized and opposite in direction, but they do *not* belong to the same action-reaction pair. Forces of an action-reaction pair always act on *different* objects, whereas the two forces shown are acting on the same object.

Figure 4b gives a more detailed picture of what is actually happening. Two pairs of forces are shown, one pair acting between the girl and the Earth, and the other pair acting between her feet and the ground. All four forces are equal, but two of them are frequently left out of force diagrams.

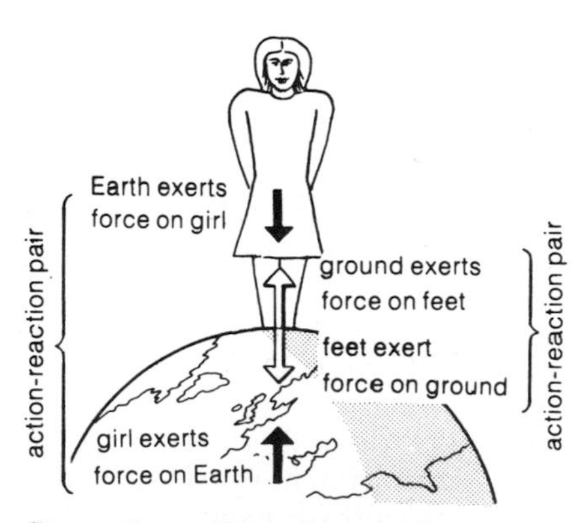

Figure 4b . . . there are actually two action-reaction pairs, as shown

Questions

1 A woman weighs 500 N when standing on the Earth's surface. What upward gravitational force does she exert on the Earth?

2 When a gun is fired, it exerts a forward force on the bullet. Why does the gun recoil backwards?

3 In figure 3b, the forces on a runner and on the ground are equal. Why does the runner move forwards, yet the ground apparently does not move backwards?

4 Figure 5 shows a stone resting on the ground. What force is the reaction force actually reacting against?

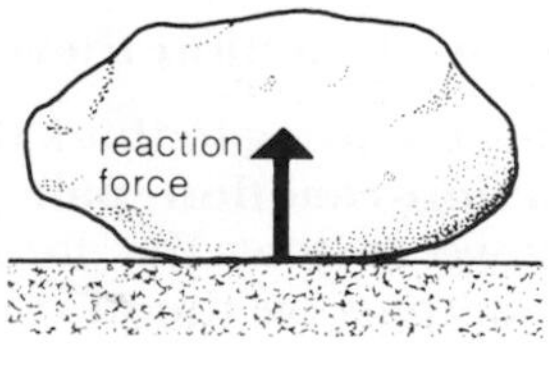

Figure 5

1.11 Conserving momentum

Take the equation force × time = change in momentum. *Combine it with Newton's third law and an important new law emerges.*

Explosions

Two trolleys springing apart might not seem much of an explosion. However, when this does happen a principle is demonstrated which applies whenever matter is pushed apart from other matter.

In the experiment shown in figure 1, one trolley has more mass than the other. The experiment is otherwise the same as that described in the previous section. After the trolleys have sprung apart, it is the trolley with less mass which has more velocity.

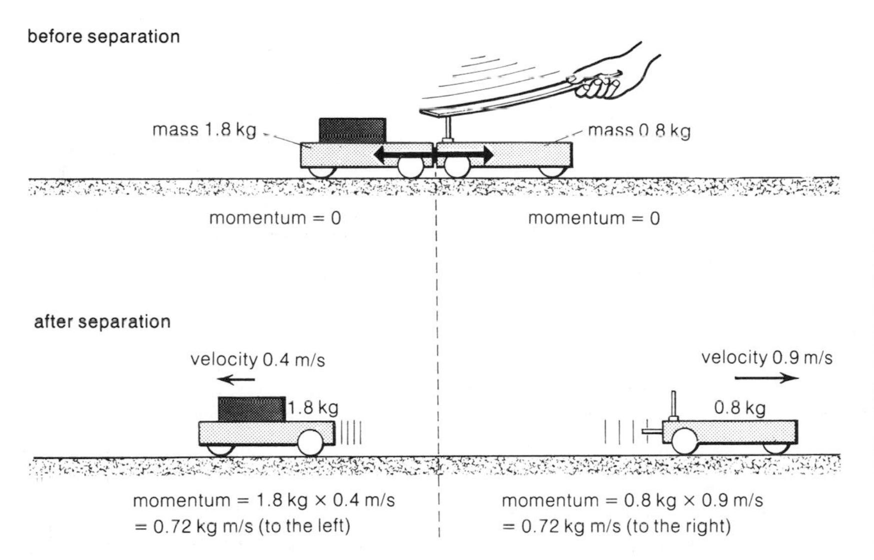

Figure 1 The momentum of both trolleys is of the same magnitude even though their velocities are different

The velocities of the trolleys after separation can be measured using ticker-tape timers. Typical mass and velocity figures are given in figure 1. The figures illustrate a rule which applies in all such experiments:

mass × velocity (trolley travelling to the left)	=	mass × velocity (trolley travelling to the right)
or, change in momentum (to the left)	=	change in momentum (to the right)

The reason each trolley gains the same amount of momentum can be found in the equation force × time = change in momentum. According to Newton's third law, equal forces act on both trolleys. The forces act for the same time, so the change in momentum must be the same in each case.

The law of conservation of momentum

Momentum is a vector quantity, so its direction must be allowed for. Using the usual convention on the figures from the experiment on the previous page:

momentum of the right-hand trolley after separation is 0.72 kg m/s

momentum of the left-hand trolley after separation is −0.72 kg m/s

The total momentum of both trolleys after separation is therefore zero, i.e. exactly what it was before they separated. This is an example of the **law of conservation of momentum**, which states:

When two or more objects act on each other, their total momentum remains constant, provided no external forces are acting.

Collisions

The law of conservation of momentum applies as much to objects colliding as it does to those pushing each other apart. Take the case of two trolleys travelling towards each other as shown in figure 2:

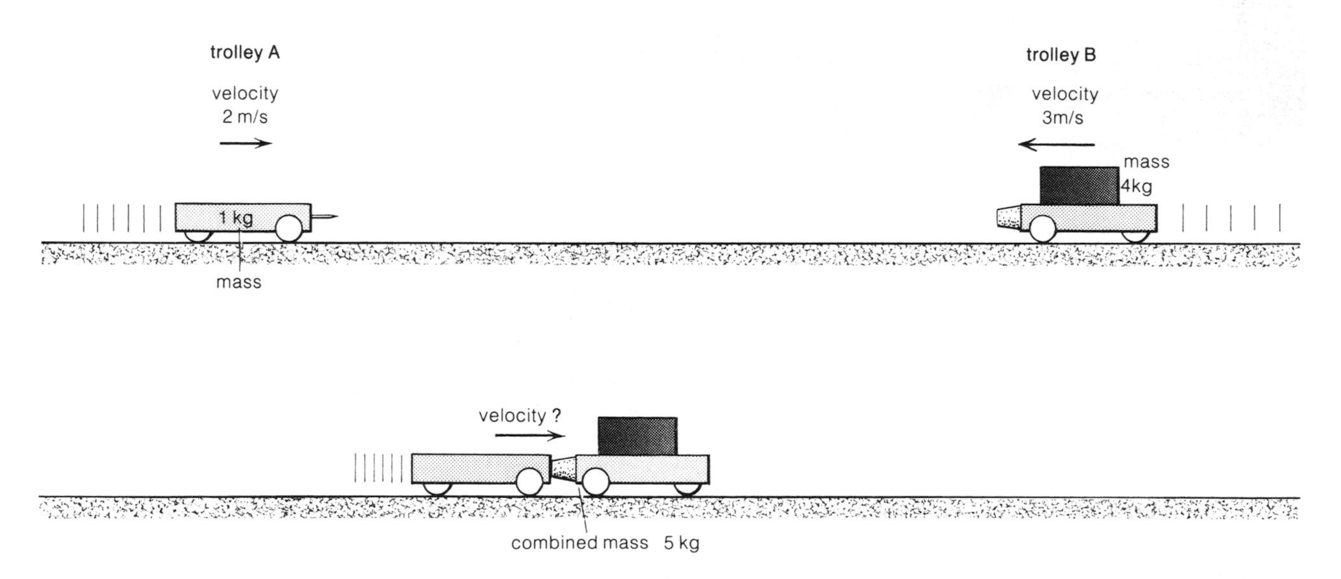

Figure 2 Conservation of momentum during a collision

When the trolleys collide, the needle attached to one trolley sticks in the cork fixed to the other. The trolleys end up with their combined masses travelling at the same velocity. If velocity and momentum to the right are regarded as positive (+):

Before the collision

the momentum of trolley $A = 1\,\text{kg} \times 2\,\text{m/s} \quad = \quad 2\,\text{kg m/s}$

the momentum of trolley $B = 4\,\text{kg} \times (-3\,\text{m/s}) = -12\,\text{kg m/s}$

$\therefore$ the total momentum of trolleys A and B $= -10\,\text{kg m/s}$

During the collision, the total momentum is unchanged, so:

After the collision

the total momentum of trolleys A and B	$= -10\,\text{kg m/s}$
combined mass × velocity	$= -10\,\text{kg m/s}$
5 kg × velocity	$= -10\,\text{kg m/s}$
∴ velocity of trolleys	$= -2\,\text{m/s}$

The trolleys have a velocity of 2 m/s to the left.

Conserving momentum in sport

In some sports, a practical understanding of the law of conservation of momentum can be useful. In golf, for example, the mass of the clubhead has a significant effect on the speed of the ball after impact, and similar ideas apply in cricket. In snooker, momentum problems tend to be more complex because the angles at which balls travel after a collision are such that momentum is conserved in all directions. Sportsmen and women of course solve such problems by experience rather than calculation.

Putting theory into practice: momentum is conserved in every collision

Questions

(Assume frictional forces are negligible in every case)

1 A trolley of mass 2 kg rests next to a trolley of mass 6 kg on a flat table. When a spring-loaded piston on one trolley is released, the trolleys are pushed apart. The lighter trolley travels at 6 m/s.
What is the momentum of each trolley? What is the velocity of the heavier trolley? What is the total momentum of the trolleys?
What average force pushed the trolleys apart if they took 0.5 s to separate completely?

2 A 16 kg mass travelling to the right at 5 m/s collides with a 4 kg mass travelling to the left at 5 m/s.
What is the momentum of each mass? What is their total momentum?
If the masses stick together when they collide, what is their final velocity?
Is this velocity to the left or to the right?

3 An object of mass 10 kg collides with a stationary object of mass 5 kg.
If the objects stick together and move forward at a velocity of 4 m/s, what was the original velocity of the moving object?

1.12

Rockets and jets

How can a rocket accelerate in space if there is nothing for it to push against? The trolley experiment in the previous section gives a clue to the answer.

Figure 1 shows an astronaut 'floating' in space. When they exert a force on the spacecraft an equal and opposite force acts on them. The spacecraft travels to the left; the astronaut travels to the right with equal momentum. To travel quickly, the spacecraft needs to have as much momentum as possible.

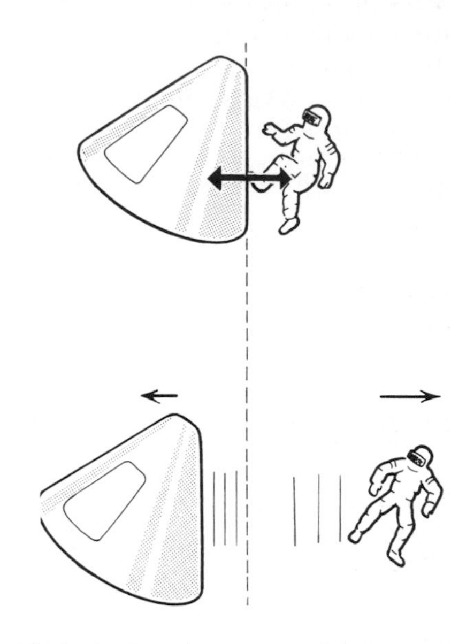

Figure 1 In order to move, the astronaut needs something to push against

The rocket engine

A rocket engine makes use of the principle just described. It doesn't push out astronauts of course, but it does push out large masses of gas at very high velocity.

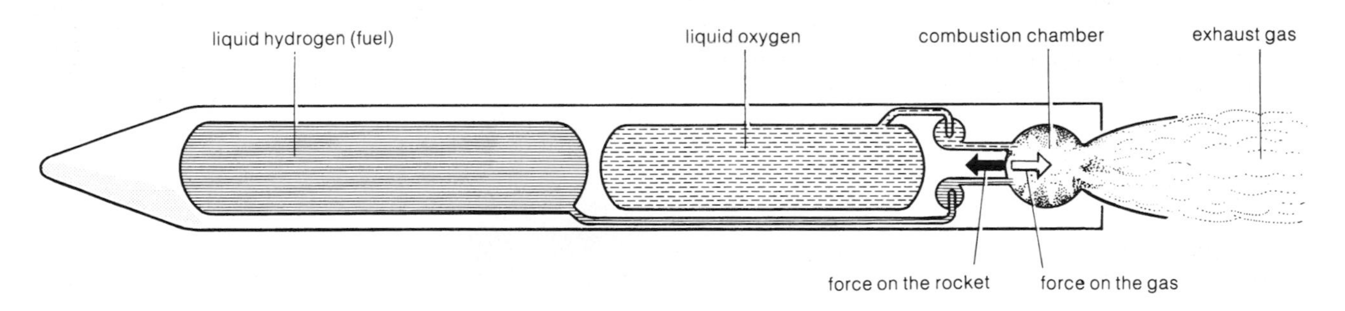

Figure 2a A rocket engine uses its own exhaust gases to push against

The rocket in figure 2a carries supplies of hydrogen and oxygen in cold liquid form. In the combustion chamber, the hydrogen fuel burns violently in the oxygen, and the gas which forms expands rapidly in the intense heat. The expansion forces the hot gas out of the exhaust nozzle at a high velocity. As the gas is given momentum in one direction, the rocket gains momentum in the opposite direction. A rocket engine *does* have something to push against – its exhaust gas.

The jet engine

Like rocket engines, jet engines produce a forward force by pushing out large masses of gas behind them. Unlike rocket engines however, they will not work in space because they need a supply of air. The air is required for two reasons: it provides the oxygen needed for the fuel to burn, and it provides mass for the engine to push out. In large quantities, air is heavy stuff. There is probably more than 80 kg of air in your sitting room at home, and a large jet engine draws in more than three times that quantity every second.

Figure 2b The engines of a Saturn Five rocket

A simple type of jet engine is shown in figure 3a. The compressor is like a series of huge fans, accepting air at the front and forcing it into the combustion chamber at high pressure. In the combustion chamber, kerosene fuel burns fiercely in the compressed air to form hot gas which expands, forcing itself out of the exhaust nozzle at high velocity and exerting a forward force on the engine in the process.

Before leaving the nozzle, the gas passes through the turbine. The turbine blades spin round in the fast-moving gas, and the rotation is used to turn the compressor. To start the engine, the compressor has to be driven by a separate motor.

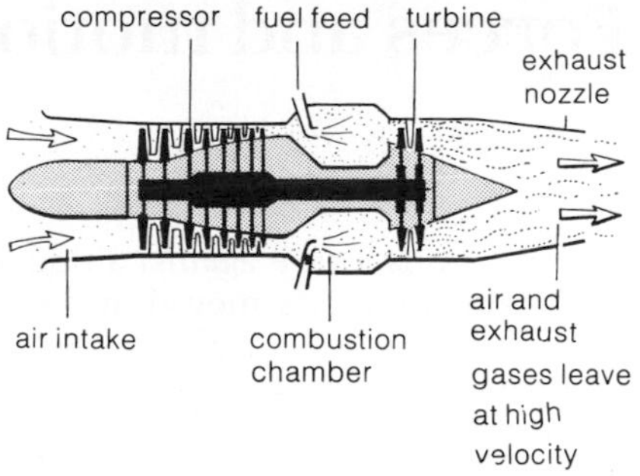

Figure 3a A jet engine uses air both for burning the fuel, and to push against

Calculating engine thrust

Example *The rocket in figure 4 is pushing out exhaust gas at the rate of* 100 kg/s. *The velocity of the gas is* 200 m/s. *Calculate the forward force, or thrust, on the rocket.*

By Newton's third law, the forward force on the rocket is equal to the force pushing out the exhaust gas to the rear. By Newton's second law, this force is equal to the momentum gained per second by the gas.

Every second, 100 kg of gas increases its velocity from 0 to 200 m/s; every second, the gain in momentum = mass × velocity change
= 100 kg × 200 m/s
= 20 000 kg m/s
the gain in momentum per second = 20 000 kg m/s per s
∴ the forward force on the rocket = 20 000 N

Figure 3b A Rolls-Royce jet engine

Solving by equation An alternative method of setting out this calculation starts with the equation $F = \frac{mv - mu}{t}$

In 1 s, 100 kg of gas is accelerated from zero velocity to a velocity of 200 m/s. In which case: $t = 1$ s; $m = 100$ kg; $u = 0$; and $v = 200$ m/s.

the force on the gas, $F = \frac{(100 \times 200) - (100 \times 0)}{1}$ N

$= 20\,000$ N

By Newton's third law, this also equals the forward force on the rocket.

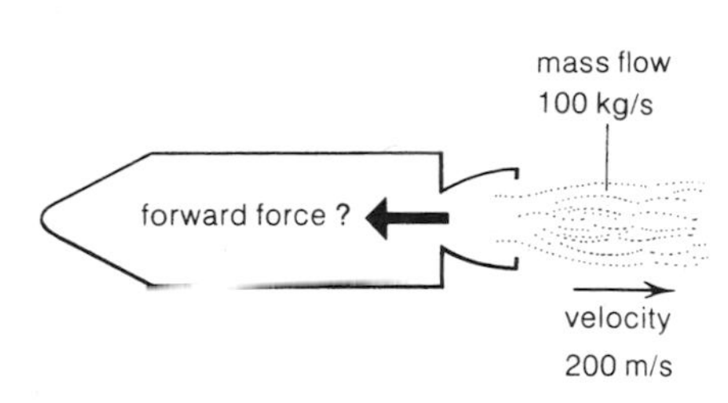

Figure 4

Questions

$g = 10\ \text{m/s}^2 = 10$ N/kg

1 Why does a jet engine need to take in air?

2 A rocket engine can work in space but not a jet engine. Explain why.

3 A jet engine pushes out 50 kg of exhaust gas every second, at a velocity of 150 m/s. What forward force does it produce?
If the engine pushed out twice the mass of gas at half the velocity, what force would it then produce?
A rocket engine produces a thrust of 16 000 N. If it pushes out exhaust gas at the rate of 80 kg/s, what is the velocity of the gas leaving the exhaust nozzle?

4 A rocket, taking off vertically, pushes out 25 kg of exhaust gas every second at a velocity of 100 m/s. What is the upward force on the rocket?
If the total mass of the rocket is 200 kg, what is the downward gravitational force acting?
What is the resultant upward force on the rocket?
What is the upward acceleration?
Calculate the acceleration of the rocket when it has burned off 100 kg of fuel.

Further questions
Forces and motion: part A

1 A man runs a race against a dog. Figure 1 is a graph showing how they moved during the race.

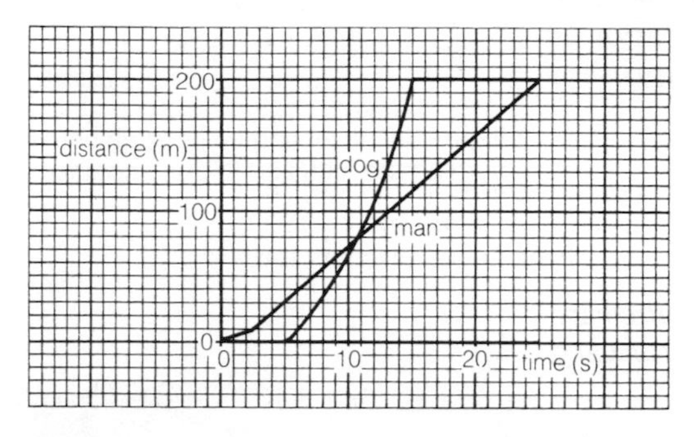

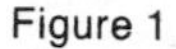
Figure 1

a) What was the distance for the race?
b) After how many seconds did the dog overtake the man?
c) How far from the start did the dog overtake the man?
d) What was the dog's time for the race?
e) Use the equation $v = \frac{s}{t}$ to calculate the average speed of the man.
f) After 8 seconds is the speed of the man increasing, decreasing or staying the same?
g) What is the speed of the dog after 18 seconds?

(NEA)

Figure 2 shows a speed-time graph for a bus as it travels from one bus-stop to the next.

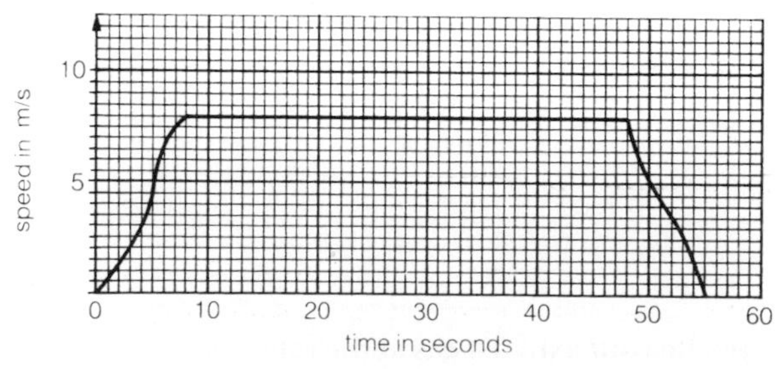

Figure 2

a) How long did the bus take for this journey?
b) For part of the journey the speed was constant.
   i) What was this speed?
   ii) For how long did the bus travel at a constant speed?
   iii) How far did the bus travel while the speed was constant?
c) *Estimate* the length of the journey. (SWEB/SEG)

3 This question is about SPEED and ACCELERATION
A cycle track is 500 metres long. A cyclist completes 10 laps (that is rides completely round the track 10 times).
a) How many kilometres has the cyclist travelled?
b) On average it took the cyclist 50 seconds to complete one lap (that is, to ride round just once).
   i) What was the average speed of the cyclist?
   ii) How long in minutes and seconds did it take the cyclist to complete the 10 laps?
c) Near the end of the run the cyclist put on a spurt. During this spurt it took the cyclist 2 seconds to increase speed from 8 m/s to 12 m/s. What was the cyclist's acceleration during this spurt? (SEG)

4 In which one of these situations must frictional forces be kept low?
A Walking along a road
B Ski-ing down a snow slope
C Leaning a ladder against a wall
D Designing brake blocks for a bicycle (SEG)

5 A solid ball is taken from the Earth to the moon. On the moon will the ball have a very different:
A volume
B density
C mass
D weight? (SEG)

6 In figure 3, a stone at the end of a light string is whirled in a horizontal circle.

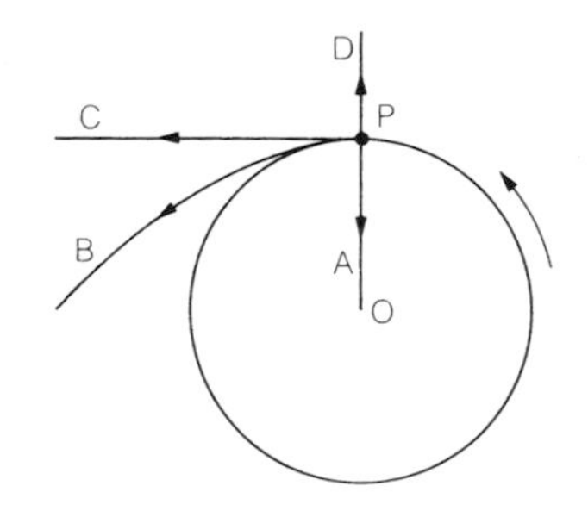

Figure 3

If the string snaps when the stone is at P, will the stone follow line
A PA **B** PB **C** PC **D** PD (SEG)

7 In figure 4 two forces act on a block of mass M. Which arrangement gives the block the greatest acceleration?

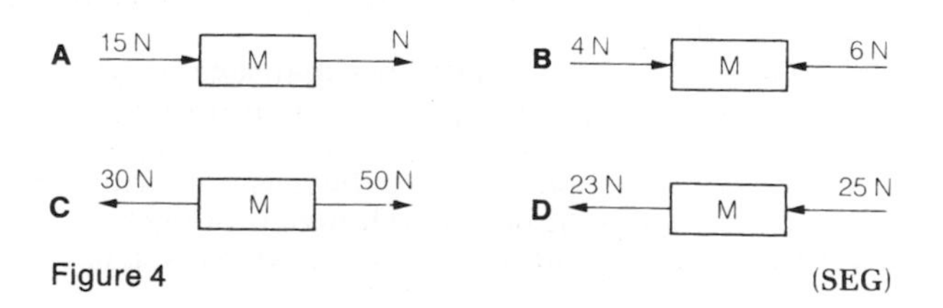

Figure 4 (SEG)

8 This question is about FORCE and ACCELERATION. The driver of a car moving at 20 m/s along a straight level road applies the brakes. The car decelerates at a steady rate of 5 m/s^2.

a) How long does it take the car to stop?
b) What kind of force slows the car down?
c) Where is this force applied?
d) The mass of the car is 600 kg. What is the size of the force slowing the car down? (SEG)

9 In figure 5, a car engine is leaking oil. The drops hit the ground at regular intervals, one every 2.5 seconds. The diagram below shows the pattern of the drops it leaves on part of its journey.

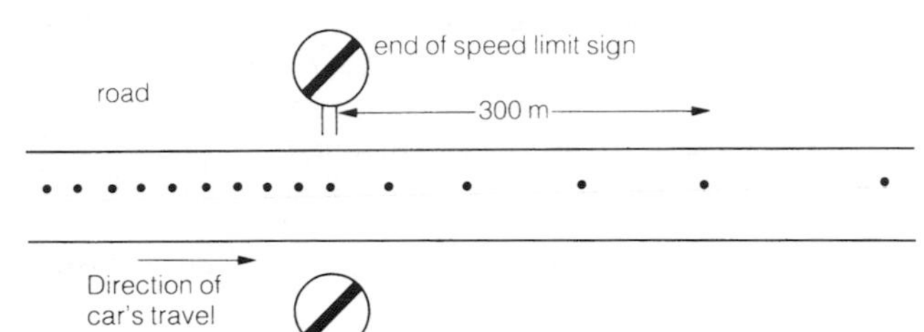

Figure 5

a) What can you say about the speed of the car before it reaches the signs?
b) If the car is travelling at 10 m/s calculate the distance between the drops on the road before it reaches the signs.
c) How can you tell that the car is accelerating after it passes the signs?
d) After the car passes the signs the fourth drop falls at a distance of 300 m past the signs. Calculate the acceleration by using the formula $s = u\,t + \frac{1}{2}\,a\,t^2$. (LEAG)

10 In figure 6, a front-wheel drive car is travelling at a constant velocity. The forces acting on the car are shown in the diagram below. *F* is the push of the air on the car and *P* is the total upward force on both front wheels

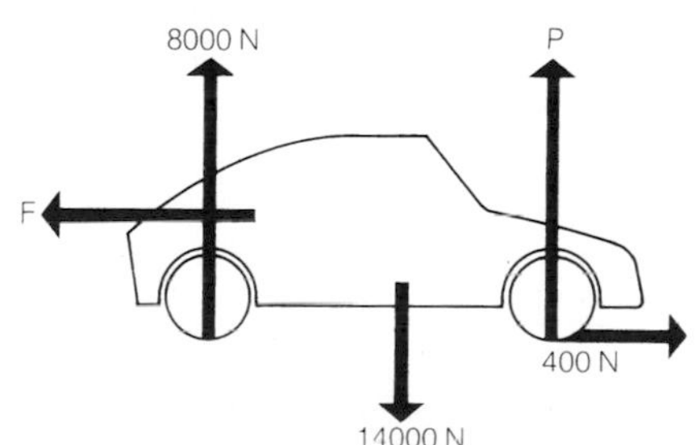

Figure 6

a) Name the 400 N force to the right.
b) Taking the weight of 1 kg to be 10 N, calculate the mass of the car.
c) The 400 N force to the right is suddenly doubled.
   i) At the instant this happens, what is now the size of the net (i.e. resultant) force moving the car forward?
   ii) Explain how this causes the car to accelerate.
   iii) Calculate this acceleration. (LEAG)

11 A girl wearing a parachute jumps from a helicopter. She does not open the parachute straight away. The table shows her speed during the 9 seconds after she jumps.

time in seconds	0	1	2	3	4	5	6	7	8	9
speed in m/s	0	10		30	40	25	17	12	10	10

a) Copy and complete the table by writing down the speed at 2 seconds.
b) Plot a graph of speed against time.
c) How many seconds after she jumped did the girl open her parachute? How do the results show this?
d) i) What force pulls the girl down?
   ii) What force acts upwards?
   iii) Which of these forces is larger:
   at 3 seconds?
   at 6 seconds?
   at 9 seconds?
e) How will the graph continue after 9 seconds if she is still falling?
f) The girl makes a second jump with a *larger area* parachute. She falls through the air for the same time before opening her new parachute.
   How will this affect the graph:
   i) during the first four seconds?
   ii) after this? (SWEB/SEG)

12 The displacements against time graphs in figure 10 shows the motion of a motor car.

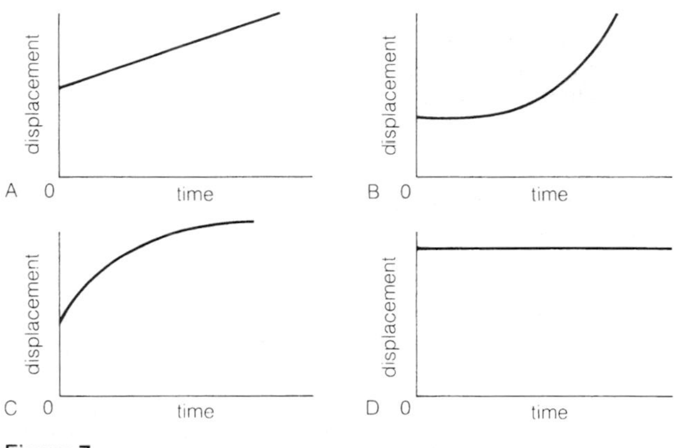

Figure 7

Which graph shows the car stopped at traffic lights? Which graph shows the car moving away from these traffic lights at a steady speed? (LEAG)

Further questions

Forces and motion: part B

Some questions require a knowledge of concepts covered in earlier units

$g = 10\,m/s^2 = 10\,N/kg \quad \sqrt{0.4} = 0.63$

1 Use the graph in figure 1 to find:
  a) the total time taken for the journey,
  b) the maximum velocity,
  c) the acceleration during the first part of the journey,
  d) the total distance travelled. (WJEC)

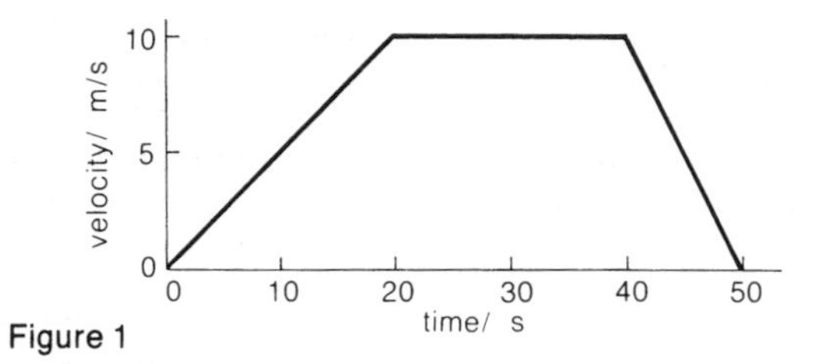

Figure 1

2 An object falls down a slope on to a friction free horizontal surface in the way shown in figure 2. The position of the object is shown every 1/8th of a second and the distances are marked in centimetres from the start.
  a) The object is accelerating whilst on the slope. Explain how you can see from the diagram that this is so.
  b) What happens to the speed of the object once it has reached the horizontal surface?
  c) Taking readings from figure 2, calculate the speed of the object on the horizontal surface
  d) Calculate the approximate value of the maximum speed reached during the whole journey.
  e) Sketch a graph to show how the velocity of the object varies with time over the whole journey. (SWEB)

3 a) Sketch a velocity-time graph for a car moving with uniform acceleration from 5 m/s to 25 m/s in 15 seconds.
  b) Use the sketch graph to find values for i) the acceleration, ii) the total distance travelled during acceleration. Show clearly at each stage how you used the graph. (JMB)

4 An electric train moves from rest with a uniform acceleration of $1.5\,m/s^2$ for the first 10 s and continues accelerating at $0.5\,m/s^2$ for a further 20 s. It continues at constant speed for 90 s and finally takes 30 s to decelerate uniformly to rest.
  a) Draw a graph of speed against time for the journey.
  b) From your graph, or otherwise, deduce the total distance travelled.
  c) What is the average speed of the train for the whole journey? (L)

5 Figure 3 represents the velocity-time graph for a ball which is rolled up a uniform slope.

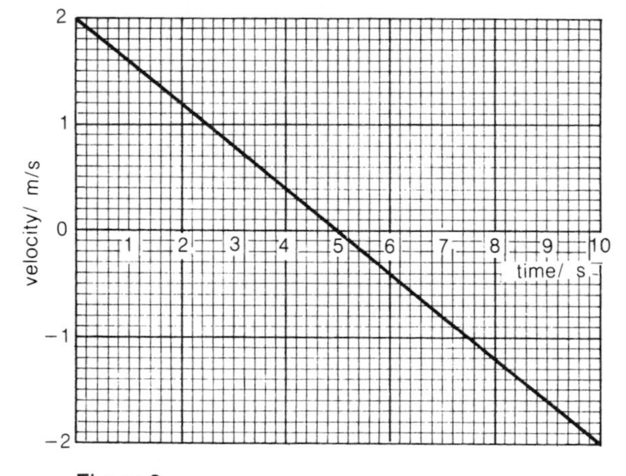

Figure 3

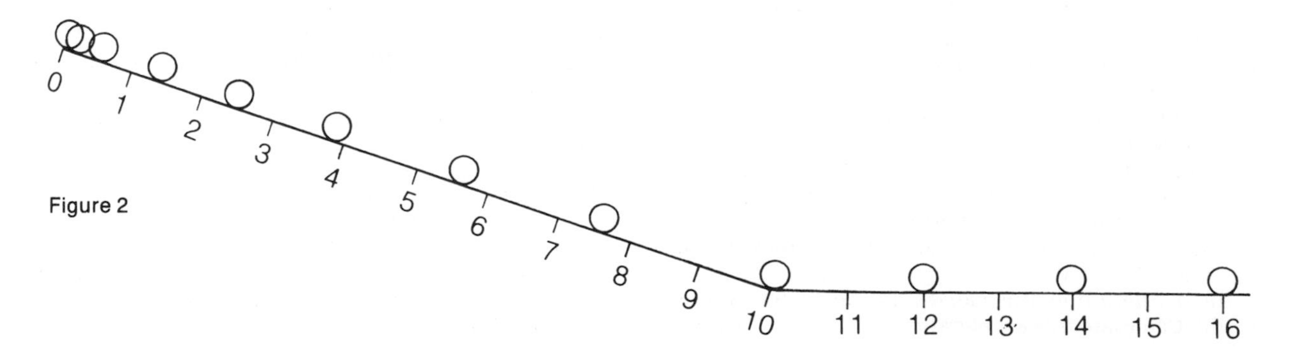

Figure 2

a) How far does the ball move in the first 5 seconds?
b) What is the velocity of the ball after 8 seconds?
c) What is the position of the ball, relative to its starting-point, after 10 seconds?
d) What is the acceleration of the ball? (O)

6 To determine the acceleration of a freely falling body, a ball bearing of mass 5 g was dropped from a height of 1.2 m above the floor and the time taken for it to reach the floor found to be 0.5 s.
a) Calculate the value of its acceleration.
b) Describe briefly how you would measure the time of fall accurately.
c) If a ball bearing of mass 10 g was used instead would the value of the acceleration be different? Give reasons for your answer. (O)

7 A simple pendulum is set swinging so that the closest distance the bob approaches the floor is 2 m. At the instant at which the bob is moving with maximum speed of 0.5 m/s the cord snaps. Calculate:
a) the time it takes the bob to hit the floor, assuming that this is equal to the time it would take to reach the floor falling vertically from rest.
b) the horizontal distance travelled by the bob between the cord breaking and the bob hitting the floor. (L)

8 The metal spheres in figure 4 are released from rest at the same time. What time is there between A striking the ground and B?

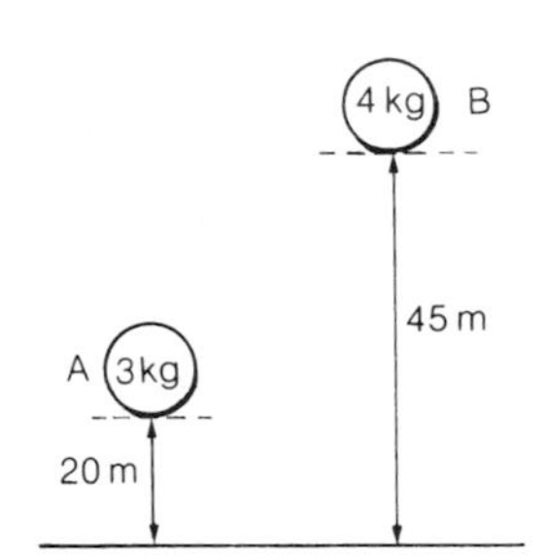

Figure 4

9 A block of low density plastic foam (e.g. polystyrene) and a block of lead each of mass 0.2 kg are released together and fall down a deep well with water at the bottom. The lead block is found to take 4.0 s to reach the water surface.
a) State and explain any difference in the times of arrival of the two masses at the water surface.
b) Calculate: (i) the speed with which the lead block strikes the water, (ii) the distance of the water surface from the top of the well.
c) Sketch the shape of a distance-time graph for the falling lead block from the instant the block is released until it reaches the surface of the water. (CLES)

10 An object 20 metres above the ground is thrown horizontally with a velocity of 2.5 m/s.
a) How long does the object take to reach the ground?
b) What horizontal distance has it then covered?

11 Describe an experiment to show how the force applied to a body is related to the acceleration it produces. Sketch the apparatus you would use, state the observations you would make and show how you would use these observations to obtain the relation between force and acceleration. (JMB)

12 A car of mass 800 kg is being towed by a truck, using a rope. The force in the horizontal rope, when the car is pulled along a level road at a constant speed, is 100 N.
a) Explain why a force is needed to tow the car at a constant speed.
b) If the truck now accelerates at 0.5 m/s^2, what is the initial value of the total force in the rope?
c) The breaking force for the rope is 1000 N. What is the maximum acceleration of the truck from this speed on a level road if the rope is not to break? (O)

13 Explain the difference between the mass and the weight of a body. State how these quantities change when the same object has its mass and weight measured, first on the Earth and then on the Moon. (JMB)

14 An object put on a beam balance gives the same reading no matter where it is on the Earth's surface but if it is on a very sensitive spring balance the reading varies from place to place. Explain why this is so. (L)

15 A projectile is fired vertically with a velocity of 20 m/s from the surface of the Moon where the acceleration of free fall is only one-sixth that on Earth. Calculate the greatest height reached and the time to reach this height. Give two reasons why g might be lower on a satellite or different planet. (SUJB)

16 A person of mass 60 kg stands on a spring weighing machine inside a lift which is accelerating upwards at 3 m/s^2. Calculate the reading (in newtons) of the weighing machine. (L)

17 A ball of mass 0.5 kg is dropped from a great height. Draw diagrams to show the force(s) acting on the ball.
a) just as it starts to fall.
b) when it is falling (through air) at its terminal velocity.
Indicate the magnitude of any forces you have shown.

18 A small satellite S moves with constant speed in a circular path round the Earth. Show on a simple sketch the directions of the velocity and acceleration of S at some instant. How does the force on S depend on a) the mass of the Earth, b) the distance from S to the Earth's centre? (SUJB)

19 A satellite is about 42 000 km from the centre of the Earth round which it travels once a day. Comment upon the truth (or otherwise) of the following four statements about it.

a) If it were not above the Earth's atmosphere, air resistance would slow it down and it would fall back to the Earth.

b) If it were not beyond the Earth's gravitational field the pull of gravity would cause it to fall back to the Earth.

c) Its solar-powered motors are continually running in order to keep it moving at a steady speed in its orbit.

d) If the satellite's mass had been larger, then it would have had to travel faster in its present orbit or be put into a more distant orbit to prevent it from falling back to the Earth. (O)

20 Figure 5 shows a small satellite S which is in orbit around the Earth.

Figure 5

a) Redraw the diagram, showing the direction of the force acting on the satellite. What causes this force?

b) A rocket on the satellite is fired so that the forward speed of the satellite is suddenly increased. Explain why the satellite cannot stay in the same orbit. (CLES)

21 The rate at which momentum changes is often more important than the size of the change. Bearing this in mind, explain how a light hammer used to drive a nail into a plank can exert a much greater force on the head of the nail than that produced by resting a large mass on the nail. (L)

22 State Newton's second law of motion, and explain how it is used to define a force of one newton.
A car of mass 1200 kg is brought to rest from a speed of 20 m/s by a constant force of 3000 N. Calculate the change of momentum produced by the force and the time taken to bring the car to rest. (WJEC)

23 Figure 6 represents the velocity-time graph for the first 50 seconds of the motion of a truck of mass 2000 kg that starts from rest.

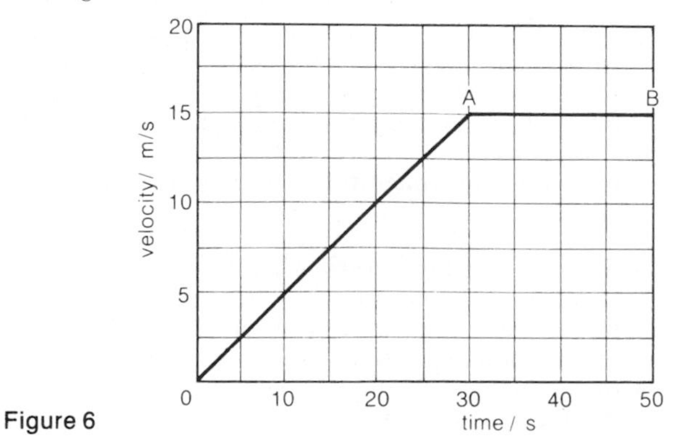

Figure 6

a) Over what part of the graph is the acceleration uniform but not zero, and what is its value?

b) What is the total distance travelled during the 50 seconds?

c) What is the average speed over the 50 seconds?

d) What is the momentum of the truck 50 seconds after the start of the motion? (O)

24 Figures 7a and 7b show a drawing pin which rests on the deck of a record player. The deck is horizontal and is rotating at a constant rate. Which of the following quantities associated with the pin are changed when the drawing pin has been carried from A to B: speed, velocity, momentum? (O&C)

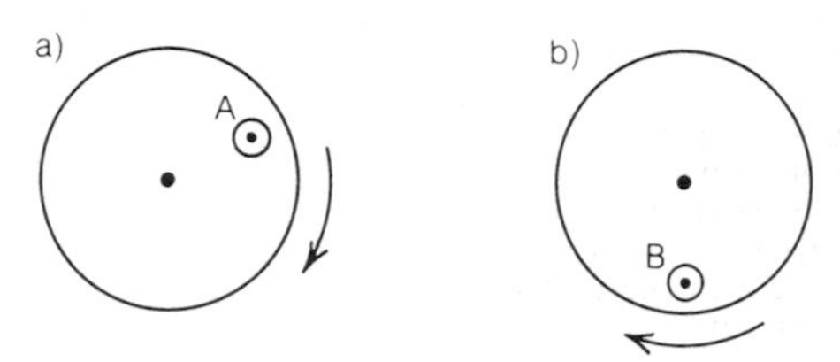

Figure 7

25 A hard ball, mass 0.50 kg, travels along a straight line with a speed 1.5 m/s. It hits a fixed sheet of metal and rebounds back along the same path with the same speed. Calculate the average force exerted on the metal during the collision if the ball remains in contact with the metal for 1.0 millisecond. (O&C)

26 An arrow of mass 0.1 kg is shot into an apple of mass 0.2 kg which is resting on the top of a wall. At the moment of impact, the arrow is travelling horizontally at 15 m/s. Calculate the common speed immediately after the impact. (AEB)

27 A stationary gun of mass 50 tonnes fires a shell of mass 100 kg with a velocity of 600 m/s. Calculate the initial velocity of recoil of the gun. (1 tonne = 1000 kg) (JMB)

28 A coal-truck of mass 5000 kg and travelling at 5 m/s couples with a stationary truck and they travel off together at 0.5 m/s. What is the mass of the truck which was stationary before impact? (O)

29 A snooker ball B makes an impact with an identical stationary ball C. Before impact B has a velocity of 2 m/s and it follows along the same path with a velocity of 0.2 m/s; calculate the velocity of C after impact. (SUJB)

30 a) State the principle of conservation of linear momentum.

b) Outline briefly (i) the similarity between the operation of a jet motor and of a rocket motor, (ii) the main difference between these two types of motor. How does this difference affect the uses which may be made of the two types of motor? (AEB)

SECTION 2

FORCE, WORK, AND ENERGY

A NEW FORCE

Uri Geller started the spoon-bending craze in the early 1970s. Since then, many others have claimed to be able to bend cutlery by gentle stroking, or even by thought alone. Several serious studies of the phenomenon have been attempted, but the results have been inconclusive. Most scientists dismiss it all as trickery, while the few who have concluded that some mysterious new force is at work have been criticized for their poor experimental methods. 'Why' writes one scientist 'do people including eminent scientists insist on being so gullible?'

Author Perplexed

The author was a strong supporter of the trickery theory – until his two children started to bend their way through the cutlery drawer . . . research continues . . .

A FUELISH PROBLEM

The world's known reserves of oil would fill a giant tank 4 kilometres long by 4 kilometres wide by 4 kilometres deep. But at the present rate of consumption, the oil level would drop 150 metres every year. So the world's oil supply will last . . . er . . .

Crunch! But everything's under control . . .

Yet another test car goes beyond the elastic limit.

Modern cars are designed to collapse! This may seem bad news – until you're in an accident.

The passenger shell of the car shown in the photograph is rigid, but **elastic**. This means that during impact, unless the stress is too great, it will quiver and shake – but will be in the same basic shape after the impact. The metal does not go beyond its **elastic limit**. The front end of the car, however, is *not* rigid – it can and does bend out of shape on impact. It goes well beyond its elastic limit. But as it does so, a large amount of energy is absorbed. This is important because by absorbing energy in this way, it helps prevent the passenger compartment from being put under an impossible stress. The strength of the passenger shell and the bendability of the rest of the car are both safety features which are deliberately engineered into the car design. The only problem comes with the repair bill . . .

Sun-powered flight

Gossamer Albatross was the first man-powered plane to cross the English Channel. Successor to Gossamer Albatross is Solar Challenger, shown below, an electrically powered craft with solar cells on its wings. On a sunny day, it should be capable of flying over 150 kilometres. Designer Paul MacCready says: 'We're trying to call attention to the fact that there is an alternate source of energy of great value. People don't think that with solar power you can actually move things.'

2.1

Combining forces

When several forces act at one point, and you start to add up their effects, you soon discover that two and two don't always make four.

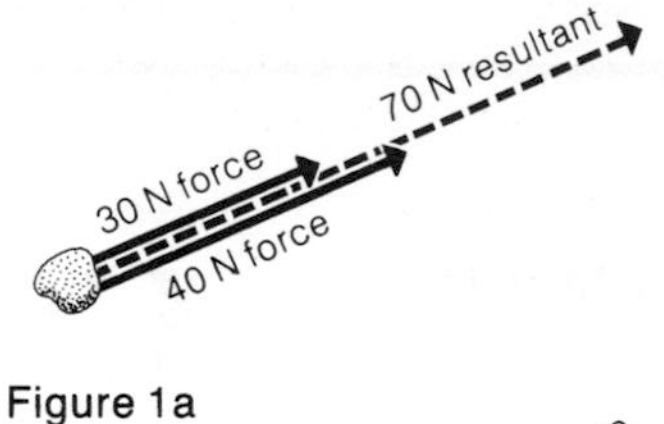

Figure 1a

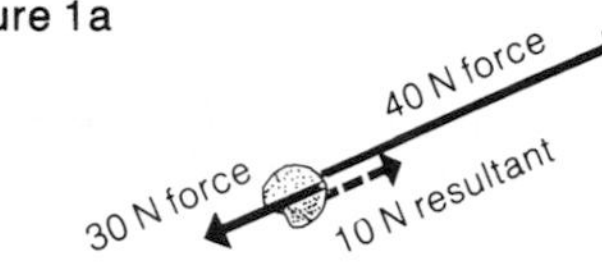

Figure 1b

Figures 1a, 1b Two forces can combine to form a single *resultant* force

Resultant of two forces

In figure 1a, an object is being pulled by two forces, one of 30 N and the other of 40 N. Both forces act through the same point and in the same direction. Together, they have the same effect as a single force of 70 N. This single force is the **resultant** of the two forces.

If the two forces act in opposite directions as in figure 1b, their resultant is a force of only 10 N.

Finding the size and direction of a resultant is not quite so simple if the forces act at an angle to each other as in figure 2. In such cases, the **parallelogram rule** must be used.

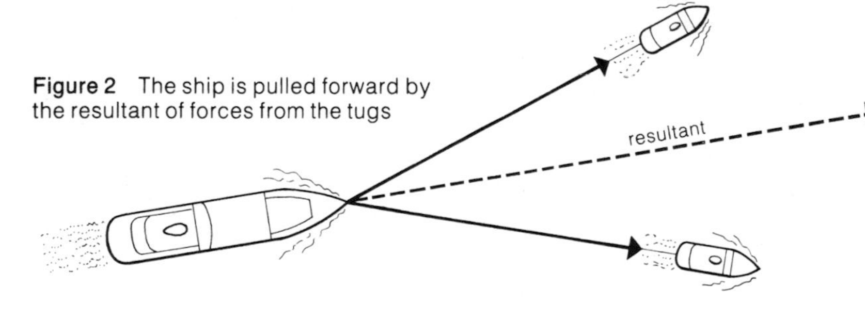

Figure 2 The ship is pulled forward by the resultant of forces from the tugs

The parallelogram rule

This is used to find the resultant of two forces acting at a named point, say point O. In figure 3a the two forces are of 30 N and 40 N, at the angle shown.

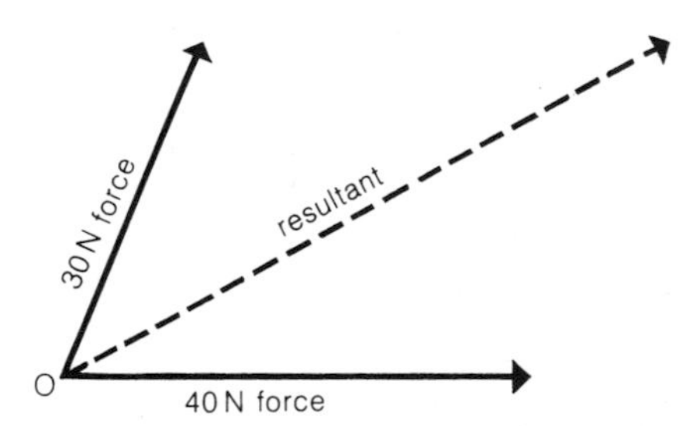

Figure 3a Two forces produce a resultant, size and direction unknown

1 On paper, draw two lines from point O to represent the two forces. The direction of each force must be shown accurately, and the length of each line must be in proportion to the size of the force. This is done in figure 3b, where one millimetre is used for each newton of force.

2 Draw in two more lines to complete a parallelogram.

3 Draw in the diagonal starting at O, and measure its length.

The diagonal represents the resultant of the two forces in both magnitude (size) and direction.

So, in the example given in figure 3, a 30 N force and a 40 N force acting in the directions shown produce a resultant of 60 N at an angle of 26° to the 40 N force.

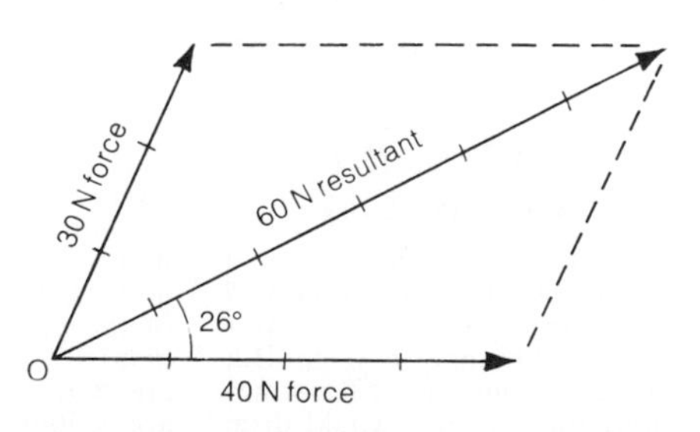

Figure 3b The parallelogram of forces is used to find the resultant

Vectors and scalars

It should be clear by now that it is important to allow for direction when adding forces together. Adding a 30 N force to a 40 N force for example gives a resultant of anything from 10 N to 70 N, depending on the direction of the forces: the actual value is given by the parallelogram rule. Like any quantity which has both magnitude and direction, force is a vector quantity. The parallelogram rule applies to all vector quantities.

Quantities such as volume or mass, which have only magnitude, are known as **scalar** quantities. Adding scalar quantities is easy: a mass of 30 kg added to a mass of 40 kg *always* gives a mass of 70 kg.

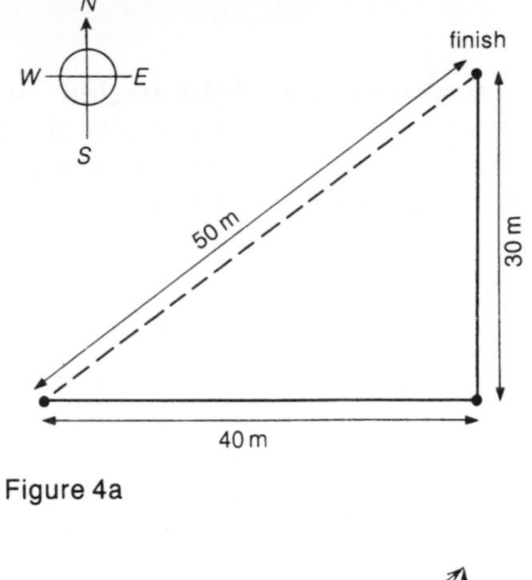

Figure 4a

Why the parallelogram rule works

To understand why the parallelogram rule works, it is best to consider a simple vector quantity such as displacement. Displacement is distance moved in a particular direction. Consider figures 4a, 4b, and 4c in turn:

Figure 4a shows the route taken by a girl who starts at a point O, walks 40 m due East, then walks 30 m due North. A quick calculation using Pythagoras' Theorem shows that she ends up 50 m away from where she started.

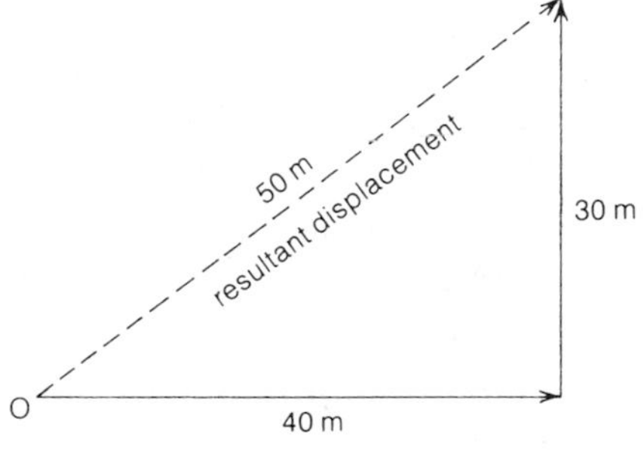

Figure 4b

Figure 4b shows how the journey can be represented using displacement vectors, the length of each line being in proportion to the displacement in each case. A displacement vector of 30 m (due North) has been added to a displacement vector of 40 m (due East) to give a resultant displacement vector of 50 m in the direction shown.

Figure 4c shows figure 4b in another form, this time as a parallelogram. The 30 m and 40 m displacement vectors have been drawn from point O. The diagonal from O gives the resultant 50 m displacement vector.

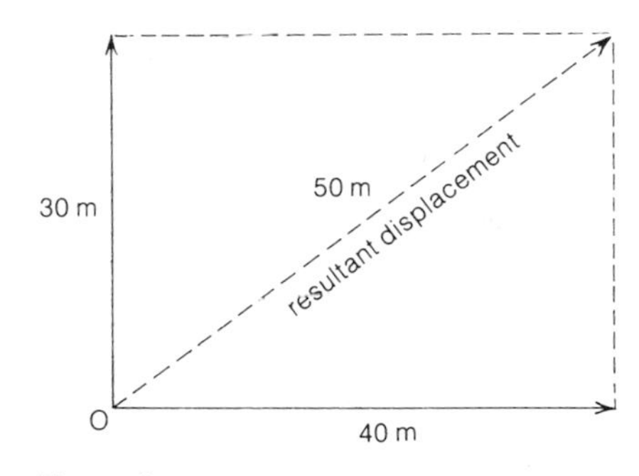

Figure 4c

Figures 4a, 4b, 4c The parallelogram rule works for *all* vectors

Components of a force

At the beginning of the section, two forces acting at a point were replaced by a single resultant. Reversing the process, a single force can be replaced by two forces having the same effect – a force can be **resolved** into two **components**. Some examples of components of a 50 N force are shown in figure 5, though any number of other examples are possible:

Figures 5a, 5b, 5c A force can be resolved into different components

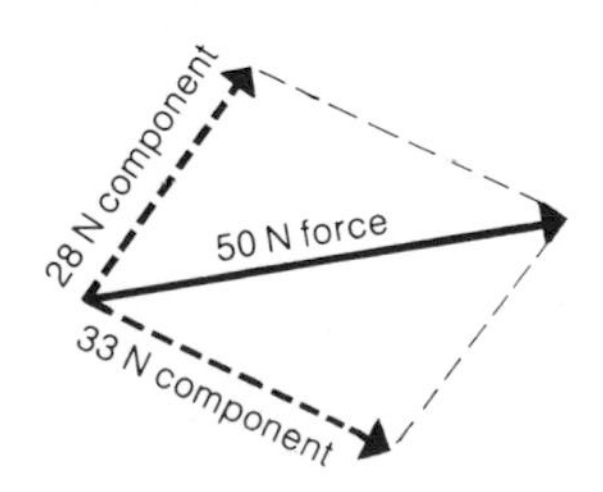

Figure 5a

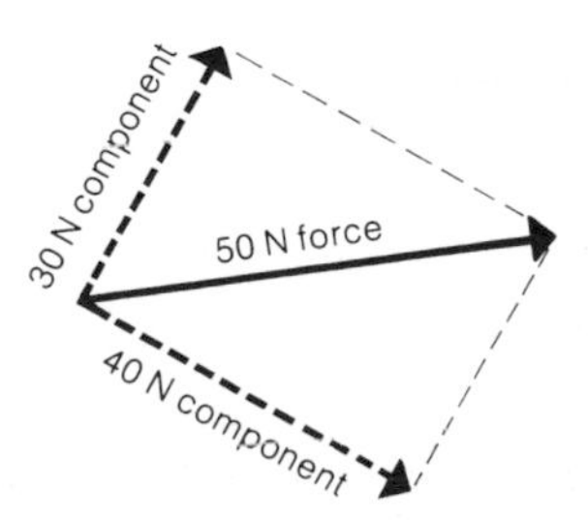

Figure 5b

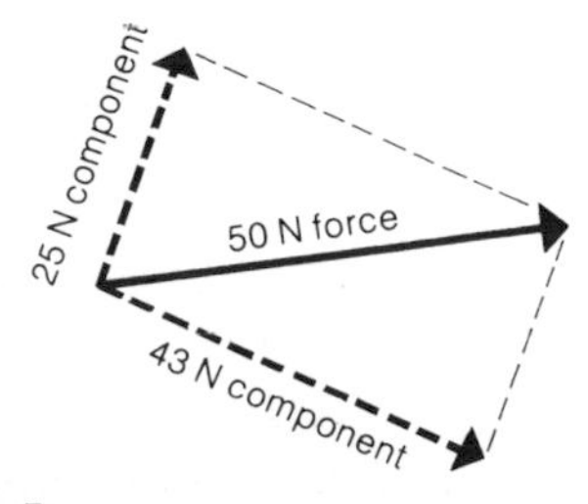

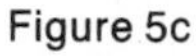
Figure 5c

Components at right angles In working out the effects of a force, it sometimes helps to resolve the force into two components at right angles to each other. The force pulling the dinghy in figure 6 has been resolved into two components. The forward component pulls the dinghy through the water, while the upward component partly lifts the bows out of the water.

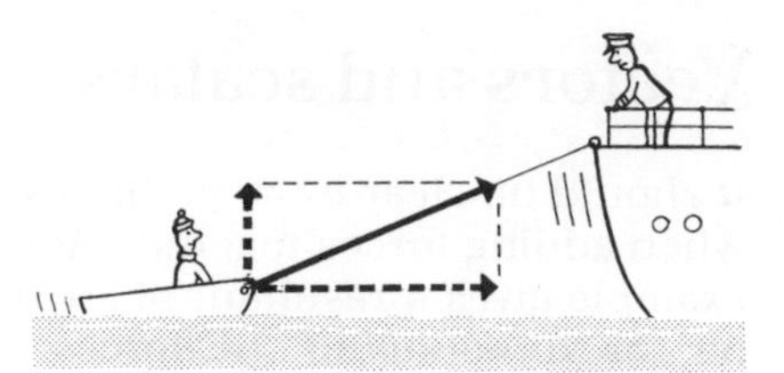

Figure 6 Components at right angles

When components are taken at right angles to each other, it is a relatively simple matter to calculate their values:

The force F in figure 7a has components of $F\cos\theta$ and $F\sin\theta$. You may find it useful to remember these results, but note carefully the position of the angle θ. Both results are easy to prove if you apply a little trigonometry to the triangles in the diagram:

Calling the components F_y and F_x,

$$\cos\theta = \frac{F_x}{F}$$

$$F_x = F\cos\theta$$

A similar argument gives

$$F_y = F\sin\theta$$

In figure 7b, two components of a 50 N force have been calculated. In the example shown, one component has been taken at an angle of 30° to the 50 N force, the other component being at right angles to the first.

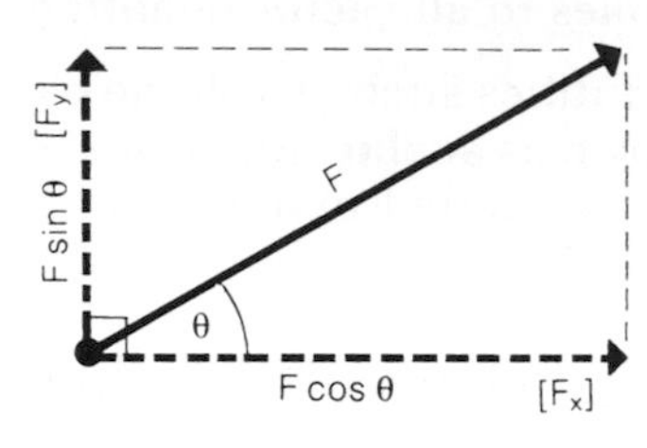

Figure 7a

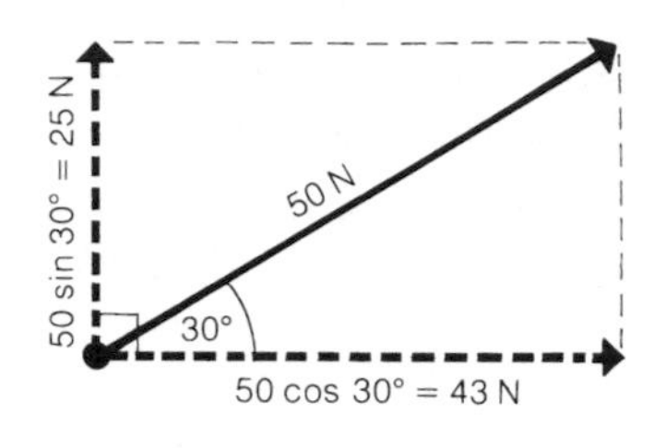

Figure 7b

Figures 7a, 7b Calculating the values of the components of forces

Questions

[$\sqrt{169} = 13$]

1 A 12 N force and a 5 N force both act at the same point.
  a) What is the greatest resultant that these forces can produce?
  b) What is the least resultant that the forces can produce?
  c) If the two forces act at right angles to each other, find by scale drawing or otherwise the size and direction of their resultant.

2 The girl in figure 8 is pushing on the handle of the lawnmower with a force of 100 N.
  a) Find the vertical and horizontal components of the force.
  b) If the lawnmower weighs 300 N, what is the total downward force on the ground when the lawnmower is being pushed?
  c) If the lawnmower is pulled, rather than pushed, using the same force on the handle, what then is the total downward force on the ground?

3 How is a vector quantity different from a scalar quantity?
  Give an example of each type.

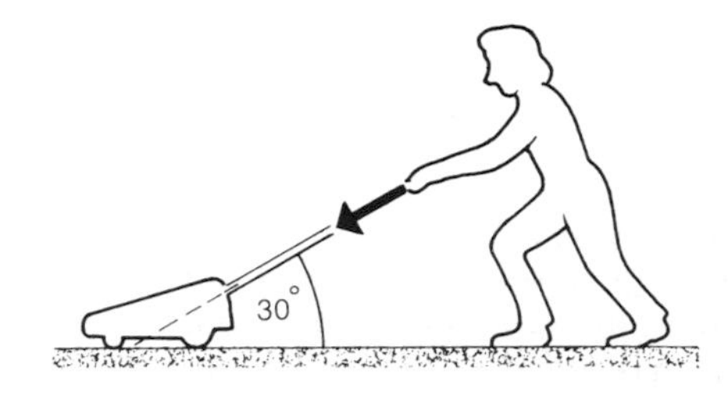

Figure 8

2.2

The turning effects of forces

When forces act at different points on an object, they may make it turn. Or they may not. It all depends on the size, direction, and position of each force.

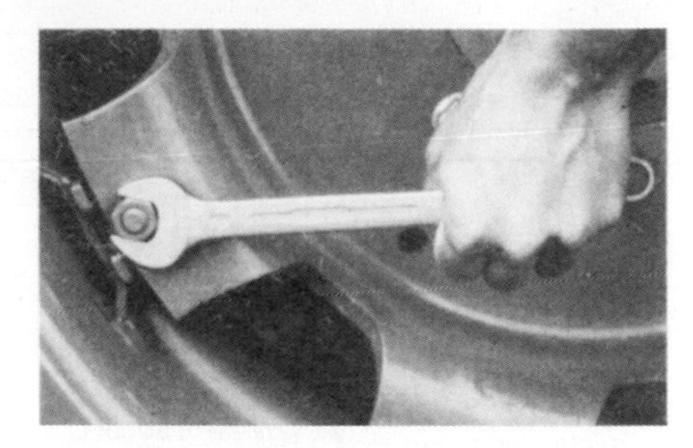

Figure 1 Using a long spanner to make a large turning force

Moment of a force

Once a nut is done up tightly, it is impossible to undo it with your fingers. Using a spanner, as in figure 1, the job becomes much easier. The greater the force, and the further from the nut it can be applied, the greater is the turning effect.

The **moment of a force** is a measure of the turning effect of the force about a particular point. It is defined as follows:

moment of a force about a point = **force** × **perpendicular distance from the point**

If force is measured in N and distance in m, the moment is measured in newton metres [N m]. For example, the force in figure 2a has a moment of 6 N m about the point O.

The force in figure 2b also has a moment of 6 N m about point O. It has a moment of only 3 N m about point P however, and its moment about point Q is zero. As the force is acting at point Q, it is zero distance from it, and has no turning effect about it.

A moment is described as **clockwise** or **anticlockwise** depending on the direction of its turning effect. The force in figure 2b for example has an anticlockwise moment about point O, and also about point P. But it has a clockwise moment about point R.

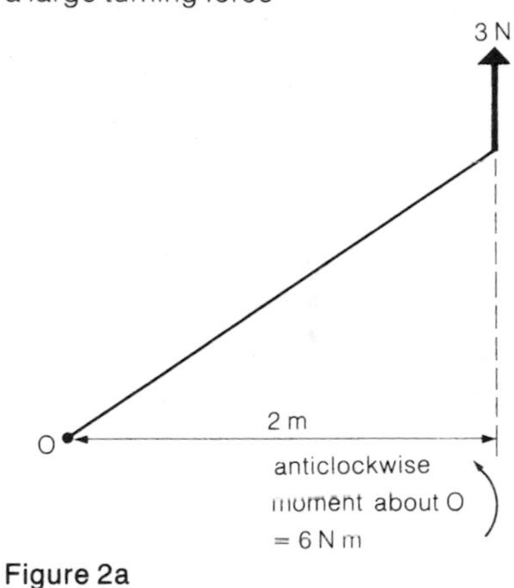

Figure 2a

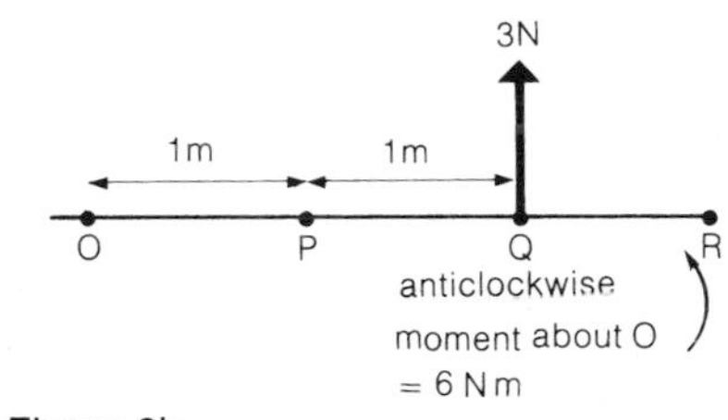

Figure 2b

Figures 2a, 2b The moment of a force varies with the size of the force and its perpendicular distance from the point in question

The principle of moments

Figure 3 shows a bar of negligible weight supported at a point O. Forces have been applied to the bar by hanging standard masses from different points. The positions of the forces have been adjusted so that the bar is in a state of balance or **equilibrium**. Note that the anticlockwise moment about O is equal to the clockwise moment about O. One turning effect balances the other.

The upward force from the support is acting at O, so its moment about O is zero.

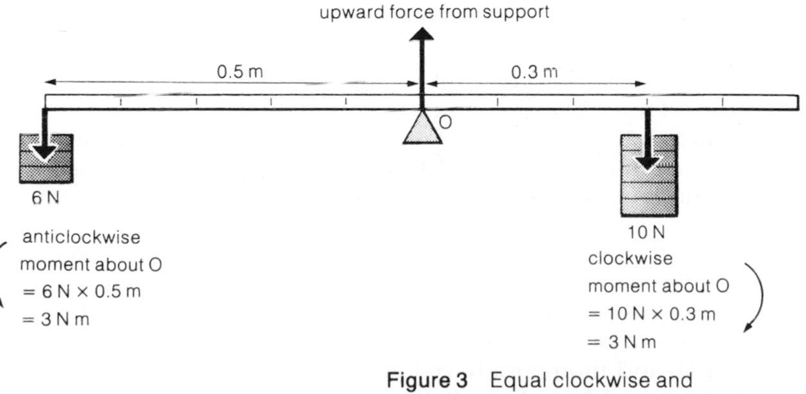

Figure 3 Equal clockwise and anticlockwise moments about O mean that the bar is balanced

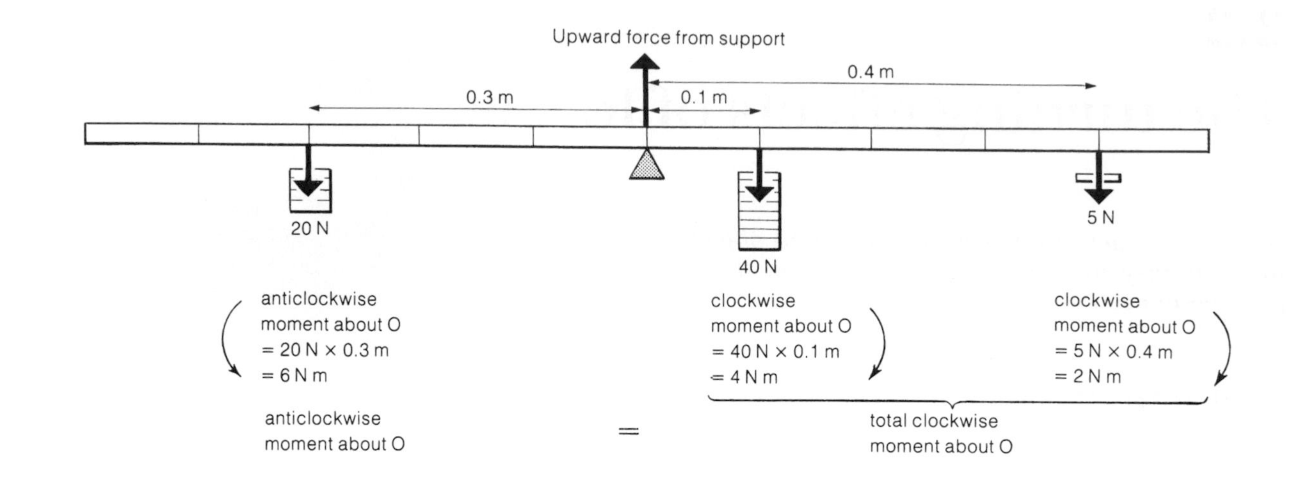

Figure 4 The clockwise and anticlockwise moments about the point O are equal. The bar is balanced.

The situation in figure 4 is a little more complicated, but once again the positions of the forces are such that the bar is balanced. The moment of each force about O is shown on the diagram. Added together, the clockwise moments about O balance the anticlockwise moment about O.

Both of these examples illustrate the **principle of moments**. This states:

When an object is in equilibrium, the sum of the anticlockwise moments about any point is equal to the sum of the clockwise moments about that point.

Conditions for equilibrium

If an object is in equilibrium, not only must the turning effects balance, the forces acting must also balance. There are therefore two conditions for equilibrium:

1 The sum of the forces in one direction must equal the sum of the forces in the opposite direction.

2 The principle of moments must apply.

Take the case of the bar in figure 4:

Fulfilling the conditions for equilibrium?

Condition 1 As the bar is in equilibrium, the upward force from the support must exactly balance the total downward force which is 20 N + 40 N + 5 N. The support must therefore exert an upward force of 65 N on the bar. If it failed to provide this force, the bar would accelerate downwards and would no longer be in equilibrium.

Condition 2 As the bar is in equilibrium, the moments about *any* point must balance. You can check this for yourself by studying the figures given in figure 5. This shows the same bar and forces as in figure 4, but distances have been measured from point P instead of point O, and moments have been calculated about point P. Note that the upward force from the support *does* have a moment about point P.

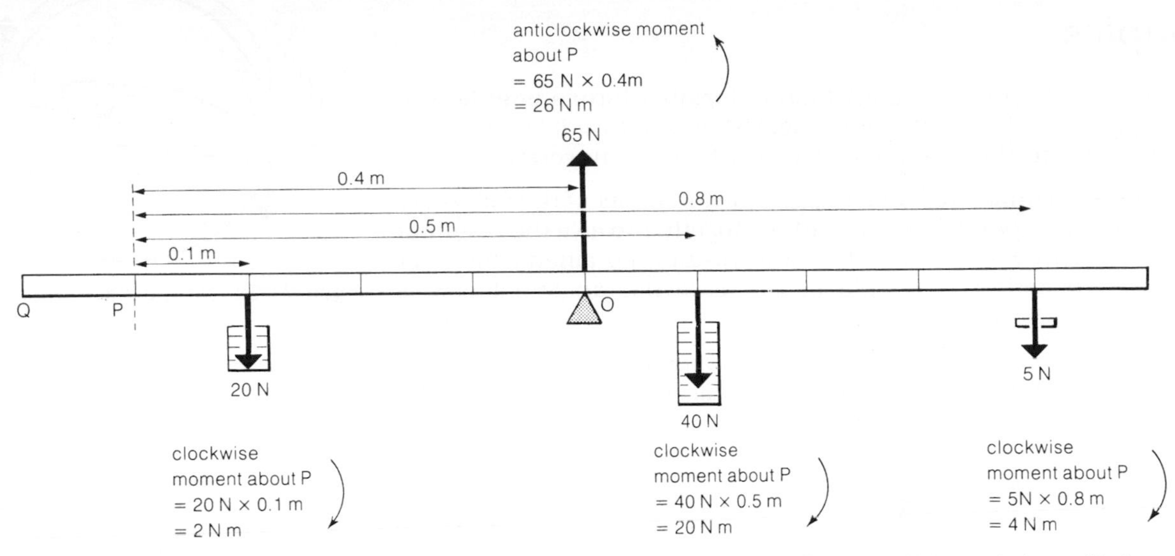

Figure 5 If the bar is in equilibrium, the clockwise and anticlockwise moments about *any* point are equal

As a further check, you could redraw the diagram and take moments about any third point Q – it doesn't necessarily have to be on the bar. Remember to take all distances from Q. If you have difficulty in deciding whether a force has a clockwise or anticlockwise moment about Q, imagine that your drawing is pinned to the table through Q and then decide which way the force arrow is trying to spin the paper.

Example *Figure 6 shows a man standing on a plank supported by two trestles. Using the information given on the diagram, calculate the upward forces X and Y exerted by the trestles on the plank.*

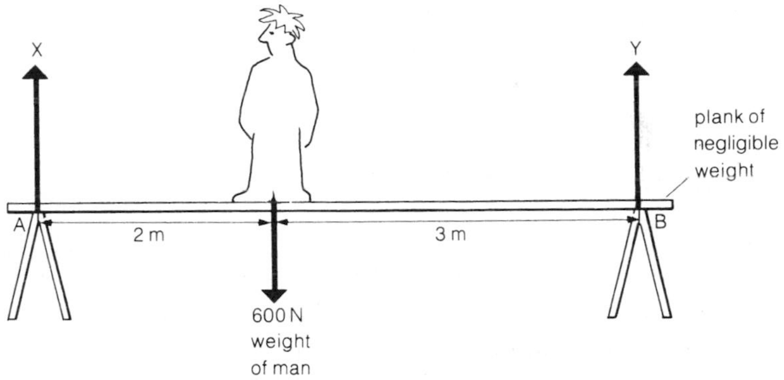

The system is in equilibrium, so the principle of moments can be used. Moments could be taken about any point, but taking moments about A or B gets rid of one of the unknowns, X or Y.

Figure 6

Taking moments about A:

clockwise moment = 600 N × 2 m = 1200 N m;
anticlockwise moment = Y × 5 m

As: anticlockwise moment about A = clockwise moment about A,

$$5Y \text{ m} = 1200 \text{ N m}$$

giving $Y = 240$ N

From here, there are two methods of finding X. You can take moments about B and proceed as before. Alternatively, you can use the fact that X and Y must add to equal the downward force of 600 N. By either method, X is found to be 360 N.

Note that the larger force is at the support nearer the man.

Couples

Two equal but opposite parallel forces together form a **couple**. You can apply a couple to a steering wheel when you turn it (as shown in figure 7), and to the top of a coffee jar when you unscrew it.

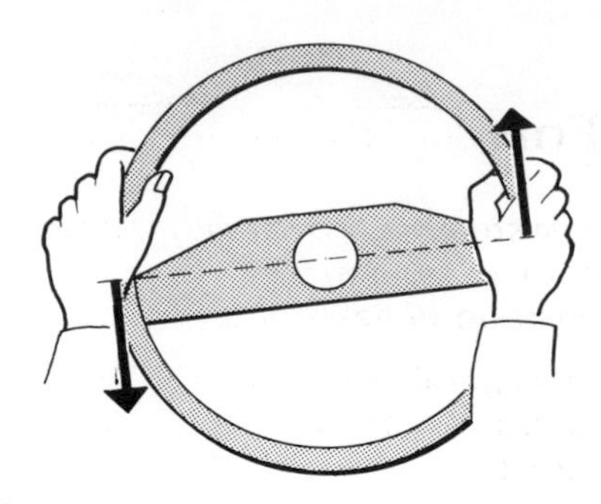

Figure 7 Couple applied to a steering wheel

To find the turning effect of a couple, the moments of the forces are taken about any point, and then added together to give the resultant moment. Figure 8 shows that the same result is obtained whichever point is selected. The couple in this case has an anticlockwise moment of 24 N m.

The simplest method of calculating the moment of a couple is to multiply one force by the perpendicular distance between the forces.

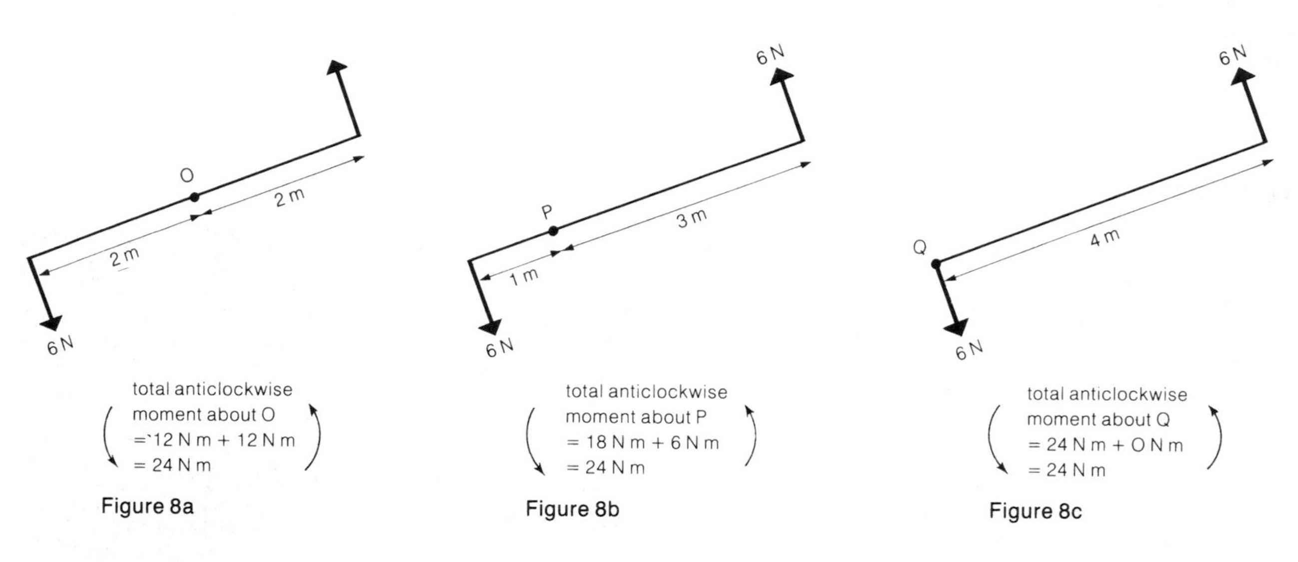

Figures 8a, 8b, 8c A couple has the same moment about all points

Anyone who has used a 'tommy-bar' type spanner as shown in figure 9 may know from experience that a couple has the same turning effect about any point. Sliding the bar to another position may make the spanner more convenient to use in cramped conditions, but it doesn't make it easier (or more difficult) to turn.

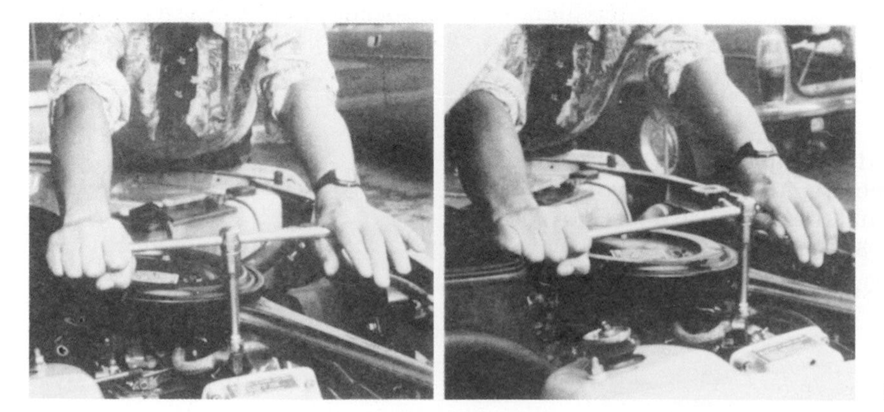

Figure 9 A 'tommy-bar' in use

Forces, couples and torques

A system of forces may make an object accelerate, or turn, or do both, or do neither. It all depends on the magnitude, direction, and position of each of the forces acting.

In general, the word 'couple' is used to describe any system of forces which will turn an object without making it accelerate in one direction. i.e. a force system in which the forces balance, but their turning effects do not. The simplest form of couple is two equal but opposite parallel forces as described above.

In engineering, the moment of a couple is called a **torque**. Typical torque values are:

torque required to turn a door handle	0.3 N m
torque produced by an electric drill	5 N m
torque required to undo a wheel nut on a car	70 N m
torque produced by a car engine	150 N m

Questions

1 The turning effect or *moment* of a force depends on two factors. What are they?

2 What is the principle of moments?
What other rule also applies if a system is in equilibrium?

3 The bar in figure 10a is in equilibrium.
What is the moment of the 6 N force about O?
What is the moment of the force F about O?
Calculate F and the reaction force X.
Using the force values you have just found, calculate the moments of F, X, and the 6 N force about P.
What is the total clockwise moment about P?
What is the total anticlockwise moment about P?

4 In figure 10b a light plank on two trestles is supporting a man and a block of concrete.
Write down the moment of each force about A.
Use the principle of moments to calculate the reaction force Y.
What is the total upward force $X + Y$?
What is the size of the reaction force X?
In figure 10c, the man walks past A towards the end of the plank.
What is the reaction force Y at the instant the plank starts to tip?
How far is the man from A as the plank starts to tip?

5 What is the moment of the couple shown in figure 10d?

The force is about 500 N and the length of the lock gate is about 3 m. What is the moment?

6 The maximum torque a bolt can withstand without breaking is 100 N m. A mechanic tightens the bolt with a spanner 0.25 m long. What is the maximum force the mechanic can use on the end of the spanner?
If he used a shorter spanner, would he be more likely or less likely to break the bolt? Give reasons for your answer.

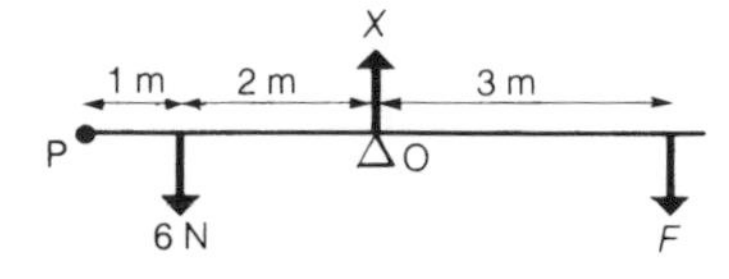

Figure 10a

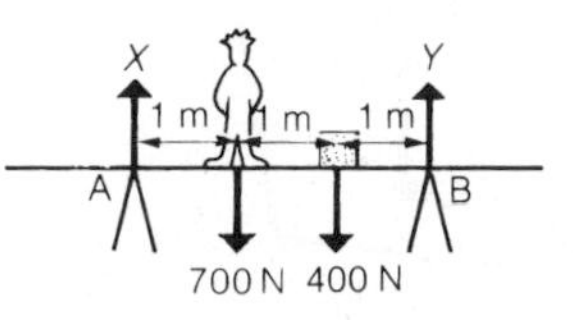

Figure 10b

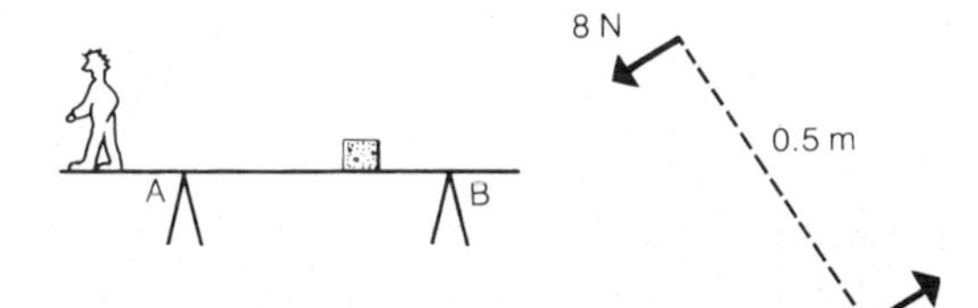

Figure 10c

Figure 10d

2.3

Centre of gravity

An object often behaves as if all its weight is acting at one point. The position of this point affects where an object will balance and how likely it is to topple.

You can think of the wooden rule in figure 1a as being made up of a large number of tiny particles of wood. Each particle has a small gravitational force acting on it. The rule balances when suspended at one particular point, G, because the gravitational forces have turning effects about G which cancel out.

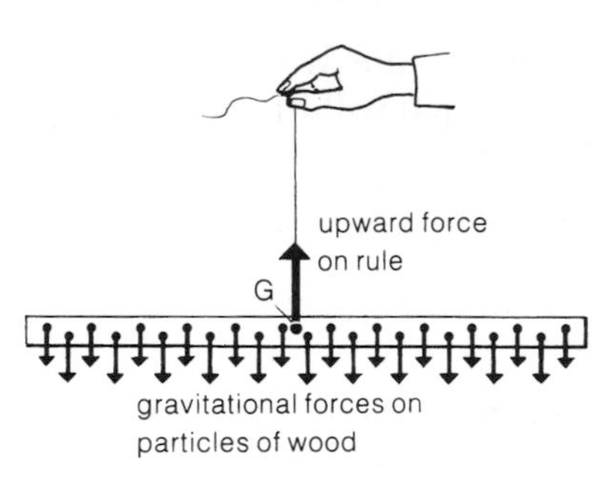

Figure 1a

The single downward force in figure 1b is the resultant of the gravitational forces in figure 1a. It represents the total gravitational force acting on the rule, and it acts at the point G. If it were to the left or right of G, it would have a turning effect about G, and the rule would not be balanced.

This downward force is the weight of the rule. The point G at which the weight acts is called the **centre of gravity** or **centre of mass**.

It is easy to find the centre of gravity of a regularly shaped object such as a metal bar, or a rectangular block of wood, or a football. The centre of gravity usually lies at the point you would immediately identify as the middle. Where an object has a more complicated shape, its centre of gravity can often be found by experiment.

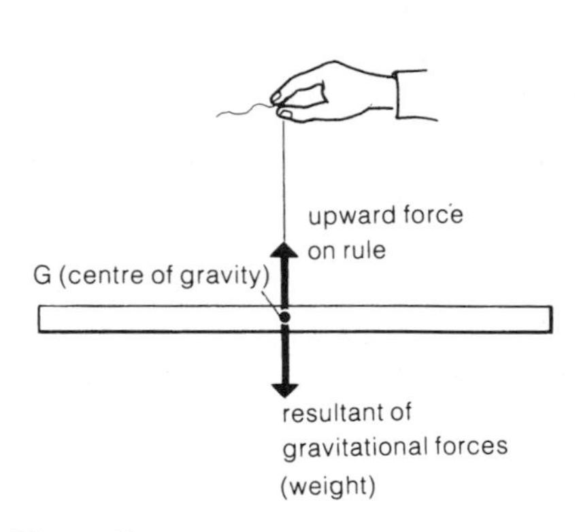

Figure 1b

Figures 1a, 1b The force on each particle adds up to make the total force on the rule

Finding the centre of gravity of a flat piece of card

When released, the card in figure 2a is able to swing freely from the pin passing through its top corner. The forces acting on the card form a couple which makes the card swing downwards, and the card finally comes to rest as in figure 2b, with its centre of gravity vertically under the pin. Whatever point the card is suspended from, it always hangs with its centre of gravity vertically under the pin. This fact can be used to find the position of the centre of gravity.

Figures 2a, 2b, 2c The card moves until it hangs with its centre of gravity vertically under the pin

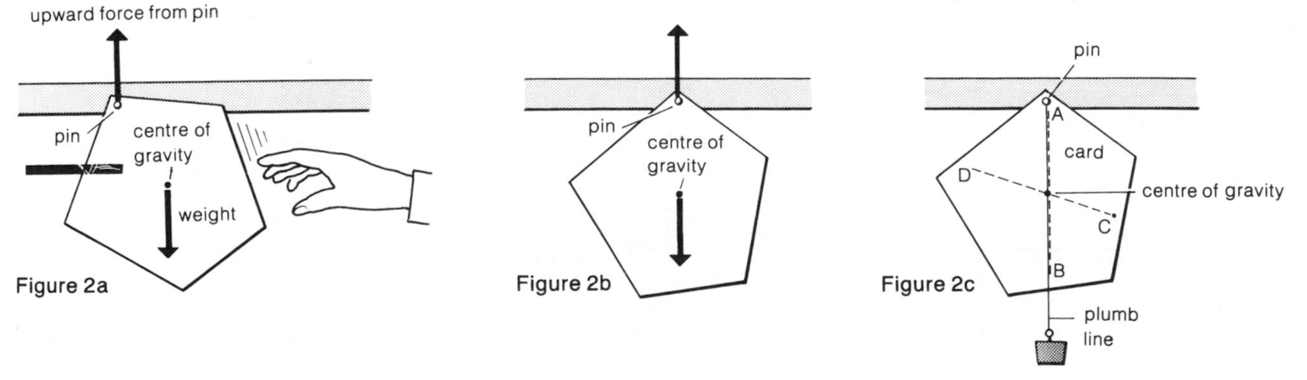

Figure 2a

Figure 2b

Figure 2c

In figure 2c, the centre of gravity of the card lies somewhere along the plumb line marked by the broken line AB. If the card is suspended at a different point, a second line CD can be marked on the card. Once again, the centre of gravity must lie along this line. The centre of gravity is therefore at the point where AB crosses CD.

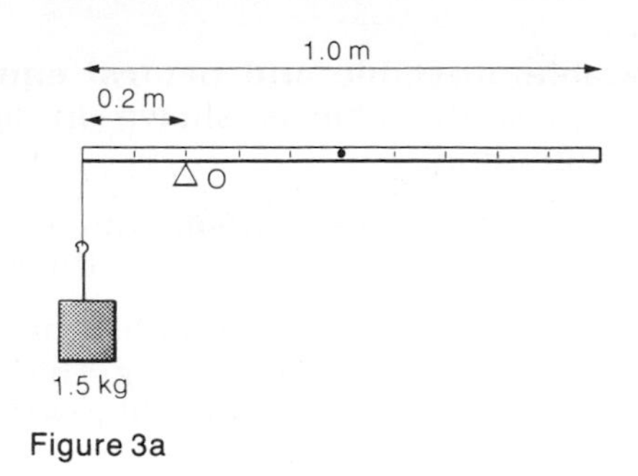

Figure 3a

More about balanced bars

In the previous unit, the balanced bars were all assumed to have negligible weight. In practice, you cannot always make this assumption. The weight of a bar often has to be included when tackling problems on moments.

Example *The bar in figure 3a balances off-centre because a* 1.5 kg *mass has been attached to one end. Use the information given in the diagram to calculate the weight of the bar. Assume that the bar is uniform and that* $g = 10\ \text{m/s}^2 = 10\ \text{N/kg}$.

'Uniform' in this case implies that the weight of the bar is evenly distributed along it. In other words, you can assume that the centre of gravity of the bar (by itself) is in the middle.

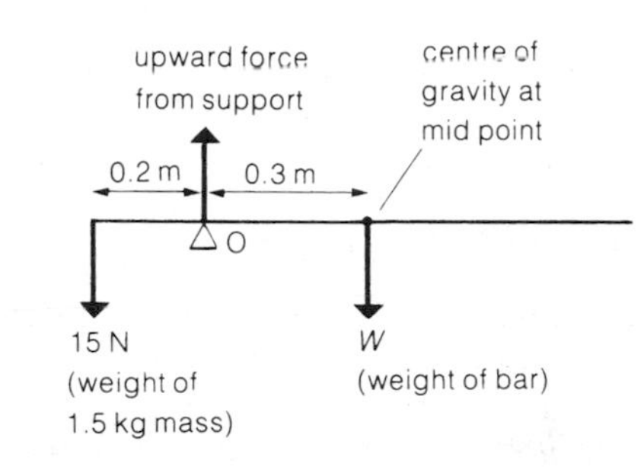

Figure 3b

Figures 3a, 3b Using moments to find the weight of a bar

The first stage in solving the problem is to redraw figure 3a to show all the forces acting on the bar. This has been done in figure 3b. Note that the 1.5 kg mass has a weight of 15 N. Note also that the weight W of the bar has been shown as a single force acting at the centre of gravity.

The problem can be simply solved by taking moments about O. The weight W, works out at 10 N.

Stability

Some things topple over more easily than others. Figures 4a, 4b and 4c show what happens when a tall, narrow box is pushed until it starts to topple.

When the box is pushed a little and then released, it falls back to its original position: the position of the box was **stable**. If the box is pushed much further, it topples. It starts to topple as soon as its centre of gravity passes over the edge of its base. From this point on, the forces on the box produce a couple which tips it even further over.

A box with a wider base and a lower centre of gravity is even more stable. Figure 4d shows that the box can be tilted to a greater angle before it starts to topple.

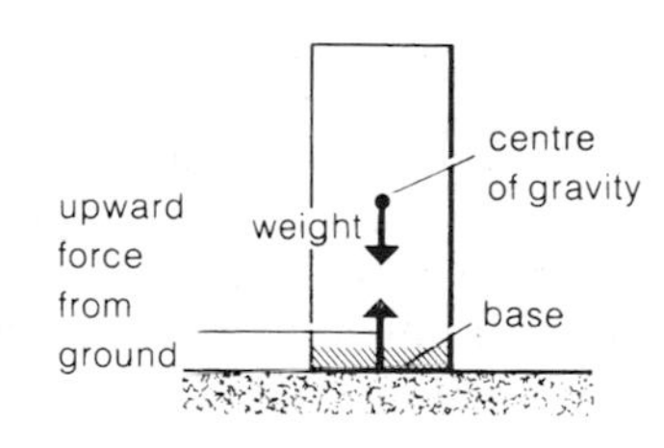

Figure 4a No tilt: no couple

Figure 4b With a small tilt, the couple turns the box back to its original position

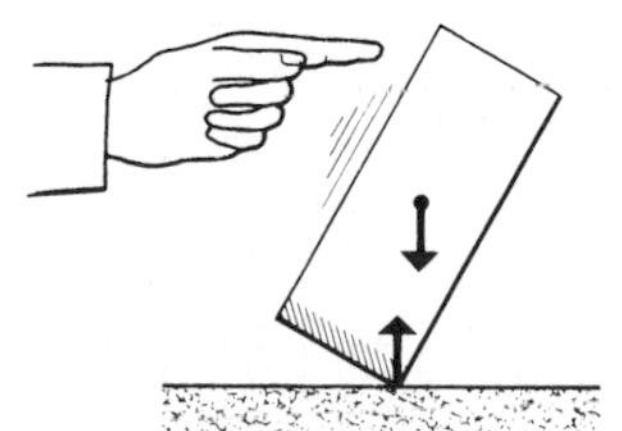

Figure 4c With a large tilt, the couple tips the box further to the right

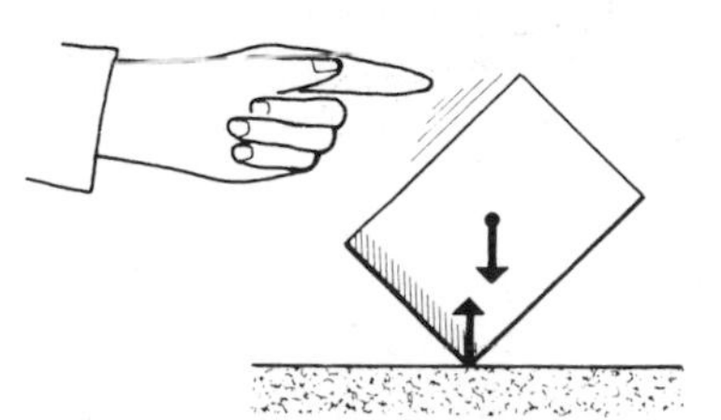

Figure 4d A box with a wider base and a lower centre of gravity can be tipped to a greater angle before it falls over

Stable, unstable and neutral equilibrium Like a box before it topples, the objects shown in figure 5 are all in a state of equilibrium.

Cone A is in **stable equilibrium**. If the cone is disturbed slightly, its centre of gravity remains over the area of its base.

Cone B is in **unstable equilibrium**. It is balanced, but it clearly isn't going to stay that way for long. The 'base' of the cone is now so small that the centre of gravity will pass beyond it immediately.

Ball C is in **neutral equilibrium**. If the ball is left alone it stays where it is; if the ball is moved, it stays in its new position. Wherever the ball lies, the centre of gravity remains over the point of contact with the bench.

Designing for stability Vehicles designed to carry large loads are often very tall; this has a bad effect upon their stability, since the centre of gravity is very high. Vehicles designed for speed, as in the racing car in figure 6, have a very low centre of gravity.

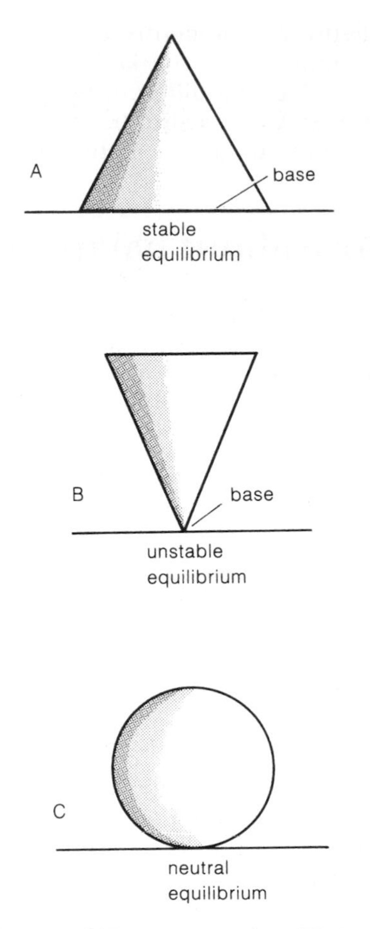

Figure 5 Different types of equilibrium

Figure 6 For high stability, a racing car has a low centre of gravity and a wide wheelbase

Questions

(Assume that bars, planks and rules are all uniform)

1 Copy and complete figure 7 to show the forces acting on the bar. Explain why the bar will tip.
2 The metre rule in figure 8 balances at O. Using information given in the diagram, calculate the weight of the rule. Calculate the upward force acting at O.
3 Redraw figure 9 to show all the forces acting on the plank. Calculate the upward forces at A and B.
4 Redraw figure 10 to show the forces acting on the cone. Explain why the cone will fall to the right. Draw another diagram showing the cone at the point of toppling, showing the positions of the forces acting on it.
5 Draw diagrams to show a drawing pin in positions of stable, unstable, and neutral equilibrium.

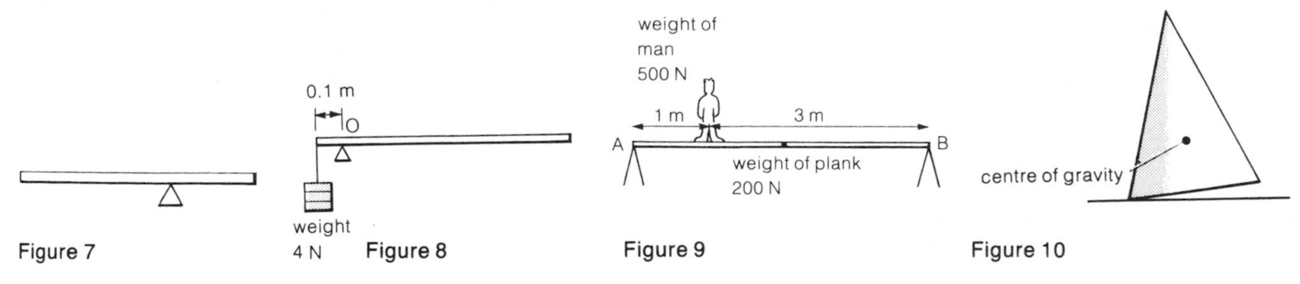

Figure 7 Figure 8 Figure 9 Figure 10

2.4
Stretching forces

Inside every spring balance is a steel spring. There is something special about the way a spring stretches that makes it very useful as a means of measuring force.

The simplest way to stretch a spring is to apply a force to each end. This can be done either by directly pulling on both ends, or by pulling on one end when the other end is fixed to a support (as shown in figure 1). As the stretching force is applied, the support distorts slightly and produces an equal pull in the opposite direction.

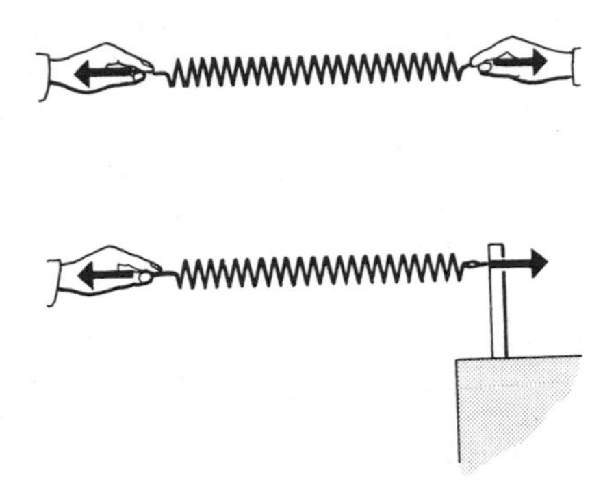

Figure 1 The spring is stretched by the force-pair acting on it

Spring stretching experiment

When a spring is stretched, the difference between its stretched and unstretched lengths is called the **extension**. To find out how the extension of a spring depends on the stretching force, you can perform an experiment similar to that shown in figure 2a.

In this experiment, the spring is fixed at one end and then stretched in stages by hanging more and more standard masses from the other end. Each kilogram of mass on Earth weighs 10 N, so every time an extra 100 g mass is hung from the spring, the stretching force is increased by 1 N. The extension in each case is measured on the millimetre scale. Typical readings are shown in the diagram.

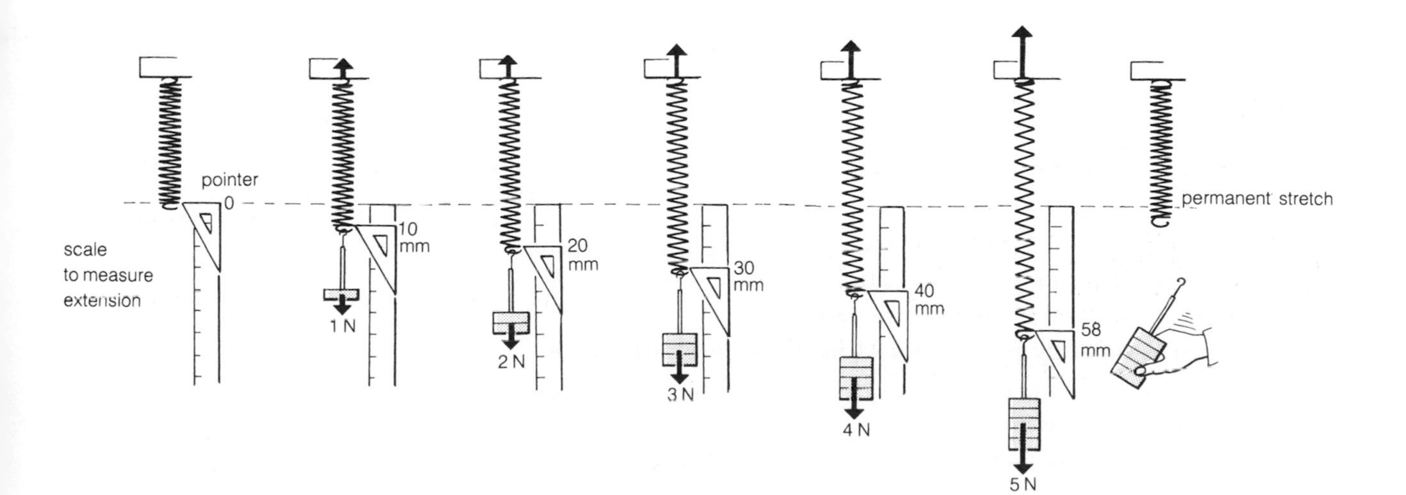

Figure 2a The spring gets longer by the same amount for each extra 10 N of force

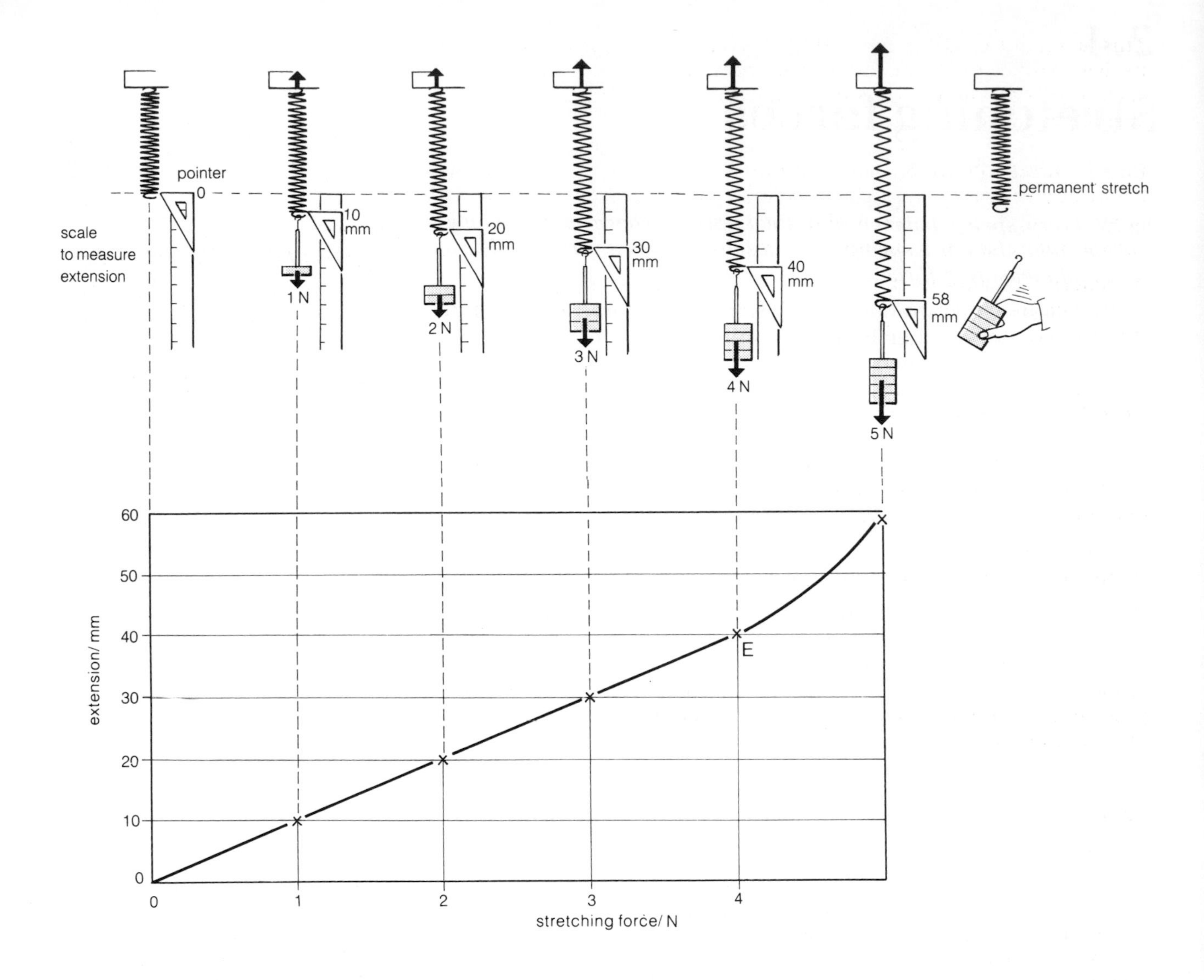

Figure 2b This shows figure 2a, along with the graph which can be drawn to demonstrate the results of the experiment

Figure 2b shows the graph which results if the extension of the spring in figure 2a is plotted against the stretching force – though new springs in particular do not always stretch in quite the same regular manner as this one.

A simple proportion The line of the graph in figure 2b passes through the origin and is straight up to point E. Mathematically, this means that over this first section of the graph:

the extension is directly proportional to the stretching force.

A simple proportion of this type has several features which you can check for yourself on the graph.

1 If the stretching force is doubled, the extension is also doubled, and so on.

2 Dividing the extension by the corresponding stretching force always gives the same result: 10 mm/N in this case.

3 Every 1 N increase in the stretching force produces the same extra extension (10 mm).

This last feature makes the spring easy to use as a force-measuring device. With a hook attached to its bottom end, and the existing scale replaced by one marked in newtons at 10 mm intervals, the spring becomes a simple spring balance.

Elastic limit Point E marks another important change in the behaviour of the spring. Up to point E, the spring returns to its original length if the stretching force is removed. Beyond E, the spring becomes permanently stretched.

A material is **elastic** if it returns to its original shape when a stretching force has been removed. The spring is elastic only up to point E. This point is called its **elastic limit**.

Figure 3 A new car design under test. The "passenger cage" should retain its shape, but the rest of the car should absorb energy by bending beyond its elastic limit

Hooke's law

More than 300 years ago, Robert Hooke carried out a series of experiments with wires and springs to discover how the extension of a material and the stretching force were related.

A material is said to obey Hooke's law if its extension is directly proportional to the stretching force.

Hooke's law applies to a great many materials including wood, glass and most metals, provided the elastic limit is not exceeded in each case. However, if the spring in the experiment in figure 2a were replaced with a piece of wood or a steel bar, any extensions produced would be far too small to measure using the existing millimetre scale.

Questions

(Assume $g = 10\ m/s^2 = 10\ N/kg$)

1 What is meant by the elastic limit of a material?
2 How does a material behave if it obeys Hooke's law?
3 A spring, fixed at one end, can be stretched by hanging standard masses from the other end. What mass is needed to achieve a stretching force of: 0.5 N, 1 N, 5 N?
4 In a spring stretching experiment, the following readings were obtained.

mass hung from spring/g	0	100	200	300	400	500	600	700	800
stretching force/N									
length of spring /mm	50	61	72	83	94	105	116	132	160
extension/mm									

a) What is the unstretched length of the spring?
b) Copy and complete the table.
c) Plot a graph of extension against stretching force.
d) Mark the elastic limit of the spring on the graph.
e) Over which region of the graph is Hooke's law obeyed?
f) What stretching force is needed to produce an extension of 25 mm?
What stretching force is needed to produce double this extension?
g) Before the elastic limit is exceeded, how much extra stretching force is needed to produce each additional 10 mm of extension?
h) What force is needed to stretch the spring to twice its unstretched length?

2.5
Work and energy

When a force moves something, work is done. And whenever work is done, energy is changed into a different form.

Work

In everyday language, work means almost anything that people do. In science, the word has a more precise meaning; work is done whenever a force produces movement. The greater the force and the greater the distance moved, the more work is done.

The SI unit of work is the **joule** [J]:

1 joule of work is done when a force of 1 newton [N] moves the point at which it acts through a distance of 1 metre [m].

The work done when any force produces movement can be calculated using the equation:

$$\textbf{work done} = \textbf{force} \times \begin{array}{l}\textbf{distance moved in the}\\ \textbf{direction of the force}\end{array}$$

in symbols: $\boldsymbol{W} = \boldsymbol{F} \times \boldsymbol{s}$.

Work is measured in J if force is measured in N and distance in m. For example, if a force of 4 N moves an object a distance of 3 m, the work done is 12 J.

Some typical work values are given in figure 2. As work quantities go, 1 joule is not a very large amount of work. Larger units of work in common use are the kilojoule and the megajoule:

1 kilojoule [kJ] = 1000 J [10^3 J]
1 megajoule [MJ] = 1 000 000 J [10^6 J]

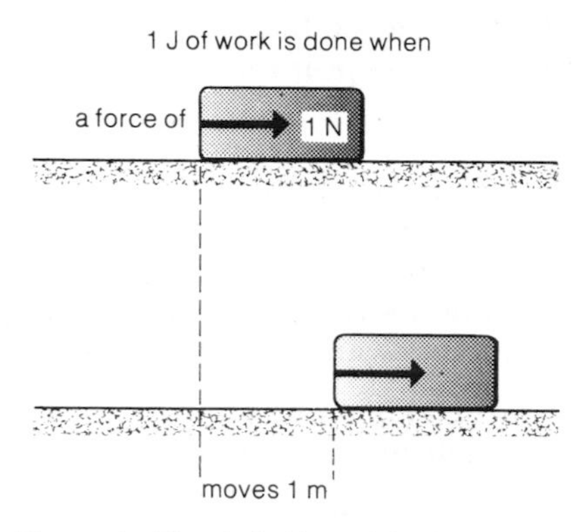

Figure 1 The definition of the joule

Typical work values	
Work done	
– in shutting a door	5J
– in throwing a cricket ball	20J
– in climbing the stairs	1500J

Figure 2 Work done in various activities

Energy

Things have energy if they are able to do work. A human body has energy; so does a tankful of petrol or a compressed spring. In each case, you can think of energy as a promise of work to be done in the future. Energy, like work, is measured in **joules** [J]. It exists in a variety of different forms.

Potential energy Objects have potential energy if they have been moved into a position from which they can do work when released, as shown in figure 3. A stretched spring has potential energy; it does work when it is allowed to spring back to its normal length. A cricket ball held up in the air has potential energy; if the ball is released, work is done as the ball is pulled to the ground by its weight.

Figure 3 Gaining potential energy in two different ways

Kinetic energy Moving objects, such as bullets, cars and cricket balls, all have kinetic energy, as shown in figure 4. When a fast-moving cricket ball is caught, it loses it kinetic energy and work is done in pushing the fielder's hand backwards. Spinning objects also have kinetic energy due to their rotation.

Figure 4 Moving objects have kinetic energy.

Thermal energy All materials are made up of tiny particles called molecules. These molecules are constantly in motion. In solids and liquids, the molecules are held close together by strong forces of attraction, and move by vibrating to and fro. In gases, the molecules have become so spaced out that they move about freely at high speed. In some cases they also spin.

Molecules have kinetic energy because they are moving. They also have potential energy because their movements keep them separated, despite the 'spring-like' attractions trying to pull them together. **Thermal energy** is the name given to the energy an object possesses because of the kinetic energy and potential energy of its molecules. The higher the temperature of an object, the faster its molecules move, and the greater is its thermal energy.

Thermal energy is commonly known as **heat energy**. However, engineers prefer to keep the term heat energy for thermal energy which is in the process of being transferred from one object to another.

Other forms of energy There are many other forms of energy. For example, batteries and generators deliver **electrical potential energy** – or **electrical energy** for short. Foods and fuels store **chemical energy** which is released in some other form when these materials combine with oxygen, during burning for example. The centres of atoms possess **nuclear energy**.

The Sun gives out vast quantities of energy in the form of **electromagnetic waves** (see page 204). Electromagnetic waves include light waves, radio waves, and X-rays. **Sound** is also a form of energy. It too travels in the form of waves, though these are different in nature from the other waves described here.

Typical energy values are given in figure 5.

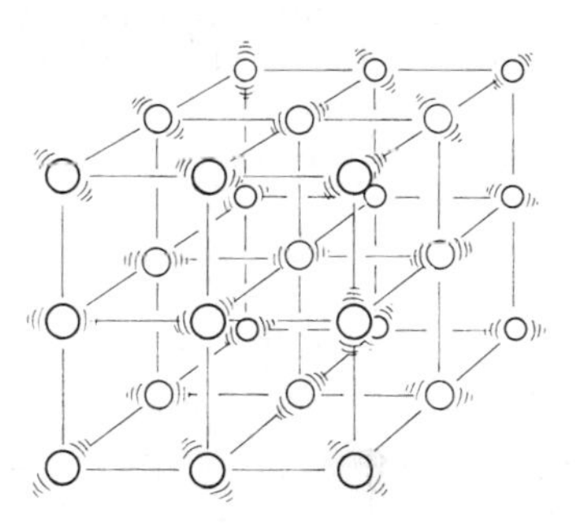

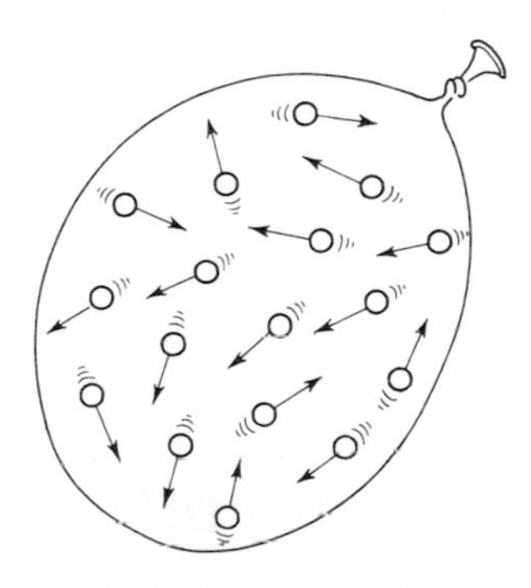

Figure 5 Solids, liquids and gases have thermal energy because of the motion of their molecules and the attractions between them.

potential energy of a house brick held at shoulder height	30 J	
kinetic energy of a brick if you throw it	75 J	
potential energy of person stood at top of stairs	1500 J	(1.5 kJ)
kinetic energy of a car travelling at 70 mph (110 km/h)	500 000 J	(500 kJ)
chemical energy stored in a pint of beer	600 000 J	(600 kJ)
heat energy needed to boil a kettleful of water	700 000 J	(700 kJ)
chemical energy in a large plateful of chips	1 000 000 J	(1 MJ)
electrical energy obtainable from fully charged car battery	2 000 000 J	(2 MJ)
kinetic energy of a heavy lorry travelling at 50 mph (80 km/h)	5 000 000 J	(5 MJ)
chemical energy stored in the food you eat in one day	11 000 000 J	(11 MJ)
chemical energy stored in one litre of petrol	35 000 000 J	(35 MJ)

Figure 5 Energy values

Changing energy from one form to another

According to the **law of conservation of energy**:

energy cannot be made or destroyed, but it can be changed from one form into another.

When a cricket ball is thrown straight up from the ground, some of the chemical energy stored in your body is transferred to the ball as kinetic energy. The ball slows down as it gains height, and its kinetic energy is changed into potential energy (as shown in figure 7).

As the ball falls to the ground, its potential energy is changed back into kinetic energy. This is changed into thermal energy and sound energy as the ball is brought to rest by the ground. The ball stops moving, but molecules of the ground, the ball, and the surrounding air all move slightly faster than before.

If air resistance is negligible, the ball returns to the ground with the same kinetic energy as it had when it left it. In practice, some of the ball's kinetic energy is changed into thermal energy as the ball collides with molecules of the air.

By the time the cricket ball finally stops moving, all its original energy has been changed into thermal energy. Even the sound produced when it strikes the ground fades away to leave the air very slightly warmer than it was before. You find similar results if you follow other energy transfer processes through to their conclusion.

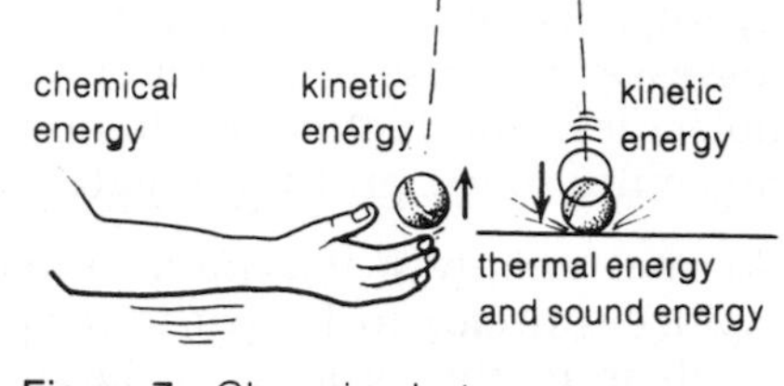

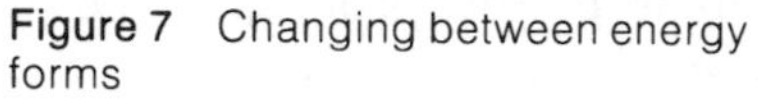
Figure 7 Changing between energy forms

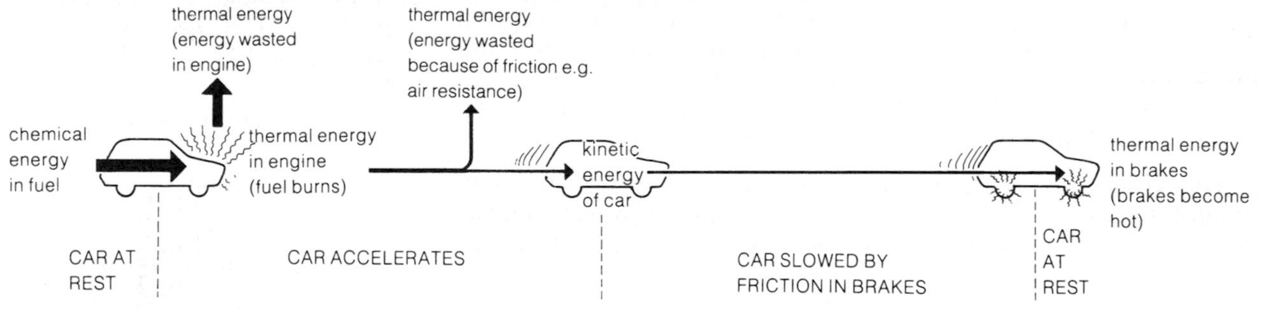

Figure 8 Energy changes in a car journey

Figure 8 shows a car which starts from rest, accelerates for several seconds, then comes to a halt when its brakes are applied. Energy changes from one form to another several times during this short journey, but all the energy released from the fuel eventually ends up as thermal energy.

Work done and energy transferred

Work is done whenever energy is changed from one form into another, as shown in figure 9. For example, if a falling stone loses 20 J of potential energy and gains 20 J of kinetic energy, the gravitational force acting does 20 J of work in accelerating the ball.

If, because of air resistance, a car loses 1000 J of kinetic energy and the surrounding air gains 1000 J of thermal energy, the car does 1000 J of work in speeding up the molecules of the air.

The quantity of energy changed from one form to another in each case is known as the energy transferred:

work done ≡ energy transferred

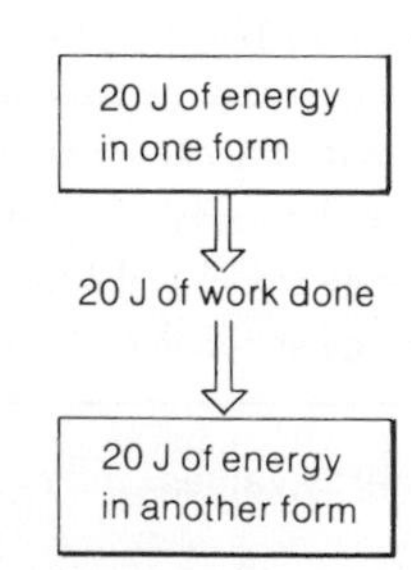

Figure 9 Work is done as the energy changes form

Finding the energy

Advanced technological societies can't operate without continuous and plentiful supplies of energy. But the world's most convenient sources of energy are running out fast. The following paragraphs explain where energy resources come from, and how they are used.

How is energy used? The chart in figure 10 shows how most of Britain's energy is used. The biggest users are industries which make things or process materials – producing steel, chemicals, plastics, and cars for example.

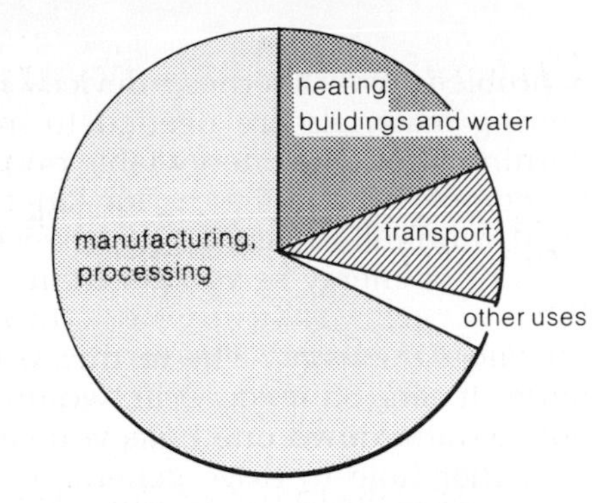

Figure 10 Use of energy in Britain

Where does the energy come from? Energy is usually supplied in the form of gas, oil-based fuels, coal, or electricity. In Britain, most of the electricity comes from power stations using other fuels as a source of energy. Heat from the burning fuel is used to change water into steam, and the steam turns the turbines which drive the generators.

Where did the energy originally come from? When fuels burn, they release energy originally radiated from the Sun.

Green plants collect energy from sunlight, as is shown in figure 11. They use it to make new plant tissue from simple materials like carbon dioxide and water. When plant tissue is combined with oxygen, during burning for example, some of this energy is released again. You get your own energy in this way, by eating plants, or animals which feed on plants.

Figure 11 Energy for plants and animals comes from the Sun

Coal, oil, and natural gas were formed from the decayed remains of plants and animals which lived on Earth many millions of years ago. Unlike the food you eat, these fuels aren't replaceable.

How long can the fuels last? There won't be much oil or natural gas left at the end of the century if consumption continues at the present rate. Coal is likely to last rather longer; perhaps 200 years or more.

Why is oil so useful? Most of the fuels used by road vehicles, ships and aircraft are extracted from oil. All these fuels store large amounts of energy in a small volume. A cupful of petrol, for example, contains enough energy to move a small car several kilometres.

Figure 12 Oil is used in the manufacture of plastics

Oil is also the basic raw material from which most plastics are made, as shown in figure 12. Oil is so valuable as a source of plastics and transport fuels that it probably makes sense to reserve it for these uses alone. Other needs would then have to be met by more bulky sources of energy.

What are the alternative energy sources? Some possibilities are outlined in the next paragraphs. All can be used to generate electricity, but none of those shown is likely to fit in the boot of a car. Storing energy in such a way that it is easily carried is quite another problem.

Wind and wave energy Generators driven by giant windmills have had a limited success. Devices for extracting energy from the waves have also shown some promise. The model shown in figure 13 uses a rocking float to drive a generator.

Figure 13 Checking to see how much energy can be gained from waves

A problem with all these devices is that such large quantities of moving materials are needed to match the energies obtained by burning fuels. The energy table on page 65 might make this clearer; try comparing the figures for kinetic and chemical energies. That giant windmill in figure 14 takes over a minute to deliver the same amount of energy as you get from a cupful of petrol.

Figure 14 A wind-powered electricity generator

Geothermal energy The further you dig into the Earth's crust, the hotter it gets. So if you drill two holes several kilometres deep and pump water down one hole, you can use the steam that comes up the other hole to drive generators or heat buildings. It is fine in principle, but technically difficult and very expensive.

Nuclear energy You can find out how a nuclear power station works on page 367. The nuclear fuel doesn't burn; it releases large amounts of thermal energy when uranium atoms within it are broken up. As much electricity can be made using a kilogram of uranium nuclear fuel as from 55 tonnes of coal. The world's known reserves of uranium are very limited, but fast-breeder reactors actually make *more* nuclear fuel as they release energy from the fuel they use. But there are serious problems. Nuclear reactors produce waste materials which will give off dangerous radiation for thousands of years. There is also concern about the risks of radiation leakage should reactor faults develop. Some people believe that we cannot afford to take the risks; others believe that we cannot afford not to.

Figure 15 A nuclear power station

Questions

1 Express the following amounts of energy in joules:
10 kJ; 0.5 MJ; 0.2 kJ; 10^6 MJ

2 What is the law of conservation of energy?

3 An object has 1 J of potential energy. Explain what this means.

4 Copy and complete the following:
In an electric kettle, ... energy is changed into ... energy.
When a car is moving and its brakes are applied, the ... energy of the car is changed into ... energy.
An object is said to possess ... energy because of the ... energy and ... energy of its molecules.

5 How much work is done when a force of 12 N moves an object a distance of 5 m?

6 10 J of work are done in throwing a stone upwards from the ground.
How much energy is given to the stone? What is the potential energy of the stone at its heighest point? (Assume air resistance is negligible.)

7 In travelling a distance of 0.5 m, an object gained 200 J of kinetic energy.
What work was done on the object, assuming no frictional forces acted?
What was the average force acting on the object?
If a force of 600 N had acted over the same distance, what would have been the gain in kinetic energy?

2.6

Energy for the world

Different countries have found different ways of solving their energy problems. But some of these have created problems of their own.

Burning coal in Britain

Much of Britain's energy comes from coal-burning power stations. Figure 1 shows how the chemical energy in the fuel is changed into electrical energy.

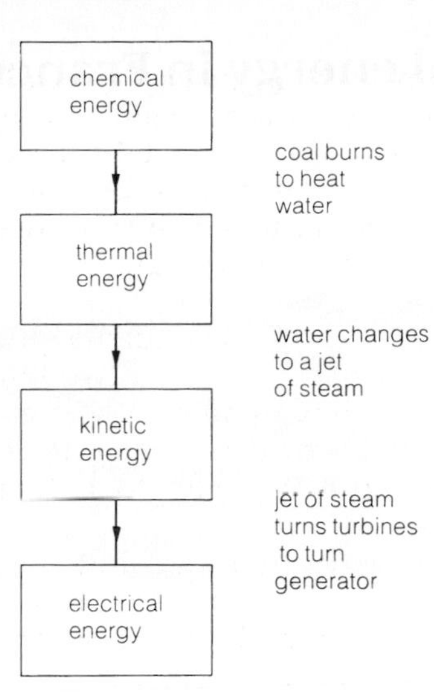

Figure 1 Energy changes in a coal-burning power station

A major problem with coal-burning power stations is the sulphur fumes they produce. The fumes make the rain acid. The acid is weak, but it kills fish in lakes, damages forests, and eats into old stone work (figure 2).

In Norway, Sweden, and Germany they blame Britain's power stations for the acid rain which is falling over Northern Europe. The Central Electricity Generating Board disagree. They say that there is no firm evidence to link their power stations with acid rain. Sulphur fumes aren't a new problem – factories and road vehicles have been producing them for years.

Figure 2 A stone carving damaged by acid rain

Hydroelectric Sweden

Much of Sweden's energy comes from hydroelectric schemes like the one in figure 3. A river is dammed to form a lake. Water rushes from the lake to turn generators at the foot of the dam. The potential energy of the water is turned into electrical energy.

Hydroelectric schemes don't cause pollution. But they alter the landscape, and they upset the balance of local animal and plant life. In Sweden, plans to build more hydroelectric schemes have been dropped because people do not want to see more of the countryside destroyed.

Figure 3 A hydroelectric dam

Tidal energy in France

Figure 4 The Rance tidal energy scheme

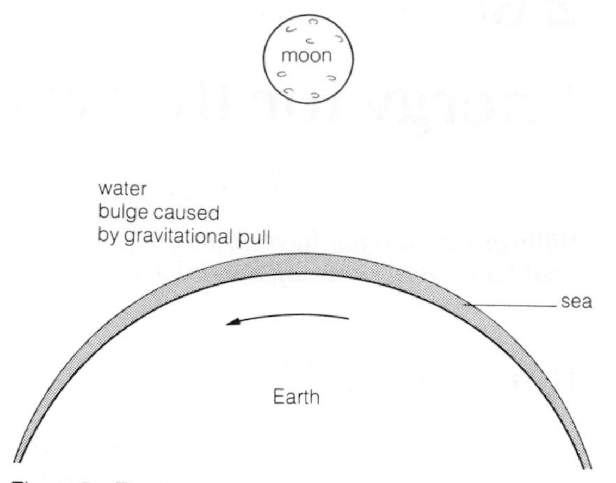

Figure 5 The Moon causes the tides on Earth

In Brittany, the River Rance has been dammed to form one of the world's largest tidal energy schemes as shown in figure 4. As high tide approaches, sea water rushes in to fill the lake, where it is trapped by the dam. At low tide, the water flows out. The flow is used to generate electrical energy as in a hydroelectric scheme.

The Moon's motion is the source of tidal energy as shown in figure 5. The Moon's gravitational attraction causes 'bulges' of sea water on the Earth's surface. As the Earth rotates, each part of the Earth passes in and out of a bulge – the tide rises and falls.

Hot Cornish rocks

At Penryn in Cornwall, they are extracting energy from hot granite rocks deep underground as shown in figure 6. Cold water is pumped down a borehole several kilometres deep. It comes up again, through a second borehole, at over 200°C. They plan to use the heat to produce steam to turn generators.

Energy from hot rocks is called **geothermal** energy. The heat is given off by radioactive atoms (see page 352) which are naturally present in the Earth's rocks.

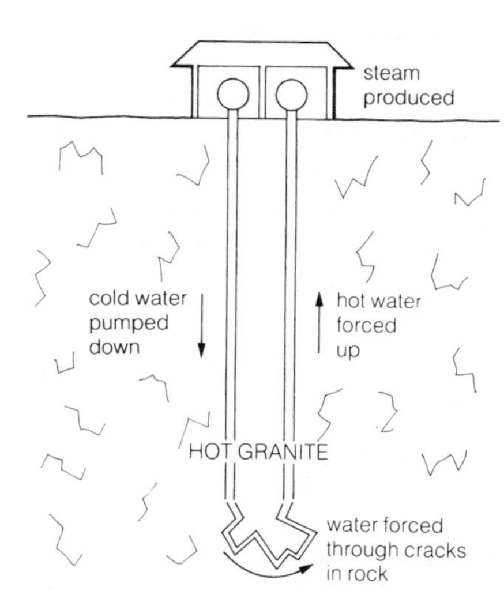

Figure 6 Geothermal energy scheme

Biomass in Brazil

Brazil's economy has suffered because of the cost of buying oil from abroad. Now, Brazil is reducing its oil bill by building cars that run on alcohol instead of petrol as shown in figure 7. The alcohol comes from sugar cane. The cane is fermented, like wine, so that the sugar is changed into alcohol. Across the country, huge areas of forest are being cut down so that large masses of sugar cane can be produced.

Sugar cane is an example of **biomass** – plants grown to 'trap' the energy in sunlight. Unlike oil, biomass is a **renewable** energy source, because new plants can be grown when the old ones have been used. But, if over-used, biomass drastically alters the landscape, and upsets the balance of nature.

Figure 7 Alcohol for fuel

Finding extra energy

The search is on for extra sources of energy. And it is producing some unusual results.

Figure 8 Electricity from your dustbin

In Edmonton, North London, the council has turned one of its rubbish incinerators into a generating station as shown in figure 8. Electricity is generated using the heat from burning household rubbish. The council gets rid of its waste, and it keeps its costs down by selling the electrical energy to the local Electricity Board.

Figure 9 Energy from the Sun: solar panels on the roof-tops

In Florida, the police have collected so much marijuana in drugs raids, that a power station has been specially converted to burn it. One tonne of marijuana gives nearly as much heat as three barrels of oil.

In sunny regions, many houses have solar panels on their roofs. Water is pumped through the panels, and absorbs energy radiated from the Sun. This helps to cut fuel bills because the water is warmed before it passes through the house's main heating system. Figure 9 shows solar panels on the roofs of English houses.

When power stations generate electricity (figure 10), much of the energy from their fuels is lost as heat. The heat warms up the cooling water that flows through the power station. Usually, the energy is wasted. But one idea is to use the warmed water to heat buildings nearby.

But there is a problem. Power station water isn't quite hot enough for room heating. To produce hotter water, each power station would have to lose some of its electrical output. And that would make it more difficult for the Electricity Board to stay in profit.

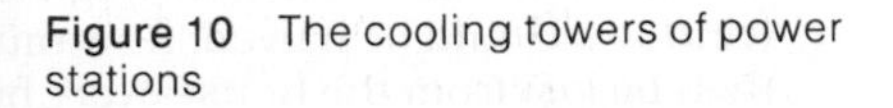

Figure 10 The cooling towers of power stations

Saving energy

All countries are looking for ways of cutting their energy bills. In Britain, large savings could be made by reducing the heat lost from buildings.

Figure 11 shows how the heat escapes from a typical house. Lost heat would cost the owner well over £500 a year in fuel bills.

Stopping draughts and air changes saves most on the fuel bills. But it can put your health – and even your life – at risk. If rooms are tightly sealed, the oxygen used by people and fires isn't replaced, and dangerous fumes can build up in the air.

In a carefully designed, energy-saving house like the one in figure 12, a one-bar electric heater would be enough to heat the whole house, even in mid-winter. Houses like this help to conserve valuable energy resources. But they are very expensive to build.

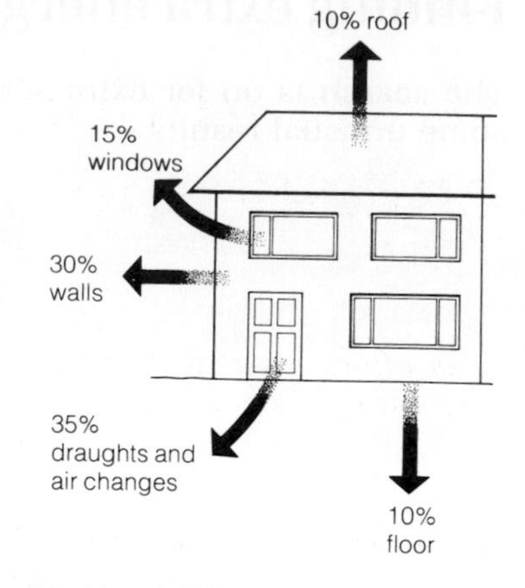

Figure 11 How the heat escapes from a typical semi

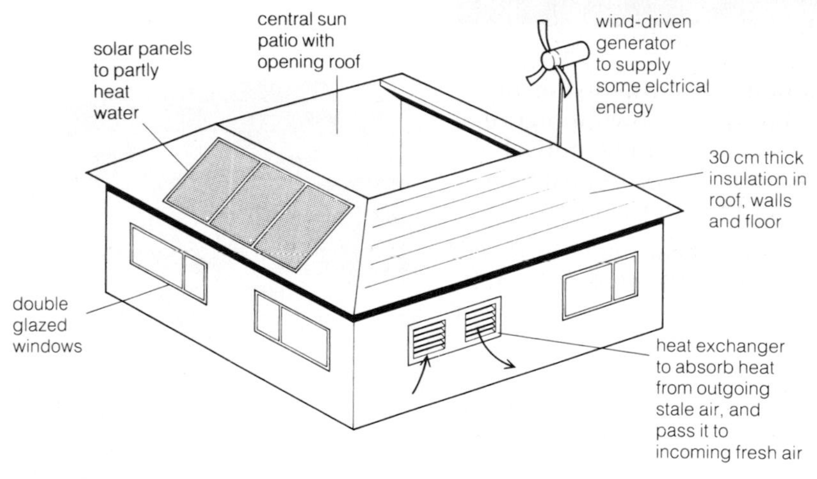

Figure 12 An energy-saving house

Questions

1 On a winter's day, this is how the heat (thermal energy) escapes from a house:

Heat escape because of	Heat lost in 1 hour
roof	6 MJ
windows	8 MJ
walls	12 MJ
floor	4 MJ
draughts and air changes	20 MJ

a) What is the total loss of heat from the house in one hour?

b) What percentage of heat is lost through the windows?

c) If double glazing and cavity wall insulation were installed, the loss of heat through the walls and windows would be halved. How much heat would then be lost from the house every hour?

d) What are the problems in trying to reduce the heat lost through draughts and air changes?

2 A. BIOMASS B. TIDAL SCHEME
C. HYDROELECTRIC SCHEME
D. GEOTHERMAL SCHEME E. SOLAR PANELS

These are all sources of energy. Which of them

a) absorb energy radiated by the Sun

b) uses the energy in hot rocks

c) depends on the motion of the moon

d) uses energy from radioactive atoms

e) use the energy carried by rushing water?

3 a) What is the difference between a *renewable* and a *non-renewable* energy source?

b) COAL OIL BIOMASS WIND TIDES NATURAL GAS

Make two lists to show which of these are renewable sources of energy and which are not.

2.7

Gravitational potential energy and kinetic energy

It is often useful to be able to calculate quantities of energy. This section shows you how.

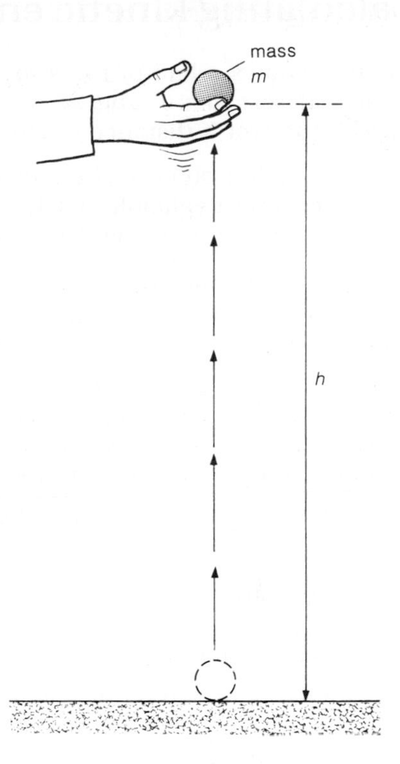

Figure 1 The work done is *force* × *distance moved* = *mgh*

Calculating gravitational potential energy

Gravitational potential energy is the energy an object possesses because of its position above the ground. The gravitational potential energy of the ball in figure 1 is equal to the work which would be done if the ball were to fall to the ground. Provided air resistance is negligible, it is also equal to the work which was done in lifting the ball up from the ground in the first place. This can be found by considering the force needed to lift the ball, and the distance the ball had to be moved:

downward force on ball = weight of ball = mg
∴ upward force needed to lift ball = mg
distance moved = h
∴ work done in lifting ball = force × distance moved
= mgh

An object of mass *m* at a vertical height *h* above the ground has a gravitational potential energy of *mgh*

If mass is measured in kg, g in m/s^2 or N/kg, and height in m, gravitational potential energy is measured in J.

For example:

If a ball of mass 0.5 kg is at a vertical height of 4 m above the ground, then:

$$\begin{aligned}\text{gravitational potential energy} &= mgh \\ &= 0.5\,\text{kg} \times 10\,\text{m/s}^2 \times 4\,\text{m} \\ &= 20\,\text{J}\end{aligned}$$

At twice this height, the gravitational potential energy of the ball would be 40 J.

Energy is a scalar quantity

Objects **A** and **B** in figure 2 have equal mass. **A** was lifted vertically from the ground, **B** was moved up the slope. Less force was used to move **B**, but the distance moved was greater. The same work was done in each case, and both objects have the same gravitational potential energy *mgh*. In calculating the gravitational potential energy of **A** or **B**, you need to know the vertical height lifted, but not the direction taken.

This example illustrates the fact that energy is a scalar quantity. Like volume or mass, it doesn't have direction.

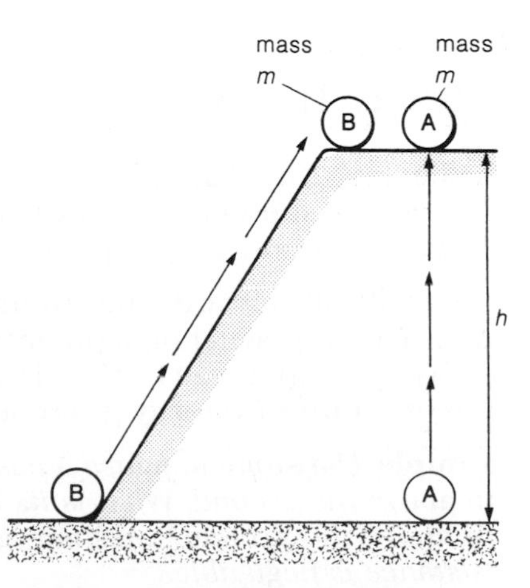

Figure 2 A and B have the same mass. The same work is done in lifting A and B to the same height. Both gain the same gravitational potential energy

Calculating kinetic energy

Figure 3 shows a ball of mass m travelling at a velocity v. The ball started at rest [$u = 0$] and gained its velocity v because a force F acted on it over a distance s. The acceleration of the ball was a.

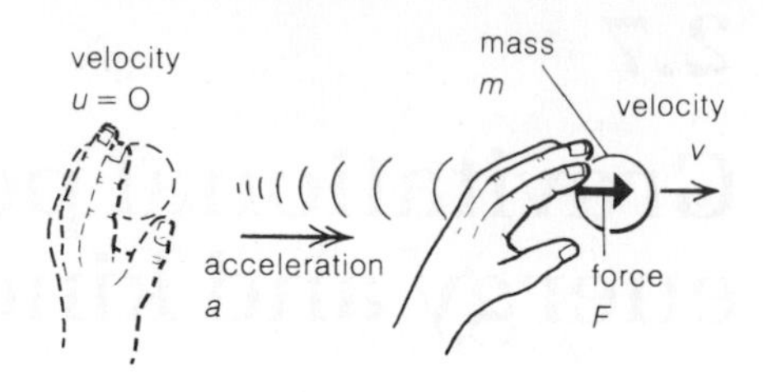

Figure 3 An object with mass m, moving with velocity v, has a kinetic energy of $\frac{1}{2}mv^2$

Like all moving objects, the ball has kinetic energy. Assuming air resistance was negligible, its kinetic energy is equal to the work which was done in giving it its velocity v:

work done on ball $= \text{force} \times \text{distance}$

$= Fs$

$\therefore$ kinetic energy of ball $= Fs$ ... *equation 1*

The kinetic energy of the ball depends on two factors: the mass of the ball, and its velocity. The problem is therefore to express the above equation *only* in terms of m and v. This can be done using $F = ma$ [Newton's second law] and $v^2 = u^2 + 2as$ [equation of motion]

As $u = 0$, the equation $v^2 = u^2 + 2as$ can be rewritten:

$$v^2 = 2as$$

Rearranged, this gives: $s = \dfrac{v^2}{2a}$

Returning to *equation 1*:

kinetic energy of ball $= Fs = ma \times \dfrac{v^2}{2a} = \frac{1}{2}mv^2$

In other words:

An object of mass m travelling at a velocity v has a kinetic energy of $\frac{1}{2}mv^2$

If mass is measured in kg and velocity in m/s, kinetic energy is measured in J. For example:

If a ball of mass 0.5 kg is travelling at a velocity of 4 m/s,

kinetic energy of ball $= \frac{1}{2}mv^2 = \frac{1}{2} \times 0.5\,\text{kg} \times (4\,\text{m/s})^2 = 4\,\text{J}$

Repeat this calculation for the same ball travelling at twice the velocity, and you find that the kinetic energy is four times its previous value.

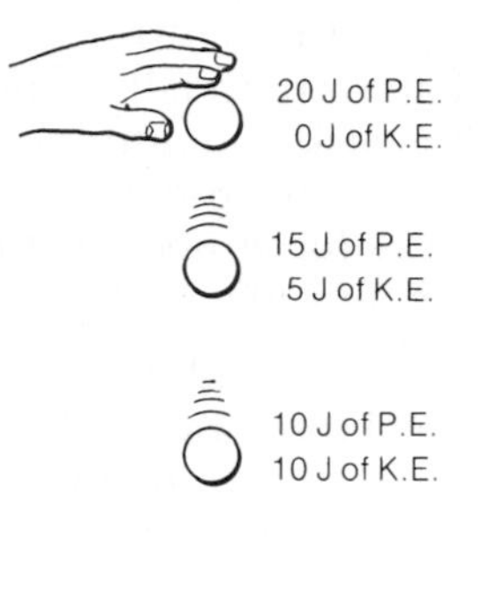

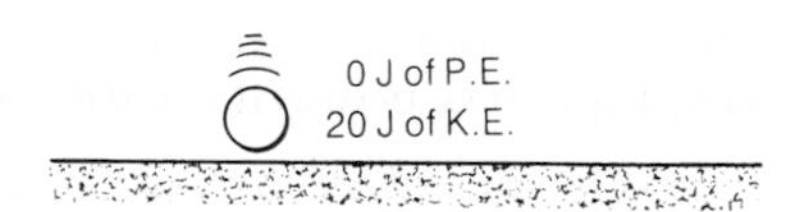

Figure 4 Transfer from gravitational potential energy (P.E.) to kinetic energy (K.E.)

Energy transfer

The ball in figure 4 starts with 20 J of gravitational potential energy. The diagram shows the energy changes which take place as the ball falls to the ground. For simplicity, air resistance has been neglected.

As the ball falls towards the ground, it gains its kinetic energy at the expense of its gravitational potential energy, so these two quantities always add up to 20 J. This is another example of the **law of conservation of energy** given on page 66.

Example *The stone in figure 5 has a mass of* 4 kg. *It is at a height of* 8 m *above the ground. What is its kinetic energy as it passes a point* 6 m *above the ground? [Assume* $g = 10\,\text{m/s}^2 = 10\,\text{N/kg}$, *and that air resistance is negligible.]*

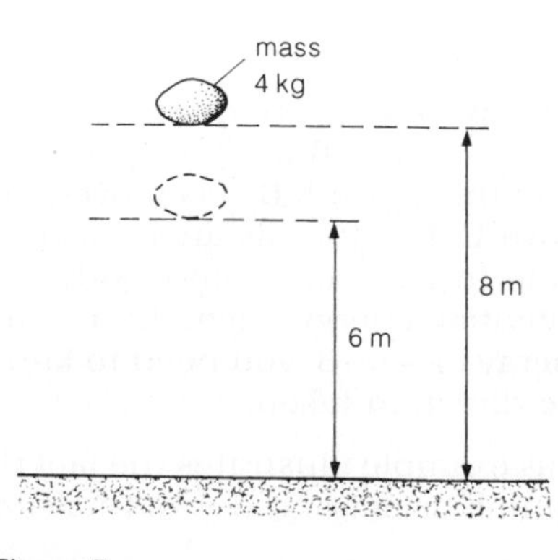

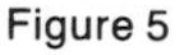

Figure 5

In problems of this type, don't immediately leap for the equation kinetic energy $= \frac{1}{2}mv^2$. It may be simpler to consider the energy changes taking place. When the stone drops from 8 m to 6 m, it loses as much gravitational potential energy as it gains kinetic energy, so try calculating that instead.

The change in height of the stone $= 2\,\text{m}$
$\therefore$ the loss of gravitational potential energy $= 4\,\text{kg} \times 10\,\text{m/s}^2 \times 2\,\text{m}$
$= 80\,\text{J}$
$\therefore$ the gain in kinetic energy $= 80\,\text{J}$

6 m *above the ground, the kinetic energy of the stone is* 80 J.

Knowing the mass of the stone and its kinetic energy the equation kinetic energy $= \frac{1}{2}mv^2$ could then be used to calculate the velocity.

Energy transfer – true or false? Problems involving energy transfer are not always numerical. Look at the diagrams in figure 6 before you read the explanations below. (In each example, the effects of air resistance and friction are neglected.)

true Both blocks start with the same amount of potential energy, so they have the same amount of kinetic energy when they reach ground level. **true** The pendulum gains the same amount of potential energy during the upward part of its swing as it loses during the downward part – despite the presence of the bench. **true** Whichever way the gun points, the shell leaves the barrel with the same amount of kinetic and potential energy. All of this energy is in the form of kinetic energy as the shell strikes the water.

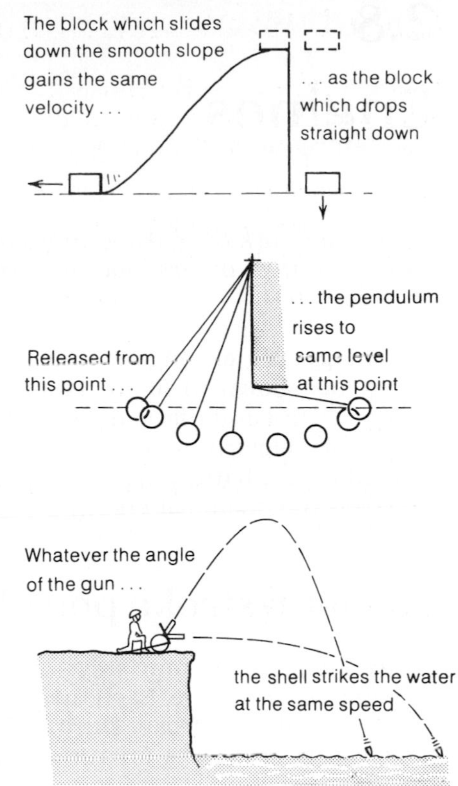

Figure 6 True or false? Consider the diagrams before looking at the text

Questions

[Assume $g = 10\,\text{m/s}^2 = 10\,\text{N/kg}$, and that frictional forces are negligible. $\sqrt{72} = 8.5$]

1 An object has a mass of 6 kg. Calculate its potential energy a) 4 m above the ground b) 8 m above the ground.
At what height above the ground will its potential energy be 360 J?

2 A stone has 100 J of potential energy. What will be its potential energy if its height above the ground is halved?

3 An object of mass 6 kg is travelling at a velocity of 5 m/s. What is its kinetic energy? What will be its kinetic energy if its velocity is doubled?

4 A ball of mass 0.5 kg has 100 J of kinetic energy. What is the velocity of the ball?

5 A trolley of mass 2.5 kg is pulled along a flat bench by a force of 5 N. When the trolley has travelled 4 m, find a) how much work has been done on it b) how much kinetic energy it has gained c) its velocity.

6 The stone in figure 7 has a mass of 2 kg. 72 J of work were done in moving the stone up the smooth slope from D to A.
a) What is the potential energy of the stone at A?
b) If the stone falls, what is its potential energy at B, the mid-point of its fall?
c) What is its kinetic energy at B?
d) What is the velocity of the stone as it passes B?
e) What is the velocity of the stone just before it strikes the ground?
f) If the stone were to slide back down the slope rather than fall, what would be its kinetic energy as it reached D? What would be its velocity at D?

7 A ball has a mass of 0.5 kg. Dropped from a cliff top, the ball strikes the sea below at a velocity of 10 m/s.
a) What is the kinetic energy of the ball as it is about to strike the sea?
b) What was its potential energy before it was dropped?
c) From what height was the ball dropped?
d) A stone of mass 1 kg also strikes the sea at 10 m/s. Repeat stages a, b, and c above to find the height from which the stone was dropped.

Figure 7

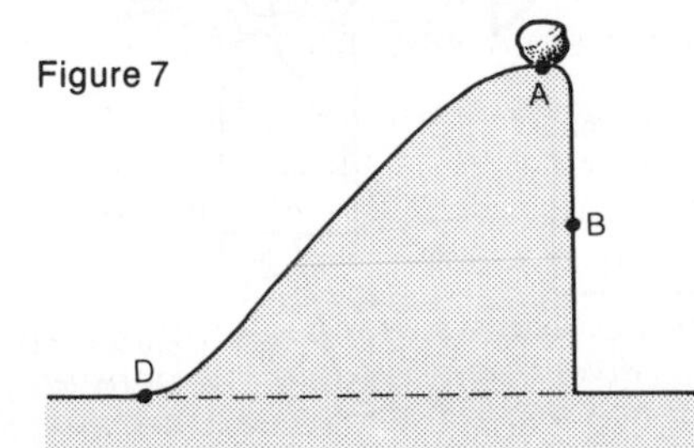

2.8
Engines

Engines do work by making things move. But some can do work at a faster rate than others, and all waste a great deal of energy in the process.

Put energy into an engine, and some of it is changed into kinetic energy [as shown in figure 1]; the job of an engine is to produce motion. Petrol and diesel engines use the chemical energy stored in their fuels; electric motors make use of the energy supplied by a generator or a battery. You are also a form of engine. Your energy source is the chemical energy stored in your food.

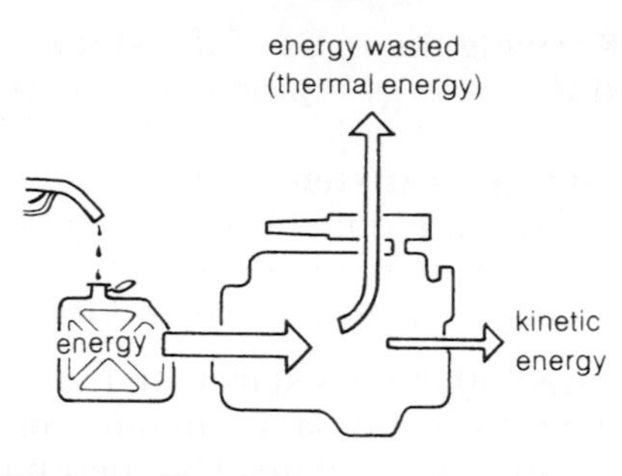

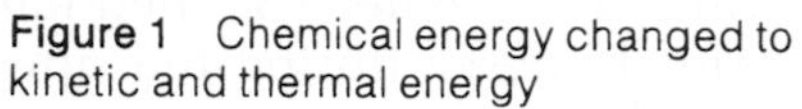
Figure 1 Chemical energy changed to kinetic and thermal energy

The four-stroke petrol engine

Figure 2 shows the four stages in the action of one type of single cylinder petrol engine. Each upward or downward movement of the piston is called a stroke. During the first stroke, the piston moves downwards to draw a mixture of air and petrol into the cylinder. During the second stroke, it moves upwards to compress it. During

Figure 2 A four-stroke engine: only one stroke in four produces power

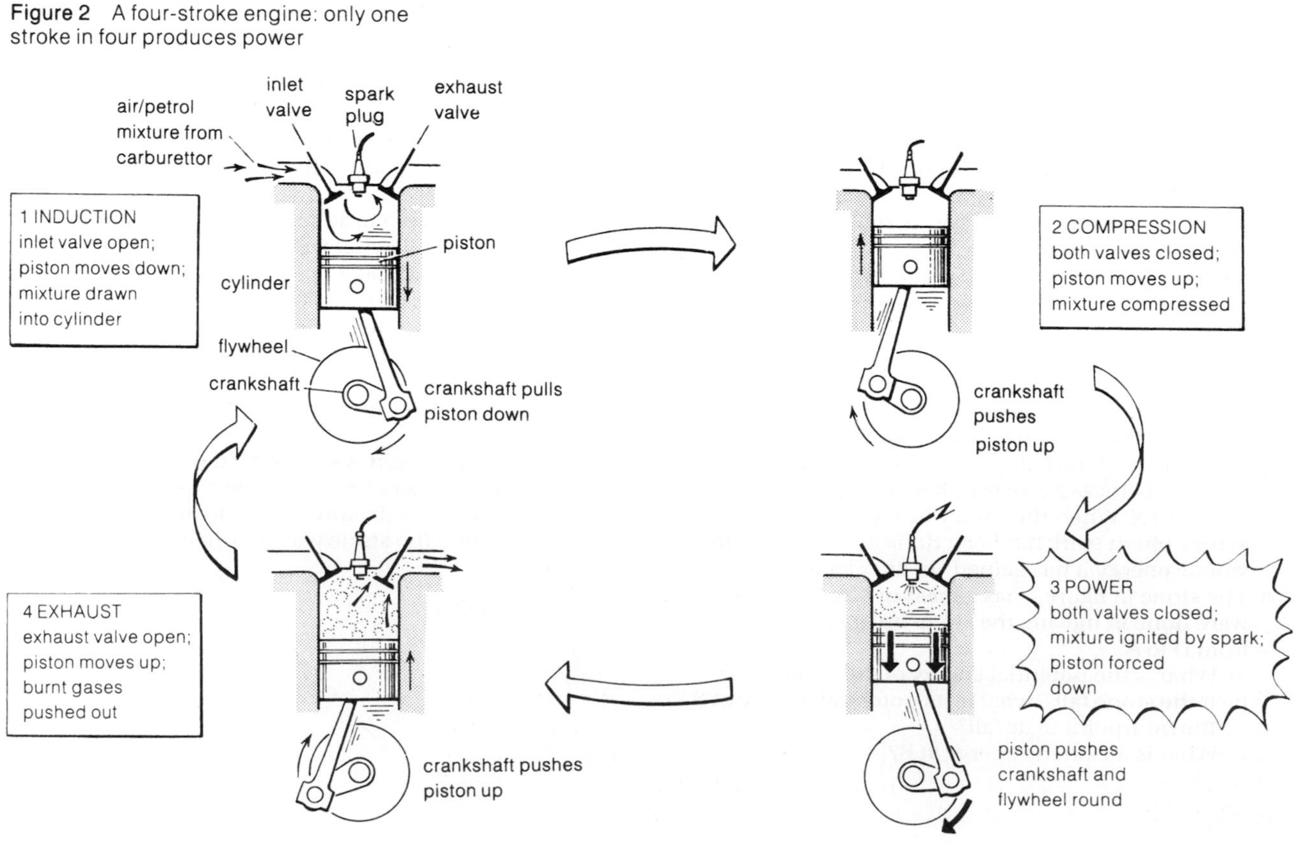

the third stroke, the piston is pushed forcefully downwards when a spark from the plug ignites the mixture. During the fourth stroke, the piston moves upwards to push the burnt gases out of the cylinder. The two valves are opened and closed at appropriate times to let in fresh mixture, seal the cylinder, or allow burnt gases out. When the engine is running fast, the whole process is repeated about fifty times every second.

The crankshaft of the engine is only pushed round during the power stroke. A heavy flywheel attached to the crankshaft keeps the engine turning through the other three strokes, these being needed to rid the engine of burnt gases and to prepare a fresh mixture for ignition. The engine shown in figure 3 is from a Metro; the valves and piston of one of the four cylinders can be seen on the left hand side; the flywheel is on the right.

Figure 3 The engine and gearbox of the 'Metro' car. The arrangement of one of the four cylinders can be seen at the top left of the diagram. The flywheel is built into the clutch, on the right hand side.

Efficiency

An engine does useful work with some of the energy supplied to it – in turning the wheels of a car for example. The rest of the energy is wasted as heat. The useful work done by an engine is called the **work output**. Figure 4 shows typical work outputs when different engines are each supplied with 100 J of energy.

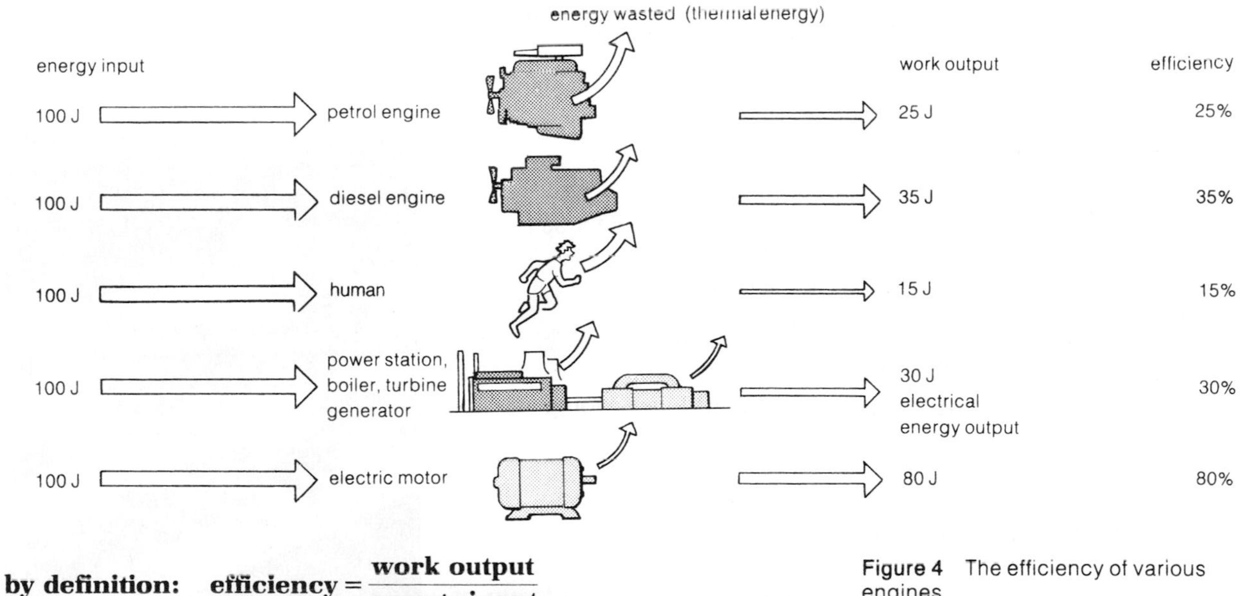

Figure 4 The efficiency of various engines

by definition: $\textbf{efficiency} = \dfrac{\textbf{work output}}{\textbf{energy input}}$

This is usually expressed as a percentage by multiplying by 100. Percentage efficiencies are given on the right-hand side of the chart. The electric motor seems to be a clear winner, but its efficiency is rather deceiving. Electricity for the motor has to be generated, and the efficiency of that process is only about 30%.

Why are engines so inefficient? Frictional losses are only part of the problem. When fuel burns in an engine, part of the thermal energy released *cannot* be changed into any other form and must stay as thermal energy. This wasted energy is carried away from the engine by the exhaust gases and the engine's cooling system.

Power

A small engine can do just as much work as a larger engine, but it takes longer to do it. The larger engine can do work at a faster rate.

The rate at which work is done is called the **power**:

$$\textbf{power} = \frac{\textbf{work done}}{\textbf{time taken}} \quad \text{or} \quad \frac{\textbf{energy transferred}}{\textbf{time taken}}$$

Power is measured in joules per second [J/s], or **watts** [**W**]. If an engine does 1 joule of useful work every second, it has a power output of 1 W. If an engine does 4000 J of work in 10 s, it has a power output of 400 W. Some typical engine powers are given in figure 5; larger powers are also given in **kilowatts** or **megawatts**.

1 kilowatt [kW] = 1000 W [1000 J/s]
1 megawatt [MW] = 1 000 000 W [1 000 000 J/s]

Sometimes, engine powers are given in **horsepower** (**hp**), a unit which dates from the time when steam engines first replaced horses as a power source. 1 hp equals 746 W, or about $\frac{3}{4}$ kW.

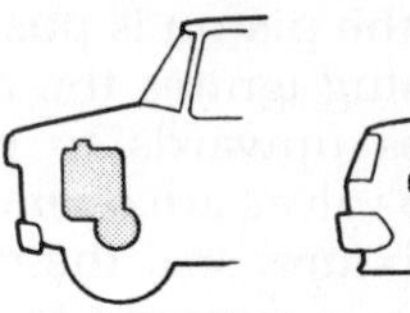

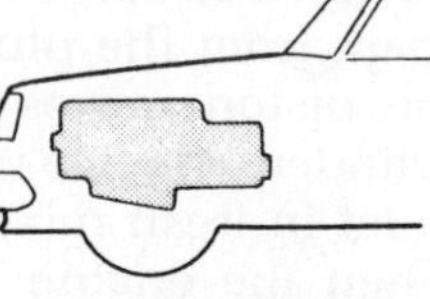

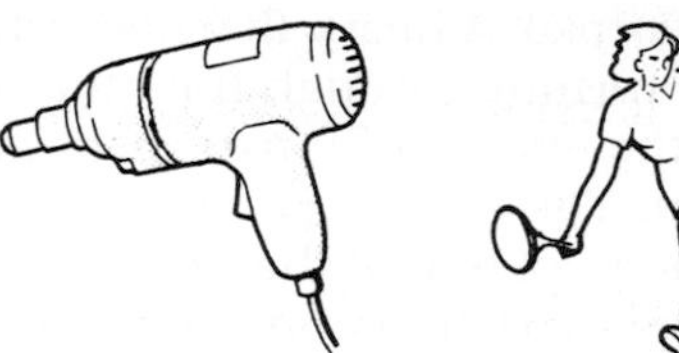

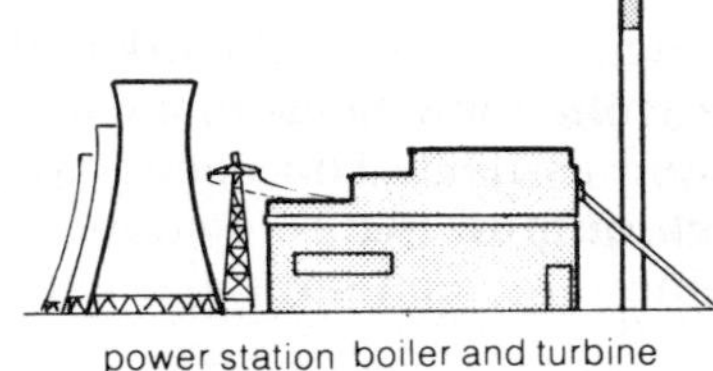

Figure 5 The power of various engines

Measuring human power

The athlete in figure 6 is doing work by running up the steps. The average power output can be calculated from the following measurements:

mass of athlete = 48 kg
vertical height from ground to top of steps = 50 m
time taken to run up steps = 60 s

$$\text{Gravitational potential energy gained} = mgh$$
$$= 480\ \text{kg} \times 10\ \text{m/s}^2 \times 50\ \text{m}$$
$$= 24\,000\ \text{J}$$

$\therefore$ work done in running up the steps = 24 000 J

$$\text{his average power} = \frac{\text{work done}}{\text{time taken}} = \frac{24\,000\ \text{J}}{60\ \text{s}} = 400\ \text{W}$$

The average power of the athlete is 400 W; the athlete is doing 400 J of work every second.

Figure 6 Running up stairs to light the Olympic flame

Power and velocity

Example *The car in figure 7 is travelling at a steady velocity of* 30 m/s *along a level road. If the total of the frictional forces acting on the car are* 700 N, *what is the power output of the engine?*

First, work out the forward force acting on the car:

As the car has a constant velocity, the resultant force on the car must be zero [Newton's first law]. The engine therefore provides a forward force of 700 N to balance the total frictional force.

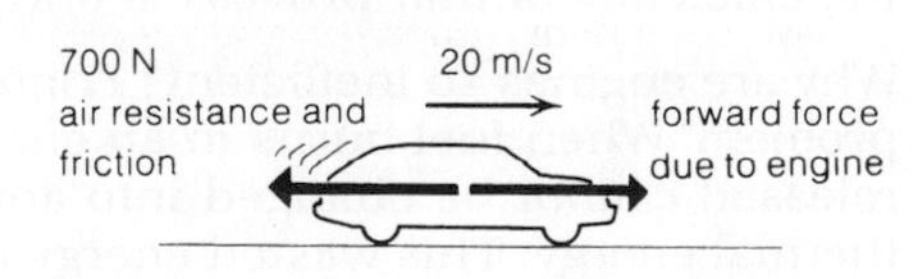

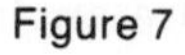

Figure 7

Next, calculate how much work the engine does in 1 second:

In 1 second, the 700 N force moves a distance of 30 m;

$\therefore$ the work done = force × distance = 700 N × 30 m = 21 000 J

As the engine does 21 000 J of work in 1 second, its power output is 21 000 W, or 21 kW.

Problems of this type can also be solved using an equation:

average power = force × average velocity

Check it using the figures given in the example.

Power and efficiency

The efficiency of an engine was originally defined as $\frac{\text{work output}}{\text{energy input}}$; efficiency is also equal to $\frac{\textbf{power output}}{\textbf{power input}}$.

Example *The crane in figure 8 lifts a* 100 kg *block of concrete through a vertical height of* 16 m *in* 20 s. *If the power supplied to the motor driving the crane is* 1 kW, *what is the efficiency of the motor?*

First, find out the power output of the motor by calculating the work done every second in lifting the concrete block:

$$\begin{aligned}\text{Gravitational potential energy gained by the block} &= mgh\\ &= 100\,\text{kg} \times 10\,\text{m/s}^2 \times 16\,\text{m}\\ &= 16\,000\,\text{J}\end{aligned}$$

$\therefore$ the work done in lifting the block = 16 000 J

$$\text{power output of the motor} = \frac{\text{work done}}{\text{time taken}} = \frac{16\,000\,\text{J}}{20\,\text{s}} = 800\,\text{W}$$

Knowing the power output and the power supplied, the efficiency can now be calculated.

$$\text{efficiency} = \frac{\text{power output}}{\text{power input}} = \frac{800\,\text{W}}{1000\,\text{W}} = 0.8 \text{ or } 80\%.$$

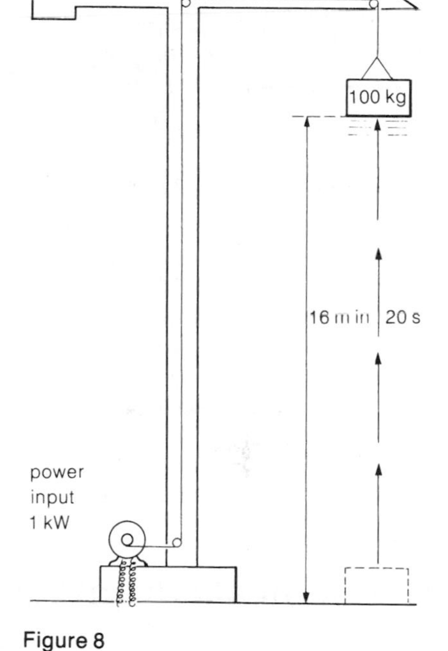

Figure 8

Questions

1 Fuel is supplied to a single cylinder, four stroke petrol engine in order to produce motion.
  a) Name the four strokes. What happens during each?
  b) Why is a flywheel necessary?
  c) The engine does 72 kJ of work in 1 minute. What is its power output?
  d) If its efficiency is 20%, what is the power input?

2 The power output of an engine is 3 kW. What does this mean?
How much work does the engine do in 20 s?
The efficiency of the engine is 30%. How much energy is supplied to the engine in 20 s?
What power is supplied to the engine?

3 A man weighing 600 N climbs a vertical distance of 10 m in 20 s. What is his average power output?

4 Because of air resistance and friction, a forward force of 2500 N is needed to keep a lorry travelling at a steady velocity of 20 m/s. What power must be provided by the engine?

5 A crane can lift a 600 kg mass through a vertical height of 12 m in 18 s. What is the power output of the motor driving the crane? If the motor has an efficiency of 80%, what power input is required?

2.9

Machines

A car jack is a machine. So is a pair of scissors or a simple lever. You don't do any less work by using a machine, but it does enable you to do work more conveniently.

The lever in figure 1 is a simple machine. You can lift a **load** on the far end of the lever by applying a force called the **effort**, at the near end. The lever turns about a point called the **fulcrum** or **pivot**.

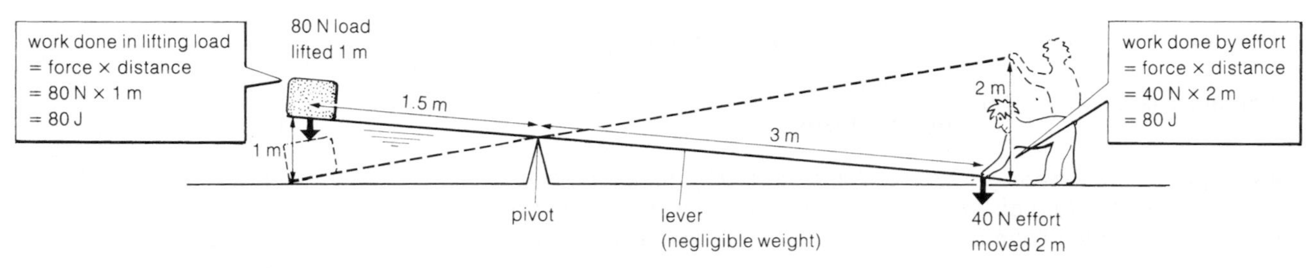

Figure 1 A simple machine in operation. The effort is smaller than the load, but the work done is the same

With this size of lever, an effort of only 40 N is required to lift the load of 80 N; this follows from the principle of moments previously explained on page 54. There is however a price to pay for this increase in lifting ability. Moving the effort 2 m downwards, only raises the load 1 m. Note from the figures in the diagram that:

1 the load is twice as much as the effort.
2 the effort moves twice as far as the load.
3 the work done by the effort is the same as the work done in raising the load. In this simple example, no energy is wasted, and the efficiency of the lever is 100%.

Mechanical advantage and velocity ratio

A machine has a high **mechanical advantage** if the load lifted is large compared with the effort. The mechanical advantage [**M.A.**] of a machine is defined as follows:

$$\textbf{M.A.} = \frac{\textbf{load}}{\textbf{effort}}$$

The lever in figure 1 has a mechanical advantage of 2. Its M.A. would be greater if the load were closer to the pivot or the effort further away.

A machine with a high mechanical advantage also has a high **velocity ratio** [**V.R.**]. This compares the distance moved by the effort with the distance moved by the load:

$$\textbf{V.R.} = \frac{\textbf{distance moved by effort}}{\textbf{distance moved by load}}$$

The lever in figure 1 has a velocity ratio of 2. The velocity ratio in this case is equal to the mechanical advantage. This is true for any machine which has an efficiency of 100%.

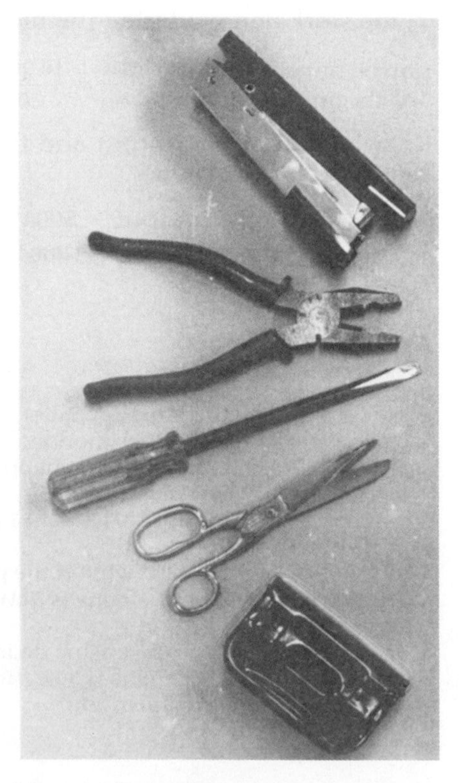

More simple machines

Common levers

Figure 2 shows several examples of levers in everyday use. The velocity ratio of any lever including those in figure 2, can be calculated as follows:

$$\text{V.R.} = \frac{\textbf{distance of effort from pivot}}{\textbf{distance of load from pivot}}$$

Check this equation by applying it to the lever in figure 1. The velocity ratio of the lever is 2.

The mechanical advantage of a lever is the same as its velocity ratio provided friction at the pivot is negligible. The wheelbarrow, scissors and hammer shown in figures 2a, 2b and 2c have mechanical advantages greater than 1. The human arm shown in figure 2d is the exception.

In this case, the load lifted at the end of the arm is much smaller than the effort produced by the muscle. The arm loses in lifting ability but it gains in moving ability; only a small movement of the muscle is needed to move the end of the arm through a large distance.

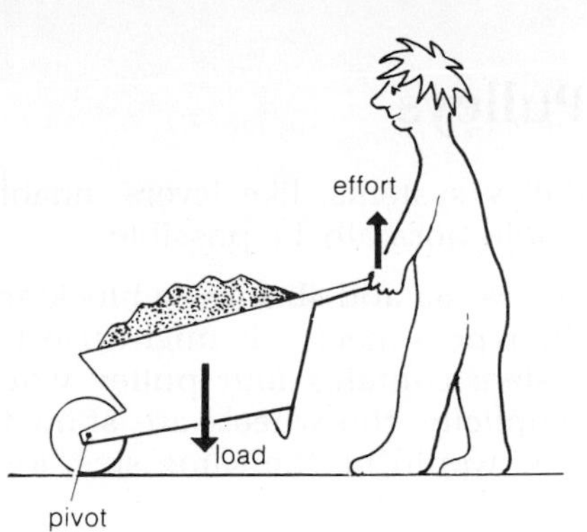

Figure 2a Lifting a wheelbarrow is easier than lifting its contents directly

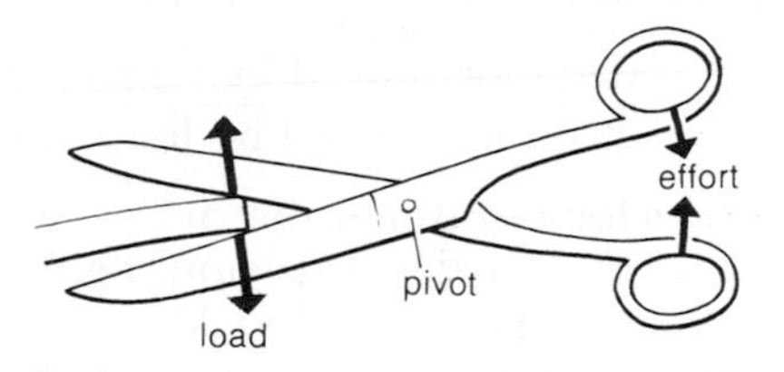

Figure 2b Scissors provide a sharp cutting edge and an increased cutting force

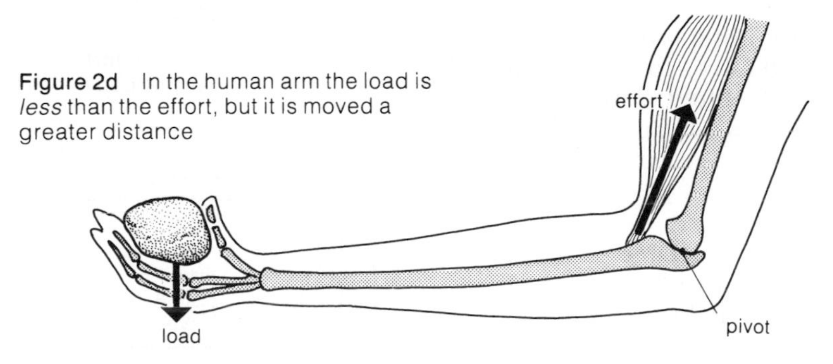

Figure 2d In the human arm the load is *less* than the effort, but it is moved a greater distance

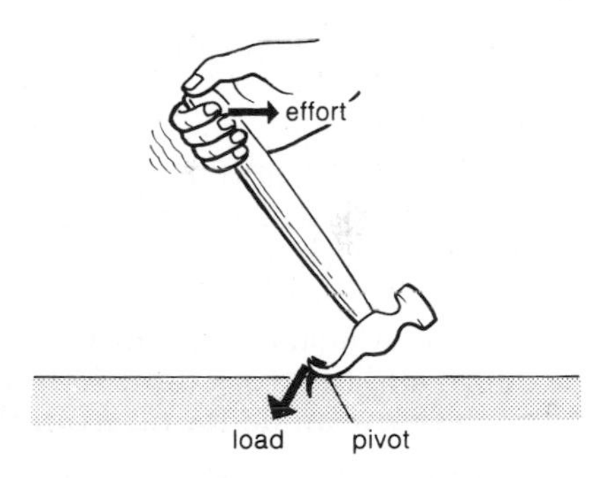

Figure 2c A small effort can provide the force needed to pull out a nail – if it doesn't pull the head off first!

The ramp [inclined plane]

A ramp counts as a machine even if it doesn't look much like one. You can pull a load up a ramp using less force than you would need to lift it vertically. In figure 3, the load is the weight of the trolley and the effort is the force needed to pull the trolley up the ramp.

$$\textbf{the V.R. of the ramp} = \frac{\textbf{distance moved by effort}}{\textbf{vertical distance moved by load}}$$

$$= \frac{\textbf{length of ramp}}{\textbf{height of ramp}} = \frac{l}{h}$$

If friction is negligible, the mechanical advantage of the ramp will also be l/h. The ramp in figure 3b for example, has a mechanical advantage of 3 if there are no frictional forces acting.

Note that l/h is equal to $\frac{1}{\sin\theta}$, where θ is the angle of the ramp.

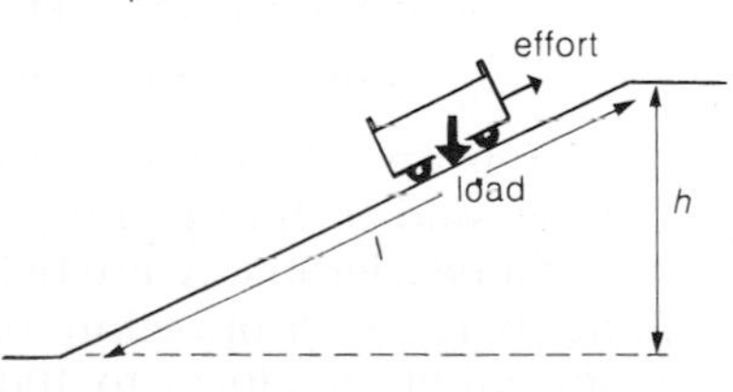

Figure 3a A ramp is also a machine. The velocity ratio is l/h

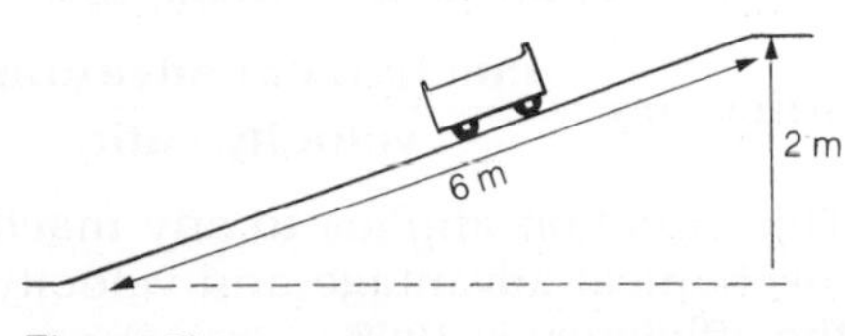

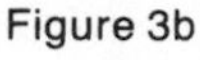

Figure 3b

Pulleys

Pulley systems, like levers, enable larger loads to be lifted than would normally be possible.

Figures 4a and 4b show a block-and-tackle pulley system similar to the type a mechanic might use to lift the engine out of a car. The system contains four pulley wheels mounted in two blocks. For simplicity, the wheels are shown above one another. In practice, they would be the same size, and mounted side by side in each block.

The velocity ratio of the pulley system can be found by comparing the distances moved by load and effort. For the load to be lifted 1 m, each section of rope, 1, 2, 3, and 4, has to be shortened by 1 m. These sections are all part of one long rope, so the end of the rope has to be pulled down a total distance of 4 m.

$$\text{V.R.} = \frac{\text{distance moved by effort}}{\text{distance moved by load}} = \frac{4\text{ m}}{1\text{ m}} = 4$$

The velocity ratio of the pulley system is 4. In practice, this can be found by counting the ropes between the blocks, or by counting the pulley wheels in the blocks.
The single downward force on the end of the rope puts the whole rope under tension and provides four upward forces to lift the load. You might therefore expect the 600 N load to be lifted with an effort of only 150 N. If this were so, the mechanical advantage of the system $\left[\text{or } \frac{\text{load}}{\text{effort}}\right]$ would be 4.

In reality, things are not so simple. Extra effort is needed to lift the lower pulley block and to overcome friction on the pulley wheels. In the example shown, a total effort of 200 N is needed to lift the 600 N load, so the mechanical advantage is only 3.

Efficiency Calculations on the diagram show the work done by the effort, and the work done in lifting the load. The efficiency of the pulley system can be found from these figures:

work output = the work done in lifting the load = 600 J

energy input = work done by the effort = 800 J

$$\therefore \text{efficiency} = \frac{\text{work output}}{\text{energy input}} = \frac{600\text{ J}}{800\text{ J}} = 0.75 \quad \text{or} \quad 75\%$$

The pulley is only 75% efficient because one-quarter of the energy input is wasted in lifting the lower pulley block and overcoming friction. If a heavier load were being lifted, the energy wasted would be a much lower proportion of the total energy input and the efficiency would be closer to 100%.

Note that the efficiency of the pulley system in the diagram is equal to the mechanical advantage divided by the velocity ratio:

$$\textbf{efficiency} = \frac{\textbf{mechanical advantage}}{\textbf{velocity ratio}} = \frac{3}{4} = 0.75$$

This equation applies to any machine. It again illustrates that the mechanical advantage and velocity ratio of a machine are equal if the efficiency is 100%.

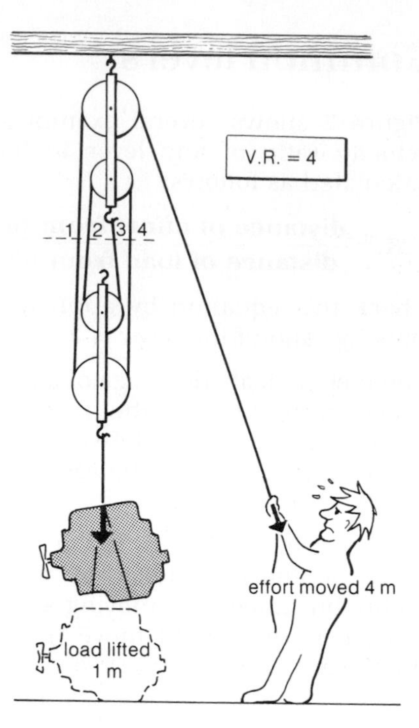

Figure 4a In this system, the effort moves four times as far as the load – the *velocity ratio* is four

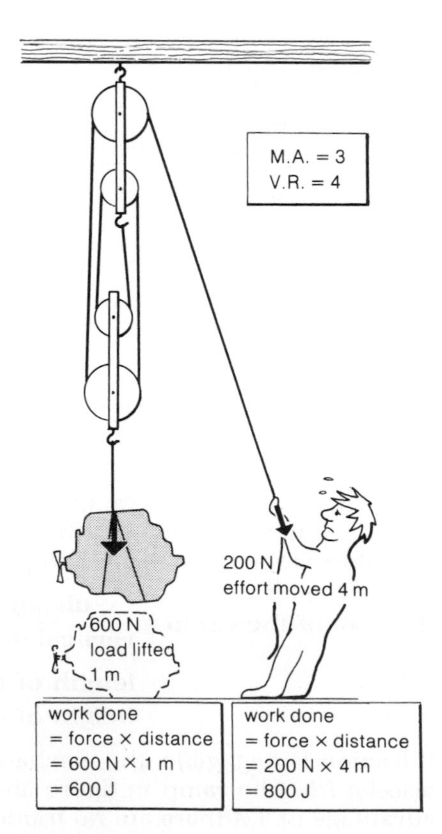

Figure 4b The velocity ratio is still four but the mechanical advantage – the ratio of the *forces* – is only three

The screw jack

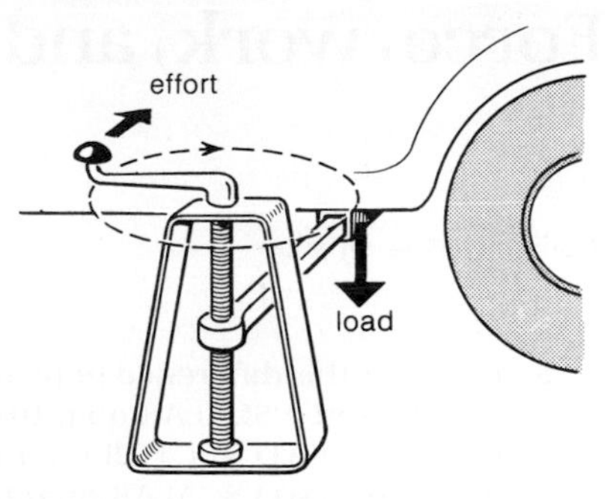

Figure 5a A car jack is likely to have a V.R. of 500 and an M.A. of 200. Would it be an advantage if it was any more efficient?

Figure 5a shows a simple screw jack of the type you might use to jack up the side of a car. As you wind the handle at the top, the nut gets carried up the thread of the screw and the car is lifted upwards. One complete turn of the handle raises the car a distance equal to the **pitch** of the screw – the distance from one thread to the next [as shown in figure 5b].

The effort is the force used on the handle. The load is the downward force from the side of the car:

$$\textbf{the V.R. of the screw jack} = \frac{\textbf{circumference of circle made by handle}}{\textbf{pitch of screw}}$$

A typical car jack might have a velocity ratio of about 500, and a mechanical advantage of about 200. Friction is usually considerable in these machines – and very useful, as it stops the screw unwinding when you let go of the handle.

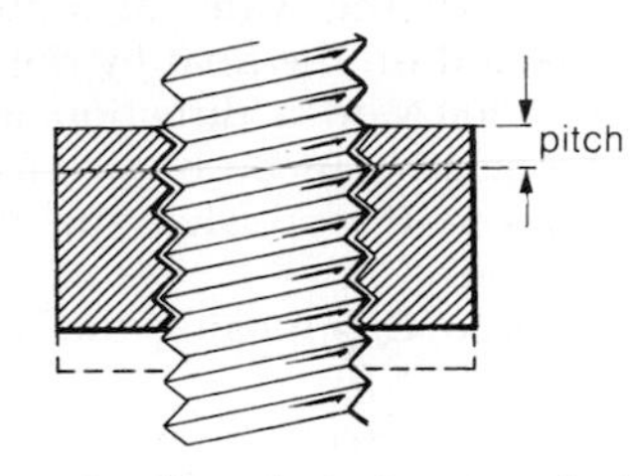

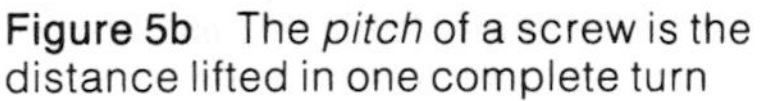

Figure 5b The *pitch* of a screw is the distance lifted in one complete turn

Gears and couples

Gears do for couples what levers do for forces: they magnify or reduce their size. The simple gearing system in figure 6 has a velocity ratio of 2; the input 'effort' shaft has to be turned twice for the output 'load' shaft to turn once.

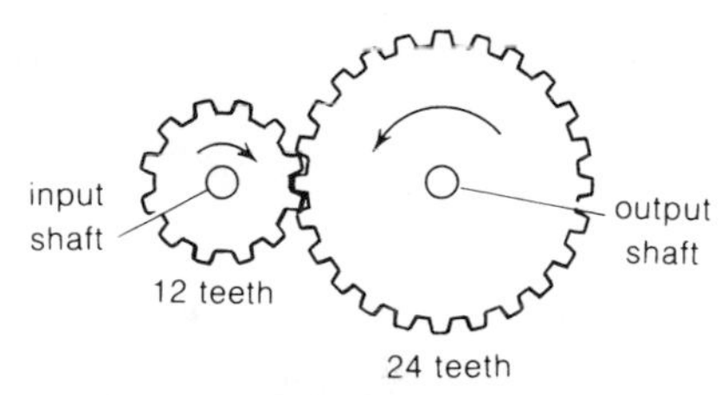

Figure 6 These gears will increase the couple. How could they be used to reduce it? What would be gained from doing this?

A car gearbox (figure 7) uses the same basic principle. It contains a number of gear wheels which can be connected together in different combinations. Bottom gear has a velocity ratio of about 4: the input shaft turns four times faster than the output shaft. When top gear is engaged, the input and output shafts are locked together so that they both turn at the same speed. Here, the velocity ratio is one.

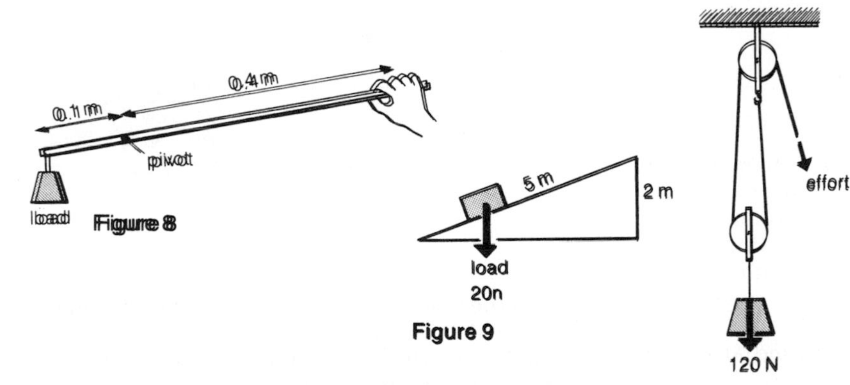

Figure 10

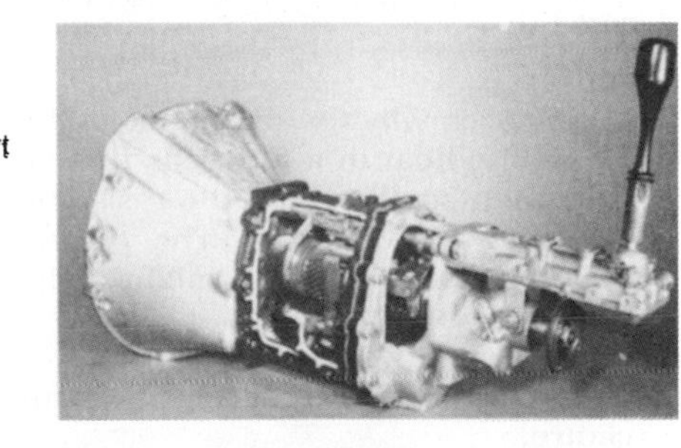

Figure 7 Car gearbox

Questions

1 What is meant by a) the mechanical advantage of a machine b) its velocity ratio?
2 What is the velocity ratio of the lever shown in figure 8? What is its mechanical advantage assuming there is no friction at the pivot? Suggest two ways in which the mechanical advantage could be increased.
3 Give an example of a lever with a M.A. less than 1.
4 Assuming friction is negligible, what is the mechanical advantage of the ramp shown in figure 9? What force is needed to pull the block up the ramp? How much work is done on the block?
5 What is the velocity ratio of the pulley system shown in figure 10?
If its efficiency is 75%, what is its mechanical advantage?
If the load is 120 N, what is the size of the effort? What work is done by the effort if the load is lifted 0.5 m? How much energy is wasted?

Force, work, and energy: part A

$g = 10\ m/s^2 = 10\ N/kg$

1 a) i) State the difference between a VECTOR quantity and a SCALAR quantity.
ii) Name TWO VECTOR quantities.
iii) Name TWO SCALAR quantities.
b) Explain the difference between adding two vector quantities and adding two scalar quantities.
c) In figure 1, a sailing boat is steered due North through the water at a steady speed. The force exerted on the boat by the wind has a magnitude of 1200 N in a direction 60° East of North. This force is balanced by two frictional forces:
force **P** opposing the forward motion of the boat,
force **Q** opposing the sideways motion of the boat.
Find, by scale drawing or calculation, the magnitudes of the forces **P** and **Q**.

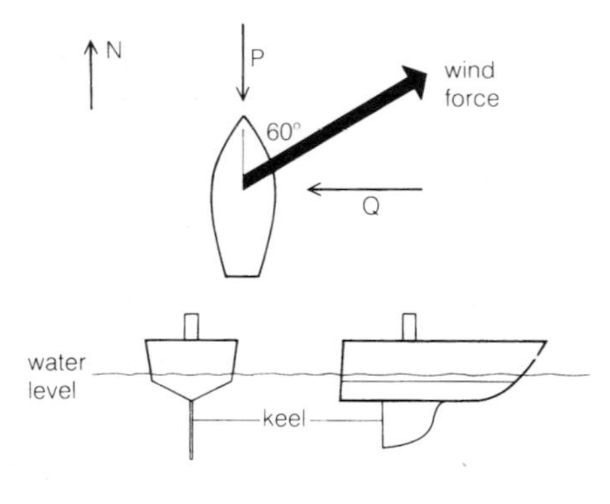

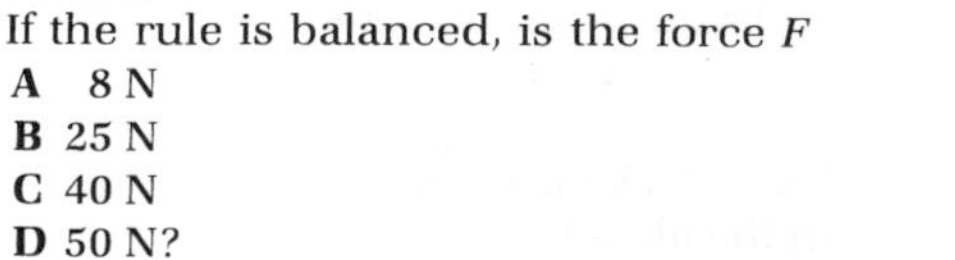

Figure 1

d) Suggest why the resistance of the water to motion of the boat in a forwards direction is likely to be much smaller than the resistance to motion in a sideways direction. The diagrams showing the shape of the boat might help you with your explanation. (SEG)

2 Figure 2 shows a uniform metre rule pivoted at its centre.
If the rule is balanced, is the force *F*
A 8 N
B 25 N
C 40 N
D 50 N? (SEG)

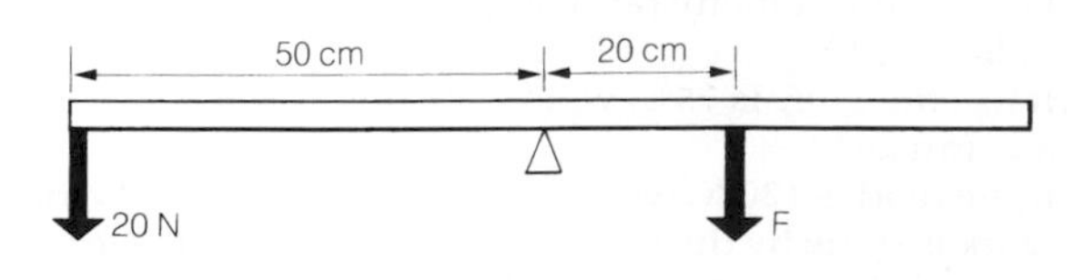

Figure 2

3 Figure 3 shows a simple bottle opener being used to remove the top from a bottle.

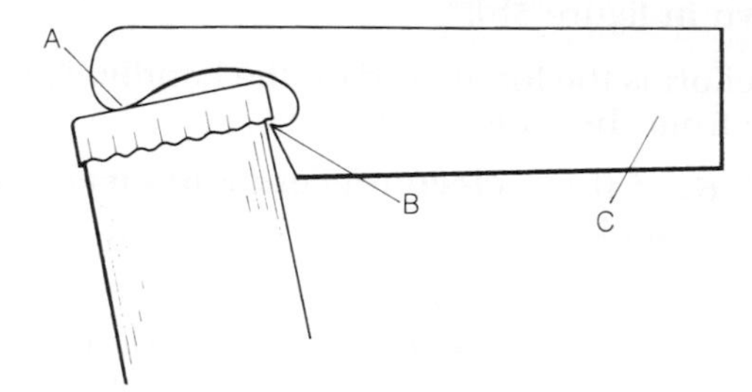

Figure 3

a) The bottle opener is being used as a lever. Which of the points A, B, or C, is the fulcrum (pivot) of the lever?
b) At which point A, B, or C, is the *load* applied to this lever?
c) Is the force on the bottle top at the point B greater or less than the force used at C? Give a reason.
d) How could you improve the design of the bottle opener so that less force is required to remove the bottle top? (SWEB/SEG)

4 In figure 4, a steam engine drives a generator which lights a lamp.

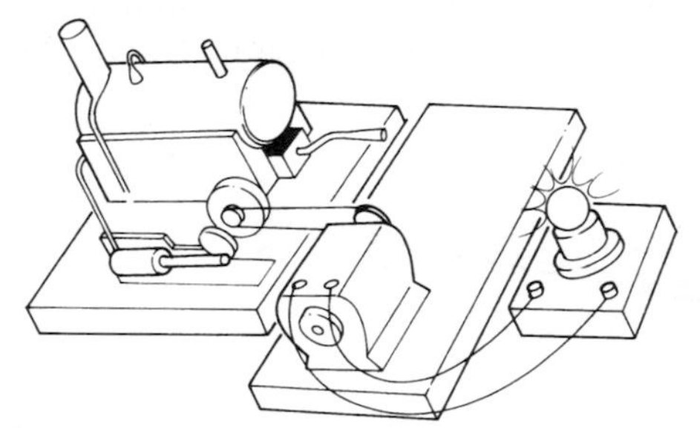

Figure 4

Which of the following lines best describes the energy changes which occur?
a) Heat — Electrical — Heat and Light
b) Electrical — Heat — Kinetic
c) Heat and Light — Kinetic — Electrical
d) Heat — Heat and Light — Electrical (LEAG)

5 Copy and complete the following sentences:
In each case state the main energy change involved.
a) An electric kettle converts electrical energy to .... energy.
b) A microphone converts sound energy to .... energy.
c) A lift raising a load converts electrical energy to .... energy.
d) A lamp converts .... energy to light energy. (SEG)

6 In a hydro-electric generating station water falls through pipes from a high reservoir to a turbine. The turbine drives an alternator.

a) Copy and label the block diagram in figure 5, with the three parts of the system. On the diagram show clearly the main energy changes in the system.

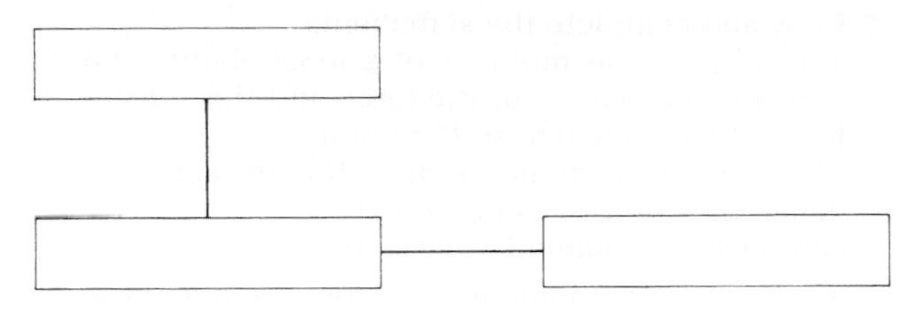

Figure 5

b) Write down the names of THREE different sources of energy which are used for driving an electrical generating system. (SEG)

7 A motor car of weight 7000 N is travelling along a level road at a constant speed of 20 m/s. When it comes to a hill, which rises vertically 100 m and is 1.0 km long, the driver increases the power output of his engine to keep the speed constant at 20 m/s.

a) How long does it take the car to climb the hill?

b) How much work does the car do against gravity as it climbs the hill?

c) What power is needed to do this work against gravity?

d) On the return journey the car crosses the top of the hill in the opposite direction at 20 m/s and the driver then disconnects the engine by pushing the clutch pedal down. Explain, in terms of energy changes, why the speed of the car increases. Include the effects of air resistance in your explanation. (MEG)

8 Is the energy supplied in 1 minute to a 3 kilowatt heater

A 50 J **C** 3000 J

B 180 J **D** 180 000 J? (SEG)

9 Figure 6 shows a mass moving up and down on the end of a spring. X and Y are the highest and lowest points reached by the mass.

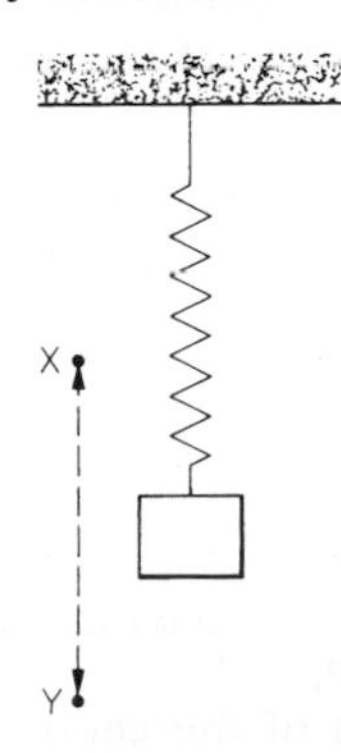

Figure 6

Which one of the following is true when the mass is at its highest point X?

a) Kinetic energy is a maximum; potential energy is a minimum

b) Kinetic energy is a maximum; potential energy is zero

c) Kinetic energy is zero; potential energy is a maximum

d) Kinetic energy is equal to potential energy (SEG)

10 Figure 7 shows a man hoisting a weight of 200N from the ground floor to the first floor of a warehouse. He is using a frictionless pulley.

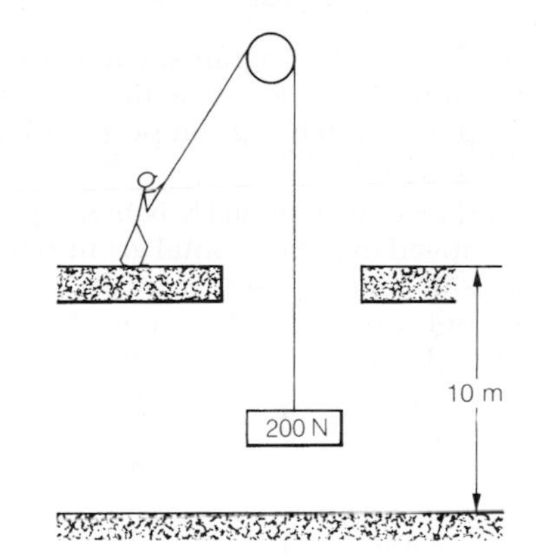

Figure 7

How has the energy of the weight changed when it is finally at rest on the first floor?

a) Its potential energy has decreased

b) Its kinetic energy has increased

c) Its kinetic energy has decreased

d) Its potential energy has increased. (LEAG)

11 In figure 8, the spring has an unstretched length of 20 cm. When a 6N weight is hung on, it stretches to 32 cm.

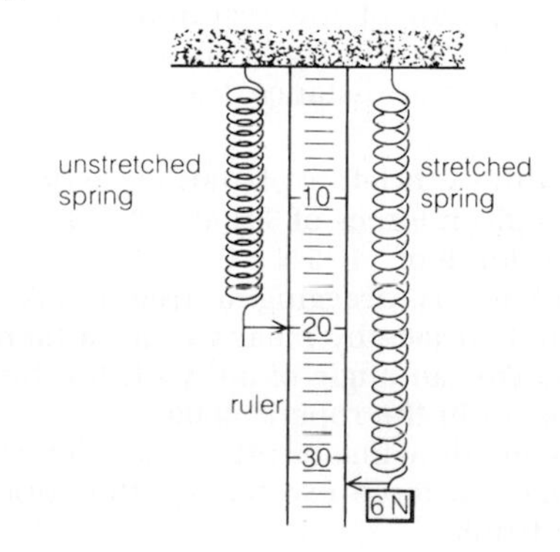

Figure 8

If the 6N weight is replaced by a 5N weight, will the new length be:

A 10 cm

B 24 cm

C 28 cm

D 30 cm? (LEAG)

Further questions

Force, work, and energy: part B

Some questions require a knowledge of concepts covered in earlier units

$g = 10\,m/s^2 = 10\,N/kg$ $\cos 36.9° = 4/5$
$\sin 30° = \cos 60° = 0.50$ $\sin 36.9° = 3/5$
$\cos 30° = \sin 60° = 0.87$ $\tan 36.9° = 3/4$
$\sqrt{60} = 7.75$ $\sqrt{0.6} = 0.775$

1 Explain, with suitable diagrams, what is meant by:
i) the resultant of two forces acting at right angles;
ii) the components in two given perpendicular directions of a single force. (O)

2 A garden roller of weight 400 N is being pulled along at a steady speed over horizontal ground by means of a force of 300 N acting parallel to the ground. If the roller does not slip, what is the value of the horizontal force exerted by the ground on the roller? (O)

3 Figure 1 shows two horizontal forces of magnitude 15 N and 20 N, acting in perpendicular directions on a small body O.

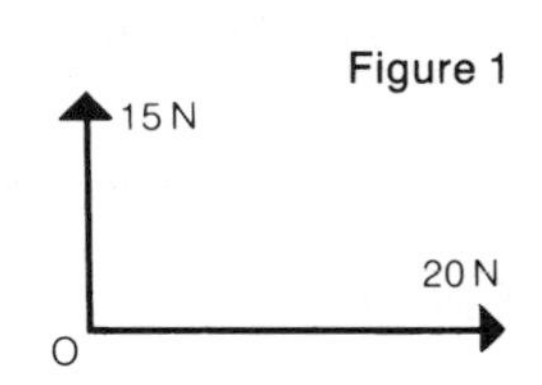

Figure 1

a) Determine the magnitude of the resultant force acting on the body.
b) The body rests on a horizontal ice surface and friction between the body and the ice can be neglected. What is the angle between the direction of motion of the body and the direction of the 20 N force?
c) The body accelerates at $4.0\,m/s^2$. What is its mass?
d) The forces continue to act so that the body moves in the direction of the resultant force for 6.0 s. Calculate i) the velocity acquired by the body in this time, ii) the distance travelled in this time. (L)

4 a) State, with a brief explanation, whether it is possible, given forces of 3 N and 8 N, to produce a resultant force of i) 5 N ii) 15 N iii) 8 N.
b) An elephant is dragging a tree trunk along a horizontal surface by means of an attached rope which makes an angle of 30° with the horizontal. The tension in the rope is 4000 N.
i) By scale drawing, or otherwise, determine the horizontal force exerted by the rope on the tree trunk.
ii) Determine also the vertical component of the force exerted by the rope on the tree trunk. Explain why it serves a useful purpose.
iii) Name the other forces which act on the tree trunk and show clearly the directions in which they act. (L)

5 Copy and complete the statements:
The value of the moment of a force about a point depends upon the * of the force and the * distance from its line of action to the point.
The principle of moments states that the sum of the * moments acting on a body which is at * is equal to the sum of the * moments, measured about any point.

6 a) Define the moment of a force about an axis. Explain what is meant by a couple.
b) The steering-wheel of a car, of radius 0.25 m, requires a couple of moment 12 N m to turn it.
i) Draw a diagram showing how this couple is provided when both hands are used to turn the wheel clockwise, and calculate the force exerted by each hand.
ii) Draw a diagram showing how this couple is provided when only one hand is used to turn the wheel clockwise, and calculate the force exerted by this hand. (O)

7 Figure 2 shows a simple form of diving board. If the diver has a mass of 40 kg, calculate the magnitude and show the direction of the forces acting on the board at A and B if the mass of the board is negligible. Describe, without calculation, what would happen to the forces at A and B if the diver were to walk towards B. (WJEC)

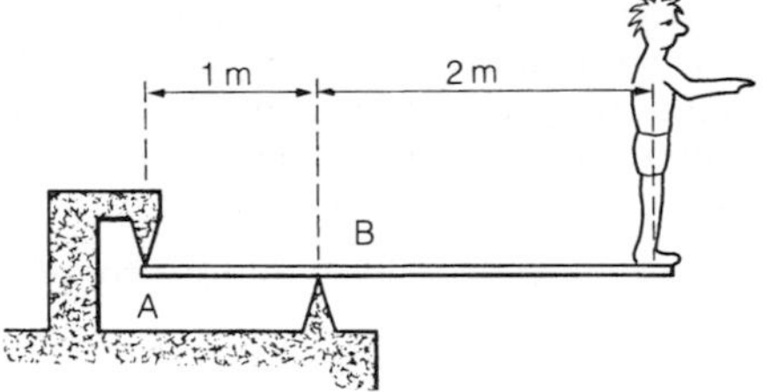

Figure 2

8 a) define the moment of a force about an axis, and explain what is meant by the centre of mass of a body.
b) A non-uniform bar AB of mass 0.60 kg balances horizontally on a pivot P 0.15 m from the end A when a 0.24 kg mass M is suspended from A, as in figure 3. Find:

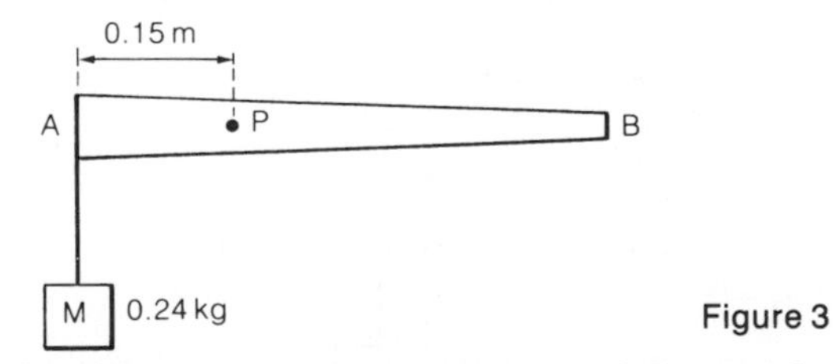

Figure 3

i) the moment (in newton-metres) of the weight of M about P;
ii) the distance of the centre of mass of the bar from the end A. (O)

9 A bridge of weight 50 000 N is supported on piers A and B 20 m apart. The weight of the bridge acts at its mid-point. A lorry, X, of weight 30 000 N stands with its centre of gravity 4 m from pier A. Find the forces acting on each pier. (WJEC)

10 The uniform rigid beam AB in figure 4 has mass 20 kg and is 4.0 m long. It rests on the fulcrum at F and FB is 1.0 m. A 5.0 kg mass hangs from A and a vertical wire attached to the beam midway between F and B prevents the beam from rotating. Calculate the tension in the wire in newtons. (O&C)

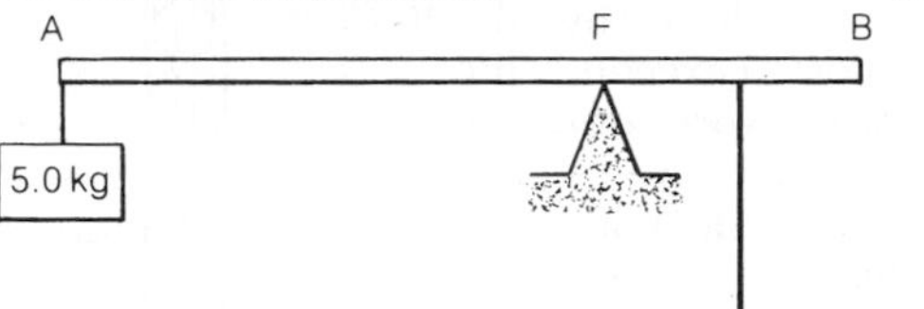

Figure 4

11 For greatest stability a table lamp should have a heavy base of large area. Draw labelled diagrams showing the lamp in tilted positions to illustrate clearly why these factors are necessary. (For simplicity the lamp may be drawn as a rectangle.) (SUJB)

12 For a light coil spring which obeys Hooke's law, a tension of 160 N would produce an extension of 72 mm. The spring is hung vertically from a fixed support, and a mass of 12 kg is attached to its lower end.

a) Calculate the extension of the spring.
b) The mass is now pulled down a further distance of 18 mm and then released. What is the resultant force acting on it immediately after release, and what is the acceleration produced? (O)

13 A spring of normal length 20 cm becomes 25 cm long when a stretching force of 15 N is applied. What is the mass of an object, which, when hung from the spring, causes the length to become 23 cm? Name the law used in your calculation. (SUJB)

14 A light spring which obeys Hooke's law is hung from a beam as in figure 5. The lower end carries a pointer which is against the 50 mm mark on a vertical scale when the spring has no load, and which moves to the 130 mm mark when a load of 20 N is carried by the spring.

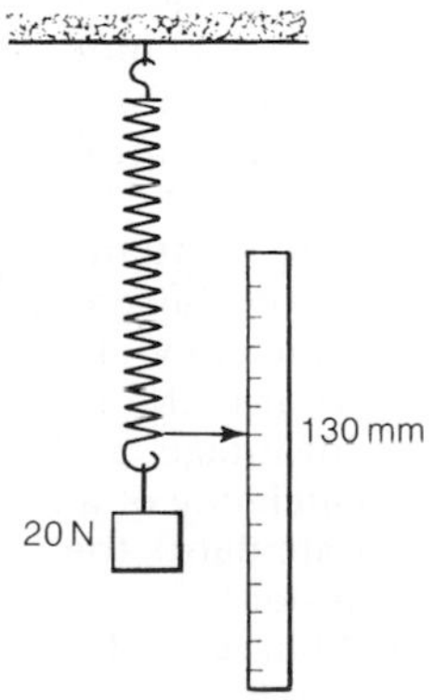

a) What load would extend the spring by 1 mm?
b) Where on the scale would the pointer be when the load was 10 N?
c) When the load is replaced by one of 12 N, what is the pull exerted on the spring by the beam? Explain.
d) Draw a graph of extension against load for the spring. (O)

130 mm
20 N
Figure 5

15 Calculate the work done against gravity by a person of mass 80 kg in walking up a flight of 12 steps each of which is 200 mm high. (L)

16 Calculate the work done in lifting a mass whose weight is 500 N through a vertical height 6.0 m.
The mass can be raised to the same height by pulling it from A to B up the ramp of length 20.0 m, shown in figure 6. If the surface of the ramp is sufficiently smooth for friction between the surface and the mass to be neglected, calculate the force acting parallel to AB required to pull the mass up the ramp at a constant speed. (L)

Figure 6
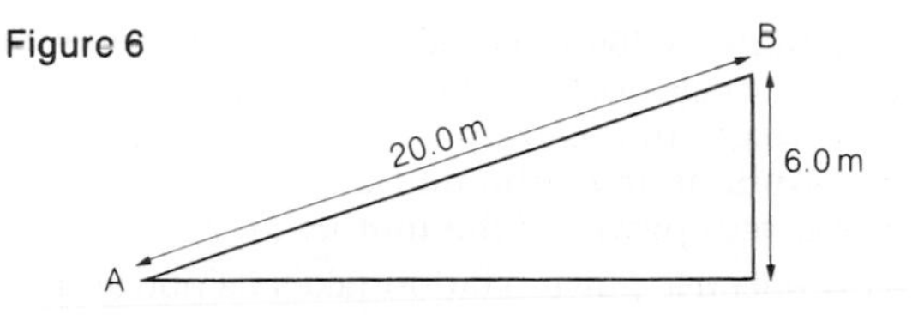

17 What kinetic energy is gained by a body of mass 3 kg on falling freely through a height of 4 m? (O)

18 An object of mass 2 kg falls freely from rest for 4 s. Calculate a) the distance fallen, b) the velocity attained, c) the gain in kinetic energy. (SUJB)

19 Masses of 30 kg and 25 kg hang on opposite ends of a light string which passes over a pulley as shown in figure 7. They are released from rest and the heavier mass falls a distance of 2 m. By how much is the total potential energy of the system reduced? (O)

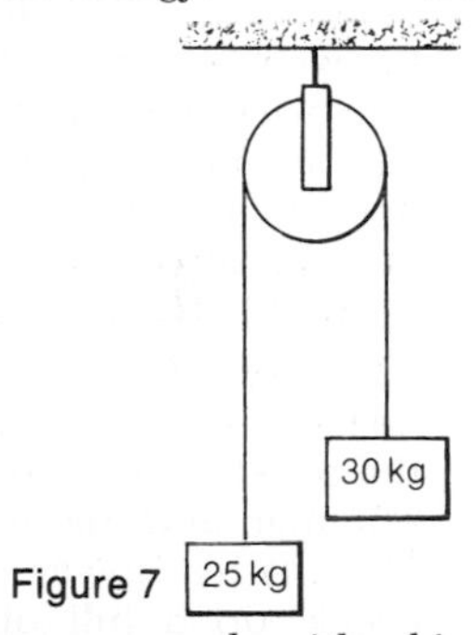

Figure 7

20 A stone of mass 3 kg is thrown upwards with a kinetic energy of 240 J. Neglecting air resistance, calculate the height to which it will rise.

21 Two wheeled trolleys, X (mass 3 kg) and Y (mass 4 kg) which can run on horizontal rails, are held together at rest against a compressed spring (figure 8). When they are released at the same instant, X moves to the left at 8 m/s. Calculate:

Figure 8
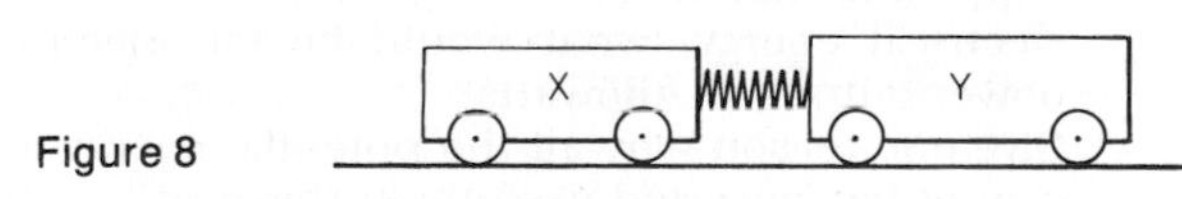

a) the momentum of X immediately after release;
b) the momentum of Y immediately after release;
c) the velocity of Y immediately after release;
d) the kinetic energy of X and the kinetic energy of Y immediately after release;

e) the potential energy that was stored in the compressed spring.

By the time the trolleys have come to rest again, all their kinetic energy has been converted 'doing work against friction'. Where does the friction occur, and what is the final form of the converted energy? (O)

22 Name and define a unit of a) force, b) work, c) power.
Describe how you would determine the average power developed by a person running up a flight of stairs. Show carefully how the result is calculated. (SUJB)

23 A crate of mass 300 kg is raised by an electric motor through a height of 60 m in 45 s. Calculate:
a) The weight of the crate.
b) The work done by the motor.
c) The useful power of the motor.

24 A hydro-electric power station takes its power from a lake whose water level is 50 m above the turbines. Assuming an overall efficiency of 50%, calculate the mass of water which must flow through the turbines each second to produce a power output of 1 MW (1 000 000 W). (JMB)

25 A builder uses a simple lift to raise bricks. The bricks are loaded on to a platform and hoisted by a cable which is wound around a rotating drum. An electric motor drives the drum.
a) The tension in the cable is 1800 N and the load is raised at a constant speed of 0.50 m/s. Calculate the work done in lifting the load for 40 s.
b) The mass of bricks in a load is 120 kg. Calculate the gain in potential energy of the bricks during the period given in a) above.
c) Account for the difference between the answers in a) and b) above. (O&C)

26 a) A man on a bicycle works at the rate of 80 W when he is travelling along a level road at 5 m/s. What is the resistance to his motion?
b) The man and the bicycle together have a mass of 60 kg. If his power output is the same when he cycles up a hill of gradient 1 in 50, and the resistance opposing motion is the same as in b), what is his speed? (O)

27 At a hydroelectric power station 600 000 kg of water fall every second from the reservoir to the turbines which are 170 m below.
a) What is the potential energy lost by the water in every second?
b) Supposing that all this energy were converted to electrical energy, what would be the electrical power output, in kilowatts?
c) Give one reason why all the potential energy lost cannot be converted usefully in this way. (O)

28 Draw a diagram showing how it is possible to raise a weight by a pulley system of velocity ratio 3. Explain your design. (O&C)

29 Figure 9 shows a pulley system.
a) What is the velocity ratio of this system?
b) The minimum effort required to lift the load is 500 N. When the load is raised through a vertical distance of 1.5 m, by how much does the work done by the effort exceed the total work done in raising the lower pulley block and the load?
c) Suggest a reason for this difference. (AEB)

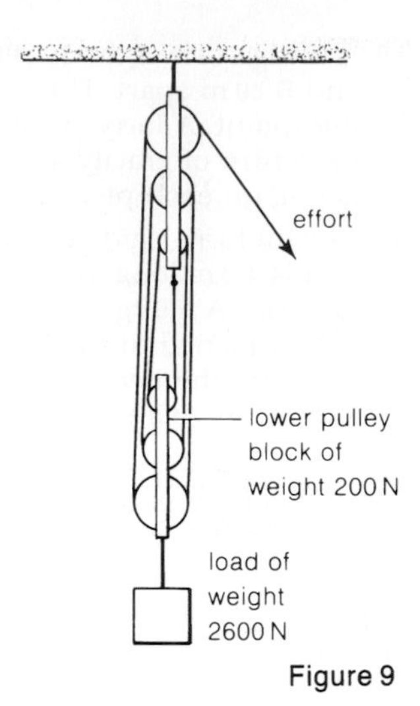

Figure 9

30 A garden roller is pulled up a ramp on to a platform as illustrated in figure 10. If the efficiency of this 'method' is 90 per cent, what is the mechanical advantage? (O)

Figure 10
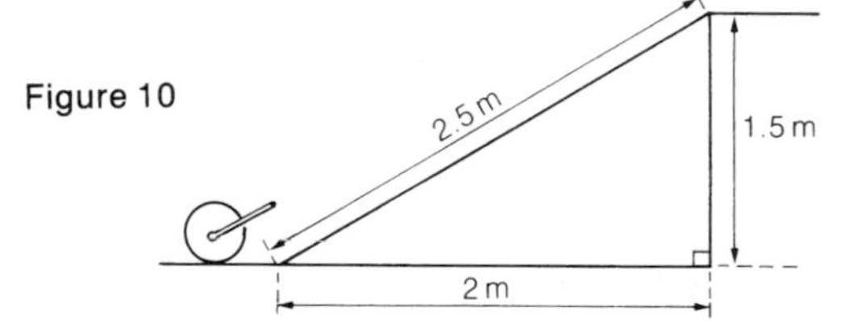

31 The simple two-pulley system shown in Figure 11 is used to lift a 3 kg mass.
a) Through what distance must the string at P be pulled to lift the mass 0.2 m? Explain your answer.
b) Define velocity ratio for a machine. What is the value of the velocity ratio of this machine?

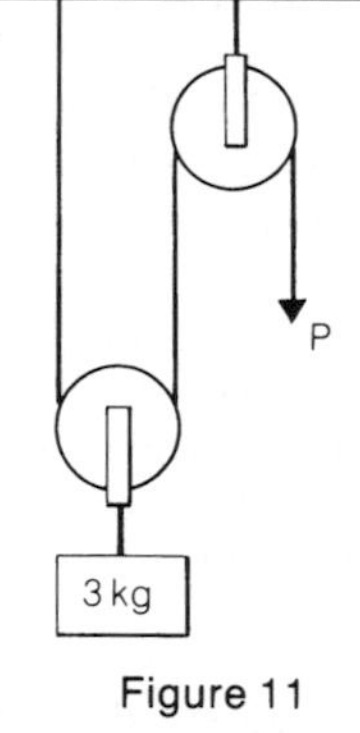

Figure 11

As the bottom pulley has to be raised at the same time as the 3 kg mass, the efficiency of the system is 75 per cent.
c) What is the relationship between the mechanical advantage (M.A.), the velocity ratio (V.R.), and the efficiency of a machine?
d) Calculate the mechanical advantage of this system.
e) What is the force at P required to lift the 3 kg mass slowly?
f) How much work is done on the machine to lift the 3 kg mass 5 m?
g) Neglecting friction and the weight of the string, calculate the mass of the lower pulley. (O)

SECTION 3

DENSITY AND PRESSURE

Suckers used in American clean-up

We refer, of course, to the domestic vacuum cleaner. In such machines, a motor-driven fan is used to lower the pressure of the air. This causes higher pressure air outside the cleaner to rush in, carrying the dust with it.

Vacuum cleaners were invented by the Englishman Herbert Booth who patented a machine in 1901. The first cleaners were horse-drawn, and parked outside people's homes – long tubes connected the machine to the room to be cleaned. Then the Americans muscled in with a radical idea. Why not make a portable cleaner that people could trundle about in their houses? 'Hoover' cleaners revolutionized domestic life in the 1920s and 1930s with their pleasantly named Grandfather, Senior and Junior machines. Modern machines are lighter, more powerful, and slightly different in appearance, but their principle remains the same.

PRESSURE RECORDS

At sea level, the Earth's atmosphere produces a pressure equivalent to the weight of ten cars pressing on every square metre. In laboratory experiments however, sustained pressures of around five million times this level have been produced – equivalent to the weight of the QE2 concentrated on an area no larger than a matchbox.

They don't make 'em like they used to.

DIVING UNDER PRESSURE

Divers hunting for pearl oysters in the Indian Ocean regularly reach depths of around 30 m. What makes this so exceptional is that they don't use breathing apparatus. Holding their breath, good divers can stay underwater for as much as 4 minutes, though this is unusual.

At a depth of 30 m, a diver experiences a pressure roughly four times atmospheric pressure. To survive the crushing effect of this pressure, he has to fill his lungs to full capacity before he dives. At 30 m, his lungs have been compressed to a quarter of their normal volume, but this makes them no smaller than they would be if he were to breathe out fully under normal conditions.

The deepest ever successful breath-held dive was a staggering 86 m by a Frenchman, Jacques Mayol, in 1973. Needless to say, this type of diving is very dangerous. Apart from the risk of drowning, pearl divers face the possibility of ruptured ear drums as well as severe heart trouble.

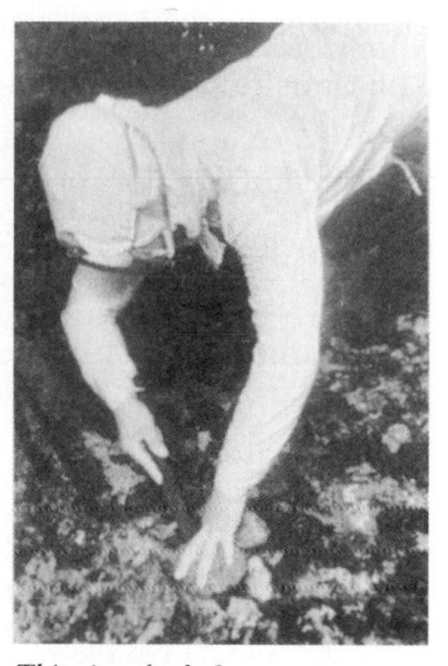

This time lucky?

KEEPING DOWN THE DENSITY

The steamroller is about the only vehicle that's actually designed to be as heavy as possible. In aircraft and cars, extra weight means higher fuel costs and lower performance, so designers are constantly searching for new materials which combine high strength with low density.

Being one of the least dense metals, aluminium has long been used by the aircraft industry. But it isn't as strong as steel. New composite materials now being developed have the strength of steel and the lightness of aluminium.

Composites can be made by setting fibres of one material in sheets of another – glass or carbon fibres in resin for example, or boron fibres in aluminium. Man's use of composites is relatively recent, but Nature has been using its own high strength low density composites for many millions of years. Wood consists of cellulose fibres set in resin, while bone is a calcium based material reinforced with tough collagen fibres.

3.1

Density

Is lead heavier than water? Not necessarily. It depends on the volumes being compared. But lead does have the greater density.

The block of lead in figure 1 has less mass than the water. The lead is the more dense of the two materials however, because it has more kilograms packed into each cubic metre.

The density of a material is given by the following equation:

$$\textbf{density} = \frac{\textbf{mass}}{\textbf{volume}}$$

If mass is measured in kg and volume in m^3, then density is measured in kg/m^3.

For example, measurements on different volumes of water show that:

1 m^3 of water has a mass of 1000 kg
2 m^3 of water has a mass of 2000 kg
3 m^3 of water has a mass of 3000 kg, and so on.

Using any of these sets of figures in the above equation, the density of water is found to be 1000 kg/m^3.

Similarly, measurements on different volumes of lead show that:

1 m^3 of lead has a mass of 11 400 kg
2 m^3 of lead has a mass of 22 800 kg
3 m^3 of lead has a mass of 34 200 kg, and so on.

Using any of the above sets of figures in the density equation, the density of lead is found to be 11 400 kg/m^3.

Approximate densities of some common substances are given in the table in figure 2.

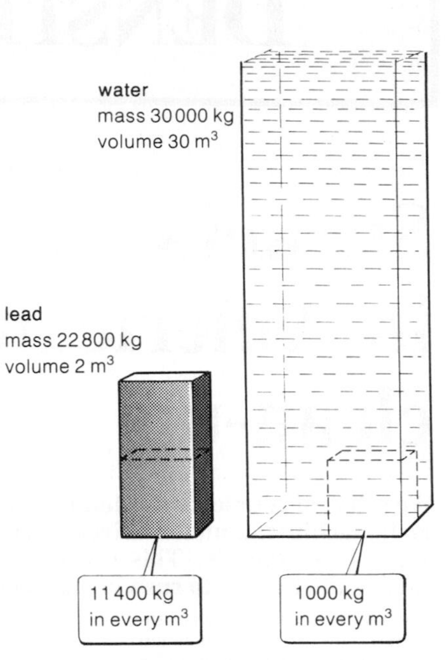

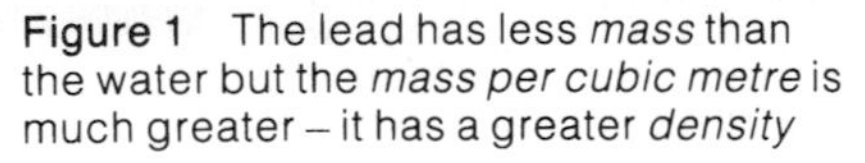

Figure 1 The lead has less *mass* than the water but the *mass per cubic metre* is much greater – it has a greater *density*

	density kg/m³	density g/cm³
air	1·3	0·0013
cork	250	0·25
wood [beech]	750	0·75
methylated spirit	800	0·80
petrol	800	0·80
polythene	920	0·92
ice [0°C]	920	0·92
water [4°C]	1000	1·00
glass [varies]	2500	2·5
granite	2700	2·7
aluminium	2700	2·7
steel [stainless]	7800	7·8
lead	11 400	11·4
mercury	13 600	13·6
gold	19 300	19·3
platinum	21 500	21·5

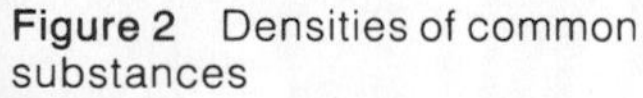

Figure 2 Densities of common substances

Expanded polystyrene: large volume but small mass: therefore, a low density

Densities of solids and liquids vary slightly with temperature. In most cases, the substances get a little bigger when they are heated, and the increase in volume causes a reduction in density. The densities of gases can vary enormously depending on how compressed they are.

Although the kg/m^3 is the basic SI unit of density, it isn't always the easiest unit to use in laboratory work. When masses are measured in g and volumes in cm^3, it is simpler to calculate densities in g/cm^3. Conversion to kg/m^3 is easy; $1\,g/cm^3 \equiv 1000\,kg/m^3$.

The table shows that the density of water is 1 g/cm^3. This simple value is no accident of nature. The kilogram [1000 g] was originally designed to be the mass of 1000 cm^3 of pure water at a temperature of 4°C. Because of imperfect early measurements however, what is now called 1 kg is not quite the same as this.

Measuring density

The density of a material can be found by calculation once the mass and the volume have been measured. You might find it useful at this stage to read again those parts of unit 1.1 dealing with mass, volume, and the measurement of both.

The mass of a small solid or of a liquid can be measured using a top-pan balance. In the case of a liquid, you must remember to allow for the mass of its container.

The volume of a liquid can be found using a measuring cylinder. In many cases, it is possible to calculate the volume of a solid from its dimensions. If the shape is too awkward for this, the solid can be lowered into a partly filled measuring cylinder as shown in figure 3. The rise in level on the volume scale gives the volume of the solid.

If the solid floats, it can be weighed down with a lump of metal. The total volume is found. The volume of the metal is measured in a separate experiment and then subtracted from this total.

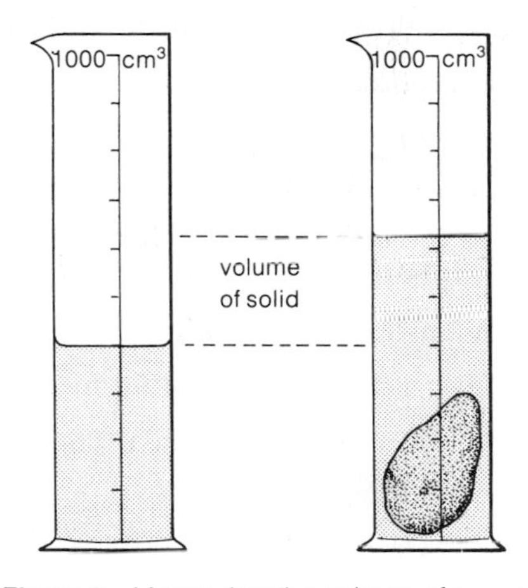

Figure 3 Measuring the volume of a small solid

Measuring the density of air

The density of air can be found by measuring the mass and the volume of the air in an otherwise empty round-bottomed flask like the one shown in figure 4. Flask, bung, tube and clip are placed on a sensitive top-pan balance, and a mass reading taken. The air is removed from the flask using a vacuum pump, and a second mass reading is taken. The difference between the two readings gives the mass of the air which was in the flask. The volume of the flask, and therefore the air, is found by filling the flask with water and measuring the volume of that.

There is just over half a gram of air in an 'empty' half-litre flask; the density of air works out at about 1.3 kg/m^3. This may be low compared with the densities of solids and liquids, but it still means that there is more than a third of a tonne of air in an average school laboratory.

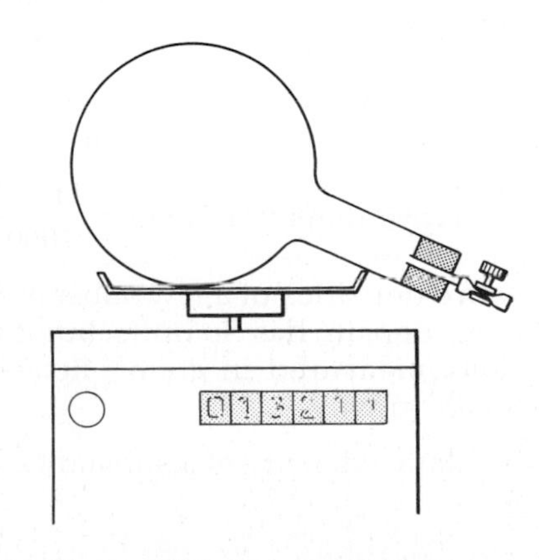

Figure 4 Measuring the mass of a flask containing air

Density calculations

The equation linking density, mass and volume can be written in symbols:

$$\rho = \frac{m}{V}$$

where ρ (Greek letter 'rho') is density, m is mass, and V is volume.

Rearranged, this equation gives: $V = \dfrac{m}{\rho}$ and $m = V\rho$

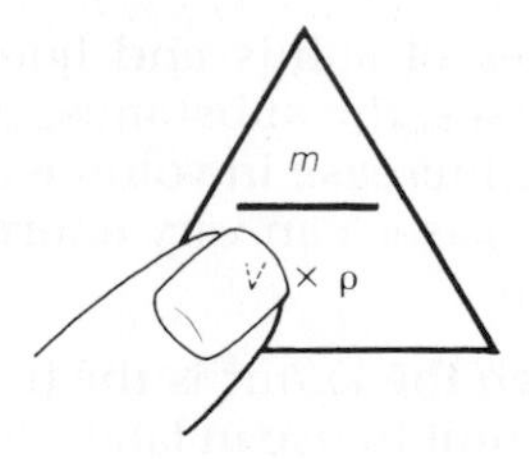

Figure 5 Cover V in the triangle and the equation for V appears. It works for m and ρ as well

These are useful if the density of a material is known, but the mass or volume is to be calculated. A method of finding all three versions of the equation is shown in figure 5. If your maths is good, you'll probably find it unnecessary.

Example *Using the density values given on page 90, calculate the mass of steel having the same volume as* 5400 kg *of aluminium.*

First, calculate the volume of 5400 kg of aluminium.

ρ is 2700 kg/m³, m is 5400 kg, V is to be calculated.

$$\therefore V = \frac{m}{\rho} = \frac{5400\text{ kg}}{2700\text{ kg/m}^3} = 2\text{ m}^3$$

This means that the volume of the steel is also 2 m³. You can now use this information to calculate the mass of the steel:

V is 2 m³, ρ is 7800 kg/m³, m is to be calculated.

$$\therefore m = V\rho = 2\text{ m}^3 \times 7800\text{ kg/m}^3 = 15\,600\text{ kg}$$

The mass of the steel is 15600 kg.

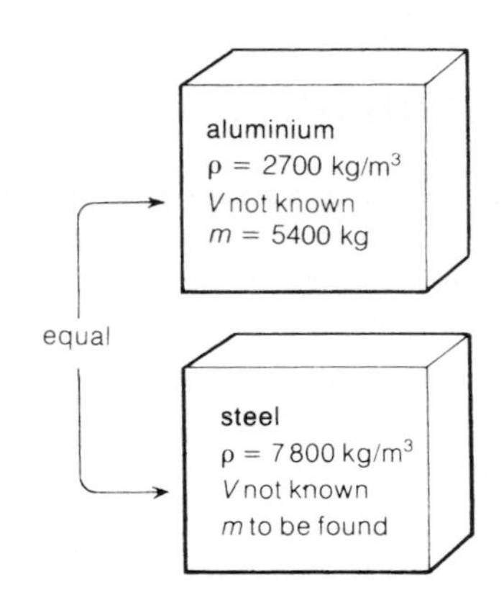

Figure 6

Relative density

The relative density of a substance tells you how many times more dense the substance is than water:

$$\textbf{relative density} = \frac{\textbf{density of a substance}}{\textbf{density of water}} \quad \textit{... equation 1}$$

For example, taking the density of lead as 11300 kg/m³ and the density of water as 1000 kg/m³:

$$\text{the relative density of lead} = \frac{11\,300\text{ kg/m}^3}{1000\text{ kg/m}^3} = 11.3$$

Relative densities of a few substances are given in figure 7. Note that relative density has no units, but it is always the same number as the density measured in g/cm³. Relative density used to be known as 'specific gravity'.

	Density		Relative density
Methylated spirit	800 kg/m³	0·80 g/cm³	0·80
Water	1000 kg/m³	1·0 g/cm³	1·0
Aluminium	2700 kg/m³	2·7 g/cm³	2·7
Lead	11 400 kg/m³	11·4 g/cm³	11·4

Figure 7 Density and relative density comparison

The relative density of a substance can also be calculated in another way:

$$\textbf{relative density} = \frac{\textbf{mass of a substance}}{\textbf{mass of the same volume of water}} \quad \textit{... equation 2}$$

This equation is particularly useful because it enables the relative density, and therefore the density, of a substance to be found from mass measurements alone. Volumes *can* be measured with great accuracy, but volume readings on a measuring cylinder are much less accurate than mass readings on a top-pan balance.

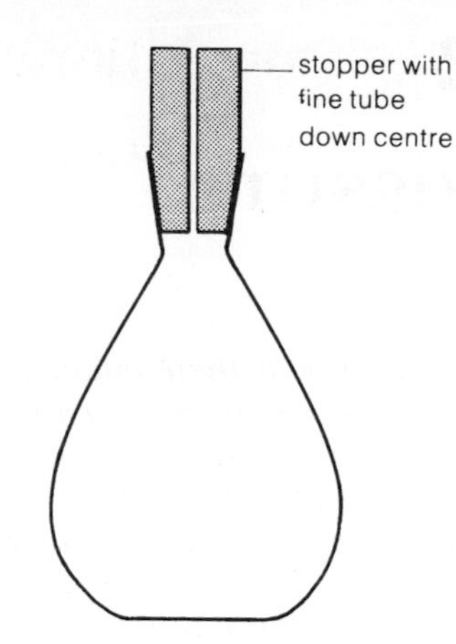

Figure 8 A density bottle

The density bottle

Figure 8 shows a density bottle, used for finding the relative density of a liquid. A narrow hole through the middle of the stopper lets out excess liquid when the stopper is pushed into the neck of the bottle. This ensures that the bottle contains exactly the same volume of liquid whenever it is filled.

To measure the relative density of a liquid, mass readings are taken with the bottle empty, then full of the liquid, then full of water. The mass of the liquid and the mass of the water are calculated from these readings, and the relative density of the liquid found using equation 2.

Figure 9 shows typical results from an experiment to measure the relative density of methylated spirit [meths].

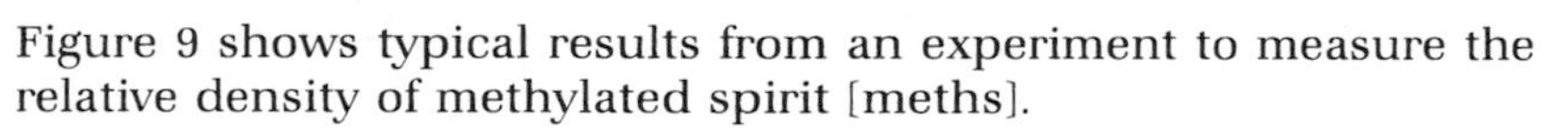

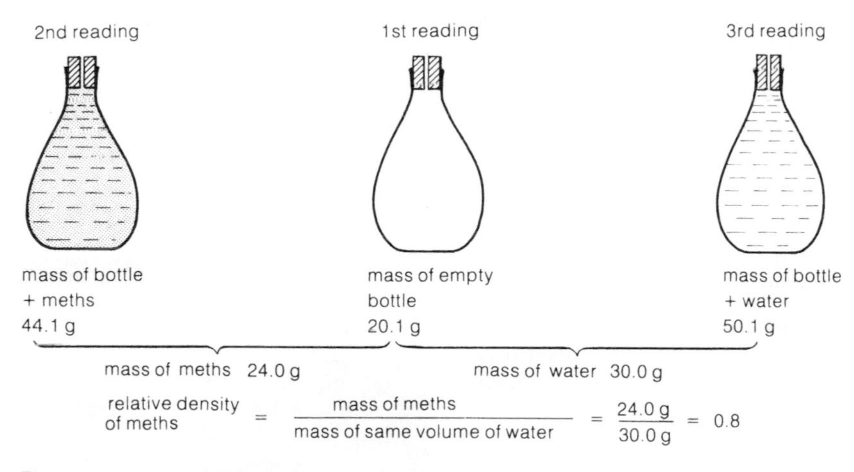

Figure 9 Using a density bottle

Questions

Where appropriate, use the density values given in the chart on page 90.

1 Water has a density of 1000 kg/m³; what does this mean? What is its density in g/cm³?
2 What is the density of mercury in kg/m³? What is its density in g/cm³? What is its relative density?
3 A small solid has an irregular shape and floats in water. Describe how you could measure its density.
4 In finding the density or relative density of a liquid, why is a method using a density bottle more accurate than one involving the use of a measuring cylinder?
5 What is the volume of 2400 kg of petrol?
6 What is the mass of air in a room measuring 5 m × 10 m × 10 m?
7 What mass of lead has the same volume as 1600 kg of petrol?
8 200 cm³ of water are mixed with 100 cm³ of meths. What is the mass of each liquid in the mixture? What is the total mass? What is the volume of the mixture? What is the average density in g/cm³?
9 An empty density bottle has a mass of 25 g. Its mass is 50 g when full of water and 45 g when full of another liquid. What is the relative density of the liquid? What is its density in kg/m³?

3.2

Pressure

When forces act, their effect is often spread out over an area. And some forces are more spread out than others.

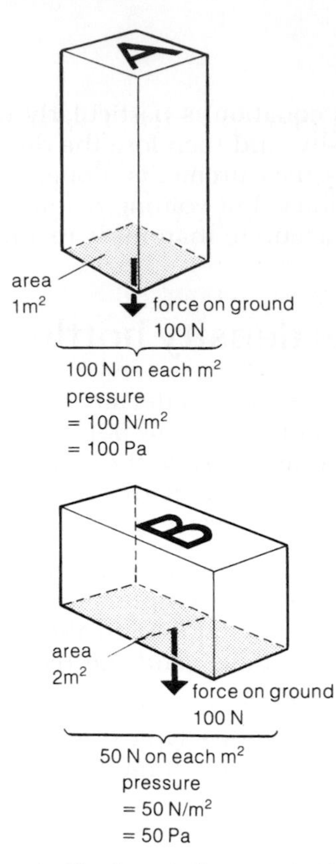

Figure 1 The force is the same but the area over which it is exerted differs. The *pressure* is therefore different

In figure 1, both blocks have the same weight. Both therefore exert the same force on the ground. The force from block A however is spread out over a larger area, so the force on each square metre of ground is reduced. Block B exerts a lower **pressure** on the ground than block A.

Pressure is calculated by dividing the force or thrust acting at right angles to a surface by the area over which it acts:

$$\textbf{pressure} = \frac{\textbf{force}}{\textbf{area}}$$

in symbols, $\boldsymbol{p} = \dfrac{\boldsymbol{F}}{\boldsymbol{A}}$

If force is measured in N and area in m^2, pressure is measured in N/m^2 or **pascal** [**Pa**]. $1\,\text{N/m}^2 \equiv 1\,\text{Pa}$.

For example,
if a 100 N force acts on an area of 1 m^2, the pressure is 100 Pa;
if a 100 N force acts on an area of 2 m^2, the pressure is 50 Pa;
if a 100 N force acts on an area of 0.2 m^2, the pressure is 500 Pa;
if a 200 N force acts on an area of 0.2 m^2, the pressure is 1000 Pa;

The larger the force and the smaller the area, the greater is the pressure. Some typical pressure values are given in figure 2.

A drawing pin can be pushed into wood because the very high pressure under the point is more than the surface can withstand. Stiletto heels, when they're in fashion, have a similar effect on some floors.

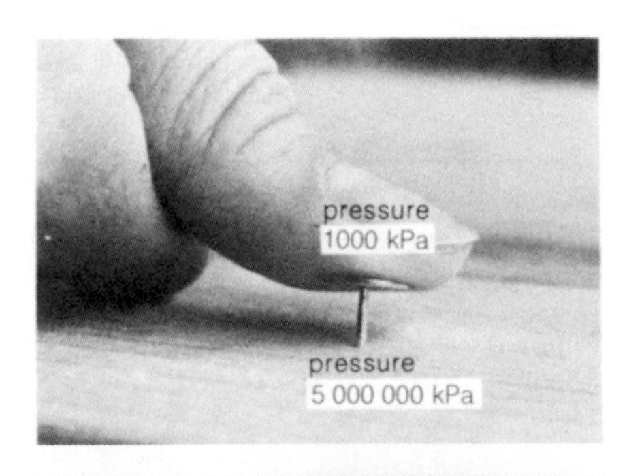

Figure 2 Typical pressure values. Note that the pressure can be very great if the area is sufficiently small

Pressure calculations: worked examples

Example 1 *The average wind pressure on the wall in figure 3 is* 100 Pa. *If the area of the wall is* $6\ \text{m}^2$, *what is the total force acting?*

This problem can be solved quite simply by rearranging the pressure equation given above:

$$\text{force} = \text{pressure} \times \text{area}$$
$$= 100\ \text{Pa} \times 6\ \text{m}^2 = 600\ \text{N}$$

The force on the wall is 600 N.

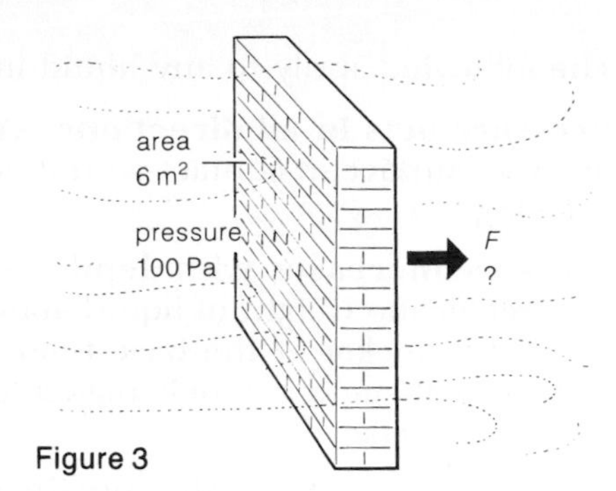

Figure 3

Example 2 *A concrete block measuring* 0.5 m × 1 m × 2 m *has a mass of* 2600 kg. *What is the maximum pressure it can exert on the ground?*

The block exerts its greatest pressure when it is resting on the end with the smallest area as shown in figure 4. To calculate this pressure, you need to know the downward force which the block exerts on the ground. Assuming that $g = 10\ \text{m/s}^2 = 10\ \text{N/kg}$:

downward force = weight of block = $2600\ \text{kg} \times 10\ \text{N/kg} = 26\,000\ \text{N}$

area over which this force acts = $0.5\ \text{m} \times 1\ \text{m} = 0.5\ \text{m}^2$

$$\therefore \text{pressure} = \frac{\text{force}}{\text{area}} = \frac{26\,000\ \text{N}}{0.5\ \text{m}^2} = 52\,000\ \text{Pa}$$

The pressure on the ground = 52 000 Pa

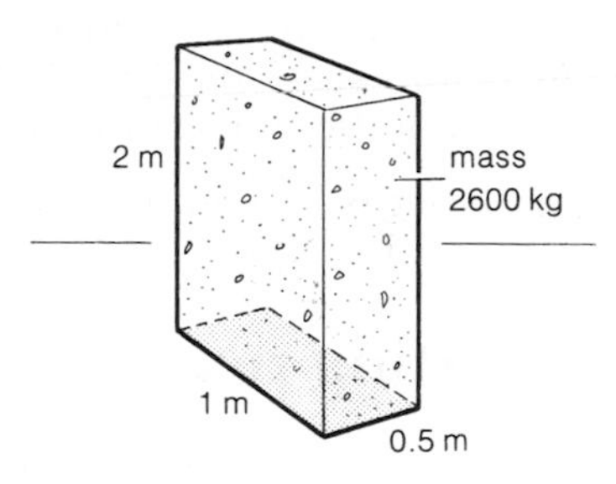

Figure 4

Pressure in liquids

Gravitational force tries to pull a liquid downwards in its container. This causes pressure on the container, and pressure on any object put into the liquid. The pressure in a liquid has several important features, some of which are illustrated by figures 5a and 5b, and the situation shown in figure 6.

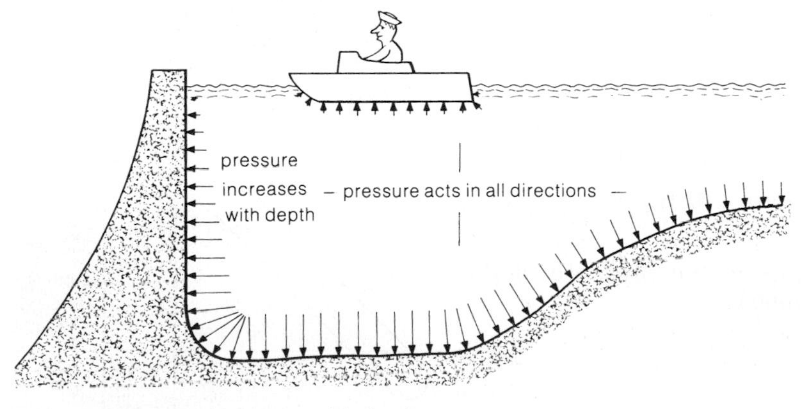

Figure 6 Pressure increases with depth but acts in all directions

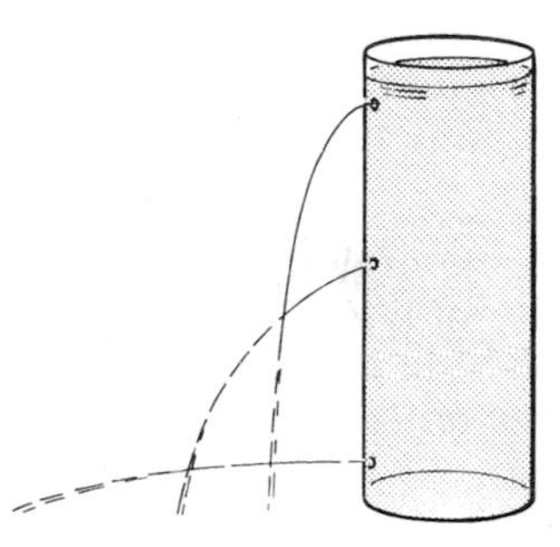

Figure 5a Pressure increases with depth

Figure 5b Pressure acts in all directions

The following apply to any liquid in an open container:

Pressure acts in all directions A liquid under pressure pushes on every surface in contact with it no matter which way the surface is facing.

Pressure increases with depth The deeper into a liquid you go, the greater the weight of liquid above and the higher the pressure. Dams are thicker at the base than at the top partly because they have to withstand a much higher pressure from the water at the bottom of a lake.

Pressure depends on the density of the liquid If water in a lake were replaced by a less dense liquid, the pressure at all points would be less.

Pressure doesn't depend on the shape of the container In the strange looking container in figure 7, the liquid stays at the same level in all four sections. Pressures at points A, B, C, and D are all the same, despite the different container shapes and widths above.

The container is sometimes known as Pascal's vases.

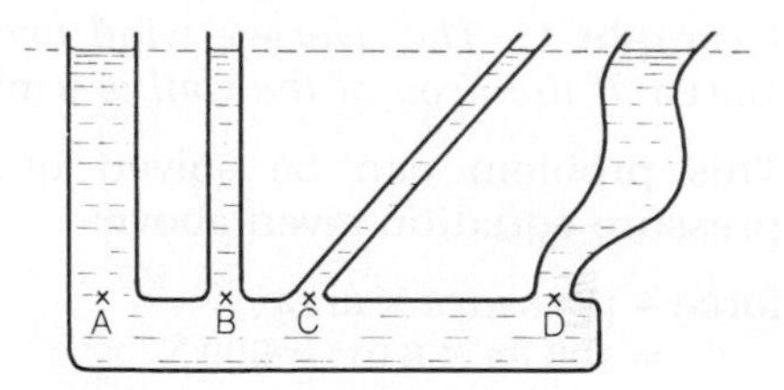

Figure 7 The pressure at A, B, C, and D is the same

Calculating pressure in a liquid

The pressure at any point in a liquid can be calculated provided you know the depth beneath the surface and the density of the liquid.

The container in figure 8 has a base area A. It is filled to a depth h with a liquid of density ρ. To calculate the pressure acting on the base, you first need to know the weight of liquid pressing down on the base.

$$\begin{aligned} \text{volume of liquid} &= \text{base area} \times \text{depth} = Ah \\ \text{mass of liquid} &= \text{density} \times \text{volume} = \rho Ah \\ \text{weight of liquid} &= \text{mass} \times g = \rho gAh \\ \therefore \text{force on base} &= \rho gAh \end{aligned}$$

This force is acting on an area A:

$$\therefore \text{pressure} = \frac{\text{force}}{\text{area}} = \frac{\rho gAh}{A} = \rho gh$$

The base area A doesn't feature in this final equation. The pressure is the same however large or small the area being considered.

The pressure at a point *h* vertically beneath the surface of a liquid of density $\rho = \rho gh$

For example, at points X and Y in figure 9, the pressure of the liquid = $1000\,\text{kg/m}^3 \times 10\,\text{N/kg} \times 2\,\text{m} = 20\,000\,\text{Pa}$.

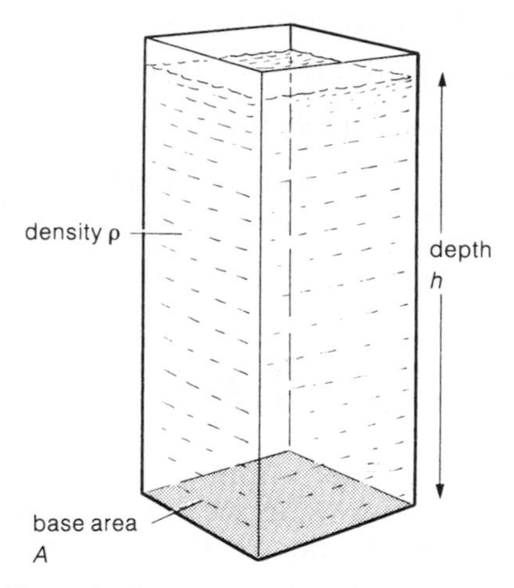

Figure 8 Pressure under a free surface is calculated from P = ρgh

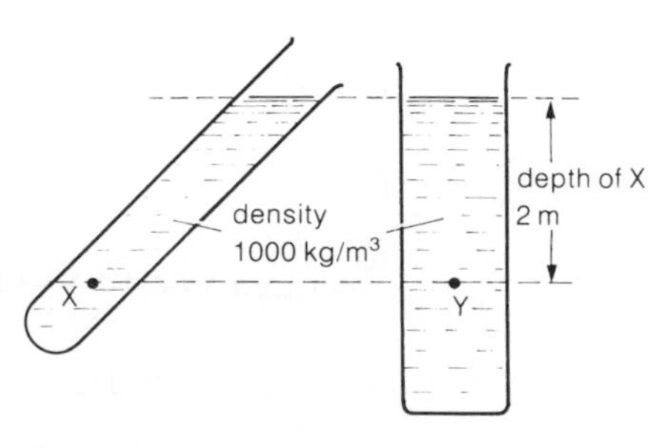

Figure 9

Hydraulic machines

Hydraulic machines work by using liquids under pressure rather than levers or wheels. They make use of two properties of liquids.

1 Liquids are virtually incompressible; they cannot be squashed.
2 If pressure is applied to a trapped liquid, the pressure is transmitted to all parts of the liquid.

Figure 10a shows a typical hydraulic machine – a simple working model of a hydraulic jack.

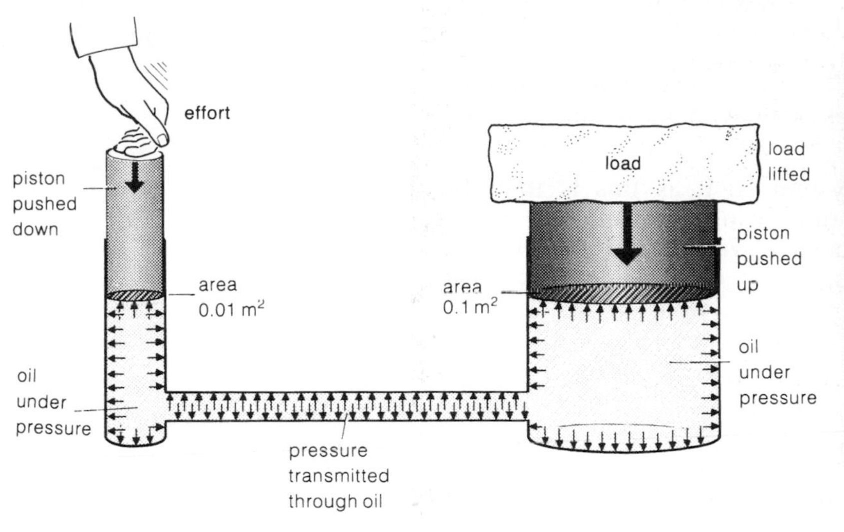

Figure 10a Pressure built up in the 'effort' cylinder is transmitted through the oil – but it acts over a larger area in the 'load' cylinder and therefore creates a larger force

A hydraulic car jack in action

The principle looks simple enough. Push down on the smaller piston; oil gets pushed through the pipe, and the larger piston is forced upwards. But the system has another important feature; the effort needed to push down the smaller piston is less than the load lifted. The downward force on the smaller piston puts the oil under pressure. This same pressure, acting on the larger area of the second piston, produces a greater upward force.

For example, if you push on the smaller piston [area = $0.01\,\text{m}^2$] with a force of 12 N as shown in figure 10b,

$$\text{the pressure applied to the liquid} = \frac{\text{force}}{\text{area}} = \frac{12\,\text{N}}{0.01\,\text{m}^2} = 1200\,\text{Pa}$$

This pressure is transmitted through the liquid. It acts on the larger piston [area = $0.1\,\text{m}^2$] as shown in figure 10c;

$$\begin{aligned} \therefore \text{ upward force on larger piston} &= \text{pressure} \times \text{area} \\ &= 1200\,\text{Pa} \times 0.1\,\text{m}^2 \\ &= 120\,\text{N} \end{aligned}$$

Therefore, assuming no frictional losses, a load of 120 N can be lifted using an effort of only 12 N. The mechanical advantage of the jack $\left[\frac{\text{load}}{\text{effort}}\right]$ is 10.

The gain in lifting ability is at the expense of the distance moved; the effort has to move ten times further than the load. The velocity ratio of the jack, 10, can also be calculated using the equation:

$$\textbf{Velocity ratio} = \frac{\textbf{area of 'load' piston}}{\textbf{area of 'effort' piston}}$$

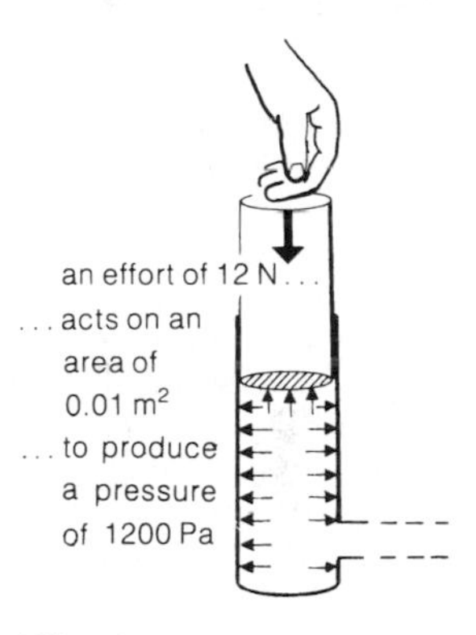

Figure 10b A pressure of 1200 Pa is produced

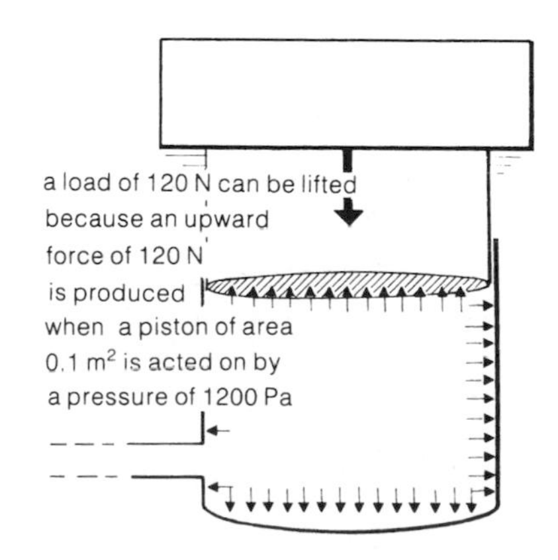

Figure 10c The same pressure acts on a greater area and develops a greater force

Using hydraulic pressure

The hydraulic principle is used in virtually all car braking systems. When you push the brake pedal down, a small piston in a master cylinder forces brake fluid through narrow pipes leading to the four wheels. Here, the fluid pushes on pistons in slave cylinders, and the pistons move brake pads or shoes against metal drums or discs attached to the wheels.

One advantage of a hydraulic system such as this is that the connecting piping can follow a route through the car body which would be very difficult to achieve using a system of levers. Some car manufacturers fit hydraulically operated clutches for the same reason.

Questions

(Assume g = 10 N/kg)

1 The pressure acting on a surface is 10 Pa. What does this mean?
2 Walking on pebbles in bare feet is more painful than walking on sand. Why?
3 A force of 200 N acts on an area of 4 m^2. What is the pressure? What would be the pressure if the same force acted on half the area?
4 A pressure of 1000 Pa acts on an area of 0.2 m^2. What force is produced?
5 A block measuring 0.1 m × 0.4 m × 1.5 m has a mass of 30 kg. Show with diagrams the positions of the block in which it produces greatest and least pressure on the ground, and calculate the pressure in each case.
6 A tank 4 m long, 3 m wide, and 2 m deep is filled to the brim with paraffin (density 800 kg/m_3). Calculate the pressure on the base. What is the thrust (force) on the base?

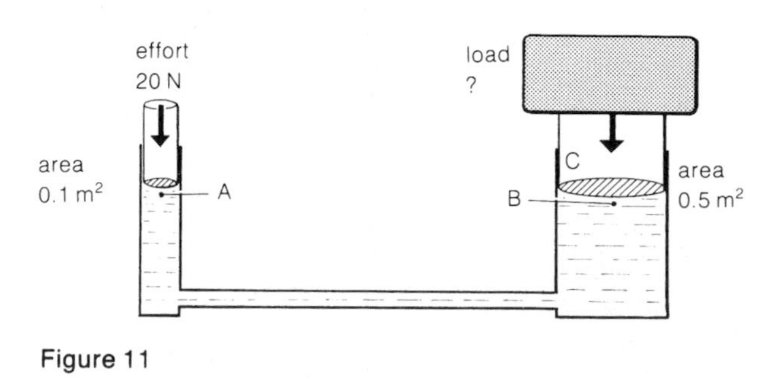

Figure 11

7 Figure 11 shows a simple hydraulic jack. The load is just being lifted using an effort of 20 N.
What is the pressure at A? What is the pressure at B?
What is the thrust acting on the piston C?
What is the load lifted? What is the mechanical advantage of the jack?
(Assume no frictional losses.)

The dam is 100 m high. What is the pressure on its base?

3.3

Pressure from the atmosphere

We live at the bottom of a deep ocean of air called the atmosphere. It may not look very dense, but it exerts a very high pressure.

In some ways, the atmosphere is like a liquid. Its pressure acts in all directions and becomes less as you rise up through it. Unlike a liquid however, the atmosphere is much more dense at lower levels. It stretches hundreds of kilometres into space, yet most of the air's bulk lies within about ten kilometres of the Earth's surface.

Down at sea level, the air pressure is about 100 000 Pa [100 000 N/m^2] – equivalent to the weight of ten cars pressing on each square metre. The effect of this intense pressure was first demonstrated by removing the air from two large hemispheres, as shown in figure 1. The same effect can be demonstrated without needing horses, by using a vacuum pump and an old oil can as in figure 2. Normally, there is air in an 'empty' oil can, and the air pressure inside the can balances the air pressure outside. If the air is removed from the can, the sides of the can are pushed in by the pressure of the atmosphere. The space in the completely empty can is called a **vacuum**.

You aren't crushed by atmospheric pressure because the pressure in your blood system is more than enough to balance it. Your ears however are very sensitive to changes in pressure. When you travel up a hill quickly in a car, the outside air pressure drops as you rise up through the atmosphere and you experience a popping sensation in your ears.

Figure 1 Count Magdeburg made a name for himself by demonstrating the effects of air pressure.

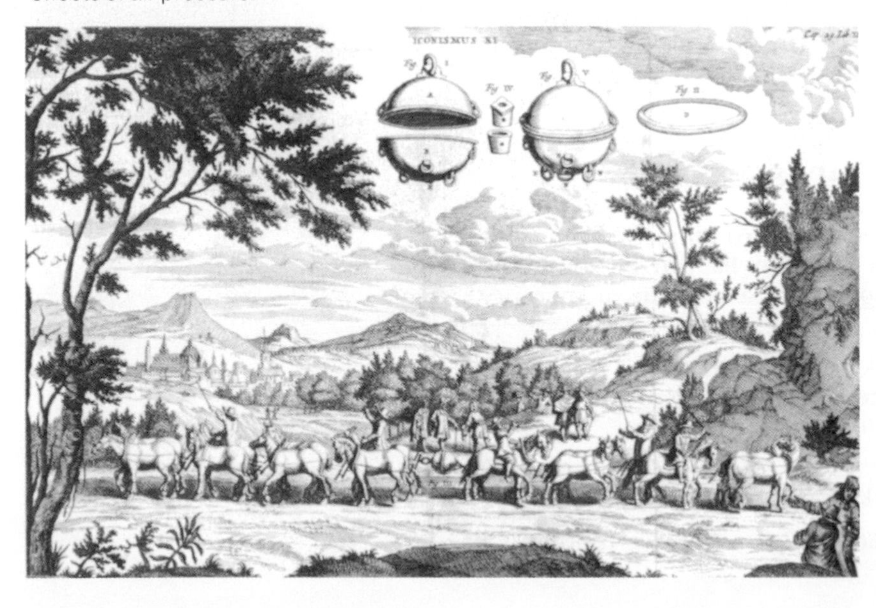

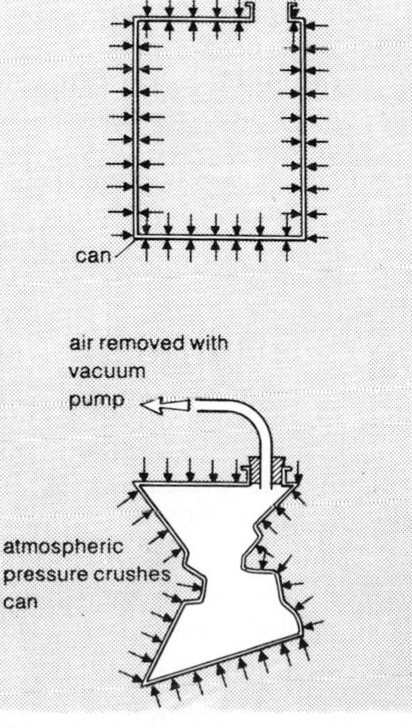

Figure 2 As the air is removed from the inside the external pressure crushes the can

The mercury barometer

Instruments which measure atmospheric pressure are called **barometers**. A simple mercury barometer is shown in figure 3. The glass tube has been sealed at one end and all the air removed from it. The bowl is open to the atmosphere and contains mercury.

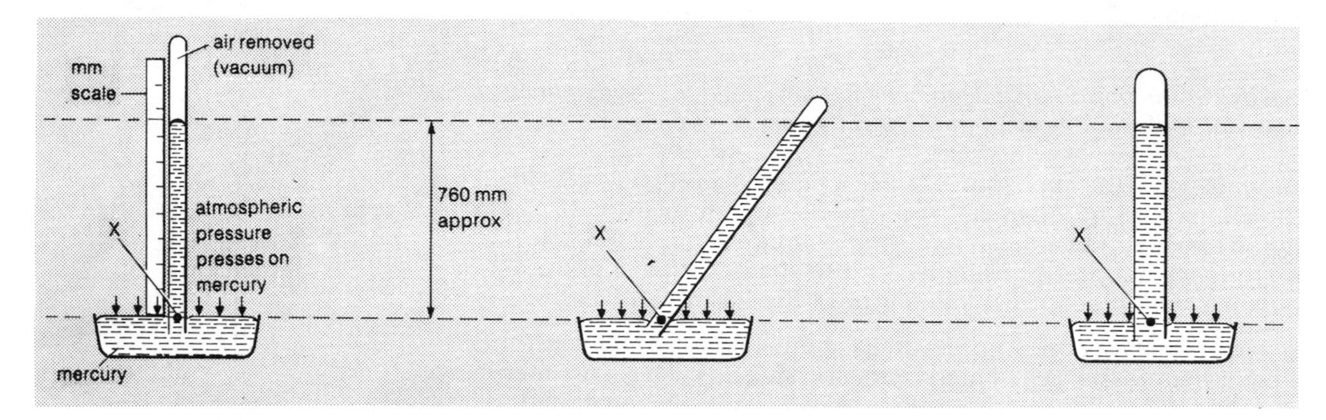

Figure 3 A simple mercury barometer: the height of the mercury column is affected neither by the width of the tube nor by the angle of tilt

With no air in the glass tube, there is nothing to stop the pressure of the atmosphere pushing mercury up the tube, and this is exactly what has happened. The mercury has risen to a height such that the pressure it produces at point X is equal to atmospheric pressure. The height of the mercury column is measured on the millimetre scale. The higher the column, the greater the atmospheric pressure.

Tilting the glass tube doesn't affect the height of the mercury column, nor does altering the width of the tube. The pressures are the same at each of the points marked X in figure 3 because the pressure in a liquid doesn't depend on the container angle, width or shape as explained on page 96.

For convenience, atmospheric pressure is often expressed in 'millimetres of mercury' (mmHg) rather than Pa. At sea level, atmospheric pressure is about 760 mmHg on average, though the actual value varies according to weather conditions. Rain clouds form in large areas of lower pressure air, called **depressions**. So a fall in the barometer reading often means that bad weather is coming.

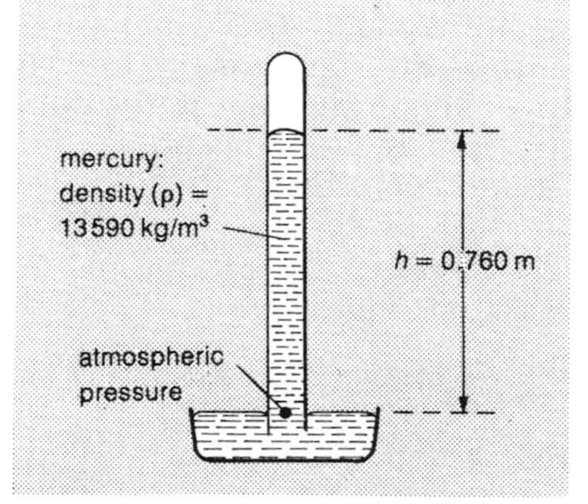

Figure 4 Standard atmospheric pressure is 760 mm Hg or 101 300 Pa

Standard atmospheric pressure

A pressure of 760 mmHg is known as standard atmospheric pressure, or 1 atmosphere (1 atm). Its value in Pa can be found by calculating the pressure at the bottom of a column of mercury 760 mm high, as shown in figure 4.

The density of mercury, ρ, is 13 590 kg/m^3: g is 9.81 N/kg if you use its more accurate value rather than the approximation of 10 N/kg; and the height of the mercury h is 0.760 m.

$$\therefore \text{pressure} = \rho g h$$
$$= 13\,590\ \text{kg/m}^3 \times 9.81\ \text{N/kg} \times 0.760\ \text{m} = 101\,300\ \text{Pa}$$

Standard atmospheric pressure, 760 mmHg or 1 atm is therefore a pressure of 101 300 Pa.

The water barometer

Water can be used in a barometer instead of mercury, but a much longer tube is needed because water is much less dense than mercury. The equation pressure = ρgh can be used to calculate the height of a column of water which air pressure will support.

Using rough values for atmospheric pressure [100 000 Pa], g[10 N/kg], and the density of water [1000 kg/m^3]:

$$\text{pressure} = \rho g h$$
$$= 1000\,\text{kg/m}^3 \times 10\,\text{N/kg} \times h$$

which gives: $h = 10\,\text{m}$

At sea level, the atmosphere can therefore support a column of water about 10 m high as shown in figure 5. Water barometers of this length have been made; von Guericke built the first one on the side of his house back in the seventeenth century.

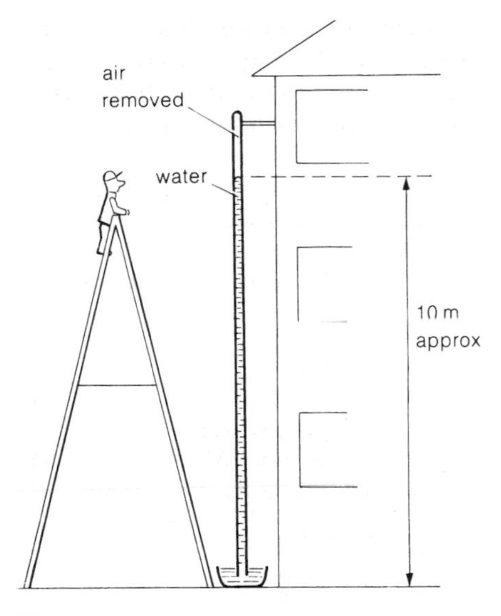

Figure 5 The water barometer is not very useful in practice

The aneroid barometer

If you have a barometer hanging in the hall at home, it will almost certainly be an aneroid barometer. Aneroid barometers are more portable, and cheaper than the mercury type. They are also easier to read because of the dial scale, as shown in figure 6.

The main feature of the barometer is the small sealed metal box containing air at low pressure. Atmospheric pressure tries to squash this box, which is corrugated to make it more flexible in the middle. If the pressure rises, the top and bottom of the box become even more squashed in. The movement of the box is magnified by a lever, and the lever pulls a chain which moves the pointer further up the scale.

An aneroid barometer is of no use until an accurate pressure scale has been marked on its dial: the instrument has to be **calibrated**. This is done by noting the needle settings for different pressures as measured on an accurate mercury barometer.

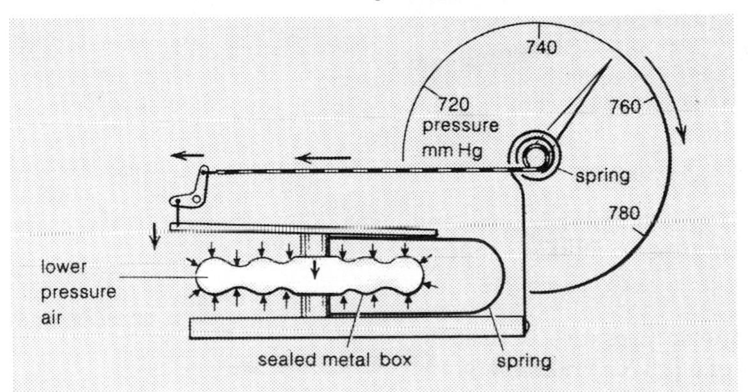

Figure 6 The aneroid barometer is much easier to use!

Figure 7 An aneroid barometer can be calibrated to measure height since the pressure drops with increased height

Parachutists can use an instrument similar to an aneroid barometer to measure their height above sea level. The pressure drops as the height above sea level increases, so the pressure scale can be replaced by an altitude scale marked in metres. Such instruments are called **altimeters** [as shown in figure 7].

Using atmospheric pressure

Figures 8a and 8b show two of the ways in which use is made of atmospheric pressure.
When you drink through a straw, you don't really suck up the liquid; the liquid is pushed up to your mouth by atmospheric pressure. You take air out of the straw by expanding your lungs. This lowers the pressure in the straw so that the pressure is no longer high enough to balance atmospheric pressure which is pressing down on the liquid in the beaker. Because of the pressure difference, the liquid is forced up the straw.

A rubber sucker works on the same principle. When a rubber sucker is pushed against a piece of glass, most of the air is squeezed out from under the sucker. If you try to pull the sucker away from the glass, the pressure of the trapped air drops as the air spreads to fill the larger space. With high pressure air outside it and low pressure air underneath it, the sucker is held firmly against the glass.

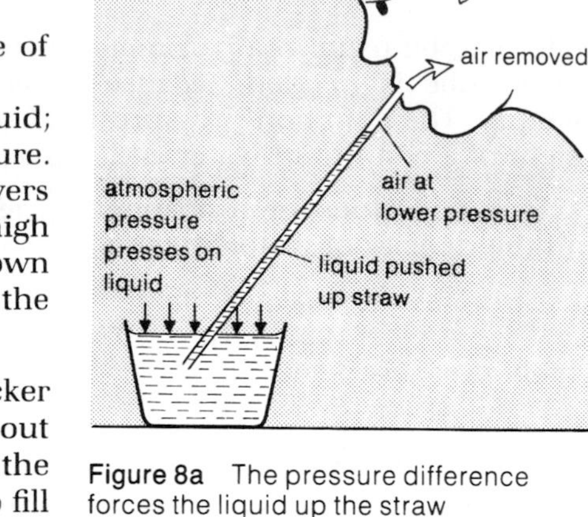

Figure 8a The pressure difference forces the liquid up the straw

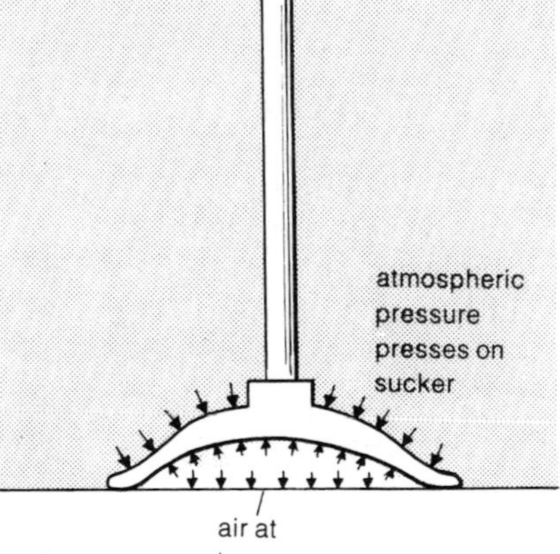

Figure 8b The pressure difference holds the sucker in place

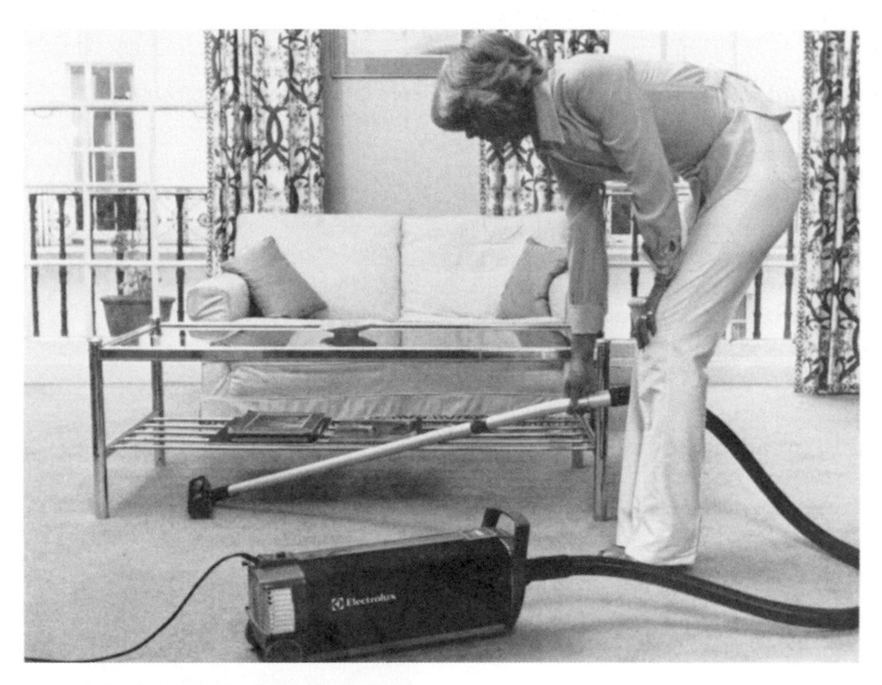

Figure 9 A domestic 'partial vacuum' cleaner

The domestic 'vacuum cleaner' shown in figure 9 works by creating a partial vacuum. Air rushes in through the tube to fill this vacuum, carrying dust with it.

Vacuum brakes In many cars, atmospheric pressure is used to apply extra force to the brakes. The heart of a power brake system is the large piston and cylinder shown in figure 10. The engine acts as a vacuum pump and removes the air on both sides of the piston. Pushing the brake pedal opens a valve which lets air back into the cylinder on the right-hand side of the piston only. The pressure difference across the piston forces the piston to the left. This applies the brakes with more force than you could produce using your foot alone.

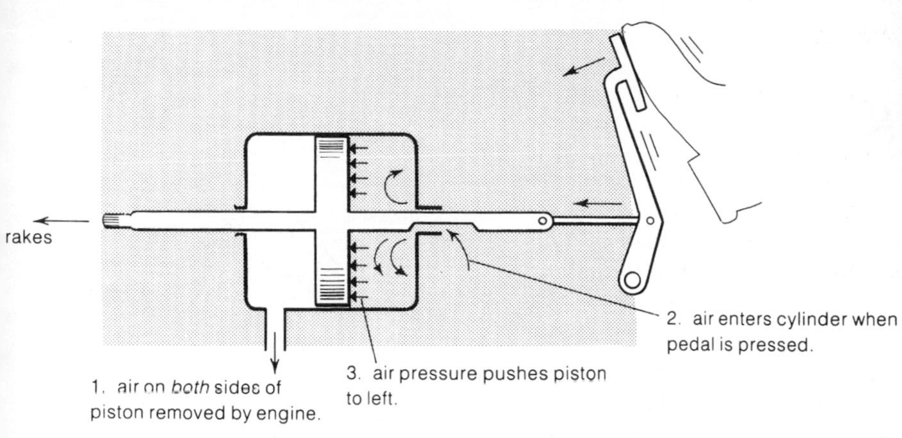

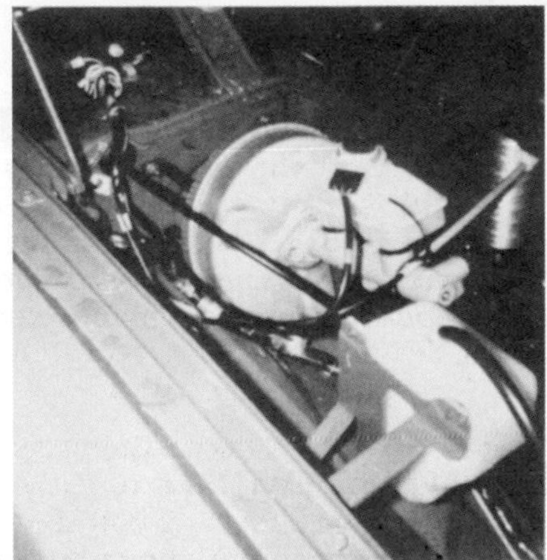

Figure 10 Power brakes: the servo unit provides extra braking force

Comparing liquid densities

The apparatus in figure 11 is used to compare the densities of two liquids. It is sometimes known as Hare's apparatus.

When air is 'sucked' out at the top, atmospheric pressure pushes the liquids up the tubes. The clip is tightened when the liquids have reached a convenient height for measurement. The liquids produce the same pressure at points X and Y because the pressure difference across the columns is the same in each case.

$$\therefore \rho_1 g h_1 = \rho_2 g h_2$$

or, density × height (liquid on the left) = density × height (liquid on the right)

If for example there is a column of meths on the left 1.0 m high, and a column of water on the right 0.8 m high;

1.0 m × density of meths = 0.8 m density of water

As the density of water is 1000 kg/m³, the density of meths works out at 800 kg/m³.

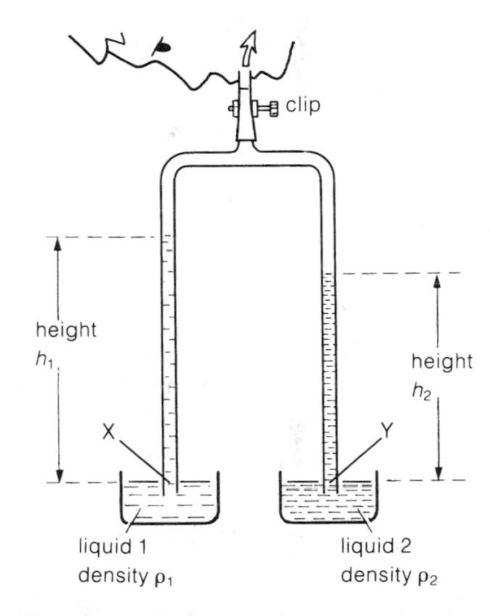

Figure 11 Apparatus for comparing the densities of two liquids

Questions

($g = 10$ N/kg)

1 At sea level, what is the approximate value of atmospheric pressure a) in Pa b) in mmHg c) in atm?
2 What is the effect on the vertical height of the mercury column in a barometer of a) using a wider glass tube b) pushing the tube further into the bowl c) tilting the glass tube at an angle d) taking the barometer to the top of a mountain?
3 Why is mercury used in a barometer rather than water?
4 Why is a mercury barometer used as a scientific standard for measuring atmospheric pressure rather than an aneroid barometer? What are the advantages of an aneroid barometer in general use?
5 Why can a barometer also be used as an altimeter?
6 Explain why a liquid travels up a straw when you 'suck' on it.
7 Using the approximate value of atmospheric pressure from question 1 a), calculate the height of a column of meths which could be supported by the atmosphere at sea level (the density of meths is 800 kg/m³).
8 If, in figure 11, h_1 is 0.9 m, h_2 is 0.72 m, and liquid 2 is water of density 1000 kg/m³, what is the density of the other liquid?

3.4

Pumps and gauges

Pumps move liquids and gases; gauges measure their pressures. This section looks at some of the more common examples of pumps and gauges.

The force pump

A typical force pump is shown in figure 1. This one is being used to squirt water onto a car windscreen. There are two valves in the pump, each allowing water to flow through in one direction only. The water is able to flow past the ball of the inlet valve by pushing it back against the spring. The water cannot return because the ball moves down and blocks the hole.

The pump is worked by moving the piston in and out. When the piston moves outwards, atmospheric pressure pushes water from the bottle through the inlet valve and into the barrel. As this happens, the outlet valve closes and stops water flowing back down the tube from the windscreen.

When the piston is pushed inwards, it forces water through the outlet valve and up to the windscreen. As this happens, the inlet valve closes and stops water being pushed back into the bottle.

Note that in this or any other force pump, the water is pushed up to the pump by atmospheric pressure. At sea level, atmospheric pressure is strong enough to push water to a height of about 10 metres, so the pump won't normally work if it is more than this height above its supply of water.

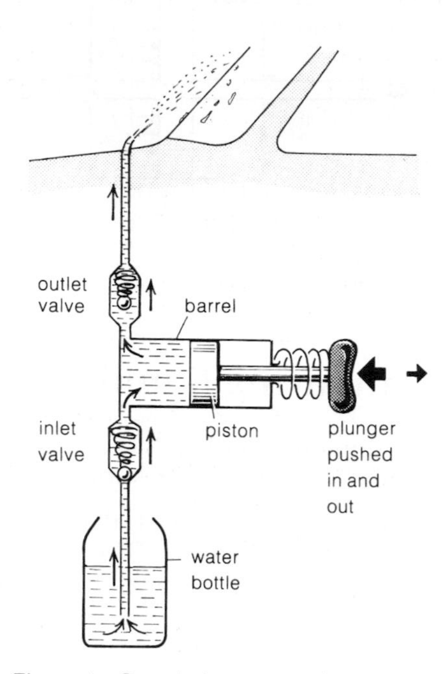

Figure 1 Car windscreen washer: a typical force pump

The bicycle pump

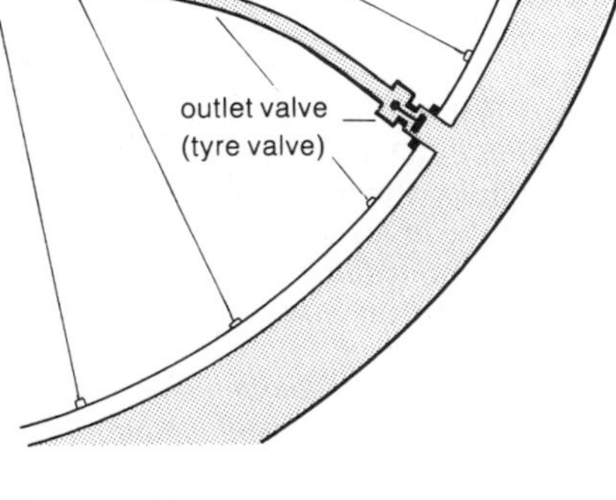

Figure 2 The bicycle pump is also a form of force pump

Like a force pump, the bicycle pump in figure 2 works because of the action of two valves and a moving piston. The cup-shaped washer acts both as inlet valve and piston; the valve fitted to the tyre is the outlet valve.

When the pump handle is pulled outwards, air moves into the cylinder past the edge of the washer. As the handle is pushed in, the washer presses tightly against the sides of the barrel and stops the air escaping. The trapped air is pushed through the tyre valve and into the tyre.

The manometer

The instrument in figure 3 is called a manometer, and it is used to measure differences in gas or liquid pressure. In the example shown, one end of the manometer U-tube is left open to the atmosphere. The other end is connected to a gas supply. The pressure difference pushes mercury round the U-bend, and there is a height difference between the mercury levels as a result.

The height difference tells you how much greater the gas pressure is than atmospheric pressure. If, for example, the height difference were 30 mmHg and atmospheric pressure were 760 mmHg, the pressure of the gas would be 790 mmHg. If the difference between the mercury levels is too small to measure accurately, water or oil can be used in the manometer instead of mercury.

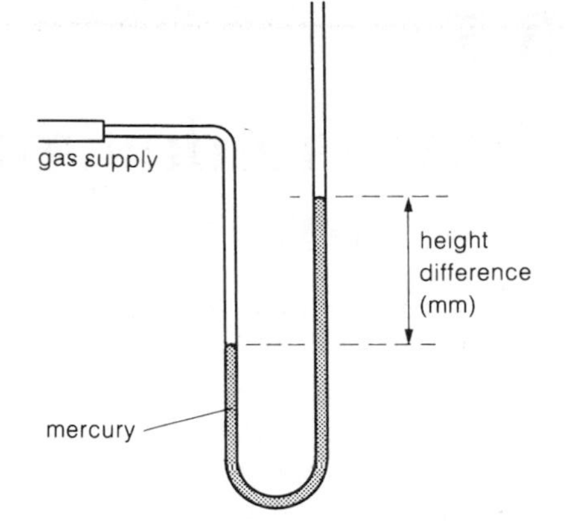

Figure 3 A manometer will indicate pressure difference

The Bourdon gauge

Figure 4 shows a Bourdon gauge. This is the type of pressure gauge that you find on the top of gas cylinders. It works along similar lines to those rolled-up paper whistles which unwind in someone's face when you blow down them. The tube in a Bourdon gauge is made of metal and it uncoils slightly when the pressure of the gas inside it increases. The movement makes a pointer move across a scale.

Bourdon gauges are more robust than manometers and they are more suitable for measuring higher pressures. But they have to be calibrated before they can be used.

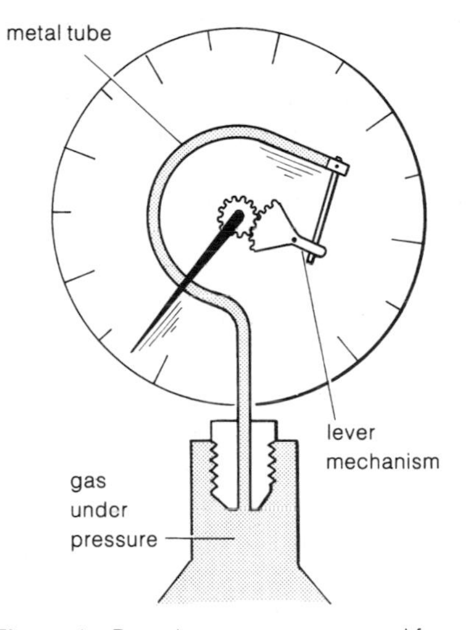

Figure 4 Bourdon gauges are used for measuring high pressures: they are very simple and not easily broken

The oil pressure gauge used in cars is a Bourdon gauge

Questions

1 What does a valve do? When does the inlet valve in a force pump open? When does the outlet valve open?
2 Where is the inlet valve in a bicycle pump? Where is the outlet valve?
3 Give one advantage that a Bourdon gauge has over a manometer. Give one disadvantage. What makes the pointer in a Bourdon gauge move across the scale?
4 What is the instrument in figure 5 called? What pressure difference is it reading in mmHg? If atmospheric pressure is 755 mmHg, what is the pressure of the gas supply?

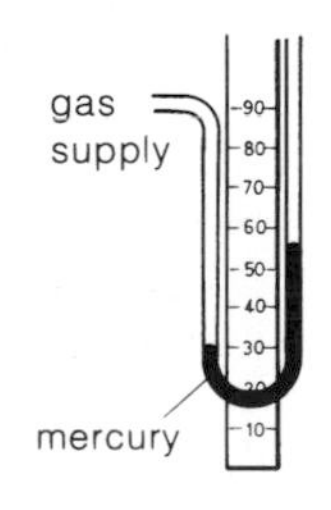

Figure 5

3.5

Archimedes' principle and flotation

More than 2000 years ago, Archimedes noticed that objects seem to weigh less when they were placed in water. This effect can be produced by any liquid or gas and it is the reason why some objects float.

When any object is placed in water, the water exerts an upward force or **upthrust** upon the object. The upthrust on a piece of wood is enough to support it, so it floats. A brick won't float but the upthrust on it makes its weight seem less than normal.

You can see the cause of the upthrust if you study the small 'pressure arrows' on the brick in figure 1. Water pressure pushes in on the brick from all sides, but the effect is greatest at the bottom where the water is deepest and the pressure highest.

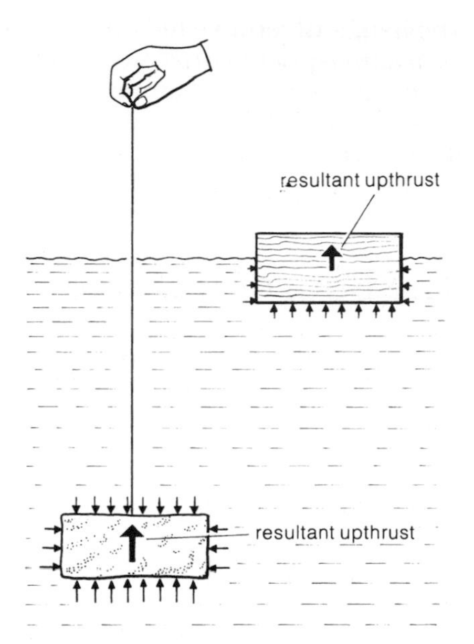

Figure 1 The fluid exerts an *upthrust* on the object in it

Archimedes' principle

The brick in figure 2 weighs 25 N in air. When the brick is immersed in water, the reading on the spring balance indicates a weight of only 15 N. The apparent loss in weight of the brick is caused by an upthrust of 10 N from the water.

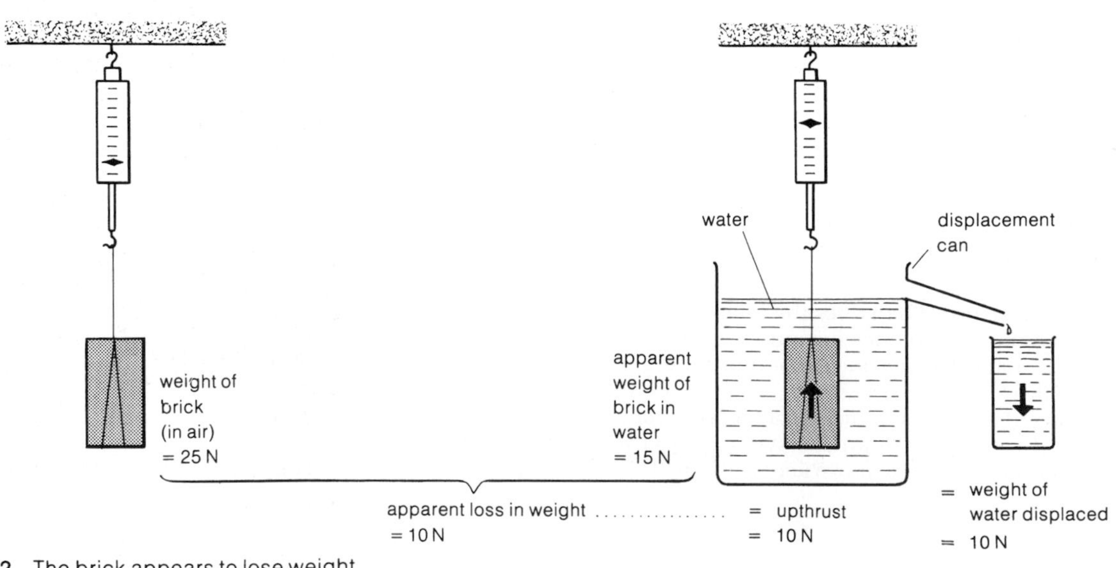

Figure 2 The brick appears to lose weight. This is caused by the upthrust on it

The brick in the water takes up space that was occupied by water. To find out how much water it has displaced, the brick is carefully lowered into a large **displacement can** which has been filled right up to the level of the spout. As the brick is immersed, the displaced water flows out through the spout and is collected in a beaker. It turns out that the displaced water weighs 10 N which is the same as the upthrust on the brick. This is an example of **Archimedes' principle** which states:

When an object is immersed in a fluid, the upthrust on the object is equal to the weight of fluid displaced.

This is illustrated in figure 3.

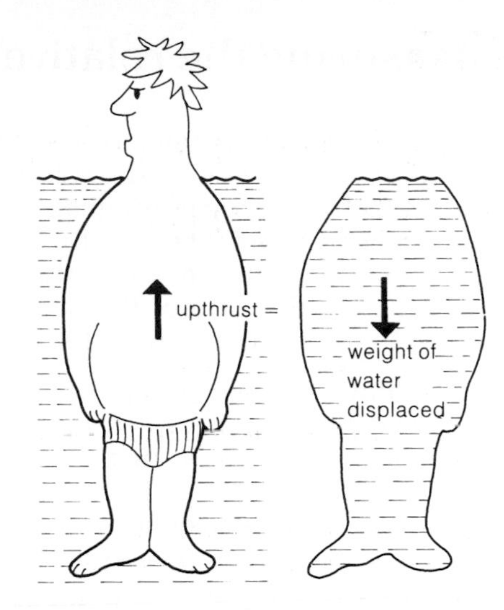

Figure 3 Archimedes' principle in action

There are several important points to note.

1 A fluid means either a liquid or a gas. There is for example an upthrust of about 0.7 N on you because of the air you displace.

2 The upthrust doesn't depend on the weight of the object. In the last experiment, a heavier brick of the same volume would still have the same upthrust acting on it.

3 Archimedes' principle applies whether an object is wholly immersed in a fluid, or only partly immersed.

Measuring the relative density of a solid

By measuring the weight of an object and its apparent loss in weight in water, its relative density can be found. In the case of the brick in the experiment shown in figure 2:

weight of brick = 25 N;
each kg of mass weighs 10 N (approx)

∴ *mass* of brick = 2.5 kg

$$\text{apparent loss in weight of brick} = \text{weight of water displaced by brick} = 10\ \text{N}$$

∴ *mass* of water displaced by brick = 1.0 kg

Remember that:

$$\textbf{relative density of an object} = \frac{\textbf{mass of object}}{\textbf{mass of same volume of water}}$$

$$\therefore \text{relative density of brick} = \frac{2.5\ \text{kg}}{1.0\ \text{kg}} = 2.5$$

This result can also be found using the equation:

$$\textbf{relative density of an object} = \frac{\textbf{weight of object}}{\textbf{apparent loss of weight in water}}$$

Lynmouth floods, 1952. The river shown used to be a road. Rocks move comparatively easily when their weight is reduced by immersion

Measuring the relative density of a liquid

When an object is immersed in meths rather than water, the upthrust on it is less. Meths is less dense than water, so the meths displaced by the object weighs less than the same volume of water as shown in figures 4a and 4b. By measuring the upthrust on an object in meths and in water, the relative density of the meths can be found. In the case of the brick, the upthrust on it was 10 N in water. In meths, the upthrust is only 8 N.

By Archimedes' principle:
the weight of meths displaced by the brick = 8 N
the weight of the water displaced by the brick = 10 N
Assuming that each kg of matter on Earth weighs 10 N,
the mass of meths displaced by the brick = 0.8 kg
the mass of water displaced by the brick = 1.0 kg

$$\textbf{relative density of a liquid} = \frac{\textbf{mass of liquid}}{\textbf{mass of same volume of water}}$$

$$\therefore \text{relative density of meths} = \frac{0.8\,\text{kg}}{1.0\,\text{kg}} = 0.8$$

This result can also be found using the equation:

$$\textbf{relative density of a liquid} = \frac{\textbf{apparent loss in weight of object in liquid}}{\textbf{apparent loss in weight of object in water}}$$

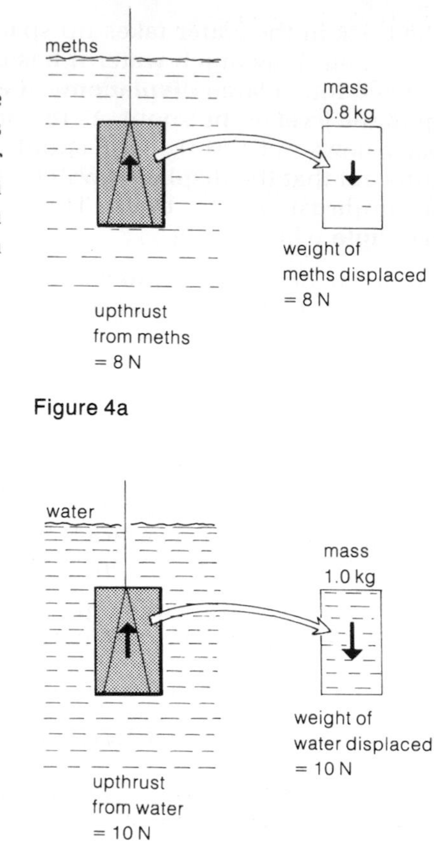

Figure 4a

Figure 4b

Figures 4a, 4b The upthrust on the block varies, depending on the density of the fluid

The law of flotation

An object will float if the upthrust on it is strong enough to support its weight. If a piece of wood weighing 10 N is lowered into water, as in figure 5, the upthrust rises as more and more water is displaced. The wood starts to float when the upthrust reaches 10 N.

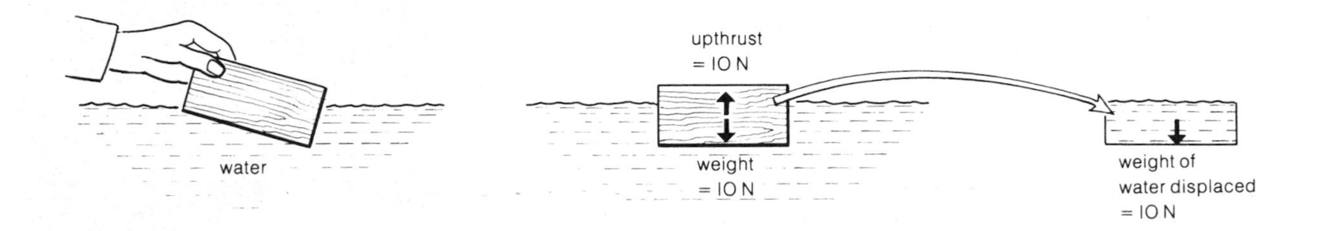

Figure 5 An object floats when the upthrust is equal to the weight of the object

If the upthrust is 10 N, the weight of water displaced is also 10 N; this follows from Archimedes' principle. This means that the weight of water displaced is the same as the weight of the wood. This is an example of the **law of flotation** which states:

A floating object displaces its own weight of the fluid in which it floats.

Levels of floating

Figure 6 shows a block of wood weighing 10 N floating in water and then in meths:

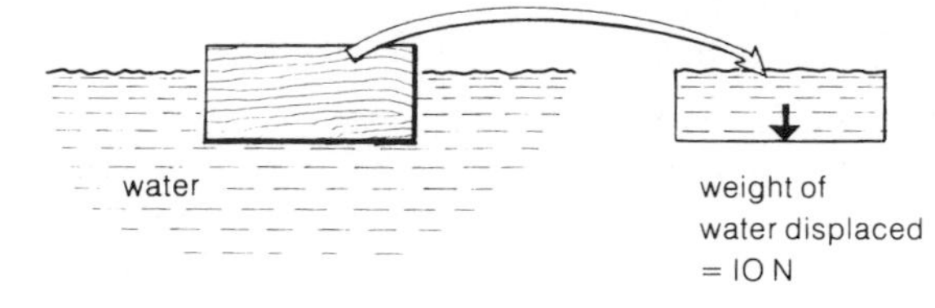

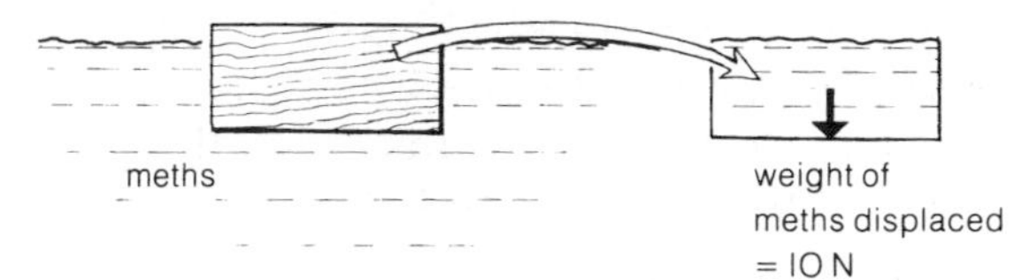

Figure 6 In less dense fluids, more fluid needs to be displaced before floating will occur

The law of flotation tells you that the weight of liquid displaced is 10 N in each case. But there must be a greater volume of meths displaced because meths is less dense than water. To displace this larger volume, the wood floats lower in the meths than it does in the water.

Any change in the density of the surrounding water affects the level at which a ship floats. Fresh water is less dense than salt water, so a ship floats lower in fresh water than it does in salt water. Warm water is less dense than cold water, so a ship floats lower in the water if the water temperature rises.

This fact is very important for shipping. A boat loaded to its maximum in cold, salt water could float dangerously low in warm, less salty water. In order to prevent this, all ships have 'maximum load' or **Plimsoll lines** marked on their side. A ship fully loaded in the North Atlantic in winter will sink to the 'tropical fresh water' line on reaching such waters, but no further.

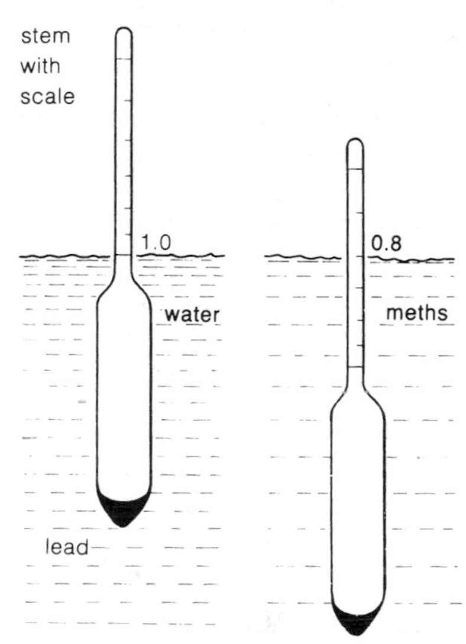

Figure 7 Hydrometers work by floating lower in liquids of lower density

Hydrometers

A hydrometer is a small float with a scale on it; they are used for measuring the relative density of liquids. They are not as accurate as some other methods, but are quicker and more convenient to use.

The hydrometer in figure 7 is shown floating in water and in meths. Meths is less dense than water, so the hydrometer has to float lower in the meths in order to displace its own weight. The level of the liquid on the scale gives the relative density in each case.

The lead in the bottom of the hydrometer helps to keep it upright when it is floating. The stem with the scale on it is made narrow so that small changes in the density of the liquid produce large changes in the level at which the hydrometer floats. This means that the scale numbers are well spaced out for greater accuracy.

Hydrometers are used for checking the relative density and therefore the quality, of beer and of milk. They are also used to test the state of charge of car batteries [as shown in figure 8]. The relative density of the acid in a fully charged car battery is about 1.25, but it is much closer to 1 for a flat battery.

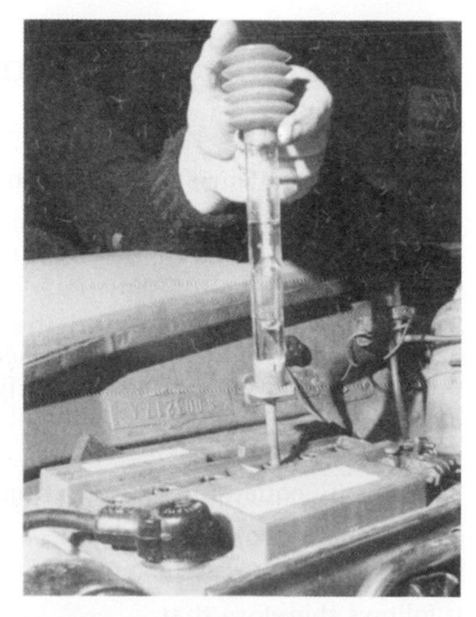

Figure 8 Using a hydrometer to check the condition of a car battery

Floating or sinking?

The lump of tar in figure 9 floats, but only just. The law of flotation tells you that it has the same weight, and therefore the same mass, as the water it displaces. Being completely immersed, it also has the same volume. This means that its density must also be the same as that of the water.

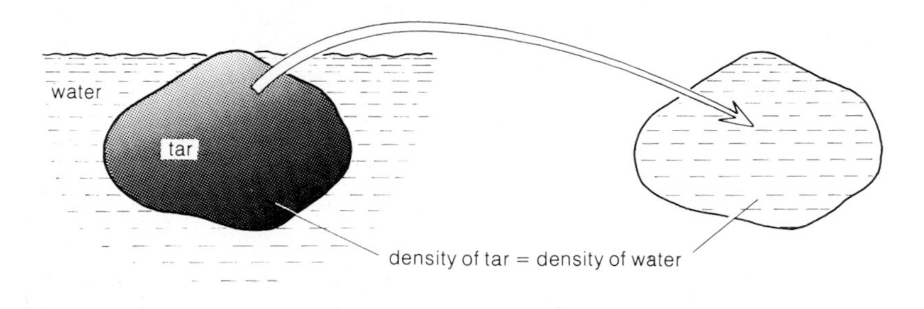

Figure 9 An object will sink if, when fully immersed, the upthrust is still not equal to its weight

You can tell whether a material will float or sink in a fluid by comparing its density with that of the fluid. A material will only float if its density is the same as or less than that of the surrounding fluid.

Densities of several substances are given in figure 10. From the values given, you can see that:

wood, petrol and ice will all float in water;
hot water will float upwards in cold water;
hydrogen gas will float upwards in air;
hot air will float upwards in cold air;
steel will float in liquid mercury but sink in water.

So how does a steel ship manage to float in water? There is far more air in a ship than steel, so the *average* density of the ship is less than that of water.

Inside the submarine are 'ballast tanks' which, when filled with water make the total weight of the submarine greater than the upthrust on it

Calculations on Archimedes' principle and flotation

In tackling problems, note the following points:

1 When an object is immersed in a fluid, Archimedes' principle applies whether the object is floating or not. The law of flotation applies only to floating objects.

2 Archimedes' principle and the law of flotation are both concerned with the weights of objects and fluids. When solving problems however, you are often dealing with volumes. There is a link between the weight of a substance and its volume, as shown below.

The density equation can be written in the form:

mass = volume × density
but: weight = mass × g

It follows therefore that:

weight = volume × density × g

	Density kg/m^3
hydrogen gas	0·09
air (760 mm Hg pressure)	
at 100°C	0·95
at 0°C	1·29
petrol	800
ice	920
water	
at 100°C	960
at 4°C	1000
steel	7800
mercury	13 600

Figure 10 Density values for common substances

Floating in air

Air may not be very dense, but it is still heavy in large volumes. The hot air balloon in figure 11 displaces more than 2.5 tonnes of air. And it will float clear of the ground if its total weight including the hot air inside it is no more than the weight of the air it displaces. This follows from the law of flotation.

On the ground, the balloon is inflated with cold air using a powerful motor-driven fan. The gas burner then heats the air in the balloon to a temperature of about 100 °C. As the air heats up, it expands and about a quarter of it is pushed out through the hole at the bottom. This reduces the weight of air that the balloon has to carry. Figure 12 shows how the total weight of a typical balloon is made up. Note that it is the weight saved by replacing the cold air by less dense hot air that gives the balloon its load-lifting ability.

If the hot air in the balloon starts to cool, it contracts and outside air flows in through the hole at the bottom to fill the space. This increases the weight of the balloon, making it sink. To maintain height, the pilot has to keep the air in the balloon hot. He does this by giving a quick blast on the gas burner every half minute or so.

Figure 11 The weight of air displaced is *just* less than the weight of the balloon and its contents

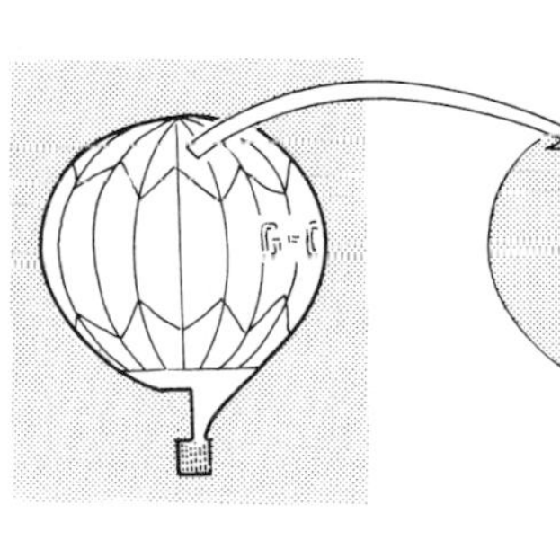

Figure 12 A hot air balloon *just* floats when its total weight is equal to the weight of air displaced

Questions

(Assume g is 10 N/kg)

1 What is a) Archimedes' principle b) the law of flotation?

2 A boat floating in water has a mass of 1000 kg. What is the weight of the boat? What is the upthrust on it? What weight of water does it displace? If the boat is floated in denser salt water, what effect does this have on a) the upthrust b) the weight of water displaced c) the level at which the boat floats?

3 A hot air balloon displaces air weighing 20 000 N. If the balloon and basket together weigh 2000 N, and the balloon contains hot air weighing 15 000 N, what is the maximum load it can lift?

4 An object weighs 24 N in air, 20 N when completely immersed in water, and 21 N when completely immersed in another liquid. Find a) the relative density of the object b) the relative density of the liquid.

5 A block of aluminium, density 2700 kg/m^3, has a volume of 0.2 m^3. What is its weight? What is its apparent weight when completely immersed in water of density 1000 kg/m^3?

6 The weighted rod in figure 13 floats with 6 cm of its length under water [density 1000 kg/m^3]. What length is under the surface when the rod floats in brine [density 1200 kg/m^3].

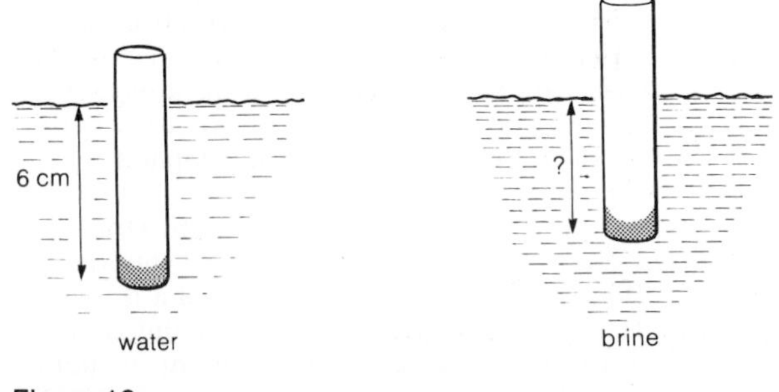

Figure 13

Density and pressure: part A

$g = 10\ m/s^2 = 10\ N/kg$

1 The container in figure 1 is filled to a depth of 4 cm with mercury.

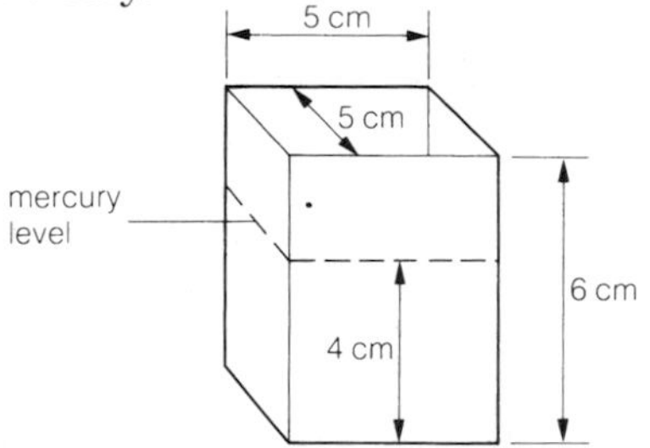

a) What is the volume of the mercury?
b) The mass of the mercury is 1360 g. What is the density of mercury?
c) What mass of water would be needed to fill completely the space in the container above the mercury? (The density of water is $1\ g/cm^3$.)
d) A small coin is gently dropped into the container completely full of mercury and water. Describe and explain what happens. (The density of coin metal is about $7\ g/cm^3$.) (MEG)

2 Jane has a motorcycle with a $125\ cm^3$ engine as shown in figure 2.

Figure 2

a) $125\ cm^3$ is often called the capacity of the engine.
   i) What word in physics would you use instead of capacity?
   ii) Jane's motorcycle engine is on a work bench. She is given some oil and a measuring cylinder. How could she check the 'capacity' of the engine?
b) The engine is made of aluminium. One reason aluminium is used is because of its low density.
   i) What is meant by the density of a material?
   ii) Describe how you would find the density of aluminium using the equipment available in your laboratory.
   iii) Why is it important to build a motorcycle with low density materials?
   iv) Suggest one *other* advantage of using aluminium for the engine. (SWEB/SEG)

3 A bath of water varies in depth from 0.2 m at the shallow end to 0.3 m at the plug-hole end. If the density of water is $1000\ kg/m^3$ and the acceleration of free fall is $10\ m/s^2$, the pressure of water acting on the plug, in Pa, which of these is:
a) 1.0×10^3
b) 2.5×10^3
c) 3.0×10^3
d) 5.0×10^3? (NEA)

4 Figure 3 shows a hydroelectric scheme. $10\ m^3$ of water flows from the top of the lake every second. The water flows down to the power station, where it turns the turbines which drive the electrical generators.

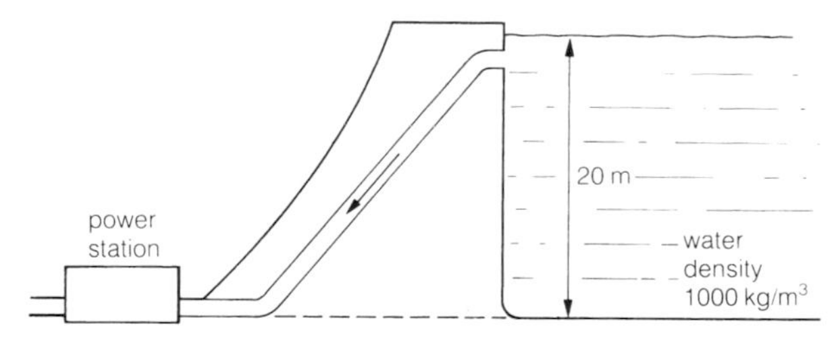

Figure 3

a) What is the pressure at the bottom of the lake?
b) Why is the dam thicker at the base than at the top?
c) How many kilograms of water flow from the lake every second?
d) How much potential energy is lost by 1 kg of water as it flows down to the power station?
e) What type of energy is this changed into?
f) If the power station has an efficiency of 80%, what is its power output?

5 A hovercraft has a mass of 1000 kg. It hovers at a constant distance above the ground.
a) Calculate the weight of the hovercraft assuming that the value of the gravitational field strength is 10 N/Kg.
b) What can you say about the upward force exerted by the air cushion? Explain your answer.
c) The hovercraft has a rectangular shape of length 5 m and width 2 m. Calculate the excess pressure (above atmospheric pressure) in the air cushion under the craft.
d) The hovercraft accelerates horizontally at $2\ m/s^2$. Ignoring air resistance, calculate the horizontal force exerted by the driving propellor on the hovercraft. (NEA)

6 Figure 4 shows a pipe with one end wider than the other. It is filled with oil. Metal cylinders A and B are fitted closely into the ends of the pipe so that they can be pushed in and out, but the oil cannot escape. If cylinder A is pushed into the pipe by a force of 100 N,

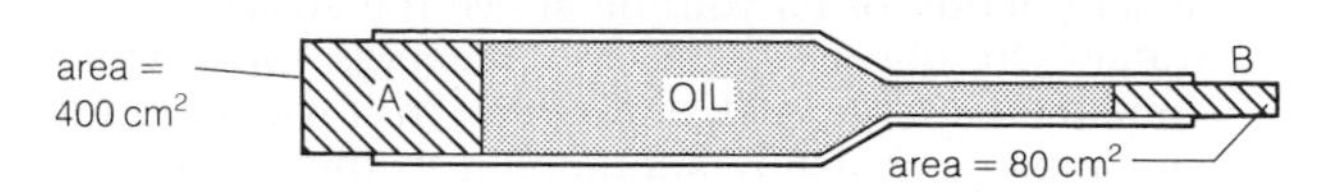

Figure 4

a) what pressure is applied to the oil?
b) what force is exerted on cylinder B? (SWEB/SEG)

7 A rectangular block measures 8 cm by 5 cm by 4 cm, and has a mass of 1.25 kg.
a) i) If the gravitational field strength is 10 N/kg, what is the weight of the block?
ii) What is the area of the smallest face of the block?
iii) What pressure (in N/cm^2) will the block exert when it is resting on a table on its smallest face?
iv) What is the least pressure the block could exert on the table?
b) i) What is the volume of the block?
ii) Calculate the density of the material from which the block is made. (NEA)

8 The area of a piston head in a car engine is $0.006\ m^2$. When the pressure in the cylinder is 3 000 000 Pa(N/m^2) is the force on the piston, in newtons:
A 18 000
B 5 000 000
C 5 000 000 000
D 18 000 000 000 (NEA)

9 Figure 5 shows a diving bell, like the one used on North Sea oil rigs. It is made for use 100 m below the surface of the sea. The pressure inside the diving bell is kept at 10 atmospheres (1 000 000 N/m^2) above atmospheric pressure.

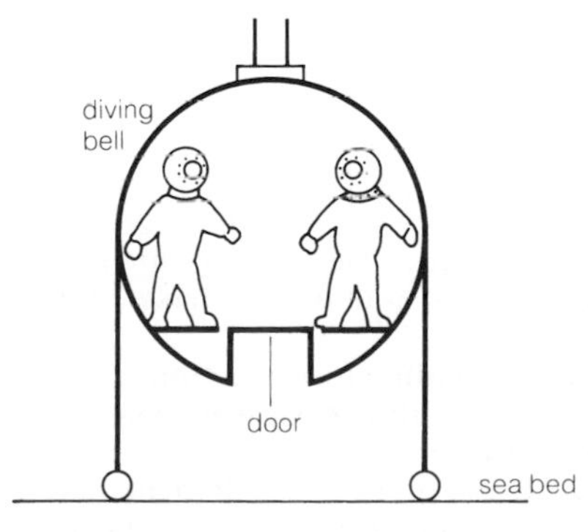

Figure 5

The pressure outside the diving bell must be equal to the pressure inside before the door is opened.
a) What is the pressure under 100 m of water?
b) Explain what would happen to the diving bell if the door was opened
i) 10 m below the surface.
ii) 200 m below the surface.
c) i) When it is under the sea, how is the pressure on top of the dividing bell different from the pressure underneath?
ii) Why does this pressure difference produce buoyancy? (SWEB/SEG)

10 Figure 6 shows two cylinders connected by a pipe. In each cylinder there is a piston and the space below each piston is full of oil.

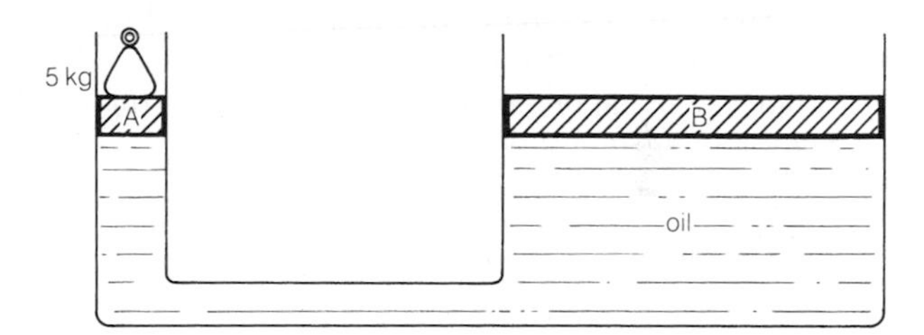

Figure 6

The area of piston A is 25 cm^2 and the area of piston B is 2000 cm^2. A 5 kg mass is placed on A.
b) i) What is the weight of this mass?
ii) What is the downward force on A?
c) i) Calculate the pressure on the oil under A?
ii) What is the pressure on the oil under B?
iii) Calculate the upward force on B?
d) George suggests that this machine can be used as a car jack.
i) Which piston (A or B) would you use to support the car? Give a reason.
ii) The car has to stay up after the force has been removed from the other piston. What device needs to be put into pipe C to do this? Say briefly how this device works.
iii) George is worried because he can lift the car with quite a small force using the jack. So he thinks that more energy comes out of the machine than he puts in. Explain why he does not get more energy out even though the force needed to raise the car is much greater than his force. (SWEB/SEG)

Further questions

Density and pressure: part B

Some questions require a knowledge of concepts covered in earlier units.

$g = 10\,\text{m/s}^2 = 10\,\text{N/kg}$

1 a) What is meant by the density of a substance? State consistent units in which the various quantities you have mentioned could be measured.

b) A tin containing $5000\,\text{cm}^2$ of paint has a mass of 7.0 kg.
i) If the mass of the empty tin, including the lid, is 0.5 kg calculate the density of the paint. (ii) If the tin is made of a metal which has a density of $7800\,\text{kg/m}^3$ calculate the volume of metal used to make the tin and the lid. (JMB)

2 A block of metal of density $3000\,\text{kg/m}^3$ is 2 m high and stands on a square base of side 0.5 m.
a) What is the base area of the block?
b) What is the volume of the block?
c) What is the mass of the block?
d) What is the weight of the block?
e) What is the pressure exerted by the weight of the block on the surface on which it stands? (O)

3 If a vehicle weighs 100 000 N, what is the force exerted on the road by one tyre if it is a four-wheeled vehicle and the weight is evenly distributed? Calculate the pressure exerted by one tyre if the area of tyre in contact with the road is $0.1\,\text{m}^2$. (SWEB)

4 The rectangular block shown in figure 1 has a mass 7200 kg and is made of material of density $1200\,\text{kg/m}^3$. The area of the base ABCD is $8.0\,\text{m}^2$.
a) Calculate the height AE of the block.
b) Calculate the pressure which the block exerts on the horizontal surface on which ABCD rests. (L)

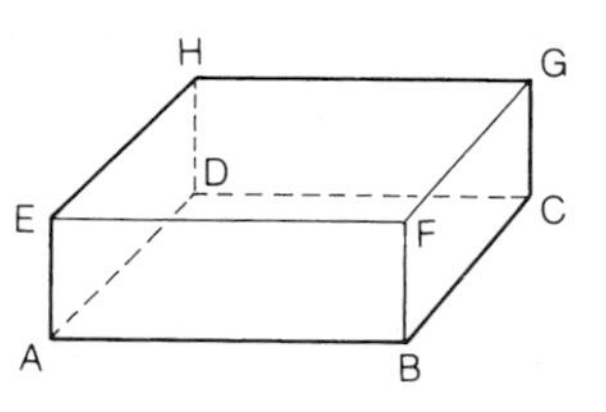

Figure 1

5 Figure 2 shows the barrel and washer of a bicycle pump. Describe what happens during one upstroke of the pump, with special reference to the action of the washer and tyre valve and the part played by atmospheric pressure. (No constructional detail of the tyre valve is expected.) (L)

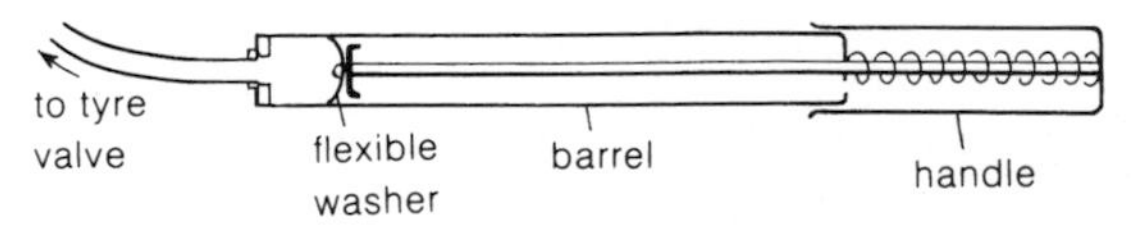

Figure 2

6 Explain why an iceberg normally floats with only a small fraction of its volume above the surface of the water. On observing several icebergs, one which contained a large rock embedded in it was found to be almost completely submerged in the water. Why was this? (L)

7 A block of metal (density $2700\,\text{kg/m}^3$) has volume $0.04\,\text{m}^3$. Calculate the mass of the block and its apparent weight when completely immersed in brine (density $1200\,\text{kg/m}^3$). (SUJB)

8 A hot air balloon is tethered to the ground on a windless day. The envelope of the balloon contains $1200\,\text{m}^3$ of hot air of density $0.8\,\text{kg/m}^3$. The mass of the balloon (not including the hot air) is 400 kg. The density of the surrounding air is $1.3\,\text{kg/m}^3$.
a) Explain why the balloon would rise if it were not tethered.
b) Calculate the tension in the rope holding the balloon to the ground.
c) Calculate the acceleration with which the balloon begins to rise when released. (O&C)

9 A solid cylinder, mass 100 kg, is suspended in water between two tight cables so that the circular faces are horizontal, as shown in figure 3, and the upper face is 0.20 m below the surface. The area of a circular cross-section of the cylinder is $0.50\,\text{m}^2$ and its height is 0.30 m. The density of water is $1.0 \times 10^3\,\text{kg/m}^3$.

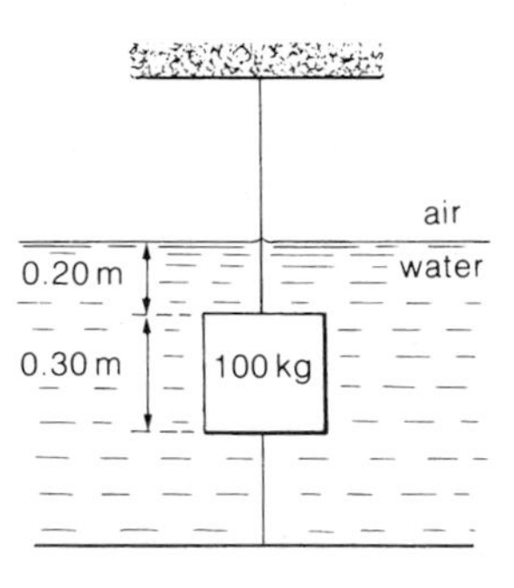

Figure 3

a) Calculate the pressure in excess of that due to the atmosphere at the top and at the base of the cylinder.
b) Calculate the corresponding forces acting on each of these faces and hence the net upward force due to the water pressure.
c) Show that the upthrust on the cylinder calculated using Archimedes' principle is equal to that calculated in b).
d) Will the cylinder sink, float or rise if the cables are removed? (O&C)

SECTION 4

MOLECULAR MOTION AND HEAT

Cool it, Stegosaurus!

Like all engines, the biological engines which power animals produce surplus heat when they are doing work. Nature has devised many methods of getting rid of unwanted heat – for example, humans sweat, dogs pant, and birds ruffle their feathers.

This may explain why stegosaurus, a dinosaur which lived on Earth around 150 million years ago, had large plates running along its back – the plates probably acted as cooling fins, transferring body heat to the outside air as blood flowed through them.

Stegosaurus: looking fierce, but keeping cool.

Riding on warm air

The hang glider pilot shown above is likely to have a very brief flight unless he can find a rising current of air to carry him upwards. Currents of rising warm air are called thermals and they can occur above hilltops, villages, or factories, or under certain types of cloud. Birds use thermals to gain height – indeed, experienced han glider pilots keep a sharp look out for birds circling in the air, as this may mean that a thermal is present.

The height gain record for a hang glider after launch stands at more than 3500 metres (over 2 miles).

Preventing bursts

Ronald Raybone's simple but effective invention may make burst pipes in winter a thing of the past. Essentially it consists of a long plastic inner tube which runs through the centre of the pipe. If the water in the pipe freezes, it expands on turning to ice – but this squashes the plastic rather than bursting the pipe. Raybone's idea has been slow to catch on in the UK, but has aroused much interest in Canada, where severe winters cause substantial damage to underground supply pipes.

Take the tension out of wash day – use soap

An average laundry basket holds about 5 kg of dirty washing. Of that, about 150 g is the dirt, made up roughly as follows:

Wax, Alcohol, bits of food	50 g
Skin, hair, and other proteins	20 g
Sweat and greasy excretions	10 g
Dust, sand, soot, and soil	70 g

Some of this dirt is jammed between the cloth fibres and some is absorbed by them. Most will not dissolve in water; nor is water attracted to it in the same way as it is to, say, glass. This tends to prevent the water passing between the fibres. At one time, the best you could hope for was to shake the dirt particles free by banging your wet washing against a stone down by the river. Nowadays, detergents make life very much easier.

Adding detergent to water lowers the **surface tension** – in other words it reduces the forces which make water molecules cling to each other rather than to other materials, and enables the water to penetrate spaces which ordinary water cannot reach. The detergent molecules can then get to work on the dirt particles themselves by clinging to them and dislodging them from the fibres.

4.1
Moving molecules

There is plenty of evidence to suggest that matter is made up of tiny particles which are constantly on the move.

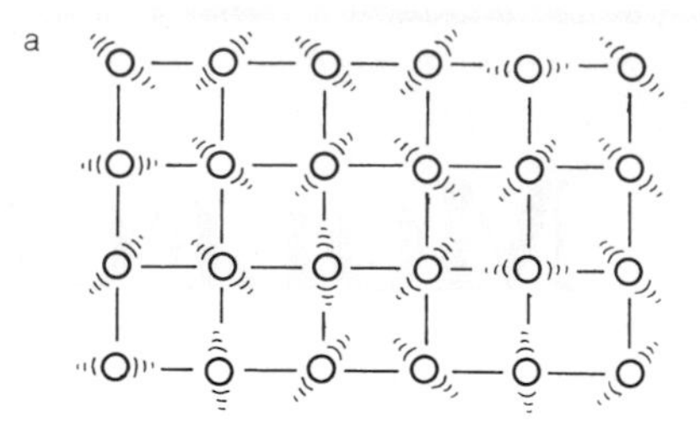

Figure 1a Molecules of a solid

The kinetic theory of matter

According to the **kinetic theory**, all matter is made up of tiny particles called molecules. These molecules are constantly in motion, and they attract each other strongly when they are close to one another. They have kinetic energy because they are moving, and potential energy because their motion keeps them separated despite the attractions which are trying to pull them together.

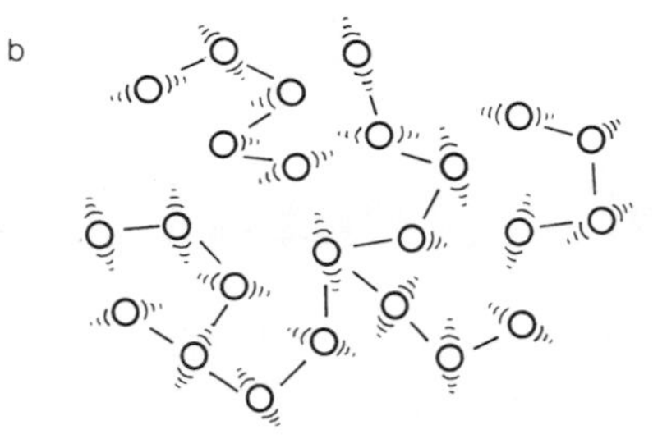

Figure 1b Molecules of a liquid

In a solid, strong forces of attraction hold the molecules close together in a regular structure. The molecules vibrate to and fro, but they are not free to change positions. A solid has a definite shape and volume, as shown in figure 1a.

In a liquid, the molecules are again close together, and they still vibrate to and fro. But they have enough energy to prevent the attractions from holding them in fixed positions. A liquid has a definite volume but no fixed shape. In other words, a liquid can flow, as shown in figure 1b.

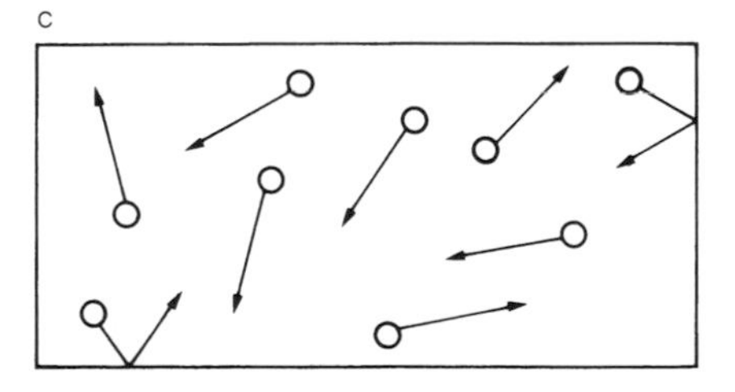

Figure 1c Molecules of a gas

In a gas, the molecules are well spaced out, and have enough energy to be virtually free of any attractions between one another. They travel at high speed and may also spin. Moving about at random, they collide with each other and the walls of any container they happen to be in. Unlike the molecules of a solid or liquid, they do not occupy a fixed volume, and quickly fill any space available, as shown in figure 1c.

Brownian motion

In 1827, Robert Brown was using a microscope to study tiny grains of pollen suspended in water, when he noticed that the grains were constantly wobbling and wandering about. The most likely explanation came some years later, when it was suggested that the motion was caused by vibrating water molecules bumping into the pollen grains. **Brownian motion**, as it came to be known, offered clear evidence in support of the kinetic theory.

Brownian motion can also be seen when you look through a microscope at smoke floating in air. Smoke from a burning straw for example is made up of millions of tiny oil droplets. These are light enough to be moved by individual molecules of air which crash into them. Bombarded by these fast moving molecules, the oil droplets wander about in zig-zag paths.

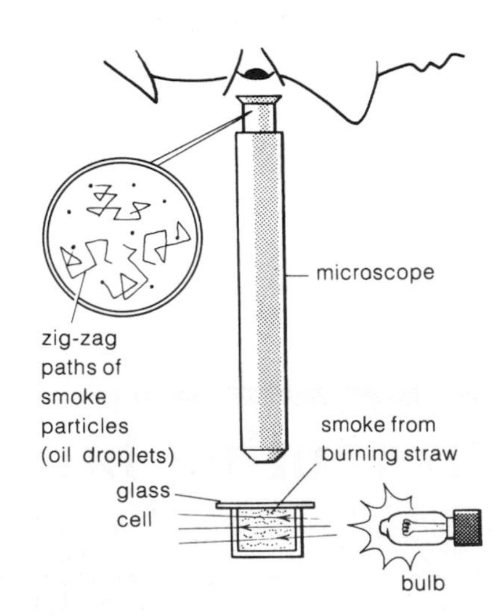

Figure 2 Experiment to demonstrate Brownian motion

Figure 2 shows the apparatus you would use to study the Brownian motion of (oil droplets) in air. Smoke from a burning straw is trapped in a small glass container or 'cell' which has a bright light shining through it from the side. Through the microscope, the smoke particles can be seen glinting in the light beam as they are jostled about by the molecules of the air.

Diffusion

When a small amount of ink is dropped into a dish of water, the colour soon spreads throughout the water. The molecules of the colouring material in the ink spread by a process called **diffusion**; they wander at random through the water as they are bumped and jostled by the water molecules around them.

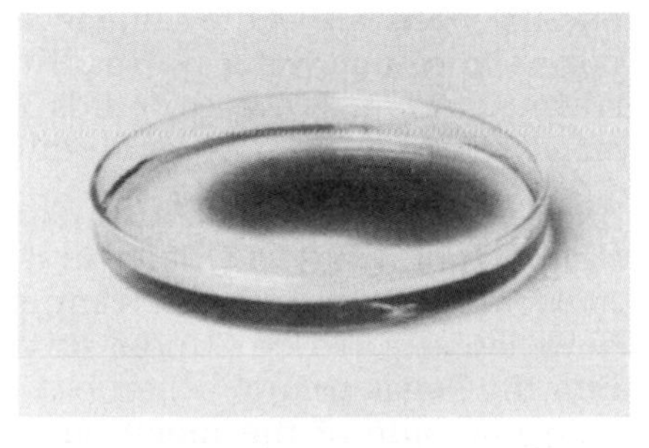

Figure 3 Diffusion in a liquid

Diffusion also occurs in gases. Diffusion causes a small amount of leaking gas to spread throughout a room. It also makes smells spread – smells are wandering gas molecules which come from whatever it is that happens to be smelling.

In the laboratory, the diffusion of gas molecules can be demonstrated using the apparatus shown in figure 4.

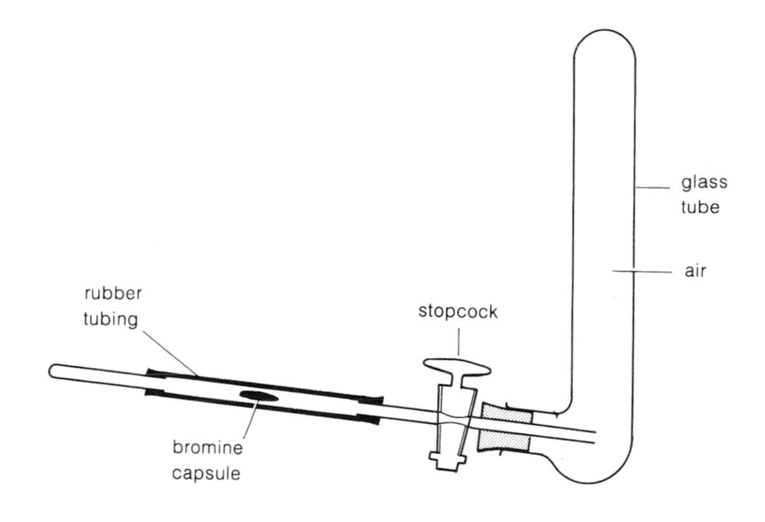

Figure 4 Diffusion in bromine

In the experiment shown, the glass tube contains air. When the rubber tubing is squeezed with a pair of pliers, the glass capsule breaks and a poisonous gas called bromine is released into the tube. The orange coloured gas can be seen slowly spreading through the air as diffusion takes place.

The experiment can also be carried out with a vacuum in the tube rather than air. In this case, a vacuum pump is used to remove the air from the tube before the start of the experiment. The tube is sealed by closing the stopcock, the apparatus is assembled, and the stopcock is then opened again. When the capsule is broken, the bromine gas fills the glass tube almost instantaneously. As no air molecules are present to collide with the bromine molecules and make them change direction, they travel freely across the tube at speeds of around 50 m/s.

Osmosis

The experiment in figure 5 shows a process called **osmosis** in action. This too is caused by wandering molecules.

The tube contains concentrated sugar solution [a mixture of sugar molecules and water molecules], but the beaker contains water only. The sugar solution and the water are separated by a thin sheet, or **membrane**, of Visking tubing. To begin with, the sugar solution and the water are at the same level. The liquid level in the tube then begins to rise as water passes through the membrane and into the sugar solution. Pressure builds up above the membrane and the sugar solution becomes more dilute.

The kinetic theory is able to explain the process of osmosis. The Visking tubing has tiny holes in it which are so small that the sugar molecules are too big to wander through. The water molecules, however, can and do wander through the membrane. More wander into the sugar solution than out of it because more are in contact with one side of the membrane than are with the other. This is shown in figure 6.

Materials such as Visking tubing, which allow some substances to pass through but not others, are described as **semi-permeable**. The pressure that builds up across such a membrane is called **osmotic pressure**.

Osmosis has importance for plant and animal life because semi-permeable membranes surround the tiny liquid-filled cells of which all living things are made. For example, plants absorb water by osmosis, and osmotic pressure keeps plant cells firm rather like balloons. Without water to maintain the osmotic pressure in their cells, plants wilt as shown in figure 7.

Prunes are wilted plums. If left soaking overnight, they absorb water by osmosis and swell up.

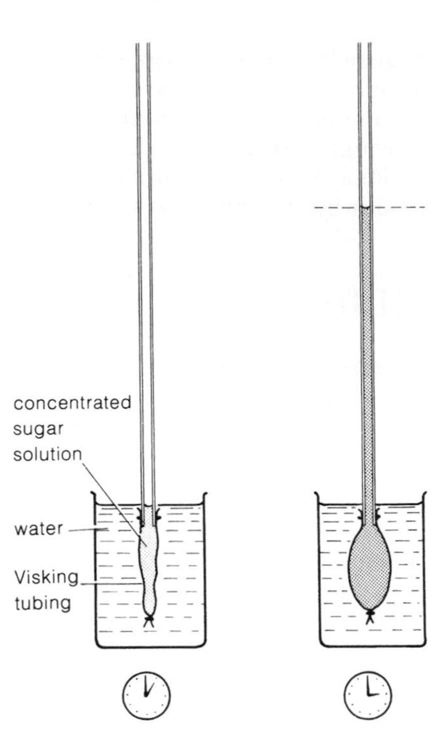

Figure 5 Experiment to demonstrate osmosis

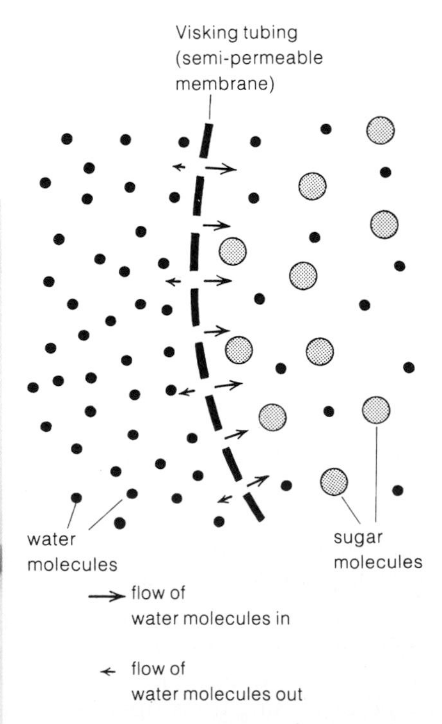

Figure 6 Small molecules pass through a semi-permeable membrane so as to even out their concentration

Figure 7 The effect of reduction in osmotic pressure

Estimating the size of a molecule

Molecules vary enormously in size. Molecules of proteins, for example, are many thousands of times larger than the smallest molecules. Figure 8a shows an experiment you could carry out to measure the size of a molecule of olive oil. It doesn't produce a very accurate result, but it does give a rough idea of molecular size.

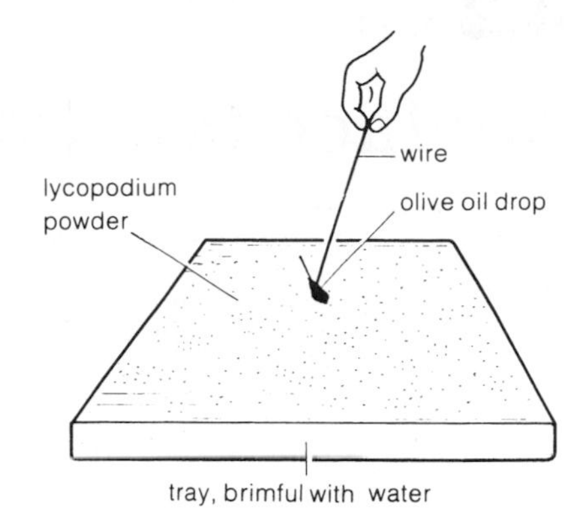

Figure 8a Experiment to estimate the size of a molecule

In this experiment, a little lycopodium powder is sprinkled over the surface of some clean water in a tray. A small drop of olive oil is then put onto the water surface, where it spreads out to form a thin circular film. The lycopodium powder enables the edge of this oil film to be seen clearly. Assuming that the oil spreads until it is only one molecule thick, the size of the molecule can be found by calculating thickness h of the film.

For h to be calculated, the radius r of the oil drop must be found, and also the radius R of the oil film it forms. In practice, it is simpler to measure the diameters of the oil drop and the oil film and halve the results. The diameter of the oil drop is measured using a millimetre scale and a magnifying glass as in figure 8b. The diameter of the oil film is also measured using a millimetre scale.

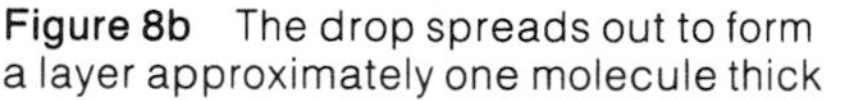

Figure 8b The drop spreads out to form a layer approximately one molecule thick

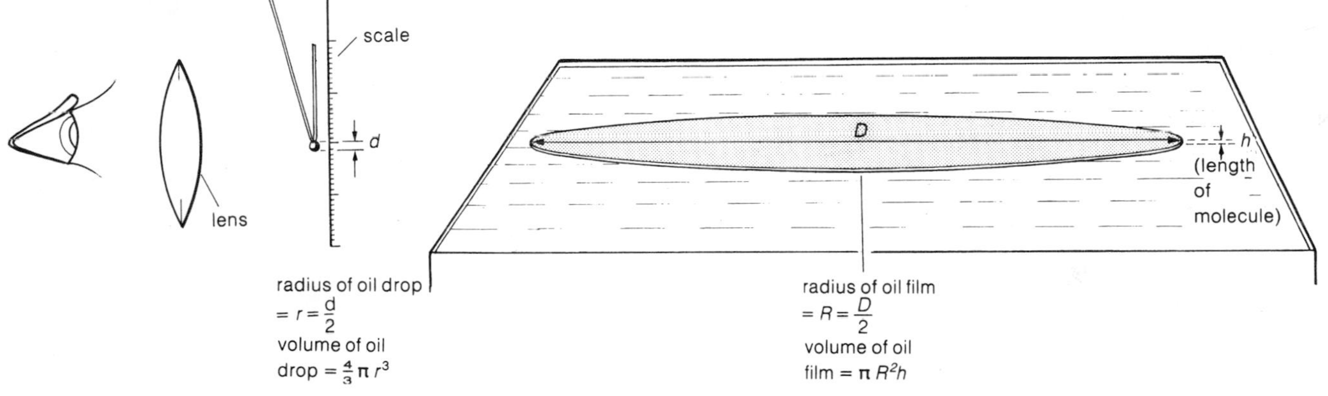

If the oil keeps a constant volume when it spreads out on the water, the volume of the oil drop, $\frac{4}{3}\pi r^3$, will equal the volume of the circular film, $\pi R^2 h$. This gives $h = \frac{4r^3}{3R^2}$

which can be used to calculate the film thickness h. This usually works out at about a nanometre [1 nm or 10^{-9} m].

Questions

1 What form [or forms] of energy is possessed by molecules in solids, liquids, and gases?

2 Describe the motion of a molecule in a) a solid b) a liquid c) a gas.

3 What do you see when you use a microscope to study illuminated smoke floating in air? What is the effect called, and how does the kinetic theory explain it? What other evidence is there to suggest that the molecules of a substance are constantly in motion?

4 What is a semi-permeable membrane? If water is on one side of such a membrane and sugar solution on the other side, what happens? What name is given to the effect?

5 A small drop of oil has a volume of 5×10^{-8} m^3. When it is put on the surface of some clean water, it forms a circular film of 0.1 m^2 in area. What is the size of a molecule of oil? State any assumptions you make in your calculation.

4.2

Surface tension

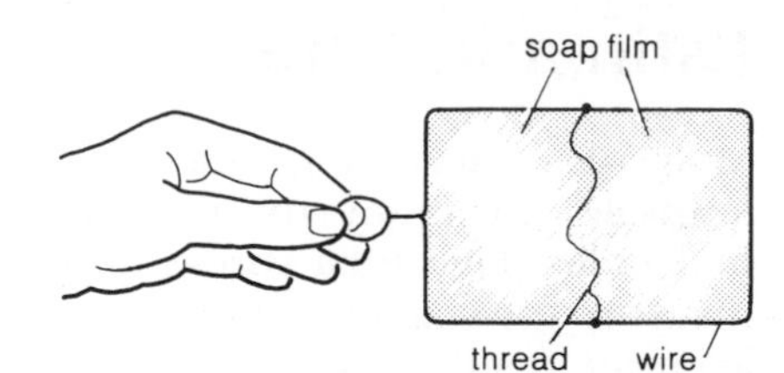

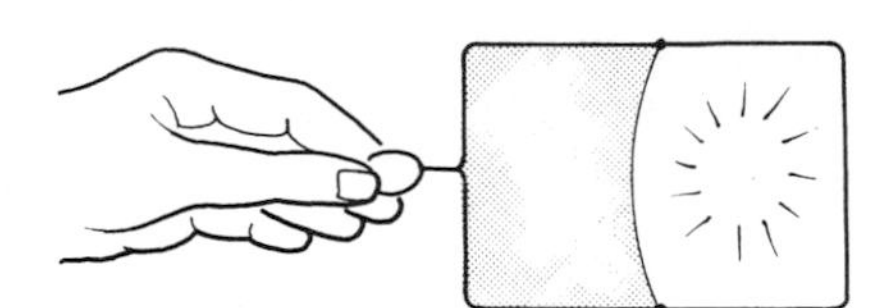

Figure 1 Demonstrating surface tension

A liquid behaves as if its surface were covered with a stretchy skin. The effect is called surface tension; this is another result of the force of attraction between molecules.

The tension in the surface of a liquid can be demonstrated using soapy water and a wire frame as shown in figure 1. The frame is first dipped in soapy water so that a thin film of liquid stretches across it. If the film to the right of the thread is popped, the remaining film pulls the thread firmly to the left.

Surface tension pulls the drips from a tap into round droplets, and it makes liquid mercury collect in drops. It enables a needle to float on the surface of water, and it allows insects to stand on a pond.

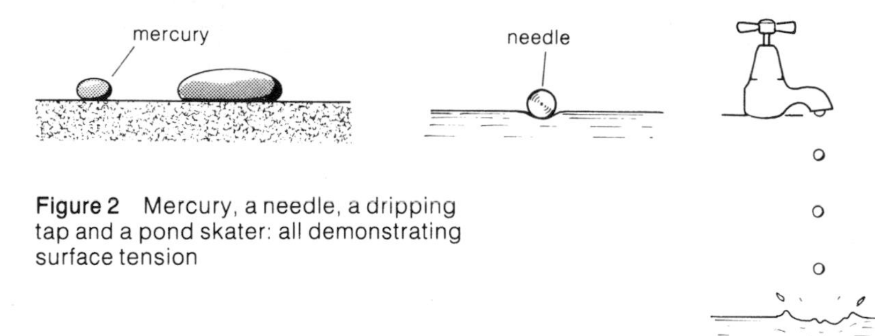

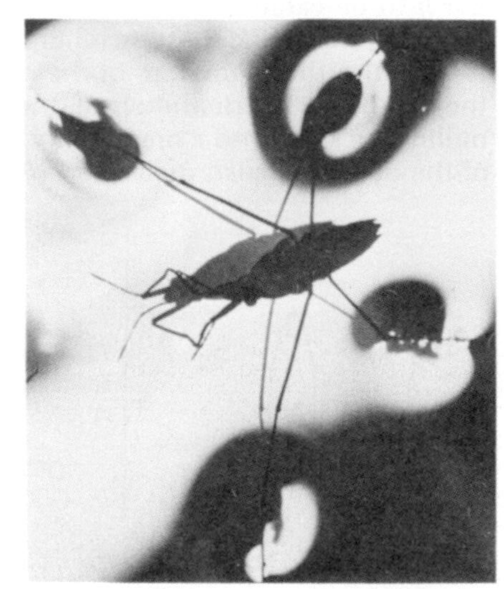

Figure 2 Mercury, a needle, a dripping tap and a pond skater: all demonstrating surface tension

Cohesion and adhesion

A liquid doesn't really have a stretchy skin. The tension in its surface is caused by the attractions which make its molecules cling together. Attraction between molecules of the same substance is called **cohesion**, and it tries to pull liquids into shapes with the smallest possible surface area. It is the cohesion of water molecules that makes small amounts of water form into round droplets, and it is cohesion that resists attempts by needles and insects' feet to push molecules apart by breaking through a water surface.

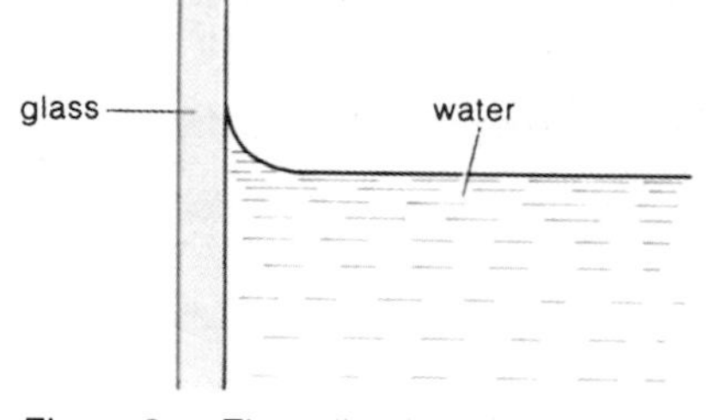

Figure 3a The adhesion of water to glass

Molecules of two different substances can also attract each other. When water is poured out of a glass beaker, for example, drops of water are left clinging to the glass because the water molecules are more strongly attracted to the glass than they are to each other. Attraction between molecules of different substances is called **adhesion**. Water 'wets' glass because the adhesion of water to glass is greater than the cohesion of water.

The strong attraction between water molecules and glass gives water an upward curving surface or **meniscus** where it meets glass, as shown in figure 3a. The meniscus of mercury, on the other hand, curves downwards where it meets glass, as shown in figure 3b because the cohesion of mercury molecules is much stronger than their adhesion to glass. When mercury is tipped out of a beaker, the glass is left completely dry.

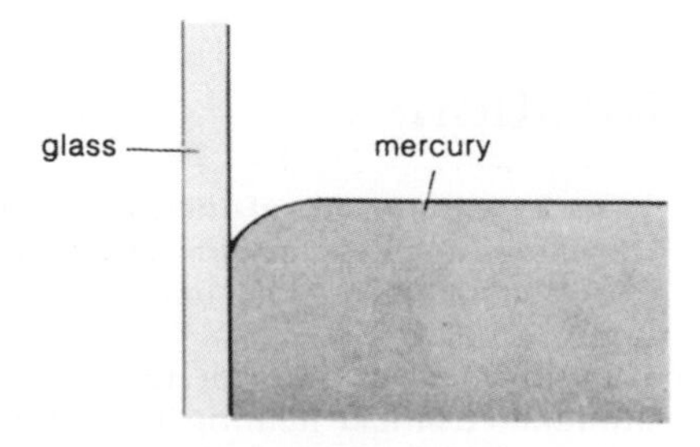

Figure 3b The strong cohesion of mercury prevents it from 'sticking' to glass

Capillary action

Glass attracts water molecules strongly enough to pull a column of water up through a very narrow tube [a **capillary** tube] made of glass. The effect is called **capillary action**. The narrower the tube, the further the water rises, as shown in figure 4a.

Mercury produces the opposite effect, as shown in figure 4b. If a narrow glass tube is dipped into mercury, the level of mercury in the tube drops below that of the surrounding liquid.

Water, alcohol and many other liquids can be drawn up through narrow gaps by capillary action. Materials like paper tissues and cotton wool 'soak up' liquids because they contain thousands of tiny air spaces through which liquids can move by capillary action, as shown in figure 5. Soil, porous rocks, bricks and concrete also contain air spaces and absorb water in the same way.

Capillary action will cause 'rising damp' through the walls and floors of houses unless steps have been taken to prevent it. In new houses, a waterproof sheet of polythene is laid in the concrete base to stop the upward movement of water from the ground. A waterproof damp course of plastic or felt and bitumen is also set into the outer brick walls just above ground level.

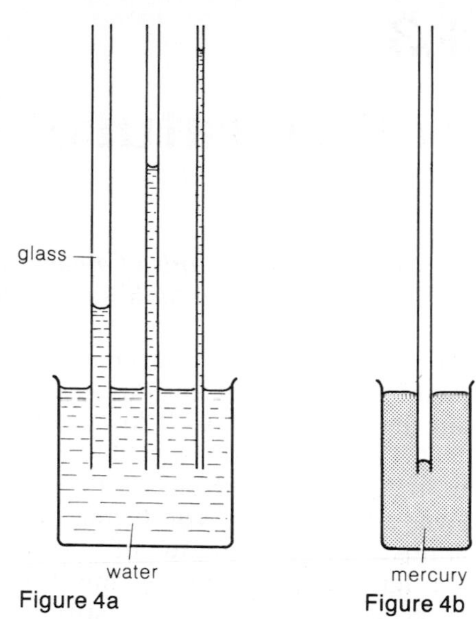

Figures 4a and 4b Adhesion causes capillary action and cohesion prevents it

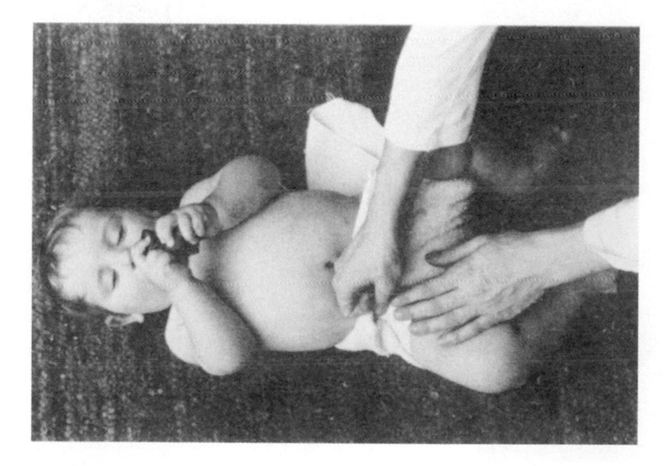

Figure 5a Using capillary action . . .

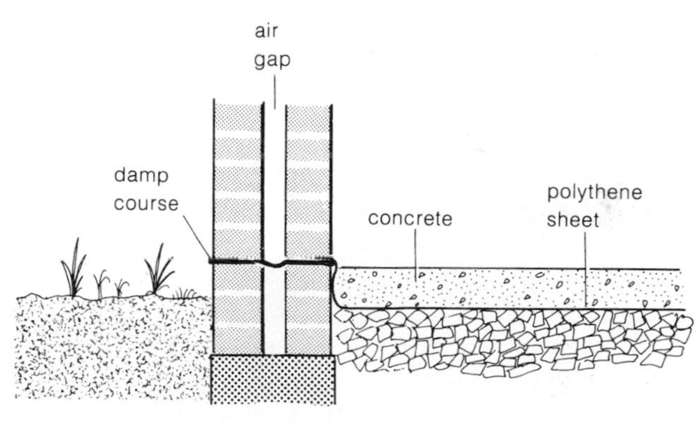

Figure 5b . . . and preventing it

Questions

1. How could you show that there is tension in the surface of a liquid? What is the cause of this tension?
2. Give three examples of the effects of surface tension.
3. What is the difference between cohesion and adhesion?
4. Why do drops of water cling to the sides of a glass beaker? Why is a glass beaker dry after mercury has been poured out of it?
5. How is the meniscus of water in a glass beaker different from that of a mercury meniscus? Why is there this difference?
6. What happens if a clean glass capillary tube is dipped into water? What would be the effect of using an even narrower glass tube?
7. What happens if a narrow glass tube is dipped into mercury?
8. Give two practical uses of capillary action.
9. What is a damp course? Why is it necessary?

4.3

Temperature

There are many methods of measuring how hot something is. But you have to decide on a suitable scale of hotness before you can use them.

A temperature scale is a range of numbers used to indicate levels or degrees of hotness. Various scales have been in use at one time or another, the most well known being the Fahrenheit and Celsius scales. The Fahrenheit scale is no longer in use for scientific work, though people still talk of hot days 'in the eighties' and fevers of 'over a hundred'. If you haven't heard of the Celsius scale, you will almost certainly know of it by its more popular name – the centigrade scale. This is now the most commonly used scale for everyday temperature measurements. Typical temperatures measured in degrees Celsius (°C) are given in figure 1.

Liquid iron. What's the temperature?

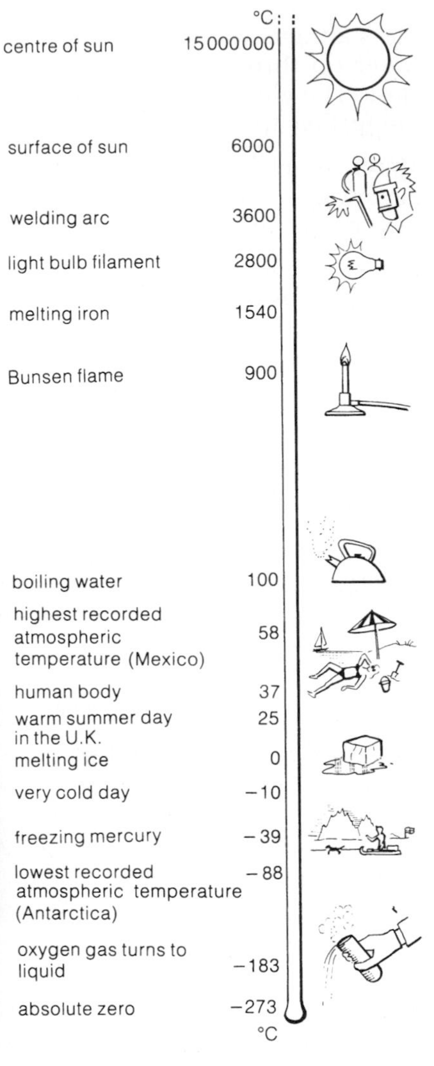

Figure 1 A range of temperatures on the Celsius scale

In the laboratory, temperatures are often measured using liquid filled thermometers like the one shown in figure 2. The bulb contains mercury or coloured alcohol which expands when the temperature rises and pushes a 'thread' of liquid along the scale. This and other types of thermometer are described in more detail later.

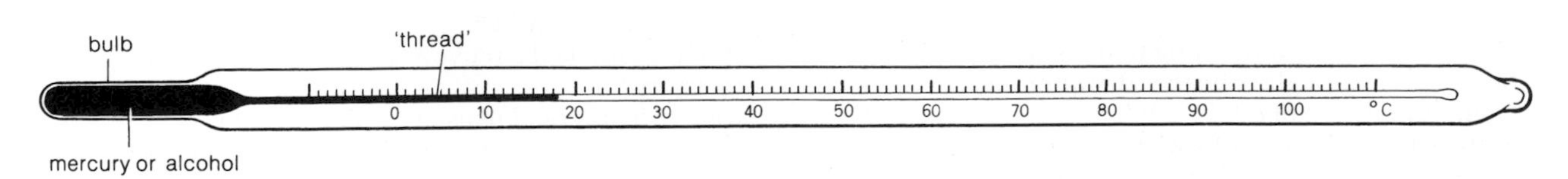

Figure 2 A liquid-filled thermometer

Fixing a scale of temperature

The numbers on a temperature scale aren't any use unless there is general agreement on the level of hotness which each number is going to represent. This isn't as difficult as it sounds. Once two standard temperatures or **fixed points** have been agreed, all other temperatures can be judged against these. On the Celsius scale:

The lower fixed point is the melting point of pure ice. This is called 0 degrees Celsius (0 °C).

The upper fixed point is the boiling point of pure water, where the water is boiling under standard atmospheric pressure (760 mmHg). This is called 100 degrees Celsius (100 °C).

By placing an unmarked thermometer in pure melting ice as in figure 3a, the 0 °C point can be fixed on its scale. The ice needs to be pure because its melting point is lowered if any impurities are present [as explained on page 151]. The 100 °C point is fixed by placing the thermometer in the steam above boiling water as shown in figure 3b. Atmospheric pressure must be the standard 760 mmHg because any change from this pressure alters the boiling point of the water. Impurities in the water also affect its boiling point, but in most cases they do not affect the temperature of the steam just above the water.

Once the 0 °C and 100 °C points have been fixed on the thermometer, the rest of the scale is made by dividing the space between the points into one hundred equal divisions or degrees; 'Centigrade' actually means 'one hundred divisions'. The idea can be extended to produce a temperature scale going above 100 °C and beneath 0 °C.

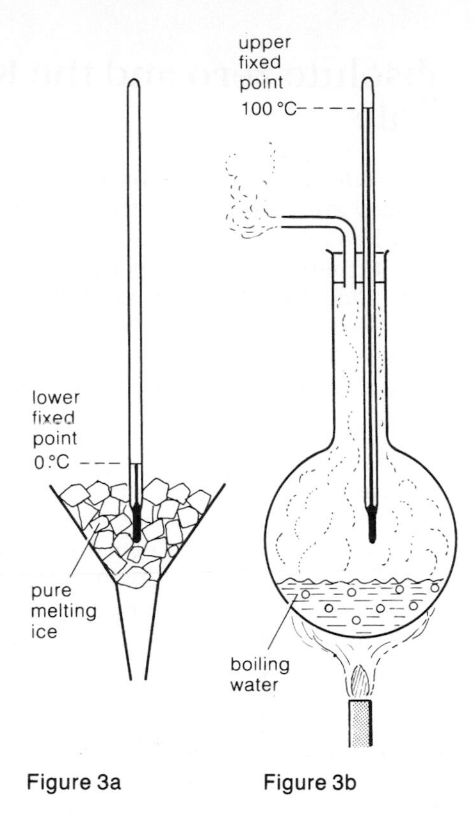

Figures 3a, 3b Calibrating a thermometer: two 'fixed points' are needed; the space between them is portioned into equal divisions

What is temperature?

The molecules of any object are constantly in motion. The hotter the object, the more kinetic energy its molecules have and the faster they move.

If a hot object is placed next to a cold object, there is a transfer of thermal energy from one to another (as shown in figure 4). One group of molecules loses kinetic energy by slowing down, the other group gains kinetic energy by speeding up. The transfer of energy stops only when molecules of both objects each have the same kinetic energy on average. The objects are then at the same temperature.

Temperature is therefore a measure of the average kinetic energy which each molecule of an object possesses. One object is at a higher temperature than another if the average kinetic energy of each of its molecules is greater.

It is important not to confuse the temperature of an object with the total quantity of thermal energy it can give out. A spoonful of boiling water, for example, has exactly the same temperature as a saucepanful of boiling water, but you would get far less thermal energy from it if you accidentally tipped it over yourself.

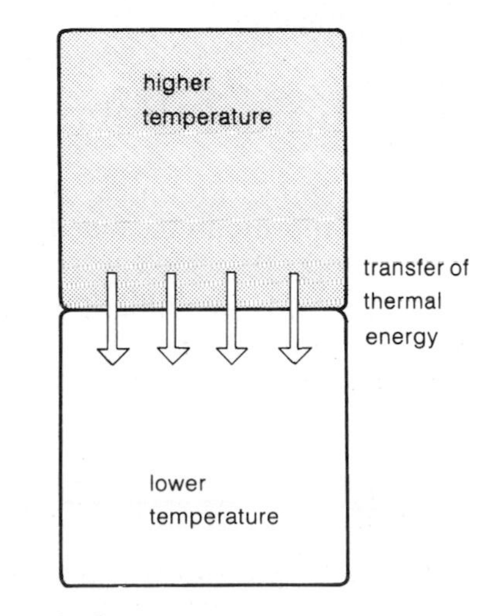

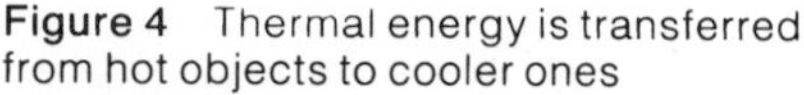

Figure 4 Thermal energy is transferred from hot objects to cooler ones

Absolute zero and the Kelvin temperature scale

The colder an object becomes, the less kinetic energy its molecules possess and the slower they move. You might expect that there would be a temperature at which they would have no energy at all. In fact, there are laws governing the behaviour of molecules which don't allow them to stop moving altogether, but at a temperature of about −273 °C they do have the minimum possible amount of energy. This temperature is called **absolute zero**; no object can reach a temperature lower than this.

In scientific work, it is often useful to measure temperatures on a scale which has its zero at absolute zero. Such a scale is called an **absolute** temperature scale. The **Kelvin** scale is the most commonly used absolute temperature scale. Each kelvin [K] on the scale is the same size as a degree Celsius [°C], so conversion from one scale to another is easy – the Kelvin temperature is found by adding 273 degrees to the Celsius temperature. Examples are given in figure 5. Note that temperatures are expressed in kelvin, and *not* 'degrees kelvin'.

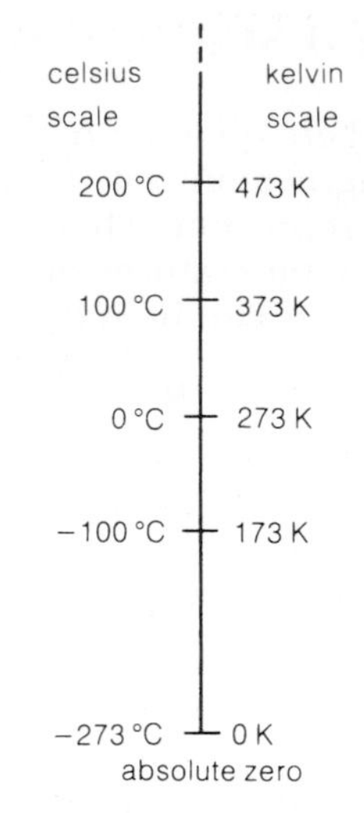

Figure 5 Absolute zero: the minimum possible thermal energy, therefore the minimum possible temperature

Liquid-in-glass thermometers

All thermometers are based on some property of a material which changes when the material becomes hotter or colder. Liquid-in-glass thermometers use the fact that most liquids expand slightly when they are heated.

When the temperature rises, the liquid in the bulb expands by moving up the tube next to the scale. The tube is very narrow, so a small increase in the volume of the liquid makes the 'thread' move a long way up the tube. The narrower the tube, the more sensitive the thermometer is to changes in temperature.

Liquid-in-glass thermometers are filled either with mercury or alcohol. Both liquids have their advantages and disadvantages, as shown in figure 6.

Mercury

Advantages

doesn't wet sides of tube
thread easy to see
conducts heat well
thermometer responds quickly to temperature changes.

Disadvantages

freezes at −39°C
thermometer not suitable for low Arctic temperatures
poisonous
thermometer hazardous if broken
expensive

Alcohol

Advantages

freezes at −115°C
thermometer suitable for low Arctic temperatures
expansion greater than mercury: wider tube can be used

Disadvantages

has to be coloured to be seen easily
clings to sides of tube
thread has tendency to break

Figure 6 Comparison of mercury and alcohol thermometers

The clinical thermometer

Figure 7 shows a clinical thermometer. This is a special type of mercury thermometer used to measure the temperature of the human body. Its scale covers only temperatures a few degrees either side of the average body temperature of about 37 °C.

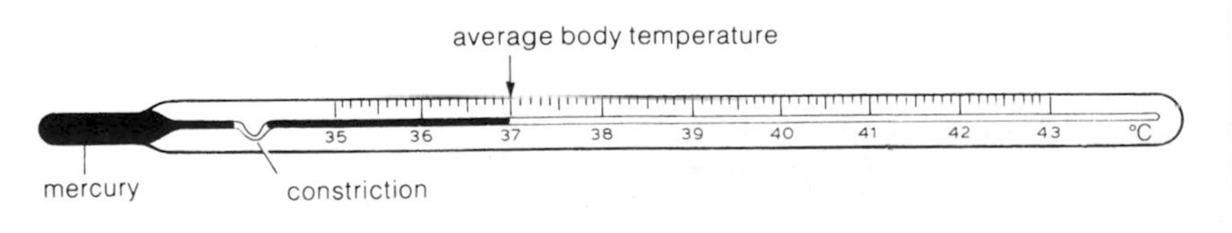

Figure 7 The clinical thermometer

The tube has a constriction (a narrowing) next to the bulb. This stops the mercury thread running back into the bulb, so the temperature reading can be taken after the thermometer has been removed from the patient's mouth. The thermometer has to be shaken to get the mercury back past the constriction.

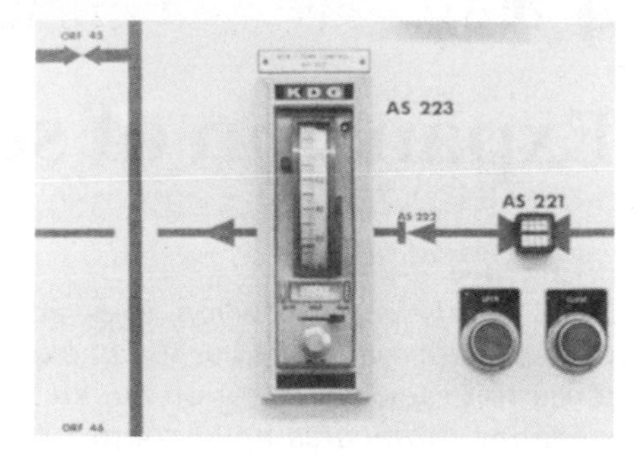

Figure 8 A computer linked temperature reading

Thermometers for industry

Many industrial processes depend on the accurate measurement and control of temperature. Liquid-in-glass thermometers aren't really suitable for industrial work. Firstly, their temperature range is often too limited. Secondly, it is usually more convenient for the operator to read the temperature on a meter or digital display placed some distance away from the source of heat. Thirdly, electrical methods of measuring temperature give readings which can be recorded automatically or fed directly to a computer controlling the heating process, as shown in figure 8.

Resistance thermometers, as shown in figure 9, are based on the principle that it becomes more and more difficult to pass an electric current through a piece of metal as its temperature rises. These thermometers usually contain a length of thin platinum wire connected to a power supply. The higher the temperature, the less the current passing through the wire and the further down the meter scale the needle moves. Resistance thermometers can be used to measure temperatures from about −200 °C to 1200 °C.

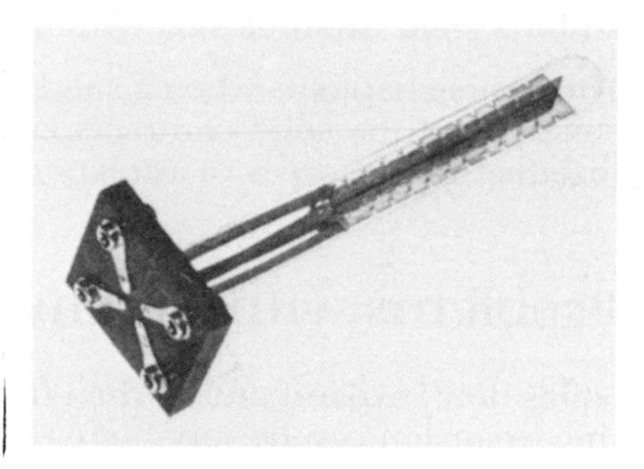

Figure 9 The temperature-sensing end of a resistance thermometer

Thermistor thermometers work along similar lines. A thermistor is a small device which offers less and less resistance to a flow of electricity as its temperature rises.

In a **thermocouple thermometer**, as shown in figure 10a, two different types of metal wire are joined together at two juctions as shown in figure 10b. A temperature difference between the junctions actually makes the metals produce a small electric current which moves the meter needle across the scale. Thermocouple thermometers are often used to measure oven and furnace temperatures. They can operate over a temperature range from about −200 °C to 1600 °C.

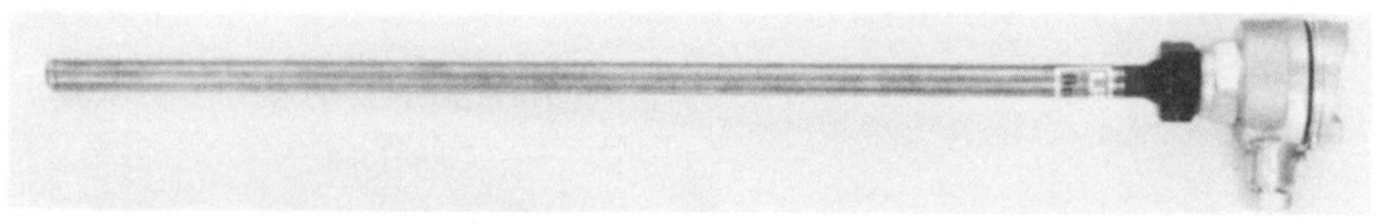

Figure 10a A thermocouple probe

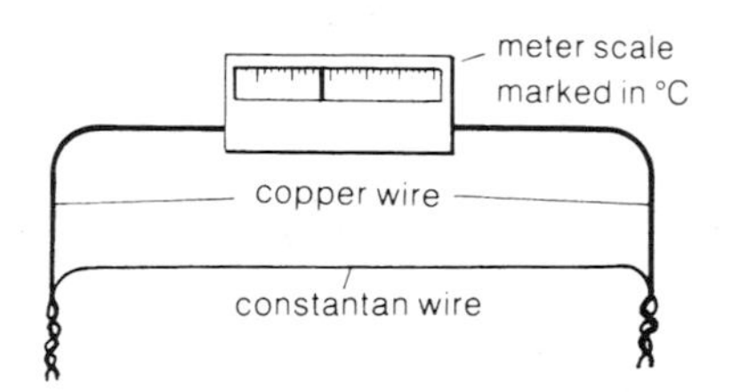

Figure 10b Thermocouple thermometer

Questions

1 What is the lower fixed point on the Celsius temperature scale? What is the upper fixed point?
2 How would you set about fixing a temperature scale on an unmarked mercury thermometer?
3 What happens to the molecules of an object when the temperature rises?
4 Two objects are at the same temperature. What conclusion can you draw about the molecules of the two objects?
5 What temperature is absolute zero on the Celsius scale? Why is this the lowest possible temperature?
6 Convert the following temperatures into kelvin: −273 °C; 0 °C; 27 °C; 100 °C.
7 Give three reasons why mercury might be used in a liquid-in-glass thermometer. Why is alcohol sometimes used instead?
8 Give two ways in which a clinical thermometer differs from the mercury thermometer which you might use in the laboratory.

4.4

Expansion of solids

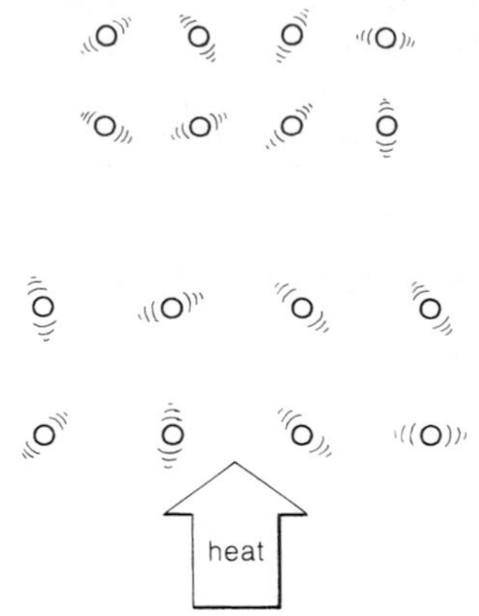

Most solids expand when they are heated – by so little that you wouldn't normally notice. Yet an expanding solid can produce enough force to crack concrete or buckle steel. And it all happens because vibrating molecules need room to move.

Heating a solid makes its molecules vibrate more vigorously. As the vibrations become larger, the molecules are pushed further apart and the solid expands slightly in all directions.

The reverse happens when a solid is cooled. The vibrations become smaller and the solid contracts as its molecules are pulled closer together by the forces of attraction between them.

Figure 1 Heating a solid causes increased vibration of the molecules – the solid *expands*

Problems with expansion

Solids don't expand much when heated, but the force produced by the expansion can be very high. Concrete and steel beams used in construction work can cause considerable damage if they do not have space to expand when the temperature rises.

Gaps are left at the ends of bridges to give room for expansion. One end of a bridge is often supported on rollers so that free movement is possible as the bridge expands and contracts, as shown in figure 2.

The problem of expansion in steel railway lines is overcome by heating the lines when they are first laid. As they cool, they try to contract, but are unable to do so because they are fastened to heavy concrete sleepers embedded in chippings. The lines are therefore permanently under tension. On a hot day, the tension is less than it would be on a cold day.

Railway lines are laid in sections of about 1 km or so in length. Some movement does take place at the ends of each section, so these are linked by overlapping joints as shown in figure 3.

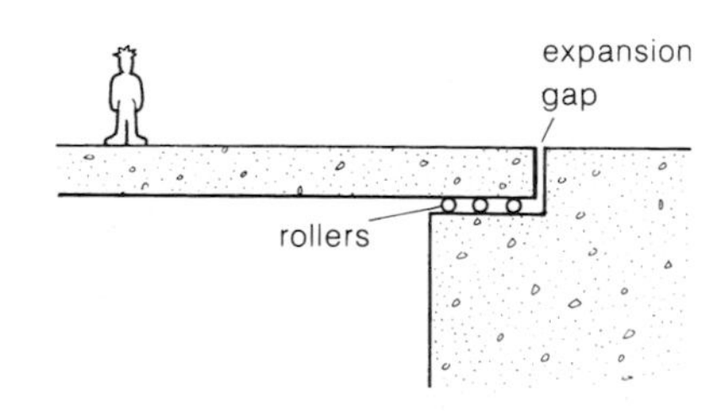

Figure 2 Allowing for expansion in a bridge

Linear expansivity

When a solid expands, it does so in all directions. For simplicity however, length increases only will be considered here.

Experiments show that for any given temperature rise, the increase in length of a solid is directly proportional to its original length. In the case of steel for example, when the temperature rises by 1 K [1 °C]:

a 1 m bar of steel increases in length by 0.000 011 m,
a 2 m bar of steel increases in length by 0.000 022 m,
a 3 m bar of steel increases in length by 0.000 033 m, and so on.

Figure 3 Overlapping joints are still used in railway lines

It is reasonable to expect this simple proportion to apply because a 3 m bar of steel, say, can be thought of as three 1 m bars joined end to end. Note in each of the cases at the bottom of the last page that:

$$\frac{\text{increase in length}}{\text{original length}} = 0.000\,011$$

this being the *fractional increase in length* of steel for a 1 K rise in temperature. Put another way, if you take any length of steel and raise its temperature by 1 K, the increase in length is always the same fraction [0.000 011] of the original length.

Further experiments show that, for small rises in temperature, the fractional increase in length of a solid is directly proportional to the temperature rise. In the case of steel for example:

If the temperature rise is 1 K, the fractional increase in length is 0.000 011.
If the temperature rise is 2 K, the fractional increase in length is 0.000 022.
If the temperature rise is 3 K, the fractional increase in length is 0.000 033, and so on.

Note in each case that dividing the fractional increase in length by the temperature rise always gives the same value, 0.000 011 /K. Steel is said to have a **linear expansivity** of 0.000 011 /K.

The linear expansivity of any solid is given by the equation:

$$\textbf{linear expansivity} = \textbf{fractional increase in length} \div \textbf{temperature rise}$$

This can be written:

$$\textbf{linear expansivity} = \frac{\textbf{increase in length}}{\textbf{original length}} \div \textbf{temperature rise}$$

though the following form is more usual:

$$\textbf{linear expansivity} = \frac{\textbf{increase in length}}{\textbf{original length} \times \textbf{temperature rise}}$$

(If this last step taxes your understanding of mathematics, satisfy yourself that, say, $\left(\frac{24}{4}\right) \div 3$ is equal to $\frac{24}{4 \times 3}$)

In the above equation, the increase in length and the original length can be measured using any length unit, provided the same unit is used for each. If the temperature rise is measured in K or °C, the linear expansivity is measured in /K or /°C.

The linear expansivity of a solid is numerically the same as the increase in length of a 1 metre length when the temperature rises by 1 K [1 °C].

For example, steel has a linear expansivity of 0.000 011 /K; a 1 m bar of steel expands by 0.000 011 m when the temperature rises by 1 K. This is illustrated in figure 4.

A 1 cm bar of steel expands by 0.000 011 cm when the temperature rises by 1 K.

A *micrometer dial gauge* is used to measure very small increases in length. This one can measure to an accuracy of one-hundredth of a millimetre

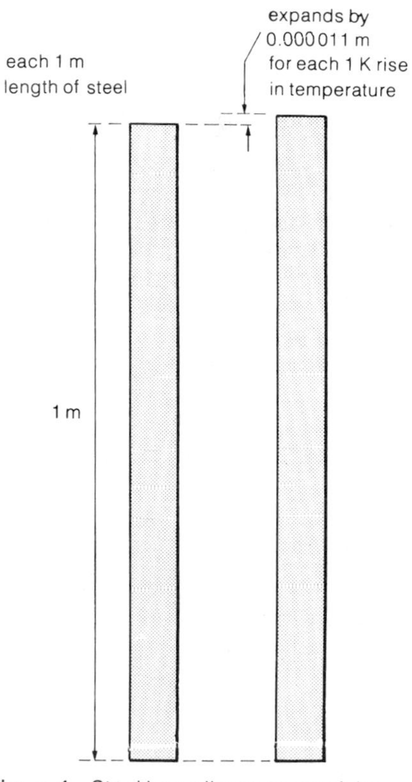

Figure 4 Steel has a linear expansivity of 0.000 011 /K

Measuring linear expansivity

The apparatus shown in figure 5 is used to measure the linear expansivity of a metal rod. The rod is heated by passing steam through the 'jacket'. Its increase in length is measured on the dial micrometer which is firmly clamped onto one end.

The rod is heated with steam to ensure that all parts reach the same temperature. The rise in temperature of the rod is found by reading the thermometer before and after the rod has been heated. As the rod expands, it pushes on the end of the micrometer. The movement is magnified in the micrometer and used to move the needle around the scale.

The linear expansivity of the rod is calculated using the equation:

$$\textbf{linear expansivity} = \frac{\textbf{increase in length}}{\textbf{original length} \times \textbf{temperature rise}}$$

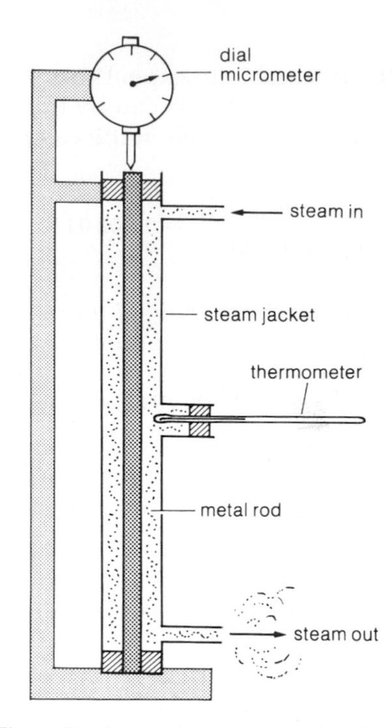

Figure 5 Apparatus for measuring the linear expansivity of a metal rod

Linear expansivities of different materials

Typical linear expansivities are given in figure 6, at the bottom of the page. The higher the value, the more any given length of material will expand for any given temperature rise.

Steel rods can be embedded in concrete beams to strengthen them because the two materials will expand equally if the temperature rises (as shown in figure 7). If the expansion of each was different, the steel might crack the concrete.

Platinum alloy wires, sealed in glass, can be used to carry the electric current into a light bulb. When the bulb heats up, the expansion of the platinum alloy is the same as that of the glass, so there isn't any risk of the glass cracking.

Pyrex glass expands less than ordinary glass. This makes it a more suitable material for dishes which have to hold hot liquids. If you tip boiling water into an ordinary glass dish, the inside starts to heat up and expand before the outside, and the stresses in the glass may make it crack. A pyrex dish is less likely to crack because it doesn't expand as much.

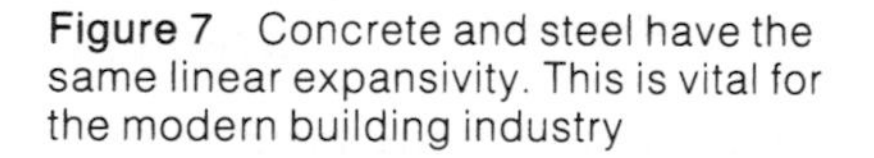

Figure 7 Concrete and steel have the same linear expansivity. This is vital for the modern building industry

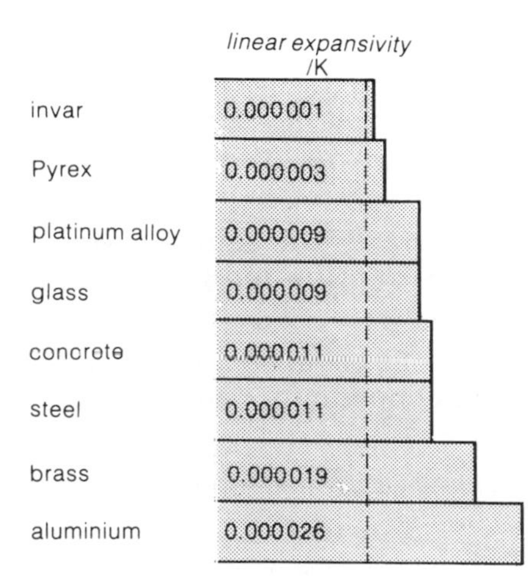

Figure 6 Linear expansivities

Calculating length increases

Figure 8

Given the length of a material, its linear expansivity, and the temperature rise, the increase in length can be calculated.

Rearranging the linear expansivity equation:

$$\textbf{increase in length} = \textbf{linear expansivity} \times \textbf{original length} \times \textbf{temperature rise}$$

In symbols: $x = \alpha l(\theta_2 - \theta_1)$

If the original length is measured in m, the increase in length is also in m. If the original length is in cm, the increase in length is also in cm, and so on. Strictly speaking, the equation is only valid where the temperature rise is small. In most cases however, the equation is a good enough approximation even when temperature rises of 200 K or more are being considered. It can also be used to calculate the decrease in length of a material when the temperature drops.

Example *The steel bar in figure 8 has a length of* 4.000 m *at* 10 °C. *If the linear expansivity of steel is* 0.000 011 /K, *what will be the length of the bar when the temperature rises to* 260 °C?

First, work out the increase in length of the bar using the equation:

$x = \alpha l(\theta_2 - \theta_1)$

substituting the values given,

$x = 0.000011 \times 4.000 \times (260 - 10) \quad \text{m} = 0.011\,\text{m}$

This can now be added to the original length to give the final length:

final length of bar at 260 °C = 4.000 m + 0.011 m = 4.011 m

Linear expansivity – of holes The linear expansivity of a hole is exactly the same as that of the material around it. All the shapes in figure 9 are made out of the same material, and the length l is the same in each case. Heat them all up together, and the increase in l is the same for all of them.

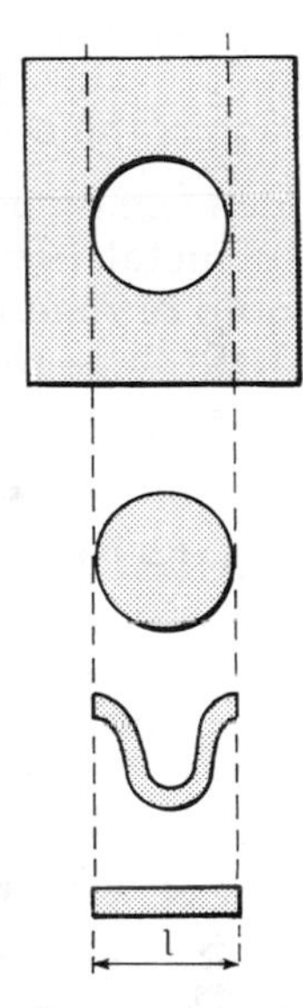

Figure 9 The linear expansivity of a hole is the same as that of the material surrounding it

Questions

1 Why do solids expand when they are heated? Give two examples of problems caused by expansion and how they are overcome.

2 The linear expansivity of brass is 0.000 019 /K. Calculate the increase in length
   a) of a brass rod 1 m long, when the temperature rises by 1 K
   b) of a brass rod 1 cm long, when the temperature rises by 1 K
   c) of a brass rod 4 m long, when the temperature rises from 20 °C to 220 °C
   d) of a brass rod 4 cm long, when the temperature rises from 20 °C to 220 °C.

3 When a metal bar is heated from 20 °C to 120 °C, its length increases from 1.500 m to 1.503 m. Calculate the linear expansivity of the metal.

4 A steel rod is 2 m long at a temperature of 15 °C. To what temperature must it be heated for it to expand 0.55 mm? The linear expansivity of steel is 0.000 011 /K.

5 Figure 10 shows a ring and a cylindrical rod. Both are made of steel and both are at a temperature of 17 °C. Calculate the temperature to which the ring must be heated so that it will just fit over the rod.

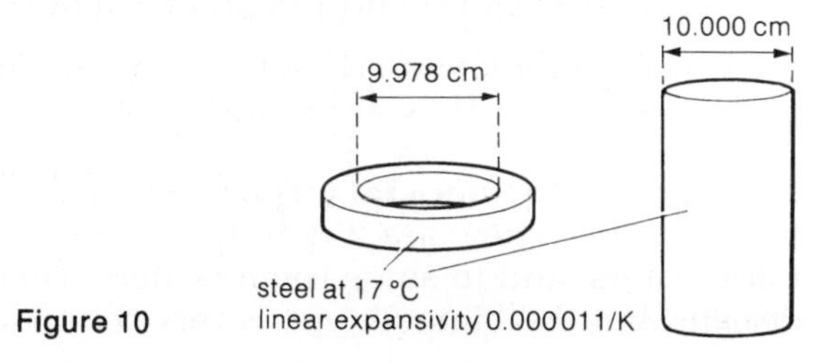

Figure 10

4.5

Uses of expanding solids

Expansion has many practical uses. One particularly useful device actually bends as it expands.

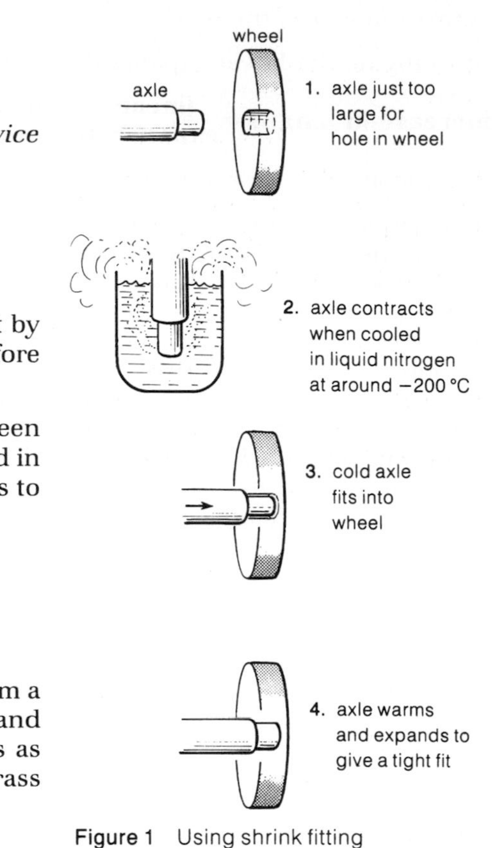

Figure 1 Using shrink fitting

Separating and joining materials

If a bottle top is too tight to unscrew, you can sometimes free it by placing it in hot water for a moment or two. The top expands before the heat reaches the bottle, which makes it a looser fit.

The same idea can be used in reverse to produce a tight fit between two pieces of metal. In figure 1, the axle shrinks when it is dipped in the cold, liquid nitrogen. It fits easily into the wheel, but expands to produce a tight fit as the axle warms up.

The bimetal strip

Two thin strips of different metals can be bonded together to form a **bimetal strip**. The bimetal strip in figure 2 is made of brass and invar. When it is heated, the brass expands about twenty times as much as the invar. This makes the bimetal strip bend with the brass on the outside (the longer side) of the curve.

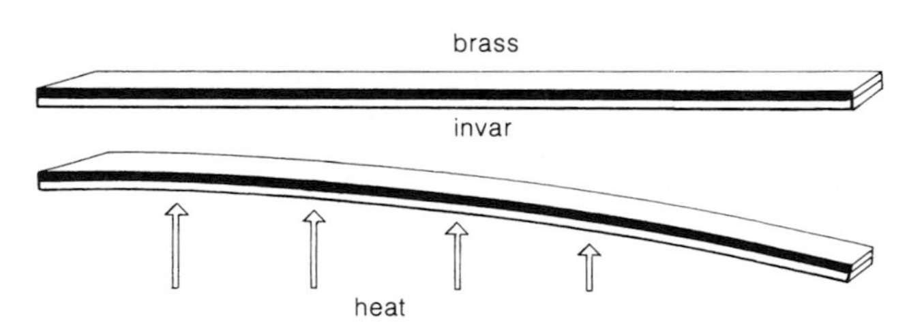

Figure 2 The bi-metal strip bends when heated because the brass expands more than the invar

If the bimetal strip is cooled rather than heated, it bends the other way. The brass contracts more than the invar, so the brass ends up on the inside (the shorter side) of the curve.

The bimetal thermometer Figure 3 shows a bimetal thermometer. This contains a bimetal strip in the form of a long spiral. The centre of the spiral is attached to a pointer; the other end is fixed. When the temperature rises, the bimetal strip coils itself into an even tighter spiral and the pointer moves across the scale.

Bimetal thermometers are not as accurate as some other types of thermometer, but they are robust and easy to read.

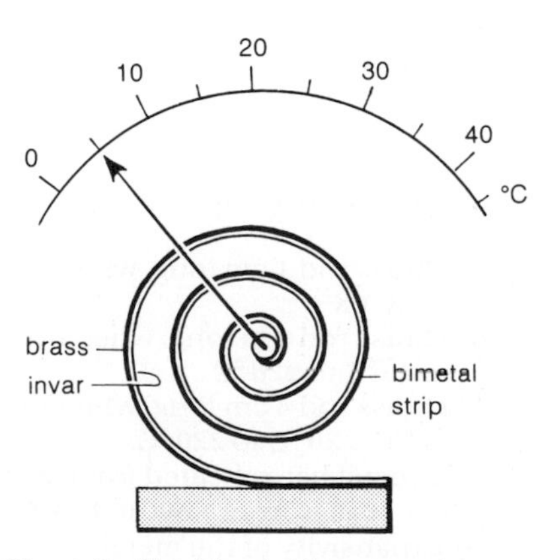

Figure 3 The bimetal thermometer

The bimetal thermostat Thermostats are devices which control temperature. They are fitted to immersion heaters, ovens and refrigerators, and to some room heaters. There are several different designs, but the bimetal type is very common.

The thermostat in figure 4 controls an electric room heater. As the room warms up, the bimetal strip bends and the two electrical contacts eventually separate. This switches off the heater. When the room cools down, the bimetal strip straightens, and the heater is switched on again as the electrical contacts touch. In this way, the thermostat switches the heater on and off to keep the room at a more or less steady temperature.

You select the temperature by turning the control knob. If the control knob is screwed inwards, the bimetal strip has to bend further before the contacts separate. This means that the room has to reach a higher temperature before the heater is switched off.

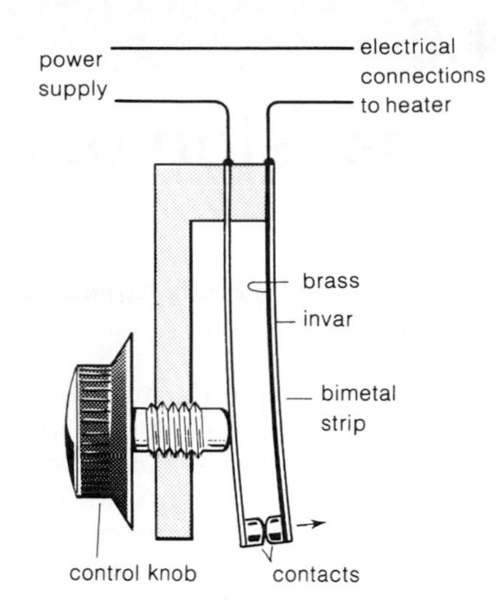

Figure 4 The bimetal thermostat – above a certain temperature, the strip bends so that electrical contact is broken and the current is switched off

Flashing indicators The indicator bulbs on a car flash on and off because of movements made by a tiny bimetal strip. The basic arrangement is shown in figure 5. When you turn on the indicator switch, a small electric current passes through the bulb and through the heating coil wound round the bimetal strip. The current is too small to light the bulb, but it does heat up the bimetal strip. This bends upwards as a result.

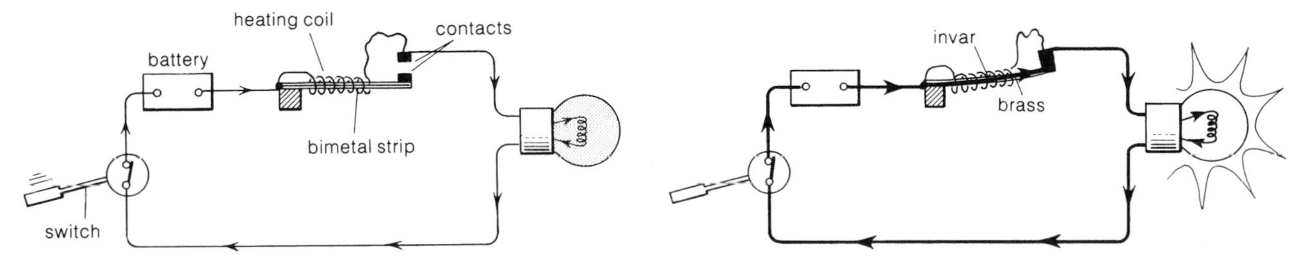

Figure 5 The same principle is used in the flashing indicators of cars

When the contacts touch, the current stops flowing through the heating coil. It takes the easier route straight along the bimetal strip. The bimetal strip now connects the bulb directly to the battery, so the bulb lights up at full brightness.

With no current passing through the heating coil, the bimetal strip cools and straightens, and the contacts separate. The current once more has to pass through the heating coil, so the bulb dims as a result. In this way, the bulb continues to flash on and off until you turn off the switch.

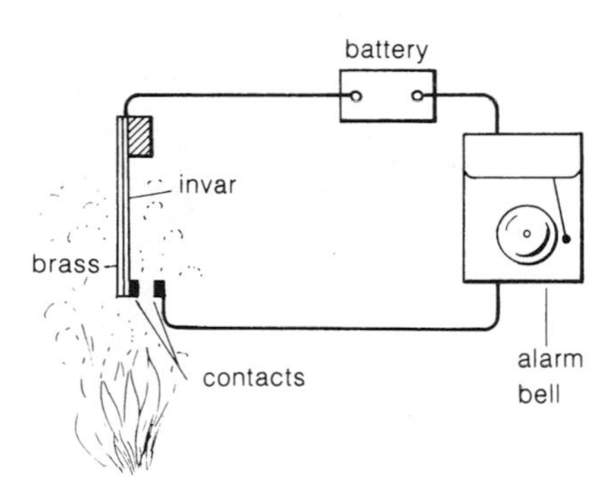

Figure 6 A bimetal strip used in a fire alarm (see question 3)

Questions

1 Why is it sometimes easier to unscrew a bottle top after you have held it in hot water?

2 Iron has a linear expansivity of 0.000 012 /K, brass has a linear expansivity of 0.000 019 /K. Draw diagrams to show what will happen to a bimetal strip made from iron and brass if it is a) heated b) cooled.
Strips of brass and iron are both 10 cm long at a temperature of 0 °C. What will be the difference in their lengths if the temperature rises to 80 °C?

3 Figure 6 shows how a bimetal strip can be used in a simple fire alarm system. How does it work?
What would be the effects of a) increasing the distance between the contacts b) using a bimetal strip of the same size, but made from brass and iron rather than brass and invar? (linear expansivity of invar = 0.000 001 /K; linear expansivities of brass and iron, see question 2).

4.6

Expansion of liquids

Like solids, most liquids expand when they are heated. But there are exceptions.

When a liquid expands, it is the increase in volume which has to be considered. It isn't possible to measure a linear expansivity because a liquid doesn't have a fixed shape. In general, liquids expand much more than solids. Mercury, for example, expands about five times as much as an equal volume of steel over the same temperature rise, yet its expansion is still small compared with that of most liquids.

The expansion of a liquid can be demonstrated using the apparatus shown in figure 1. The flask is heated by placing it in hot water. At first, the level of the liquid in the tube may fall slightly because the flask starts to expand outwards before the heat gets through to the liquid. Eventually, the liquid does warm through and it expands up the tube to the level shown. The same basic principle is used in liquid-in-glass thermometers (see page 124).

Whenever a liquid is stored in a closed container, an air space must be left for expansion to take place. As shown in figure 2, car radiator systems have space in them for the coolant to expand, and bottles containing liquids are never completely full when you buy them.

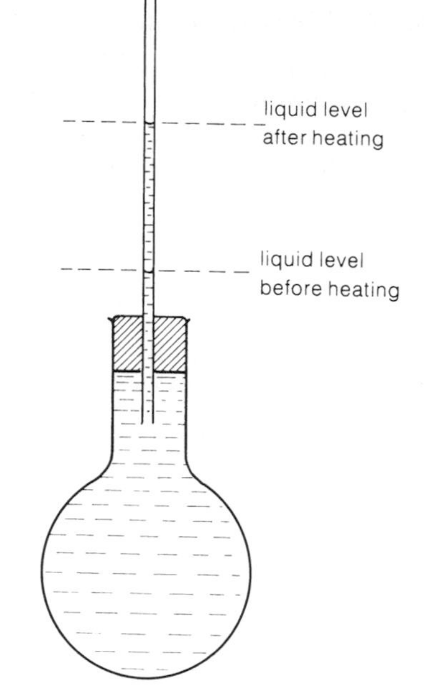

Figure 1 Liquids expand on heating

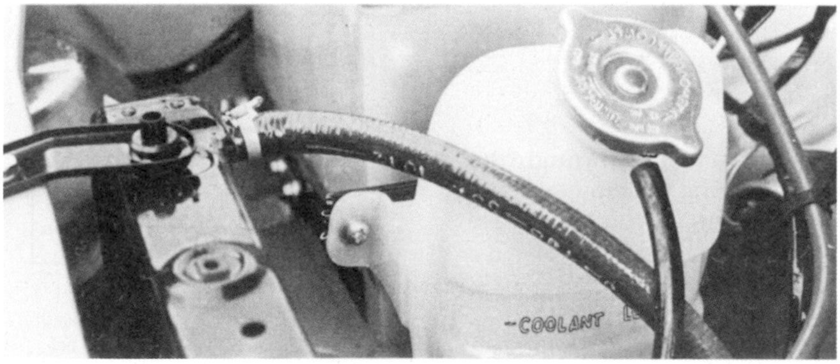

Figure 2 Radiator expansion chamber in a modern car. When the engine is cold, the liquid should reach the 'coolant level' indicated. It will be almost full when the engine is hot.

The unusual behaviour of water

Unlike most other liquids, water doesn't always expand when it is heated and contract when it is cooled. The unusual behaviour of water is illustrated by the graphs in figures 3a and 3b.

By following the line of the graph in figure 3a from right to left, you can see how the volume of 1 kg of water changes as the temperature drops from about 100 °C. As the water cools, it contracts until it reaches 4 °C, but from 4 °C down to 0 °C, it actually expands slightly. When the water freezes at 0 °C to form ice, it shows a considerable expansion. With further cooling, ice contracts in the normal way.

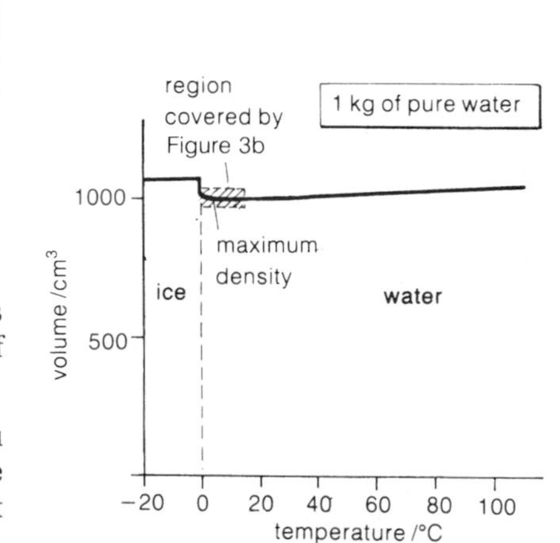

Figure 3a Hot water is less dense than cold water; ice is less dense than either

The change in volume of the water can be seen more clearly in figure 3b. This shows on a much larger scale the shaded area of the graph in figure 3a.

As the volume of the water changes, so does its density. Water has its least volume at 4 °C, so it has its greatest density at this temperature. For this reason, water at 4 °C will sink through colder or warmer water around it.

When water changes to ice, its volume increases by about 9%, which is why water pipes burst if they freeze up. This increase in volume means that ice is less dense than water, so any ice which forms in water will float.

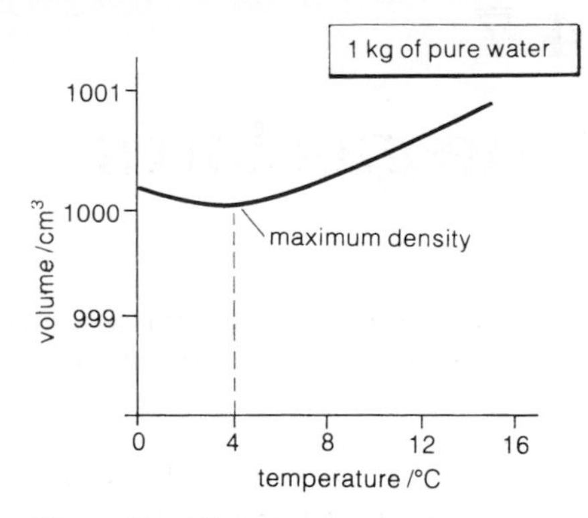

Figure 3b Water has its maximum density at 4 °C

Freezing ponds

If the temperature above a pond falls and then stays below freezing point, the pond will eventually freeze over. The unusual expansion of water affects this process in two ways:

1 *The surface of the pond won't start to freeze until all the water has cooled to* 4 °C. As water on the surface cools, it becomes more dense and starts to sink. Warmer, less dense water is pushed upwards to take its place and this in turn is also cooled down. The water only stops circulating in this way when it has all cooled to 4 °C and is at its maximum density.

2 *The pond freezes over with the deepest water staying at a temperature of* 4 °C. When the water on the surface of the pond cools below 4 °C, it stays on the surface because it is less dense than the water underneath. In time, ice forms on the surface, and the water temperatures become as shown in figure 4. The denser, warmer water at the bottom of the pond is unlikely to cool any further because the water does not circulate, and the ice and upper layers of water act as an insulating blanket. Fish can survive a harsh winter by staying in this warmer water.

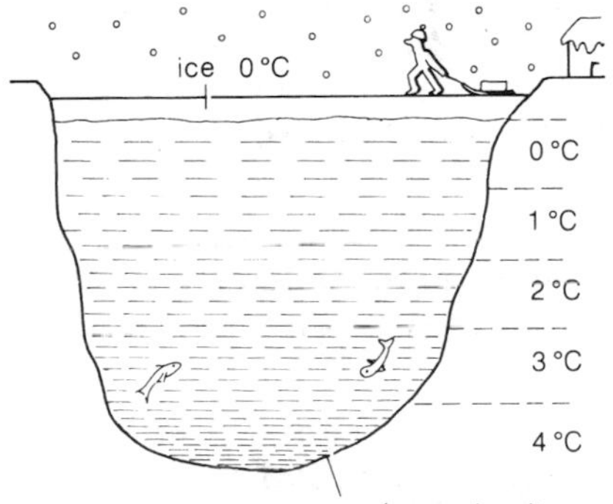

Figure 4 The 'warm' water stays at the bottom of the pond because it is most dense

Questions

1 In general, which expand the most, liquids or solids?
2 How would you demonstrate the expansion of a liquid?
3 What practical use is made of liquid expansion?
4 In what way is the expansion of water different from that of most other liquids?
5 At what temperature does water have its maximum density?
6 When a quantity of water freezes, what happens to its volume? Where can this cause problems?
7 In winter, a pond doesn't start to freeze over until all the water has cooled to 4 °C. Why not?
8 Why can fish survive a harsh winter by staying in the water at the bottom of a pond?

4.7

The gas laws

The expansion of a gas isn't as simple as the expansion of a liquid or solid. There are more factors to consider because a gas is so much more compressible.

Studying the behaviour of a gas is complicated by the fact that you have to consider its pressure as well as its volume and temperature. The pressure, volume and temperature of any fixed mass of gas are all related, so a change in one of these factors always produces a change in at least one of the other two. Sometimes, all three factors may change at once. This happens, for example, when air rises in a thundercloud or gases expand in the cylinders of a car engine.

To find the laws linking pressure, volume and temperature experimentally, each factor is kept constant in turn while the relation between the other two factors is investigated. Three such experiments are described below. In each case, the gas being studied is a fixed mass of dry air.

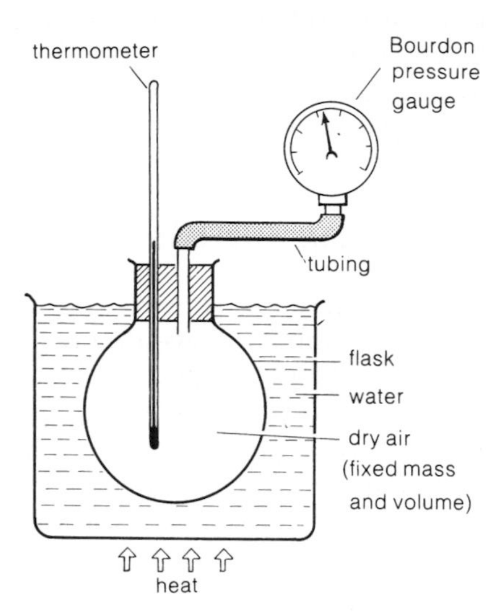

Figure 1 At constant *volume* the pressure of gas goes up in proportion to the absolute temperature

Relation between pressure and temperature at constant volume

In this experiment, the air is enclosed in a sealed flask in order to keep its volume constant (as shown in figure 1). The temperature of the air is increased in stages by heating the water around the flask. At each stage, the pressure of the air is measured on the Bourdon gauge which is attached to the flask by a short length of tubing. The tubing needs to be as short as possible because the air it contains is at a different temperature from the rest of the air in the flask.

The readings are used to plot a graph showing how the pressure of the air varies with temperature. A typical graph is shown in figure 2. Note that temperatures have been measured on an absolute temperature scale, in this case the Kelvin scale. Zero on the temperature axis is therefore at absolute zero. The graph obtained from experimental results is shown by the solid line. The dotted line shows the result of extending the solid line backwards. It meets the temperature axis at 0 K [−273 °C]. The graph shows all the usual signs of a simple proportion:

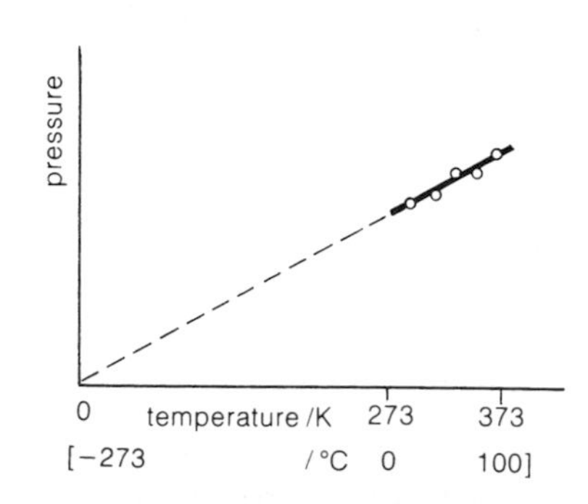

Figure 2 The pressure of the gas varies with the temperature. The pressure is *directly proportional* to the absolute temperature

1 It is a straight line passing through the origin (0 K).

2 Doubling the absolute (Kelvin) temperature of the air doubles its pressure.

3 Dividing the pressure of the air by its absolute (Kelvin) temperature always produces the same value.

Straight line graphs are also obtained if the experiment is repeated with different masses of air and with other gases. In all cases, the lines pass through absolute zero when extended backwards.

These findings can be summed up in a law which states:

The pressure of a fixed mass of gas is directly proportional to its absolute temperature, provided the volume of the gas is kept constant.

or, for a fixed mass of gas of volume V, pressure p, and absolute temperature T (measured in kelvin for example)

$\frac{p}{T}$ = **constant**, provided V is constant

This is known as the **Pressure law**. The value of the constant depends on the particular mass, volume and type of gas being considered.

The Pressure law and the kinetic theory According to the kinetic theory, gas molecules are constantly bombarding the sides of any container they happen to be in (as shown in figure 3). There is an outward force on the container whenever a molecule strikes it and bounces off. Many billions of molecules hit the container every second and this produces a steady outward pressure.

If the temperature rises, the molecules move faster. They strike the container with greater force, and the pressure increases as a result. A fall in temperature produces the opposite effect. The molecules move more slowly and the pressure drops.

If the pressure continued to drop as shown on the graph, a gas would exert no pressure at all at absolute zero. It was this feature of a pressure-temperature graph that first suggested to experimenters that there might be an absolute zero of temperature at −273 °C. In practice, all gases turn liquid before absolute zero is reached. Oxygen for example liquefies at about 90 K.

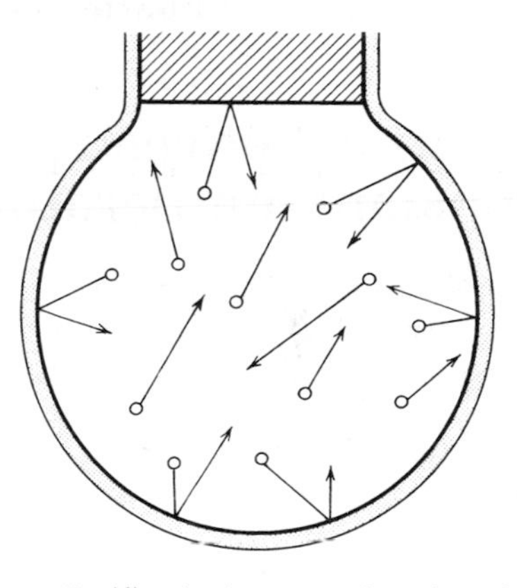

Figure 3 Kinetic theory explanation of the pressure law

Relation between volume and temperature at constant pressure

In the experiment shown in figure 4, the air is trapped in a capillary tube by a bead of concentrated sulphuric acid (dampness in the air would upset the results, and the acid acts as a drying agent). Inside the tube, the air is free to expand or contract. Its pressure is always constant and equal to atmospheric pressure plus the pressure due to the weight of the acid.

As in the previous experiment, the temperature of the air is increased by heating the water around it. The water is kept well stirred and heated slowly to ensure that the temperature indicated on the thermometer is the same as the temperature of the air column.

The length of the air column is measured for several different temperatures. The length of the column is not of course equal to its volume, but it is *proportional* to the volume provided the width of the capillary tube is constant. Length measurements can therefore be used to represent air volumes.

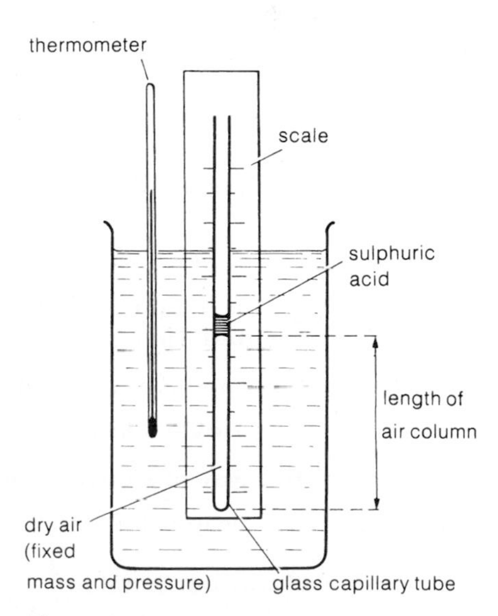

Figure 4 At constant *pressure* the volume goes up in proportion to the absolute temperature

A graph of volume against temperature is as shown in figure 5. Like the previous graph, this meets the temperature axis at 0 K when extended backwards, and a similar law applies:

The volume of a fixed mass of gas is directly proportional to its absolute temperature provided the pressure of the gas is kept constant.

or, for a fixed mass of gas $\frac{V}{T}$ = **constant**, provided p is constant and T is measured on an absolute temperature scale (e.g. the Kelvin scale).

This is often called **Charles' law**.

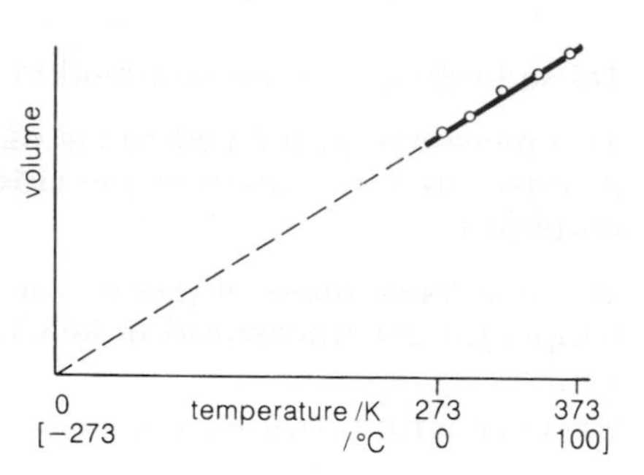

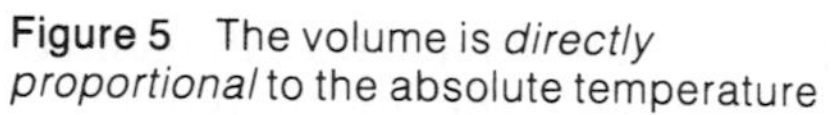

Figure 5 The volume is *directly proportional* to the absolute temperature

Relation between pressure and volume at constant temperature

In the experiment shown in figure 6, the air is trapped above oil in a glass tube. The volume of the air is reduced in stages by pumping outside air into the oil reservoir. This forces more oil up into the glass tube. Changing the volume of the air affects its temperature slightly, so you have to wait a few moments after each adjustment to give the air time to settle down to its original temperature.

The pressure of the air is measured for several different volumes. The Bourdon gauge actually measures the air pressure above the oil reservoir, but this is the same as the pressure of the trapped air because the pressure is transmitted through the oil.

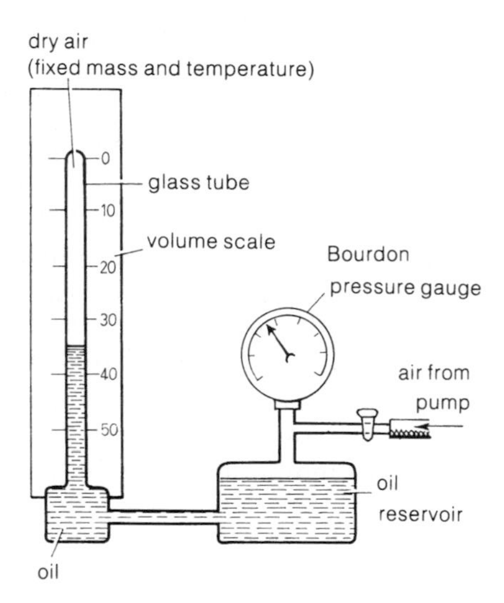

Figure 6 At constant *temperature*, when the pressure goes up, the volume goes down

Figure 7a shows the result of plotting pressure against volume. A simple law linking the two quantities is not immediately obvious. However, a detailed study of the points on the graph shows that the pressure and the volume have an *inverse* proportion to one another. This inverse proportion can be recognized in two ways:

1 When the volume is reduced, the pressure increases, but pressure × volume keeps the same value;

2 When the volume is halved, the pressure is doubled, and so on.

These findings can be expressed in the form of a law which states:

The pressure of a fixed mass of gas is inversely proportional to its volume provided the temperature of the gas is kept constant.

or, for a fixed mass of gas: pV = **constant**, provided T is constant.

This is known as **Boyle's law**.

There is another way of showing the relationship between p and V:

If: p is *inversely* proportional to V, then

p is *directly* proportional to $\frac{1}{V}$

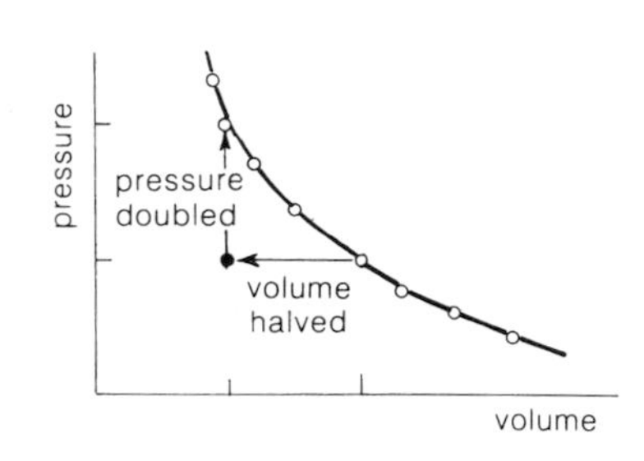

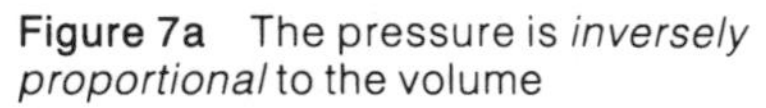

Figure 7a The pressure is *inversely proportional* to the volume

This proportion is illustrated by the graph in figure 7b. Here, pressure has been plotted against $\frac{1}{\text{volume}}$. The result is a straight line passing through the origin.

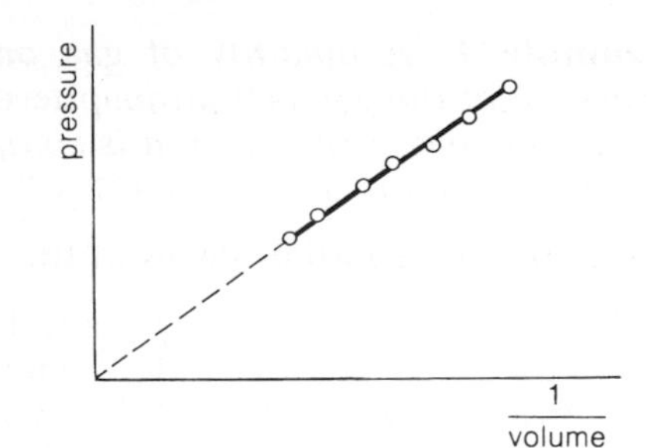

Figure 7b The pressure is directly proportional to $\frac{1}{\text{volume}}$

Boyle's law and the kinetic theory The kinetic theory explains Boyle's law as follows. If the volume of a gas is reduced to half its value, each cubic metre of a container will hold twice as many molecules as before. Every second, there will therefore be twice as many impacts with each square metre of the container sides. The pressure is doubled as a result (as shown in figure 8).

The combined gas equation

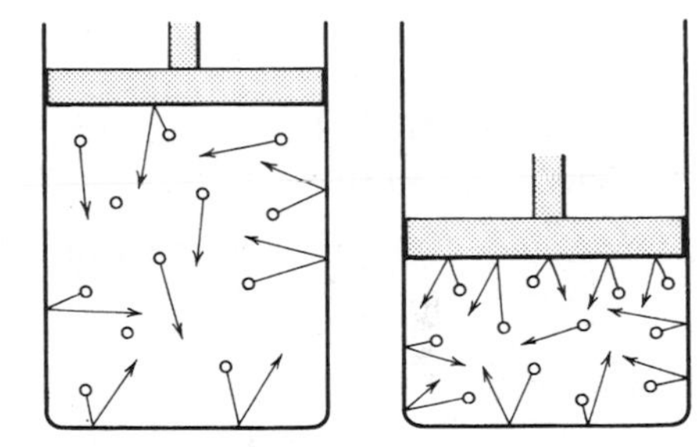
Figure 8 Kinetic theory explanation of Boyle's law

The results of the three experiments just described can be expressed in the form of a single equation:

For a fixed mass of gas, $\boldsymbol{\frac{pV}{T}}$ **= constant**

This is sometimes known as the combined gas equation. The three gas laws can all be obtained from this equation:

If V is constant, p/T is constant (the Pressure law)
If p is constant, V/T is constant (Charles' law)
If T is constant, pV is constant (Boyle's law)

A gas which obeys the gas laws exactly is known as an **ideal gas**. In reality, no gases are ideal though most can be regarded as such at low or medium pressures and medium temperatures. If a gas is near its liquefying temperature, attractions between its molecules begin to influence its behaviour and it no longer acts like an ideal gas. If a gas is highly compressed, the size of its molecules restricts the space available for movement and this too affects the way the gas behaves.

p, V, and T calculations

Figure 9 illustrates a fixed mass of gas before and after changes in pressure, volume and temperature have taken place. From the combined gas equation above, it follows that:

$$\frac{p_2V_2}{T_2} = \frac{p_1V_1}{T_1}$$

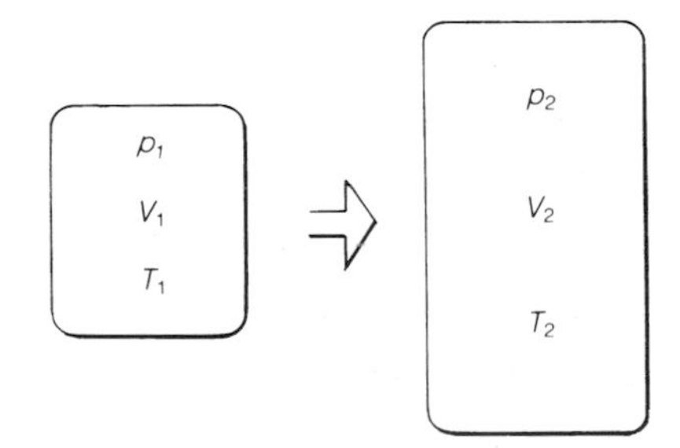

Figure 9 For a fixed mass of gas, $\frac{p \times V}{T}$ stays constant

Examples of problems which can be solved using this equation are given on the next page. Note that:

1 p and V can be measured in any appropriate unit provided the same unit is used on both sides of the equation.

2 T must always be measured on an *absolute temperature scale* (e.g. the Kelvin scale) (number of kelvin = 273 + number of °C).

Example 1 *A quantity of gas occupies a volume of* 4 m³. *The pressure of the gas is* 3 atmospheres *when its temperature is* 27 °C. *What will be its pressure if it is compressed into half the volume and heated to a temperature of* 127 °C*?* (As shown in figure 10).

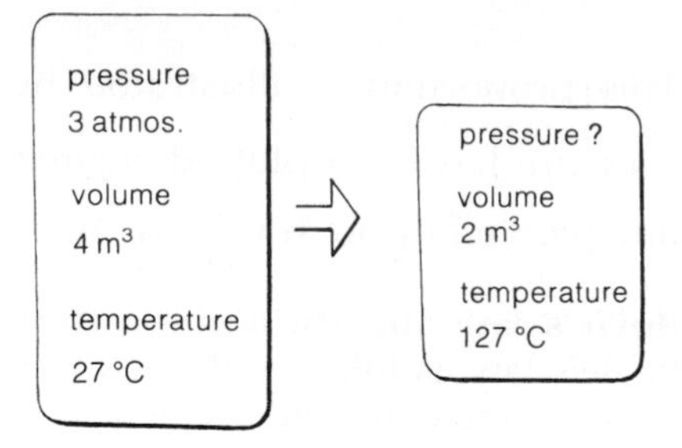

Figure 10

Using the information given in the question:

$p_1 = 3\text{ atm}$	p_2 is to be found
$V_1 = 4\text{ m}^3$	$V_2 = 2\text{ m}^3$; half the original volume
$T_1 = (273 + 27)\text{ K}$	$T_2 = (273 + 127)\text{ K}$
$= 300\text{ K}$	$= 400\text{ K}$

$$\frac{p_2V_2}{T_2} = \frac{p_1V_1}{T_1}$$

$$\frac{p_2 \times 2\text{ m}^3}{400\text{ K}} = \frac{3\text{ atm} \times 4\text{ m}^3}{300\text{ K}}$$

$$\frac{p_2 \times 2}{400} = \frac{3\text{ atm} \times 4}{300}$$

Rearranged, this gives $p_2 = 8\text{ atm}$. The final pressure of the gas is therefore 8 atmospheres.

Example 2 *When an upturned beaker is placed on the surface of water, it contains* 300 cm³ *of trapped air at atmospheric pressure. What will be the volume of the air when the beaker is taken* 20 m *beneath the water surface? Assume that the temperature is constant, and that atmospheric pressure can support a column of water* 10 m *high.*

The problem is illustrated in figure 11.

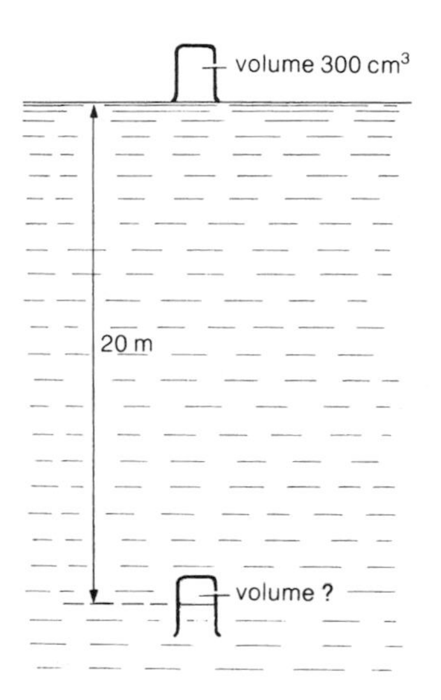

Figure 11

Using the information given in the question:

p_1 = atmospheric pressure	p_2 = atmospheric pressure
= 10 m of water	+ pressure due to 20 m of water
= 1 atm	= 1 atm + 2 atm
	= 3 atm
$V_1 = 300\text{ cm}^3$	V_2 is to be found

T_1 isn't given but is equal to T_2

$$\frac{p_2V_2}{T_2} = \frac{p_1V_1}{T_1}$$

Substituting numbers, and cancelling T_1 and T_2 because they are equal,

$$3\text{ atm} \times V_2 = 1\text{ atm} \times 300\text{ cm}^3$$

Rearranged, this gives $V_2 = 100\text{ cm}^3$. The final volume of the trapped air is therefore 100 cm³.

The constant volume gas thermometer

Figure 12 shows a simple form of constant volume gas thermometer. Basically, the instrument is the same as the apparatus used to establish the Pressure law except that the Bourdon gauge has been replaced by a more accurate mercury manometer. In this case, the pressure of the gas (in mmHg) is equal to the height difference h (in mm) plus atmospheric pressure (in mmHg).

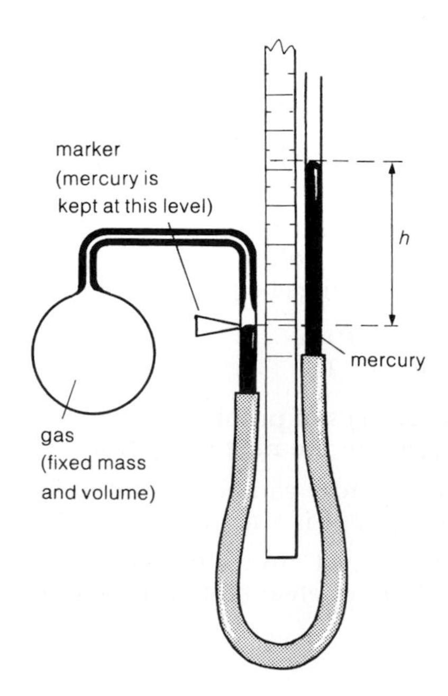

Figure 12 The constant volume gas thermometer

It isn't possible to mark a temperature scale on a constant volume gas thermometer. To find some unknown temperature θ, you measure the pressure of the gas at 0 °C, 100 °C, and θ, and deduce the value of θ from a pressure-temperature graph. The 0 °C and 100 °C points are plotted first and a straight line drawn between them. Knowing the pressure at θ the value of θ can be read off on the temperature axis.

This process is clearly far too involved for everyday temperature measurements in the laboratory. Constant volume gas thermometers are important however, for reasons which are given below.

Thermodynamic temperatures

All types of thermometer agree with each other at the lower and upper fixed points (0 °C and 100 °C). They don't necessarily agree at other temperatures however. Water, for example, which has a temperature of 50.0 °C measured on a constant volume gas thermometer will have an indicated temperature of 49.8 °C on a mercury thermometer. This doesn't mean that either thermometer is wrong, just that the pressure of a gas doesn't vary with hotness in quite the same way as the volume of mercury. As with any thermometer, the temperature indicated depends on the particular properties of the material being used to detect the change in hotness.

Figure 13 Lord Kelvin

To overcome this problem, Lord Kelvin (shown in figure 13) proposed a **thermodynamic** temperature scale based on the average kinetic energies of molecules rather than on some property of a particular substance. This is the scale now known as the Kelvin scale. It can be shown theoretically that temperatures measured on a constant volume gas thermometer are exactly the same as temperatures on the thermodynamic scale, provided the thermometer contains an ideal gas. No gas is ideal, of course, but hydrogen comes very close to it over a wide temperature range. For this reason, constant volume hydrogen thermometers are used as standard thermometers against which other thermometers are calibrated.

Questions

1 What experimental evidence is there to suggest that there might be an absolute zero of temperature at −273 °C?

2 Explain in terms of the kinetic theory why the pressure of a gas increases when its temperature rises. Assume that the volume of the gas is constant.

3 What is meant by an ideal gas?

4 A gas in a fixed container is at a pressure of 1200 mmHg and a temperature of 27 °C. What will be its pressure if it is heated to a temperature of 177 °C?

5 An air bubble has a volume of 2.5 cm^3 when released at a depth of 40 m in water. What will its volume be when it reaches the surface? Assume its temperature is constant, and that atmospheric pressure = 10 m of water.

6 A gas occupies a volume of 2 m^3 when its pressure is 1140 mmHg and its temperature is 27 °C. What volume would it occupy at standard temperature and pressure (0 °C and 760 mmHg)?

4.8

Conduction

Whenever there is a temperature difference in a material, heat flows to try to reduce it. Vibrating molecules can't keep their vibrations to themselves!

If you put one end of a metal rod in a bunsen flame, the other end eventually becomes too hot to hold. Thermal energy, commonly called heat, is transferred from the hot end of the bar to the cold, and the molecules at the cold end begin to move more quickly as a result, as shown in figure 1. The process is called conduction.

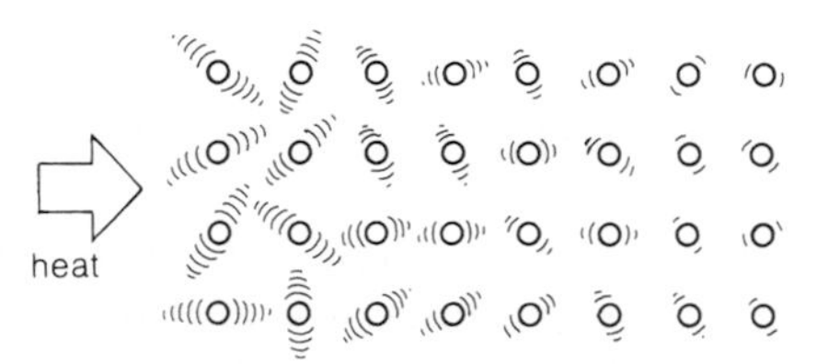

Figure 1 Thermal energy conducts easily through metals

Good and bad conductors

Some materials are much better at conducting heat than others. Examples of good and bad conductors are given in figure 2. Bad conductors of heat are known as **insulators**.

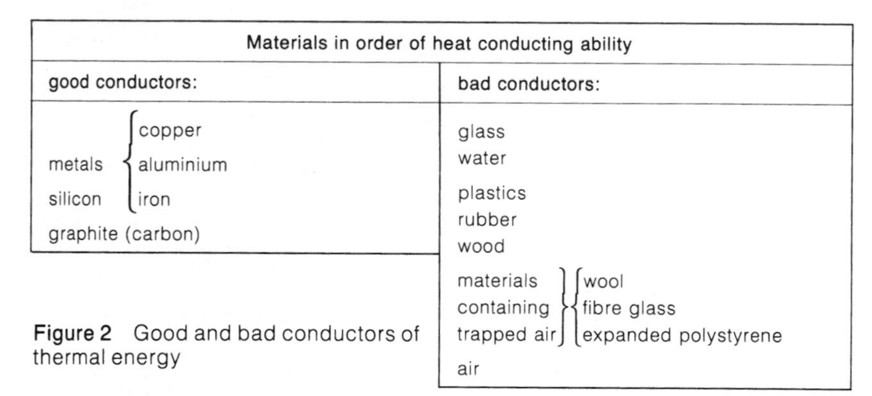

Materials in order of heat conducting ability	
good conductors:	bad conductors:
metals { copper, aluminium, iron }	glass
silicon	water
graphite (carbon)	plastics
	rubber
	wood
	materials containing trapped air { wool, fibre glass, expanded polystyrene }
	air

Figure 2 Good and bad conductors of thermal energy

Metals are the best conductors of heat, and figure 3 shows a simple method of comparing their conducting abilities. The metals in this case are in the form of rods of equal size. Before the start of the experiment, each is coated with a thin layer of wax. Boiling water is tipped into the metal tank, and the apparatus left for ten minutes or so. A greater length of wax melts on the copper rod than any other, showing that copper is the best of the conductors present.

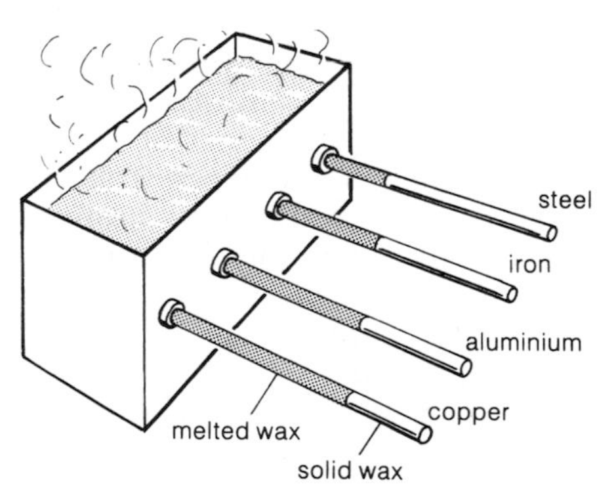

Figure 3 Comparing four good conductors

Non-metal solids tend to be poor conductors of heat, as do most liquids. Figure 4 shows that water is a poor conductor; the water at the top of the tube can be boiled without the ice melting.

Gases are much poorer conductors than liquids. Air has about one-twentieth of the conducting ability of water. Many materials are poor conductors because they contain tiny pockets of trapped air.

An iron bar at room temperature feels cold because it quickly conducts heat away from your hand. A polystyrene tile feels warm compared with other materials because it conducts hardly any heat at all from your hand when you touch it.

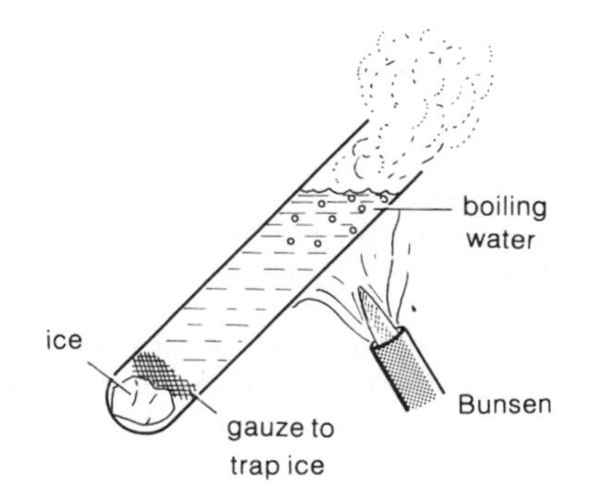

Figure 4 Water is a bad conductor of heat

Uses of good and bad conductors

Wire gauze is often placed over a bunsen to conduct heat outwards from the flame. A glass beaker can safely be heated on the gauze because this protects it from the concentrated heat of the flame.

Saucepans are normally made of aluminium, copper or iron, because these materials will readily conduct the heat from a gas ring or hotplate. Saucepan handles, on the other hand, are made of plastic or wood to keep conduction to a minimum.

Figure 5 Using the insulating properties of air

Insulating the house

Nowadays, insulating materials are extensively used in houses to reduce heat losses as much as possible. This not only keeps down fuel bills; on a national scale, it helps to save limited energy resources. Figure 6 shows how different insulating materials are used in a house. All the materials shown owe their excellent insulating properties to the trapped air they contain.

To calculate likely heat losses from a house, architects need to know the **U-values** of different materials. For example:

the single brick wall in figure 7 has a U-value of 3.6 W/(m^2 °C)

This means that the 1 square metre wall, with a 1°C temperature difference across it, will conduct heat at the rate of 3.6 joules every second.

The heat flow would be greater if
a) the temperature difference was higher
b) the area was greater
c) the thickness of the wall was less

U-values compared:	U-value W/(m^2 °C)
Single brick wall	3.6
Double wall, with air cavity	1.7
Double wall, with insulating foam in cavity	0.5
Glass window, single layer	5.7
Double glazed window	2.7

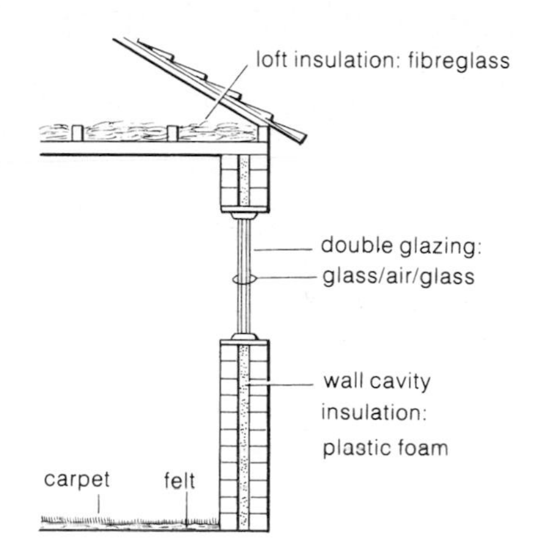

Figure 6 Preventing conduction of thermal energy in the home

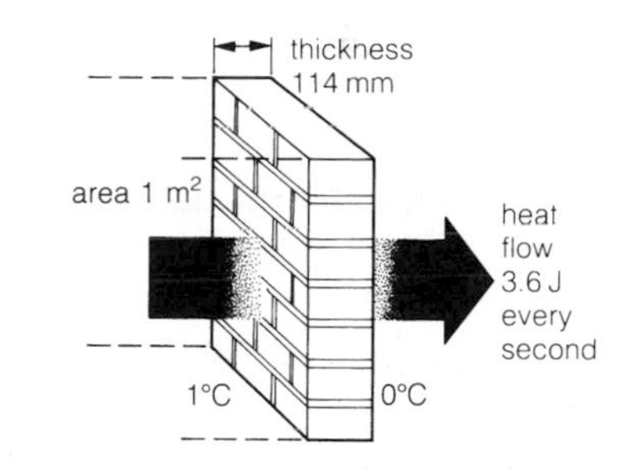

Figure 7 Wall with a U-value of 3.6 W/(m^2 °C)

Questions

1 Which of the materials mentioned in this section is the best conductor of heat? Name two other good conductors.
2 Which are the worse conductors of heat, liquids or gases?
3 Why are materials such as wool and fibreglass good insulators?
4 A piece of iron and a piece of wood are both at room temperature. Why does the iron feel colder than the wood?
5 Why is wire gauze often placed over a bunsen flame?
6 Give three ways in which insulating materials can be used to reduce heat energy losses from a house.
7 A glass window has a U-value of 5.7 W/(m^2 °C).
 a) Explain what this means.
 b) Look at the table of U-values on this page. Then explain why houses with small windows are likely to lose less heat than those with larger windows. Does this apply if the windows are double glazed and the wall cavities are insulated?

4.9

Convection

Liquids and gases may be poor conductors of heat, but they can rapidly carry heat from one place to another if they are free to circulate.

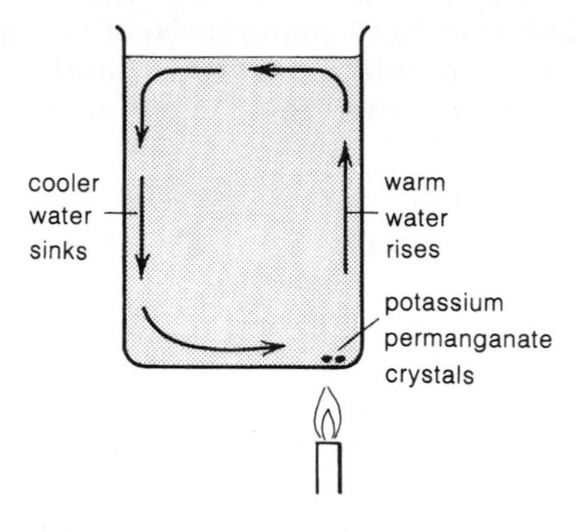

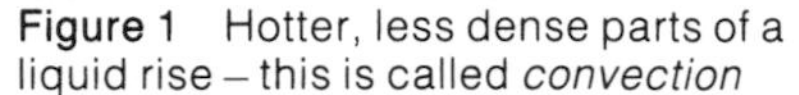

Figure 1 Hotter, less dense parts of a liquid rise – this is called *convection*

Convection in a liquid

In figure 1, the bottom of the beaker of water is being gently heated at one point only. As the water above the flame becomes warmer, it expands, becomes less dense, and is pushed upwards by the cooler, more dense water around it. This cooler water sinks to take its place. The circulating stream of water which is set up is called a **convection current**, and it rapidly transfers heat energy from the bottom to the top of the beaker. It is possible to see the path of the convection current by dropping a few crystals of potassium permanganate into the water to colour it. Convection doesn't occur if the water is heated at the top rather than at the bottom. The warmer, less dense water simply stays at the top.

The domestic hot water system

Convection is used in many domestic hot water systems to circulate hot water around the pipes. A simplified system is shown in figure 2; the storage tank is often in an upstairs cupboard and the header tank in the loft.

Water heated in the boiler rises by convection to the top of the storage tank. Cooler water in the bottom of the tank sinks down to the boiler where it is heated. In this way, a supply of hot water collects in the storage tank from the top downwards. The hot tap is supplied with water from the top of the storage tank.

The header tank provides the pressure needed to push water out of the taps, and also replaces water in the system as it is used. The tank is linked to the main water supply by a valve which opens to refill the tank whenever the water level drops. The expansion pipe serves as an overflow should steam or air bubbles build up in the system.

Actual systems are usually more complex than that shown. There may be separate circuits for taps and for radiators, and a pump to assist the flow of water.

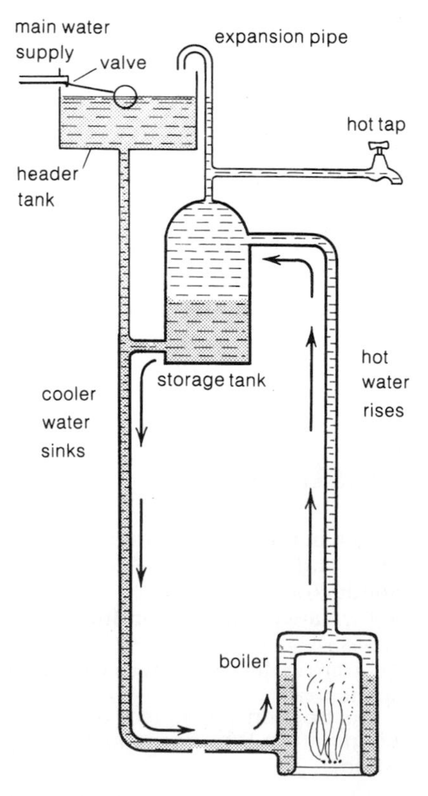

Figure 2 Convection currents move hot water from the boiler to the storage tank

Convection in air

Convection can occur in gases as well as in liquids (as shown in figure 3). Air convection currents are set up when warm air is pushed upwards by cooler, more dense air around it.

Many room heaters, including 'radiators', warm the air in a room by means of convection. Figure 4 shows a convector heater in use. Warm air rises from the top of the heater and cooler air flows in at the bottom to replace it. In time, all the air in the room passes through the heater.

In a refrigerator, convection is used to circulate cold air around the food. Air is cooled by the freezer compartment at the top of the refrigerator. As it sinks, it is replaced by warmer air rising from below. The circulating air carries heat energy away from all the food in the fridge.

Convection causes the onshore and offshore winds which sometimes blow on the coast during the summer. In hot sunshine, the land heats up more quickly than the sea. Warmer air rises above the land as cooler air blows in from the sea, as shown in figure 5a. At night, the reverse happens. The land loses heat more rapidly than the sea. Warmer air now rises above the sea, as cooler air blows out from the shore, as shown in figure 5b.

Figure 3 Convection currents in air

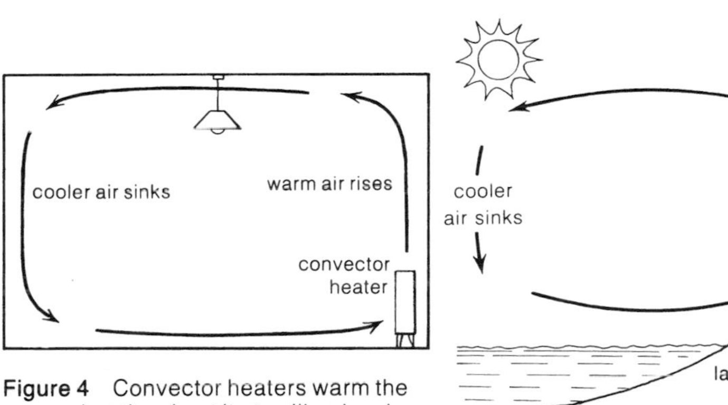

Figure 4 Convector heaters warm the room – but they heat it at ceiling level more than floor level!

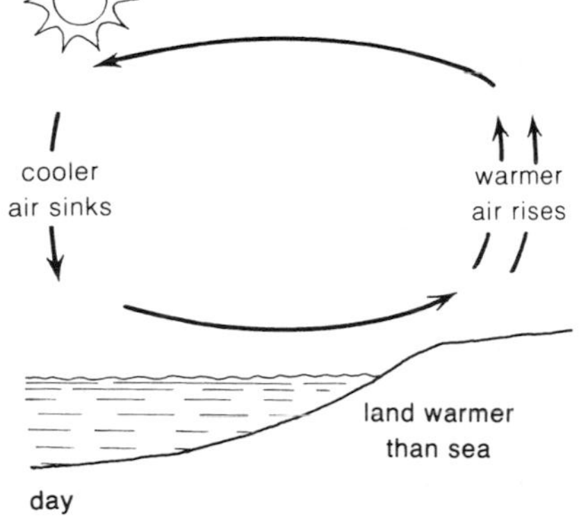

Figure 5a During the day, the breeze is towards the shore . . .

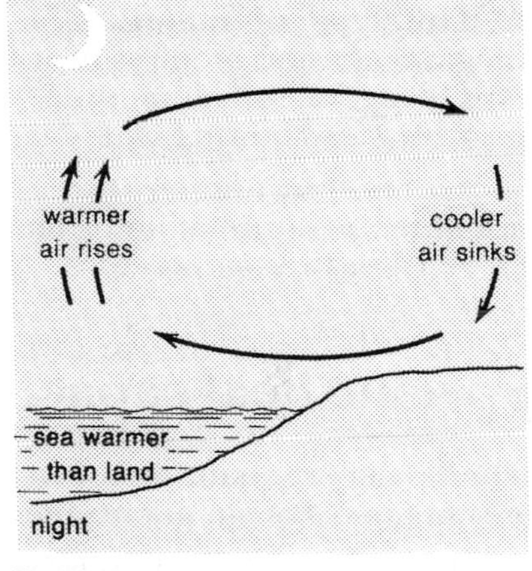

Figure 5b . . . during the night, the breeze is towards the sea

Figures 5a, 5b The direction of sea breezes varies with the relative temperature of the land and sea

Questions

1 In figure 6, what will happen to the water at A and at B? Give reasons in each case.

2 Why doesn't convection occur when a beaker of water is heated at the top?

3 Why is a header tank fitted to a hot water system? Why is an expansion pipe necessary? Explain in detail what happens in the system shown on the opposite page after hot water has been drawn off from the tap.

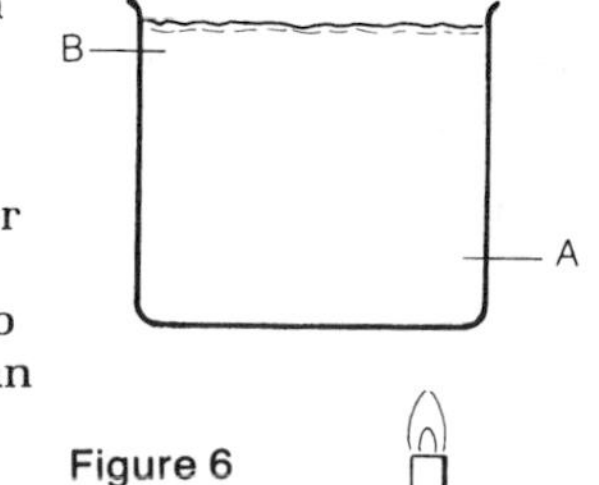

Figure 6

4 How does a 'radiator' in a house distribute most of its heat.

5 Why is the freezer compartment placed at the top of a refrigerator?

6 On a hot summer's day, coastal breezes often blow in from the sea. What causes these breezes, and why do they change direction at night?

4.10

Specific heat capacity

Some materials are more difficult to heat up than others. They have a greater capacity for absorbing thermal energy.

When an object is placed over a bunsen flame or an electric heating element, it starts to gain thermal energy. This input of energy can have several effects, (as explained in units 4.11 and 4.12). The most likely effect however is that the molecules will move faster. In other words, the temperature will rise.

The thermal energy needed to produce a temperature rise depends on three factors:

1 The mass of the material. The greater the mass, the more molecules there are to speed up, and the more thermal energy is needed for any given temperature rise. It takes more energy to warm up a kettleful of water than a cupful of water.

2 The type of material. For any given temperature rise, a kilogram of water requires nearly five times as much thermal energy as a kilogram of aluminium. Water has a greater capacity for absorbing and storing heat energy.

3 The rise in temperature. For any given object, a 10 K [10 °C] rise in temperature requires ten times as much thermal energy as a 1 K [1 °C] rise in temperature.

Specific heat capacity J/kg K	
water	4200
sea water	3900
meths	2500
ice	2100
aluminium	900
concrete	800
granite	800
glass	700
steel	500
copper	400
mercury	150

Figure 1 The specific heat capacity of some common substances

Specific heat capacity

When different quantities of water are heated, the following results are obtained. For simplicity, the energy values are approximate only.

To produce a 1 K temperature rise in a mass of 1 kg, 4200 J of thermal energy are required;
to produce a 1 K temperature rise in a mass of 2 kg, 8400 J of thermal energy are required;
to produce a 10 K rise in temperature in a mass of 2 kg, 84 000 J of thermal energy are required.

In each case, the gain in thermal energy ÷ mass ÷ temperature rise has the same value, 4200 J/(kg K). Water is said to have a **specific heat capacity** of 4200 J/(kg K).

The specific heat capacity of any substance is given by the equation:

$$\textbf{specific heat capacity} = \textbf{gain in thermal energy} \div \textbf{mass} \div \textbf{temperature rise}$$

though this is usually written in the following form:

$$\textbf{specific heat capacity} = \frac{\textbf{gain in thermal energy}}{\textbf{mass} \times \textbf{temperature rise}}$$

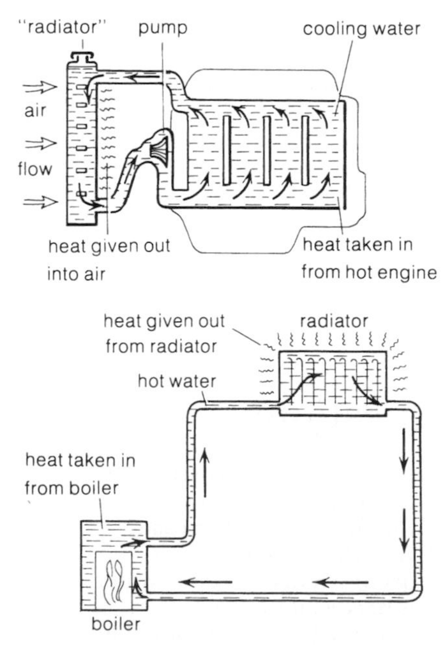

Figure 2 The high specific heat capacity of water makes it useful in various ways

(to convince yourself of this last step, try using simple figures:

$24 \div 4 \div 3$ does equal $\frac{24}{4 \times 3}$)

The specific heat capacity of a substance is numerically the same as the thermal energy required to produce a 1 K rise in temperature in a mass of 1 kg.

For example, water has a specific heat capacity of 4200 J/(kg K); 4200 J of thermal energy are required to produce a 1 K temperature rise in water of mass 1 kg.

Specific heat capacities of different substances are given in figure 1. Water has a particularly high specific heat capacity, which makes it a useful substance for storing and carrying thermal energy. In each of the examples in figure 2, water is used to take in thermal energy. This energy is released again when the water cools down.

Concrete has a lower specific heat capacity than water, but being more dense, the same mass takes up less space. Concrete blocks are used in night storage heaters to store thermal energy (as shown in figure 3). Electric heating elements heat up the blocks overnight when electricity is cheaper to buy. The hot blocks continue to release their thermal energy through the day as they cool down.

Figure 3 A night storage radiator contains concrete blocks which store thermal energy

Thermal energy calculations

Knowing the mass of a substance and its specific heat capacity, it is possible to calculate the thermal energy required to produce any given temperature rise.

The specific heat capacity equation can be rearranged to give:

thermal energy required = mass × specific heat capacity × temperature change

In symbols, $\boldsymbol{E = mc(\theta_2 - \theta_1)}$

Thermal energy is measured in J, if mass is measured in kg, specific heat capacity in J/(kg K), and the temperature rise in K or °C.

The same equation can also be used to calculate the thermal energy given out by a substance when its temperature falls.

Example *How much thermal energy is required to raise the temperature of* 3 kg *of aluminium from* 15 °C *to* 25 °C*?* [*The specific heat capacity of aluminium is* 900 J/(kg K)]

The problem is illustrated in figure 4.

The thermal energy required is calculated using the equation:

$E = mc(\theta_2 - \theta_1)$

Substituting the values given in the question:

$E = 3 \times 900 \times (25 - 15)$ J
$= 27000$ J

27 000 J of thermal energy are required.

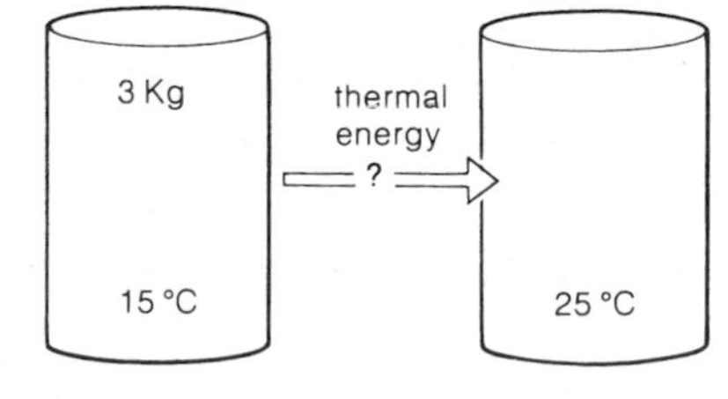

Figure 4

Thermal energy gains and losses

Mix roughly equal quantities of hot and cold water, and you end up with warm water. During the mixing process, the hot water loses thermal energy while the cold water gains it (as shown in figure 5). Assuming that no thermal energy is lost to the surroundings, it follows from the law of conservation of energy that:

thermal energy lost = **thermal energy gained**
(hot water) (cold water)

The same principle can be applied whenever substances at different temperatures are mixed.

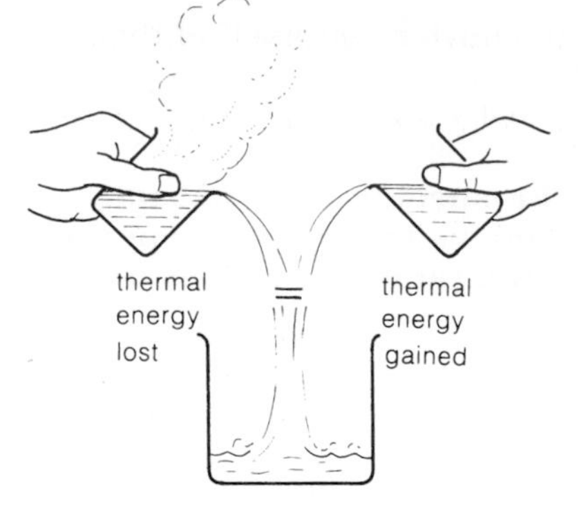

Figure 5 Thermal energy may be transferred but not lost

Example *A* 0.5 kg *block of aluminium at a temperature of* 100 °C *is placed in* 1.0 kg *of water at* 20 °C. *Assuming that no thermal energy is lost to the surroundings, what will be the final temperature of the aluminium and water when they come to the same temperature?*

The problem is illustrated in figure 6.

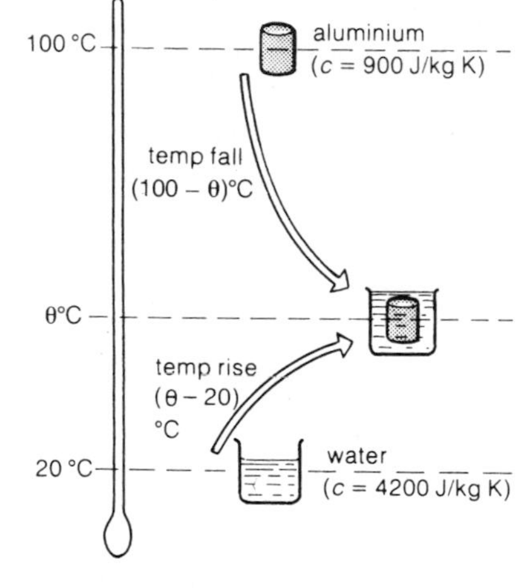

Figure 6

Call the final temperature θ °C. The temperature of the aluminium falls by (100 − θ) °C while the temperature of the water rises by (θ − 20) °C.

The thermal energy lost by the aluminium and gained by the water is calculated using the equation

$E = mc(\theta_2 - \theta_1)$

Substituting the values given above:

thermal energy lost by aluminium = 0.5 × 900 × (100 − θ) J

thermal energy gained by water = 1.0 × 4200 × (θ − 20) J

as thermal energy gained = thermal energy lost,

$0.5 \times 900 \times (100 - \theta) = 1.0 \times 4200 \times (\theta - 20)$

multiplying out: $45\,000 - 450\,\theta = 4200\,\theta - 84\,000$

rearranging, and collecting 'θ' terms on the left hand side,

$4650\,\theta = 129\,000$

giving: $\theta = 27.7$

The final temperature is therefore 27.7 °C.

Calculations involving power

It is possible to calculate the thermal energy supplied by an electric heater if its power rating is known. An electric kettle with a power rating of 2000 watts [W], for example, supplies 2000 J of thermal energy every second (as shown in figure 7). In 2 seconds, it would supply 4000 J, in 3 seconds it would supply 6000 J, and so on. In general,

thermal energy supplied by heater = power × time

In symbols: $E = Pt$

Thermal energy is measured in J, if power is measured in W [J/s] and time in s.

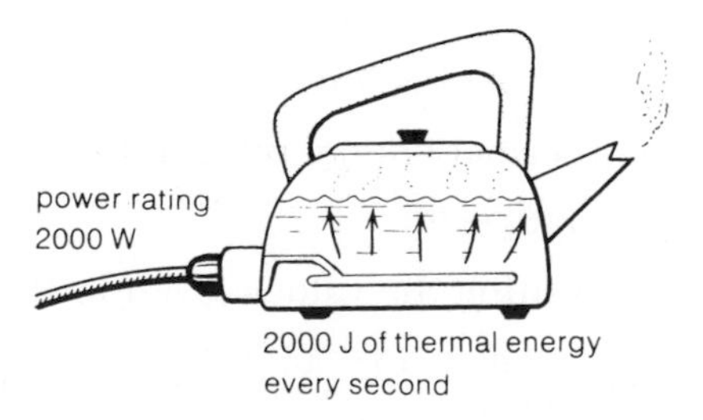

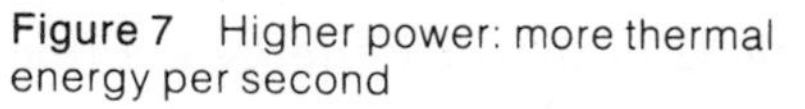

Figure 7 Higher power: more thermal energy per second

Example *A 2 kg block of aluminium (as shown in figure 8) is heated by an electric heater rated at 90 W. If the temperature of the block rises by 5 °C and no heat energy is lost to the surroundings, for how long is the heater switched on?*

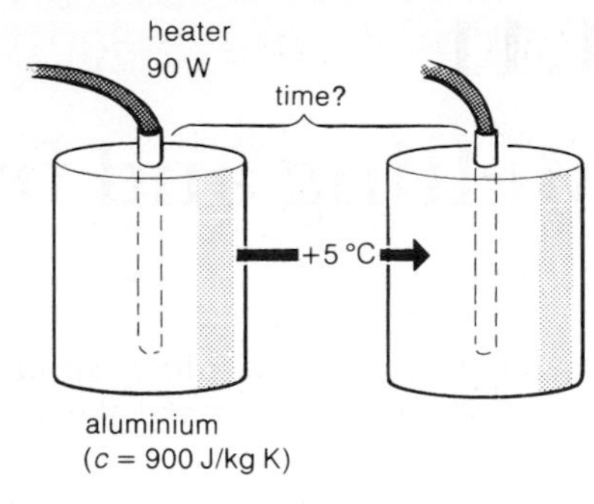

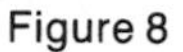

Figure 8

Call the unknown time t seconds.

thermal energy supplied by heater = thermal energy gained by aluminium

$$\therefore \quad Pt = mc\,(\theta_2 - \theta_1)$$

Substituting the values given, $90t \text{ J} = 2 \times 900 \times 5 \text{ J}$

which gives: $t = 100$

The heat is switched on for 100 seconds.

Measuring specific heat capacity

Figure 9 shows an experiment for measuring the specific heat capacity of water. The pan contains a measured mass of water. The electric heater is switched on for several minutes and the rise in temperature of the water noted. Assuming that:

thermal energy supplied by heater = thermal energy gained by water

then: $$Pt = mc\,(\theta_2 - \theta_1)$$

rearranged, this gives $$c = \frac{Pt}{m(\theta_2 - \theta_1)}$$

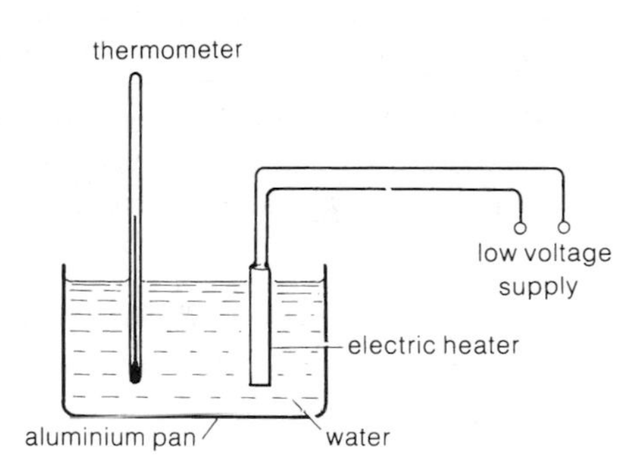

Figure 9 Apparatus for measuring the specific heat capacity of water

The specific heat capacity of the water (c) is calculated from this equation by substituting the values of m, t, $(\theta_2 - \theta_1)$ and P. If the power rating of the heater isn't known, it can be found as explained in unit 6.9.

As described, this experiment only gives an approximate value for the specific heat capacity of water. To obtain a more accurate value, allowance must be made for thermal energy losses during the heating process, and for the thermal energy used in raising the temperature of the pan.

A similar experiment can be carried out to measure the specific heat capacity of a metal – aluminium for example. In this case, the pan of water is replaced by a block of aluminium with holes drilled to take the electric heater and the thermometer.

Questions

In questions 3, 4, and 5, assume that no heat energy is lost to the surroundings. Where appropriate, use the values for specific heat capacity given in the chart on page 144.

1 Aluminium has a specific heat capacity of 900 J/(kg K). What does this mean? A 10 kg block of aluminium cools from 100 °C to 50 °C. How much thermal energy does it give out? What thermal energy would be given out by the same mass of water over the same temperature fall?

2 Water has a very high specific heat capacity. In what ways is this useful?

3 A 210 W heater is placed in 2 kg of water. What temperature rise is produced if the heater is switched on for 200 s?

4 An electric kettle has a power rating of 2.1 kW. The kettle is filled with 1.5 kg of water at a temperature of 20 °C. How long after the kettle is switched on will the water start to boil?

5 A lump of metal of mass 0.2 kg and temperature 100 °C is placed in water of mass 0.4 kg and temperature 16 °C. If the final temperature of the metal and water is 20 °C, what is the specific heat capacity of the metal?

4.11
Melting and freezing

Putting heat energy into a solid does not necessarily increase its temperature. It may make the solid change into a liquid.

The container of crushed ice shown in figure 1 is being heated by a small electric heater. Although the ice is absorbing thermal energy, frequent checks show that the temperature in the container remains at 0°C until virtually all the ice has melted. This is an example of thermal energy producing a change of *state* rather than a change in temperature. In this case, all the thermal energy supplied is being used to change water from a solid to a liquid state.

Figure 1 Thermal energy is needed to change ice at 0°C to water at 0°C

The thermal energy absorbed during the melting process is called **latent heat of fusion**. 'Latent' means hidden; the water ends up with more energy as a liquid than as a solid, but this extra energy isn't immediately obvious because of the absence of any temperature change. If any solid is to become a liquid, it must gain the necessary latent heat. Equally, if a liquid is to change back into a solid, it must lose this latent heat.

Latent heat of fusion: the Kinetic Theory

In a solid, the vibrating molecules have kinetic energy because they are moving, and potential energy because of the 'spring like' attractions which try to pull them close together. If the molecules can be separated sufficiently, the attractions weaken, the molecules become free to change positions, and the solid becomes a liquid.

The latent heat of fusion represents the work done in separating the molecules during the melting process. As thermal energy is absorbed, the molecules move further apart and their potential energy is increased. There is no change in the average kinetic energy of the molecules during melting, so the temperature stays the same.

In ice, the molecules are held in a regular ring structure, as shown in figure 2a. When ice melts, work is done in separating some molecules so that the rings are broken, and the regular structure is lost, as shown in figure 2b. The molecules now have more freedom of movement, but are more closely packed: water in the liquid state takes up less space than ice.

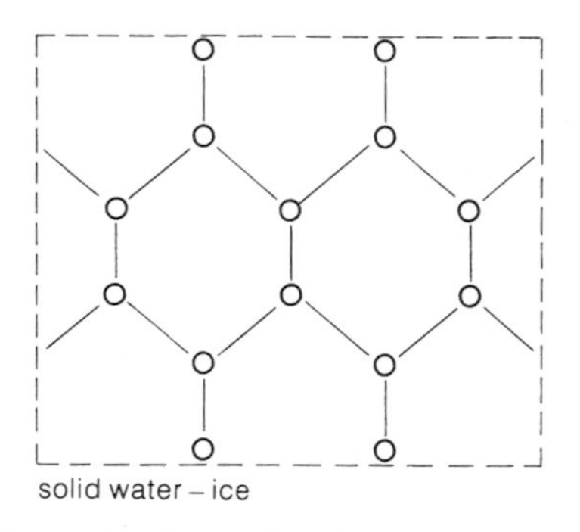

Figure 2a The molecular structure of ice

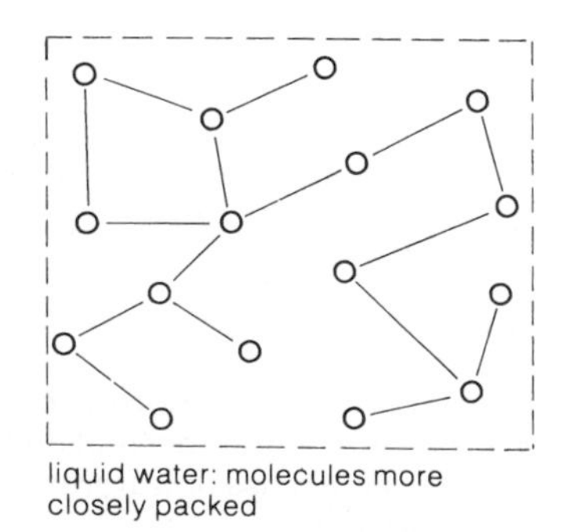

Figure 2b The molecular structure of water

Specific latent heat of fusion

The thermal energy required to change a solid material into a liquid depends on the mass of the material. In the case of water:

334 000 J of thermal energy are required to change 1 kg of ice at 0 °C into water at the same temperature;
668 000 J of thermal energy are required to change 2 kg of ice at 0 °C into water at the same temperature; and so on.

In each case, dividing the thermal energy by the mass gives the same value, 334 000 J/kg. Water is said to have a **specific latent heat of fusion** of 334 000 J/kg.

The specific latent heat of fusion of any substance is given by the equation:

$$\textbf{specific latent heat of fusion} = \frac{\left[\begin{array}{c}\textbf{thermal energy required to change a mass}\\ \textbf{of a substance from a solid into a liquid}\\ \textbf{without change in temperature}\end{array}\right]}{\textbf{mass of substance}}$$

Specific latent heat is measured in J/kg, if energy is measured in J and mass in kg.

The specific latent heat of fusion is numerically equal to the thermal energy required to change 1 kg of a substance from a solid into a liquid without change in temperature.

For example, water has a specific latent heat of fusion of 334 000 J/kg; 334 000 J of thermal energy are required to change 1 kg of ice at 0 °C into water at the same temperature, as shown in figure 3. This means incidentally that it takes nearly as much thermal energy to melt 1 kg of ice as it does to heat 1 kg of tap water right up to boiling point.

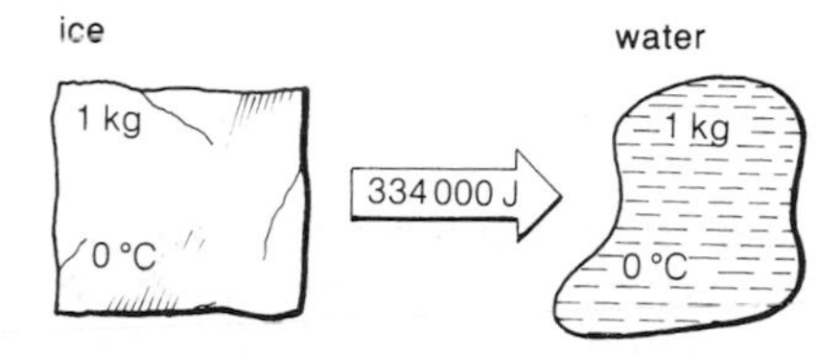

Figure 3 The specific latent heat of fusion of ice is 334 000 J/kg

Specific latent heat of fusion calculations

It is possible to calculate the thermal energy required to change any mass of a substance from a solid into a liquid, provided the mass and the specific latent heat of the fusion are known.

Rearranging the specific latent heat of fusion equation:

thermal energy required = mass × specific latent heat

in symbols: $E = mL$

Thermal energy is measured in J if mass is measured in kg and specific latent heat in J/kg.

Example *How much heat energy is required to change 2 kg of ice at 0 °C into water at 20 °C? [Specific latent heat of fusion of water = 334 000 J/kg; specific heat capacity of water = 4200 J/(kg K).]*

The problem is illustrated in figure 4.

The thermal energy required to change ice at 0 °C into water at 0 °C is calculated using the equation:

$$E = mL$$

in this case: $E = 2 \times 334\,000\ \text{J} = 668\,000\ \text{J}$

The 2 kg of water formed now has to be heated from 0 °C to 20 °C. The thermal energy required for this is calculated using the equation:

$$E = mc(\theta_2 - \theta_1) \quad \text{(as explained on page 137)}$$

in this case: $E = 2 \times 4200 \times 20\ \text{J} = 168\,000\ \text{J}$

The total energy required is therefore 668 000 J plus 168 000 J, or 836 000 J.

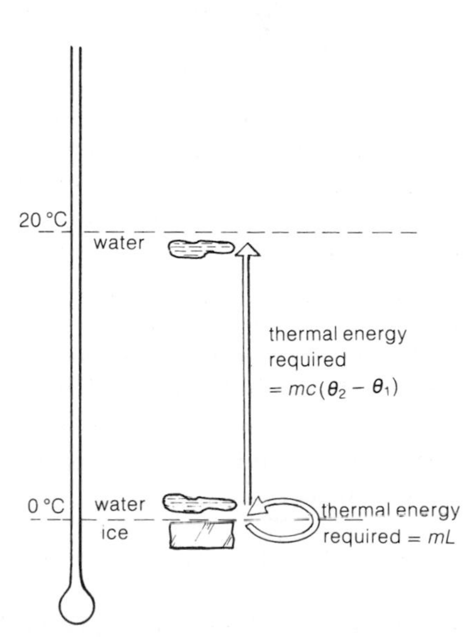

Figure 4

Measuring the specific latent heat of fusion of ice

Figure 5 shows the apparatus used. Ice at 0 °C is heated by a small electric heater which is left switched on for several minutes. Some of the ice melts to form water which runs down through the funnel and is collected in the beaker. The mass of ice melted is found by measuring the mass of water collected. Then:

the thermal energy supplied by the heater = thermal energy used to melt ice

therefore: $Pt = mL$

rearranging: $L = \dfrac{Pt}{m}$

Knowing the power of the heater P, the time t, and the mass of ice melted m, the specific latent heat of fusion of the ice L can be calculated. The experiment gives only a rough result unless allowance is made for extra heat absorbed from the surroundings.

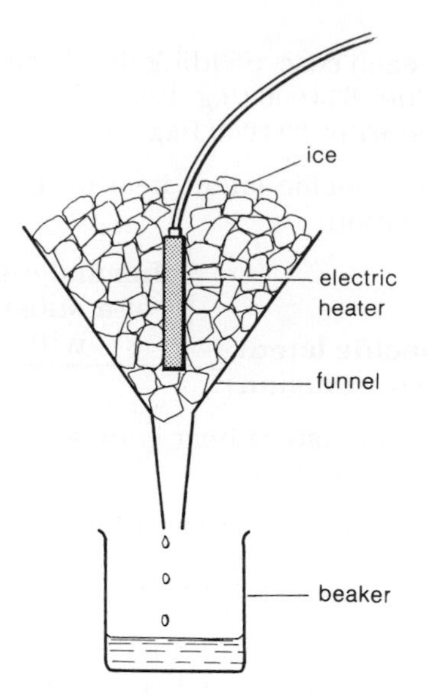

Figure 5 Apparatus for measuring the specific latent heat of fusion of ice

Melting and freezing points

When a pure substance melts, it does so at one particular temperature, called its **melting point**. When it turns solid, it does so at exactly the same temperature, its freezing point. In either case, it is more usual to refer to the melting point.

Figure 6 shows an experiment to find the melting point of naphthalene. The experiment should be carried out in a fume cupboard because naphthalene gives off a poisonous vapour.

The naphthalene is first melted and heated beyond its melting point by placing the test tube in near boiling water for a few minutes. The test tube is then removed from the water; the temperature of the naphthalene is measured every half minute as it cools. A graph of temperature against time gives a 'cooling curve' as shown in figure 7.

The cooling liquid naphthalene reaches its melting point, 80 °C, at point B on the graph. Between B and C, the naphthalene gradually changes back into a solid. It continues to lose heat energy over this region, but its temperature doesn't change; it is losing its latent heat. At point C, the naphthalene is completely solid and its temperature starts to fall again.

Only pure substances have clearly defined melting points. Many materials are a mixture of several substances which melt at different temperatures.

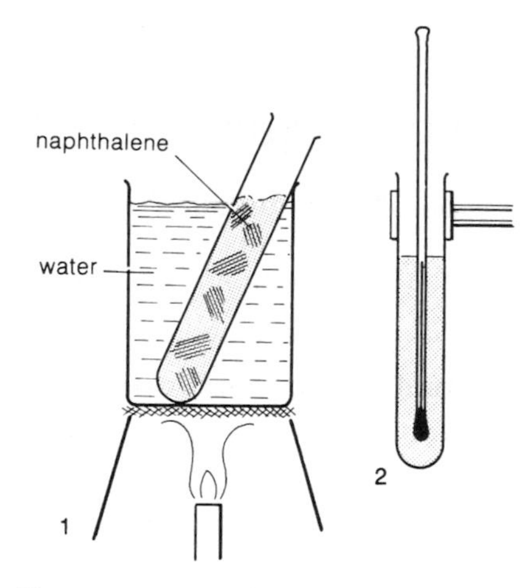

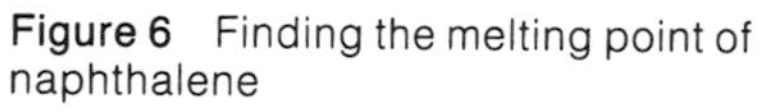

Figure 6 Finding the melting point of naphthalene

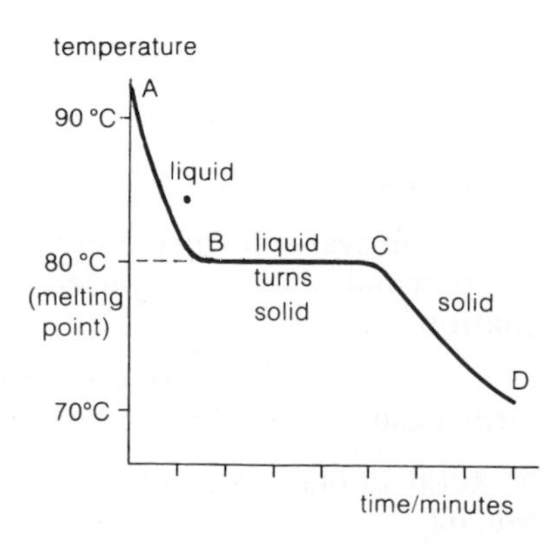

Figure 7 Cooling curve for naphthalene

Factors affecting melting point

If a substance is put under pressure, its melting point is changed slightly. Applying pressure to ice, for example, lowers its melting point. The change in this case is very small, being less than 0.01 °C for each atmosphere increase in pressure.

The experiment shown in figure 8a demonstrates the effect of pressure on the melting point of ice. Heavy objects are hung from the ends of a length of thin copper wire which exerts a high pressure on the ice underneath it.

Figure 8a A high pressure will lower the melting point of ice. As the pressure is removed behind the wire the ice re-forms

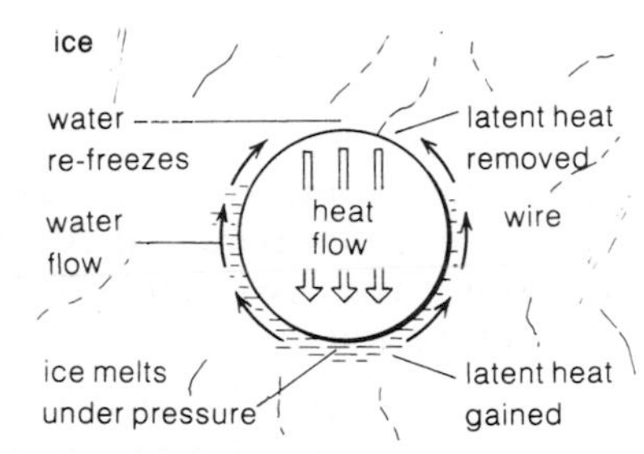

Figure 8b The complex flow of thermal energy as the wire moves down through the ice

This lowers the melting point and the ice melts. As the wire moves downwards, the water from the melted ice flows up round the wire and refreezes above it where the pressure is normal. In time, the wire passes right through the ice, though the block remains in one piece.

The experiment only works quickly if the wire is made of a material which conducts heat well. Underneath the wire, latent heat must be given to the ice to enable it to melt. Above the wire, latent heat must be removed from the water to enable it to freeze. The wire must rapidly conduct this thermal energy from one side to the other (as shown in figure 8b).

The melting point of a substance is also affected by the presence of impurities. Adding salt to melting ice, for example, can reduce its melting point to as low as −18 °C. Adding antifreeze has a similar effect.

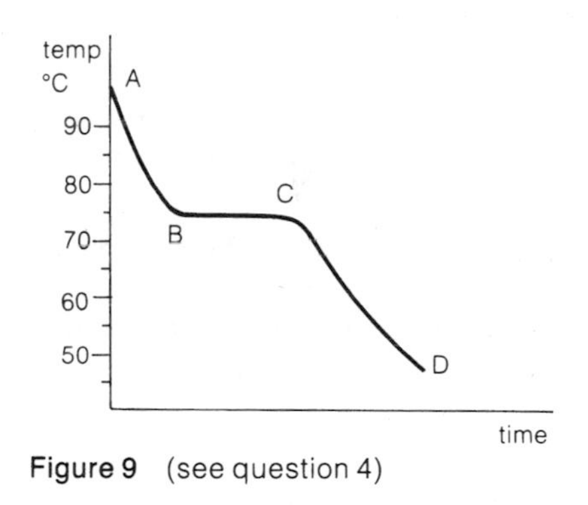

Figure 9 (see question 4)

Questions

Where necessary, use the value of the specific latent heat of fusion of ice given in question 2.

1 Why is thermal energy needed to turn a solid into a liquid?

2 The specific latent heat of fusion of ice is 334 000 J/kg. What does this mean?

3 How much thermal energy must be removed from 5 kg of water at 0 °C to turn it into ice?

4 A substance is melted and then allowed to cool. A cooling curve is plotted as shown in figure 9. What is the melting point of the substance? What is happening to the substance between B and C? Over which of the sections AB, BC and CD, is the substance losing heat?

5 What is the effect on the melting point of ice of a) exerting a high pressure on the ice b) adding salt to the ice?

6 How much thermal energy must be removed from 2 kg of water at 100 °C to turn it into ice at 0 °C? [The specific heat capacity of water is 4200 J/(kg K)].

7 A panful of crushed ice at 0 °C is heated by a 167 W heater. What mass of ice will melt if the heater is switched on for 100 s? State any assumptions you make in your calculation.

4.12

Latent heat of vaporization

Thermal energy is also needed to change a liquid into a gas. Liquids have a tendency to turn into gases whether you heat them or not. If the thermal energy isn't supplied, they will extract it from somewhere.

Figure 1 An electric kettle can vaporize steam at about 50 g/min

When an electric kettle is switched on, the temperature of the water rises until boiling point is reached. From that point on, nearly all the heat supplied by the element is used to change the water into a gas called water vapour or steam, as shown in figure 1, and there is no further rise in temperature. A full kettle boils dry in about twenty minutes if it isn't switched off. All the water has then been turned into water vapour.

The thermal energy needed to turn a liquid into a gas is called **latent heat of vaporization**. It represents the work done in forcing the molecules far enough apart for the attractions between them to be almost non-existent. The molecules are then free to travel at random through any space available to them, and their potential energy is greatly increased.

When a gas changes back into a liquid, the molecules are pulled close together again by the forces of attraction between them, and the potential energy lost is released as heat. Steam gives out latent heat when it condenses to form water, which is why a scald from steam is much worse than one from boiling water.

Specific latent heat of vaporization

The **specific latent heat of vaporization** of a substance is given by the equation:

$$\textbf{Specific latent heat of vaporization} = \frac{\left[\begin{array}{c}\textbf{thermal energy required to change a}\\ \textbf{mass of a substance from a liquid into}\\ \textbf{a gas without change in temperature}\end{array}\right]}{\textbf{mass of substance}}$$

Specific latent heat is measured in J/kg, if energy is measured in J and mass in kg.

The specific latent heat of vaporization is numerically equal to the thermal energy required to change 1 kg of a substance from a liquid into a gas without change in temperature.

For example, water has a specific latent heat of vaporization of 2 260 000 J/kg (at 100 °C); 2 260 000 J of thermal energy are required to change 1 kg of water at 100 °C into steam at the same temperature, as shown in figure 2. The value varies depending on the temperature at which the change of state is taking place.

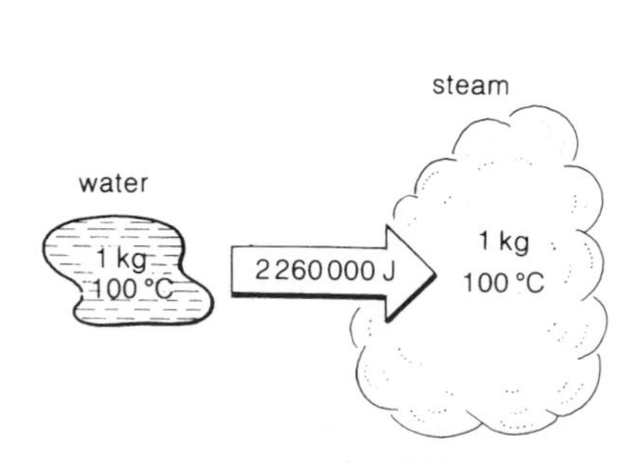

Figure 2 The specific latent heat of vaporization of steam is 2 260 000 J/kg

Note that the specific latent heat of vaporization of water is much greater than its specific latent heat of fusion. It takes nearly seven times as much thermal energy to change each kg of water into steam as it does to change each kg of ice into water. This is understandable when you think how much more the molecules increase their separation when steam is formed.

Practical use of latent heat theory: using a blast of steam to heat up coffee

Specific latent heat of vaporization calculations

The thermal energy required to change any mass of a substance from a liquid into a gas can be calculated using the equation:

thermal energy required = mass × specific latent heat

in symbols: $E = mL$

Energy is measured in J, if mass is measured in kg and specific latent heat in J/kg.

Example *A pan contains* 2.0 kg *of water at* 0 °C. *A jet of steam at* 100 °C *is passed through the water. What is the temperature of the water when* 0.10 kg *of steam have condensed in it? Assume that no heat is lost, or absorbed by the pan.*

[*Specific latent heat of vaporization of water* = 2 260 000 J/kg; *specific heat capacity of water* = 4200 J/(kg K)].

The problem is illustrated in figure 3.

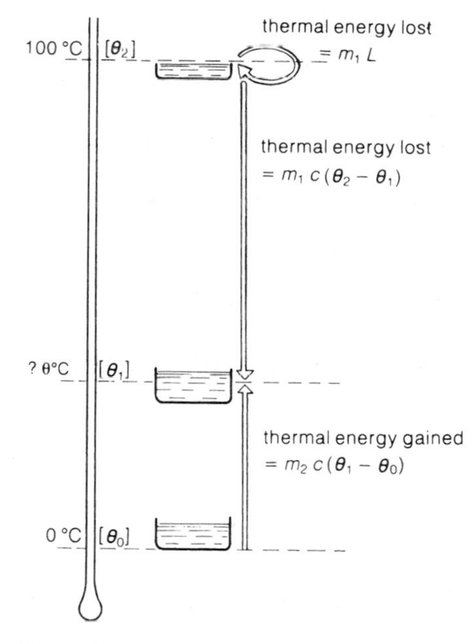

Figure 3

Call the final temperature of the water θ °C.

The steam loses thermal energy in two ways; firstly by condensing to form water at 100 °C, secondly by cooling, as water, from 100 °C to θ °C. The water in the pan gains thermal energy when its temperature rises from 0 °C to θ °C. By writing down expressions for the total thermal energy lost by the steam and gained by the water, and putting them equal, the value of θ can be found.

thermal energy lost by steam in condensing $= m_1 L = 0.10 \times 2260000 \text{ J} = 226000 \text{ J}$

thermal energy lost by condensed steam in cooling to θ °C $= m_1 c(\theta_2 - \theta_1) = 0.10 \times 4200 \times (100 - \theta) \text{ J}$
$= 420(100 - \theta) \text{ J}$

$\therefore$ total thermal energy lost $= [226000 + 420(100 - \theta)] \text{ J}$

thermal energy gained by water in pan $= m_2 c(\theta_1 - \theta_0) = 2.0 \times 4200 \times (\theta - 0) \text{ J}$
$= 8400\,\theta \text{ J}$

But remember that: thermal energy gained = thermal energy lost

$$\therefore 8400\,\theta = 226000 + 420(100 - \theta)$$

rearranged, with 'θ' terms collected on the left hand side, this gives

$$\theta = \frac{268000}{8820} = 30.4$$

The final temperature of the water (plus condensed steam) is 30.4 °C.

Measuring the specific latent heat of vaporization of water

Figure 4 shows the apparatus used. When the water in the can is boiling vigorously, the mass reading on the balance is noted and a stopwatch started. Several minutes later, the stopwatch is stopped and the mass reading taken again. The difference in the mass readings gives the mass of water which has been changed into steam during the time measured.

Thermal energy supplied by heater = thermal energy used to change water into steam

therefore: $Pt = mL$

giving: $L = \dfrac{Pt}{m}$

Knowing the power rating of the heater P, the mass of steam produced m and the time taken t, the specific latent heat of vaporization of the water L can be calculated. Allowances must be made for heat losses if an accurate result is required.

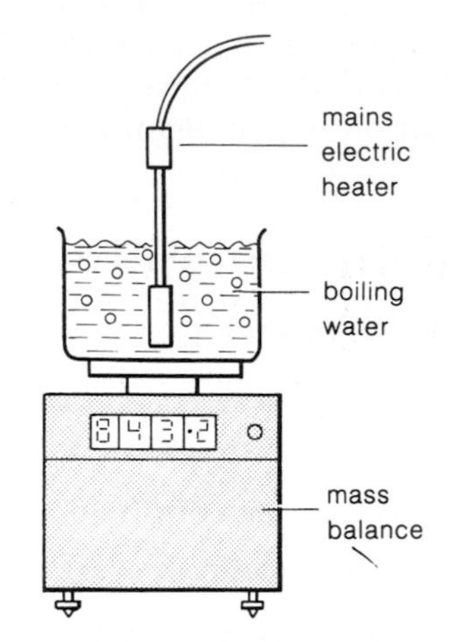

Figure 4 The rate of water loss and the power of the heater can be used to calculate the specific latent heat of vaporization

Evaporation

A liquid doesn't have to boil in order to change into a gas. The change of state occurs most rapidly at the boiling point, but a liquid starts to turn into a gas well below this temperature. The process is called **evaporation**, and it explains why wet clothes dry and rain puddles disappear even on a cold day.

Liquids evaporate because some of their molecules move very much faster than others. Although most of the molecules remain in the liquid, the faster ones have enough energy to overcome the attractions of other molecules and are constantly escaping from the liquid surface.

There are several ways of making a liquid evaporate more rapidly:

1 Increase its temperature This makes the molecules move faster, so that more of them have enough energy to escape from the liquid.

2 Increase its surface area This gives the faster molecules a greater chance of escaping. Wet roads dry out quickly because the rainwater is spread over such a large area.

3 Pass air through it or across its surface When air moves across a liquid surface, it carries away molecules escaping from the liquid and reduces their chances of returning to it. This is why wet washing dries best on a windy day.

4 Make the liquid into a fine spray A spray is made up of millions of tiny liquid droplets with a very large total surface area. The highly curved surfaces make it easier for molecules to escape. The spray principle is used in a carburettor where a spray of petrol quickly evaporates as it is drawn into the air entering a car engine, as shown in figure 5. The petrol vapour formed is burnt in the engine.

Evaporation at work

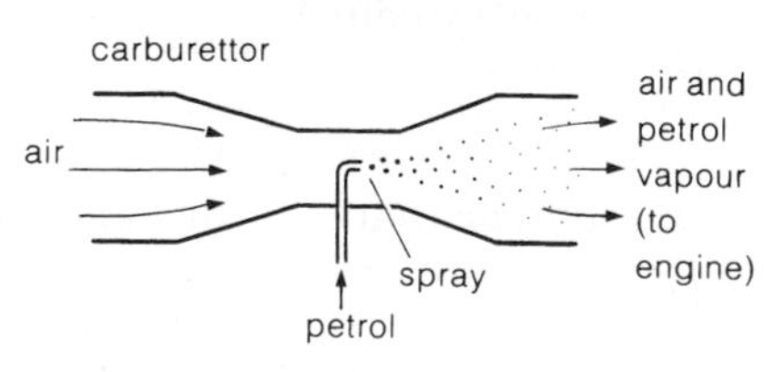

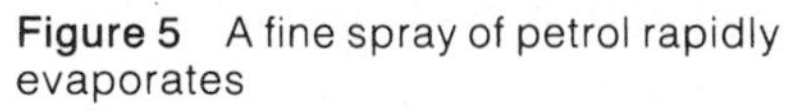

Figure 5 A fine spray of petrol rapidly evaporates

Cooling by evaporation

A liquid needs latent heat in order to evaporate. If the liquid isn't being heated, the thermal energy must come from some other source. Meths tipped on the back of your hand quickly evaporates, and your hand feels cold as a result. The evaporating meths extracts the necessary latent heat from the liquid meths remaining. This cools, and extracts heat from your hand.

Ether evaporates more readily than meths; it is more **volatile.** If air is bubbled through ether as in figure 6, the latent heat needed for evaporation is drawn from the liquid ether and the surrounding glass. As the ether evaporates, the glass may become so cold that frost begins to form on the outside.

Water evaporates much more slowly than ether or meths, but it is still possible to be badly chilled by the evaporation of water if you are wearing wet clothes. This is particularly so if there is a wind blowing to speed up the process.

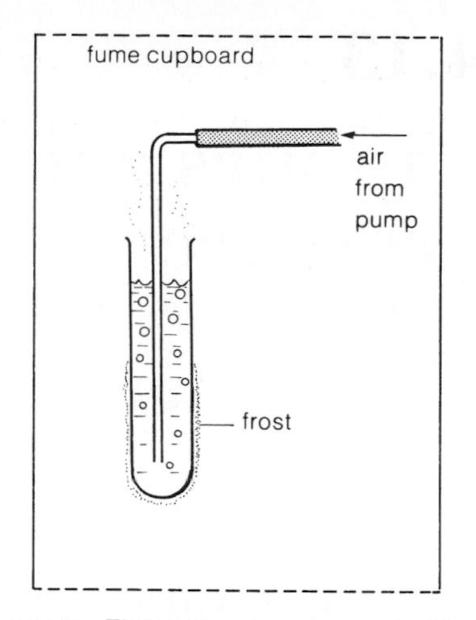

Figure 6 Thermal energy is needed for evaporation. Its loss causes cooling

Cooling and the kinetic theory The kinetic theory explains the cooling caused by evaporation in the following way. During evaporation, it is the faster molecules which escape from the liquid surface. As the liquid is losing molecules with greater than average kinetic energies, the average kinetic energy of the remaining molecules is reduced. The temperature of the liquid is lowered as a result.

Refrigerators The cooling effect in many refrigerators is produced by the evaporation of a volatile liquid called Freon, as shown in figure 7. The liquid Freon evaporates rapidly in the pipes in the freezer compartment as more and more of its vapour is drawn away by the electric pump. As the Freon evaporates, it draws the necessary latent heat from the food inside the refrigerator.

The pump compresses the vapour which turns liquid again on being forced through the zig-zag pipe at the back of the refrigerator. The latent heat released is given off through the cooling fins.

In this way, thermal energy is extracted from the food inside the refrigerator and given out at the back. A refrigerator actually makes your kitchen warmer.

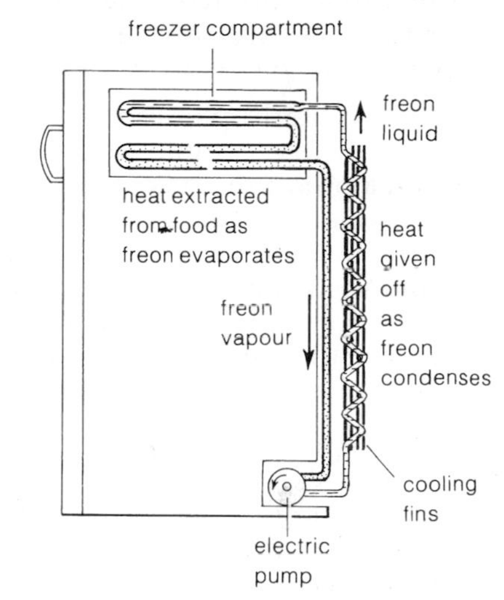

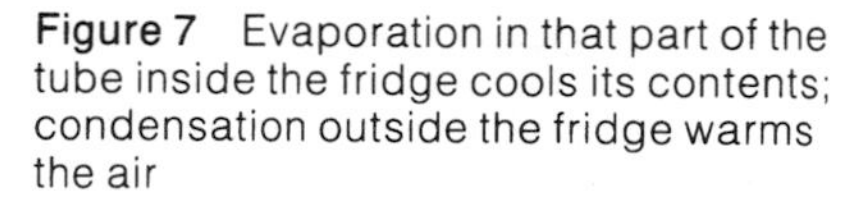
Figure 7 Evaporation in that part of the tube inside the fridge cools its contents; condensation outside the fridge warms the air

Questions

In all calculations, assume there are no heat losses to the container or to the outside. Where necessary, use the values of the specific heat capacity of, and specific latent heat of vaporization of, water given in questions 1 and 3.

1 At 100 °C, water has a specific latent heat of vaporization of 2 260 000 J/kg. What does this mean? How much thermal energy would be needed to change 2 kg of water at 100 °C into steam at the same temperature? How much thermal energy would be needed to change 10 kg?

2 Explain in terms of the kinetic theory why a) thermal energy is needed to turn a liquid into a gas b) evaporation produces cooling.

3 How much thermal energy is needed to turn 3.0 kg of water at 50 °C into steam at 100 °C? (The specific heat capacity of water is 4200 J/(kg K).)

4 Give two ways, other than by direct heating, in which a liquid may be made to evaporate more rapidly.

5 Describe two practical applications of the cooling effect produced by evaporation.

6 The heating element of a kettle has a power rating of 2.26 kW. If the kettle contains boiling water, what mass of steam will be produced in 10 minutes?

7 0.9 kg of water at 0 °C is heated by bubbling through it a jet of steam at 100 °C. What is the temperature of the water when 0.1 kg of steam have condensed?

4.13

Vapours and vapour pressure

Put a liquid in a closed container and there is a limit to the amount that can evaporate. There is also a limit to the pressure exerted by the vapour which forms.

Saturated vapour pressure

The barometer shown in figure 1 contains mercury. The pipette is being used to put water into the space at the top of the tube one drop at a time.

The first water drop evaporates completely, and the fall in the level of the mercury column shows that the water vapour formed is exerting a pressure on the mercury surface (as shown in figure 2). As more drops are added, the vapour pressure increases and the mercury level continues to fall until water starts to appear above the mercury in liquid form. Further drops cause more water to appear but produce no further increase in vapour pressure. The space at the top of the tube contains the maximum mass of water vapour possible under these conditions and the vapour is said to be **saturated**. The pressure exerted by the vapour when liquid is also present is known as the **saturated vapour pressure**, or **SVP** for short.

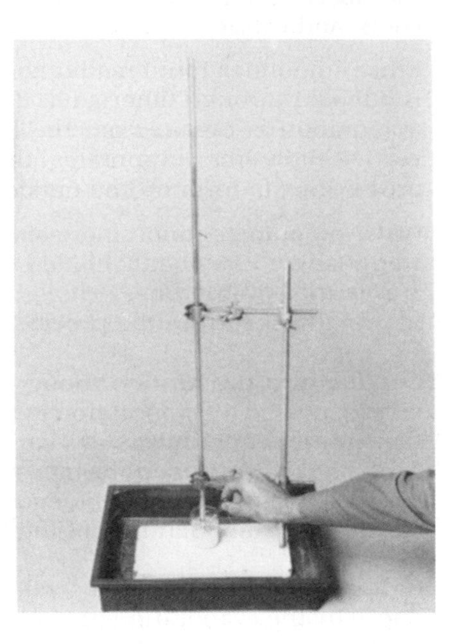

Figure 1 Adding water to the vacuum above a mercury column

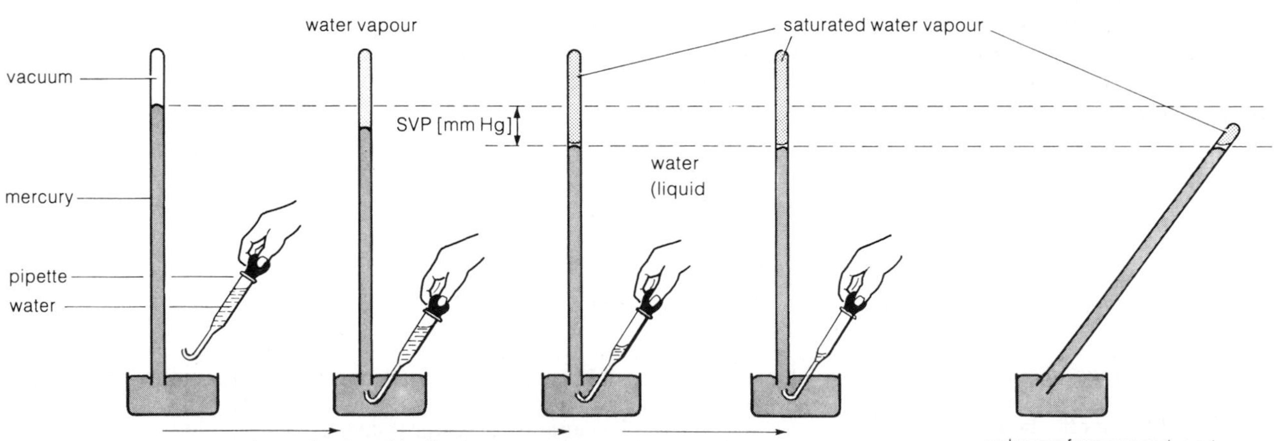

Figure 2 As the water is added, it evaporates. Eventually, no more will – the vapour is *saturated*

Provided the vapour remains saturated, changing its volume has no effect at all on the pressure it exerts. If the tube is tilted, the volume of the water vapour reduces. Some of the vapour returns to liquid form so that the pressure returns to the same value as before. The vertical height of the mercury column stays the same however, showing that there is no change in the saturated vapour pressure:

The SVP does not depend on the volume occupied by a saturated vapour.

Gases and vapours

If the volume of the water vapour in the last experiment is reduced sufficiently, all the vapour is changed into liquid. A gas can only be liquefied in this way if it is below a temperature known as its **critical temperature**. In the case of water, this is 374 °C and the water used in the experiment is obviously well below this temperature. Oxygen on the other hand has a critical temperature of −118 °C, so it isn't possible to liquefy oxygen by reducing its volume at room temperature. The gas has to be cooled first.

A gas is described as a **vapour** if it is below its critical temperature, i.e. if it can be liquefied by reducing its volume. Above the critical temperature, the gas molecules have enough energy to overcome the attractions between them even when closely pushed together.

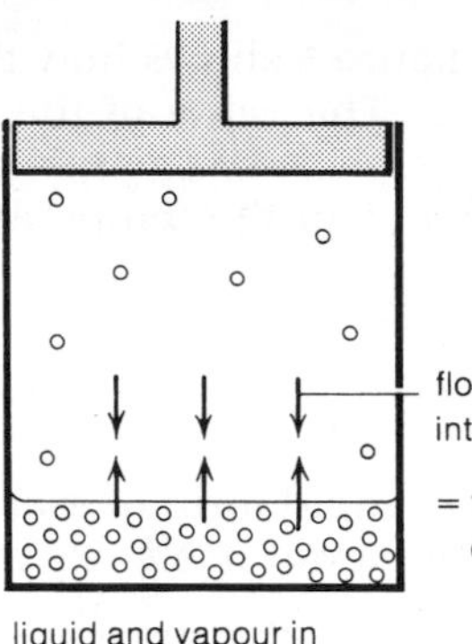

Figure 3a Liquid and vapour in dynamic equilibrium

Dynamic equilibrium

When water vapour is present above a liquid, there is a continuous exchange of molecules between the two. Molecules are constantly escaping from the liquid to form vapour and returning from the vapour to form liquid. If the vapour is saturated, the rate of flow of molecules in each direction is the same, and liquid and vapour are said to be in a state of **dynamic equilibrium**. Under these conditions, liquid is evaporating to form vapour at the same rate as vapour is condensing to form liquid, as shown in figure 3a.

A vapour is saturated if it is in a state of dynamic equilibrium with its liquid.

If the volume of the vapour is reduced (as shown in figure 3b), the number of molecules in each cubic metre rises and the rate at which molecules return to the liquid is increased. As a result, more return to the liquid than leave it, and the number of molecules in the vapour falls until dynamic equilibrium is restored. When the vapour is again saturated, there are as many molecules in each cubic metre as before, so the vapour pressure is unchanged.

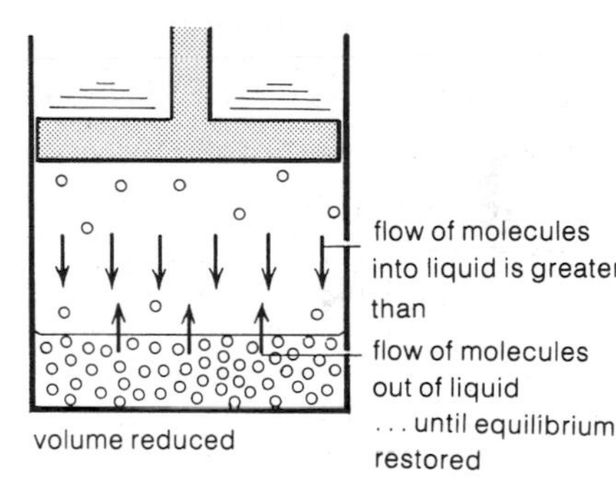

Figure 3b If the volume is reduced, more molecules return to the liquid than leave it, until equilibrium is restored.

Effect of temperature on a saturated vapour

If a liquid and vapour are in dynamic equilibrium, an increase in temperature increases the rate at which molecules escape from the liquid, so that equilibrium is temporarily upset. With more molecules leaving the liquid than returning to it, the number of molecules in the vapour rises until dynamic equilibrium is restored (as shown in figure 4). The vapour is once again saturated, but there are now more molecules in each cubic metre than there were before and they are moving faster on average because of the increased temperature. From this it follows that:

1 increasing the temperature increases the mass of vapour that a container can hold before saturation is reached.

2 increasing the temperature increases the SVP of a liquid. The effect can be seen in the experiment in figure 2 by gently heating the top of the tube and noting the fall in the mercury levels.

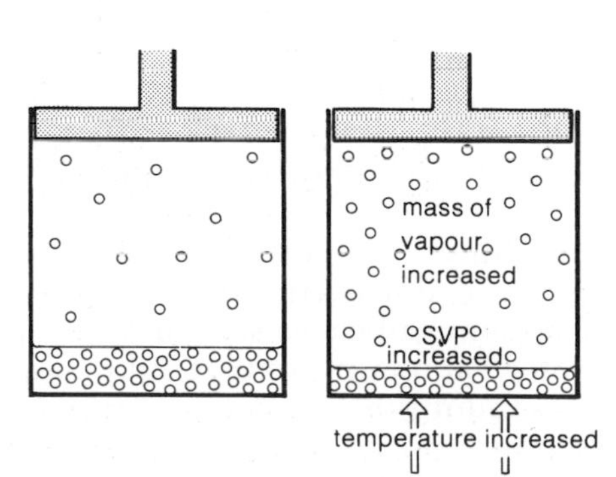

Figure 4 If the temperature is increased, more molecules leave the liquid than return to it, until equilibrium is restored.

The graph in figure 5 shows how the SVP of water changes between 0 °C and 120 °C. The value of the SVP of water at 100 °C may look familiar. This is because there is a close link between SVP, atmospheric pressure, and the temperature at which a liquid boils.

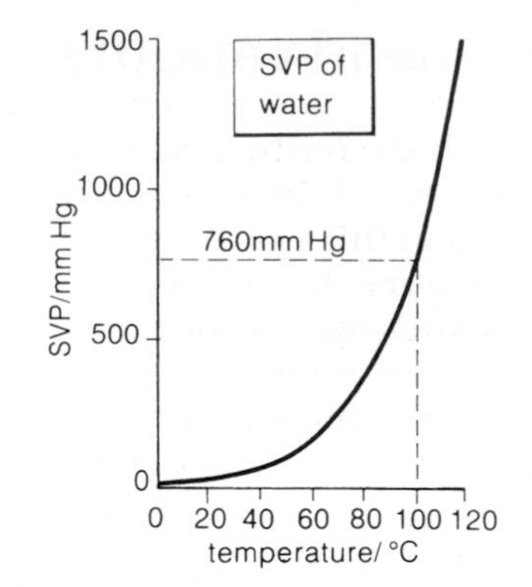

Figure 5 A graph of SVP against temperature, for water

Boiling

Boiling is a very rapid form of evaporation in which vapour bubbles form within the liquid itself, expand, rise to the surface, and burst.

Below boiling point, the pressure of the atmosphere pressing on a liquid stops vapour bubbles forming and evaporation only takes place on the surface of the liquid. As the temperature rises, the SVP of the liquid increases. When the SVP is equal to atmospheric pressure, vapour bubbles are able to form within the liquid and expand by pushing back the atmosphere (as shown in figure 6).

A liquid boils when its SVP is equal to atmospheric pressure.

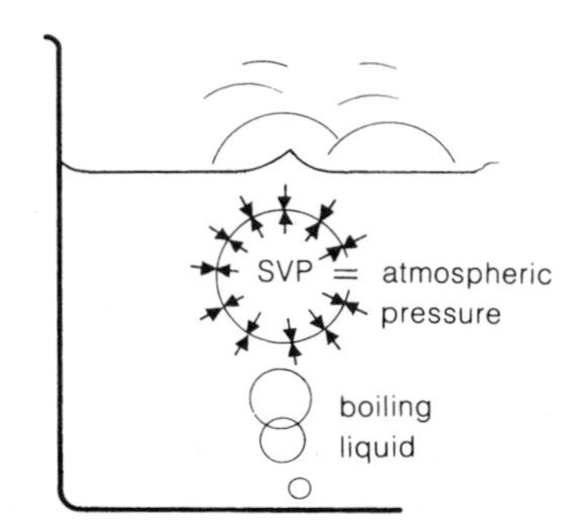

Figure 6 A liquid boils when its SVP is equal to atmospheric pressure

Water boils at exactly 100°C if atmospheric pressure is the standard 760 mmHg because water has a SVP of 760 mmHg at 100 °C. A fall in atmospheric pressure lowers the boiling point. At the top of Mount Everest for example, where the air pressure is only about a third of the pressure at sea level, water boils at around 70 °C. Atmospheric pressure changes slightly from day to day, but in the UK it is rare for the boiling point of water to fall below 98.5 °C.

Increasing the outside pressure raises the boiling point. The principle is used in the pressure cooker (as shown in figure 7) in which pressure is allowed to build up to about twice normal atmospheric pressure, so that the water in the cooker boils at around 120 °C. Food cooks much faster at this temperature than it does at 100 °C.

The boiling point of a liquid is also raised if impurities are present. Salt water, for example, has a slightly higher boiling point than pure water, and the water in a car radiator boils at a higher temperature than normal if mixed with antifreeze.

Figure 7 Water boils at a temperature above 100°C when the pressure is greater than atmospheric pressure

Water vapour in the atmosphere

The atmosphere acts as a huge container with water at the bottom in the form of seas, lakes and rivers. Its size however is such that there are many variations in pressure and temperature between different regions, and also variations in the amount of water vapour present. It is rare to find conditions of dynamic equilibrium.

Humidity Humidity is a term used to describe how close the water vapour in one particular part of the atmosphere is to saturation. The local atmosphere is said to have a **relative humidity** of 100% if the water vapour present is fully saturated – it is then holding the maximum possible mass of water vapour for that particular temperature.

The greater the humidity, the more slowly any surface water will evaporate. Sweat evaporates slowly when the humidity is high, which is why you feel so hot and uncomfortable on a 'close' or 'muggy' day. A breeze makes you feel more comfortable because it aids evaporation.

Cloud and condensation A warm atmosphere can hold a greater mass of water vapour than a cold one, as shown in figure 8. If warm, saturated water vapour in the atmosphere is suddenly cooled, some of the vapour must condense to form liquid. This water may appear as dew on the ground, or as millions of tiny droplets in the air which are seen as mist, fog or cloud. The so-called 'steam' which you see coming out of a kettle is in fact a cloud of tiny water droplets produced when hot water vapour condenses on being cooled by the air. Water vapour (true steam) is invisible.

Cold surfaces also cool any water vapour which touches them, and the vapour may condense as a result. Condensation on cold windows, taps and pipes is caused in this way. If condensation freezes, it forms frost.

a warm atmosphere can hold more water vapour . . .

. . . than a cold one.

Figure 8

Water vapour and the weather

Bad news from the weatherman in figure 9: rain threatens as a depression approaches the British Isles. Clouds have already formed over the west side of the country, where a large 'tongue' of warm damp air has a colder mass of air around it. The warm air is cooled by the colder air and water vapour condenses to form cloud.

The likelihood of rain increases as the warm damp air rises through the cold air (because of it being less dense). As it rises the pressure decreases, and the warm air expands, doing work as it pushes back the surrounding atmosphere. The energy required comes at the expense of its own internal energy. In other words, the expansion causes cooling. As the water vapour in the air is cooled, it condenses and may well cause rain.

Figure 9 Warm, damp air rising, cooling, and forming rain over the west coast of Britain

Questions

1 When can a gas be described as a vapour?

2 A liquid and its vapour are in dynamic equilibrium. What does this mean?

3 The container in figure 10 holds a liquid and its vapour. The vapour is saturated. What is the effect on the SVP of a) increasing the volume of the container b) increasing the temperature?

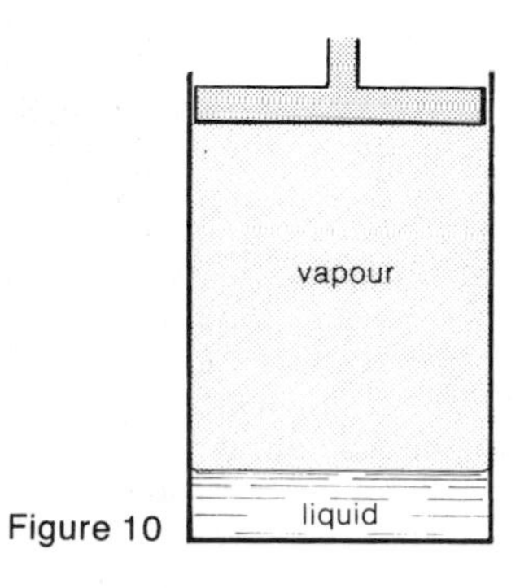

Figure 10

4 What condition is necessary for a liquid to boil? Give two ways in which the boiling point of a liquid can be raised.

5 Explain why a liquid boils at a lower temperature if atmospheric pressure falls.

6 Explain why a) wet roads dry more slowly when the humidity is high b) in a warm room, condensation forms on a cold window c) an overnight fall in temperature causes dew to form on the ground d) people tend to feel hot and uncomfortable on a humid day.

Molecular motion and heat: part A

1 Which of the following describes particles in a solid at room temperature?
A close together and stationary
B close together and vibrating
C close together and moving around at random
D far apart and moving at random (LEAG)

2 Jane couldn't unscrew a metal bottle top because it was too tight but, after she ran it under a hot tap for a few minutes, she found that she could unscrew it.
Was this because:
A the hot water acted like oil between the glass and the bottle.
B the increased pressure of the air in the bottle caused the cap to expand
C the glass in the neck of the bottle contracted
D the metal cap expanded more than the glass? (LEAG)

3 In cold weather, the metal handlebars of a bicycle, as shown in figure 1, feel colder to the hands than the plastic handgrips.

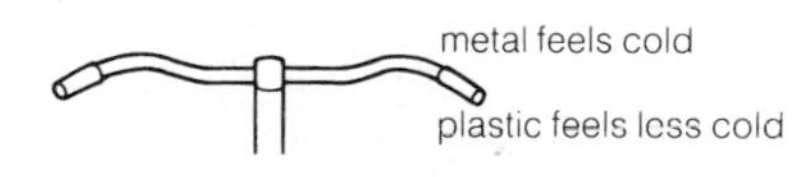

Figure 1

This is because:
A the metal is at a colder temperature than the plastic
B the plastic contains more heat energy than the metal
C the metal conducts heat better than the plastic
D the plastic is a good radiator of heat (LEAG)

4 A metal block is heated. Which line in the following table shows what happens to its volume, density and mass?

Result	*Volume*	*Density*	*Mass*
A	increases	decreases	same
B	increases	increases	increases
C	increases	increases	same
D	same	decreases	same

(LEAG)

5 Figure 2 represents a section through a particular type of building board.
Which line in the following table shows why such boards provide such good heat insulation?

	Aluminium foil is	*Expanded Polystyrene is*
A	a poor conductor	a good reflector
B	a poor reflector	a poor conductor
C	a good reflector	a poor conductor
D	a good conductor	a good reflector

(LEAG)

Figure 2

6 The table below gives information about the rate of heat flow through different surfaces of a room in very cold weather.

SURFACE	RATE OF HEAT FLOW in kJ/h
single glazed window	1200
wall between room and outside	3460
wall between room and inside rooms	860
door	200
ceiling	480
floor	1820

a) What is the meaning of 'k' in kJ?
b) Which surface gives the lowest rate of energy loss?
c) Suggest why less energy is lost through the walls between inside rooms than is lost to the outside.
d) Find the total heat loss per hour from the room.
e) If one bar of an electric fire supplies 2700 kJ/h, how many bars would be needed? (SEG)

7 Figure 3 shows the main parts of a solar heating system designed to provide hot water for a house. Heat energy from the sun warms the water in the solar panel. This water is then pumped through a spiral of copper tube inside the hot water tank so that it can transfer its heat to the water in the tank.

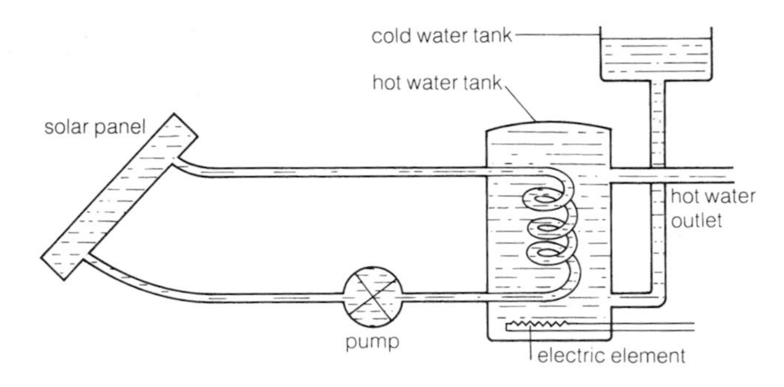

Figure 3

a) Why is copper a good material from which to make the spiral?
b) Give ONE reason why the tube is bent into a spiral rather than being straight.
c) The pipes between the solar panel and the hot water tank are lagged.
i) Name a suitable material for the lagging.
ii) Give ONE reason why this should be done.
d) Explain why the hot water outlet pipe is at the top of the hot water tank. (SEG)

8 Figure 4 represents a piston and a cylinder. The trapped air cannot pass the piston.

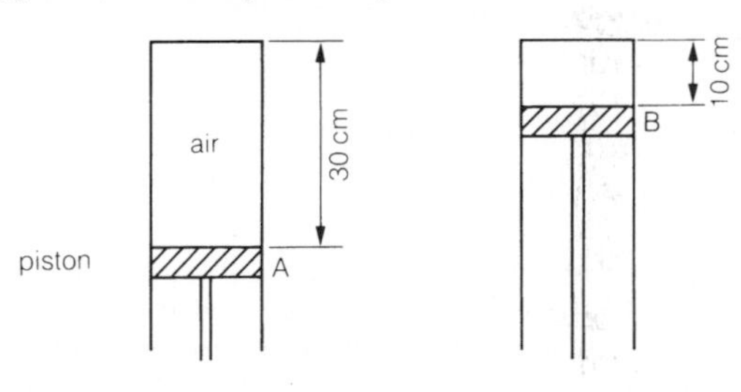

Figure 4

The top of the piston moves from position A to position B without changing the temperature of the enclosed air. Does this cause the air pressure in the cylinder to be:
a) reduced to a third
b) unchanged
c) doubled
d) trebled? (LEAG)

9 a) Why do air-filled cavity walls, as shown in figure 5, keep a house warmer in winter than solid brick walls?

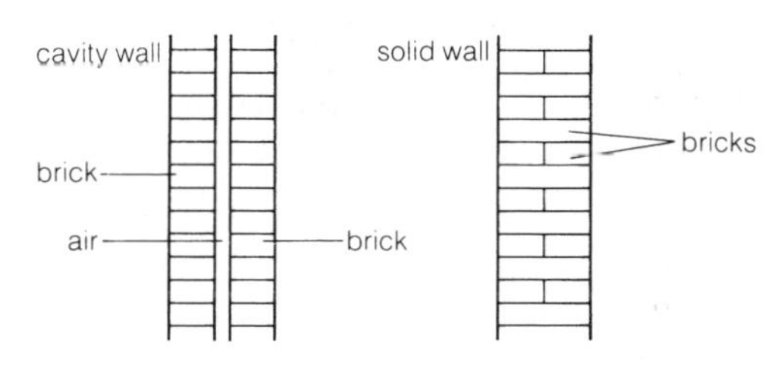

Figure 5

b) Why does filling the cavity with plastic foam keep the house even warmer?
c) Explain how a hot water radiator heats a room. (MEG)

10 Copy, and write in the missing words. a) to h) are about a petrol engine; some of the following words may help you: air, carbon dioxide, nitrogen, oxygen, petrol vapour, pressure, temperature, volume, clutch, connecting rod, crank.
a) In a petrol engine cylinder a mixture of .... and .... is burned.
b) This burning raises the .... of the mixture.
c) Raising the .... of the mixture increases the .... on the piston.
d) Increasing the .... on the piston exerts a force on the ....
e) to f) are about molecules and their movement; some of the following words may help you: air, conduction, convection, diffusion, jerky, molecules.
e) Scent moves through still air by a process called ....
f) In 'Brownian motion' smoke particles move in a .... manner. This movement is caused by .... hitting the smoke particles. (SEG)

11 a) Many domestic oil-fired central heating systems operate by pumping water through a boiler and circulating the heated water through pipes to radiators. The same water is recirculated continuously through the system.
In one such system the water flows at a rate of 0.6 kg/s. Water enters the boiler at a temperature of 35°C and leaves the boiler at a temperature of 75°C. Each kg of oil consumed provides 3×10^7 J to heat the water. The density of the oil is 850 kg/m^3. The specific heat capacity of the water is 4200 J/(kg K).
Calculate
i) the energy absorbed by the water per second as it passes through the boiler,
ii) the mass of oil which would provide this amount of energy,
iii) the time required to consume 1 m^3 of oil if the system runs continuously.

b) In practice the action of the boiler is a little more complicated. The water inlet and outlet temperatures are not constant and when the outlet temperature exceeds a given preset value the burner is switched off. The burner is switched on again when the outlet temperature falls below a second, and lower, preset value.
i) State and explain the factors which would determine the fraction of the time in which oil would be burned.
ii) State and explain the steps a householder might take to keep this fraction to a minimum.
iii) Outline the principle of operation of a device which might be used to switch the burner on or off. (SEG)

12 In a room at 20°C, liquid naphthalene (melting at 80°C) is allowed to cool from 100°C. Which one of the graphs in figure 6 best represents this cooling process?

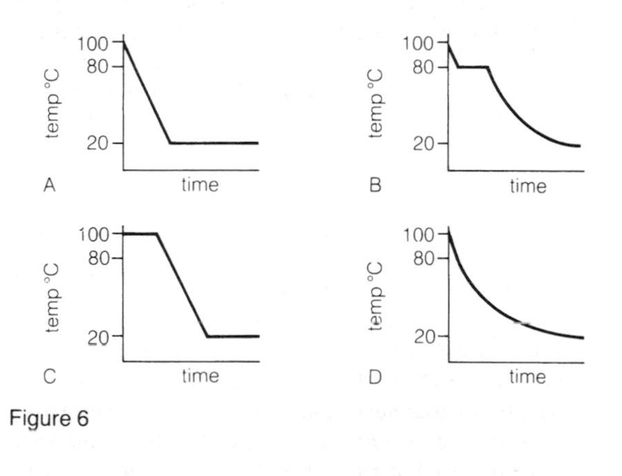

Figure 6

Further questions

Molecular motion and heat: part B

Some questions require a knowledge of concepts covered in earlier units

Specific heat capacity of water = 4200 J/kg

g = 10 N/kg $\pi = 3.14$

Specific heat capacity of ice = 2100 J/(kg K)

Specific latent heat of fusion of ice = 334 000 J/kg

Specific latent heat of vaporization of water = 2 260 000 J/kg

1 The sealed jar of figure 1 is completely evacuated and contains a small thin-walled glass capsule containing the liquid bromide.

a) Describe what is seen when the capsule is broken.

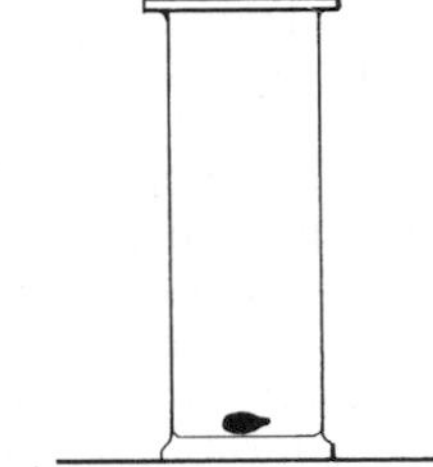

Figure 1

b) What does this tell us about the speed with which the molecules of bromine vapour travel?

c) What is the cause of the pressure which is now exerted on the inside walls of the jar?

d) If, instead of being evacuated, the jar contained air at atmospheric pressure, what difference would this have made to your answer to a). Explain this difference. (O)

2 When a drop of olive oil is placed on the surface of water in a large dish, the oil spreads out to form a circular film.

a) Describe the measurements you would make to find the area of the film.

b) The volume of the oil drop is 4×10^{-11} m^3 and the circular film is 0.2 m in diameter. Calculate the thickness of the film.

c) What conclusions about the size of a molecule of olive oil can you draw from this experiment? Explain. (O)

3 An oil drop of volume 8×10^{-12} m^3 forms a circular patch of area 4×10^{-3} m^2 when spread on water. Estimate the size of an oil molecule. (JMB)

4 a) Draw a labelled diagram of the apparatus you would use to measure the linear expansivity of copper in the form of a rod.

b) A copper rod 750 mm long is placed in such an apparatus at 20 °C, and the expansion-measuring device reads 0.30 mm. When the temperature is raised to 100 °C, the reading is 1.26 mm.

i) What is the expansion of the rod?

ii) What is the linear expansivity of copper? (O)

5 Calculate the length of a steel cable on a day when the temperature of the air is 20 °C if it is known that the cable has a length of 100 m when the temperature is 10 °C. The linear expansivity of steel is 0.000 011/K. (O&C)

6 The linear expansivity of copper is 1.7×10^{-5}/K. By how much will a copper bar 3.0 m long expand when it is heated from 0 °C to 60 °C? (O&C)

7 Two marks on a metal bar are 50.00 cm apart at a temperature of 293 K. This distance increases by 0.096 cm when the temperature rises to 373 K. Calculate the linear expansivity of the metal. (SUJB)

8 A football pitch 90 m long is marked out in the winter using a steel measuring tape. The same tape makes it appear 2.7 cm shorter when a check is made at midsummer. Explain why this happens, and estimate the difference in temperature, taking the linear expansivity of steel to be 0.000 011/K. (O)

9 The temperature of the water inside an aquarium can be controlled by a thermostat which switches an electric heater on and off. Draw a diagram showing how this may be done using a bimetallic strip. (Your diagram must clearly show the construction of the bimetallic strip.) How may different constant temperatures be achieved using your arrangement? (L)

10 A bimetal strip of brass and invar is straight at 300 K. Sketch and explain the appearance of the strip at a) 270 K, b) 400 K. (N.B. The alloy invar has negligible linear expansivity.) (SUJB)

11 The kinetic theory describes the behaviour of a gas in terms of the properties of its molecules. Explain the following:

a) A gas in a container at room temperature exerts a pressure on the walls of the container.

b) When more of the same gas, at the same temperature, is introduced into the same container, the pressure increases. (O&C)

12 a) On aerosol cans there is a warning not to leave them in strong sunlight or to throw them on to a fire when empty. Give the physical reasons for this warning.

b) Early in the morning the pressure in a car tyre was 2.00×10^5 Pa when the temperature was 7 °C. What would you expect the pressure in the tyre to be when the temperature has risen to 27 °C in the afternoon? (Assume that the volume of the tyre does not change.) (L)

13 A metal globe will withstand a pressure 7 times that of the atmosphere. It was sealed containing air at atmospheric pressure and at a temperature of 200 K. To what maximum temperature may it be heated

safely? (You should ignore the expansion of the metal but say why this assumption is reasonable.) (SUJB)

14 In figure 2, dry air is trapped in the closed end of a horizontal glass tube by mercury which occupies a 250 mm length of the tube. The other end of the tube is open to the atmosphere, the atmospheric pressure being 750 mmHg. The temperature is 290 K.

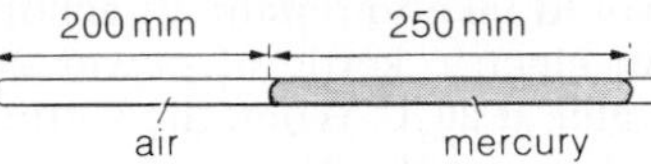

Figure 2

a) The open end is now raised until the tube is vertical. What is now the pressure on the trapped air?
b) What length of the tube does the trapped air now occupy?
c) With the tube still vertical, the temperature is raised to 348 K. What length of the tube does the trapped air now occupy?
d) With the tube returned to the horizontal position at 348 K, what length of the tube does the trapped air now occupy? (O)

15 Dry gas is collected at a pressure of 800 mm of mercury and temperature 300 K. The volume is found to be 280 cm^3. What would this volume have been if the pressure was 700 mm of mercury and the temperature 270 K? (SUJB)

16 Figure 3 shows a fixed mass of dry gas enclosed in a tube. The temperature is initially 290 K.
a) Given that the atmospheric pressure is 760 mmHg, what is the pressure of the gas in the tube?
b) What will the pressure become if the temperature is raised to 320 K, and the volume V of gas is kept constant by raising the right-hand tube?
c) What is now the difference between the mercury levels in the two tubes? (O)

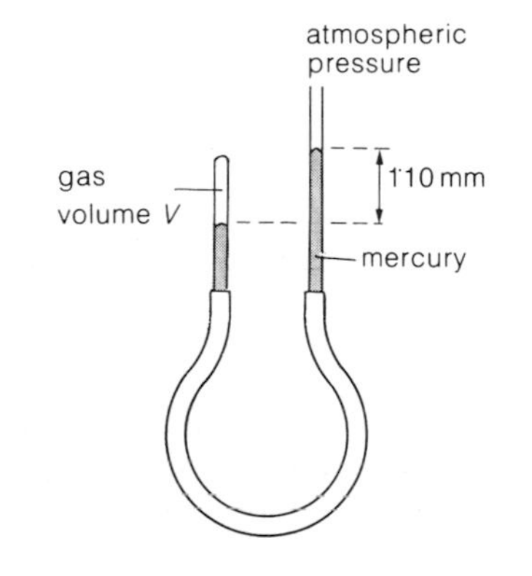

Figure 3

17 Boyle's law states that for a fixed mass of gas at constant temperature the product pressure × volume is constant.
a) Describe how you would test the law experimentally at room temperature.
b) The density of oxygen at 1 atmosphere pressure and 288 K is 1.3 kg/m^3. What is the mass of oxygen contained in a cylinder of volume 0.0004 m^3 at 50 atmospheres pressure at this temperature?
c) If some of the gas is used from the cylinder, explain how a pressure gauge, fitted to the cylinder, may be used to help you to estimate the mass of gas that remains. (Assume that the temperature stays constant.) (O)

18 a) As the surface of a pond freezes it is found that each equal increase in the thickness of the ice takes longer to form, even when the air above the ice remains at the same low temperature. Explain why this is so.
b) In the experiment shown in figure 4 the ice remains intact for several minutes as heating progresses. Explain how this can be so. (L)

water
HEAT
ice, weighted with copper gauze

Figure 4

19 a) Draw a labelled diagram of an apparatus you could use to show that copper conducts heat more readily than iron.
b) Describe what happens and how this shows that copper is the better conductor. (O)

20 a) Define the term specific heat capacity.
b) Describe how you would attempt to measure the specific heat capacity of aluminium and point out two sources of experimental error in your method. What steps could you take to reduce the magnitude of the errors you have mentioned?
c) If a 3 kW immersion heater takes 40 minutes to heat 30 kg of water from 10 °C to 60 °C, how much heat is lost to the tank holding the water and its surroundings? (O&C)

21 A 125 W heater and a thermometer were immersed in 0.6 kg of oil in a vessel of negligible heat capacity. The following observations were noted:

Temperature/K	294	302	313	324	334
Time/minutes	2	4	6	8	10

Plot a suitable graph and use it to find i) the average rise in temperature per minute, ii) the temperature at which heating started. Hence calculate the specific heat capacity of the oil; what precautions should be taken in such an experiment to ensure as accurate a result as possible? (SUJB)

22 An electrical heater marked 200 W is immersed in 1.50 kg of oil [specific heat capacity 2000 J/(kg K)] in an open vessel of negligible thermal capacity. Estimate the time required for a temperature rise of 40 K. State the energy losses which must be neglected in your calculation and say how these could be reduced in practice. (SUJB)

23 The hot water tap of a bath delivers water at 80 °C at a rate of 10 kg/min. The cold water tap of the bath delivers water at 20 °C at a rate of 20 kg/min. Assuming that both taps are left on for 3 minutes, calculate the final temperature of the bath water, ignoring heat losses. (L)

24 A waterfall is 210 m high. The temperature of the water at the top is 10.0 °C and the temperature of the water at the bottom is 10.5 °C. Use this information to obtain a value for the specific heat capacity of water. State any assumption you have made regarding energy in your calculation. (AEB)

25 The graph in figure 5 shows the variation of temperature with time for a pure metal cooling from 300 °C. In what state is the metal in stage a) AB, b) BC, c) CD? If the average rate of heat loss during stage BC is 120 J/min and the mass of metal is 80 g, what is the specific latent heat of fusion of the metal? (L)

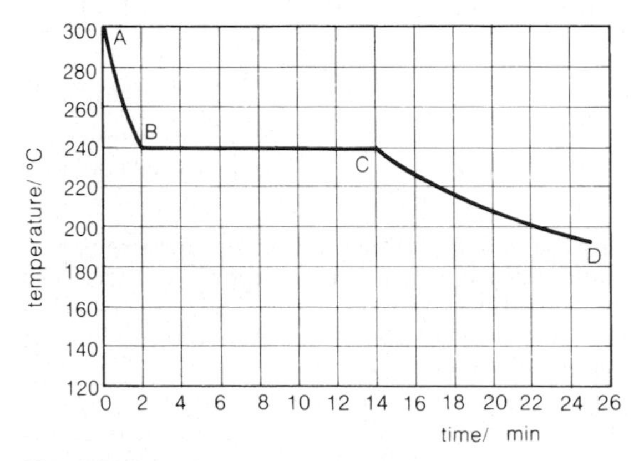

Figure 5

26 A vacuum flask contains water at 15°C, with a thermometer and an electric heater immersed in the water. 0.02 kg of dried ice at 0 °C is added to the water, the heater is connected to a power supply and switched on. The heater gives an output of 24 J/s.
a) How much heat is required to melt the ice?
b) How much heat is required to raise the temperature of this melted ice to 15 °C?
c) For how long must the heater be used before the thermometer begins to rise above 15 °C, assuming there is no heat exchange between the vacuum flask and the surroundings, and the contents of the flask are thoroughly mixed?
d) Why would the value for the time be different if the ice had not been dried before it was added?
e) Why is it not necessary to know either the heat capacity of the vacuum flask or the initial mass of water it contained in order to calculate the time to reach 15 °C? (C)

27 A few small ice cubes, of total mass 10 g, which are at a temperature of −10°C are added to a drink to cool it. When all the ice cubes are melted and the drink and the melted ice thoroughly mixed, the temperature is 6 °C.
Assuming no heat interchange with the surroundings, calculate
a) the heat removed from the drink to raise the temperature of the ice to 0 °C.
b) the heat removed from the drink to melt all the ice at 0 °C.
c) the total heat removed from the drink by the addition of the ice. (C)

28 Calculate the total quantity of heat required to change 0.01 kg of ice at −10 °C completely into steam at 100 °C. (JMB)

29 Describe an experiment you could perform to find the specific latent heat of vaporization of water. State the readings you would take and how you would use them to obtain your result. List any precautions you would take to obtain an accurate result.
An electric kettle of power 2000 W contains 2 kg of water at 20 °C. When the kettle has been switched on, determine the minimum time required for the water to reach its boiling point of 100 °C.
Give two reasons why the actual time would be longer than your calculated value.
Show whether nor not the kettle will boil dry if it is left on for a further 15 minutes after the water has reached its boiling point. (L)

30 Air is pumped through some ether in a copper can or calorimeter resting upon a thin layer of water on a wooden block. Very soon the water freezes. Explain this and give reasons why the water will not freeze unless air is pumped through. (Assume that the air is at the same temperature as the ether.) (L)

31 a) Define (i) a saturated vapour; (ii) saturated vapour pressure.
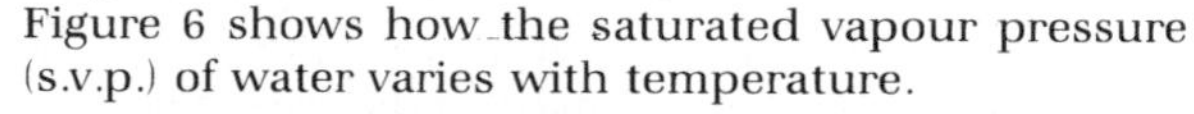
Figure 6 shows how the saturated vapour pressure (s.v.p.) of water varies with temperature.

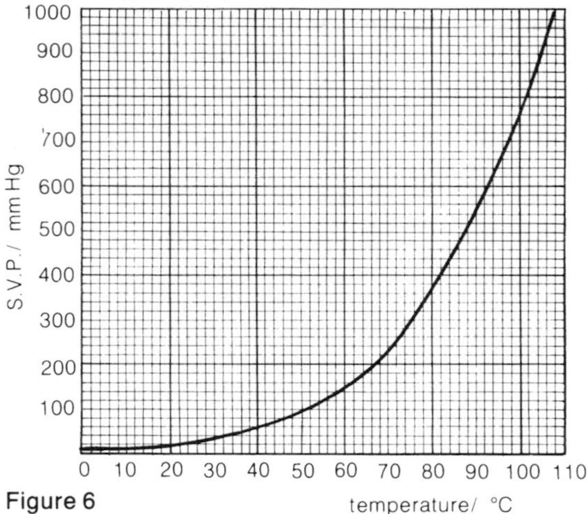

Figure 6

b) Why does water normally boil at about 100 °C?
c) Find from the graph: i) the temperature at which the saturated vapour pressure of water is 500 mmHg; ii) the saturated vapour pressure of water at 50 °C. (O)

32 Some water is heated at a constant rate in a kettle. How, if at all, does the average kinetic energy of the water molecules change a) before the water boils, b) at the boiling point? State also how the boiling point is affected if c) the water is not pure, d) the atmospheric pressure is much lower than normal. (SUJB)

33 State and explain, briefly, how the rate of evaporation of a pool of water is affected by a) the area of surface exposed to the atmosphere and b) the humidity of the atmosphere. (SUJB)

SECTION 5

WAVES: LIGHT AND SOUND

What the eye does not see . . .

Scientists now understand that the image on the retina is just one factor in determining the view your brain gives you of the outside world.

What you see also depends on past knowledge and experience, and anticipation of what you expect to see.

The simplest example is the 'blind spot' – there are no light-sensitive cells on the retina at the place where the optic nerve joins onto it. But usually you cannot detect this. To do so, close your *left* eye, and, holding the book at arms length, focus on the cross below. Then bring the book gradually closer. At a certain place, the spot 'disappears'. What happens as you bring the book closer? Does the same thing happen with your other eye closed?

Now, hold the open book against a well-lit, blank wall. Stare at the dot in the middle of the shape on the right for about two minutes, with your face about 20 cm from the page. Remove the book, stare at the wall, and wait until something appears.

+ ●

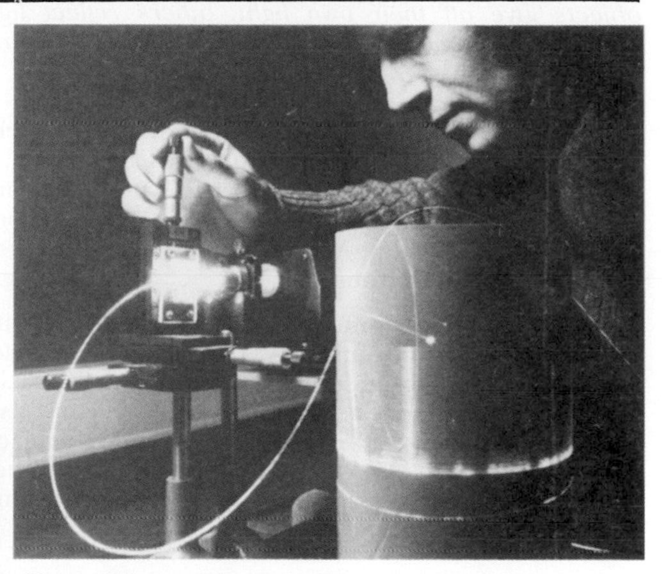

Sending light down tubes

Put light into one end of a glass rod and it comes out of the other, however twisty the rod. The light zig-zags its way from one end to the other as it is totally internally reflected from the sides of the rod.

Many applications have been found for this simple idea, particularly with the development of optical fibres. These are hair-like bendy rods of glass which can be arranged together in bundles. Provided the fibres are packed side-by-side in an orderly arrangement, a light image put into one end of such a bundle emerges at the other end.

As an experiment, optical fibres have been used instead of cables to carry telephone calls. A transmitter converts an electrical 'copy' of speech patterns into pulses of laser light which travel along the fibres. A receiver at the other end converts the light pulses back into electrical signals which can be handled by a telephone exchange in the normal way. Development of the system is still at an early stage, but optical fibres are proving lighter and easier to handle than conventional cables, and they don't need so many signal boosters along the route between telephone exchanges.

Birds defend territory by changing their tune

According to one group of bird experts, many birds defend their territory by changing the length and frequency of their songs. This gives likely intruders the impression that the territory is being defended by a large number of birds, rather than just one. Frequent change of perch also adds to the effect. The theory – called the Beau Geste effect, after the old Hollywood movie in which Beau Geste of the Foreign Legion successfully defends a desert fort by propping up the bodies of his dead comrades against the parapets – has been given strong support by recent results from the Rockefeller University Field Research Centre. Scientists from the Centre replaced resident male birds in a territory with loudspeakers playing single songs, a range of songs, or nothing at all, and then measured the rates at which non-resident male birds trespassed into the territory in each case. Trespass rates were lowest when song patterns were being varied, but only for birds normally resident well away from the territory. Near neighbours apparently weren't so easily fooled.

5.1

Light and shade

You can see objects only if light from them enters your eyes. Some objects give off their own light, others reflect light from other sources. But wherever light comes from, it normally travels in straight lines.

Figure 1 A laser beam in use

Light is given out or **emitted** by very hot objects such as the Sun, or the filament of a bulb, or the hot gases in a flame. It may also be emitted from much cooler materials when electrons lose energy. This happens in a fluorescent tube, or a TV screen, or a laser like the one shown in figure 1. Any object which produces its own light is said to be **self-luminous**.

Most of the objects around you do not produce their own light and are said to be **non-luminous**. They are only visible because they reflect light from some other sources. Some surfaces are better at reflecting light than others; the white surface of this page reflects a high proportion of the light striking it, while the black letters reflect hardly any at all. Light which is not reflected is either absorbed, or, in the case of a transparent material like glass, transmitted, i.e. it passes right through.

The nature of light

What actually enters your eyes when you see something? The list below gives a general outline of the nature of light. Many of the points mentioned are examined in more detail in later units.

Light transfers energy from one place to another Energy is needed to produce light. Materials gain energy when they absorb light. Mostly, this causes an increase in their thermal energy. The solar cells in figure 2 however, change some of the energy in sunlight directly into electrical energy.

Figure 2 Light energy can be changed into other forms of energy

Light is a form of radiation This is a useful label, but it doesn't tell you very much. Radiation is a general term applied to almost anything that travels outwards from its source but can't immediately be identified as solid, liquid or gas like the more familiar forms of matter.

Light is a form of wave motion The way in which light radiates from its source is similar in many ways to the way in which ripples spread outwards across a pond when a pebble is dropped into the water. In the case of light, the 'ripples' are electric and magnetic in nature. They can travel through empty space, and do so at a speed of about 300 000 km/s. This speed is equivalent to travelling seven times round the Earth in less than a second.

Light is something detected by the human eye This may sound obvious, but it is really a matter of definition. Objects emit many types of radiation, most of which are not detectable by the human eye. Light is the name given to radiation which the eye can detect.

Beams and rays

It is often possible to see the path of a beam of light from a projector or a laser because smoke and dust particles in the air reflect some of the light into your eyes. The edges of the beam show clearly that **light travels in straight lines**.

In diagrams, arrowed lines called **rays** are used to show the direction in which light is travelling, as shown in figure 3. In practice, it is often convenient to think of a ray as a very narrow beam of light, and a wider beam as a large number of individual rays. Only a few of these rays are normally shown in a diagram.

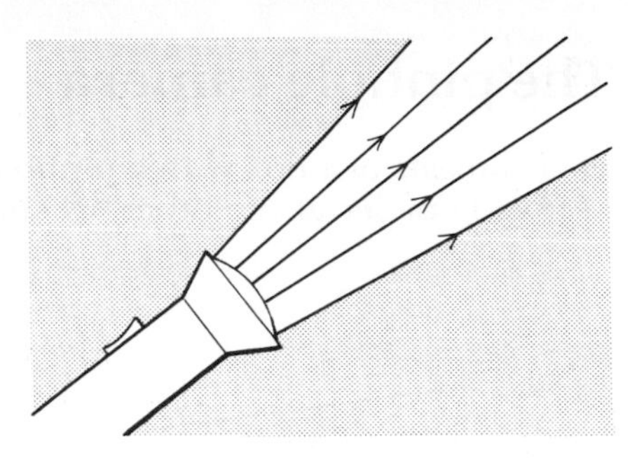
Figure 3 A beam of light is represented by several rays

Shadows

These are formed because light travels in straight lines.

Figure 4a shows the shadow formed when a football is placed between a screen and a point source of light such as a small light bulb. The shadow has a sharp edge; this marks the region on the screen light cannot reach because it has been stopped by the ball.

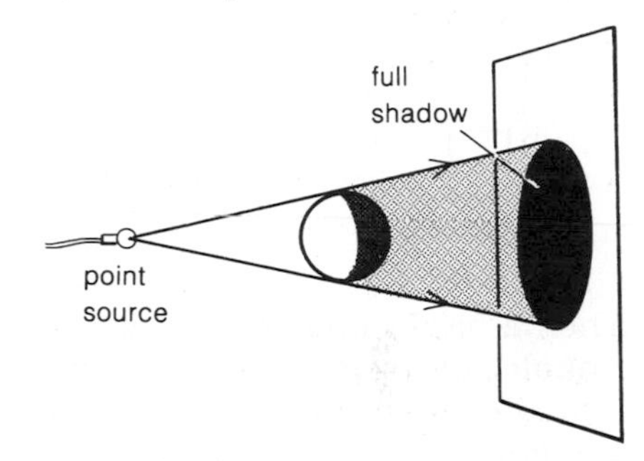

Figure 4a When light comes from a point source, the shadow of an object has a sharp edge

If the point source of light is replaced by a spread-out or **extended** source such as a table lamp, the edge of the shadow becomes fuzzy and indistinct. Around the area of full shadow, there is a region of part shadow where only some of the light from the lamp has been stopped. This is shown in figure 4b; the only rays included in the diagram are those marking the boundaries between regions of full shadow, part shadow, and no shadow.

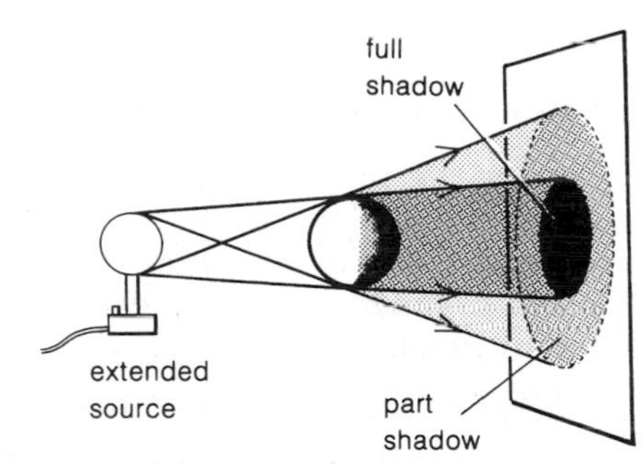

Figure 4b When the source is *extended*, the edge of the shadow is indistinct

Eclipse of the Sun

Like the football, the Moon is non-luminous and is visible only because of the light it reflects.

On the rare occasions when the Moon passes exactly between the Sun and the Earth, the Sun's rays form a shadow of the Moon with the Earth acting as a screen. This is shown in figure 5. In the region of full shadow, the face of the Sun appears to be completely covered or 'eclipsed' by the Moon. In the region of part shadow, only part of the Sun's face appears to be covered.

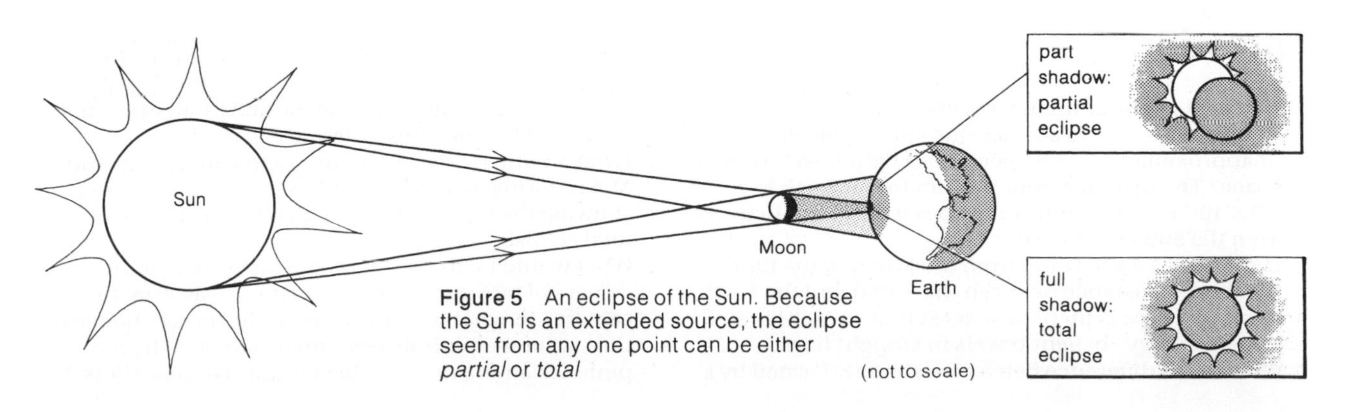

Figure 5 An eclipse of the Sun. Because the Sun is an extended source, the eclipse seen from any one point can be either *partial* or *total*

The pinhole camera

This consists of a box with a pinhole at one end and a screen made of tracing paper at the other, as shown in figure 6.

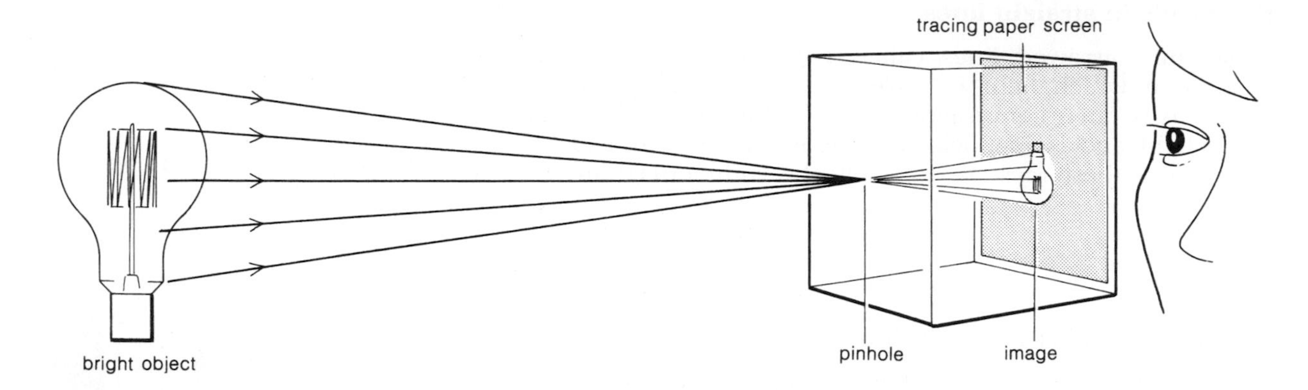

Figure 6 A pinhole camera. Note that the image inside the camera is upside-down

When a bright object such as a light bulb is placed in front of the pinhole, an upside-down **image** appears on the screen. This image is smaller and dimmer than the original object, but otherwise looks the same.

Most of the rays which leave the light bulb do not pass through the pinhole. The diagram shows just five of the rays that do. Each is made up of light of a different intensity and colour depending on the point it was emitted from, and each illuminates one point on the screen. The complete image is built up by these and the other rays, not shown on the diagram, which also pass through the pinhole.

Enlarging the pinhole makes the image brighter because a greater amount of light energy then reaches the screen. Unfortunately, it also makes the image more blurred. A large pinhole acts like a number of small pinholes, each producing an image in a slightly different position.

Photographs can be taken with a pinhole camera if the tracing paper is replaced by photographic film (as shown in figure 7). It is a slow process however. The pinhole has to be uncovered for several minutes to allow sufficient light to fall on the film for the image to be recorded satisfactorily.

Figure 7 Photo taken with a pinhole camera?

Questions

1 Name four self-luminous objects.
2 Why is it possible to see non-luminous objects?
3 At approximately what speed does light travel through space? The distance from the Sun to the Earth is 1.5×10^{11} m. About how long does it take light to travel from the Sun to the Earth?
Light takes 1.3 s to travel from the Moon to the Earth. What is the distance between the Moon and the Earth?
4 What evidence is there to suggest that a) light is a form of energy b) light travels in straight lines?
5 What is the difference between a shadow formed by a point source of light and one formed by an extended source? Why does this difference occur?
6 What are the relative positions of the Sun, Earth and Moon during an eclipse of the Sun? Draw a diagram showing the region of the Earth which experiences a total eclipse.
7 What would be the effect on the image in a pinhole camera of moving the object further away from the camera? In what two ways would the image change if the size of the pinhole were made larger? Why is a pinhole camera not suitable for normal 'snapshots'?

5.2

Reflection of light

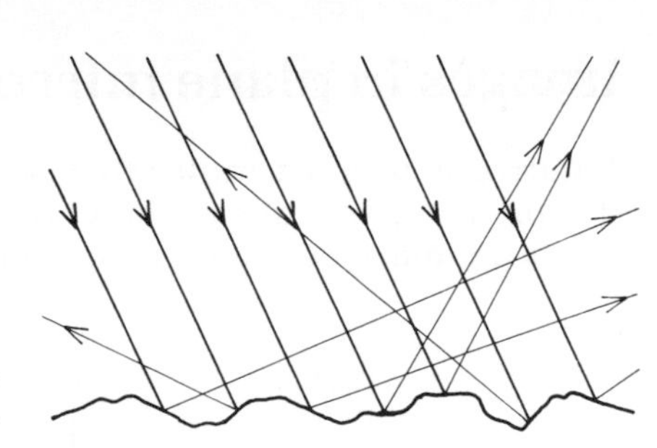

Figure 1a Diffuse reflection

Most surfaces reflect light in all directions. Very smooth surfaces like mirrors reflect light in a regular manner, and the eye can see images in them as a result.

Like most surfaces, this page is nothing like as smooth as it looks. Under a microscope, many pits and bumps would be visible. Light reflected from a rough surface such as this is scattered in all directions and the reflection is described as **diffuse**. On the other hand, when light strikes a very smooth surface such as polished metal or glass, the reflection is **regular**. This is shown in figures 1a and 1b.

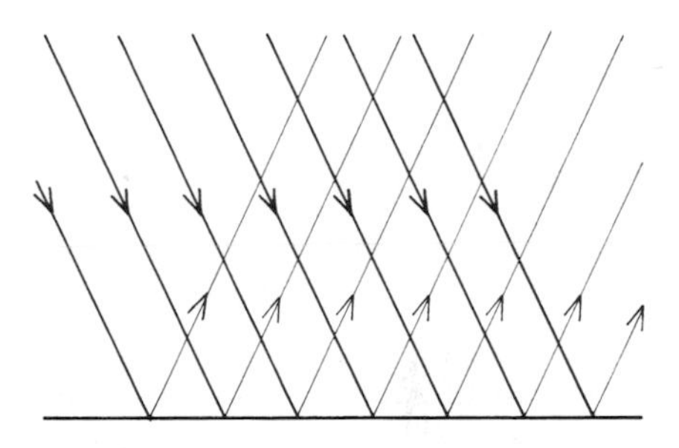

Figure 1b Regular reflection

Mirrors can be made by coating a piece of glass with a thin metallic layer. The surface that results not only gives regular reflection, it also reflects a high proportion of the light striking it, with very little being absorbed.

The laws of reflection

When a ray of light strikes a reflecting surface, it is reflected as shown in figure 2. The line drawn at right angles to the surface is called a **normal** and it is from this line that the angles of the rays striking and leaving the surface are usually measured. These are the **angle of incidence** and the **angle of reflection** and they are illustrated in the diagram. The connections between them are summarized by the **laws of reflection**:

1 The angle of incidence is equal to the angle of reflection; the ray leaves the surface at the same angle as it arrives.

2 The incident ray, the reflected ray and the normal all lie in the same plane; all three could be drawn on the same flat piece of paper.

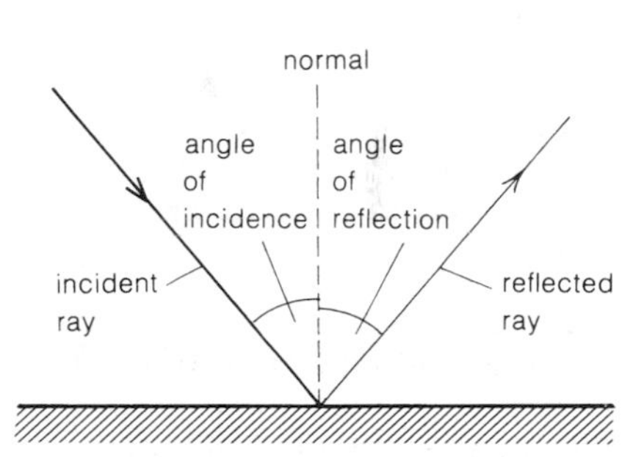

Figure 2 The angle of incidence is equal to the angle of reflection

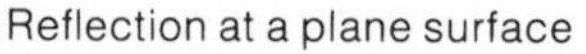

Reflection at a plane surface

Images in plane mirrors

A mirror with a flat rather than a curved surface is called a **plane** mirror. Figure 3a shows how, by reflecting light, a plane mirror forms an image of a point source of light such as a small light bulb.

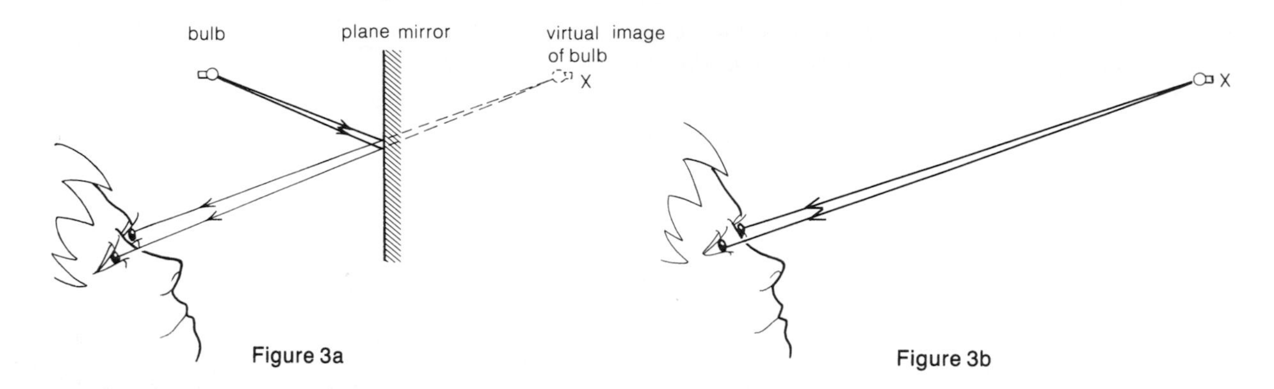

Figure 3a

Figure 3b

Figures 3a, 3b Light rays from the object are reflected by the mirror. The observer sees an image 'behind' the mirror

For simplicity, two rays only have been shown in the diagram. To the girl, both rays seem to come from the point X and this is where she sees an image of the bulb.

The reason the girl sees an image of the bulb at X may be clearer if figure 3a is compared with figure 3b. Here, the mirror has been removed and the light bulb has actually been placed at X. To the girl, little seems to have changed because the rays are entering her eyes exactly as before. Either way, she sees a bulb at X, but in one case it isn't really there.

The image the girl sees in the mirror is known as a **virtual** image; it cannot be formed on a screen, nor do the light rays pass through it. The dashed lines are only construction lines used to fix the position of the image in the diagram.

Finding an image position by experiment

Figure 4 shows a simple method of finding the position of an image formed by a plane mirror. The object in this case is a pin placed upright about 10 cm in front of the mirror. A ruler is placed at A so that its edge lines up with the image of the pin, and a line is drawn on the paper to mark the direction of the image. The ruler is then moved to B and the process repeated. When the two lines drawn are extended, the point at which they meet gives the image position.

The image position can be checked by placing a second pin at this point. The image of the first pin (which you see in the mirror) should line up with the top of the second pin (which you can see above the mirror). They should stay in line even if you move your head from side to side.

In this case, there is **no parallax** between the image and the second 'image' pin. This indicates that they are in the same position.

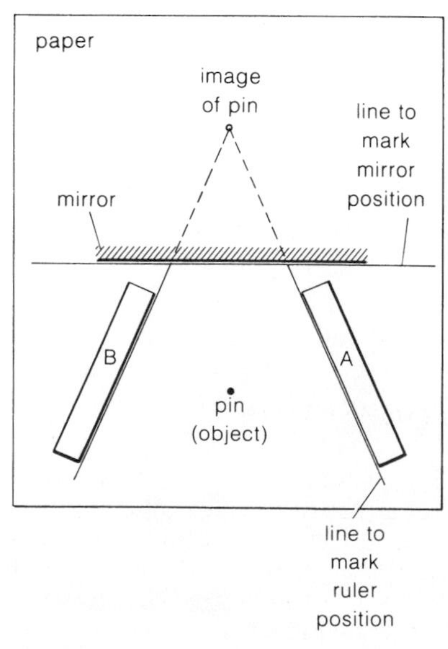

Figure 4

Image of real objects

In reality, objects do not occupy single points. Figure 5 shows how a plane mirror forms an image of an **extended** object. As in any ray diagram, an infinite number of rays could have been drawn, but two rays from any point on the object are sufficient to establish the position of the image of that point:

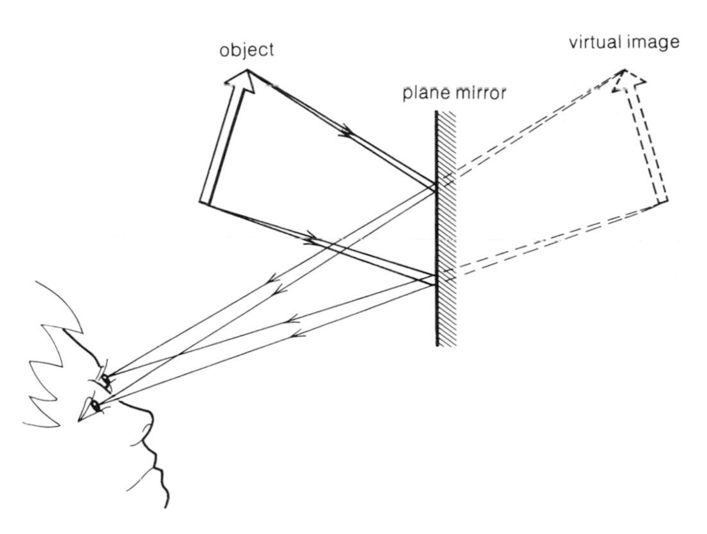

Figure 5 An image of the whole object is formed by both eyes. The images are slightly different. This information is used by the brain to create a three dimensional image

Figure 6 The word 'AMBULANCE' is laterally inverted, so that it reads correctly when seen in a driving mirror

Rules for image position Figure 5 illustrates several features of image formation in plane mirrors. When an object is placed in front of a plane mirror:

1 **The image formed is the same size as the object.**

2 **The image is as far behind the mirror as the object is in front.**

3 **A line joining any point on the object to the equivalent point on the image cuts the mirror at right angles.**

4 **The image is virtual.**

5 **The image is laterally inverted** ('back to front'). This is also illustrated in figure 6.

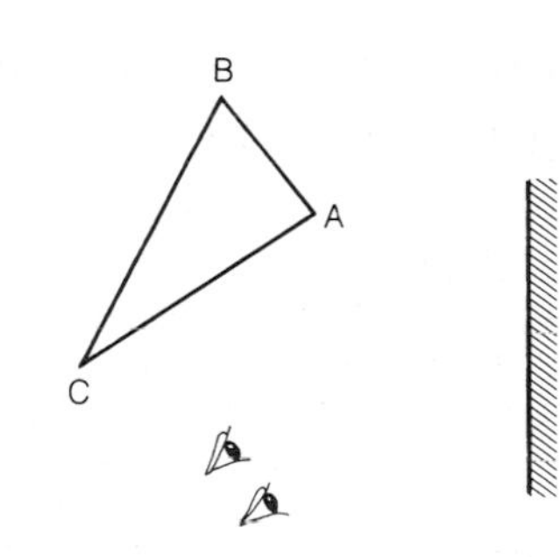

Figure 7 (see question 4)

Questions

1 Give an example of a surface which gives a) diffuse reflection b) regular reflection.

2 A ray of light strikes a mirror at an angle of 30° *to the mirror surface*. What is the angle of reflection?

3 The image seen in a plane mirror can't be formed on a screen. What name is given to an image of this type?

4 Copy and complete figure 7 to show the position of the image. Draw two rays from point A which reflect from the mirror and enter the observer's eyes. Is it possible for the observer to see the reflections of the corners B and C in the mirror? If not, why not?

5 A man stands 10 m in front of a large plane mirror. How far must he walk before he is 5 m away from his image?

5.3

Curved mirrors

When a mirror has a curved surface, those simple rules for image position and size no longer apply. The image may even be out in front of the mirror and upside-down.

Curved mirrors can be **concave** or **convex** in form. The reflecting surface of a concave mirror 'caves' inwards, whereas that of a convex mirror bulges outwards. Small mirrors are usually made with **spherical** reflecting surfaces; they curve like small pieces taken from a large hollow sphere.

Several terms are used in connection with curved mirrors. The **centre of curvature** marks the centre of the sphere of which the mirror forms a part; the **pole**, the **principal axis**, the **aperture**, and the **radius of curvature** of a mirror are all illustrated in figure 1.

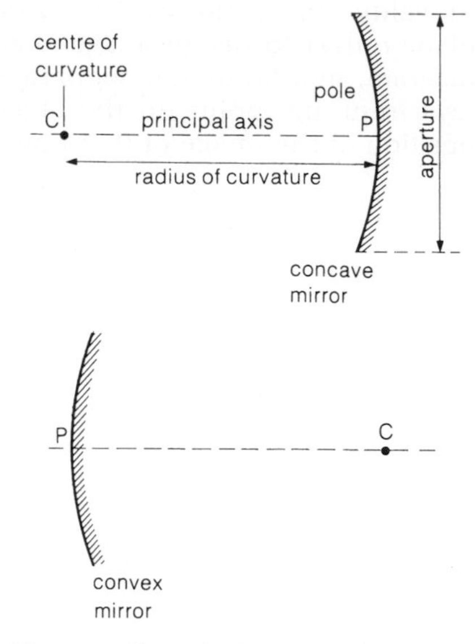

Figure 1 Curved mirrors: terms used

Principal focus

Figures 2a and 2b at the bottom of the page show how concave and convex mirrors reflect rays of light which are travelling parallel to the principal axis.

The rays of light reflected from the concave mirror converge on a single point F; the rays reflected from the convex mirror diverge as if travelling outwards from a single point F. In each case, F is the **principal focus** of the mirror, and the distance from F to P is called the **focal length**. In both cases, the principal focus is exactly midway between the centre of curvature of the mirror and its pole:

focal length $= \frac{1}{2} \times$ **radius of curvature**

In symbols: $f = \frac{1}{2}r$

so the more highly curved the mirror, the shorter is its focal length. This is explained in figure 3. A single ray is shown striking the concave mirror. Like all other rays, it is reflected such that the angles of incidence and reflection are equal. If the angles are small, the distances CF and FP are also equal.

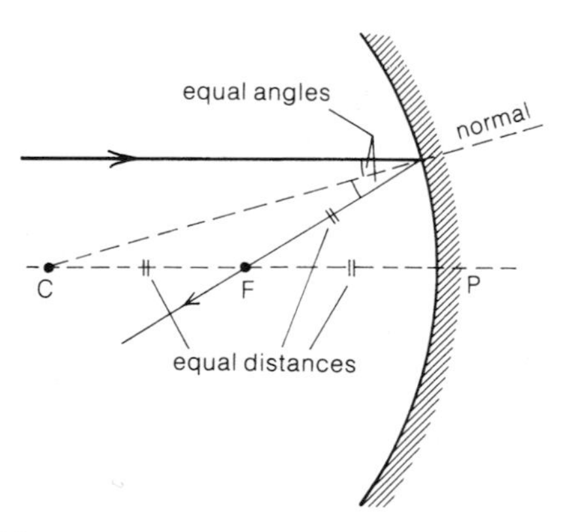

Figure 3 At any reflecting surface, the angles of incidence and reflection are always equal

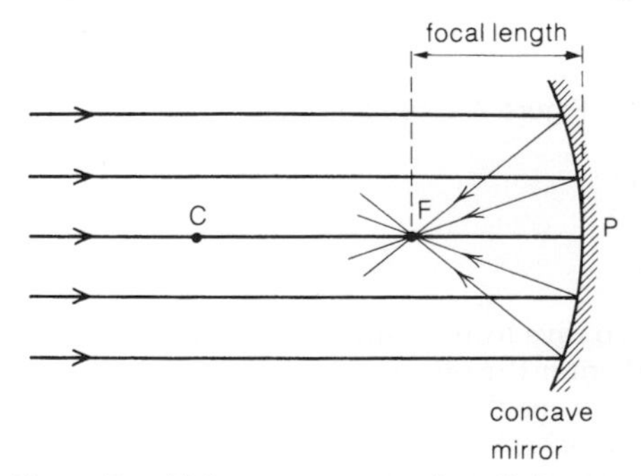

Figure 2a Using a concave mirror light rays may be brought to a *focus* in front of the mirror

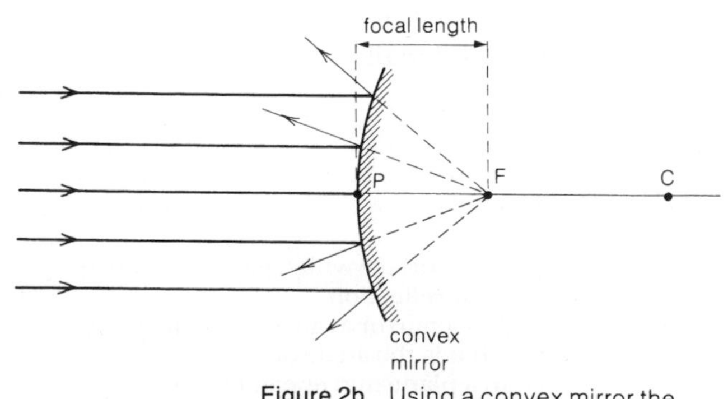

Figure 2b Using a convex mirror the reflected rays appear to come from a point behind the mirror

Strictly speaking, only rays parallel and very close to the principal axis are reflected precisely through F. Further out from the axis, parallel rays strike the mirror at greater angles of incidence, and are reflected through points on the axis closer to the mirror than F.

When the aperture of the mirror is large compared with the radius of curvature, the reflected rays meet along a **caustic curve** rather than at one single point. You may have seen a caustic curve on the surface of tea in a cup as shown in figure 4. Such complications are going to be ignored in the rest of this section. You can assume that all rays shown are **paraxial** (close to the principal axis), and that spherical mirrors are of small aperture only.

Figure 4 Tea-drinkers may have noticed a 'caustic curve'

Images formed by concave mirrors

distant object

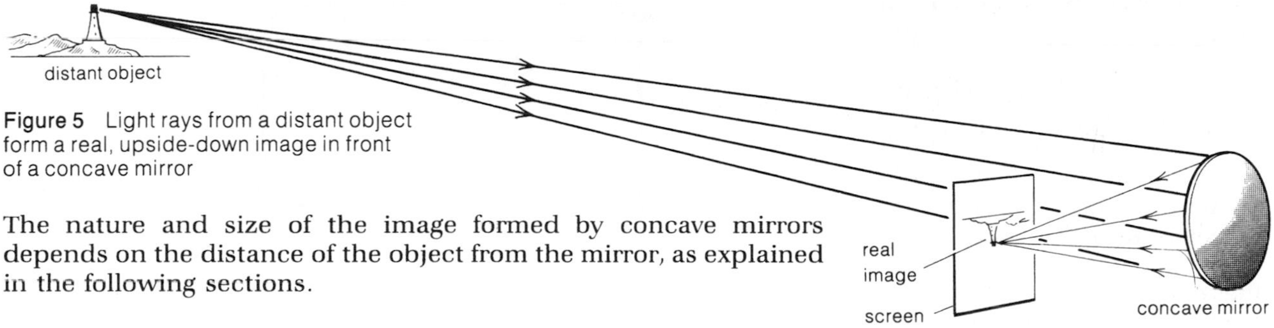

Figure 5 Light rays from a distant object form a real, upside-down image in front of a concave mirror

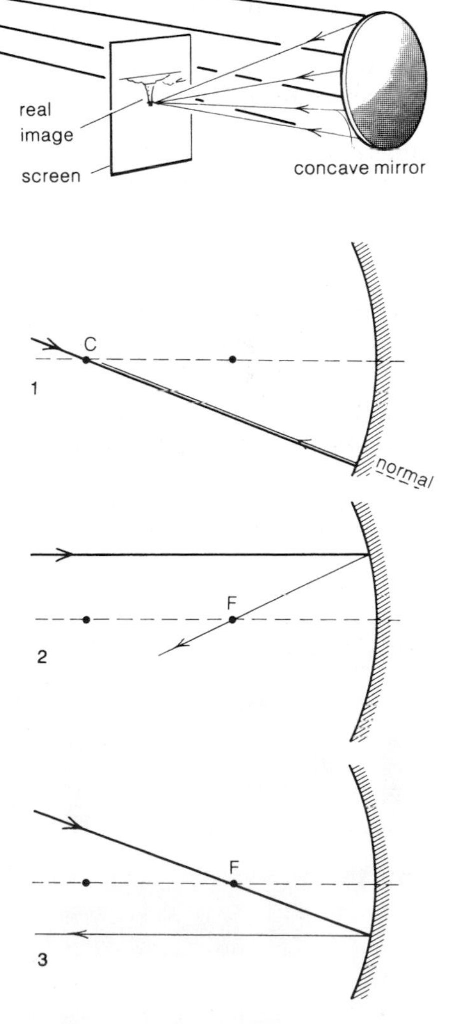

The nature and size of the image formed by concave mirrors depends on the distance of the object from the mirror, as explained in the following sections.

Distant object (beyond C) A concave mirror forms a smaller, inverted (upside-down) image of any object placed beyond its centre of curvature. The image appears to 'float' out in front of the mirror when you look towards it. Reflected rays of light actually meet to form this image, which can be picked up on a screen. The image is known as a **real** image, as distinct from a virtual image which can't be picked up on a screen. Figure 5 illustrates the formation of a real image by a concave mirror. For simplicity, rays have been drawn from one point on the object only.

If an object is very distant, the rays of light arriving at the mirror are nearly parallel, and the real image is formed very close to the principal focus. A quick method of finding the approximate focal length of a concave mirror is to pick up the image of a bright, distant object on a screen and then measure the distance between the screen and the mirror.

As a distant object is brought towards a concave mirror, the image moves further away from the mirror and becomes larger. It is possible to work out the size and position of an image by drawing a ray diagram. Rays are drawn from a point on the object which does not lie on the principal axis. Any number of rays could be drawn but it is simplest to choose from three rays whose paths you can predict without resorting to angle measurements at the surface of the mirror. Figure 6 shows the three rays, which are:

1 The ray of light through C. This is reflected back through C.

2 The ray of light parallel to the principal axis. This is reflected through F.

3 The ray of light through F. This is reflected parallel to the principal axis. (This is equivalent to rule 2 in reverse.)

Figure 6 Three rules for light rays meeting concave mirrors

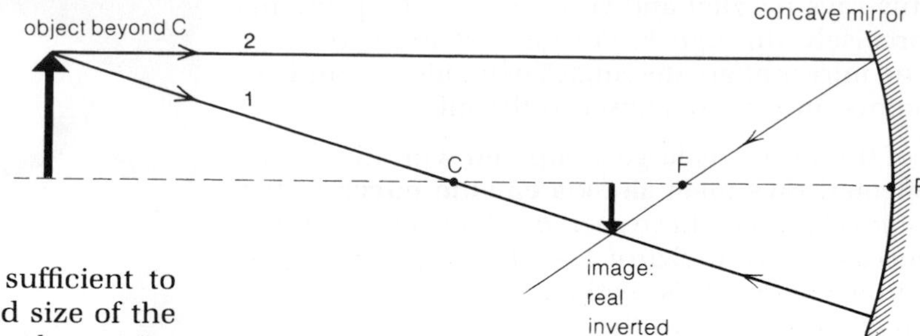

Any two rays are sufficient to fix the position and size of the image. Look for the point where the rays cross after reflection from the mirror, as shown in figure 7:

Figure 7 Object beyond C

Object between C and F If an object is placed at the centre of curvature of a concave mirror, the image is again real and inverted, and it is the same size and distance from the mirror as the object. This is shown in figure 8. Moving the object towards F makes the image move further away from the mirror and become magnified, as shown in figure 9.

Close object (between F and P) A concave mirror forms a magnified, upright and virtual image of any object placed very close to it (as shown in figure 10). Figure 11 shows how this comes about. The rays leaving the object aren't brought to a focus after reflection. They diverge, but appear to come from a point behind the mirror.

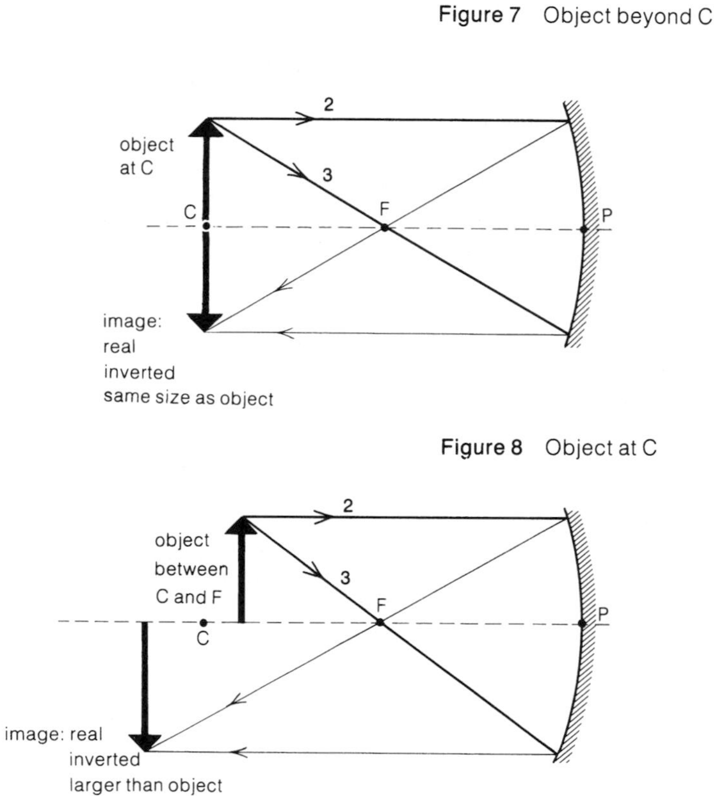

Figure 8 Object at C

Figure 9 Object between C and F

Figure 10 A concave mirror can be used to magnify

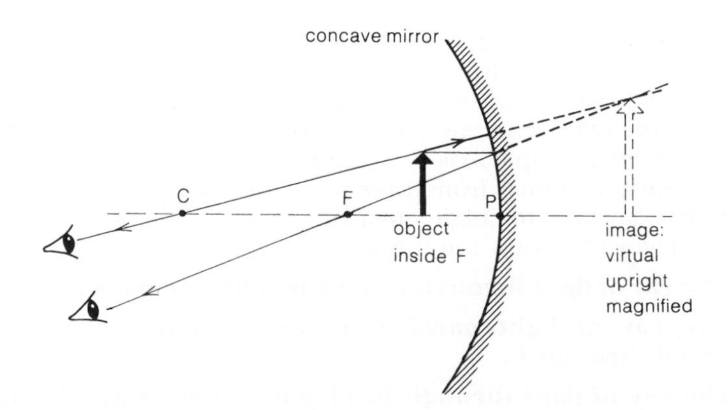

Figure 11 Object between F and P

Accurate ray diagrams

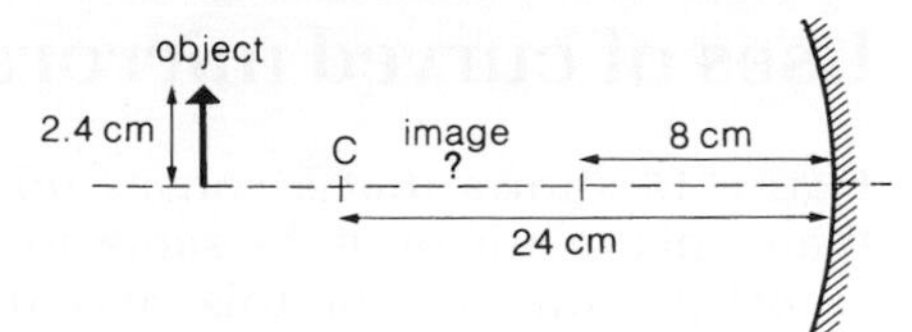

Figure 12

Example *An object* 2.4 cm *high stands on the principal axis at a distance of* 24 cm *from a concave mirror. If the focal length of the mirror is* 8 cm, *what is the position, height, and nature of the image?*

Problems of this type can be solved by drawing an accurate ray diagram as shown in figure 13. For accuracy, it is best to use graph paper, choosing a scale that enables the diagram to fit on the paper conveniently. In the example given, 1 cm on the graph paper represents 2 cm actual distance. Note that $r = 2f = 16$ cm.

In drawing accurate ray diagrams, the following points should be noted:

1 The mirror is represented by a straight line, not a curve. Every ray drawn then counts as a paraxial ray, no matter how far out from the principal axis it lies. In effect, the straight line represents the centre part of the mirror only, but drawn on a very stretched-out vertical scale.

2 The scales used vertically and horizontally needn't be the same. Changing the height of the object arrow on the above diagram doesn't change the position of the image arrow. You can check this by redrawing the diagram with an object of, say, twice the height.

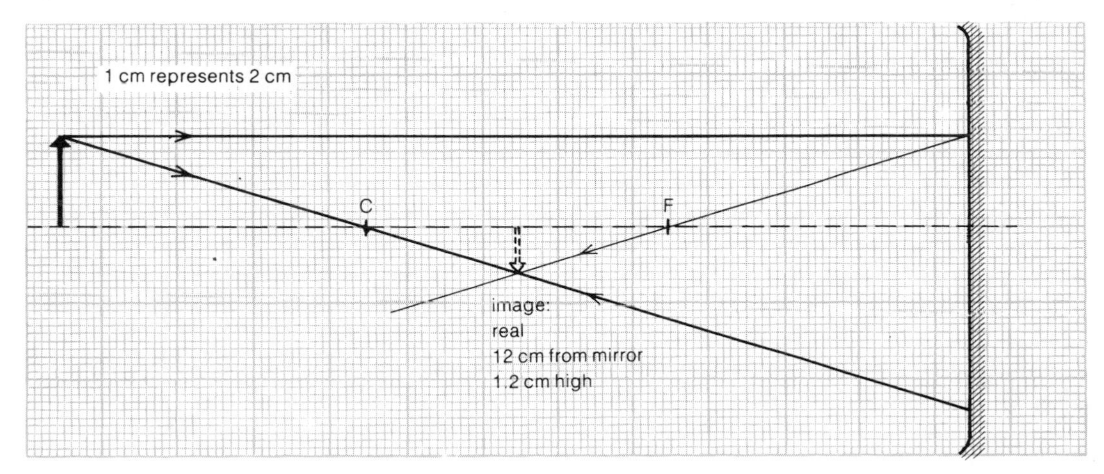

Figure 13 Drawing a ray diagram

Images formed by convex mirrors

A convex mirror forms an upright, virtual image of any object placed in front of it (as shown in figure 14). The image is always smaller than the object and closer to the mirror. Changing the position of the object changes the position and size of the image.

Figure 14 A convex mirror produces an image smaller than the object

Uses of curved mirrors

Figure 15 shows that a convex mirror gives a wider angle of view than a plane mirror of the same size. Convex mirrors are often used as driving mirrors for this reason, though by making the image smaller, they give the driver a false sense of distance.

Shaving and make-up mirrors are often concave in shape because of the magnification they give close up. They are also used in some optical instruments as an alternative to lenses (see page 190).

Figure 16 shows a car headlight. Here, a concave reflector is used to produce a parallel beam of light from a small bulb placed at the principal focus of the reflector. The surface used in the reflector is parabolic in shape because this avoids, at least for parallel rays, 'caustic curve' problems. The satellite tracking 'dish' shown at the bottom of the last page also has a parabolic reflecting surface. The concave dish is used to bring microwave signals from satellites to a focus. Microwaves obey the same laws of reflection as light when they strike a metallic surface (figure 17).

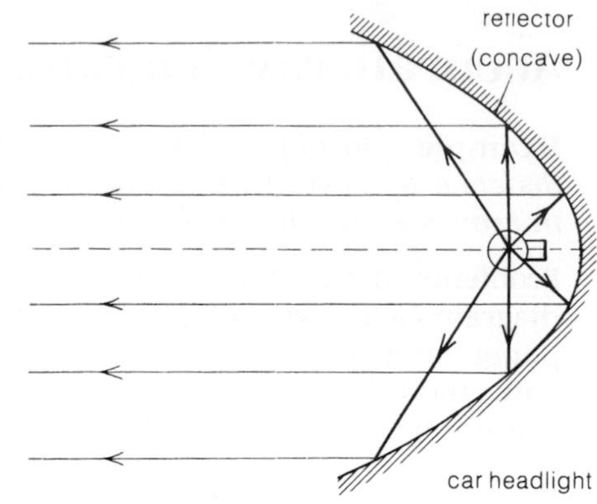

Figure 16 A parallel beam created by a concave mirror

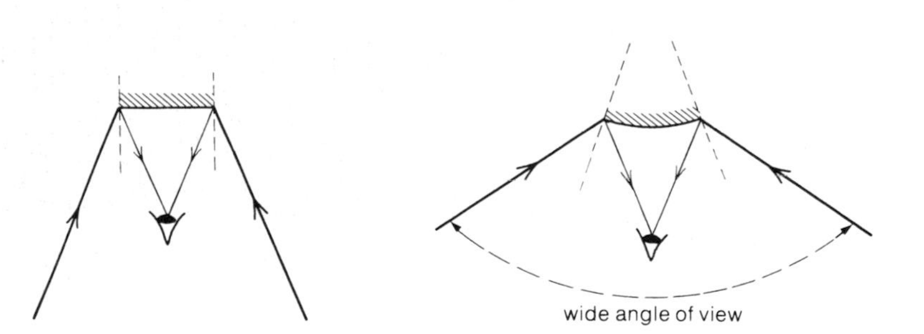

Figure 15 Convex mirrors give a wide angle of view

Figure 17 A parallel beam of microwaves can be brought to a focus by a concave metal surface

Questions

1 Explain the meaning of the following terms with reference to a concave mirror: *radius of curvature*, *principal focus*, *aperture*.
2 Small concave and convex mirrors are often *spherical* in form. What does this mean? What disadvantages do such mirrors have if they are of wide aperture?
3 A small upright object is placed just outside the principal focus of a concave mirror. Draw a ray diagram to show how the image is formed, and describe its size, position and nature. If the object is moved further away from the mirror, what changes are there in the position and size of the image?
4 Give two circumstances in which a concave mirror can form a magnified image of an object placed in front of it. Illustrate your answers with ray diagrams.
5 Briefly describe how you would find the focal length of a concave mirror.
6 An object 3 cm high is placed 30 cm away from a concave mirror of focal length 12 cm. Find by scale drawing the position and height of the image. If the object is moved so that it is only 6 cm away from the mirror, what is the new position and height of the image?

5.4

Refraction

Figure 1 Refraction: terms used

Light rays are bent when they pass at an angle in or out of materials such as glass and water. The effect is called refraction, and it can drastically affect the way you see things.

A swimming pool looks less deep than it really is, and a stick appears to bend when you place one end under water. These effects and many others are caused by the bending of light rays when light passes from one material or **medium** into another.

Figure 1 shows the path of a ray of light as it passes from air into glass. In this case, the ray is bent or **refracted** towards the normal, and the **angle of refraction** of the ray is less than the angle of incidence. Similar results are obtained for light passing from air into other media such as water and paraffin.

Figure 2 A light ray refracted twice by a glass block (and partly reflected as well)

Materials such as glass, water and paraffin are said to be **optically more dense** than air. Optical density is only a descriptive term however and does not necessarily relate to the actual density of a material. Paraffin, for example, is optically more dense than water because it has a greater refracting effect on light, but it is less dense as a liquid.

Light passing into an optically more dense medium is bent towards the normal; light passing into an optically less dense medium is bent away from the normal.

Both cases can be seen in figure 2. A ray of light is passing from air into glass and then back into air again. Note that light passing through a parallel-sided block emerges parallel to the direction in which it arrived.

The laws of refraction

Early experimenters noticed that the angle of refraction (r) increased when the angle of incidence (i) was increased, but they were unable to find a simple proportion between the angles. In 1620, the Dutch scientist Snell discovered that it was the sines of the angles which were in proportion rather than the angles themselves. This is illustrated in figure 3, where three rays are striking a glass block at different angles. Dividing sin i by sin r always produces the same number whatever the angle of the incident ray. Experiments show that there are two laws of refraction:

1 The incident and refracted rays are on opposite sides of the normal at the point of incidence, and all three lie in the same plane.

2 The value of $\frac{\sin i}{\sin r}$ is constant for light passing from one given medium into another. This is known as **Snell's law**.

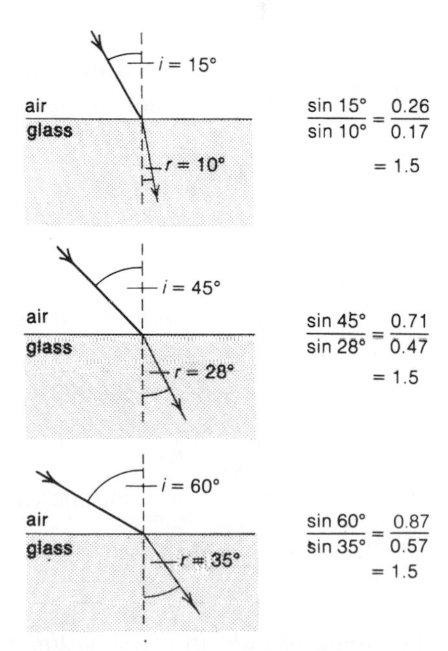

Figure 3 sin *i* divided by sin *r* stays constant

Refractive index

Figure 4 shows the values of $\frac{\sin i}{\sin r}$ obtained when light rays pass from a vacuum into different media. The value of $\frac{\sin i}{\sin r}$ in each case is called the **refractive index** of the medium and it gives you an indication of its light-bending ability. For example, light is refracted more by glass than by paraffin, and more by paraffin than by water.

The refractive index of a medium = $\frac{\sin i}{\sin r}$ **for light passing from a vacuum into the medium.**

In practice, it isn't necessary to make measurements of refractive index in a vacuum, except in very accurate work. Results obtained for rays passing from air into the different media are not significantly different from those given in the table.

Refractive Index		
water	1·33	(4/3 approx)
paraffin	1·44	
Perspex	1·49	
glass (soft crown)	1·52	(3/2 approx)
diamond	2·42	

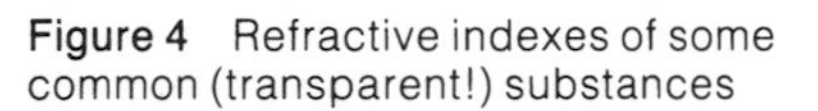

Figure 4 Refractive indexes of some common (transparent!) substances

Refractive index and the speed of light

If a fast moving car is driven at an angle over a kerb, the speed of the car is reduced and its direction of travel is changed. The change in direction occurs because one wheel strikes the kerb and is slowed before the other. Light entering a glass block also changes its direction because its speed is reduced, though for different reasons from those in the car example. The causes of refraction are examined on page 198, but you may find figure 5 useful as a means of remembering that a reduction in speed produces refraction towards the normal.

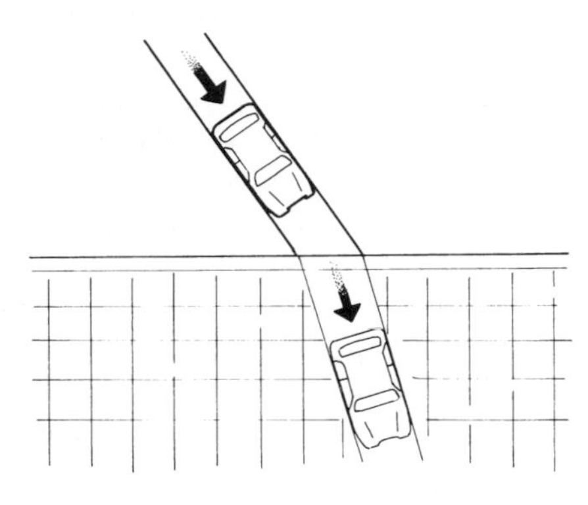

Figure 5 The speed of one side of the car is reduced *before* the other; therefore the direction changes

In a vacuum, light travels at about 300 000 km/s. In glass, its speed drops to 200 000 km/s (as shown in figure 6). The ratio of these two speeds gives the refractive index of glass, which is 3/2 approximately.

$$\textbf{refractive index} = \frac{\textbf{speed of light in vacuum}}{\textbf{speed of light in medium}}$$

In symbols, $\boldsymbol{n = \frac{c}{c'}}$

Note that the greater the refractive index of a medium, the lower is the speed of light. The more light is slowed, the more it is bent.

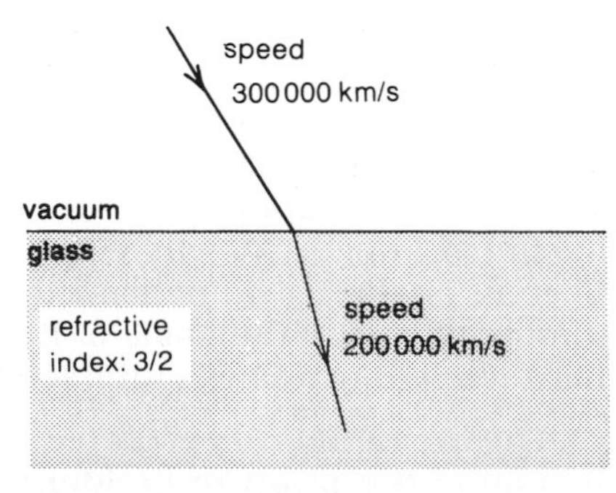

Figure 6 The speed of light reduces in the more dense medium

Example *If the refractive index of water is 4/3, what is the speed of light in water?*

In this case, $n = 4/3$, $c = 300\,000$ km/s, and c' is to be found.

$$n = \frac{c}{c'}$$

$$\frac{4}{3} = \frac{300\,000 \text{ km/s}}{c'}$$

which gives: $$c' = \frac{300\,000 \text{ km/s} \times 3}{4} = 225\,000 \text{ km/s}$$

The speed of light in water is therefore 225 000 km/s.

Real and apparent depth

The bending of light can give you a false impression of depth. Figure 7 shows two rays of light leaving a point on the bottom of a swimming pool. The rays are refracted as they leave the water. To the observer, the rays seem to come from a higher position, and the bottom looks closer to the surface than it really is.

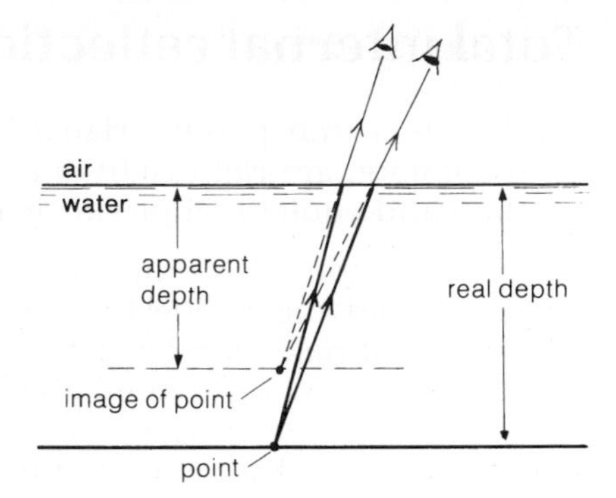

Figure 7 Refraction makes water look shallower than it really is

The real depth of the water and its apparent depth are marked on the diagram. These are related to the refractive index of the water by the following equation:

$$\textbf{refractive index} = \frac{\textbf{real depth}}{\textbf{apparent depth}}$$

provided rays entering the observer's eyes are close to the normal.

Figure 8 shows how the apparent depth of a glass block differs from its real depth. To find the apparent depth of a glass block, you can position a pin against its furthest face and then locate the image position by the method described on page 170.

Figure 8 Real and apparent depth in a glass block

Example *A glass block appears to be* 6 cm *thick when viewed from above. If the refractive index of the glass is* 3/2, *what is the actual thickness of the block?*

Applying the equation given above: $\frac{3}{2} = \frac{\text{real depth}}{6\,\text{cm}}$

∴ *the thickness of the block (real depth) is* 9 cm.

Total internal reflection and the critical angle

Figure 9 shows three of the rays leaving a small lamp placed underwater.

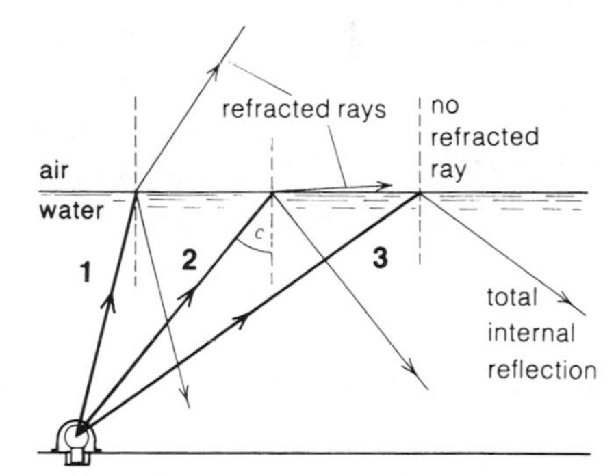

Figure 9 Total internal reflection

Ray 1 splits into two parts at the surface of the water. One ray passes into the air and is refracted; the other weaker ray is reflected.

Ray 2 also splits into two parts, but the ray passing into the air is refracted so much that it is only just able to leave the water. The angle marked c on the diagram is called the **critical angle**. For angles of incidence greater than c, no refraction is possible.

Ray 3 strikes the surface at an angle of incidence greater than c. There is no refracted ray; the surface of the water acts like a perfect mirror, and the ray is said to have been **totally internally reflected**.

It is possible to calculate the critical angle using the information given in figure 10. Here, ray 2 has been reversed so that the angle of incidence is 90° and c now becomes the angle of refraction:

$$\text{refractive index } (n) = \frac{\sin i}{\sin r} = \frac{\sin 90°}{\sin c} = \frac{1}{\sin c}, \quad \text{giving} \quad \sin c = \frac{1}{n}$$

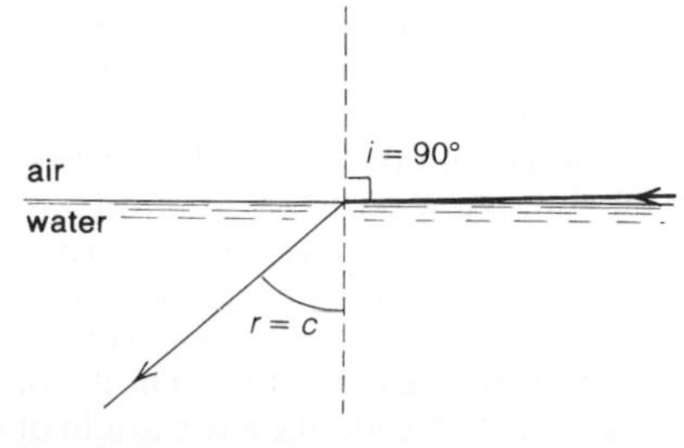

Figure 10

Water has a refractive index of about 4/3, and the critical angle for water works out at about 49°. Glass has a refractive index of about 3/2, and the critical angle in this case is about 42°.

Total internal reflection in prisms

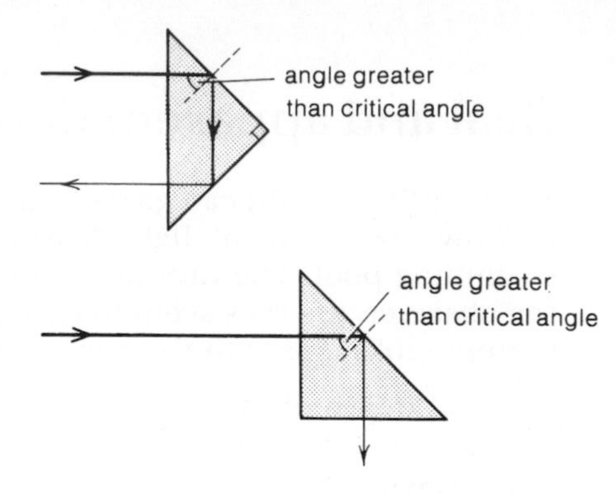

A prism is a transparent triangular block, usually made of glass. Most common are right right-angled prisms (corner angles 45°, 45° and 90°) and equilateral prisms (each corner angle 60°).

Figures 11a and 11b show rays of light being totally internally reflected from the inside faces of two right-angled glass prisms. Total internal reflection occurs because each ray striking an inside face does so at an angle of incidence of 45°, and this is greater than the critical angle of glass. Prisms are used as alternatives to mirrors in several types of optical instrument (as shown in figure 12).

Figures 11a, 11b The path of a light ray can be altered by internal reflection inside a prism

Optical fibres

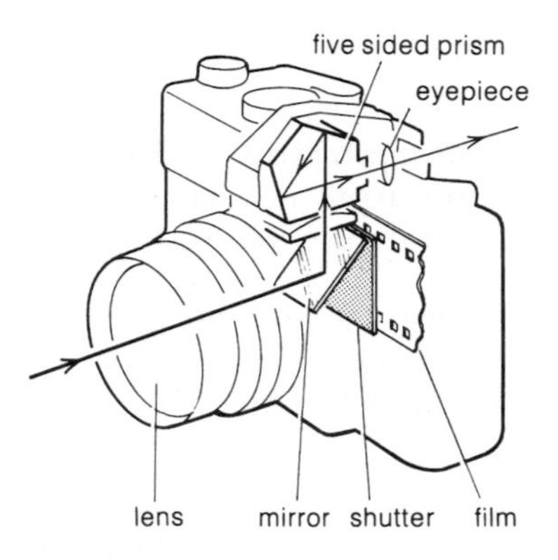

Figure 12 A five-sided prism is used in the view-finder system of many cameras

Optical fibres are very thin, flexible glass rods, used for carrying light. When a ray enters a fibre at one end, it is totally internally reflected off the sides until it comes out of the other end, as shown in figure 13.

A bundle of optical fibres is called a light pipe. Surgeons use them to see inside the body as shown in figure 14. British Telecom use them to carry telephone messages – a single pipe can carry more than a thousand at once. The messages are coded, and sent along the fibres as 'pulses' of laser light. Even if the pipe is several kilometres long, the light comes out almost as bright as it goes in. So less 'booster' stations are needed than with ordinary telephone cables.

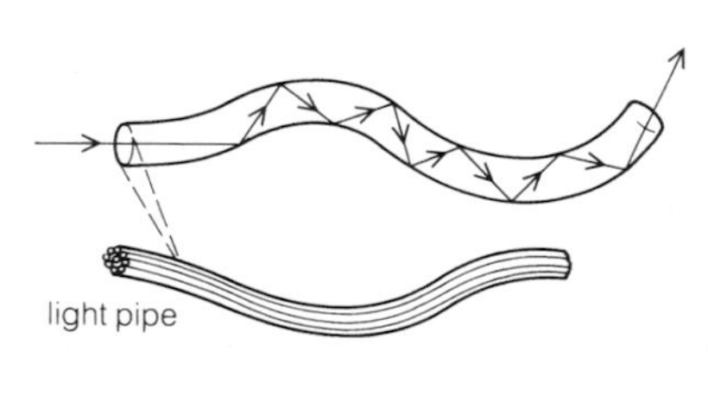

Figure 13 Optical fibres

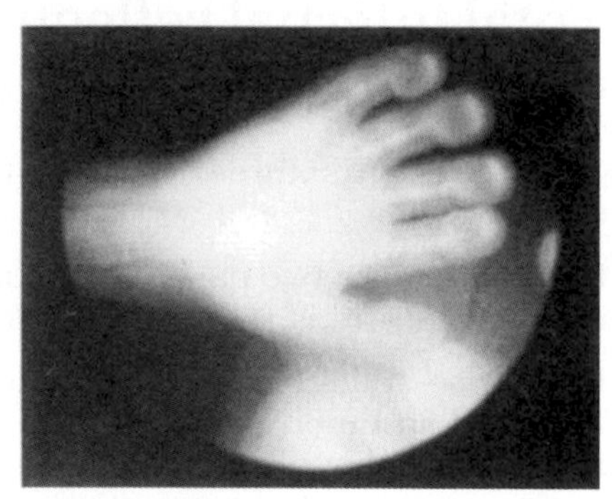

Figure 14 Light pipe photo of a baby inside the womb

Questions

In the following questions, assume that where refraction occurs, light is passing either out of or into air. Take the refractive index of glass as 3/2 and the refractive index of water as 4/3.

1 Draw a diagram to show how a ray of light is refracted when it passes into a) an optically denser medium b) an optically less dense medium.

2 Light enters water at an angle of incidence of a) 24° b) 53°. Calculate the angle of refraction in each case.

3 Light entering glass has an angle of refraction of a) 24° b) 37°. Calculate the angle of incidence in each case.

4 Draw a diagram to show what happens to a ray of light travelling in water when it strikes the surface of the water at the critical angle.

5 A transparent material has a refractive index of 2. Calculate the value of the critical angle. If the refractive index were less than 2, would the critical angle be greater or less than before?

6 A liquid is 25 cm deep. What is its apparent depth when viewed from above if the refractive index of the liquid is 1.25? What is the critical angle for the liquid?

7 Give two practical applications of total internal reflection.

5.5

Lenses

Lenses produce images similar to those formed by curved mirrors, but they do so by refracting light rather than reflecting it.

Like curved mirrors, lenses are either convex or concave in form. Convex lenses are thickest through the middle, concave lenses are thickest around the edge, but several variations on these basic shapes are possible, as shown in figure 1. Most lenses are made of glass and have spherical surfaces (as explained on page 172).

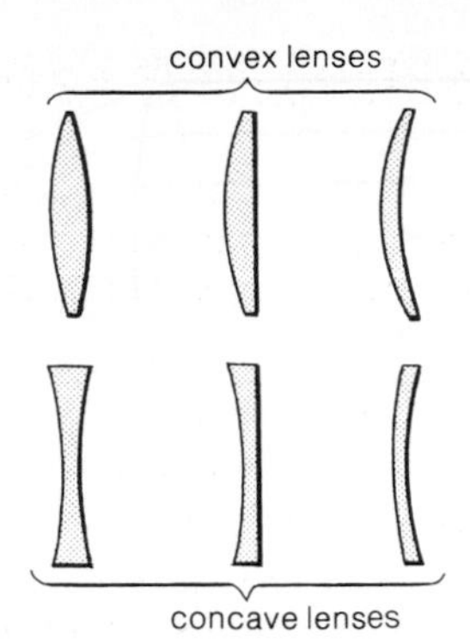

Figure 1 Convex lenses are thicker in the middle than the outside. Concave lenses are thinner in the middle

Light rays passing through a convex or **converging** lens are bent towards the principal axis, whereas rays passing through a concave or **diverging** lens are bent away from the principal axis. It is possible to see why this bending takes place if you imagine a lens to be a series of glass blocks as in figure 2. Each ray is refracted towards the normal as it passes into the glass and away from the normal as it leaves it. Light passing through the central block emerges in the same direction as it arrives because the faces of this block are parallel. P marks the **optical centre** of the lens.

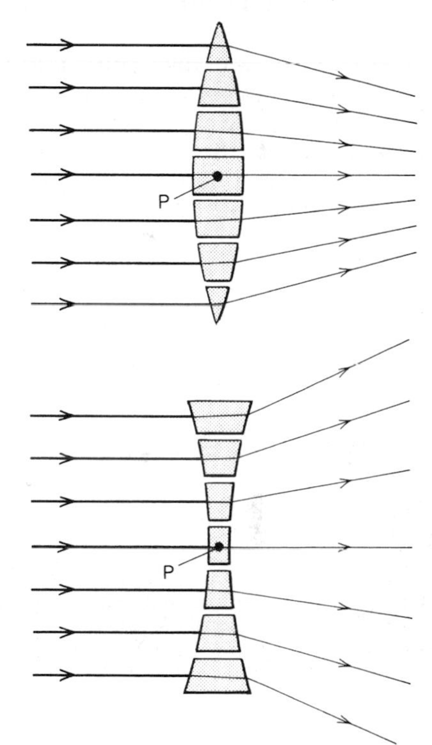

Figure 2 Each part of the lens acts like a prism

Principal focus

Figures 3a and 3b show how rays travelling parallel to the principal axis are refracted by a convex and a concave lens. The rays passing through the convex lens converge to a point F; the rays passing through the concave lens diverge as if travelling outwards from a point F. In each case, F is the **principal focus** of the lens and the distance from F to P is called the **focal length**.

Rays of light can pass through a lens in either direction, so every lens has two principal foci, one on each side of the optical centre. Whatever the shape of the lens, the distances FP and F′P are equal.

The focal length of a lens depends on the curvature of each surface, but the connection isn't as simple as it is in the case of a curved mirror. In general, the more highly curved the surfaces, the shorter is the focal length. The thick lens in figure 4 has a shorter focal length than the thin lens.

Like the mirrors described on page 172 spherical lenses bring parallel rays of light to an exact point focus only if the rays are paraxial (close to the principal axis). For this to happen in practice, a lens must have a large focal length compared with its aperture – like the thin lens shown in the diagram.

An additional problem with lenses is that they refract light of different colours by slightly different amounts. The principal focus for, say, red light isn't in quite the same position as the principal focus for blue light. These various complications are going to be ignored in the rest of this section. You can assume that all lenses shown are thin, all rays paraxial, and all colours refracted equally.

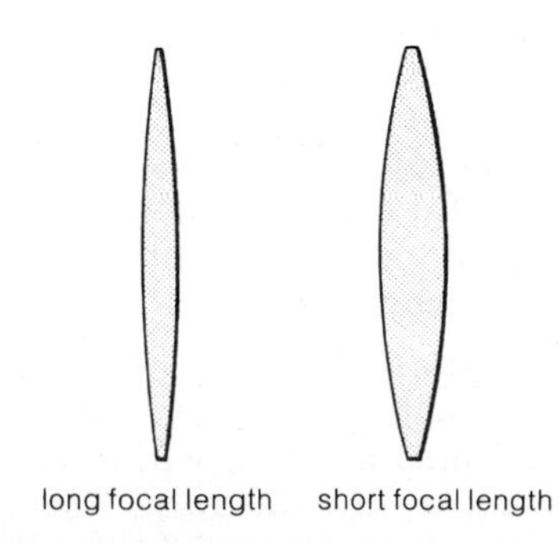

Figure 4 Thick lenses have a shorter focal length than thin ones

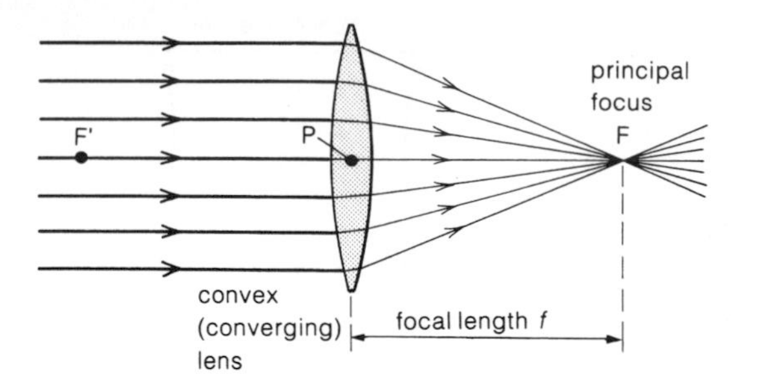

Figure 3a Parallel rays are brought to a focus on the far side of a convex lens

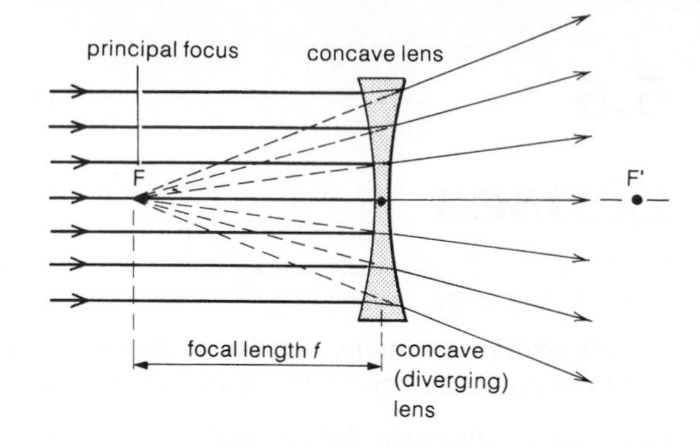

Figure 3b Parallel rays leave a concave lens as if they had come from the focal point on the near side of it

Images formed by convex lenses

Convex lenses are very similar to concave mirrors in their image-forming properties.

Distant objects (beyond 2F′) Figure 5 illustrates the formation of a real image by a convex lens. For simplicity, rays have been drawn from one point on the object only:

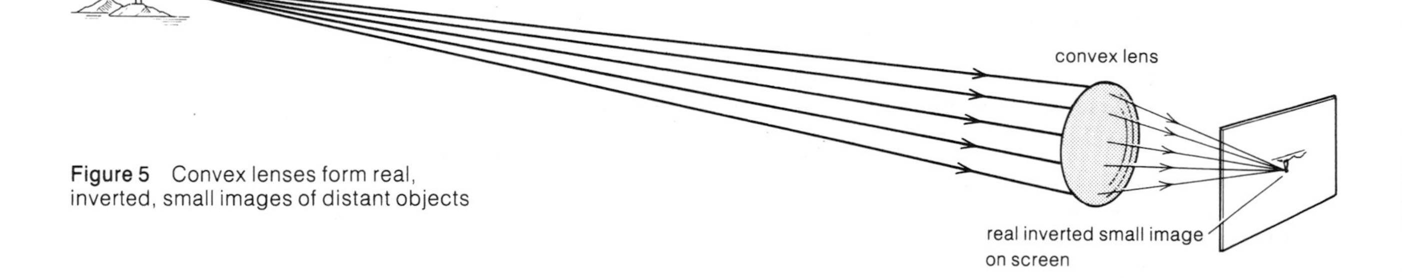

Figure 5 Convex lenses form real, inverted, small images of distant objects

A convex lens forms a small, inverted image of any object placed more than $2f$ along its principal axis. The image is real and can therefore be picked up on a screen.

If an object is very distant (as in figure 5), rays of light arriving at the lens are nearly parallel, so the image formed is very close to the principal focus. A quick method of finding the approximate focal length of a convex lens is to pick up the image of a bright, distant object on a screen and then measure the distance from the screen to the lens.

As a distant object is brought towards a convex lens, the image moves further away from the lens and becomes larger. As with a curved mirror, the position and size of an image can be found by drawing a ray diagram. *For simplicity, rays are usually shown bending along the upright line through the middle of the lens, though in reality bending takes place at each surface.*

A convex lens forms a real image of the Sun on a piece of paper . . .

Any two of the following three rays are sufficient to fix the position and size of the image:

1 A ray of light through the optical centre of the lens, P. This passes through the lens unbent.

2 A ray of light parallel to the principal axis. This passes through F when it leaves the lens.

3 A ray of light through F′. This leaves the lens parallel to the principal axis. It is equivalent to ray number 2 in reverse.

These rays are illustrated in figure 6. A ray diagram including all three rays is shown in figure 7.

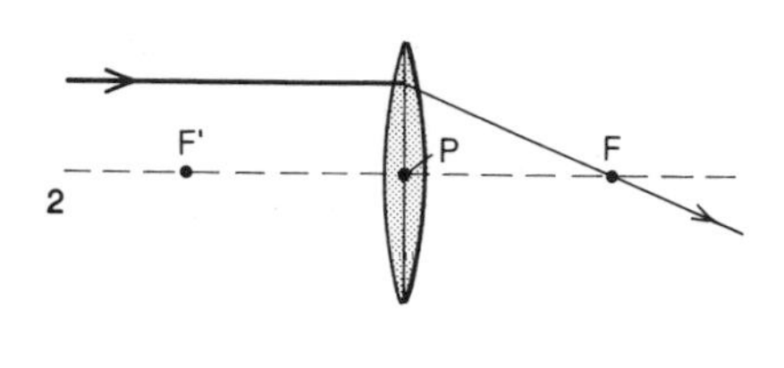

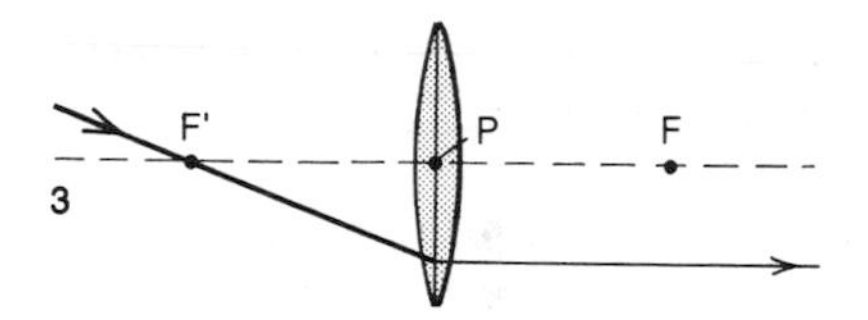

Figure 6 Three rules for light-rays passing through convex lenses

Object at 2F′ If the object is placed exactly *2f* from the lens, the image distance is also *2f*, and the image is the same size as the object, as shown in figure 8. When an object and its real image are *4f* apart, they are at the minimum possible distance from each other.

Object between 2F′ and F′ The image is still real and inverted, but figure 9 shows that it is now larger than the object and further away from the lens. Moving the object towards F′ increases the image size and distance still further. If the object is bright and very close to F′, a large image can be picked up on a projector screen placed several metres away.

If the object is at F′, the rays leave the lens parallel to one another, and do not meet.

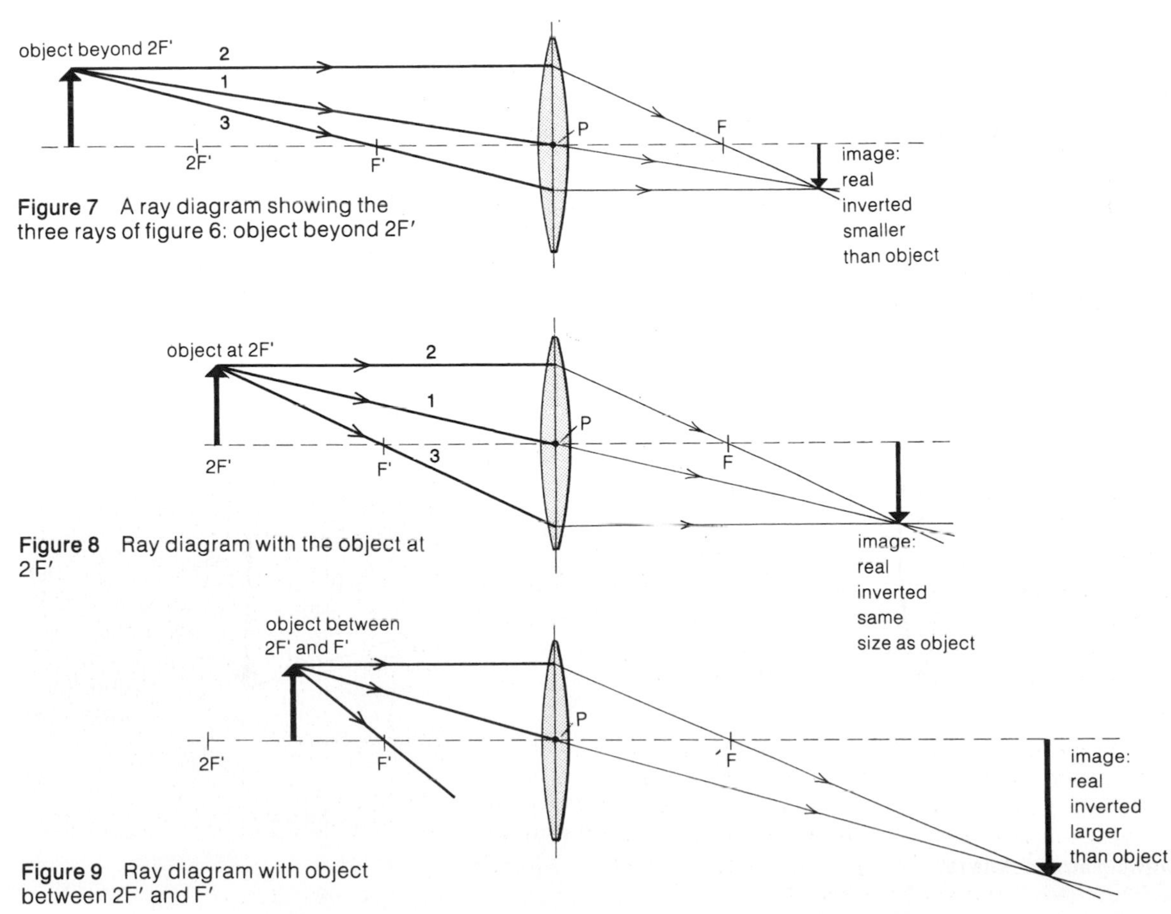

Figure 7 A ray diagram showing the three rays of figure 6: object beyond 2F′

Figure 8 Ray diagram with the object at 2F′

Figure 9 Ray diagram with object between 2F′ and F′

Close object (between F′ and P) A convex lens forms a magnified upright and virtual image of any object placed very close to it. Figure 10 shows how this comes about. The rays leaving the object aren't brought to a focus; they diverge, but appear to come from a point behind the lens. Used in this way, a convex lens is commonly known as a magnifying glass.

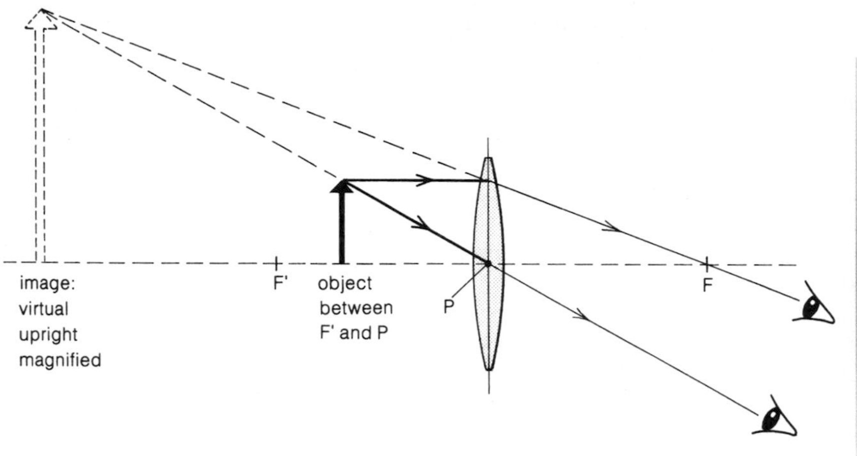

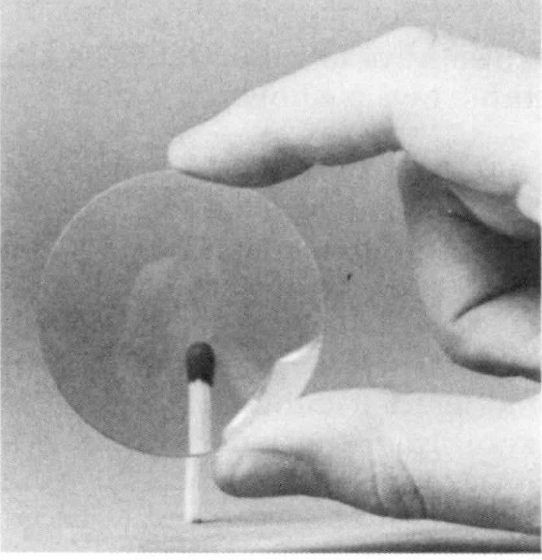

Accurate ray diagrams

Example *An object* 2 cm *high stands on the principal axis at a distance of* 9 cm *from a convex lens. If the focal length of the lens is* 6 cm, *what is the position, height, and nature of the image?*

Problems of this type can be solved by drawing an accurate ray diagram, as shown in figure 11. The comments on scale drawing on page 175 apply as much to lenses as they do to mirrors.

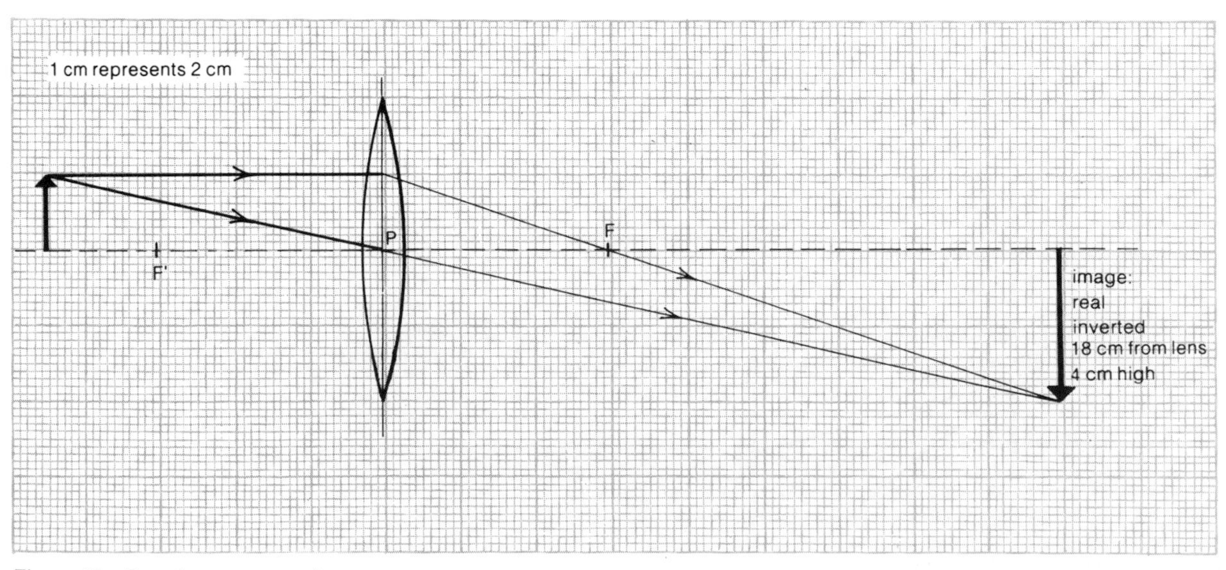

Figure 11 Drawing an accurate ray diagram

Measuring the focal length of a convex lens

The experimental arrangement is shown in figure 12. A plane mirror is placed behind the lens and a bright object in front. The object in this case is a set of illuminated cross-wires, and its position is adjusted until its image appears alongside on the card. In this position, rays of light from the object are parallel when they strike the mirror and are reflected back to the lens as a parallel beam. The cross-wires and card mark the position of the principal focus of the lens.

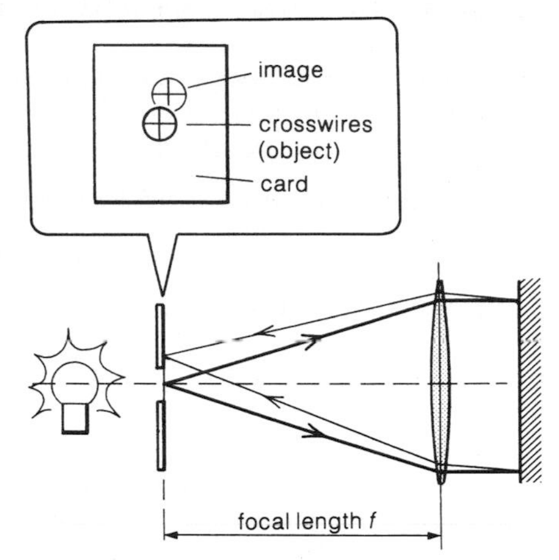

Figure 12 When the object is at the focal point of the lens, the image is clearly superimposed on the object

Images formed by concave lenses

Concave lenses are very similar to convex mirrors in their image-forming properties.

A concave lens forms an upright, virtual image of any object placed in front of it. The image is always smaller than the object and closer to the lens. In figure 13, two standard rays have again been used to show how the image is produced. Changing the position of the object changes the position and size of the image, but the basic form of the diagram is unchanged.

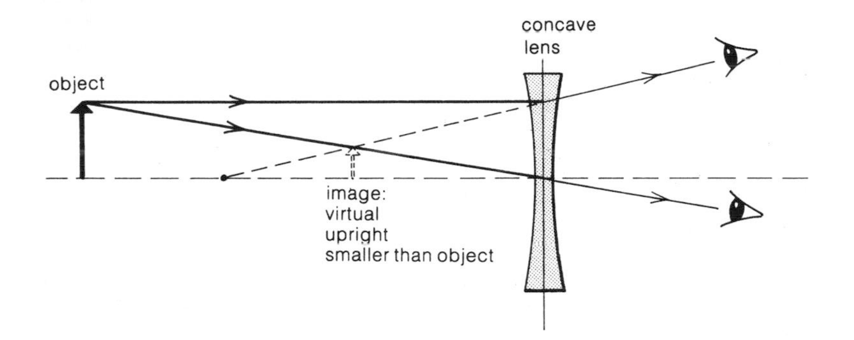

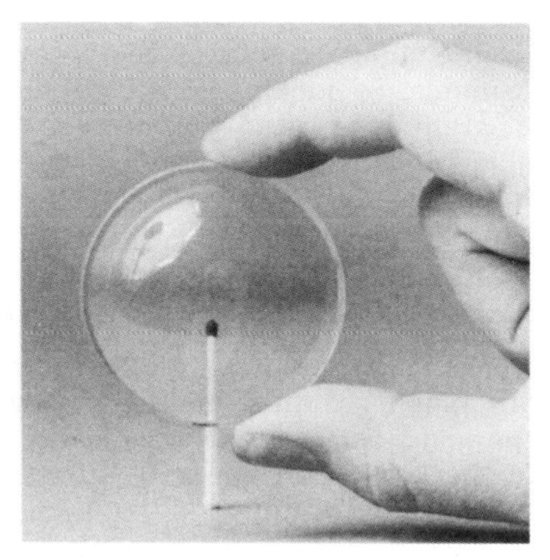

Figure 13 The image formed by concave lenses is always virtual, upright, and smaller than the object

Calculating image positions

The following equation may be used to find the position and nature of the image:

$$\frac{1}{u}+\frac{1}{v}=\frac{1}{f}$$

Where u is the distance from the object to the lens, v is the distance from the image to the lens, and f is the focal length. u, v, and f can be measured using any length unit, provided the same unit is used in each case.

In using the equation, note the following:

1 If the lens is convex, f is taken as positive (e.g. $f = 6$ cm)

2 If the lens is concave, f is taken as negative (e.g. $f = -6$ cm)

3 When v is positive, the image is real

4 When v is negative, the image is virtual.

The equation can also be used for curved mirrors as well as lenses. Note that converging lenses and mirrors have positive focal lengths, whereas diverging lenses and mirrors have negative focal lengths.

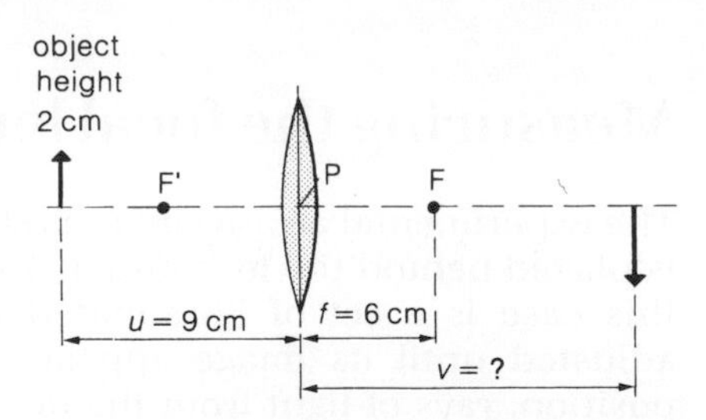

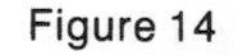

Figure 14

Example *An object* 2 cm *high stands on the principal axis at a distance of* 9 cm *from a convex lens. If the focal length of the lens is* 6 cm, *what is the position and nature of the image?*

The problem is illustrated in figure 14.

In this case, u is 9 cm, f is 6 cm and v is to be found.

Applying the equation given at the bottom of page 185:

$$\frac{1}{9\text{ cm}} + \frac{1}{v} = \frac{1}{6\text{ cm}}$$

giving: $$\frac{1}{v} = \frac{1}{6\text{ cm}} - \frac{1}{9\text{ cm}} = \frac{1}{18\text{ cm}}$$

therefore: $v = 18$ cm

The image is real and 18 cm away from the lens – a result previously found by scale drawing.

The first known use of lenses: as simple spectacles during work on Medieval manuscripts

Linear magnification

In the above problem, the image distance is twice the object distance. By looking at the similar triangles in the simplified ray diagram in figure 15, you can see that the image height is twice the object height. The lens in this case is said to be producing a **linear magnification** of 2:

$$\textbf{linear magnification} = \frac{\textbf{image height}}{\textbf{object height}} = \frac{\textbf{image distance}}{\textbf{object distance}} = \frac{v}{u}$$

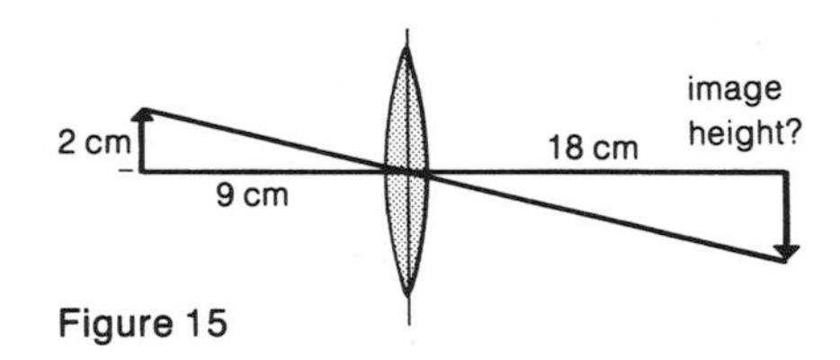

Figure 15

Note the simple proportions that these equations imply. An image and object are the same size if they are equal distances from a lens; the image is, say, ten times larger than the object if it is ten times further away from the lens, and so on. This principle applies to *all* thin lenses and mirrors.

Questions

1 What is meant by the principal focus of a) a convex lens b) a concave lens?

2 How could you find the focal length of a convex lens a) rapidly, but approximately b) accurately?

3 Give two reasons why in practice a convex lens may not bring parallel rays to an exact point focus.

4 A small upright object is placed well outside the principal focus of a convex lens. Draw a ray diagram to show how the image is formed, and say whether the image is real or virtual. What is the effect on the size and position of the image, of moving the object towards the lens?

5 State where an object should be placed if the image formed by a convex lens is to be

a) large, upright and virtual
b) real and smaller than the object
c) real and larger than the object
d) real and the same size as the object
e) real, and at the minimum possible distance from the object?

What is this minimum distance in part e)?

6 An object 3 cm high is placed 24 cm away from a convex lens of focal length 8 cm. Find by scale drawing or calculation the position, height and nature of the image. If the object is moved to a point only 3 cm away from the lens, what is the new position, height and nature of the image?

5.6

The camera and the eye

Both contain a converging lens system, and both pick up an image of the outside world on a special type of screen.

Figure 1 A modern 'single lens reflex' camera

The camera

In a camera (as shown in figures 1 and 2), a convex lens is used to form a small, inverted, real image on a piece of photographic film. The film, which is normally kept in total darkness, contains a light-sensitive chemical called silver bromide. When you press the camera button, a **shutter** in front of the film opens then shuts again, exposing the film to light for a brief moment only. Different intensities and colours of light across the image cause varying chemical changes in the film, which can later be developed, 'fixed', and used in printing a photograph.

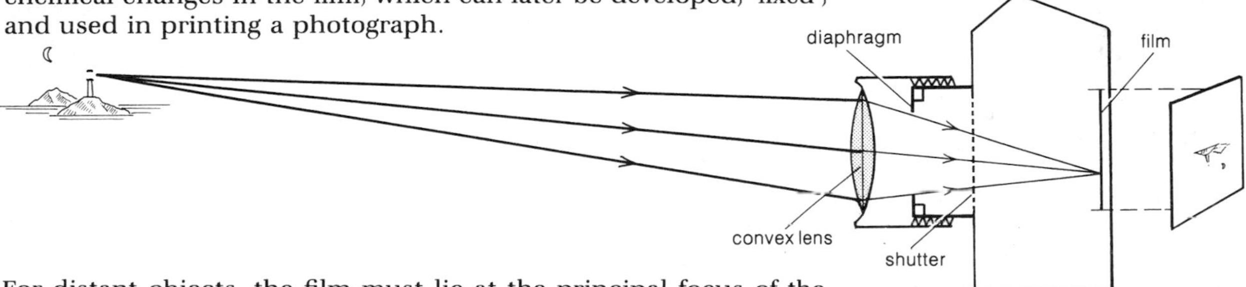

Figure 2 To obtain a clear photo of a distant object, the film must be located at the principal focus of the lens. To focus close objects, the lens is moved forward

For distant objects, the film must lie at the principal focus of the lens if the image is to be in sharp focus. For closer objects, the distance between lens and film must be increased. Accurate focusing of the image is achieved by screwing the lens backwards or forwards in its holder to suit the particular object distance.

On many cameras, the shutter speed can be varied, with exposure times ranging from perhaps 1/15 s to 1/1000 s. Shortening the exposure time cuts down the amount of light reaching the film, and reduces blurring if moving objects are being photographed.

The amount of light reaching the film is also controlled by an adjustable ring of sliding plates called the **diaphragm** (as shown in figure 3). This alters the diameter or **aperture** of the hole through which the light passes. Increasing the aperture lets more light into the camera, but it can also lead to some of the focusing problems mentioned on page 181. High quality cameras have expensive 'multi-element' lenses which produce sharply-focused images even at wide apertures.

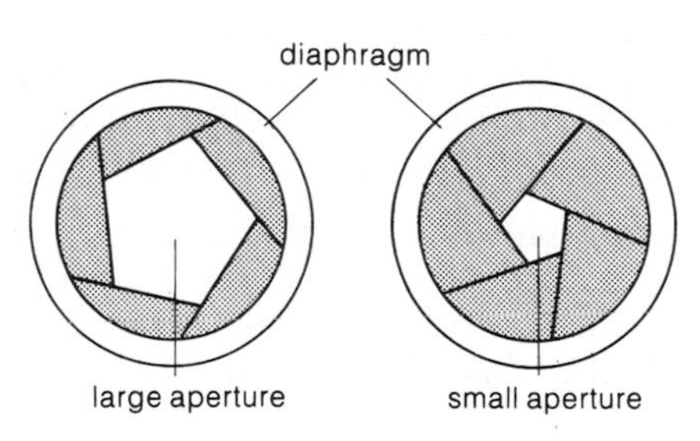

Figure 3 The diaphragm alters the amount of light reaching the film

The human eye

Like a camera, the eye (as shown in figures 4 and 5) uses a convex lens system to produce a real inverted image of an object in front of it. The 'screen' in this case is called the **retina** and it contains nearly 130 million light-sensitive cells. These cells send electrical impulses along the optic nerve to the brain. The brain uses them to form a view of the outside world.

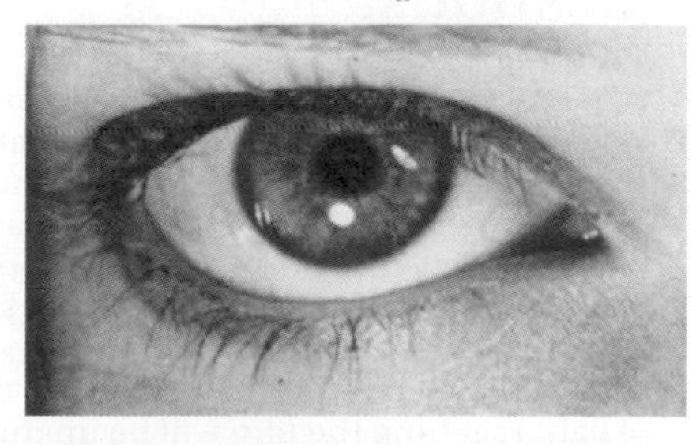

Figure 4 The iris is similar to the diaphragm of the camera

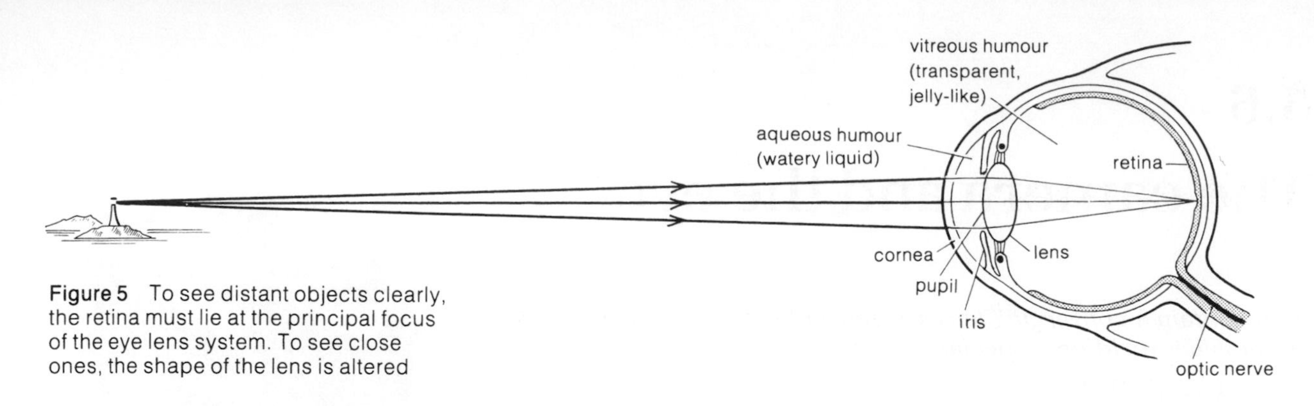

Figure 5 To see distant objects clearly, the retina must lie at the principal focus of the eye lens system. To see close ones, the shape of the lens is altered

The amount of light reaching the retina is controlled by the **iris** – the bit that makes your eyes brown, blue or hazel. If you walk into a dark room, the hole in the middle of the iris (the **pupil**) automatically grows larger to let in more light.

Light entering the eye is converged mainly by the **cornea** and the watery liquid behind it. The lens itself is used to make focusing adjustments – a process called **accommodation**. There is no backwards or forwards movement of the lens as in a camera. Instead, the shape of the lens is changed by the ring of muscles around it. When an object is brought closer to the eye, the lens thickens, and the reduced focal length enables the image to be kept in focus on the retina. For normal eyes, 25 cm is about the closest an object can be brought for comfortable viewing for any length of time, though this distance varies with age.

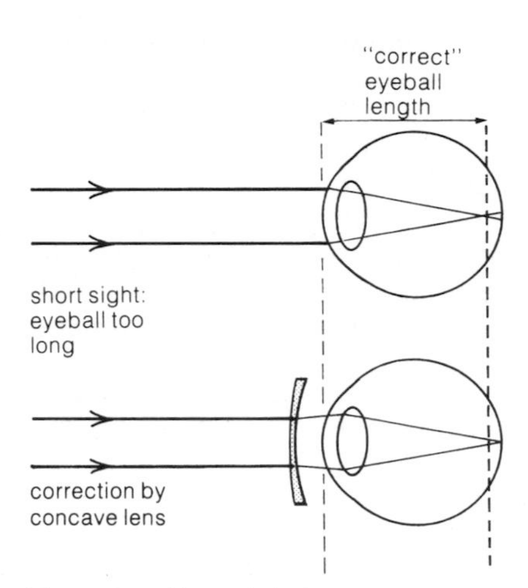

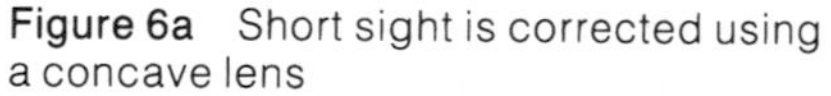

Figure 6a Short sight is corrected using a concave lens

Short and long sight

Changes in the shape of the lens may not be enough to produce sharp focusing on the retina. Spectacles or contact lenses must then be worn to overcome the problem.

Short-sighted people cannot see distant objects clearly. The eyeball is too long for the lens system and rays from any distant object come to a focus just in front of the retina. A concave lens is placed in front of the eye to correct the fault (as shown in figure 6a).

Long-sighted people cannot see close objects clearly. The eye-ball is too short for the lens system and rays from a close object converge towards a point just beyond the retina. A convex lens is placed in front of the eye to correct the fault (as shown in figure 6b).

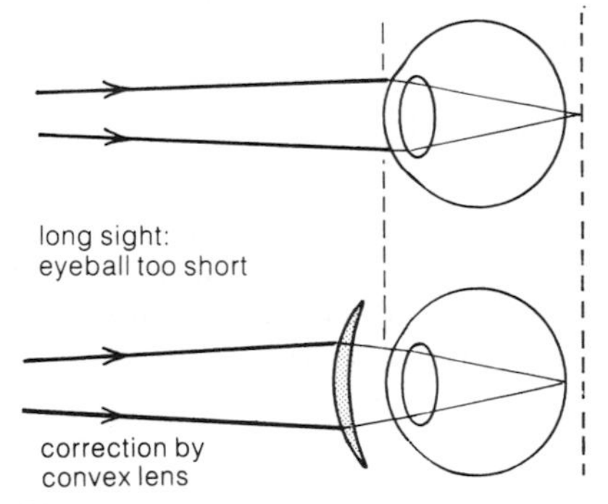

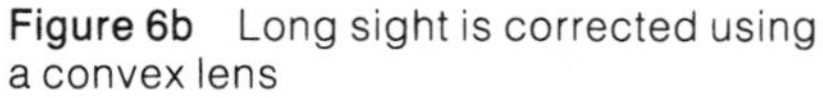

Figure 6b Long sight is corrected using a convex lens

Questions

1 Where is the image formed in a camera?
2 Where is the image formed in a human eye?
3 What is the function of the lens in a human eye?
4 An object is moved closer to a camera and to an eye. What changes must take place in a) the camera b) the eye in order to keep the image in sharp focus?
5 If the exposure time on a camera is increased, what other change must be made in order that the amount of light reaching the film will be unchanged?
6 How is the amount of light entering a human eye controlled?
7 Give two ways in which a camera is similar to the human eye, and two ways in which it is different.
8 What type of lens is used to correct a) short sight b) long sight?
9 A camera lens has a focal length of 5 cm. How far away from the film must it be positioned if it is to form a sharply focused image of a distant object? How far must the lens be moved if it is to form a sharp image of an object only 1 m from the lens?

5.7

Telescopes

A single lens cannot produce a magnified image of a distant object. But two lenses can ...

The astronomical telescope

A simple astronomical telescope is shown in figure 1. The lenses in this case are both convex:

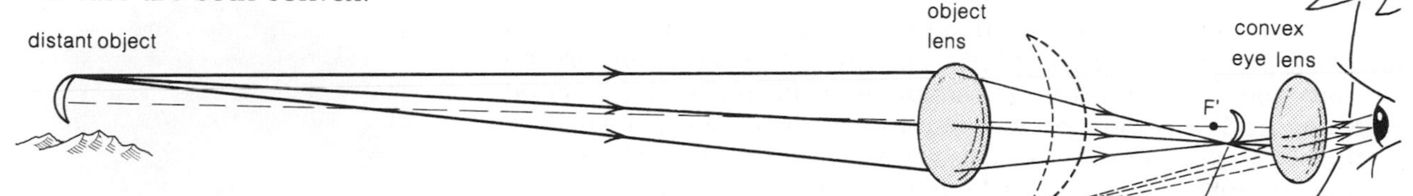

Figure 1 A simple telescope uses the small 'eye piece' lens to look at the real image formed by the large 'object' lens

The **object lens** (or **objective**) forms a small, inverted real image of a distant object – in this case the Moon – just inside the principal focus of the **eye lens** (or **eyepiece**). This image acts as a close object to the eye lens which forms a virtual magnified image of it. The eye lens is being used as a magnifying glass, but it is magnifying an image of the object rather than the object itself. The final image is upside-down, but this doesn't matter in astronomical observations.

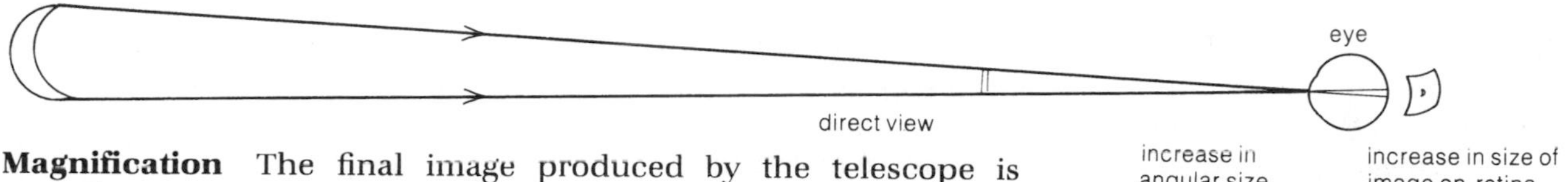

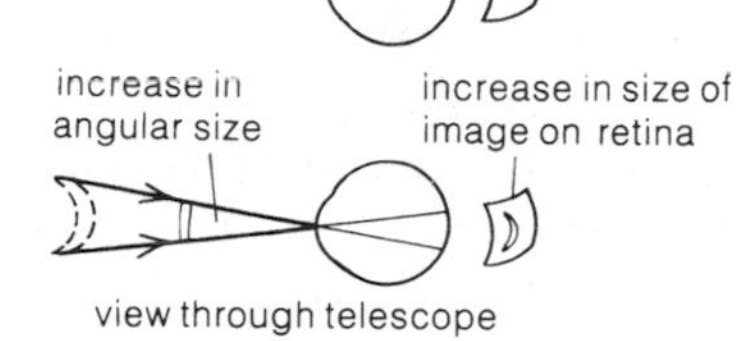

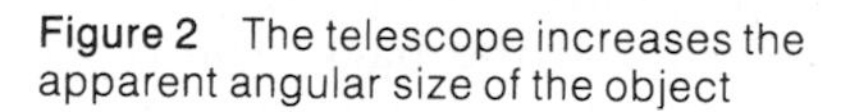

Figure 2 The telescope increases the apparent angular size of the object

Magnification The final image produced by the telescope is actually smaller than the original object. It looks larger because it is very much closer to the observer's eye. Figure 2 shows that the telescope increases the apparent **angular size** of the Moon, and it is this which causes the magnified image on the retina of the observer's eye.

The magnification produced by a telescope depends on the size of the real image formed by the object lens, and the extent to which this image is magnified by the eye lens. The two diagrams in figure 3 show that the object lens must have a long focal length to produce a relatively large real image. The eye lens on the other hand must have a short focal length if it is to act as a powerful magnifying glass.

To produce a high magnification, a telescope must have a long focal length object lens and a short focal length eye lens.

Light-gathering power For astronomers viewing very faint stars and galaxies, the light-gathering power of a telescope is as important as its magnification. The larger the diameter of the object lens, the greater is the light-gathering power, and the brighter and more detailed is the final image. In practice, the limit for the diameter of a glass object lens is about a metre. Above this, the lens would sag under its own weight.

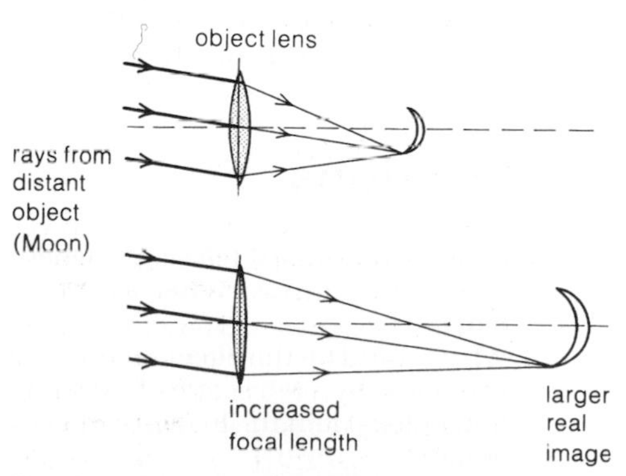

Figure 3 To get a high magnification you need a long focal-length 'object' lens (and therefore a long telescope!)

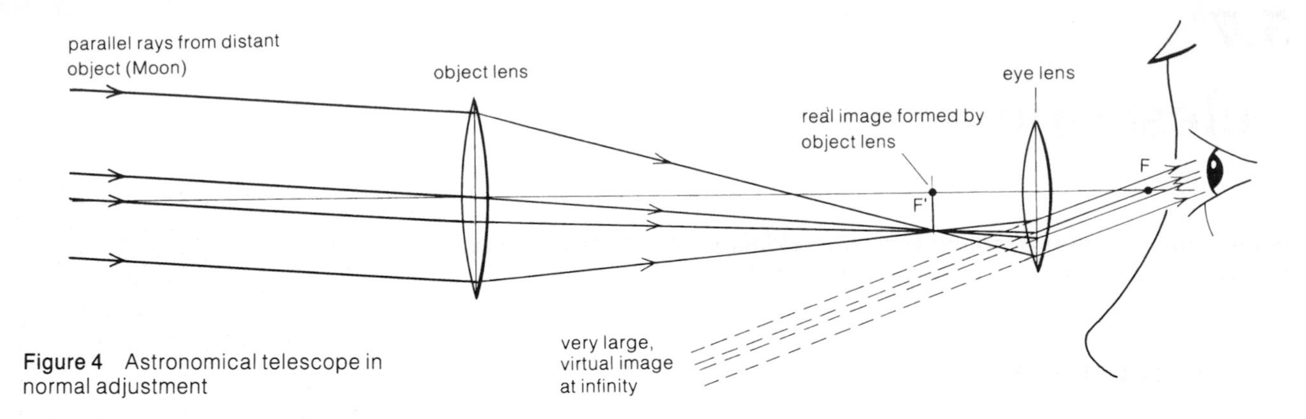

Figure 4 Astronomical telescope in normal adjustment

Binoculars: two telescopes side by side, with prisms to turn the image the right way up

Normal adjustment In figure 4, the lens positions have been adjusted so that the image formed by the object lens lies exactly at the principal focus of the eye lens. The rays entering the observer's eye are parallel, and the telescope is said to be in **normal adjustment**. For prolonged viewing, this arrangement is less tiring for the eye as the focusing muscles are fully relaxed. To the observer, the view of the Moon seems little different than before, though he is actually looking at an infinitely large image of the Moon an infinite distance away from his eye!

Astronomical reflecting telescopes

In a reflecting telescope, a concave mirror is used instead of a convex lens to form a real image in front of the eye lens. Figure 5 shows a typical arrangement; the eye lens is on the side of the telescope, and a small plane mirror is used to reflect light from the concave mirror towards it. The small mirror stops a little of the incoming light from reaching the concave mirror, but it doesn't block the observer's view in any way. A concave mirror (or a convex lens for that matter) forms a complete real image even when partly covered.

The main advantage of reflecting telescopes is that very large concave mirror diameters are possible. Unlike lenses, mirrors can be supported at the back. The world's largest reflecting telescope is sited on Mount Semirodriki in the USSR. It has a 6 metre diameter mirror and a light-gathering power which should enable it to detect a lighted candle more than 20 000 km away!

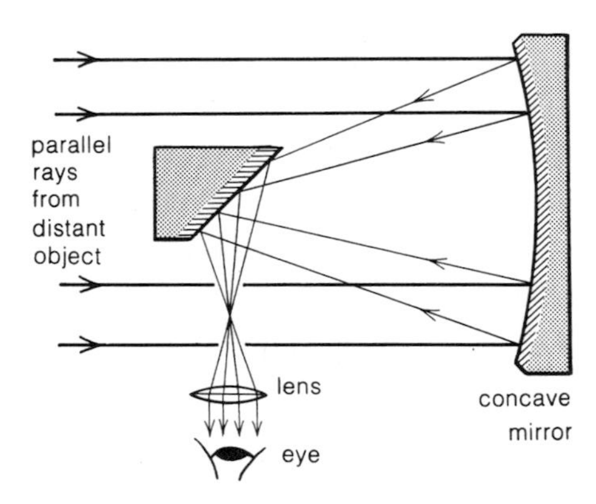

Figure 5 A reflecting telescope

Questions

1 In an astronomical telescope, where does the object lens form an image? What is the function of the eye lens?
2 What would be the effect on the magnification produced by a telescope of a) using an object lens of longer focal length b) using an eye lens of longer focal length?
3 In a telescope, what would be the effect of replacing the object lens by one of the same focal length but larger aperture (diameter)?
4 Astronomical telescopes are usually used in normal adjustment. What does this mean? What is the advantage of this arrangement?
5 How does a reflecting telescope differ from a lens telescope?
6 What is the principal advantage of a reflecting telescope over a lens telescope?

5.8
Colour

White light is made up of a whole range of colours, yet a TV set can give you a full colour picture using three colours only.

Figure 1 Newton and his original separation of white sunlight into a spectrum

Dispersion

Figures 1 and 2 illustrate an experiment first performed by Sir Isaac Newton in 1666. As the narrow beam of white sunlight passes through the glass prism, the light splits into a range of colours. White light is actually a mixture of colours rather than a single colour, and the prism refracts these different colours by different amounts.

The effect described above is called **dispersion**, and the colour range produced is known as a **spectrum**. You can see the colours of the spectrum on the rear cover of this book; most people seem to see seven colours in the spectrum, though in reality the spectrum is a continuous change of colour from beginning to end.

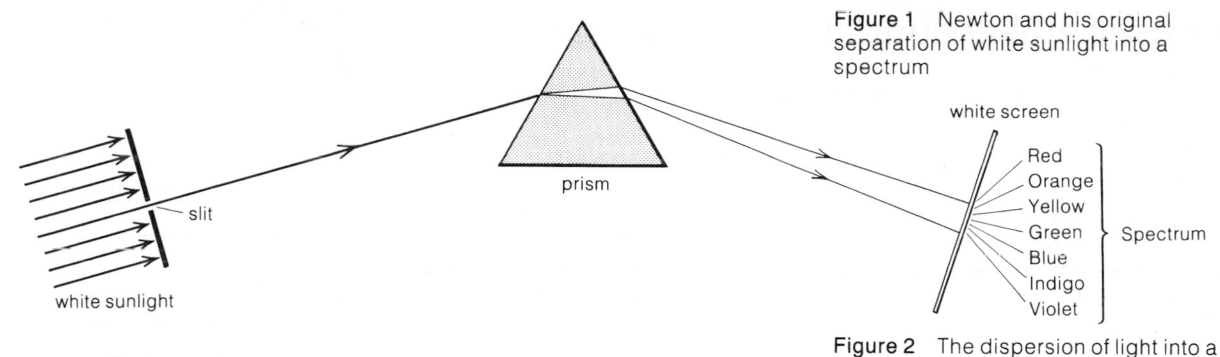

Figure 2 The dispersion of light into a spectrum

Recombination

The colours of the spectrum can be combined again to produce white. Figures 3 and 4 show two methods of doing this experimentally.

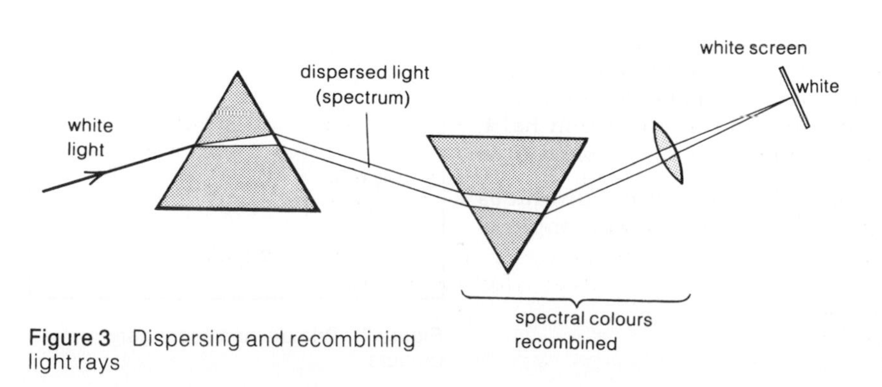

Figure 3 Dispersing and recombining light rays

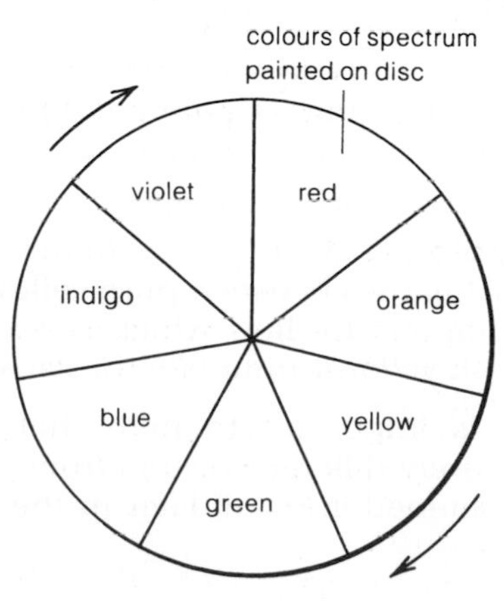

Figure 4 When rotated, the coloured disc appears white

Colour vision

It isn't necessary to provide all the colours in the spectrum to give the eye the sensation of white. If red, green and blue light beams from three spotlights are overlapped on a white screen as in figure 5, the bright patch in the middle of the screen appears white even though not all the spectral colours are present. To the human eye:

red + green + blue = white

Why does it happen? The retina of the human eye contains three main types of colour-sensitive cell. One type responds to a range of colours around red, another type to a range of colours around green, and the third type to a range of colours around blue. White light from the Sun or a light bulb stimulates all three types of colour-sensitive cell, and a mixture of red, green and blue light has exactly the same effect.

You experience *all* colours by the extent to which the three types of colour-sensitive cell are stimulated, and it is possible to give the eye almost any colour sensation by mixing red, green, and blue light in the right proportions. Colour television makes use of this principle. The screen of a colour TV (as shown in figure 6) is covered with thousands of tiny red, green and blue strips which glow in different combinations to produce a full colour picture.

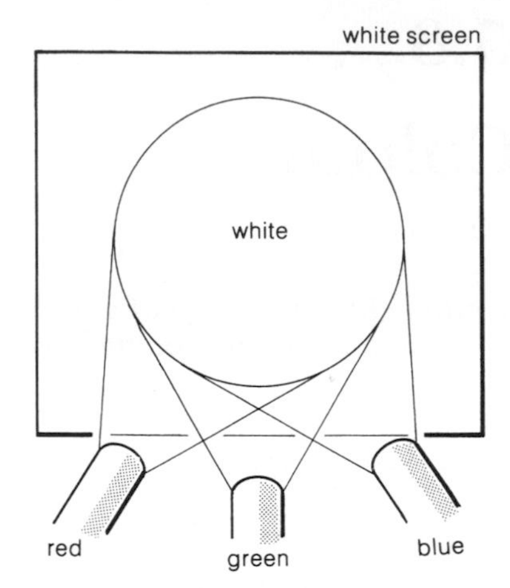

Figure 5 When red, green and blue lights are combined, the result looks white

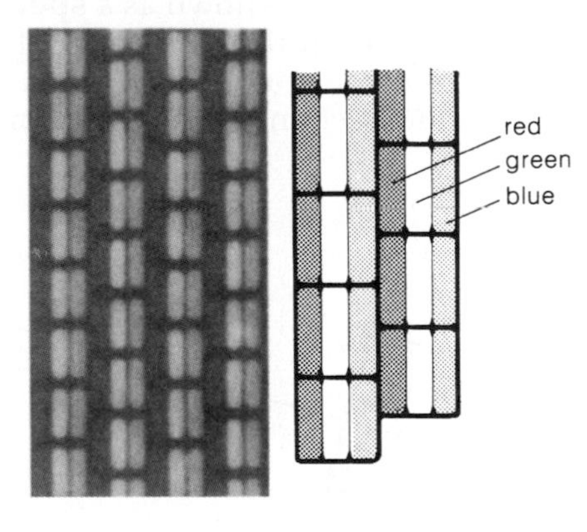

Figure 6 A colour TV screen

Primary and secondary colours – colour addition

Red, **green** and **blue** are known as the **primary colours** because it isn't possible to produce them by mixing light of any other colours together. Artists use different 'primary' colours, which will be explained later.

If the red, green and blue light beams in the experiment described above are moved apart slightly, new colours appear on the screen where any two of the beams overlap, as shown in figure 7. These are the **secondary colours**: **yellow**, **peacock blue** and **magenta**, and you can see them on the rear cover of this book. Producing new colours by mixing light of other colours together is known as **colour mixing by addition**. To the human eye:

red + green = yellow
green + blue = peacock blue (turquoise or cyan)
red + blue = magenta

The secondary colours aren't really single colours at all; they only appear as such to the human eye. The eye cannot for example distinguish between pure yellow in the spectrum and secondary or compound yellow which is really a mixture of red and green light. Both yellows stimulate the same colour-sensitive cells in the retina.

It is important to note that mixing coloured paints produces entirely different results from mixing coloured lights. This topic is examined in detail later in the unit.

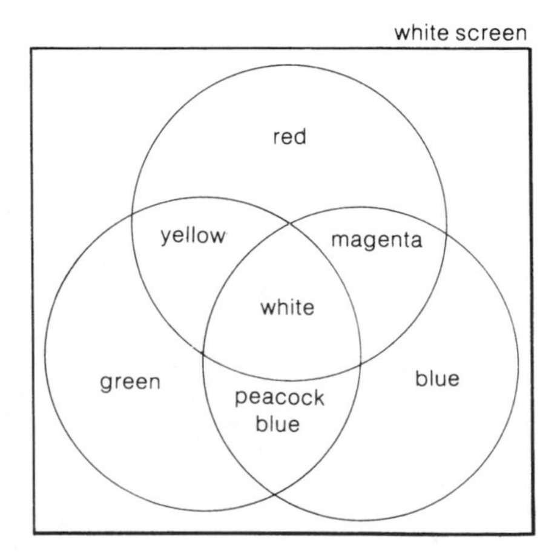

Figure 7 Primary and secondary colours

Complementary colours

As: red + green + blue = white
and: red + green = yellow

it follows that: yellow + blue = white

similarly: red + peacock blue = white
and: green + magenta = white

Any two colours which produce white light when mixed are called **complementary** colours. For example, yellow and blue are complementary. To find complementary colours in the colour-circle diagram shown on the last page in figure 7, look for colours which lie opposite each other on either side of the central white patch.

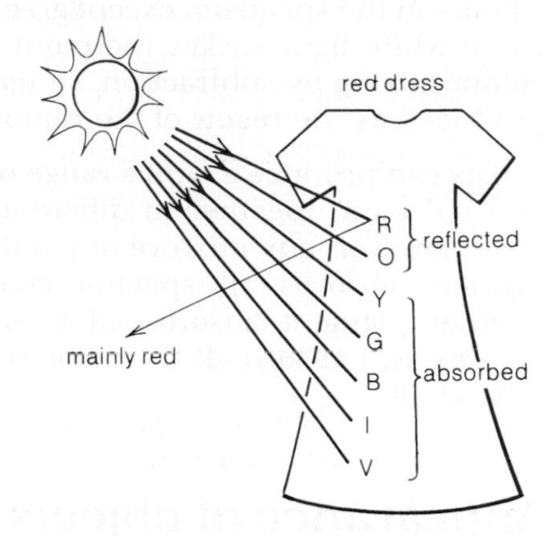

Figure 8 An object of a certain colour reflects that colour and absorbs all others

Colour of objects

A spotlight gives out its own light, so does a TV screen. Most objects however are only visible because they reflect light which has come from the Sun or some other source. Objects appear coloured when they reflect only some of the colours in the spectrum and absorb the rest. For example:

If a dress appears white in white light (e.g. sunlight), it is reflecting all the colours in the spectrum.

If a dress appears black in white light, it is absorbing all the colours in the spectrum and reflecting no light at all.

If a dress appears red in white light, it is reflecting mainly red light and absorbing the other colours in the spectrum, as shown in figure 8.

Filters are pieces of plastic or glass which only allow light of certain colours to pass through. A red filter for example lets through mainly red light but absorbs the other colours in the spectrum, as shown in figure 9. If a red filter is placed in front of a white spotlight, a red beam of light is produced.

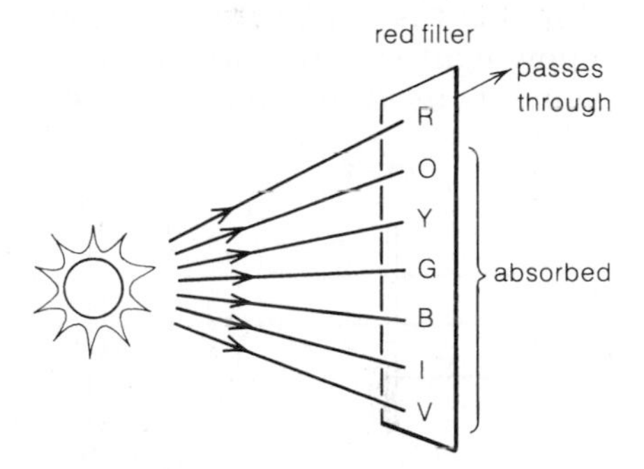

Figure 9 A filter of a certain colour allows that colour to pass through; it absorbs all others

Mixing pigments – colour subtraction

Substances used to colour paints, filters, plastics and other materials are known as **pigments**. In the past they have been made from some fairly exotic materials – for example, purple from crushed sea snails, and yellow from the urine of cows fed on mangoes! Nowadays, the chemical industry produces a whole range of pigments, many of which come from coal products.

Most of the pigments in common use are impure, and reflect more than one colour. The pigment in yellow paint for example reflects red, orange and green light as well as yellow, though this mixture appears as yellow to the human eye. The rest of the colours in the spectrum are absorbed.

Our attempts to create yellow pigments proved inconclusive

Figure 10 shows the effect of mixing together yellow and blue paint. Between them, the yellow and blue pigments present absorb all the colours in the spectrum except green, so only green light is reflected when white light strikes the paint mixture. This is an example of colour mixing by subtraction, so named because the final colour is produced as the result of the removal of other colours.

Artists can produce a whole range of colours by mixing yellow, blue and red paint together in different proportions. If pigments were pure however, any mixture of paints would always be black. If one pigment absorbs all spectral colours except, say, yellow, and another pigment absorbs all colours except blue, the resultant mixture will absorb all the colours in the spectrum and reflect no light at all.

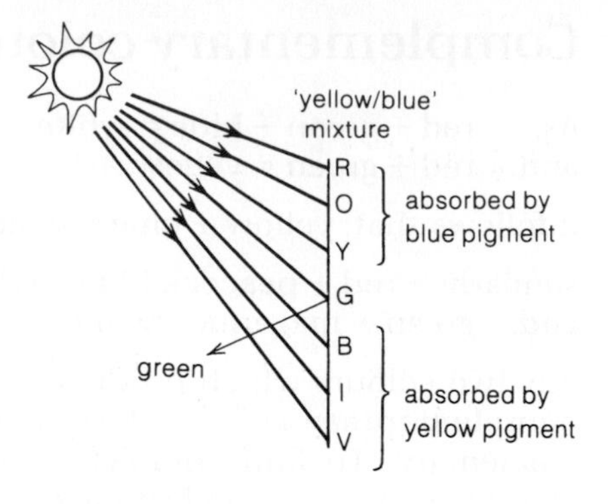

Figure 10 The selective reflection of a yellow/blue mixture of pigments

Appearance of objects in coloured light

When coloured light strikes an object, the object may appear to be an entirely different colour from its colour in white light. This is illustrated by the examples in figures 11a and 11b. For simplicity, the following assumptions are made:

white light is a mixture of red, green and blue light;
yellow light is a mixture of red and green light;
pigments used in the illuminated materials are pure.

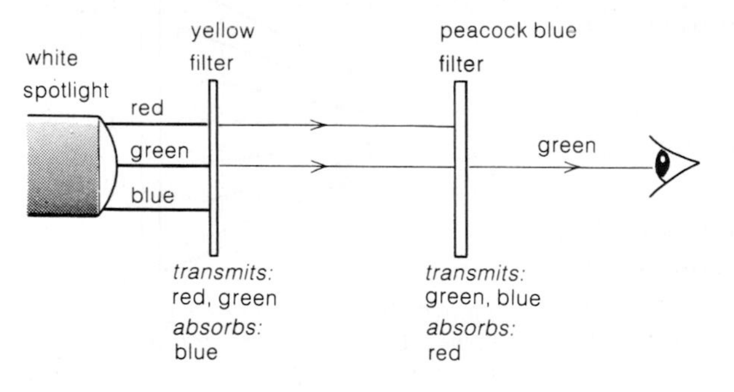

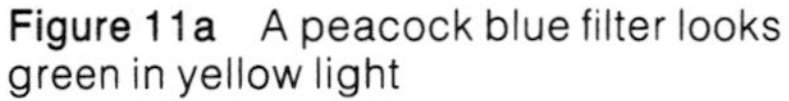
Figure 11a A peacock blue filter looks green in yellow light

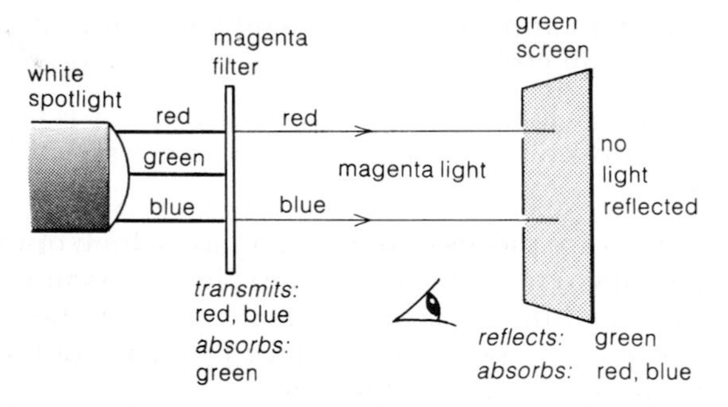

Figure 11b A green screen looks black in magenta light

Questions

In questions 9 to 12, assume that white is a mixture of the three primary colours only, and that yellow is secondary or compound yellow.

1 What is meant by dispersion?
2 How could you show that the colours of the spectrum combine to give white?
3 What colours are sufficient to give the eye the sensation of white?
4 What are the three primary colours?
5 Draw a diagram to show beams of red, green and blue light overlapping on a white screen. Indicate the colours seen, in the different areas of overlap.
6 Redraw the diagram in question 5 to show the colours which would be seen in the different areas if the surface of the screen were green.
7 What are complementary colours?
8 Yellow (secondary) light mixed with blue light strikes a white screen. What colour is seen?
Yellow paint mixed with blue paint is viewed in white light. What colour does the mixture appear?
9 If white light shines on a black surface, which colours are reflected and which absorbed?
10 If white light strikes a yellow filter, which colours pass through and which are absorbed?
11 Light, initially white, passes through a yellow filter then a peacock blue filter before striking a white screen. What colour is seen on the screen?
12 What colour will a red dress appear in a) yellow light b) green light?

5.9

Wave motion

There are many kinds of waves. This section looks at waves travelling along stretched springs and across the surface of water. Similarities between the way these waves behave and the way light behaves seem to suggest that light may itself be a form of wave motion.

Figure 1 Ripples on a pond: wave energy moving

Transverse progressive waves

Figure 1 shows ripples moving out from the place in a pond where a stone has been dropped in. Figure 2 shows how a similar wave effect can be produced in a spring. One end of a stretched 'slinky' spring is moved rapidly from side to side, and waves are produced which travel along the spring:

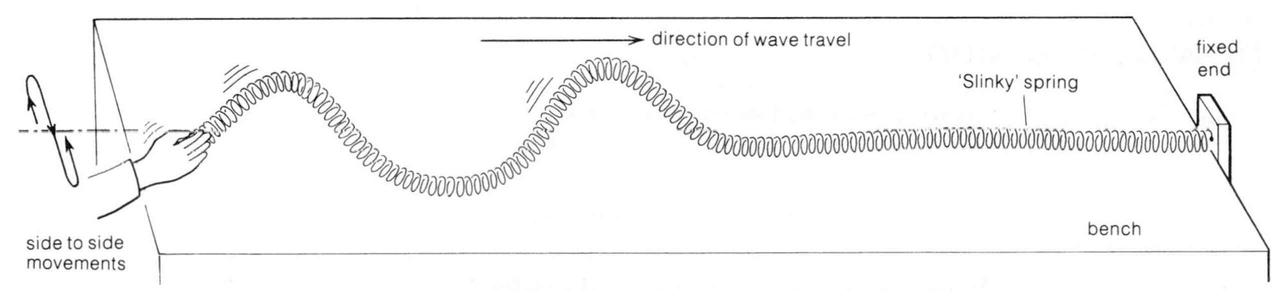

Figure 2 Wave energy moving in a slinky spring

When the end turn of the spring is moved sideways, it pulls the next turn sideways a fraction of a second later. This pulls the next turn, and so on. In this way, sideways movements are passed from turn to turn, and the travelling wave effect is produced. As the waves pass along the spring, turns initially at rest are set in motion, so there is a transfer of energy from one end of the spring to the other.

The turns themselves do not move in the direction of travel but **oscillate** (move to and fro) about the positions they would occupy if there were no movement at all. The waves are **progressive** because they are moving, and **transverse** because the oscillations are at right angles to the direction of travel, i.e. from side to side.

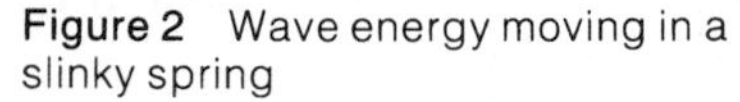

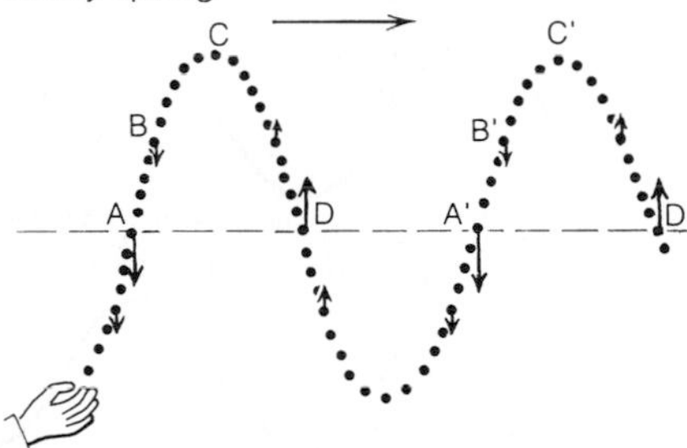

Figure 3a

Definitions

Figure 3a represents transverse progressive waves travelling along a 'slinky' spring: each dot is a point along the middle of the spring. The diagram shows the waves at one instant. A fraction of a second later, the waves will be further to the right.

Frequency Figure 3b shows a single wave made by one complete oscillation of the hand. The number of complete waves produced per second is called the frequency and it is measured in **hertz (Hz)**. If the hand makes 4 oscillations every second, 4 complete waves are produced per second, and the frequency is 4 Hz.

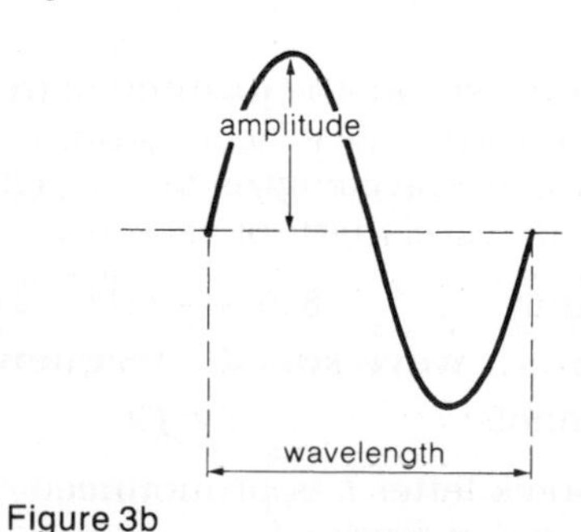

Figure 3b

Figures 3a, 3b The spring moves sideways and the waves move forwards: A transverse, progressive wave

Wavelength The distance occupied by one complete wave is called the wavelength and it is measured in metres. It is equal to the distance AA′ in figure 3a, also to BB′, CC′, and so on.

Amplitude Also measured in metres, this is the maximum distance a point moves away from its rest position when a wave passes.

Phase The small arrows shown in figure 3a indicate the movements of the points A, B, C etc. Points A and A′ have the same speed and direction of movement and are said to be in phase. B and B′ are also in phase, so are the other pairs of points marked. On the other hand, the points A and D are exactly out of phase because their movements are opposite.

Wavefronts Waves in a 'slinky' spring travel along one line only, whereas waves on the surface of water may spread out in all directions. For convenience, such waves are often represented in diagrams by a series of lines called wavefronts (as shown in figure 4). You can think of these wavefronts as the 'crests' or the 'troughs' of the waves, though they can be lines joining any points which are in the waves; they can be lines joining any points which are in phase.

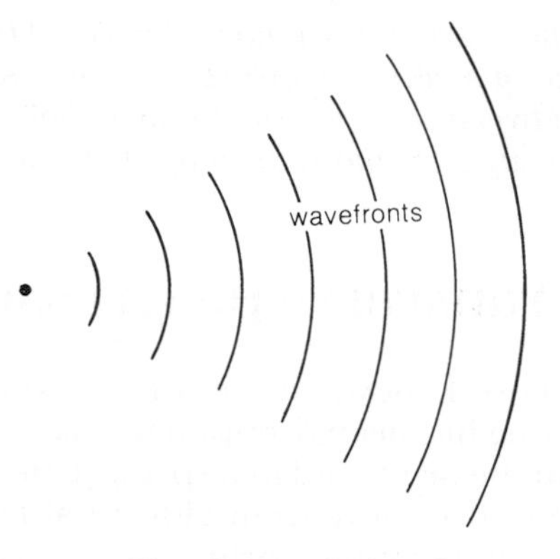

Figure 4 Wavefronts travel out from a source

The wave equation

The speed of a wave and its frequency and wavelength are linked by a simple equation. Consider the case of the transverse progressive waves shown in figure 5.

The frequency of the waves is 4 Hz; 4 complete waves are produced every second.

The wave speed is 8 m/s; these waves move 8 m further to the right in 1 second:

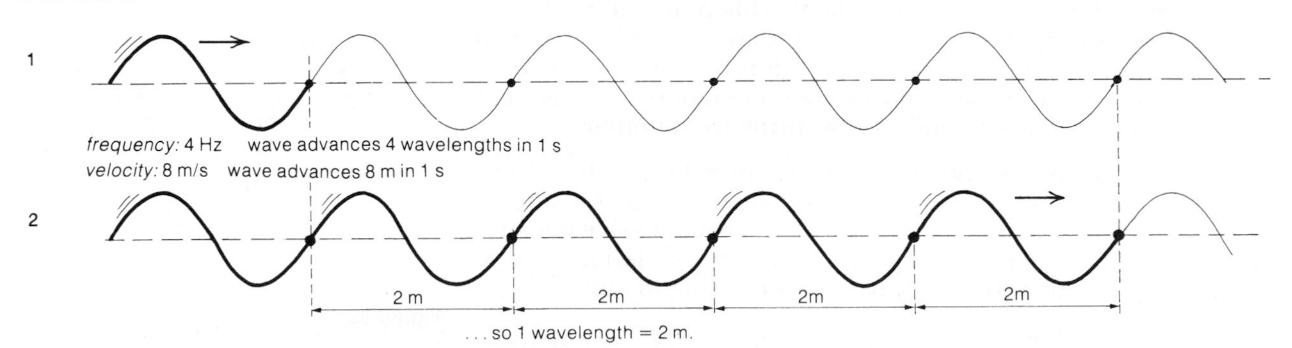

Figure 5 Four waves are produced every second. The wavelength is 2 metres; the velocity is 8 metres per second i.e. $v = f\lambda$

Diagram 2 shows the position of the waves 1 second after diagram 1. 4 more waves have been produced; the wave pattern has moved 4 complete wavelengths to the right and has covered a distance of 8 m. The wavelength of each wave is therefore 2 m.

Note that: $8\ \text{m/s} = 4\ \text{Hz} \times 2\ \text{m}$

in words: **wave speed = frequency × wavelength**

in symbols: $\boldsymbol{v = f\lambda}$

(the Greek letter λ is pronounced 'lamda')

This equation applies to any wave. The wave speed is measured in m/s, if frequency is measured in Hz and wavelength in m.

The ripple tank

The properties of progressive waves can be studied in the laboratory using a ripple tank as shown in figure 6a. The waves in this case are ripples travelling across the surface of shallow water in a tray, and they are produced by a small electric vibrator. A bar attached to the vibrator gives plane (straight) waves, as shown in figure 6b; circular waves are produced by fixing a small metal ball to the bar, as shown in figure 6c. The bottom of the tray is transparent so that a lamp can be used to cast an image of the ripples on a white screen placed underneath.

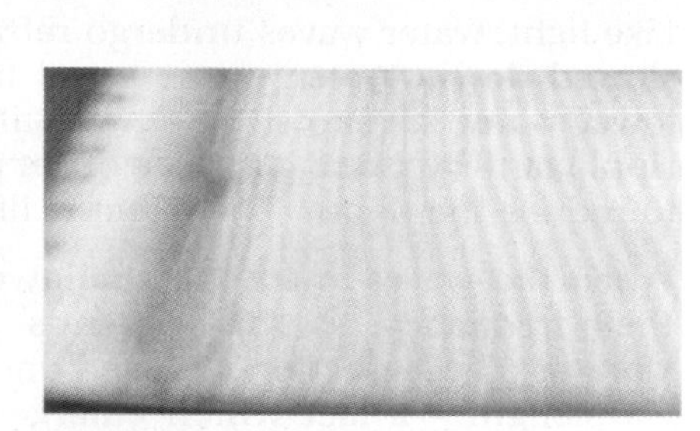

Figure 6b Plain waves

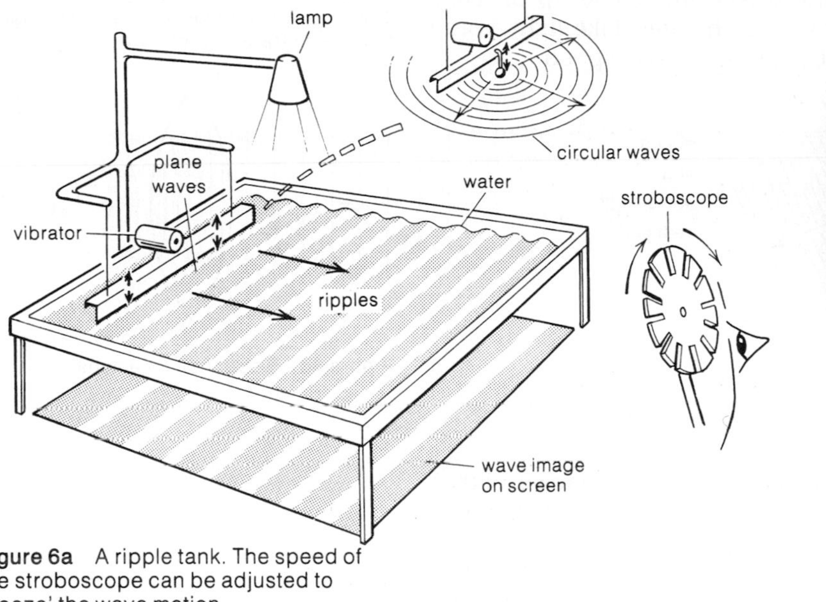

Figure 6a A ripple tank. The speed of the stroboscope can be adjusted to 'freeze' the wave motion

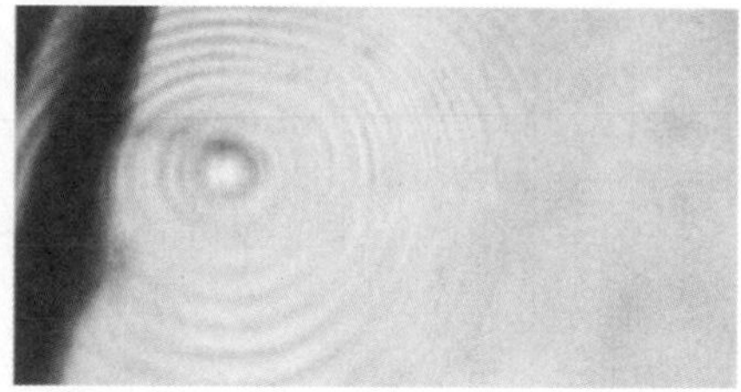

Figure 6c Circular waves

The waves are easier to study if their motion is 'frozen' by means of a **stroboscope**. The stroboscope shown in figure 6a consists of a disc with slits cut in it, and it is spun round using a finger. When you look through the spinning disc, the waves are only visible when a slit passes in front of your eye. If the speed of the disc is such that the waves have moved forward exactly one wavelength each time they are seen, they will appear motionless.

Four wave effects which can be demonstrated using a ripple tank are described below. Look for similarities between the way water waves behave and the way light behaves.

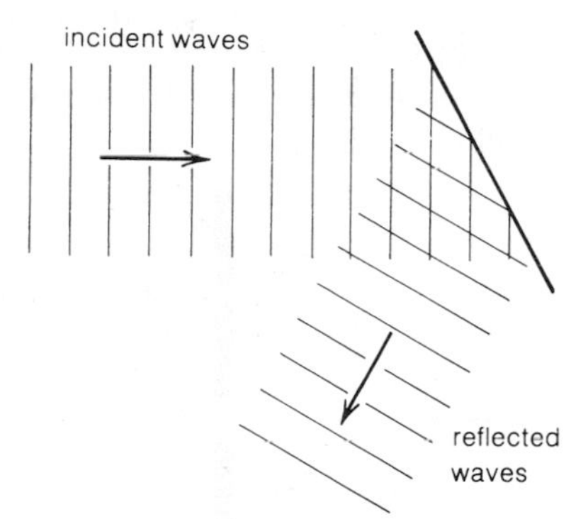

Figure 7a Reflection at a flat surface

Reflection

Like light, water waves are reflected when an obstacle is placed in their path, and the same laws of reflection apply. In a ripple tank, reflection can be demonstrated by placing an upright surface in the water. Figure 7a shows reflection from a plane surface; figure 7b shows how waves are brought to a focus when they are reflected from a concave surface.

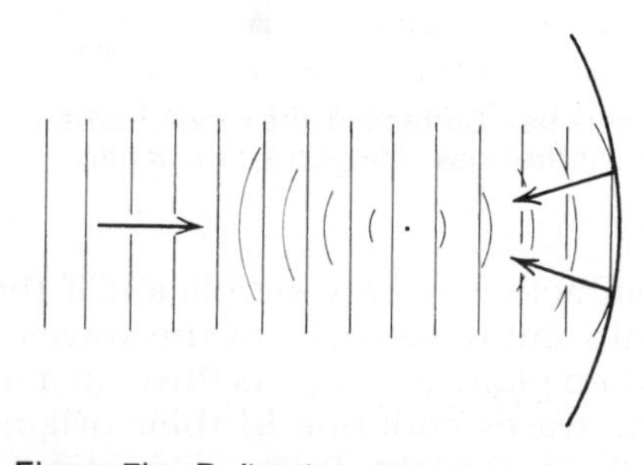

Figure 7b Reflection at a concave surface

Refraction

Like light, water waves undergo refraction (bending) when they are slowed down. Water waves travel more slowly if they enter shallower water (as shown in figure 8a), so they can be refracted in a ripple tank by placing a piece of flat plastic in the bottom of the tray to reduce the depth. The effect is illustrated in figures 8b and 8c.

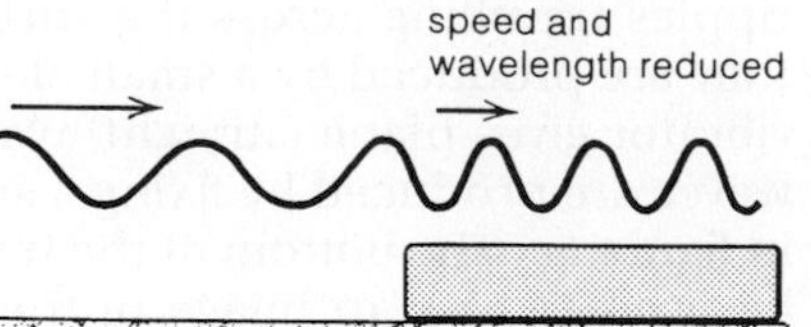

Figure 8a Water waves travel more slowly in shallow water. This can be used to demonstrate refraction

When the waves enter the shallower water there is no change in their frequency, as this depends only on the frequency of the vibrator. The reduction in speed does however cause a reduction in wavelength – a fact which emerges if you consider the equation $v = f \times \lambda$ with f kept constant. As the wavefronts close up on each other, the direction of travel of the waves changes. Like light, water waves bend in towards the normal when they lose speed.

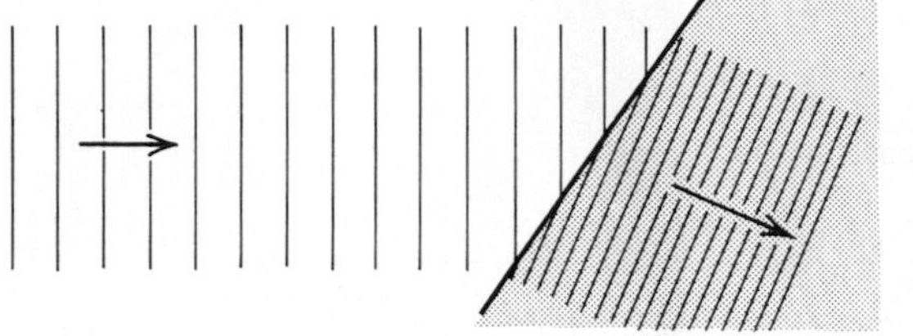

Figure 8b Refraction at a plane surface

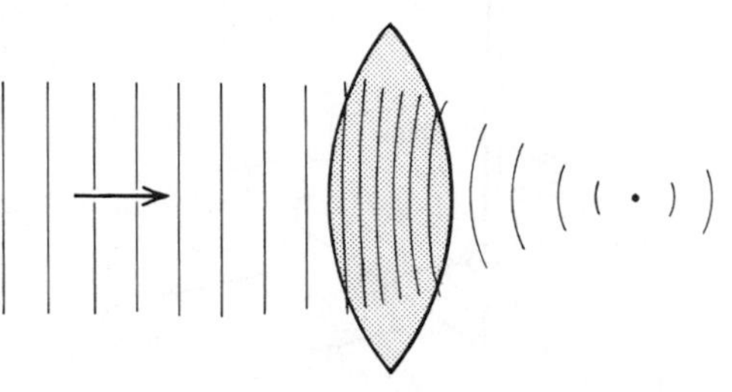

Figure 8c Refraction by a 'lens'

Diffraction

Figure 9a below shows the effect of placing two obstacles with a narrow gap between them in the path of plane waves in a ripple tank. Waves passing through the gap spread out in all directions and the wavefronts produced are circular. This is an example of an effect known as **diffraction** – the bending of waves as they pass around obstacles.

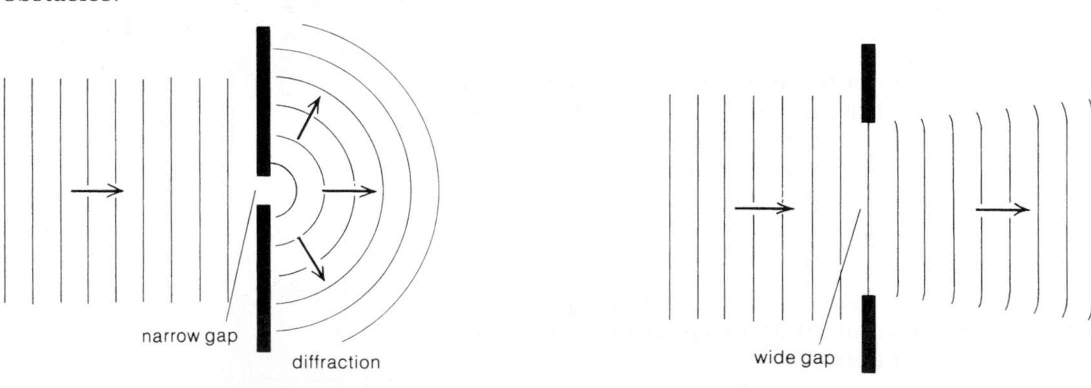

Figure 9a Diffraction: the waves spread out as they pass through a narrow gap

Figure 9b Diffraction is insignificant if the width of the gap is much more than the wavelength of the waves

Diffraction is only significant if the size of the gap is comparable with the wavelength of the waves. Figure 9b shows what happens when plane waves pass through a much wider opening. In the main, the waves continue in their original direction and the wavefronts remain straight. Some diffraction does occur at the edges of the wave 'beam', but the effect is slight.

Interference

If two identical sets of waves travel through the same region of water in a ripple tank, they may, depending on their phase, reinforce each other or cancel each other out. The effect is known as **interference**.

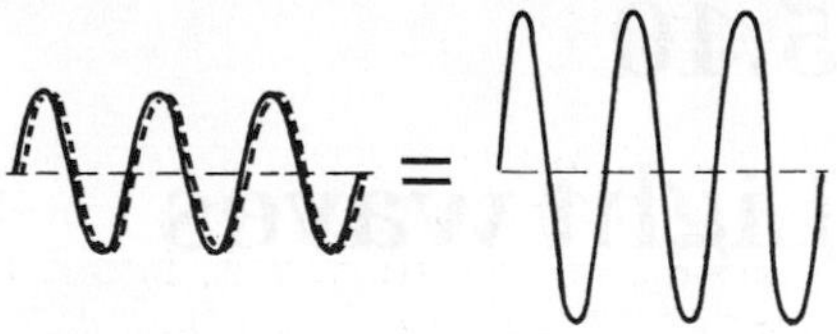

Figure 10a illustrates **constructive interference**. The waves are in phase, both are moving the water surface in the same direction, and the amplitude is doubled. Figure 10b illustrates **destructive interference**. The waves are exactly out of phase and try to move the water surface in opposite directions. So there is no movement at all.

The vibrator in a ripple tank can send out two sets of circular waves at once by fixing two metal balls to the bar as shown in figure 11. Both sets of waves have the same frequency, and are exactly in phase as they leave points S_1 and S_2. Interference occurs where the waves cross; the wave pattern formed is an **interference pattern**.

Figure 10a Constructive interference: two waves reinforce one another's effect

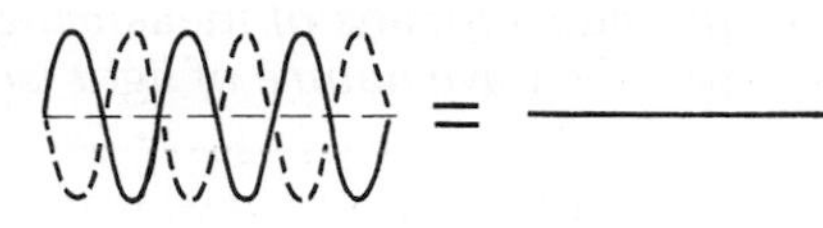

Figure 10b Destructive interference: the waves cancel one another's effect

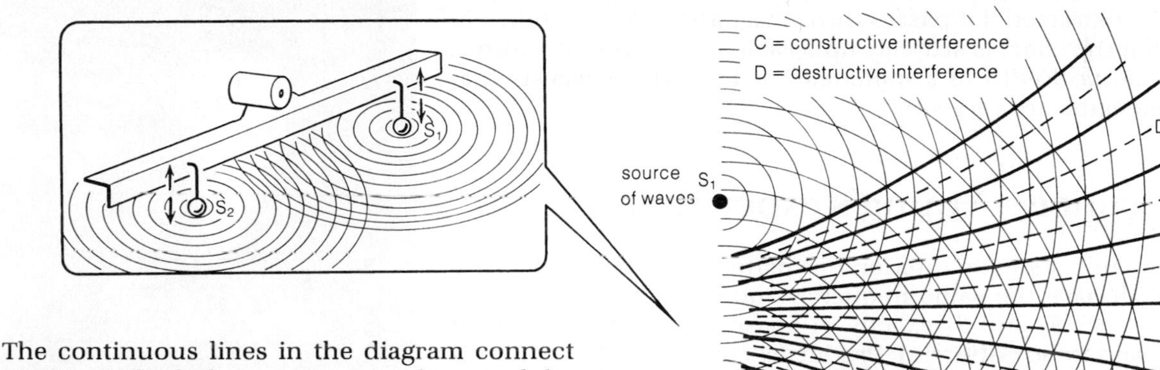

Figure 11 Two sources can produce a complex pattern of simultaneous constructive and destructive interference

The continuous lines in the diagram connect points at which the waves are in phase and the oscillation of the water surface at a maximum. The dotted lines connect points at which the waves are exactly out of phase and the water surface is therefore undisturbed.

If the wave sources S_1 and S_2 are moved further apart, the interference pattern becomes less spread out. If the distance between S_1 and S_2 is many times larger than the wavelength of the water waves, the lines of constructive and destructive interference are too close together to be visible.

Questions

1 If a wave is *transverse*, what does this mean?
2 Explain the meaning of the terms *frequency*, *wavelength*, *amplitude*, *wavefront*.
3 What equation links the speed, frequency, and wavelength of a wave?
4 The wave crests seen in a ripple tank are 5 mm apart, and the frequency of the vibrator is 10 Hz. What is the wave speed? If the frequency of the vibrator is doubled, what then is the spacing between the crests?
5 How can the waves in a ripple tank be made to travel more slowly? When this happens, what is the effect on a) the frequency b) the wavelength?
6 What happens when plane waves in a ripple tank pass through a narrow opening? What is the effect called? What happens when plane waves pass through a much wider opening?
7 What can be said about the phase of two identical sets of waves if they a) constructively interfere b) destructively interfere?

5.10 Light waves

Experiments which demonstrate the diffraction and interference of light offer convincing evidence that light travels in the form of waves. They provide a means of measuring light wavelengths and give clues as to why the wave nature of light isn't more immediately apparent to the eye.

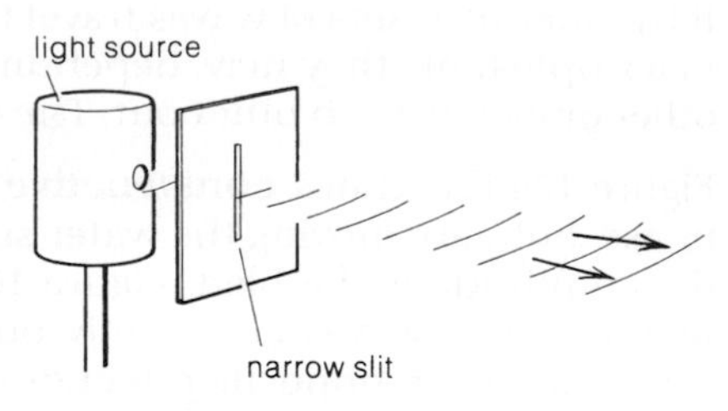

Figure 1 Diffraction of light waves through a very narrow slit

Diffraction

Water waves in a ripple tank are diffracted through wide angles if they pass through a gap comparable in size to their wavelength. Light is also diffracted if it passes through a narrow slit (as shown in figure 1), but the effect is scarcely noticeable unless the slit width is around one-hundredth of a millimetre or less. This suggests that light wavelengths are very short.

Thomas Young in 1810

Interference – Young's experiment

In a ripple tank, a steady interference pattern is obtained because the waves all have the same frequency and wavelength, and leave the two sources S_1 and S_2 exactly in phase. The matching sets of waves are said to be **coherent**. Producing coherent light sources is much more difficult. Most light sources emit a whole range of wavelengths, and nearly all send out light waves in short random bursts, each out of phase with the next. This makes it impossible to obtain a steady interference pattern using two separate lamps as wave sources.

These problems were neatly overcome in a series of experiments devised by Thomas Young in 1801. The modern version of one of his experiments is shown in figure 2:

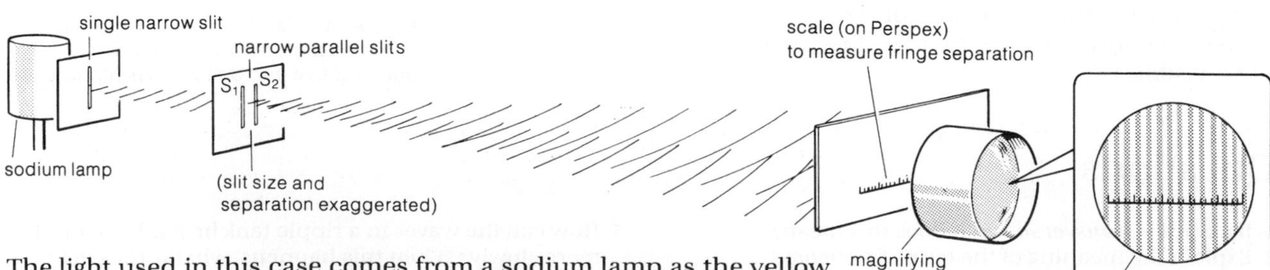

Figure 2 Light waves from the sources S_1 and S_2 cause interference

The light used in this case comes from a sodium lamp as the yellow light it emits is effectively of one wavelength only. The wave sources S_1 and S_2 are two narrow parallel slits about 0.5 mm apart which diffract light received from a single slit placed in front of the lamp. The arrangement produces two matching sets of waves from the same light source and ensures that waves leaving S_1 and S_2 are always coherent whatever phase changes occur in the bursts of light from the sodium lamp.

Interference occurs where the waves from S_1 and S_2 cross, and a series of bright and dark bands called **fringes** are seen in the eyepiece (as shown in figure 3). The bright fringes are lines of constructive interference where waves from the two slits arrive in phase and reinforce each other; the dark fringes are lines of destructive interference where the waves cancel each other out. Note the similarities between this interference pattern and that produced in a ripple tank.

Figure 3 The interference pattern from Young's experiment

Measurement of wavelength – Young's experiment

To find the wavelength λ of the light in the above experiment, three quantities must be measured; the distance s between the centres of the parallel slits, the distance D from the slits to the eyepiece scale, and the distance w between the centre of one bright (or dark) fringe and the next. w is known as the **fringe separation**. It is found by measuring the distance occupied by ten or so fringes on the scale, and calculating from this figure the distance between the centres of two neighbouring fringes.

The relationship between λ, s, D, and w can be found from figure 4. For clarity, distances in the diagram are not to scale. In an actual experiment, D might be set at about a metre, while the fringe separation w would be only about a millimetre at the most.

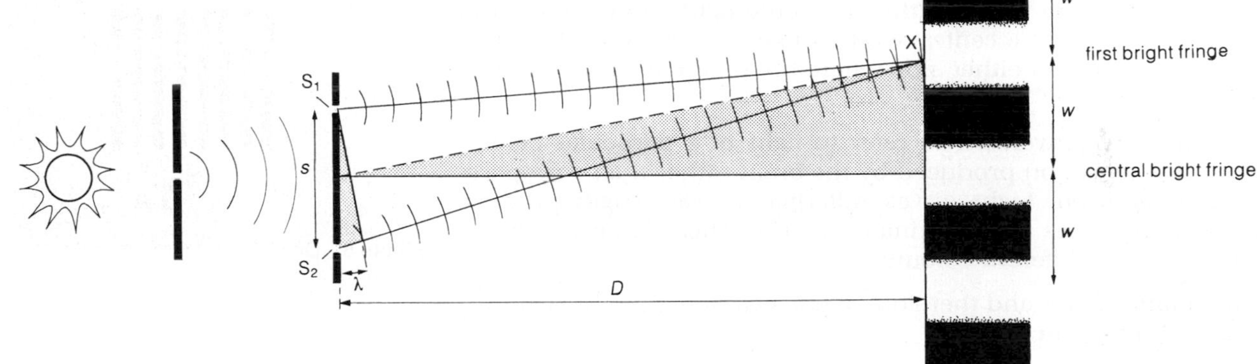

Figure 4 The first bright fringe occurs when one wave travels exactly one wave length further than the other, so they arrive in phase. Using the two similar (shaded) triangles, the wavelength of the light, λ, can be found

The first bright fringe is formed where light travelling from S_2 to X has exactly one wavelength further to travel than light travelling from S_1 to X. As waves leave S_1 and S_2 in phase, they will then arrive at X in phase.

The shaded triangles are approximately similar, so corresponding sides are in the same ratio. Assuming that the length of the dotted line is approximately equal to D, it follows from the similar triangles that $\frac{\lambda}{s} = \frac{w}{D}$

Rearranged, this gives $\lambda = \frac{sw}{D}$, from which the wavelength can be calculated.

Keen mathematicians, worried by all those approximations, should remember the exaggerated vertical scale of the diagram. Try drawing the diagram to an accurate scale and you will see that the approximations are quite acceptable.

Typical readings from an experiment might be:

slit separation $= s = 1.0\,\text{mm}$
fringe separation $= w = 0.6\,\text{mm}$
distance from slits to scale $= D = 1000\,\text{mm}$

Substituting these values in the equation given above:

$$\lambda = \frac{1.0\,\text{mm} \times 0.6\,\text{mm}}{1000\,\text{mm}}$$

$$= 0.0006\,\text{mm}$$

The wavelength of the yellow light from the sodium lamp is 0.0006 mm [6×10^{-7} m], which means that there are about 1700 of these light waves to the millimetre.

0.0007 mm
red light
0.0004 mm
violet light

Figure 5 Different colours of light have different wavelengths

Wavelength and colour

Experiments with a white light source and a series of coloured filters show that different colours have different wavelengths. Red light has the longest wavelength, about 0.0007 mm, and produces the most widely spaced fringes (as shown in figures 5 and 6). Violet light has the shortest wavelength, about 0.0004 mm, and gives more closely spaced fringes. The other colours of the spectrum have wavelengths between these values, while white light contains the full range of visible wavelengths. If a white light source is used in Young's experiment, a central white fringe is produced with bands of mixed colours on either side where the different fringe patterns overlap.

Although it is convenient to refer to light of a particular colour, colour is a sensation produced by the brain rather than a property of the light itself. Light waves differing in wavelength produce different responses in the retina of the eye. These are experienced by the brain as different colours.

Light of one colour, and therefore of one wavelength only, is known as **monochromatic** light.

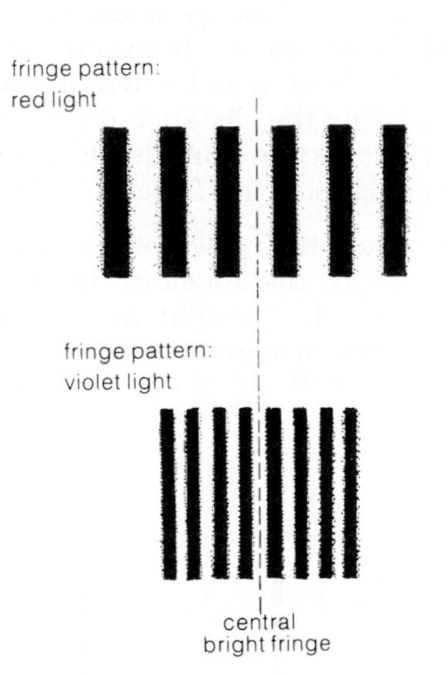

Figure 6 Different colours produce different fringe patterns

Laser light

Lasers as shown in figure 7 give off a narrow beam of intense light. They are particularly useful for producing interference effects, because their light is coherent – the waves are regular, and of just one wavelength.

A laser can be used to make a special type of photographic image called a hologram. Figure 8 shows the basic principle. The hologram is formed on the photographic film. It doesn't look like a 'real' picture. It is an interference pattern, produced when waves direct from the laser meet waves reflected from the chess piece. If, later, laser light is shone back through the hologram, a three-dimensional image of the chess piece can be seen.

Figure 7 laser in use

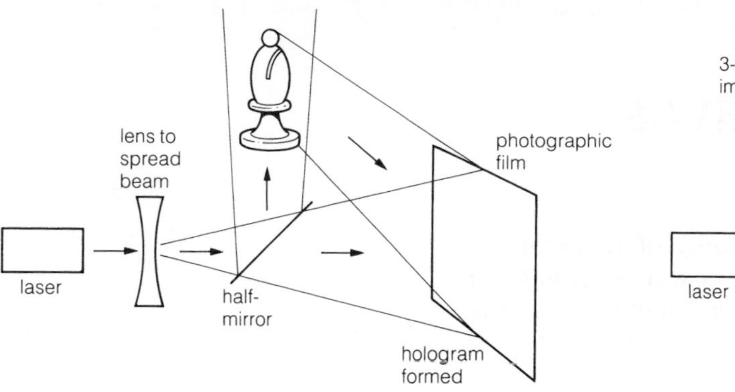

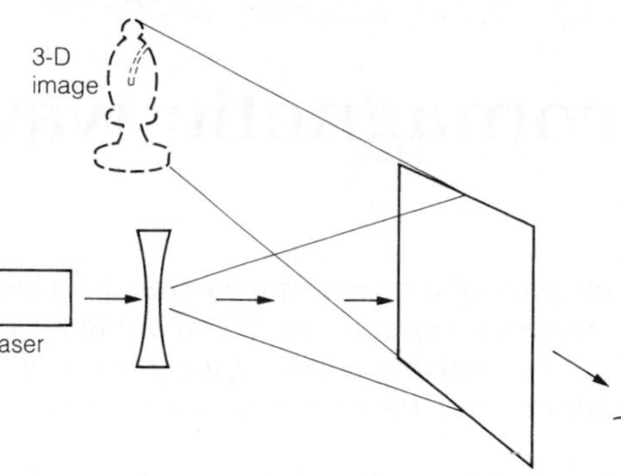

Figure 8 Making a hologram .... and using one

Not all holograms need laser light to give an image. Some work using reflected daylight. Credit cards often have reflection holograms printed on them as shown in figure 9. They make the cards very difficult for forgers to copy.

Figure 9 Hologram on a credit card

Lasers are used in compact disc players as shown in figure 10. Like records, compact discs 'store' music in coded form. But they don't have a groove for a stylus to travel along. Instead, a tiny laser beam travels over thousands of tiny pits on the surface of the disc. A detector picks up the reflected beam, and the light 'pulses' are changed into sound.

In surgery, lasers are used in delicate operations on eyes and nerves. Extra-fine beams give a concentrated heat which can seal blood vessels and cut tissue with great precision.

Figure 10 Compact disc player and disc

Questions

1 When light passes through a gap, diffraction is not always apparent. Why not?

2 Why is it not possible to obtain a steady interference pattern when two lamps are used as sources of light waves?

3 In a Young's slits experiment, the light source emits monochromatic light. What does monochromatic mean?
What would be the effect on the fringe pattern obtained, of
a) using slits which were slightly further apart?
b) moving the screen further away?
c) using light of a shorter wavelength?
d) using white light instead of monochromatic light?

4 In a Young's slits experiment, the slits are 0.5 mm apart and the eyepiece side is 1.0 m away from the slits. If the bright fringes are 0.8 mm apart on the scale, what is the wavelength of the light?

5 Which colour light has the longest wavelength? Which has the shortest wavelength? About how many light waves are there to the millimetre on average?

5.11
Electromagnetic waves

Light waves are part of a much more extensive range of waves known as the electromagnetic spectrum. Electromagnetic waves come from a whole variety of sources. They differ greatly in their wavelengths and in their effects, but they have certain fundamental properties in common.

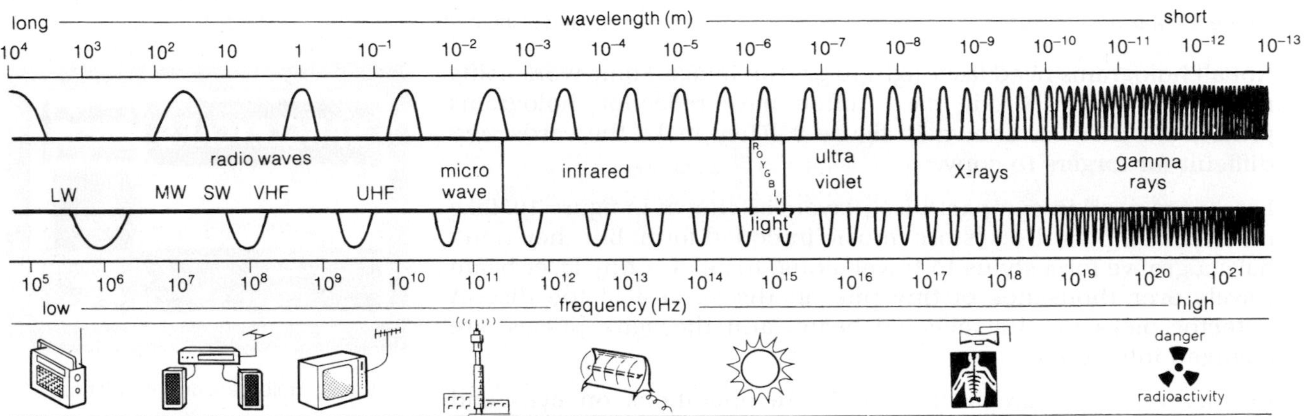

Figure 1 The electromagnetic spectrum

Properties of electromagnetic waves

Figure 1 shows the different types of waves which make up the electromagnetic spectrum. Features common to all these waves include the following:

They travel through free space in straight lines at a speed of approximately 300 000 km/s. Free space means a vacuum in which there are no nearby materials to influence the behaviour of the waves. In free space, all the electromagnetic waves travel at the speed commonly known as the speed of light (3×10^8 m/s).

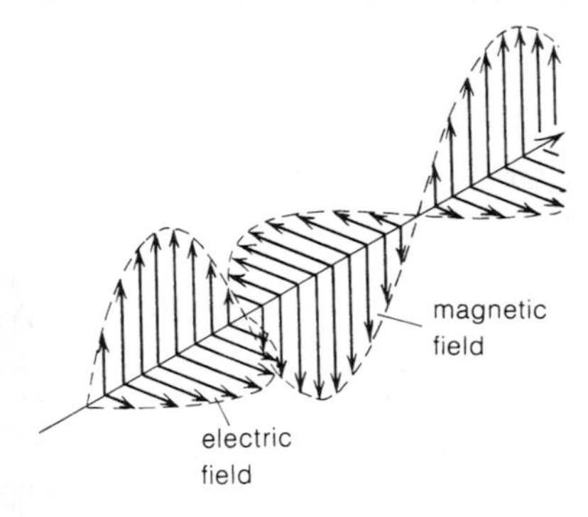

Figure 2 Electromagnetic waves

They are electric, magnetic and transverse in nature. The moving wave effect is produced by oscillating electric and magnetic **fields** (more details are given on pages 236 and 298). These fields oscillate at right angles to the direction of travel and at right angles to each other, as illustrated in figure 2.

Waves emitted from a point source in free space obey an inverse square law. The intensity of an electromagnetic wave is a measure of the wave energy passing every second through each square metre at right angles to its path. A wave has an intensity of 1 watt per square metre (1 W/m^2) if 1 joule of wave energy is passing through each square metre every second.

As a wave travels outwards from a point source, its intensity falls because the wave energy becomes spread over an increasingly large area, as shown in figure 3. At twice the distance from the source, the intensity is only a quarter of its previous value; at three times the distance, it is reduced to a ninth, and so on.

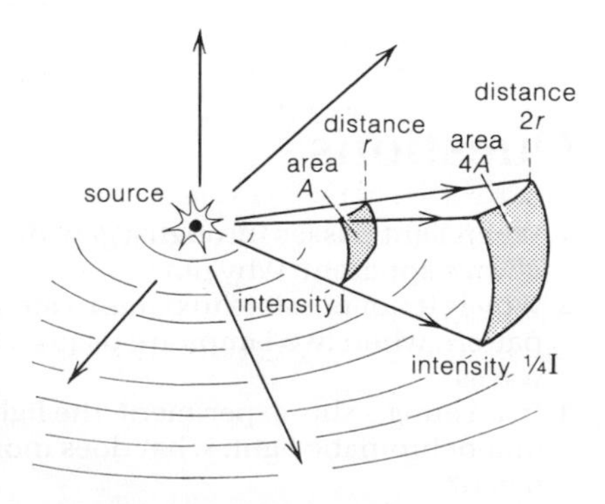

Figure 3 At twice the distance from a source, the intensity is a quarter of its original value

Expressed mathematically:

$$I \propto \frac{1}{r^2}$$

where I is the intensity and r is the distance from the source. This is known as the **inverse square law**.

They obey the equation $c = f \times \lambda$, where c is the speed of light (3×10^8 m/s), f is the frequency of the wave, and λ is the wavelength. As c is a constant, it follows that the higher the frequency of a wave, the shorter is its wavelength. This fact is also illustrated by the frequency and wavelength scales in figure 1.

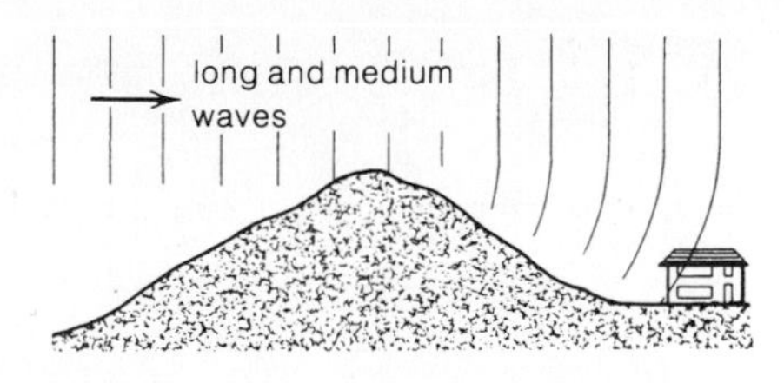

Figure 4 Diffraction of radio waves

Where electromagnetic waves come from

All matter, non-living and living, is made up of molecules, each containing one or more atoms. Atoms are themselves made up of a central nucleus around which tiny particles called electrons are in orbit, as explained in unit 8.11. Electrons and nuclei all carry a small electric charge.

Electromagnetic waves are emitted when electrically charged particles change energy in some way. This happens, for example, when an electron changes to a lower orbit around a nucleus. It also occurs when electrons or nuclei oscillate; their kinetic energy is then constantly changing. The greater an energy change, or the more rapid an oscillation, the higher is the frequency and the shorter the wavelength of the waves produced. The energy changes causing X-rays, for example, are much greater than those giving rise to radio waves.

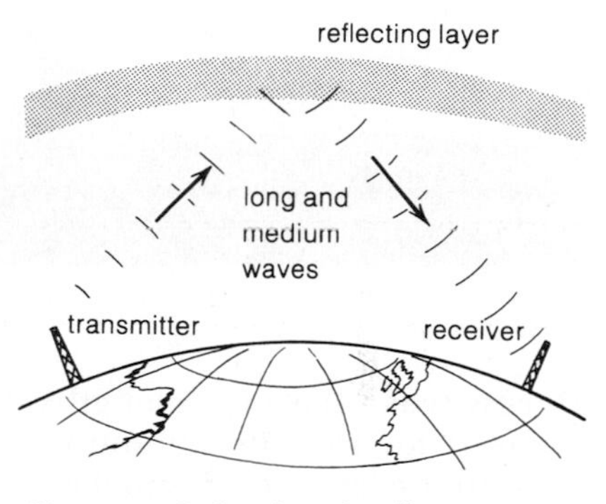

Figure 5 Reflection of radio waves

Radio waves

These have the longest wavelengths in the electromagnetic spectrum. They can be produced by making electrons (commonly called 'electricity') oscillate in an aerial or **antenna**, and are used to transmit sound and picture information over long distances. **Long** and **medium waves** will diffract around a hill, so you can still pick up a signal on a transistor radio even if a hill blocks the direct route from the transmitting aerial, as shown in figure 4. Long and medium waves are also reflected by layers of charged particles in the Earth's upper atmosphere, as shown in figure 5, so long distance reception is possible despite the curvature of the Earth's surface.

VHF and UHF (TV) waves have shorter wavelengths. They are not reflected by upper atmospheric layers and are diffracted only slightly by hills. For good reception, there normally needs to be a straight path from the transmitting aerial to your TV or stereo radio receiving aerial, as shown in figure 6.

Figure 6 A local U.H.F. transmitter

Figure 7 Goonhilly Down receiving dish is used for microwave communication with satellites

Microwaves have wavelengths of a few centimetres or less. They are used for satellite communication, as shown in figure 7, and radar. In the laboratory, small microwave transmitters and receivers can be used as an alternative to a ripple tank to demonstrate the reflection, refraction, diffraction and interference of transverse waves. Like all electromagnetic waves, microwaves produce a heating effect when they are absorbed. Food absorbs certain microwave frequencies very strongly; the principle is used in microwave ovens.

Infrared waves and light

When an electric fire is first switched on, you can detect the infrared radiation emitted by the heating effect it produces in your skin. All objects radiate infrared waves, mostly as a result of the continuous motion of their molecules. The waves may come from molecules which are oscillating or rotating, or from electrons which are displaced from their 'orbits' when fast moving molecules collide. Most objects radiate a wide range of wavelengths.

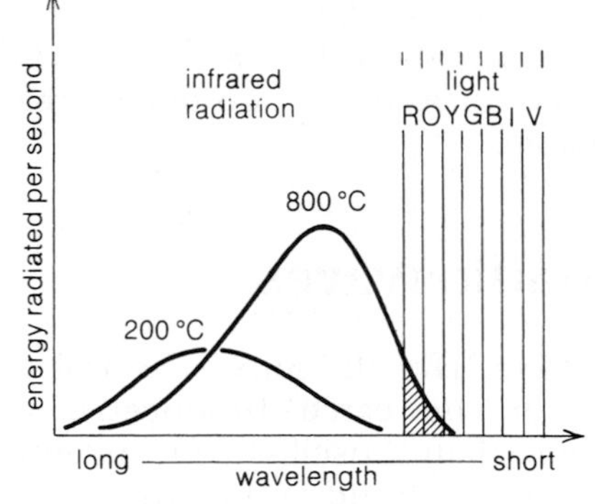

Figure 8 Hotter objects give out more infrared radiation than cooler ones

An object gives out more and more infrared radiation as its temperature rises, as shown in figure 8, the wavelengths becoming shorter as the molecules vibrate more vigorously. At about 700°C, the shortest waves present can be detected by the eye and now become known as light waves. These waves lie at the red end of the visible spectrum, so the object appears red-hot. If the temperature rises still further, more wavelengths are produced in the visible part of the spectrum and the object eventually becomes white-hot.

In the experiment shown in figure 9, a small phototransistor is being used to detect the presence of infrared in the radiation emitted by the white-hot filament of a light bulb. A phototransistor is placed just beyond the red end of the visible spectrum where it picks up short wavelength infrared radiation refracted by the prism. Infrared and light can also come from cooler sources, such as gases through which an electric current is passing.

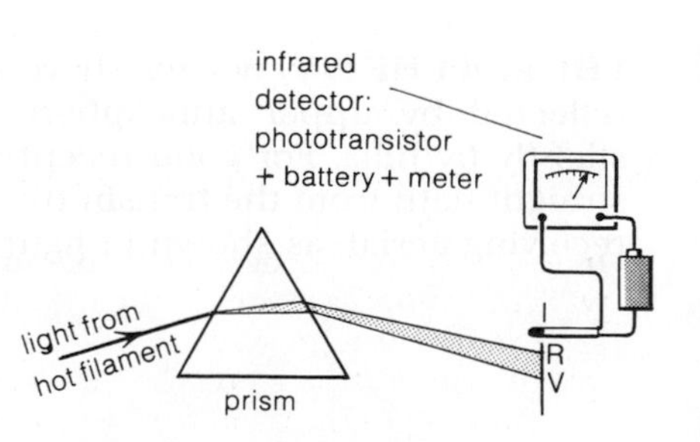

Figure 9 Detecting infrared radiation beyond the red end of the spectrum

Ultraviolet waves

At very high temperatures, objects begin to emit shorter wavelengths which can't be detected by the eye. These are ultraviolet waves. Ultraviolet radiation from the Sun is used by your skin in producing vitamin D, and it will also give you a sun tan. But an excess is harmful, and can cause skin disease and damage to the retina.

Some materials **fluoresce** when they absorb ultraviolet radiation; they convert its energy into visible light, and can be seen to glow. Fluorescent paints glow because of the ultraviolet present in sunlight, and washing powders contain chemicals which fluoresce for the same reason, as shown in figure 10. They make your shirts or blouses look whiter than white in daylight and glow when disco lights strike them. The inside of a fluorescent light tube is coated with powders which glow when they absorb ultraviolet radiation, this being produced by passing an electric current through the gas in the tube.

Ultraviolet waves can also cause **ionization**, a process in which electrons absorb enough energy to break free of the atoms which hold them. Ionization can be harmful to living things, and ultraviolet radiation is frequently used to kill off bacteria.

Figure 10 Turning ultraviolet radiation into visible light makes your washing 'whiter than white'

X-rays and gamma rays

X-rays are given off when fast moving electrons lose energy very rapidly. Gamma rays are emitted by some radioactive materials when large energy changes take place within the nuclei of their atoms. Both types of radiation are very penetrating, and the shorter wavelengths can pass through dense metals like lead.

Gamma rays tend to have shorter wavelengths than X-rays because energy changes within the nucleus are normally much larger than those that take place outside it. There isn't however any difference between X-rays and gamma rays of the same wavelength and intensity, and the names are used only in order to identify the nature of their source.

The two types of radiation are examined in more detail on pages 350 and 352.

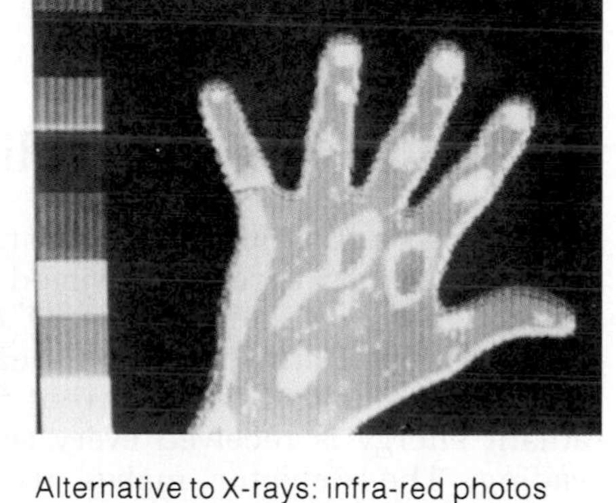

Alternative to X-rays: infra-red photos can be used to detect cancer, and blocked blood-circulation

Questions

1 List the different types of waves in the electromagnetic spectrum, putting longest wavelengths at the top and shortest at the bottom. Which of the radiations shown result from the highest energy changes?

2 Write down three properties common to all electromagnetic waves.

3 In each case write down a type of radiation which a) will cause fluorescence b) is emitted by a hot object c) can pass through lead d) can be detected by the eye e) can be diffracted by a hill f) causes damage to the retina g) is used for satellite communications h) can be used to kill bacteria.

4 2 m away from a point source of infrared waves, the intensity is $4\,W/m^2$. What is the intensity a) 1 m from the source b) 4 m from the source?

5 Give two reasons why a long wave radio can pick up a signal despite the presence of hills and the curved surface of the Earth between it and the transmitting aerial.

6 A VHF radio station emits radio waves at a frequency of 100 MHz.
a) What is the wavelength of the waves?
b) How long will it take the waves to travel a distance of 100 km?
(Speed of electromagnetic waves = 3×10^8 m/s; 1 MHz = 10^6 Hz.)

5.12

Thermal radiation

Figure 1 The surface which absorbs radiation most easily will lose its coin first

All objects give out infrared waves. Hot objects may emit light and ultraviolet as well. These waves warm up anything that absorbs them and are known as thermal radiation – or radiation for short. Radiation, like conduction and convection, is a process by which thermal energy (heat) can be transferred from one place to another.

Good and bad absorbers

Some surfaces are better absorbers of thermal radiation than others. The inside of a black car warms up more rapidly than a white one on a hot sunny day; white clothes are often worn in hot countries to reduce the amount of radiation absorbed.

A simple method of comparing the absorbing abilities of different surfaces is shown in figure 1; the better the absorber, the more rapidly the wax melts and the sooner the coin falls off. Figure 2 shows how different surfaces compare. Dull black surfaces are the best absorbers. Silvery mirror-like surfaces are the poorest absorbers, reflecting nearly all of the thermal radiation that strikes them.

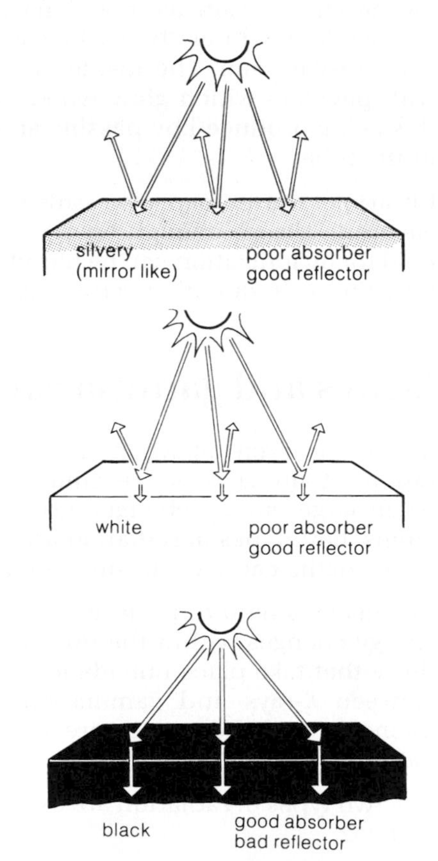

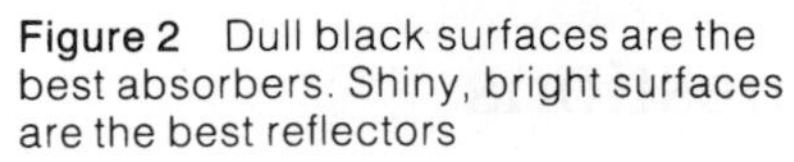

Figure 2 Dull black surfaces are the best absorbers. Shiny, bright surfaces are the best reflectors

Detecting thermal radiation

A simple form of thermal radiation detector is shown in figure 3. Copper and iron wires are twisted together at one end to form a junction, and the other ends of the wires are connected to a sensitive meter. If the junction is heated in any way, by radiation for example, a small electric current flows in the circuit. The more radiant energy is received every second, the greater is the meter reading. The detector makes use of an effect known as the **thermoelectric effect**, a current being produced whenever the ends of two different metals in a circuit are at different temperatures.

Figure 4 shows a more sensitive form of thermal radiation detector. Known as a **thermopile**, it consists of a large number of junctions joined in the same circuit to give a greater current. The metal cone reflects radiation on to the blackened surface next to the junctions.

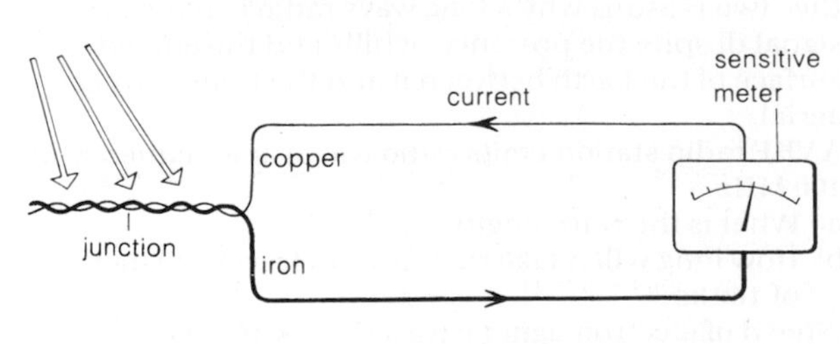

Figure 3 A thermal radiation detector

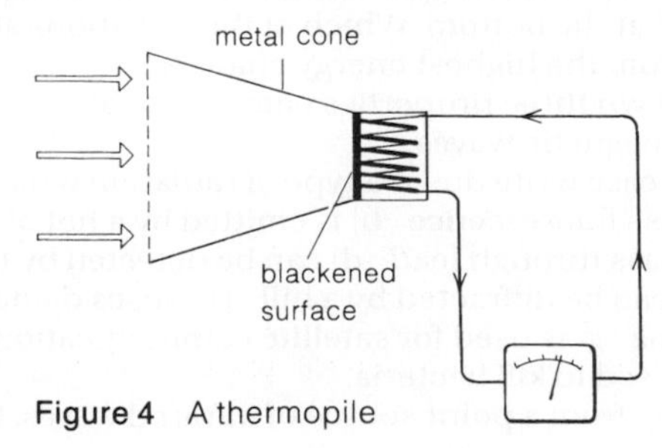

Figure 4 A thermopile

Good and bad emitters

Some surfaces are better emitters of thermal radiation than others. One method of comparing different surfaces is shown in figure 5. The metal cube is filled with boiling water which heats all the surfaces of the cube to the same temperature. The thermopile is placed in turn at the same distance from each surface and the meter readings compared.
Good emitters of thermal radiation are also good absorbers, while poor emitters are poor absorbers (and therefore good reflectors). Dull black surfaces are the best emitters, silvery surfaces the poorest. This fact is used in 'space blankets':

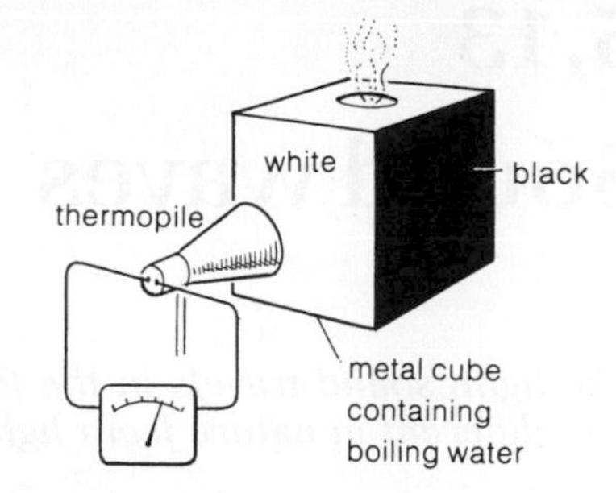

Figure 5 A thermopile can detect the difference between good and bad emitters of thermal radiation

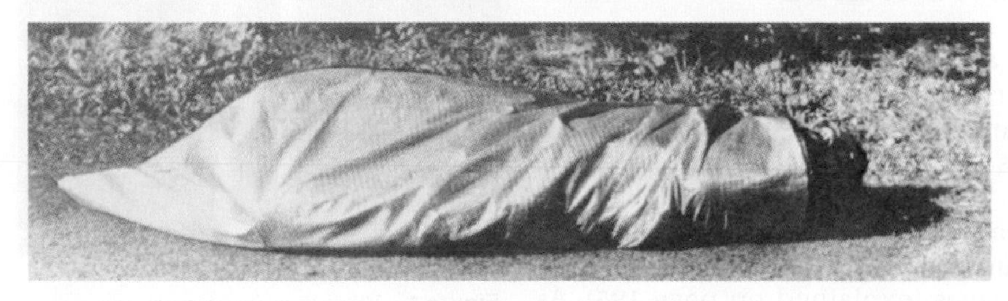
Figure 6 Emergency 'space' blankets reflect and conserve body heat

The vacuum flask

A 'vacuum flask' has two thin glass walls with a narrow space between them. As shown in figure 7, it has several features to reduce flow of thermal energy, and will keep liquids hot (or cold) for several hours.

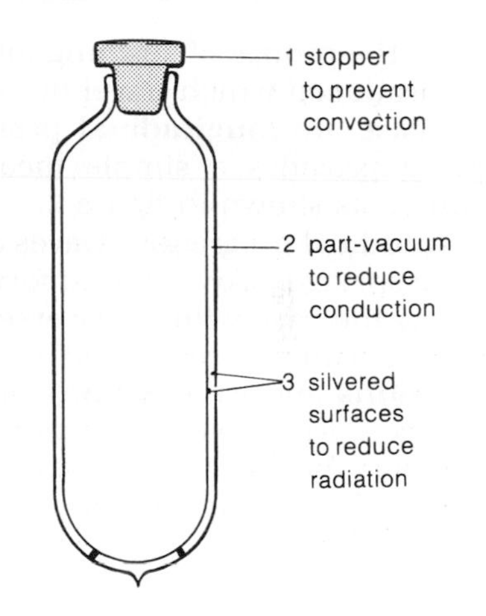

Figure 7 The vacuum flask reduces heat loss by conduction, and radiation

The greenhouse effect

Glass is transparent to the light and short wavelength infrared radiation emitted by very hot objects. On the other hand, glass won't transmit the long wavelength infrared given out by cooler objects. The radiation is either reflected or absorbed. These properties of glass are used in a greenhouse to 'trap' heat on a sunny day.

Light, and short wavelength infrared radiation from the Sun passes easily through the glass roof of a greenhouse and warms the air and other materials inside, as shown in figure 8. These warmed materials also give off thermal radiation. The wavelengths are longer however, and much of the radiation is reflected back into the greenhouse when it strikes the glass.

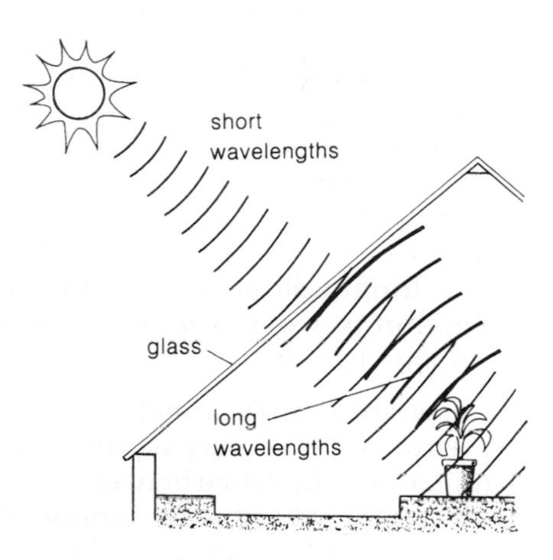

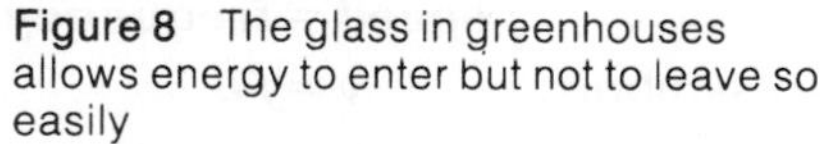
Figure 8 The glass in greenhouses allows energy to enter but not to leave so easily

Questions

1 What type of surface is best at a) absorbing b) emitting radiation c) reflecting thermal radiation?
2 Explain why the clothes worn in hot countries are often white.
3 What instrument can be used to detect thermal radiation?
4 How does the radiation emitted by a very hot object differ from that emitted by a warm one?
5 Give three ways in which a hot object may lose heat.
6 How are heat losses reduced in a Thermos flask?
7 Why does thermal radiation pass more easily into a greenhouse than out of it?

5.13

Sound waves

Like light, sound travels in the form of waves. But sound waves are very different in nature from light waves.

Longitudinal progressive waves

When the prongs of a tuning fork are vibrated, as in figure 1, waves are produced which travel outwards through the air. The waves in this case are **longitudinal** progressive waves. A stretched 'slinky' spring provides a simple means of studying this type of wave motion, as shown in figure 2.

Longitudinal progressive waves are made by rapidly moving the end of the spring backwards and forwards, rather than from side to side as was the case with transverse waves (explained on page 195). As the oscillations are passed from turn to turn, a series of **compressions** and **rarefactions** are seen travelling along the spring. The compressions are regions where the turns are bunched together; the rarefactions are regions where the turns are more stretched out than normal.

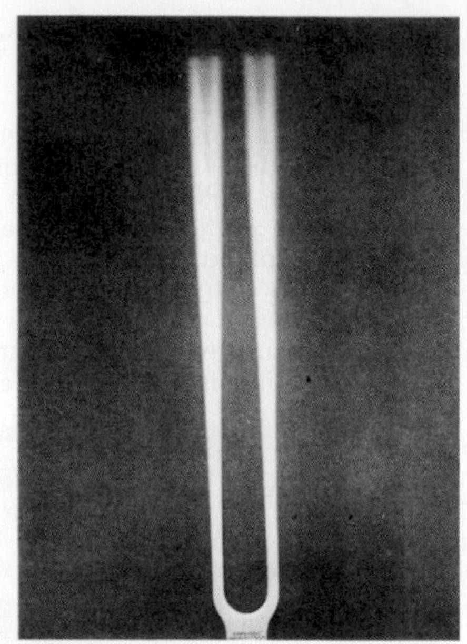

Figure 1 Vibrating objects usually produce sound. This is a *tuning fork*

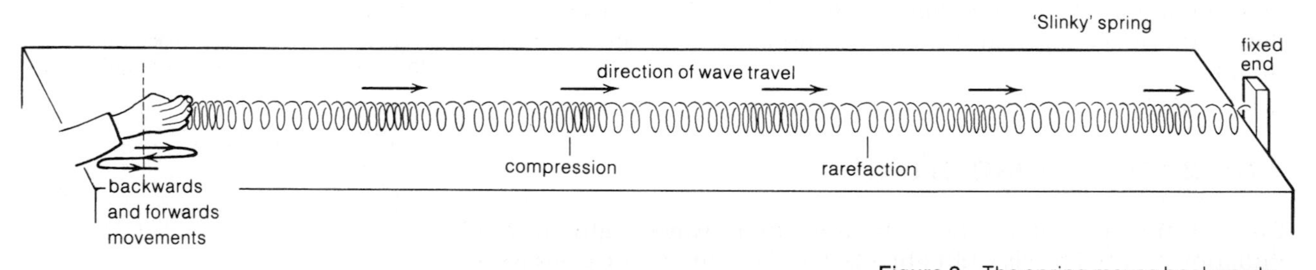

Figure 2 The spring moves backwards and forwards; the compressions move forward: longitudinal progressive waves

In a longitudinal wave, the oscillations take place in the direction of wave travel, i.e. backwards and forwards rather than from side to side.

In diagrams, longitudinal waves are usually drawn as a series of wavefronts, these marking the regions of compression (as shown in figure 3). The distance between wavefronts is the wavelength, and as with other forms of wave motion, the following equation applies:

speed = frequency × wavelength

In symbols: $v = f\lambda$

Speed is measured in m/s if frequency is measured in Hz and wavelength in m.

If the 'slinky' spring is stretched out further so that the tension is greater, the oscillations are passed from turn to turn more rapidly and the speed of the waves is increased.

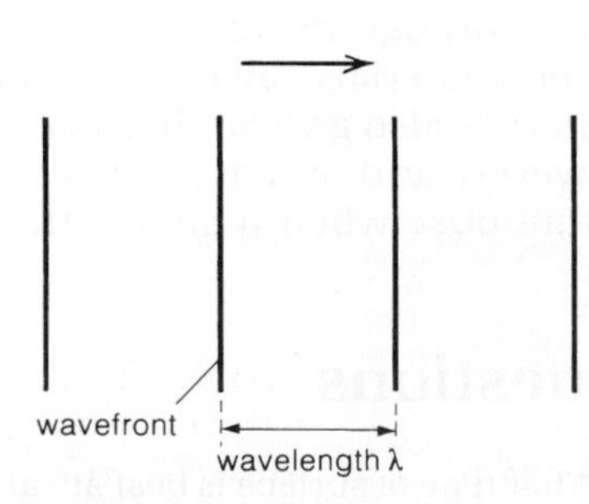

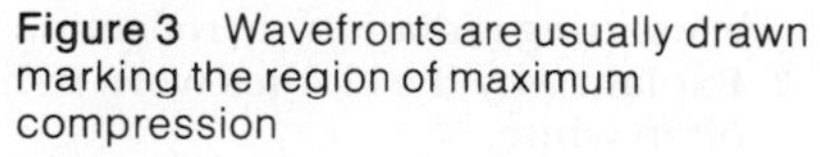

Figure 3 Wavefronts are usually drawn marking the region of maximum compression

The nature of sound waves

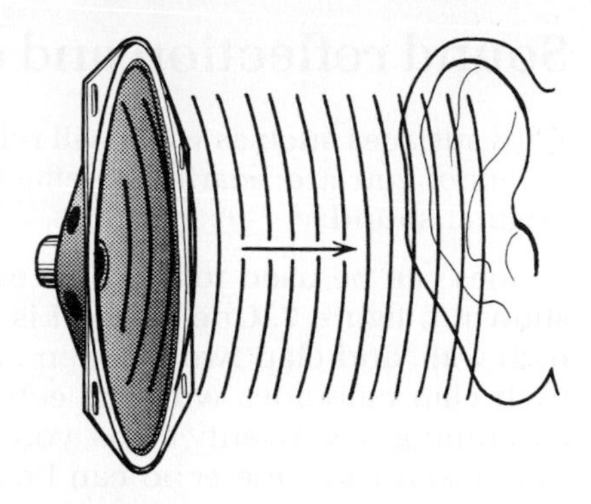

Figure 4 Sound waves move out from a loudspeaker. These can be detected by the ear

Sound waves are caused by vibrations When a loudspeaker cone vibrates, it alternately compresses then 'stretches' the air next to it, as shown in figure 4. The backwards and forwards movements cause a series of compressions and rarefactions to travel out through the air – the air transmits longitudinal progressive waves. The waves in this case are known as sound waves. When they enter your ear, they cause small but rapid pressure changes on a thin sheet of skin and muscle called the ear-drum; from this, you experience the sensation of sound.

Sound waves are produced when any vibrating object compresses and stretches the material around it. The prongs of a tuning fork give out sound waves when vibrated, as was shown in figure 1. The strings on a guitar, or the mass of air in a whistle or a flute will do the same. If two solid materials are banged together, sound waves are produced by the vibration of each surface.

Sound waves need a material to travel through Sound waves can only exist if there is a material present to pass on oscillations, so it isn't possible for sound to travel through a vacuum. Figure 5 shows how this can be demonstrated; less and less sound is heard from the electric bell as the air is pumped out of the glass jar.

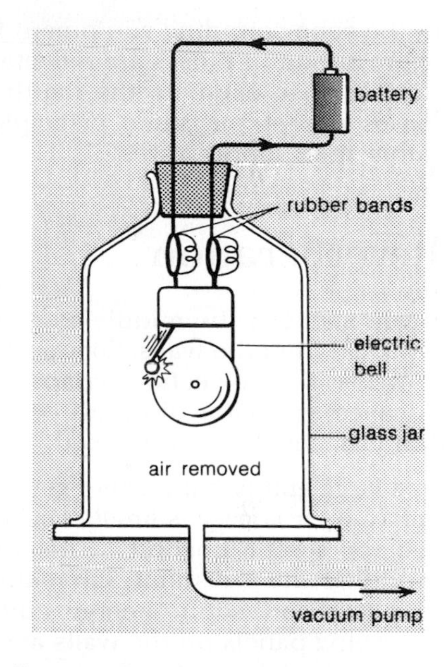

Figure 5 By removing the air from the jar, it is possible to show that sound cannot travel through a vacuum

Sound waves can travel through solids, liquids and gases Most of the sound waves reaching the ear do so by travelling through air, but it is also possible to hear sounds when swimming underwater. Walls, windows, doors and ceilings will all transmit sound.

The speed of sound

The speed of sound varies considerably depending on the material through which the waves are travelling. In general, sound travels more rapidly through liquids than through gases, and fastest of all through solids. Higher speeds result partly from stronger forces between molecules; as in a tightly stretched 'slinky' spring, the oscillations are passed on more rapidly:

speed of sound in air ______ 330 m/s (dry air, at 0 °C)
speed of sound in water ______ 1400 m/s (at 0 °C)
speed of sound in concrete ______ 5000 m/s

In the case of gases, for example air:

The speed of sound does not depend on the pressure; if atmospheric pressure changes for example, there is no change in the speed of sound.

The speed of sound increases with temperature At high altitudes, the speed of sound is less than it is at sea level, because the temperature is lower and *not* because the pressure is less.

Figure 6 shows how the speed of sound in air relates to other typical speeds. It is worth noting that the speed of sound in air is only about one three-millionth of the speed of light.

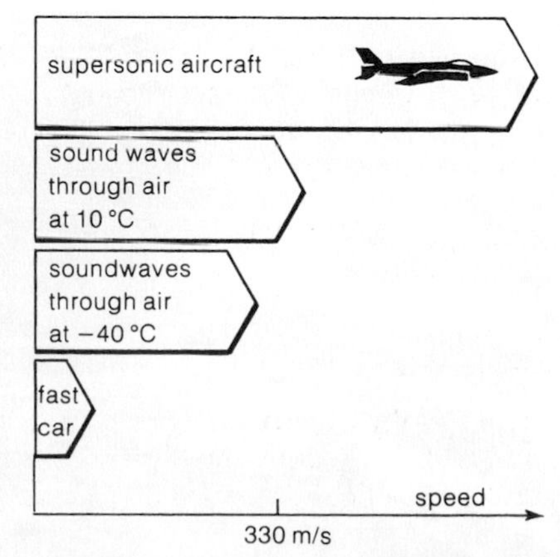

Figure 6 The speed of sound

Sound reflection and echoes

Hard surfaces such as walls will reflect sound waves. When you hear an echo, you are hearing a reflected sound a short time after the original sound.

Echoes can be used to make an estimate of the speed of sound, as shown in figure 7. One method is to stand 100 metres or so from a high wall, and clap two wooden blocks together at such a rate that each clap coincides with the echo of the one before. By counting and timing, say, twenty claps, you have timed twenty echoes, so the time taken t for one echo can be calculated. If the distance to the wall is d:

$$\textbf{speed of sound} = \frac{\textbf{distance travelled by sound}}{\textbf{time taken}} = \frac{2d}{t}$$

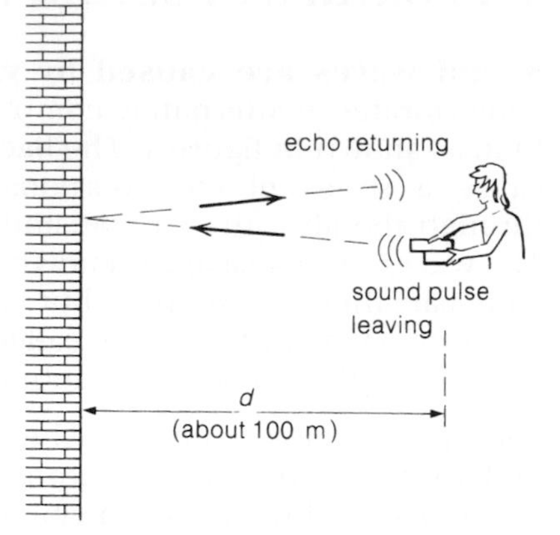

Figure 7 The blocks are clapped again as each echo is heard. The time for say twenty claps is found; the time for one echo is thus one twentieth of this

In the echo-sounding equipment fitted to some boats (as shown in figure 8), sound pulses are reflected from the sea bed, and the echo time used to estimate the depth of water under the boat. Radar makes use of the same principle, though microwaves are used rather than sound waves.

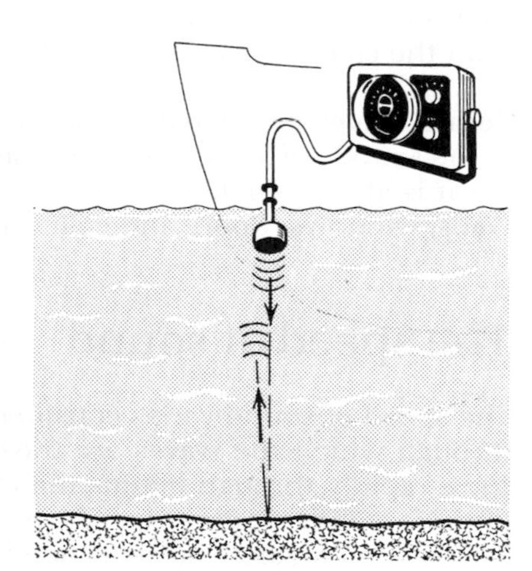

Figure 8 Echo-sounding machines time the echo to find the distance

Reverberation

If you stand in the middle of a room and shout, sound waves are reflected from the walls, floor and ceiling. The echo time is so short however, that the echo overlaps with the original sound. The effect is called **reverberation**; an echo isn't detected, but the original sound seems to be prolonged.

In a large empty hall, where sound waves are reflected many times before their energy is finally absorbed, the reverberation effect may last for several seconds. This makes music and speech sound muddled and indistinct. Soft materials such as curtains and carpets help to absorb sound energy, and many concert halls have sound-absorbing panels on the walls and ceilings to reduce reverberation (as shown in figure 9).

In practice, it isn't desirable to cut out reverberation altogether as this makes a concert hall sound 'dead'.

Figure 9 Inside the Albert Hall, London, saucer-shaped sound absorbers have been hung from the ceiling to stop unwanted echoes

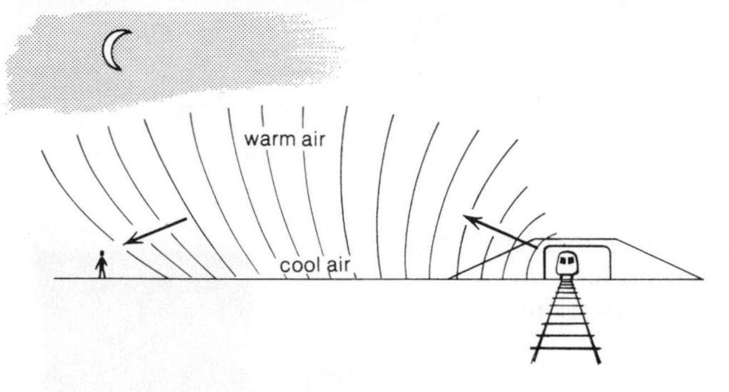

Figure 10 Refraction of sound waves by air of differing temperature

signal generator
loudspeaker

Figure 11 Coherent sound waves also 'interfere' with one another to produce areas of louder and softer sound

The wave nature of sound

Like other types of wave, sound waves can be diffracted and refracted, and they can also give rise to interference effects:

Sound waves detected by the human ear have wavelengths ranging from about 1 cm up to 15 metres, and are diffracted appreciably by everyday objects of comparable size. The diffraction of sound waves makes it possible to hear round corners.

Distant traffic and trains often sound much louder (and closer) at night. The effect is caused by the refraction of sound waves, as shown in figure 10. During the night time, the air layers near the ground become cooler than those above, so sound waves travel more slowly through them. As a result, sound waves leaving the ground tend to be refracted back towards the ground rather than being lost in the upper air layers.

Two loudspeakers emitting sound waves of the same constant frequency, will produce a steady interference pattern, as shown in figure 11. Walking along a line in front of the speakers, you can hear the sound become alternately louder and quieter as waves from the two speakers reinforce then cancel each other out.

Light does not diffract round corners . . . but sound does

Questions

1 What is the difference between a longitudinal progressive wave and a transverse progressive wave? What type of wave motion is sound?
2 Why is it not possible for sound waves to travel through a vacuum? How could you show this experimentally?
3 What evidence is there to suggest that sound waves can travel through a) a solid b) a liquid c) a gas?
4 Why is it possible to hear round corners?
5 What happens to the speed of waves along a 'slinky' spring if the tension is increased?
6 What happens to the speed of sound in air if a) the air pressure rises b) the temperature rises?
7 What evidence is there to suggest that sound is a form of wave motion?
8 Echo-sounding equipment on a ship receives sound pulses reflected from the sea bed 0.02 seconds after they were sent out. If the speed of sound in sea water is 1500 m/s, what is the depth of water under the ship?
9 A woman stands 120 m away from a high wall. She claps two blocks of wood together at a steady rate such that 40 claps are made in 30 seconds. If each clap coincides with the echo of the one before, what is the speed of sound?
10 What evidence is there that sound waves interfere?

5.14

Hearing sounds

Through music, speech and noise, the human ear experiences a wide variety of sound sensations. Yet all these sensations depend only on differences in frequency and amplitude of the sound waves entering the ear.

In laboratory experiments, a signal generator and loudspeaker are often used as a source of sound waves, as shown in figure 1. The signal generator produces an oscillating electric current, which makes the loudspeaker cone vibrate backwards and forwards. The size of the vibration can be increased by turning up the gain ('volume') control on the signal generator. The frequency of the vibration, and therefore of the sound waves, can be set to any particular value by turning a second control knob which moves a pointer across a frequency scale.

Figure 1 A signal generator and loudspeaker produce a continuous note at an adjustable frequency

Amplitude, loudness and intensity

When a loudspeaker cone vibrates, the **amplitude** of the oscillation is the maximum distance the loudspeaker cone moves backwards or forwards from its rest position, as shown in figure 2. If the amplitude of the cone oscillations is increased, the amplitude of the sound waves is also increased and more sound energy travels out through the air every second. To the ear, the sound becomes louder.

The **intensity** of a sound wave is a measure of the wave energy passing every second through each square metre at right angles to its path. A sound wave has an intensity of 1 watt per square metre ($1\,W/m^2$) if 1 joule of wave energy is passing through each square metre every second. The loudest sounds the ear can endure without damage have an intensity of about $1\,W/m^2$.

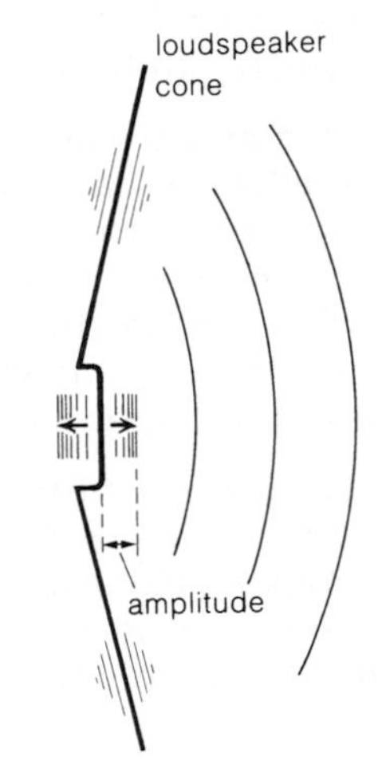

Figure 2 Increasing the *amplitude* of the vibration increases the *loudness* of the note

Frequency and pitch

Tests using a signal generator and a loudspeaker show that the human ear can detect sound waves with frequencies ranging from about 20 Hz (20 waves every second) up to about 20 kHz (20 000 waves every second), though the ability to hear high frequencies becomes less with age.

Different frequencies sound different to the ear. High frequencies are heard as notes said to be of **high pitch**; low frequencies are heard as notes said to be of **low pitch**. Examples of sound frequencies are given in figure 3.

Some notes give a harsh and discordant sound when played together, others blend well and produce a sound which is easy to listen to. To the ear, two notes blend best of all if one frequency is exactly twice the other. On a musical scale, such notes are said to be one **octave** apart in pitch.

pitch		frequency
high	upper limit of hearing	20 000 Hz
	whistle	10 000 Hz
	high note (soprano)	1000 Hz
	low note (bass)	100 Hz
low	drum note	20 Hz

Figure 3 High frequency vibrations produce high-pitch notes

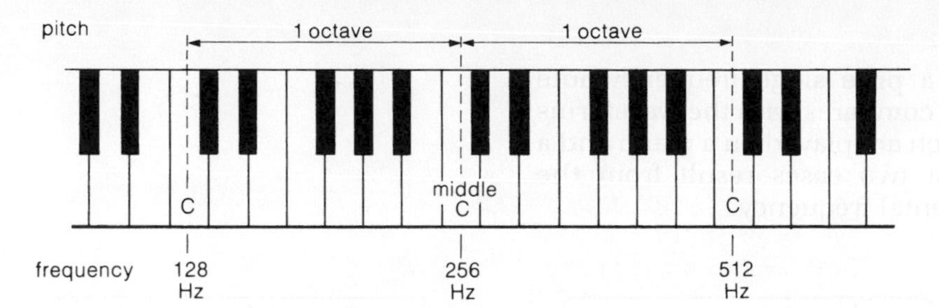

(piano tuned to scientific pitch)

Figure 4 Adding a note which has double the frequency of the first produces a pleasing blend of sound. This is the basis of musical scales

Figure 4 illustrates the keyboard of a piano tuned to **scientific pitch** in which the note called Middle C corresponds to a frequency of 256 Hz. Frequencies of notes one octave above and below Middle C are also shown. To overcome certain tuning difficulties, pianos are normally tuned to frequencies slightly different from those given in the diagram.

Waveforms

The instrument with the display screen in figure 5 is called an oscilloscope. Connected to a microphone, an oscilloscope can be used to display incoming sound waves in graphical form.

The microphone contains a thin sheet of metal called a diaphragm which oscillates rapidly backwards and forwards as the compressions and rarefactions of sound waves strike it. These movements are changed into electrical oscillations which are used to move a white spot upwards and downwards on the oscilloscope screen. At the same time, the spot is pulled rapidly across the screen from left to right. The combination of the two movements produces a line called a **waveform** on the screen. The waveform is repeated many times every second.

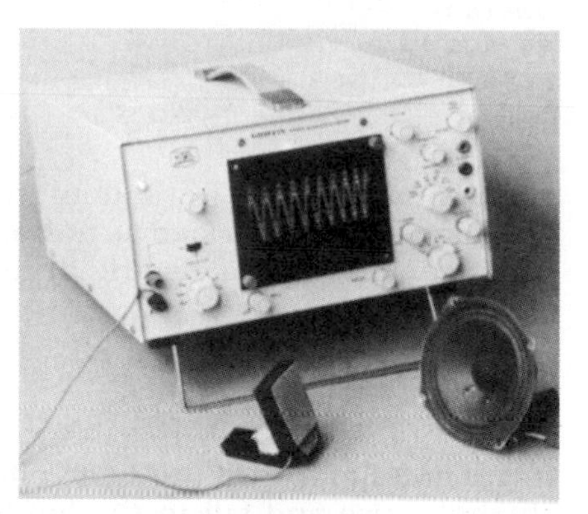

Figure 5 An oscilloscope can produce a sound waveform

The waveform looks like a series of transverse waves. In reality, it is a graph showing how the pressure of the air next to the microphone diaphragm varies with time. The crests correspond to sound wave compressions and the troughs to rarefactions.

Waveforms on an oscilloscope screen can be used to measure the frequency and amplitude of sound waves. They can also be used to study the characteristics of the sounds produced by different musical instruments, as discussed below.

Quality

Middle C played on a piano has a different characteristic sound from Middle C played on a guitar. The notes are said to have a different **quality** or **timbre**.

Both notes contain a strong **fundamental frequency** which corresponds to Middle C, but many other weaker frequencies called **overtones** are also present. It is the number, frequency, and relative amplitude of these overtones which give each sound its particular quality (as shown in figure 6). Overtones are usually exact multiples of the fundamental frequency: 2, 3, or 4 times this frequency for example.

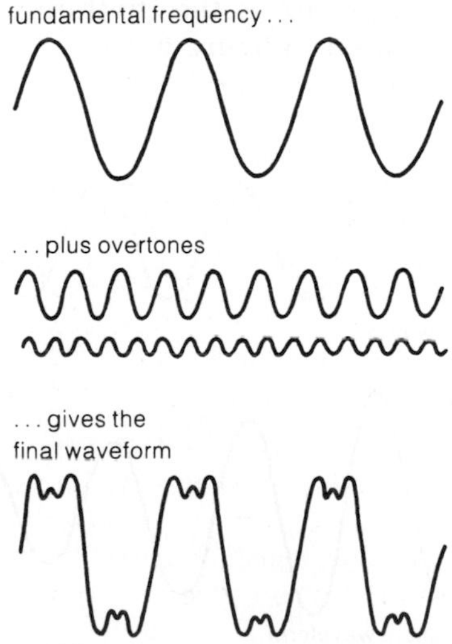

fundamental frequency . . .

. . . plus overtones

. . . gives the final waveform

Figure 6 The overtones to a basic note give the instrument its particular quality

Figure 7 shows how the waveform of a 'pure' single frequency note from a signal generator or tuning fork compares with the waveforms produced when notes of the same pitch are played on a piano and a guitar. The waveforms in these last two cases result from the addition of overtones to the fundamental frequency.

waveforms

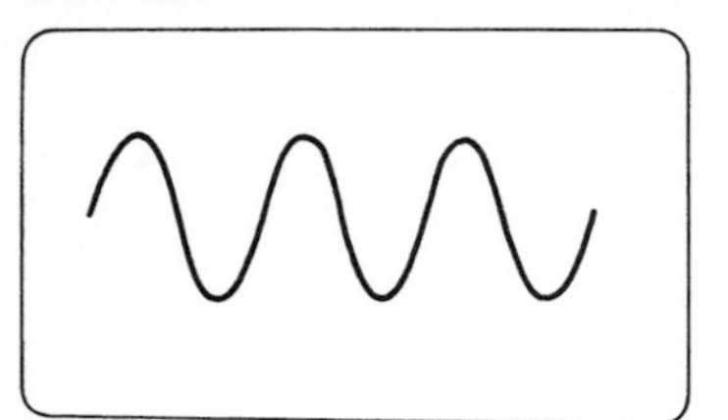

pure note:
signal generator,
tuning fork

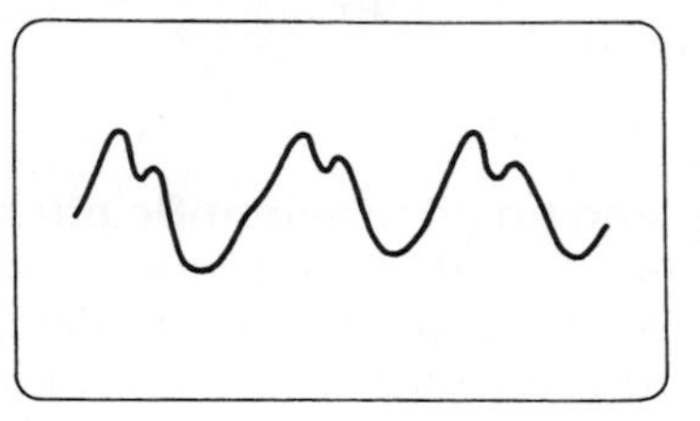

guitar

piano

Figure 7 Wave forms from different sources

Sounds which are a random mixture of constantly changing frequencies are described as noise, as shown in figure 8. Jet engines produce noise, so do electric drills.

noise

Figure 8 Noise is a jumble of many frequencies together

Beats

If two notes of almost the same pitch are played together, there is a rhythmic rise and fall in the loudness of the sound heard. These pulsations in sound are known as **beats**. They can be demonstrated by connecting two loudspeakers to separate signal generators, each set to a slightly different frequency.

As the frequencies of two sets of sound waves differ, the waves go alternately in and out of phase with each other. As a result, the waves alternately reinforce and cancel each other out, and it is this which produces the rhythmic variations in loudness. The effect is illustrated in figure 9.

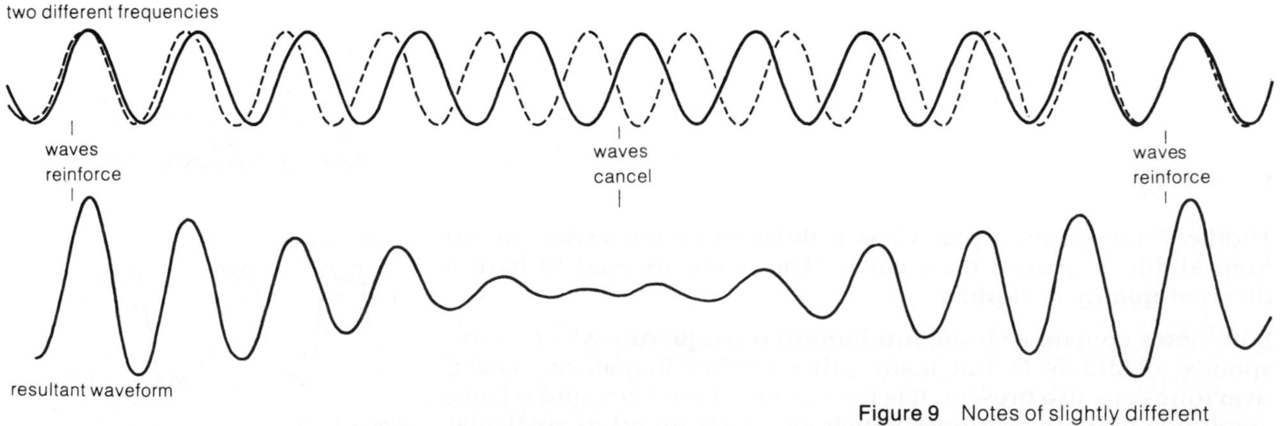

Figure 9 Notes of slightly different frequencies, sounded together, produce beats. At some times the waves reinforce one another, at other times they cancel one another's effect

If two sound sources emit notes of frequencies f_1 and f_2,

beat frequency $= f_1 - f_2$

For example, if the frequency of one source is 300 Hz and the frequency of the second source is 304 Hz, the beat frequency is 4 Hz: 4 beats are heard every second. The closer the frequencies of the two sources, the fewer beats are heard every second. The sources are exactly 'in tune' if no beats are heard.

Beats provide a useful means of tuning musical instruments. A guitar can be tuned by adjusting each string so that no beats can be heard when the same note is played on it and the string next to it.

Waveforms on disc

No wider on average than a human hair, the grooves on a stereo LP record actually form part of one continuous groove a kilometre long which spirals inwards from the edge of the disc. The groove has two walls with a different wave-like shape along each, as shown in figure 10. These are the waveforms of the sounds which emerge from the left and right hand speakers of the record player when the record is played. As the record rotates, the groove walls make a small stylus oscillate in the two directions shown in figure 11. Each oscillation is used to generate a small oscillating electric current which is then amplified (made bigger) and fed to the appropriate speaker.

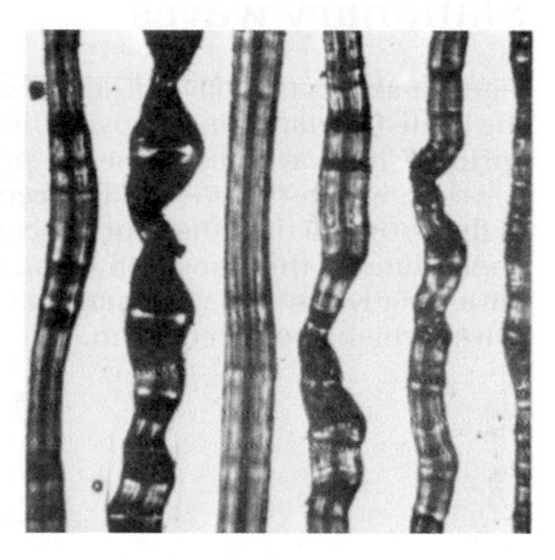

Figure 10 Grooves of a stereo record

When an LP is first made, the grooves need to be 'cut' as close together as possible in order to keep playing time to a maximum. The cutting stylus is controlled by a machine which spaces out the grooves for loud music passages when waveform amplitudes are high, and bunches up the grooves for quieter music. Loud bass (low) notes have the largest amplitudes of all, so the bass content of the recording is reduced when the record is cut in order to save groove space. During playback, the record player amplifier compensates for this by boosting the bass.

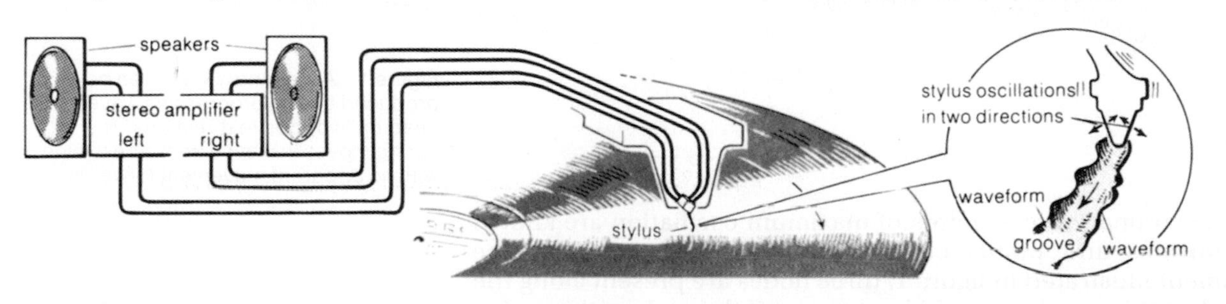

Figure 11 A stereo system: two oscillations at right angles are used to produce two separate sounds

Questions

1 Two sounds have the same frequency but one is louder than the other. In what way do the sound waves differ?

2 What happens to the pitch of a note if the frequency is increased?

3 A note has a frequency of 200 Hz. What does this mean? What is the frequency of a note a) one octave lower b) two octaves higher?

4 What are overtones?

5 Why does a note played on a guitar sound different from a note of the same pitch played on a piano?

6 One loudspeaker emits a note of frequency 200 Hz. A second loudspeaker emits an equally loud note of frequency 205 Hz. What would you hear when stood near the loudspeakers? What change would occur if the frequency of the note from the second loudspeaker were lowered to a) 202 Hz b) 200 Hz c) 195 Hz?

5.15

Vibrating strings and air columns

Most musical instruments produce sound when a stretched string or air column is made to vibrate. The vibrations give rise to waves known as stationary waves in the string or air column itself, and cause progressive waves of the same frequency to travel out from the instrument.

Stationary waves

Figure 1 shows the effect of sending two sets of progressive waves of the same frequency in opposite directions along a stretched 'slinky' spring. The waves combine to produce **stationary** or **standing** waves in which there is no apparent wave movement from one end of the spring to the other but the coils oscillate by different amounts depending on their position along the spring. The stationary waves have exactly the same frequency and wavelength as the progressive waves which produced them.

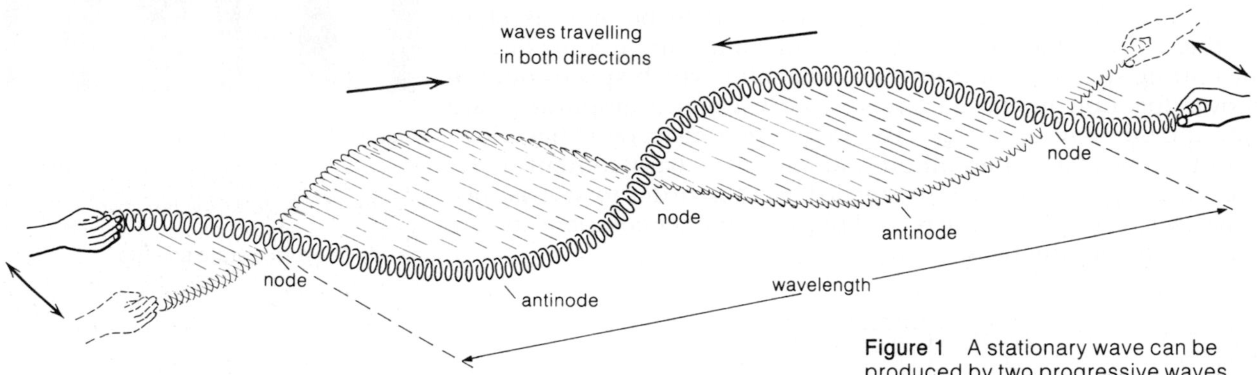

Figure 1 A stationary wave can be produced by two progressive waves heading in opposite directions. The spring moves in waves from side to side with no apparent movement forwards

Along a stationary wave, points of maximum oscillation are known as **antinodes** and points of zero oscillation as **nodes**. In the experiment illustrated in figure 1, three nodes are present along the spring, but this number would be different if the ends of the spring were made to oscillate at a different frequency or the spring were stretched to a different extent.

Stationary waves in a stretched string

When a guitar string is plucked, transverse waves are sent along the string in both directions (as shown in figure 2). The waves are reflected at the fixed ends, and as they pass through each other they combine to produce a stationary wave.

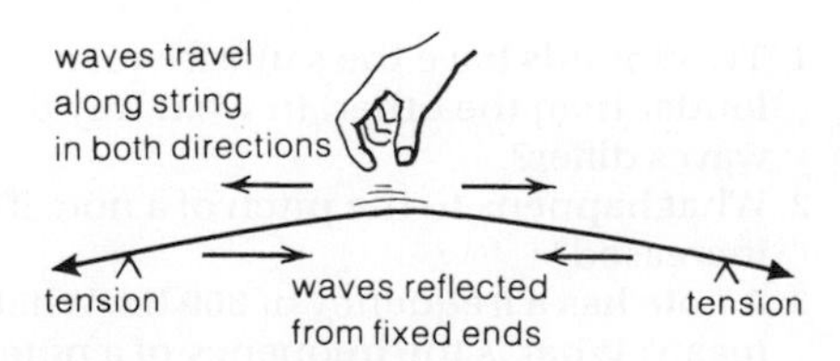

Figure 2 A plucked guitar string produces a stationary wave

A stretched string can vibrate in various ways or **modes** depending on the way it is plucked. The simplest mode of vibration is shown in figure 3a; only one antinode is present, and there is a node at each end of the string where motion is not possible. The string in this case is said to be vibrating at its fundamental frequency, and the length of the string is half the wavelength of the stationary wave.

Other possible modes of vibration occur at 2, 3, and 4 times the fundamental frequency, and so on. The first two of these other modes are shown in figures 3b and 3c. Whatever the mode of the vibration, there is always a node at each end of the string. A guitar string may vibrate in several modes at once when plucked, and it is the combination of the fundamental note and overtones which gives the sound its particular quality (see page 215).

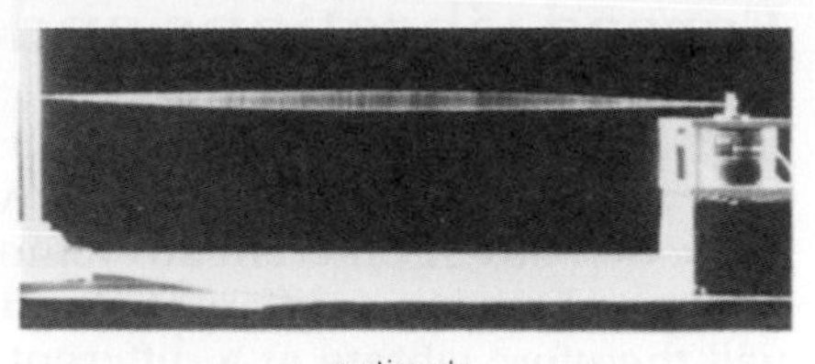

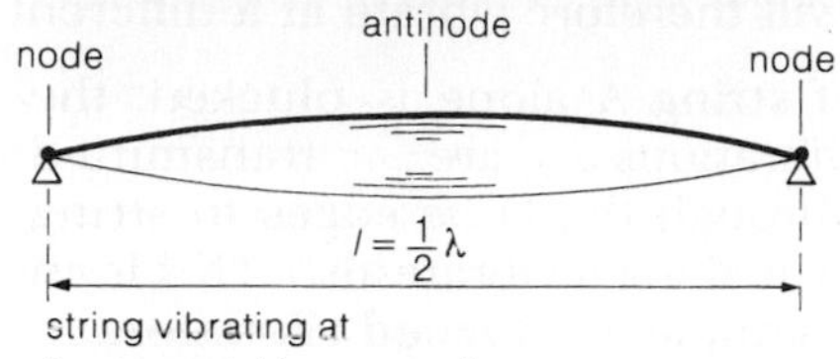

Figure 3a

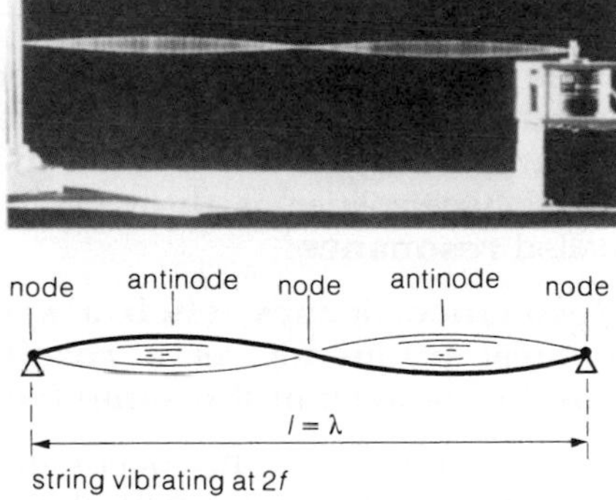

Figure 3b

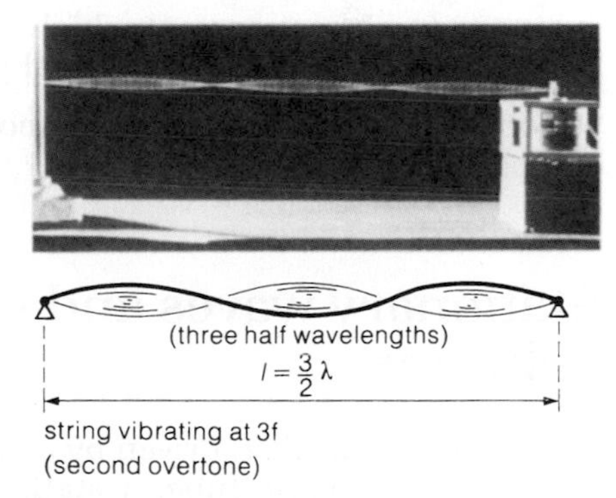

Figure 3c

Figures 3a, 3b, 3c Fundamental note with first and second overtone

Frequency of a vibrating string

The frequency of a vibrating string depends on several factors. Experiments show that, if the fundamental frequency of a string is f:

***f* is proportional to 1/length** Halving the length of a string doubles the frequency of vibration providing the stretching force on the string (called the **tension**) stays the same (as shown in figure 4). On a guitar, you push the strings against metal strips called frets in order to change the length of string which vibrates.

***f* is proportional to $\sqrt{\text{tension}}$** If the tension in a stretched string is increased to four times its previous value, the frequency of vibration doubles. In a guitar, small tension changes are used to make tuning adjustments to the strings. Tightening a string raises the pitch; slackening a string lowers it.

***f* is proportional to $1/\sqrt{\text{mass of the string per unit length}}$** Interpreting this in the simplest possible way – heavy strings vibrate more slowly than light ones. Bass strings on a guitar are usually in the form of thin coiled springs around a nylon core, as these give a high mass per unit length while allowing smaller tension changes to be made than would be possible with a completely solid string.

All of the above can be expressed in the form of an equation:

$$f = \frac{1}{2l}\sqrt{\frac{T}{m}}$$

where f is the fundamental frequency, l is the length of the string, T is the tension (stretching force), and m is the mass per unit length.

Figure 4 Halving the length of a string *doubles* the frequency

Forced vibrations and resonance

In the experiment shown in figure 5, two identical guitar strings have been stretched between the two knife edges which are fixed to a wooden box. A different stretching force has been applied to each string by hanging a different standard mass from each. Each string will therefore vibrate at a different frequency when plucked.

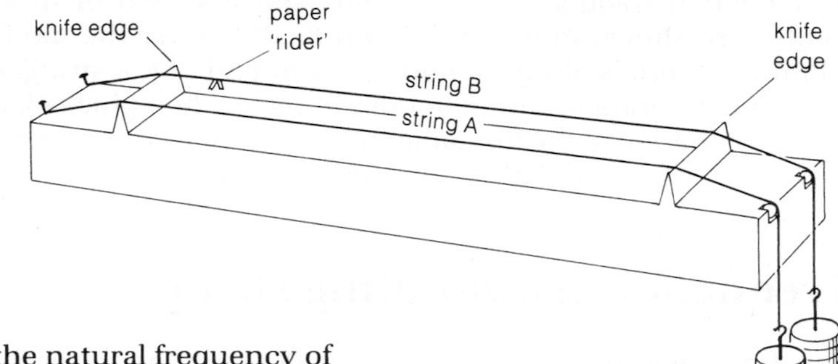

Figure 5 The vibration of string A is transmitted to string B. If the vibration is at the *natural frequency* of string B the vibration will be intense. This is *resonance*

If string A alone is plucked, the vibrations are transmitted through the knife edges to string B, making it vibrate also. This is an example of a **forced vibration**. In this case, string B is being forced to vibrate at a frequency other than its natural frequency, and the amplitude of the vibrations is small.

If the tension in string B is changed so that the natural frequency of vibration of B becomes the same as that of string A, the forced vibrations become much larger in amplitude – so much so that the small paper 'rider' is thrown off. This is an example of an effect called **resonance**:

Resonance occurs when a system is made to vibrate at its natural frequency as a result of vibrations received from another source of the same frequency.

Resonance doesn't only occur in stretched strings. The sides of a bus vibrate strongly when the engine frequency matches their natural frequency vibration. And if you ever used a playground swing, you made your pushes at the same frequency as the natural frequency of the swing in order to go high. Some electrical circuits have a natural frequency at which an electric current will oscillate; use is made of this fact in a signal generator.

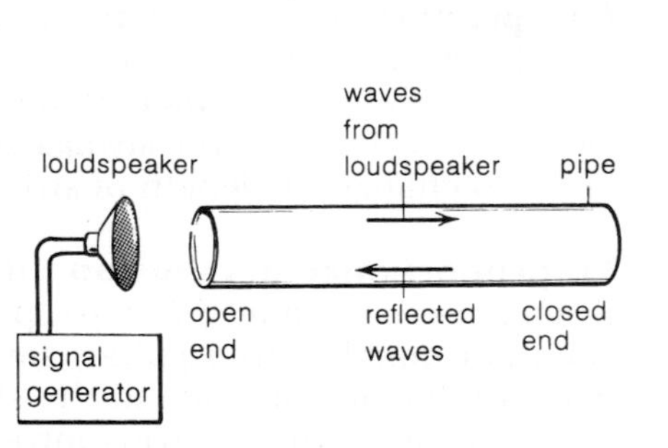

Figure 6 Sound waves can produce resonance

Resonance can also occur when a column of air is made to vibrate in a tube.

Stationary waves and resonance in air columns

If a small loudspeaker, driven by a signal generator, is held near the open end of a glass tube, a stationary wave is set up in the air column. The stationary wave is produced as waves travelling along the tube combine with reflected waves travelling in the opposite direction (as shown in figure 6).

Like a stretched string, an air column has certain natural frequencies at which it will vibrate. If the frequency of the loudspeaker matches any of these exactly, resonance occurs and a loud note is heard coming from the glass tube.

The sounds produced by wind instruments are caused by resonance in a column of air. When you blow into a recorder (as shown in figure 7), a wedge breaks up the air flow and causes the air to vibrate over a wide range of frequencies. Some of these match the

air flow

wedge

Figure 7 The recorder

natural frequencies of vibration of the air column in the recorder and a loud note is heard as a result.

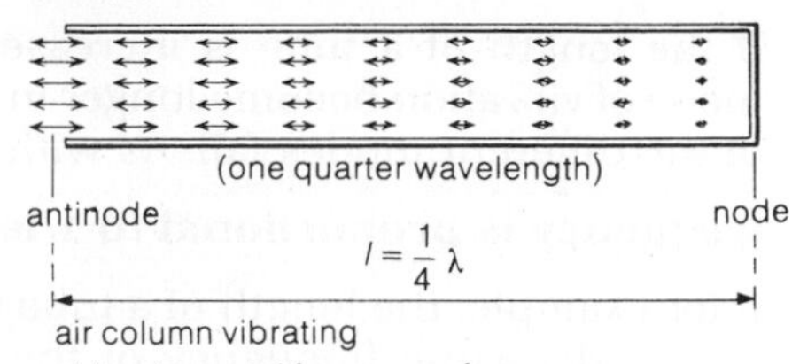

Figure 8 An air column vibrating at its fundamental frequency

Modes of vibration in an air column

The vibrations which cause stationary waves in a column of air are longitudinal (backwards and forwards) rather than transverse as they are in a guitar string. Figure 8 shows the simplest natural mode of vibration which can occur in a tube open at one end only. The air column in this case is vibrating at its fundamental frequency. There is an antinode at the open end of the tube where the air vibration is a maximum, and a node at the closed end where the air is unable to vibrate backwards and forwards. The tube is about a quarter of a wavelength long, though not exactly so because the antinode occurs slightly beyond the open end.

It is difficult to represent longitudinal stationary waves in diagram form, so they are usually drawn as transverse waves instead. Figure 9a shows how the fundamental mode is drawn using this convention. It is important to remember that the stationary wave is not really transverse and that the node exists right across the column and not just at one point.

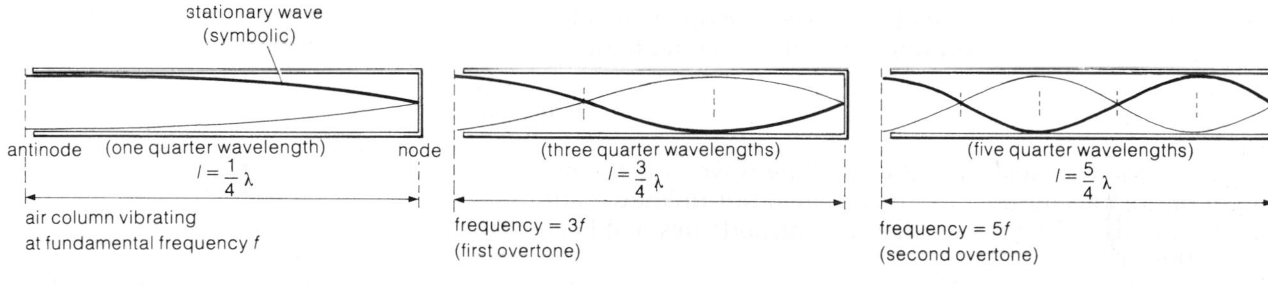

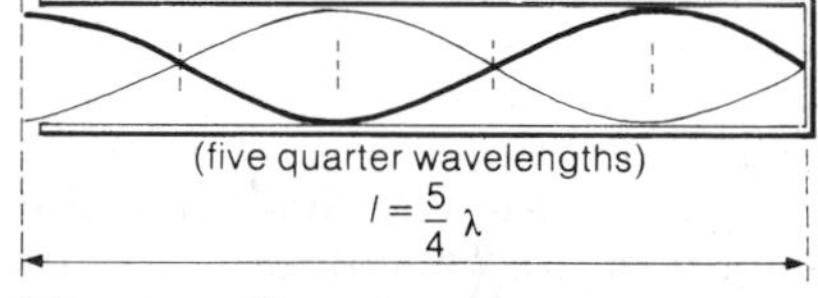

Figure 9a The stationary wave of figure 8 is usually drawn this way

Figure 9b The frequency is 3 f; this is the first overtone

Figure 9c The frequency is 5 f; this is the second overtone

Figures 9b and 9c show other natural modes of vibration of the air column, all of which occur at frequencies higher than the fundamental. In each case, there is a node at the closed end of the tube and an antinode at the open end. When you blow into a wind instrument like a recorder, several modes may be present at once, and the combination of the fundamental and overtones gives the instrument its particular sound quality.

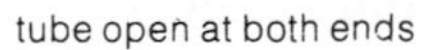

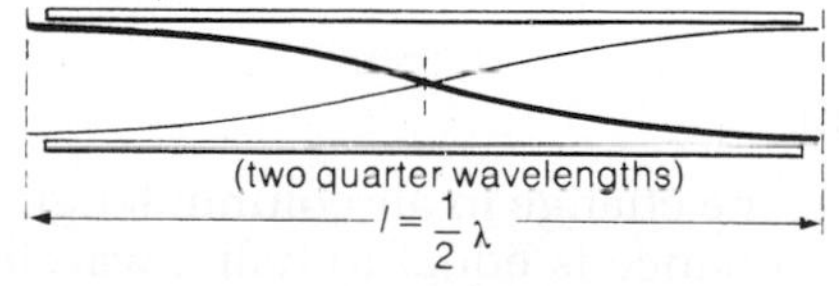

Figure 10 In an open tube the fundamental frequency is 2 f

Stationary waves can also be produced in a tube which is open at both ends (as shown in figure 10). Whatever the mode of vibration, there is an antinode at each end of the tube and the frequency is double that in a tube which is open at one end only.

Effect of pipe length on frequency

If the length of a tube is increased, the standing waves for each mode of vibration become longer in wavelength, and the frequencies of the different modes fall. As with a guitar string;

frequency is proportional to 1/length

If for example, the length of a tube is halved (as shown in figure 11), the fundamental frequency of the air column is doubled and the note emitted rises by one octave. To produce different notes on a recorder, you change the effective tube length by blocking and unblocking air holes with your fingers.

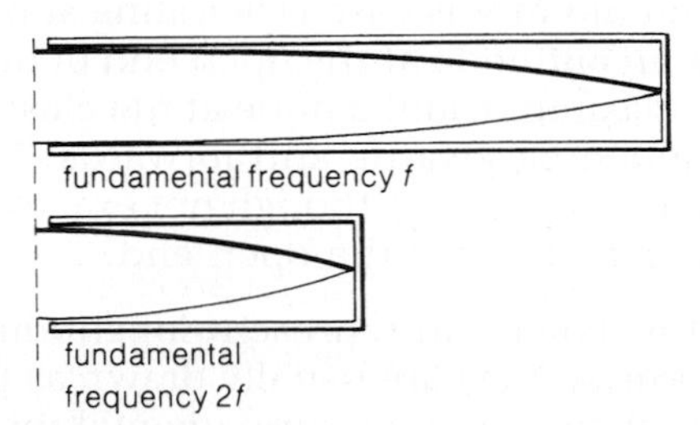

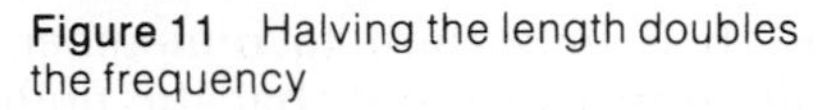

Figure 11 Halving the length doubles the frequency

Measuring the speed of sound using a resonance tube

The wavelength of the sound waves emitted by a tuning fork can be measured using the apparatus shown in figure 12. Knowing the frequency f of the tuning fork – normally marked on the side – the speed of sound in air can be calculated using the equation $v = f\lambda$.

The length of the air column in the tube can be varied by raising or lowering the water reservoir. Starting with a very short air column, the column length is slowly increased until resonance occurs when the vibrating tuning fork is held just above the open end of the tube. The air column is now vibrating at its fundamental frequency and:

$$\frac{\lambda}{4} = l_1 + c \qquad \textit{... equation 1}$$

where λ is the wavelength of the standing wave, l_1 is the measured length of the air column, and c is a small but unknown additional length to correct for the fact that the antinode lies just beyond the end of the tube.

The air column is now lengthened until resonance occurs again. In this case,

$$\frac{3\lambda}{4} = l_2 + c \qquad \textit{... equation 2}$$

where the same end correction, c, applies as in the first case. Subtracting (1) from (2) gives

$$\frac{\lambda}{2} = l_2 - l_1$$

i.e. the change in air column length betwen the first two positions of resonance is equal to half a wavelength.

In this way, λ is calculated. f is known from details printed on the tuning fork, so:

speed of sound $\quad v = f\lambda$
or $\quad v = 2f(l_2 - l_1)$

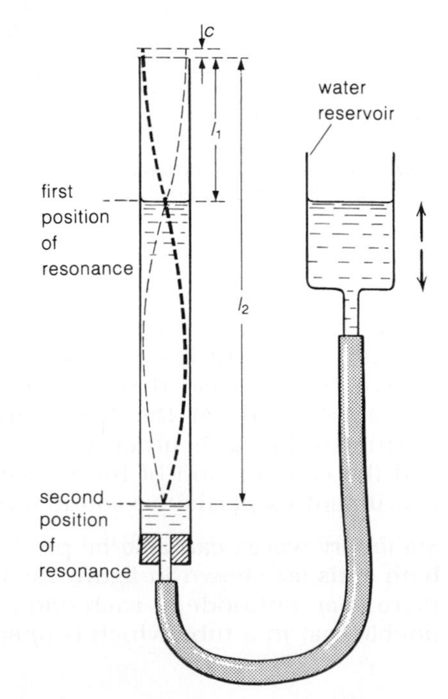

Figure 12 A resonance tube can be used to measure the wavelength of a sound. This can be used to calculate the speed of the sound

An open-ended horn 3 m long. What would be its fundamental frequency? What is the note produced if it actually sounds at the third overtone?

Questions

1 longitudinal and progressive; transverse and progressive; longitudinal and stationary; transverse and stationary.
Which of the above describes the waves found in each of the following cases?
a) a vibrating guitar string b) the vibrating column of air in a flute c) the sound waves emitted by a flute.

2 Figure 13 shows a stretched string vibrating at a frequency of 600 Hz.

Figure 13

Copy the diagram. Mark in the wavelength, the maximum amplitude of the oscillations, and the positions of the nodes and antinodes.
Draw another diagram to show the string vibrating at its fundamental frequency. What is this frequency?
What is the difference in pitch between the notes emitted in the two cases?
What would be the effect on the fundamental frequency of a) increasing the tension in the string b) using a heavier string?

3 A stretched string 60 cm long vibrates at a frequency of 100 Hz. At what frequency would it vibrate if its length were reduced to 15 cm but the tension were unaltered? How would this affect the pitch of the note produced?

4 Draw diagrams to represent the standing wave in a tube open at one end when the air column is vibrating at a) its fundamental frequency b) a frequency corresponding to the first overtone.
If the speed of sound is 330 m/s and the fundamental frequency is 110 Hz, what is the approximate length of the tube?
What is the frequency of the first overtone?
What would be the fundamental frequency and the frequency of the first overtone if the tube were open at both ends?

5 In an experiment to measure the speed of sound using a resonance tube, a tuning fork of frequency 640 Hz is used. When the length of the air column is increased from zero, resonance occurs at lengths of 12 cm and 37 cm. What is the speed of sound?

Further questions

Waves: light and sound: part A

1 a) Figure 1 shows two rays of light leaving an object O and striking a plane mirror.

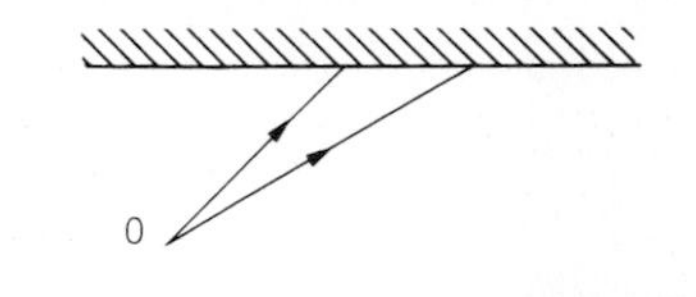

Figure 1

Copy the diagram. Draw the two reflected rays and use them to find the position of the image.

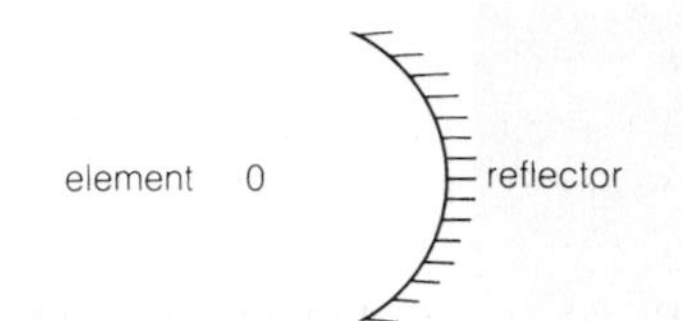

Figure 2

b) Figure 2 shows a side view of an electric fire.
i) What types of electromagnetic waves are given out by the element?
ii) What name is given to the shape of the reflector?
iii) The reflector is made of metal. Describe its surface, and explain why metal is used. (NEA)

2 a) Figure 3 shows a ray of red light incident on a glass prism. Copy and complete the diagram to show the path of the light through and out of the prism.

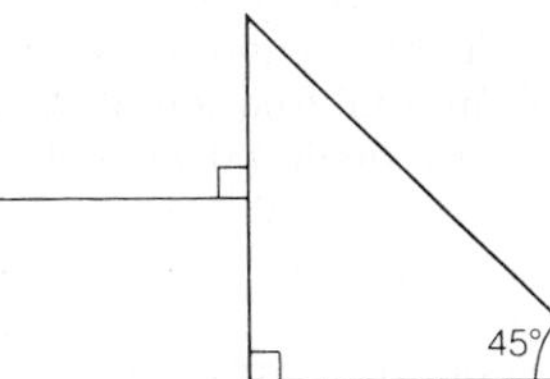

Figure 3

b) If white light were used instead of red light on this prism, what difference, if any, would you notice?

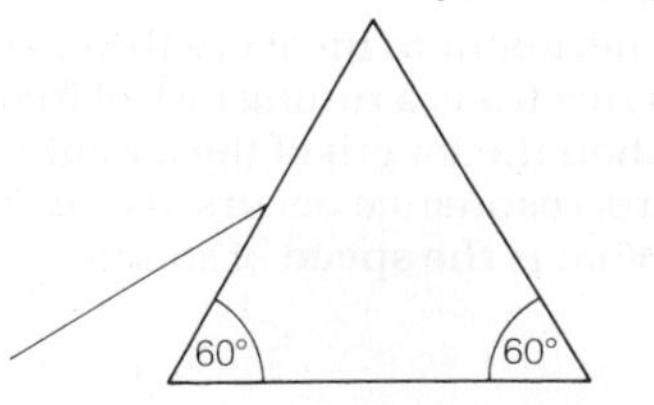

Figure 4

c) Figure 4 shows a ray of red light incident on a different prism. Copy and complete the diagram to show the path of the light through and out of the prism.
d) If white light were used instead of red light on this prism, what difference, if any, would you notice? (MEG)

3 This question is about solar panels, devices that are sometimes seen on the roofs of houses and are used to provide hot water. An example is shown in figure 5.

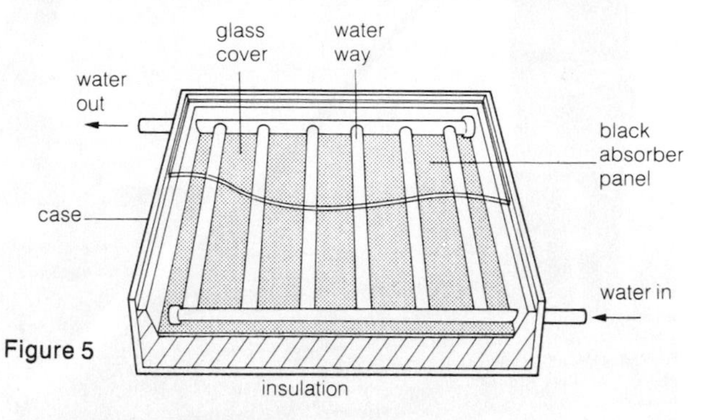

Figure 5

a) State the purpose of the following:
i) the insulation behind the solar panel.
ii) having the absorber panel painted black.
iii) having a glass cover on the top of the panel.

b) i) Name suitable materials for making the absorber panel and water-ways. (Do not use brand names)
ii) Give your reasons for the choice of such materials.

c) The pipe connecting the water outlet from the panel to the hot water storage tank should be kept short. Why is this desirable?

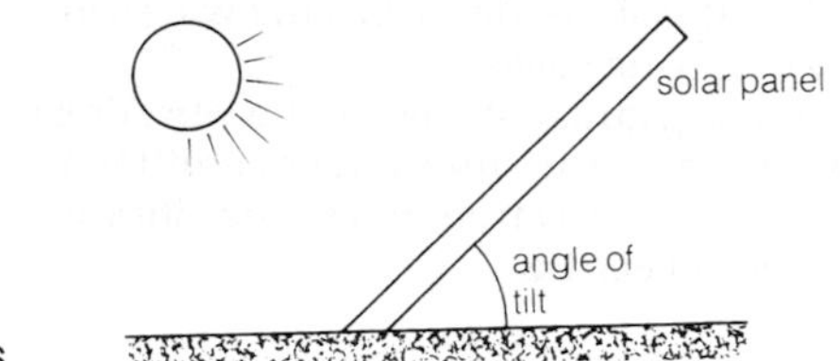

Figure 6

d) The angle of tilt of a solar panel greatly affects the amount of energy it receives at different times of the year. Figure 6 shows what is meant by the angle of tilt. The table of data shows the effect of different angles of tilt for the summer months.

Maximum daily input of energy in megajoules to a 1 m² panel

Month	Angle of tilt of panel to the horizontal					
	20°	30°	40°	50°	60°	70°
Apr	23.8	24.9	24.8	24.1	22.7	20.5
May	28.4	28.8	27.4	25.2	23.0	19.8
Jun	29.2	29.2	27.4	25.2	22.3	19.1
Jul	28.8	29.2	27.4	25.6	23.0	20.2
Aug	25.6	25.9	26.3	24.8	22.7	20.5
Sept	20.5	21.6	22.3	22.7	21.6	20.5

Use the table above to answer the following questions.

(i) What angle of tilt would be ideal for a solar panel in April?

(ii) Is it better to have the solar panel tilted at an angle of 40° or at an angle of 50° for all the months shown in the table. Give reasons for your answer.

(iii) What is the maximum amount of energy that a 4 m^2 panel could receive during a day in July?

4 Figure 7 shows four lenses (drawn in cross-section).

a) Which *two* of these lenses are converging lenses?

b) What does the word 'converging' mean when used to describe a lens?

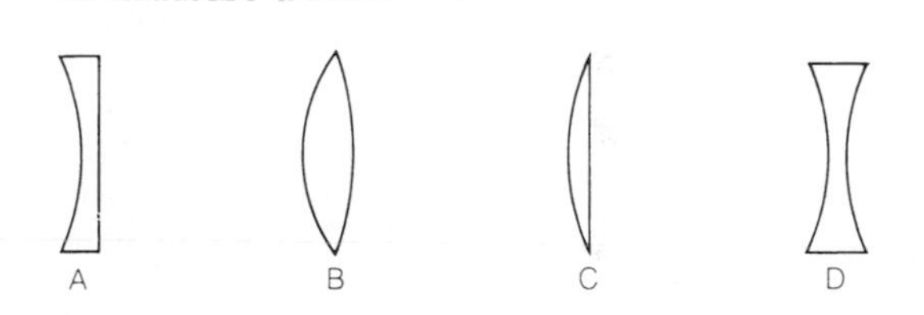

Figure 7

c) What word might be used to describe the other two lenses?

d) The lenses are drawn to scale. Which *one* of these would make the most powerful magnifying glass? (SWEB/SEG)

5 A teacher showed his class how to measure the speed of sound as shown in figure 8. He set up two microphones A and B, in line with a balloon. Each microphone was connected through a sound-operated switch to an electronic timer. The connections were made so that when a loud noise reached A the timer started, and when it reached B the timer stopped. After the balloon was burst the timer showed a reading of 5 milliseconds.

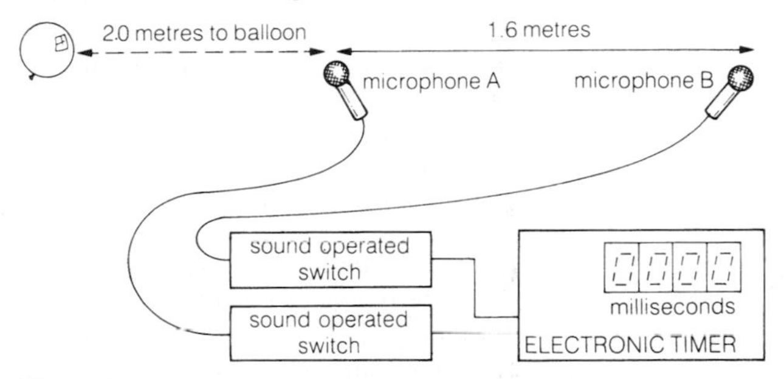

Figure 8

a) Write 5 milliseconds in seconds.

b) How far did the sound travel in 5 milliseconds?

c) Calculate the speed of sound in air (in metres per second). Show your working. (SWEB/SEG)

6 Figure 9 shows the electromagnetic spectrum.

Radio		Visible	Ultra Violet		γ-rays

Figure 9

a) Copy figure 9 and fill in the names of the two missing regions.

b) Which region

i) has the longest wavelength.

ii) has the highest frequency,

iii) causes a sun tan,

iv) is used in burglar alarms?

c) some washing powders contain a chemical which is sensitive to ultra-violet radiation. State and explain what you see when clothes washed in such a powder are put in sun-light. (NEA)

7 Figure 10 shows water waves approaching a harbour wall.

a) Copy the diagram. Draw the four waves in front of those shown. Take care and keep your drawing to scale.

b) The water waves travel 20 m in 10 s. Calculate their speed.

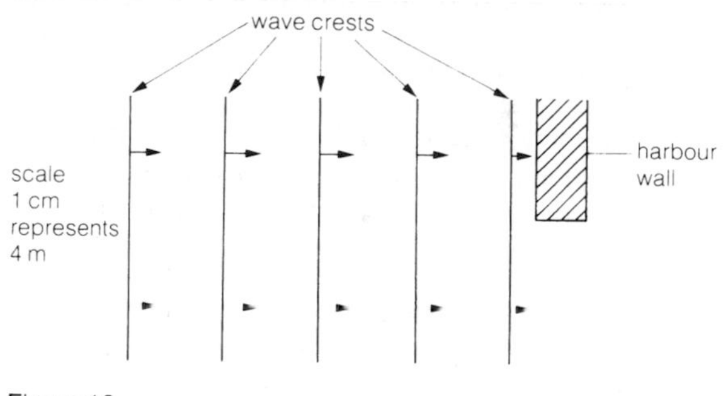

Figure 10

c) Measure the wavelength of the waves in figure 10. (Remember that 1 cm on the diagram represents 4 m).

d) Use your answers to b) and c) to calculate the frequency of the waves.

e) Sandra is sitting on the harbour wall. Write down how she could measure the frequency of the waves passing the harbour wall. What equipment does she need? How should she use it? (SWEB/SEG)

8 A microphone is connected to a cathode ray oscilloscope. Three sounds are made in turn in front of the microphone. The traces A, B and C produced on the screen are shown in figure 11. (The controls of the oscilloscope are not altered during the experiment.)

a) Which trace is due to the loudest sound? Explain your answer.

b) Which trace is due to the sound with the lowest pitch? Explain your answer. (MEG)

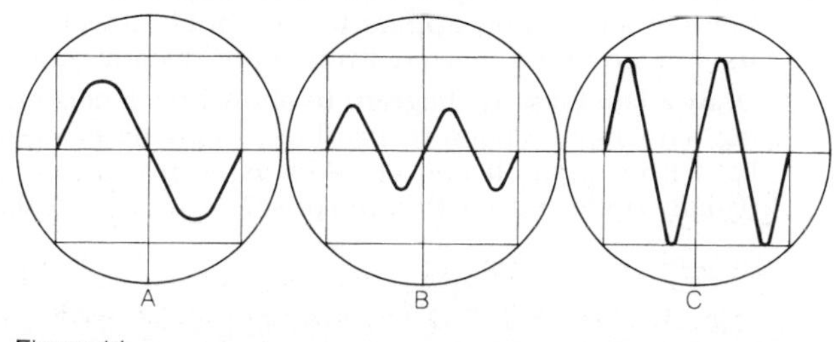

Figure 11

Further questions

Waves: light and sound: part B

Some questions require a knowledge of concepts covered in earlier units

sin 60° = 0.87; sin 49° = 0.75; sin 35.5° = 0.58

1 Draw a ray diagram to show how a concave (converging) spherical mirror may be used to give a magnified virtual image of an object. (SUJB)

2 An object is placed 15 cm in front of i) a plane mirror, ii) a concave mirror of radius 60 cm, and iii) a convex mirror of radius 60 cm.

a) Describe how you would locate the image in case i) by experiment. State clearly the nature and relative size of this image, as well as its position.

b) Draw scale diagrams for both ii) and iii) to show where the images are produced. State the nature and position of each image. State a practical application of each case. (SUJB)

3 An object 4.0 cm tall is placed 18.0 cm from a concave mirror of focal length 6.0 cm so that it is perpendicular to and has one end on, the principal axis of the mirror. Either by calculation or by full-scale ray drawing, determine: a) the distance of the image from the mirror b) the height of the image c) whether the image is erect or inverted, d) whether the image is real or virtual. (AEB)

4 The speed of light in air = 3×10^8 m/s. Calculate the speed of light in glass of refractive index 1.5. Rays of light, in air, are incident on a plane air/glass boundary. Show on clear diagrams the directions the rays would take in glass when the angles of incidence are i) 0°, ii) 60°. Calculate the angle of refraction in case ii). (SUJB)

5 a) Draw a clearly-labelled ray diagram to show why a tank of water viewed vertically downwards appears less deep than it really is. Describe in detail an experiment to measure the apparent depth of a glass block. State how you would expect this depth to relate to the real value.

b) To a person under water, looking upwards, all objects above and outside the water appear to be within a certain cone of vision. Explain this, using a clear diagram. If the refractive index of water is 1.33, calculate the vertical angle of the cone. (SUJB)

6 Explain with the help of a suitable diagram why, on a hot day, a road may appear to have pools of water on its surface when viewed from some distance away.

7 Draw a labelled ray diagram to show how a converging lens may be used to produce a parallel beam of light from a small source. Why is a curved mirror usually preferred for this purpose? (SUJB)

8 a) Define the terms principal focus and focal length as applied to a thin converging lens.
Describe how you wold find the focal length of such a lens by the plane mirror method.

b) A lamp and screen are fixed 120 cm apart, and a thin converging lens placed between them forms a real image on the screen, 3 times as long as the lamp filament itself.

i) What are the distances of object and image from the lens, and what is its focal length?

ii) For what other position of the lens can a real image be formed, and what is its length as compared with that of the filament? (O)

9 A convex lens of focal length 6 cm is held 4 cm from a newspaper which has print 0.5 cm high. By calculation or scale drawing, determine the size and nature of the image produced. (WJEC)

10 Figure 1 shows a lens, of focal length 10 cm, being used as a simple magnifier. By a ray diagram drawn to scale, or otherwise, find the value of the height, h, of the lens above the object if the virtual image is to be formed 25 cm from the lens.

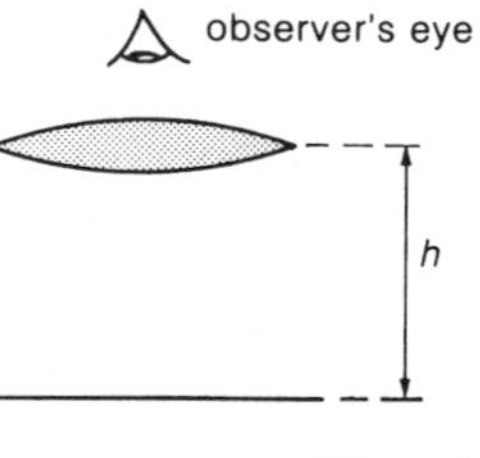

Figure 1

If the object is 7.0 mm wide, what is the width of the image? (O&C)

11 A thin converging lens of focal length 30 cm is used to form a real image on a screen 90 cm from the lens.
either: find, by a graphical construction, the object distance from the lens and the magnification, and explain your construction.
or: calculate the value of the object distance from the lens and the magnification using the appropriate formulae, and draw a diagram (not necessarily to scale) showing how the image is formed. (O)

12 An illuminated object is set up 2 m from a white screen. Where should a converging lens be placed in order to give a clear image, four times the height of the object, on the screen? What focal length lens is necessary? (SUJB)

13 A normal eye viewing a distant object adjusts to see clearly an object at a distance of 0.5 m. State the effect, if any, which this has on a) the shape of the cornea, b) the shape and the focal length of the eye lens.
The object at 0.5 m is then strongly illuminated so that it is much brighter. What is the effect of this change on the diameter of the pupil of the eye? (C)

14 a) A camera is used to photograph a small statue 80 cm high which is 500 cm away from the camera lens. The distance between the lens and the film is

5 cm. Without making any calculation, what can be deduced about the focal length of the lens? Account for your answer.
Determine the height of the image produced on the film. If the camera is now used to photograph a distant scene, what adjustment would need to be made?
Draw a ray diagram to show how a lens similar to that in the camera may be used as a magnifying glass. Mark on the diagram the position of the observer's eye.

b) A person finds that in order to read a newspaper easily he must hold it at arms' length although he can see distant objects quite normally. Name the defect from which he is suffering. Suggest a possible cause of the defect. Explain how this defect may be overcome by the use of suitable spectacle lenses. (L)

15 Describe the arrangement of a simple ripple tank for demonstrating wave behaviour, and explain how a steady pattern is obtained and made visible. Draw simple diagrams to show the ripple patterns produced in experiments to demonstrate; a) plane waves being reflected at a plane boundary; b) plane waves being refracted at a plane boundary; c) diffraction at a slit aperture. (O)

16 A continuous progressive transverse wave of frequency 8 Hz moves across the surface of a ripple tank.
a) With reference to the frequency, describe the movement of the water surface.
b) If the water waves have a wavelength of 32 mm, calculate the speed with which the waves travel across the water surface. (L)

17 a) Figure 2 represents the fringe pattern obtained in a double-slit experiment when monochromatic red light was used.

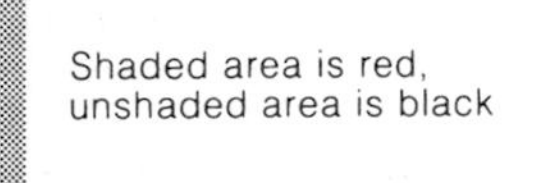

Shaded area is red, unshaded area is black

Figure 2

i) Explain clearly, using the wave theory of light, why dark and red fringes occur.
ii) State clearly how the patterns would change if monochromatic blue light were used, the rest of the apparatus remaining unchanged.
iii) What deduction could be made about the difference between red and blue light from the two fringe patterns?

b) In the experiment referred to in a) the two slits were 0.0002 m apart and the distance from the double-slit to the screen on which the fringes were formed was 4 m.
i) Sketch and label the arrangement of the apparatus, showing the positions of the source, slits and screen.
ii) On the screen, the distance between the first red (bright) fringe and the eleventh red fringe was 0.13 m. Calculate the fringe separation.
iii) Calculate a value for the wavelength of the red light used in the experiment. (JMB)

18 A filament lamp emitting white light, a narrow slit, a converging lens and a glass prism are used to focus a spectrum of light onto a white screen, the spectrum being spread between X and Y. (See figure 3.)

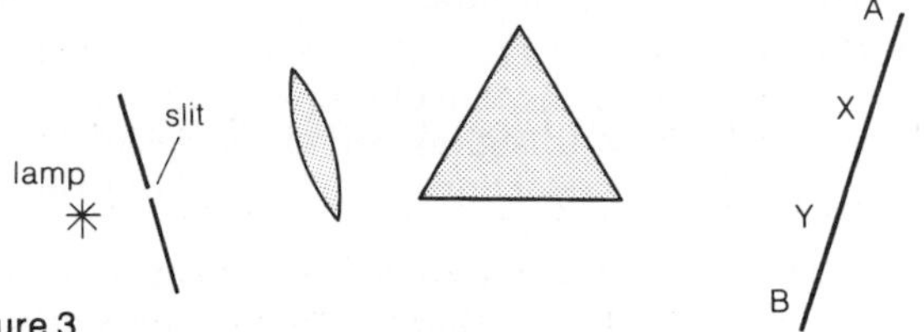

Figure 3

a) List in order the colours to be expected between X and Y.
b) What differences would it make to the spectrum between X and Y, if the voltage across the lamp was reduced from 12 V to 10 V?
c) What would be the effect on the spectrum of covering the slit with a red filter?
d) Discuss whether there might be any radiation from the lamp so as to reach the screen between A and X.
e) Discuss whether there might be any radiation from the lamp so as to reach the screen between B and Y. (O)

19 Light and gamma rays are both examples of electromagnetic radiation.
a) Name three other types of electromagnetic radiation.
b) State two differences between light and gamma rays.
c) The speed of light is 3×10^8 m/s. Calculate the frequency of yellow light of wavelength 6×10^{-7} m. (O)

20 Describe an experiment which you would carry out to show how the nature of a surface affects the heat radiated from that surface in a given time. State any precautions which you would take and state your findings for two named surfaces.
How would you then show that the surface which was the better radiator was also the better absorber of radiation? (L)

21 Two similar cans are equally filled with a light oil at 110 °C (see figure 4). Each contains a thermometer and is fitted with a lid, but one can has a dull black surface whilst the other has a bright silver surface. Both are left to cool on a wooden bench in the same draught-free room where the air is at 10 °C.
a) Why do the cans not lose much heat by conduction?
b) Explain the processes by which the cans do lose heat.
c) Sketch a cooling curve for the black can.
d) How would the cooling curve for the silver can differ from that for the black can? Explain this difference. (O)

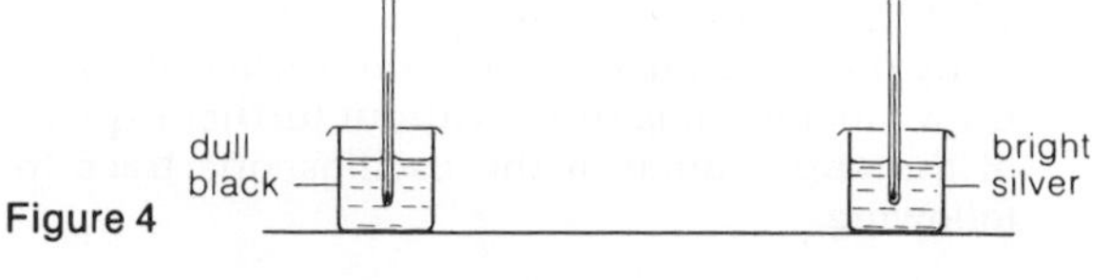

Figure 4

22 Two similar loudspeakers, connected to the same generator, are set up at one end of a laboratory. Describe how you would expect to observe an interference pattern from these sources of waves. Explain why this pattern occurs. (SUJB)

23 The diaphragm of a sound generator is made to move to and fro along the line of propagation of the sound.
   a) Explain how this gives rise to a series of compressions and rarefactions which travel outwards as a longitudinal wave.
   b) Draw a diagram representing such a wave and indicate the distance that represents one wavelength.
   c) If the frequency of the oscillator is 400 Hz and the speed of sound in air is 320 m/s, what is the wavelength of the sound? (O)

24 Two vertical walls A and B (figure 5) are 55 m apart. A man standing at P, 22 m from A, claps his hands once.

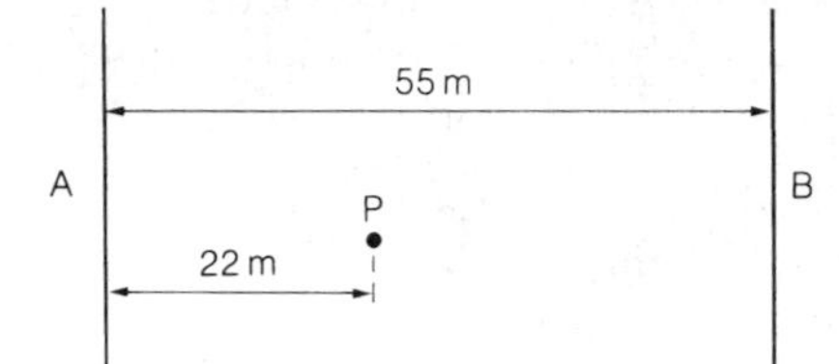

Figure 5

   a) What is the time-interval between the clap and the first echo that he hears?
   b) What is the time-interval between the clap and the second echo?
   c) Why does the man standing at P hear a sequence of echos which gradually dies away? (Take the speed of sound in air to be 330 m/s.) (O)

25 A person claps his hands at approximately $\frac{1}{2}$-second intervals in front of a wall 90 m away. He notices that each echo produced by the wall coincides with the next clap.
   a) Calculate an approximate value for the speed of sound.
   b) If you were using the above as a basis for an experimental method to determine the speed of sound, what procedure would you adopt to obtain high accuracy in the timing part of the experiment? (L)

26 Figure 6 shows the trace on an oscilloscope produced when an audible pure tone is sounded and the sound picked up by a microphone connected to the oscilloscope. In the following, assume that the oscilloscope controls are not altered.

Figure 6

Copy the diagram. Using the same scale as your copy, draw labelled diagrams (without further explanation) of the appearance of the oscilloscope trace for the following:

   a) a pure tone of the same frequency as the original but louder,
   b) a pure tone of the same amplitude as the original but having twice the frequency,
   c) a sound from a musical instrument, such as a trumpet or clarinet, having the same pitch as the original. (O&C)

27 With reference to a thin wire under tension which is vibrating between two fixed clamps, explain the terms: node, transverse wave, stationary wave, amplitude, frequency.
How, if at all, are the pitch and loudness of the note emitted by the wire affected if i) the amplitude is increased, and ii) the vibrating length of the wire is decreased? (SUJB)

28 a) Explain what you understand by resonance. Describe briefly how this phenomenon may be demonstrated by two methods involving different aspects of physics.
   b) A thin wire under tension is 60 cm long and fixed at both ends. The wire emits a note of frequency 255 Hz when plucked at its mid point. What would you expect to observe if a sound generator of frequency 256 Hz is set up near the vibrating wire? If the vibrating length is reduced to 50 cm what frequency note would you then expect from the wire? (SUJB)

29 A vibrating tuning fork of frequency 256 Hz is held over the top of a tube full of water, as shown in figure 7. The water is then allowed to flow slowly from the tube, and there is a first loud resonance when the air column is 300 mm long.

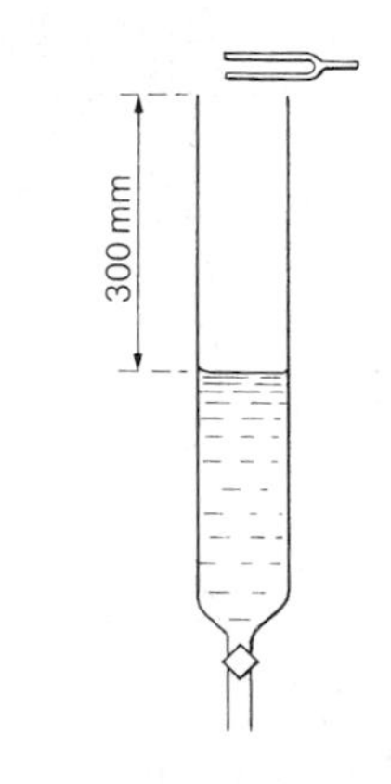

Figure 7

   a) What is meant by 'frequency 256 Hz'?
   b) What estimate does this experiment give for the wavelength in air of the note from the fork?
   c) What is the relationship between the frequency of the fork and the wavelength in air of the sound it produces?
   d) What estimate does this experiment give for the speed of sound in air?
   e) Why is the loudness of the note heard much less when the length of the air column is less than 300 mm? (O)

SECTION 6

ELECTRICAL ENERGY

Lightning never strikes twice?

They say that lightning never strikes the same spot twice, but some unfortunate people have discovered otherwise.

In 1918, a Major Summerford was wounded in Flanders, not by the enemy, but by a flash of lightning which knocked him off his horse and left him paralyzed from the waist down. He was invalided out of the army, retired to Vancouver and took up fishing. In 1924 he was by a river with three fellow anglers when lightning hit the tree beneath which he was sitting and paralyzed his right side. Within two years he had more or less recovered from these shocks and was able to take walks in the Vancouver park, where, in the summer of 1930, during a sudden thunderstorm he was again struck by lightning. This time he was permanently paralyzed and died two years later. Even then he had not finished his career as an involuntary lightning conductor. In June 1934 there was a storm over Vancouver. Lightning struck the cemetery and shattered a tombstone. It was Major Summerford's.

Major Summerford's case is not unique. Ex-ranger Mr Roy C. Sullivan of Virginia earned a place in the Guinness Book of Records as the only man to have survived seven attacks by lightning. In 1942, lightning destroyed his big toe nail; in 1969 it took away his eyebrows. A year later it seared his left shoulder, and three years after that it set his hair on fire. At this point, Mr Sullivan decided to carry a five gallon can of water around with him in his car as a precaution. But it didn't help much – in 1973 his (newly grown) hair caught fire again during a lightning strike. And in 1977, he was taken to hospital with chest and stomach burns after being struck while fishing.

The professor of natural history at the University of St Petersburg: famous for being the first (recorded) man to be killed by lightning.

The All-electric Alternative?

2 million joules of energy are available from a fully charged car battery. That may *sound* impressive but the fact remains that it takes about ten such batteries to provide as much useable energy as a gallon of petrol. So far, the great weight, high cost, and limited performance of battery-powered cars have made them an unacceptable alternative to petrol driven vehicles. Research promises some improvements in the performance of lead-acid batteries – perhaps three times the energy storage for the same weight – but practical storage batteries may eventually be based on other materials such as sodium and sulphur or zinc and chlorine. Development costs are likely to be very high.

ELECTRIFYING QUOTES?

Genius is one percent inspiration and ninety-nine percent perspiration.
(Thomas Edison (1847–1931), inventor of the filament light bulb.)

Lord Finchley tried to mend the
Electric Light Himself. It
struck him dead: and serve him
right!
It is the business of the wealthy
man
To give employment to the
artisan.
(Hilaire Belloc)

6.1

Electric charge

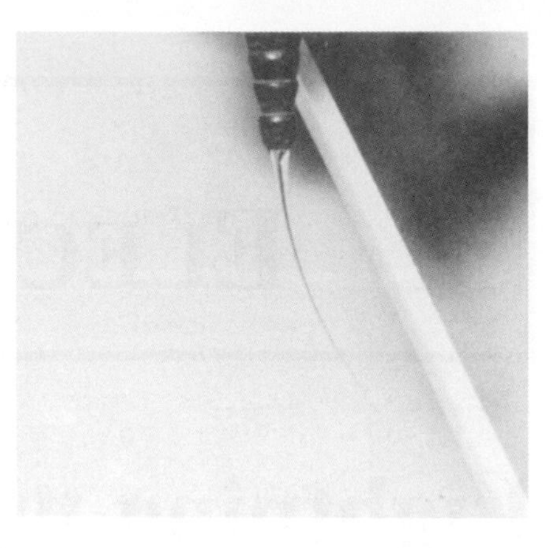

Figure 1 A rubbed polythene rod will attract a stream of running water

Some materials take on mysterious new properties when rubbed. They exert forces on other materials nearby, and can cause lightning-like sparks and crackles. When materials behave in this way, they are said to possess an electric charge. There is evidence to suggest that there are two types of electric charge, and that both exist in all atoms.

Negative and positive charges

Polythene and Perspex are both materials which become charged with 'static electricity' when rubbed with a dry, woollen cloth.

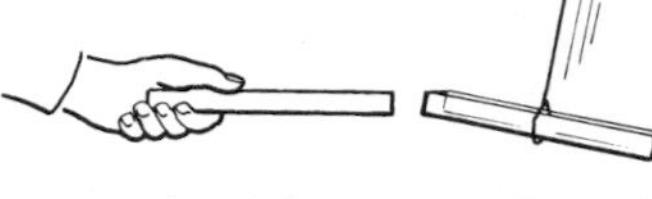

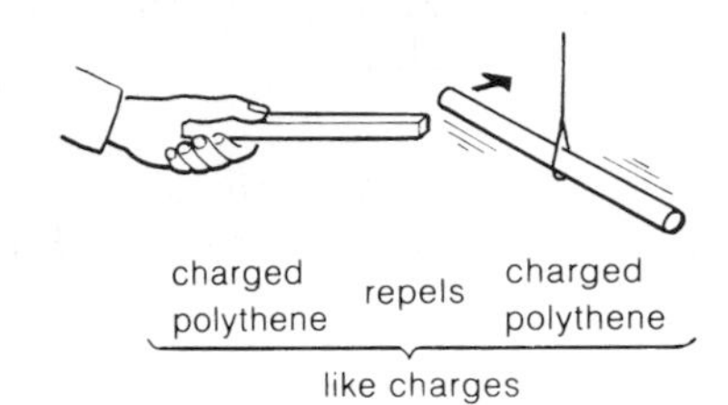

Figure 2 Like charges repel; unlike charges attract

If a charged polythene rod is brought near a charged Perspex rod as in figure 2, it will attract it. On the other hand, a charged polythene rod will push away or **repel** another charged polythene rod. This suggests that the electric charges on the polythene and Perspex rods are of different and opposite types. Experiments with other materials confirm that there are two types of electric charge; these are called negative (−) charge and positive (+) charge. When rubbed with a dry, woollen cloth, a polythene rod becomes negatively charged, and a Perspex rod becomes positively charged.

Figure 3 Repulsion between hairs carrying the same type of charge

Experiments like those in figure 2 show that:

like charges repel, unlike charges attract;
the closer the charges, the greater the force between them.
The repulsion between like charges is illustrated in figure 3.

Where charges come from

All materials are made up of tiny particles of matter called atoms. Atoms are thought to be made up of smaller particles, some of which are electrically charged.

Unit 8.11 describes some of the experimental results which have enabled scientists to build up a description or **model** of the atom. At the centre of each atom is a nucleus made up of particles called **protons** and **neutrons**. Surrounding this nucleus are very much lighter particles called **electrons**. This is illustrated in figure 4.

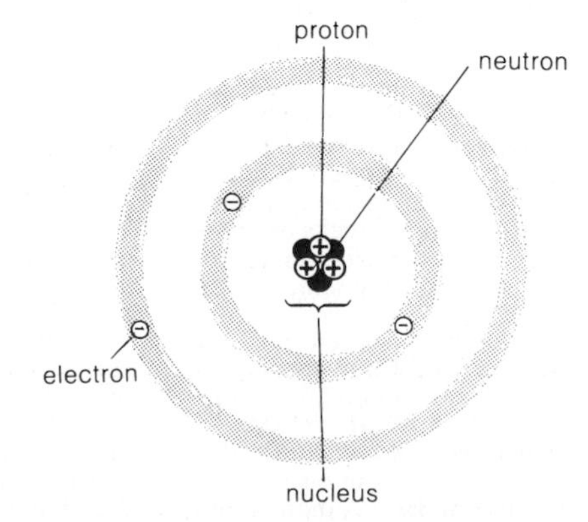

Figure 4 A model of the atom: neutral neutrons and positive protons in the centre; negative electrons in orbit

Electrons have a negative (−) charge;
protons have an equal positive (+) charge;
neutrons have no charge.

Normally, atoms have equal numbers of electrons and protons, so the total amounts of negative and positive charge within a material are the same. The overall or **net** charge on the material is zero. However, when two materials are rubbed together, electrons may be transferred from one to another. This upsets the balance between the opposite charges within each material, so that each is left with a net negative or positive charge.

When a polythene rod is rubbed with a dry cloth as shown in figure 5, the polythene pulls electrons away from atoms on the surface of the cloth. This leaves the polythene with more electrons than normal and the cloth with less. The polythene therefore ends up with a negative charge while the cloth is left with a positive charge.

When a Perspex rod is rubbed with a dry cloth as shown in figure 6, the opposite happens. In this case, it is the cloth which pulls electrons away from the rod. With less electrons than normal, the Perspex rod is left with a positive charge while the cloth gains an equal amount of negative charge.

Note that rubbing materials together doesn't make charge; it simply separates negative and positive charges which already exist within the materials.

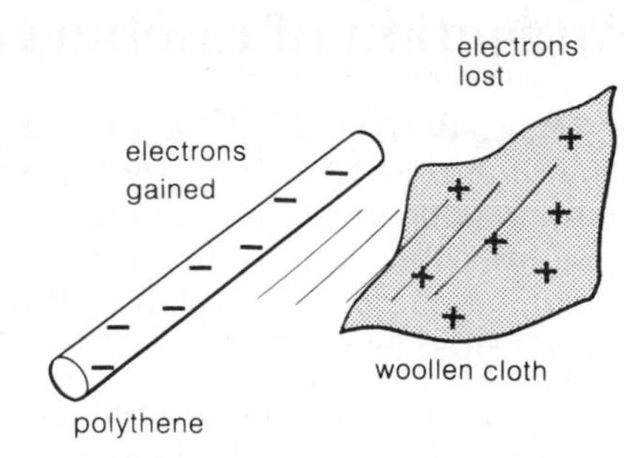

Figure 5 Rubbing can transfer electrons: in this case, from the cloth to the polythene

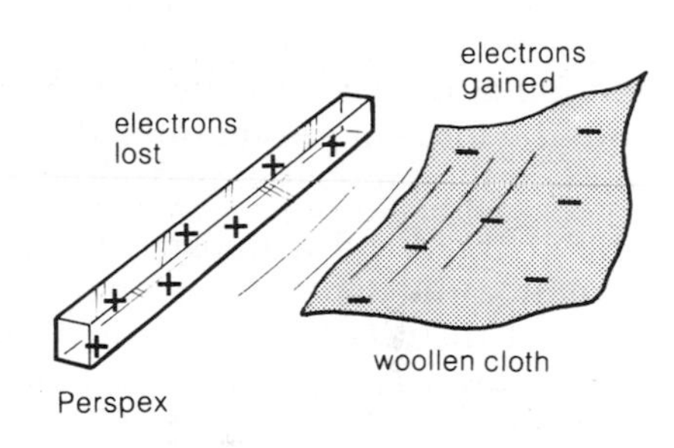

Figure 6 Using Perspex, the electrons transfer from the Perspex to the cloth

Conductors and insulators

When rubbed, some objects lose charge almost as soon as they gain it. This happens because electrons flow through the object or surrounding materials until the balance of negative and positive charge is restored.

Materials which allow electrons to flow through them are called **conductors**, as shown in figure 7. Metals are the best conductors of all. The outermost electrons in each atom are so loosely held that they are able to move freely between atoms. These **free electrons** also make metals very good conductors of heat (see page 141). Most non-metals conduct charge poorly or not at all, though carbon is an important exception.

Materials which do not conduct charge are called **insulators** (see figure 7). Their electrons are all tightly held to atoms and are not normally free to move – though they can of course be disturbed if a material is rubbed. Because of this, insulators are relatively easy to charge by rubbing because any electrons which are transferred tend to stay where they are. Conductors **can** be charged by rubbing – but only if held in insulating handles. If you rub a hand-held metal rod, any electrons transferred are immediately replaced by electrons which flow through the rod and your body, and the rod remains uncharged.

When you switch on a light, the 'electricity' passing through the cable is actually a flow of electrons. The cable has copper conducting wires through its centre as shown in figure 8. These are enclosed in an insulating material – usually the plastic PVC.

Conductors	
Good	*Poor*
metals especially silver copper aluminium carbon	water human body earth semiconductors silicon germanium
Insulators	
rubber plastics e.g. PVC polythene Perspex	glass dry air

Figure 7 Good conductors, poor conductors, and insulators

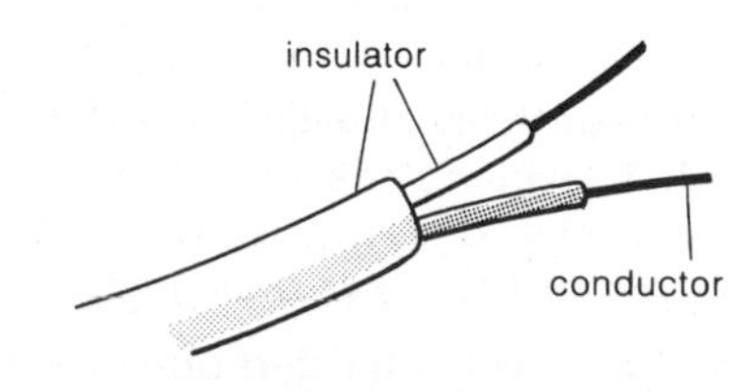

Figure 8 A cable contains conducting and insulating materials

Attraction of uncharged objects

A charged object will attract any uncharged object close to it. Charge a rubber balloon by rubbing it against your sleeve and the balloon will cling to a wall; charge a comb by pulling it through your hair and the comb will pick up small pieces of paper. Records become charged when you pull them out of their sleeves, and will attract dust as a result.

You can see why such attractions occur by considering the effect of a charged Perspex rod on a small piece of aluminium foil placed just underneath it as shown in figure 9. Free electrons in the aluminium are pulled towards the positively charged rod, the top end of the foil becoming negatively charged while the bottom end is left with a net positive charge. The charged rod therefore attracts the top end of the foil and repels the bottom end. As the top end is closer to the rod, the force of attraction is the stronger of the two forces and the foil is pulled towards the rod as a result.

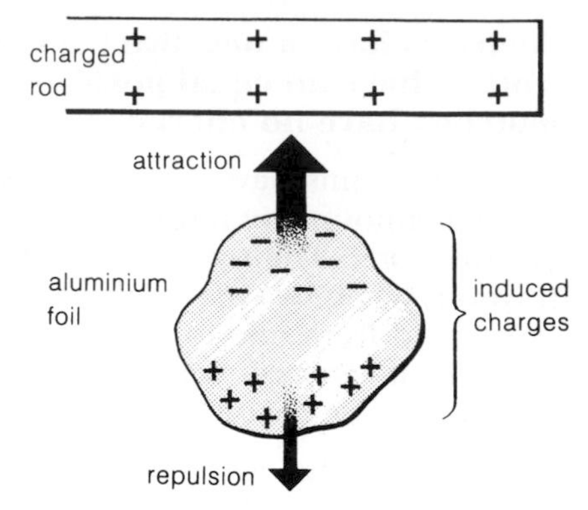

Figure 9 A charged rod will attract uncharged objects by causing induced charges

The charged rod will also attract a small piece of paper. Being an insulator, the paper doesn't contain free electrons but the charge on the rod distorts the atoms in the paper, pulling electrons a little closer towards the rod and pushing the nuclei a little further away. The end result is the same; the paper behaves as if it has a negative charge at the top end and a positive charge at the bottom.

In both of these examples, charges appear on the foil and the paper because of the presence of a nearby charged object. Charges caused in this way are known as **induced** charges.

Sometimes, the force of attraction between charges on different objects may be so strong that the air separating them ceases to act as an insulator. A flash is seen, as moving electrons collide with molecules in the air and cause them to give out light as shown in figure 10; sparks, on a smaller scale, share the same cause.

Figure 10 A huge-scale electric spark: moving electrons collide with air molecules and make them give out light

The leaf electroscope

Detecting charge Small charges can be detected using a leaf electroscope as shown in figure 11. If a charged object touches the metal cap at the top of the electroscope, some of the charge is transferred to the gold 'leaf' and metal plate at the bottom. Charges on the leaf and plate repel each other and the leaf rises as a result.

Testing for negative or positive charge Once charged, an electroscope can be used to find out whether the charge on an object is positive or negative. Figure 12a shows an electroscope which has been given a negative charge; cap, plate and leaf all contain more electrons than normal.

If a negatively charged object is brought towards the cap as in figure 12b, free electrons are pushed away from the cap and down into the leaf and plate. This increases the repulsion between the leaf and the plate, and the leaf rises even more.

If a positively charged object is brought towards the cap as in figure 12c, the reverse happens. Free electrons in the leaf and plate are attracted upwards towards the cap, and the leaf falls.

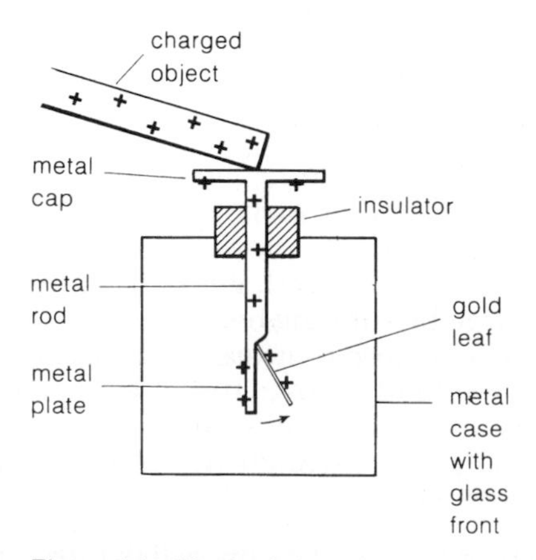

Figure 11 The leaf electroscope. When a charged object touches the electroscope, the leaf rises because like charges repel

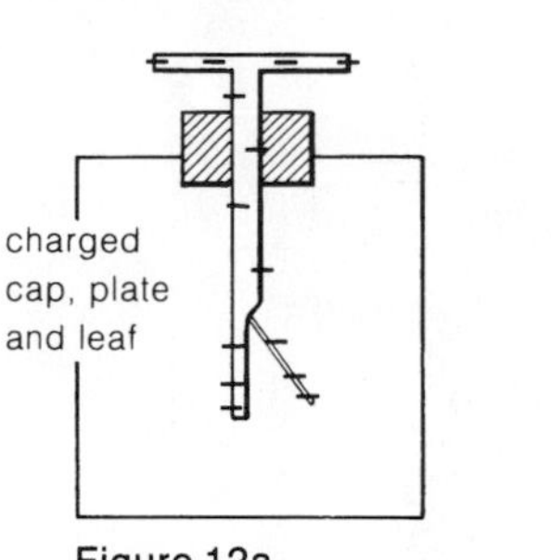

Figure 12a

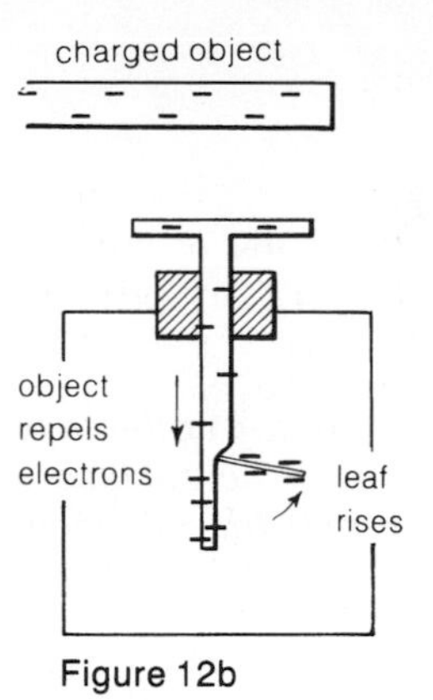

Figure 12b

charged object

object
attracts
electrons

leaf
falls

Figure 12c

In general, the leaf of a charged electroscope rises if the object brought towards the cap carries the same type of charge, and falls if the object carries the opposite type of charge.

Figures 12a, 12b, 12c A negatively charged electroscope will respond differently to nearby negative and positive charges

Charging by induction

Figure 13 shows how a charged rod can be used to give a nearby conductor a charge of the opposite type. The process is called **charging by induction**. The rod in this case is made of Perspex and is carrying a positive charge. The conductor is a metal can and it is mounted on an insulating stand.

Figure 13 Charging by induction. Notice that the retained charge on the conductor is opposite to that of the charged insulator.

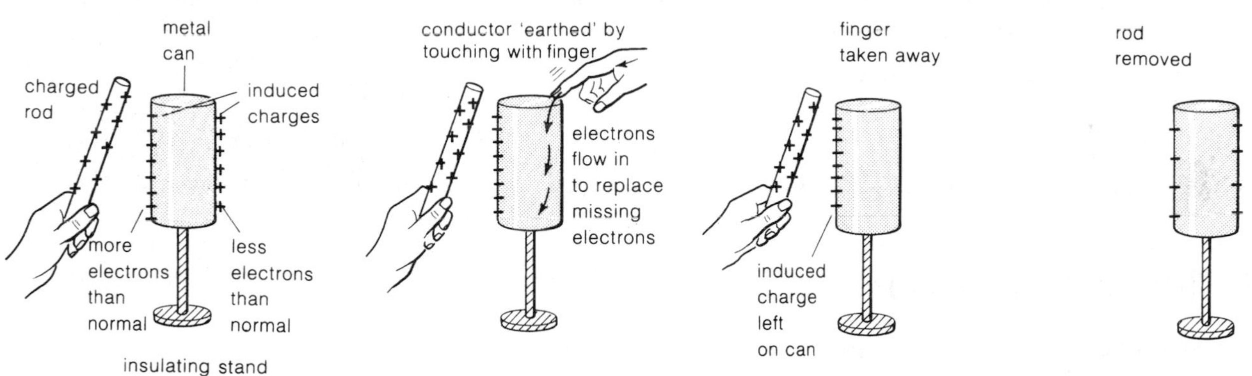

Questions

1 Name a material which, when rubbed with a dry cloth, becomes a) negatively charged b) positively charged. Explain in terms of electron transfer what happens in each case.

2 In an atom, what type of charge is carried by a) protons b) neutrons c) electrons?

3 Why are metals such good conductors of charge? Name a non-metal which is also a good conductor.

4 Why doesn't a conductor become charged when it is held in the hand and rubbed with a cloth? How can a conductor be charged by rubbing?

5 A positively charged rod is brought towards the cap of an uncharged electroscope. Is there any effect on the leaf before the cap is touched? If so, explain why.

6 A leaf electroscope is given a negative charge. When a charged rod is brought towards the cap of the electroscope, the leaf rises even more. What type of charge is there on the rod? Give reasons for your answer.

7 In figure 14, a charged rod has been brought close to an uncharged metal can. Copy the diagram and add any induced charges you would expect to find on the can. What would you need to do in order to leave the can with a negative charge?

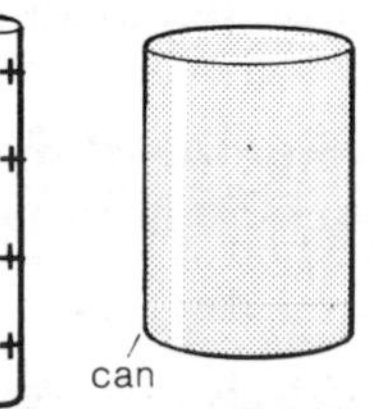

Figure 14

6.2
Charged conductors

Experiments show that any charge gained by a conductor collects on its outside surface and is most concentrated where the surface is most highly curved. A van de Graaff generator provides a useful source of charge for such experiments.

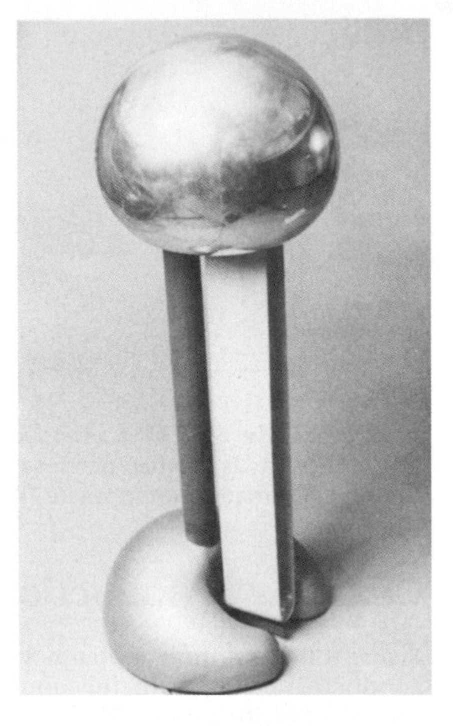

Figure 1 A van de Graaff generator: negative charges generated on the belt are stored on the dome

The van de Graaff generator

A van de Graaff generator gives a continuous supply of charge, and in rather larger quantities than you can get by rubbing an insulator with a cloth.

A small van de Graaff generator is illustrated in figure 1. The rubber belt gains extra electrons because of slight slipping which occurs between it and the roller driving it. These electrons are carried by the moving belt up to the metal dome, where they collect. A supply of negative charge therefore builds up on the dome.

A conductor on an insulating stand can be given a negative charge by connecting it to the dome by a length of conducting wire. If a positive charge is required instead, the conductor can be placed half a metre or so away from the dome and then charged by induction as explained on page 233.

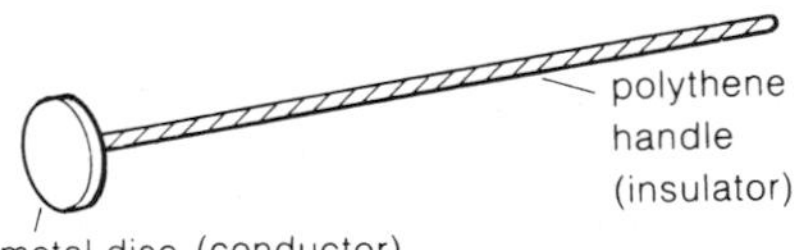

Figure 2 A proof plane is used for transferring small amounts of charge

Distribution of charge on a conductor

A **proof plane**, as shown in figure 2, is used for transferring small amounts of charge from a charged object to an electroscope.

If a proof plane is touched against the inside surface of a charged hollow conductor as in figure 3, and then placed against the cap of an electroscope, the leaf of the electroscope does not rise. It will however rise if the proof plane is touched against any part of the outside surface of the conductor, showing that:

the charge on a conductor collects only on its outside surface.

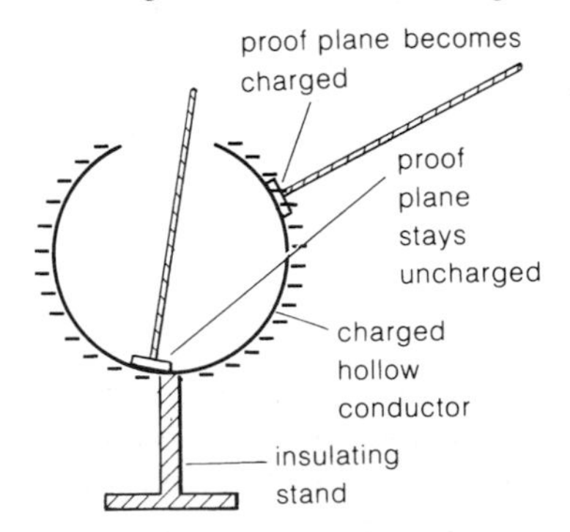

Figure 3 Using a proof plane to find the charge distribution

Further tests with a proof plane on different parts of the outside surface of a charged conductor as shown in figure 4 show that in most cases the charge is not evenly distributed. The greatest deflection of the electroscope leaf occurs when the proof plane has been touched against the most highly curved part of the conductor surface. This shows that:

the concentration of charge on a conductor is greatest where the surface is most sharply curved.

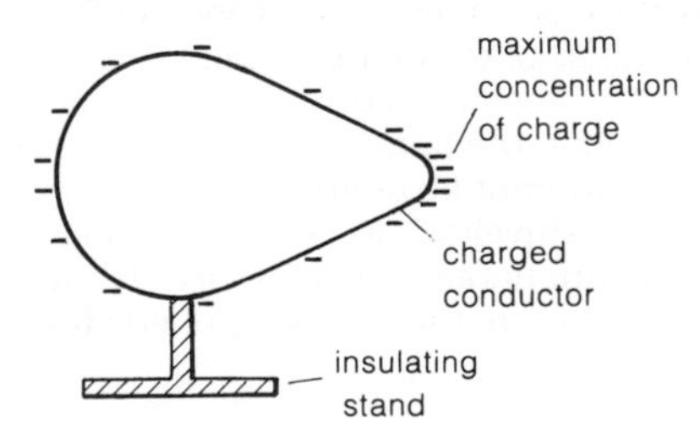

Figure 4 Distribution of charge on differently shaped surfaces

The action of a point

If an upturned drawing pin is placed on top of a van de Graaff generator, any charge which collects on the dome immediately 'leaks' away from the sharp point. The charge seems to be carried away by a stream of air, sometimes called an **electric wind**, which you can feel if you hold your hand a few centimetres above the point.

The action of the point is illustrated in figure 5. The metal surface is so sharply curved, and the charge so highly concentrated, that the forces on nearby air molecules are strong enough to strip electrons from them. Some of these electrons become attached to other molecules. As a result, the point is surrounded by a large number of molecules which have either lost or gained electrons. These charged molecules are called **ions**; the air is said to be **ionized**. The charge on the dome falls rapidly as positive ions are attracted to the point, strike it, and collect electrons to replace those which they have lost. At the same time, a fast-moving stream of air is produced as negative ions are pushed rapidly away from the point.

A sharp point with a positive charge on it loses charge by a similar process. In this case however there is a flow of positive ions away from the point, which is gaining electrons rather than losing them.

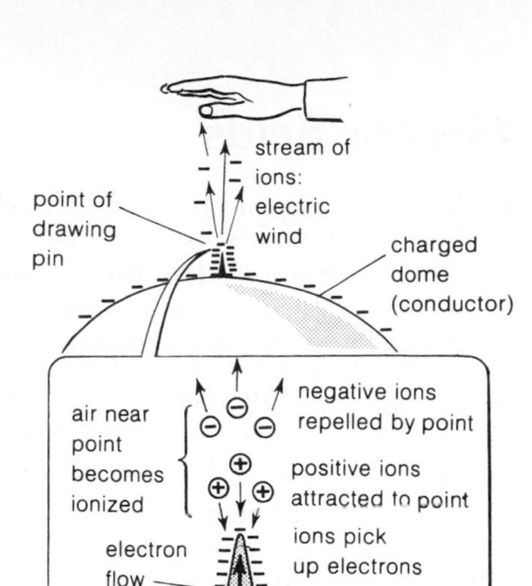

Figure 5 Air molecules are given a negative charge, and then are repelled from the point where electrons are concentrated – creating an *electric wind*

The lightning conductor

Tall buildings usually have a strip of copper called a lightning conductor attached to at least one side as shown in figure 6. One end of the strip is fixed to a metal plate buried in the ground; the other end is attached to a sharp spike or spikes which point upwards above the highest part of the building.

Thunderclouds carry electric charges. If, say, a negatively charged thundercloud passes over a building as in figure 7, a positive charge is induced on the roof, and the force of attraction between these opposite charges may be strong enough to produce a sudden flow of electrons from cloud to roof. In other words, the roof may be struck by lightning.

The lightning conductor reduces the risk to the building in two ways:

1 The flow of ions from the spikes lowers the induced charge on the roof and cancels out some of the charge on the cloud. This reduces the chances of lightning striking.

2 If lightning does strike, the lightning conductor provides a route for electrons to pass into the ground without damaging the building. The ground or **earth** has an almost infinite capacity for absorbing extra electrons.

Figure 6 A lightning conductor on top of St Paul's Cathedral, London. Do you think he ever received flashes of inspiration?

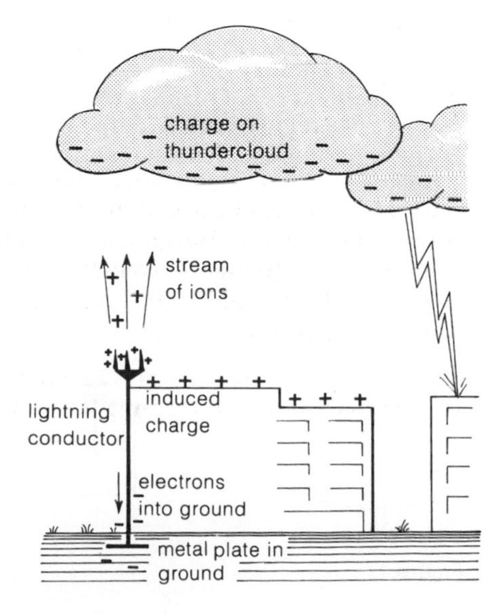

Figure 7 How a lightning conductor works.

Electric fields

Figure 8 shows what happens when pieces of thread are sellotaped to the dome of a van de Graaff generator. The threads become negatively charged like the dome, and the force of repulsion between these like charges makes the threads stand outwards at right angles to the dome surface.

The region around the charged dome where other charges will experience a force on them is called an **electric field**, and it is represented by a series of arrowed lines as shown in figure 9.

An electric field line shows the path which would be taken by a positive charge free to move in the field, with the arrow giving the direction of the force acting.

A method of finding electric field patterns experimentally is shown in figure 10. In each case, oppositely charged conductors were dipped into a dish of castor oil, and grass seed sprinkled on the oil surface. Charges induced in the seeds made them line up in the direction of the electric field.

Figure 8 Demonstrating field lines

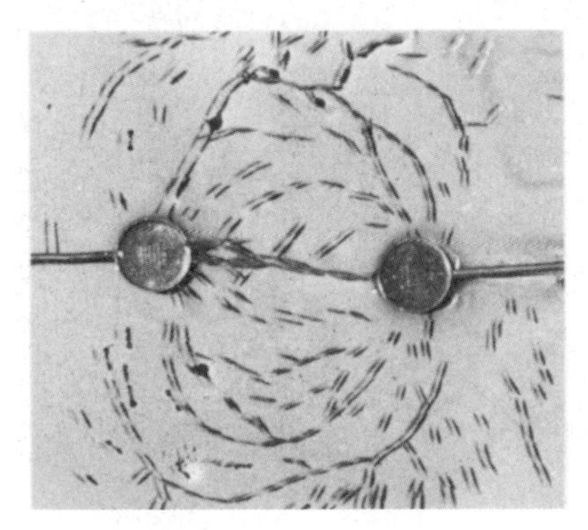

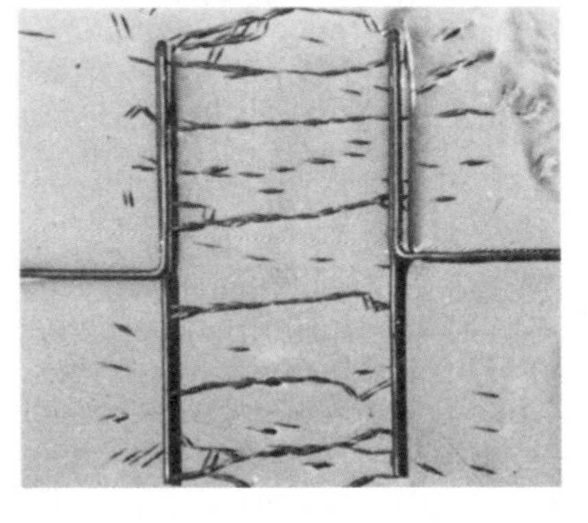

Figure 10 Electric fields between oppositely charged conductors

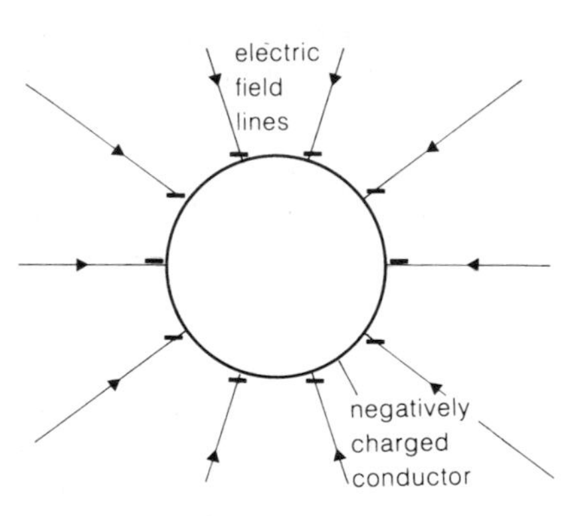

Figure 9 Electric field lines show the path of a 'free' positive charge

Questions

1 Figure 11 shows a positively charged hollow conductor. A proof plane is touched at point A and then placed in contact with the cap of an electroscope. The experiment is repeated at point B and then at point C. The lower diagrams show the possible effects on the electroscope leaf. Which diagram corresponds to each of the points of contact A, B, and C? What two properties of charged conductors do these experiments demonstrate?

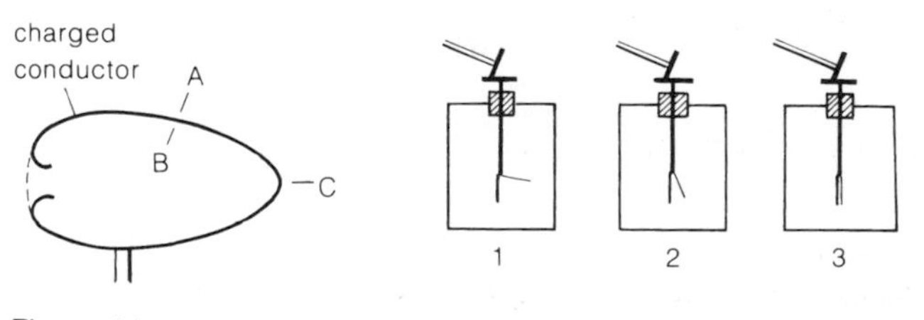

Figure 11

2 When a van de Graaff generator is switched off, the charged dome eventually loses all its charge. Why? Why is the loss of charge much more rapid if an upturned drawing pin is placed on top of the dome?

3 Give two functions of a lightning conductor.

4 Copy and complete figure 12 to show the electric field around a positively charged metal sphere. Redraw the diagram to show the electric field around a negatively charged sphere. What would happen in each case to a positive charge placed at point X if the charge were free to move?

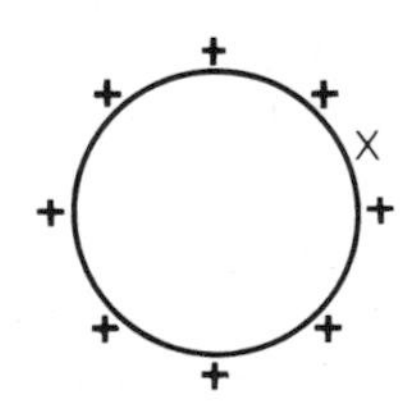

Figure 12

6.3

Charge, potential and capacitance

Work must be done to charge a conductor, and energy is gained by the charge as a result. But some conductors are easier to charge than others.

Quantities of charge

An uncharged object becomes negatively charged if it gains electrons and positively charged if it loses them. All electrons carry the same quantity of charge, so the more electrons an object gains or loses the greater is the overall negative or positive charge upon it.

Charge is measured in coulombs [C], a coulomb being equal to the charge on about 6 million million million electrons. The coulomb is not defined in this way however (see page 247).

A coulomb is a very much larger quantity of charge than is normally produced by rubbing, and it is often more convenient to measure charge in **microcoulombs**:

1 microcoulomb [μC] = 10^{-6} C [one millionth of a coulomb]

Before charge is lost as in figure 1a, the charge on the dome of a small van de Graaff generator is around 5 μC. The charge on a rubbed polythene rod is thousands of times smaller again. Even so, a negatively charged polythene rod carries something like 1000 million extra electrons.

Figure 1a A van de Graaff generator can lose about 5 μC of charge . . . in a flash

Electrical potential energy

Usually, work must be done to increase the charge on an object, and energy is released when charge is lost.

In a van de Graaff generator for example, the motor driving the rubber belt has to do work to move charge up to the dome because the charge being carried on the belt is repelled by the charge already on the dome (see figure 1b). You can tell that work is being done as the dome charges up, because the motor noticeably slows down as the belt becomes more difficult to move. When there is no charge on the dome, the motor revolves freely.

When the dome charges up, free electrons are pushed closer and closer together on its surface. As a result, they gain potential energy – rather like the turns of a spring being compressed. This energy is often referred to as **electrical potential energy**, or just **electrical energy**.

When a spark jumps between the dome of a van de Graaff generator and Earth, the dome loses electrons. The potential energy lost is changed into heat, light and sound energy.

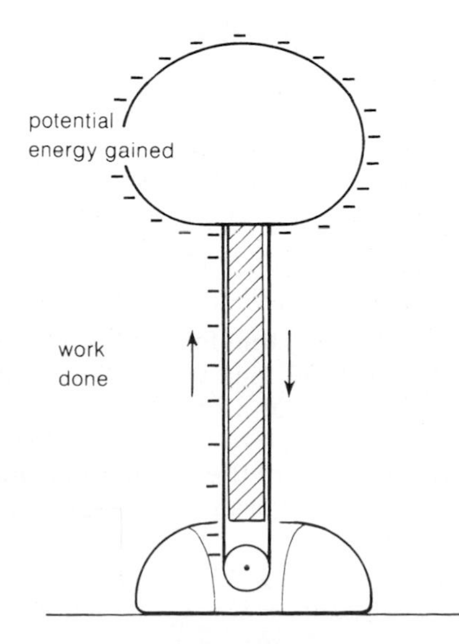

Figure 1b Work must be done to cause an increase in electrical potential energy

Potential

If work has to be done to move charge from one point to another, the points are at different **potentials**.

For convenience, *earth potential is taken to be zero*. The potential of any given point is then found by considering the work which must be done to move a small positive charge from earth up to that point, using the equation:

$$\textbf{potential} = \frac{\textbf{work done}}{\textbf{charge moved}}$$

in symbols:

$$V = \frac{W}{Q}$$

If work is measured in joules [J], and charge in coulombs [C], potential is measured in joules/coulomb or **volts [V]**.

For example:
If a point is kept at a steady potential of 1 volt:
1 joule of work must be done to move 1 coulomb of positive charge from earth to the point (as shown in figure 2);
2 joules of work must be done to move 2 coulombs of positive charge from earth to the point;
and so on.

If a point is kept at a steady potential of 1000 volts,
1000 joules of work must be done to move 1 coulomb of positive charge from earth to the point;
2000 joules of work must be done to move 2 coulombs of positive charge from earth to the point;
and so on. In other words:

the potential in volts is the work done per coulomb in bringing positive charge from earth to the point.

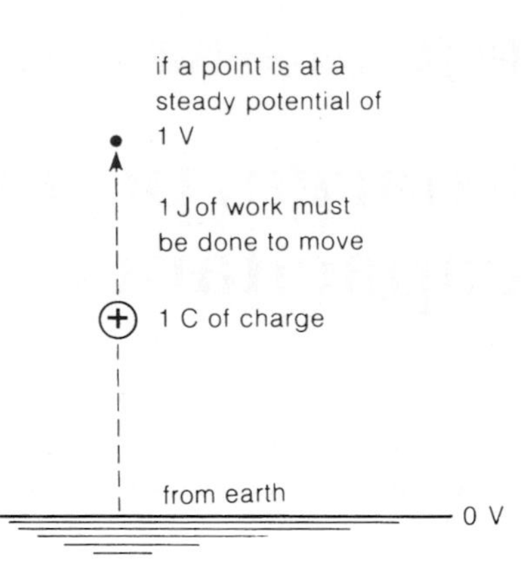

Figure 2 One volt is one joule per coulomb

Positive and negative potential

In figure 3, work must be done to move positive charge from earth to conductor A. Conductor A is at a *higher* potential than earth, and its potential is taken as *positive*.

On the other hand, work must be done to move positive charge from conductor B to earth. B is at a *lower* potential than earth; its potential is *negative*.

It is quite possible for a negatively charged object to have a positive potential (and vice versa). An example is given in figure 4. Despite the presence of a negative charge on the smaller conductor, work would still have to be done to bring a positive charge up to it from earth because of the repulsion from the positive charge on the larger conductor. Overall, the potential of the smaller conductor is positive.

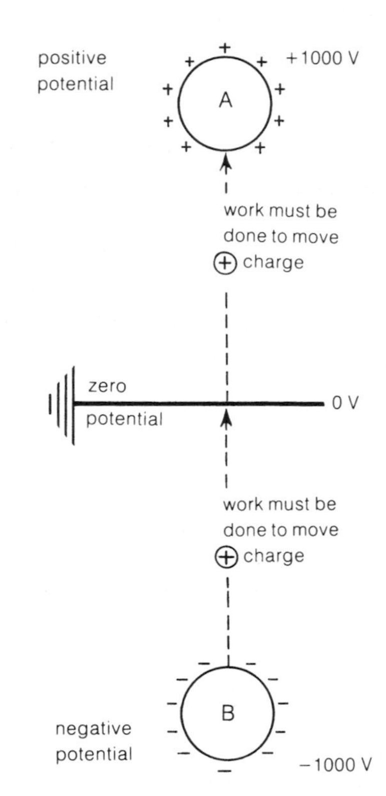

Figure 3 Negative and positive potential the object

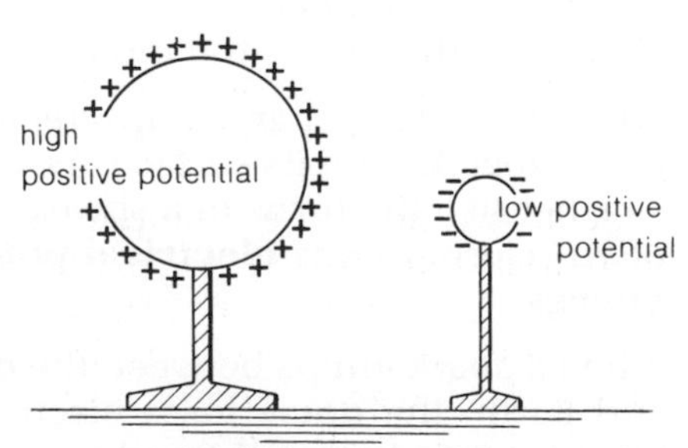

Figure 4 A negatively charged object can have a positive potential under special circumstances

Potential difference and electron flow

If two conductors are at different potentials, there is a **potential difference** or **p.d.** between them. In figure 5 for example, there is a p.d. of 10 000 V between the conductors. A leaf electroscope can be used to measure the p.d. One conductor is connected to the cap of the electroscope, the other to the metal case. The greater the p.d., the greater the deflection of the leaf.

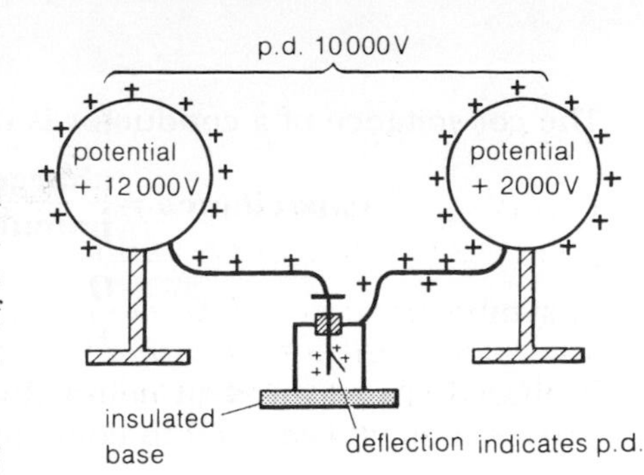

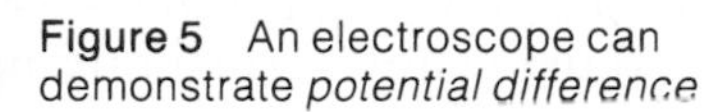

Figure 5 An electroscope can demonstrate *potential difference*

If two conductors at different potentials are joined by a conducting wire, electrons will flow from one to the other until both are at the same potential. Free electrons in both conductors then have the same average potential energy.

Electrons always flow from lower to higher potential, i.e. towards the more positive potential.

When electrons move in this way, there is a **current** flowing as shown in figure 6.

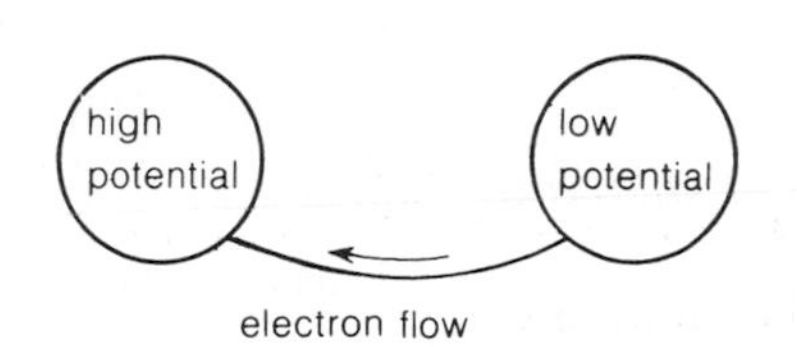

Figure 6 Electrons flow from regions of low potential to regions of high potential

Potential at different points on a charged conductor

If there is a p.d. between any two points on a charged conductor, electrons will flow until both points were at the same potential:

All points on a charged conductor are at the same potential.

This can be demonstrated using a leaf electroscope as shown in figure 7. The deflection of the leaf is a measure of the p.d. between the wire and earth, and the deflection stays the same whatever the point of contact between the wire and the charged conductor.

Don't confuse this experiment with the experiment to study charge distribution described on page 234. In that experiment, small quantities of charge were *transferred* from the charge conductor to the electroscope cap, using a proof plane. In the experiment just described, the cap is *connected* to the charged conductor.

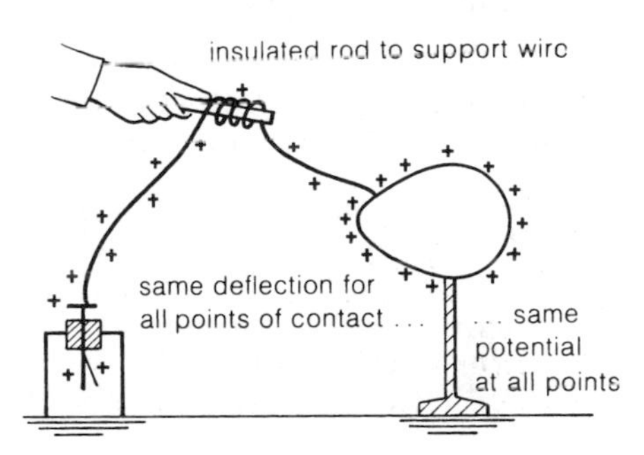

Figure 7 All points on a charged conductor are at the same potential

Capacitance

When a conductor is given more charge, its positive or negative potential rises in value. This happens because the increased charge repels any incoming charge more strongly than before, so larger quantities of work have to be done to increase the charge on the conductor still further.

Two van de Graaff generators are illustrated in figure 8, the figures showing how the potential of each dome rises as more and more charge collects on the dome. Comparing the two sets of figures, you can see that the larger dome has to be given twice as much charge as the smaller dome for its potential to rise to the same extent. The larger dome has twice the **capacitance** of the smaller dome.

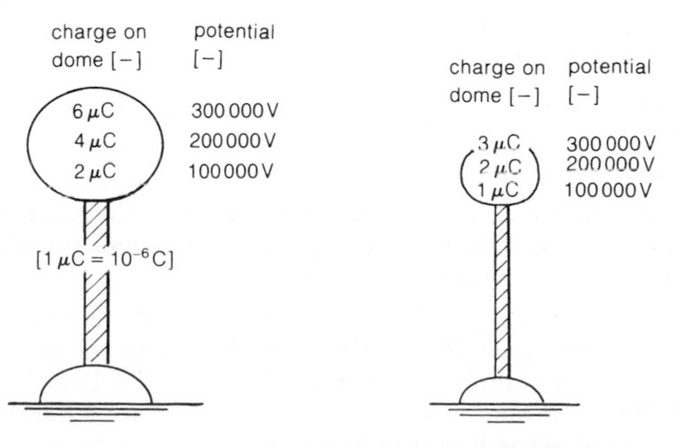

Figure 8 The larger dome needs twice the charge to reach the same potential: it has twice the *capacitance*

The capacitance of a conductor is defined as follows:

$$\textbf{capacitance} = \frac{\textbf{charge of conductor}}{\textbf{potential of conductor}}$$

in symbols: $$C = \frac{Q}{V}$$

If charge is measured in coulombs [C] and potential in volts [V], capacitance is measured in coulombs/volt or **farads [F]**.

Figure 9 A variety of different capacitors

For example, the capacitances of the domes in figure 8 can be calculated using any corresponding values of charge and potential:

For the smaller dome:

$$\text{capacitance} = \frac{\text{charge}}{\text{potential}} = \frac{5 \times 10^{-6}\,\text{C}}{500\,000\,\text{V}} = 10^{-11}\,\text{F}$$

For the larger dome:

$$\text{capacitance} = \frac{\text{charge}}{\text{potential}} = \frac{10 \times 10^{-6}\,\text{C}}{500\,000\,\text{V}} = 2 \times 10^{-11}\,\text{F}$$

In carrying out capacitance calculations of this type, be careful not to confuse C (printed in *ITALIC*), the symbol for capacitance, and C, the abbreviation for a coulomb of charge. Note also that:

the capacitance of a conductor in farads is numerically equal to the charge in coulombs needed to increase the potential by one volt.

For practical capacitance measurements, the farad is far too large a unit. The domes above for example have capacitances of less than a ten thousand millionth of a farad. The most commonly used unit of capacitance is the microfarad:

1 microfarad [μF] = 10^{-6} F [one millionth of a farad]

Capacitors

Devices designed to store small quantities of charge are called **capacitors**. They have many different applications, some of which are described at the end of the section.

The dome of a van de Graaff generator is a simple form of capacitor, though it has a very low capacitance compared with some of the capacitors shown in figure 9, and is much more bulky. Capacitors with a high capacitance are able to store charge at a relatively low potential, and their design is normally based on the parallel plate capacitor shown in figure 10.

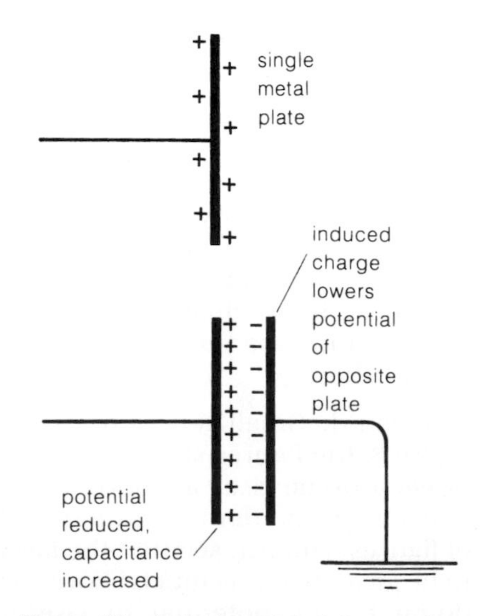

Figure 10 A single metal plate can store charge; a nearby earthed plate greatly increases its capacity to do so.

By itself, a single metal plate has a relatively low capacitance, though this can be increased by increasing the area of the plate. If however a second earthed plate is placed close to the first, the charge induced on the second plate considerably lowers the potential of the first one. Much less work has to be done to put extra charge on the first plate because the repulsion from charge already on it is offset by the attraction from opposite charge from the second one. With the charge on the first plate now at a lower potential, the capacitance of the plate is greatly increased. The closer the two plates, the larger the capacitance becomes.

The capacitance can be increased still further by sandwiching an insulating material called a **dielectric** between the plates. The electric field between the plates distorts the atoms of the dielectric, and the displaced electrons and nuclei further lower the potential of the first plate. Dielectric materials include aluminium oxide, Mylar, mica, and plastics such as polyesters and polypropylene.

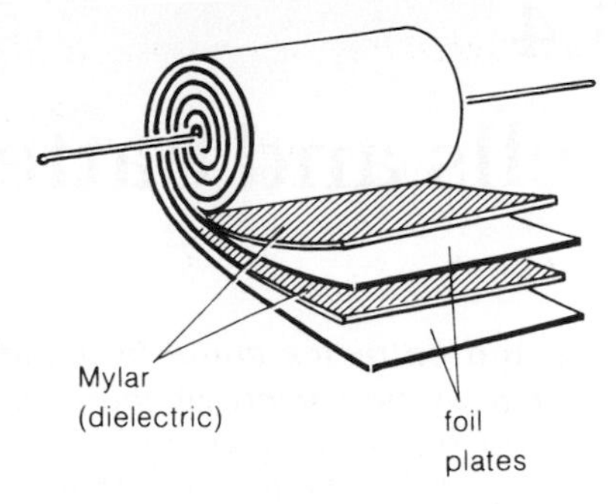

Figure 11. Plates separated by dielectric material and rolled up, form very good capacitors

Practical capacitors often have parallel plates and dielectric rolled up 'swiss roll' style as shown in figure 11. This arrangement gives a large plate area in a small vol·ıme. Typical capacitance of such a capacitor might be 5μF or so – more than ten thousand times the capacitance of the dome of a small van de Graaff generator.

Using capacitors Capacitors are extensively used in the electrical circuits in radios, TVs, record players, and other electronic equipment, as shown in figure 12.

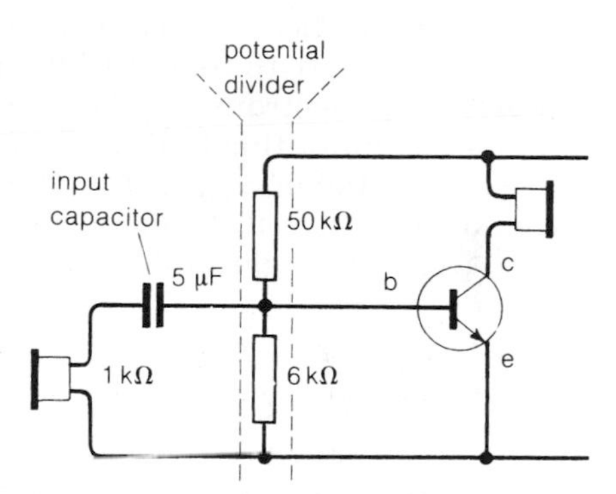

Figure 12 Part of a circuit, showing a capacitor in use. (The circuit is explained in unit 8.5)

A capacitor does not allow electrons to flow through it, but any change of charge on one plate produces an equal change in the induced charge on the opposite plate. A capacitor can therefore be used to transfer a *change* of current from one section of a circuit to another, while stopping a *continuous* current flowing between the sections.

In some circuits, electrons are made to flow backwards and forwards from one plate of a capacitor to the other. Oscillating currents of this type are used in a signal generator, and also in radios and TVs to enable them to be tuned in to different frequencies.

Questions

1 What quantities are measured using the following units? a) coulombs b) volts c) farads.
2 Why does work have to be done to increase the charge on the dome of a van de Graaff generator?
3 A conductor is at a potential of +1000 V. What does this mean?
4 What is the p.d. between the two conductors in figure 13? If the conductors are connected by a length of wire, which way will electrons flow? When will this flow of electrons cease?

Figure 13

5 A conductor is positively charged. What will happen to its potential if it loses electrons?
6 A conductor well away from the influence of other charges carries a charge of 0.0002 C and is at a potential of +1000 V. What would be its potential if the charge were doubled? What is the capacitance of the conductor?
7 A capacitor consists of two parallel metal plates with an air gap between them. Give three changes which could be made to increase the capacitance. Why does the arrangement just described have a greater capacitance than one of the plates by itself?
8 Figure 14 shows a charged conductor. How does the concentration of charge at A compare with that at B? How does the potential at A compare with that at B?

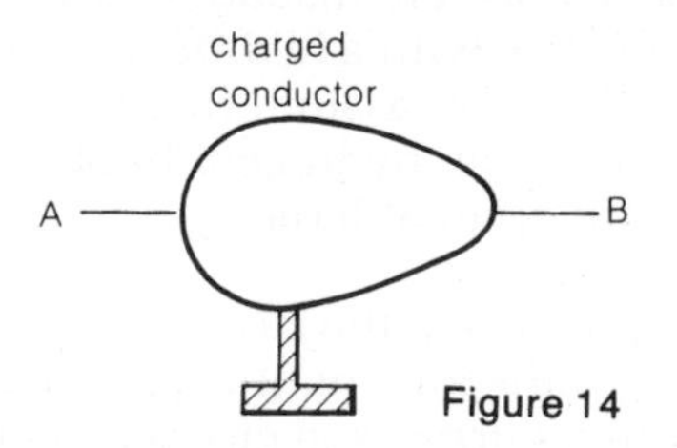

Figure 14

6.4

Cells and batteries

Cells and batteries come in a wide range of sizes and are used to power a whole variety of things from wrist watches to submarines. They push out charge in the form of electrons when chemical reactions take place inside them.

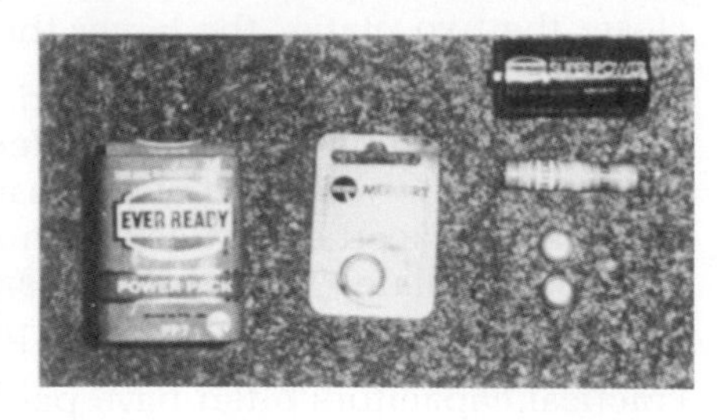

Figure 1 Different types of cells and batteries

All the cells in figure 1 have two terminals. Chemical reactions inside a cell help to create a small potential difference, usually a volt or so, between these terminals and this makes electrons flow along any conducting path that links them, as shown in figure 2. The flow or **current** of electrons may last for many hours, and some cells give out charge amounting to thousands of coulombs before the current ceases. Potential energy lost by the electrons as they flow from one terminal around to the other is changed into other forms – heat and light energy in a light bulb for example.

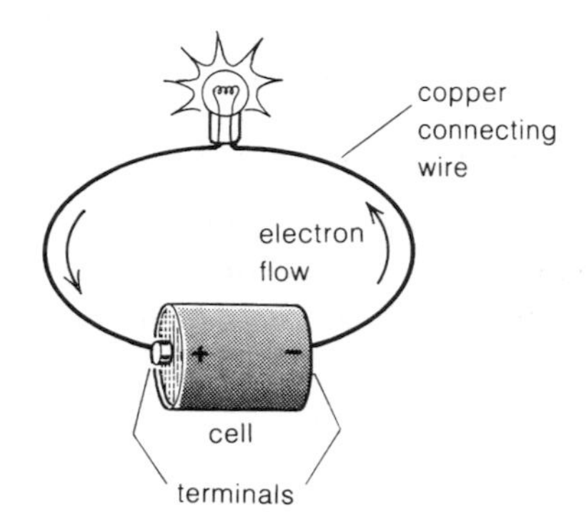

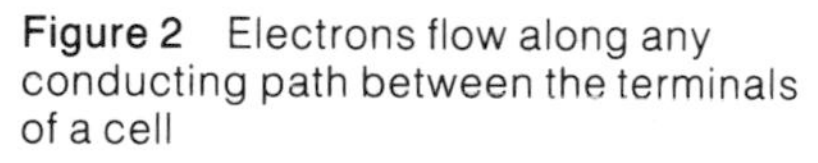

Figure 2 Electrons flow along any conducting path between the terminals of a cell

The simple cell

Figure 3 shows a simple cell consisting of a zinc plate and a copper plate in a dilute solution of sulphuric acid. The two metals behave differently in the presence of the acid, and this causes a p.d. of about 1 V between them.

When the plates are linked by a conducting path, the zinc slowly dissolves in the acid solution. The zinc goes into the solution in the form of positive ions rather than complete atoms however, so electrons get left behind on the plate. The copper plate on the other hand doesn't dissolve in the acid solution. Instead, it loses electrons to positively charged ions of hydrogen in the acid, turning them into uncharged molecules of hydrogen gas which bubbles up round the plate. The overall effect of these reactions is that the dissolving zinc plate becomes negatively charged, the copper is left with a positive charge, and electrons flow from the zinc round to the copper.

A simple cell has two main defects:

Polarization The cell only works for a short time before a blanket of bubbles builds up on the copper plate and drastically reduces the electron flow. The effect is called **polarization**. It occurs partly because the bubbles act as an insulator, and partly because a positive plate is less effective with a hydrogen surface than a copper one. A **depolarizer** such as potassium dichromate is added to the acid solution to remove the hydrogen bubbles chemically and restore the current to its original level.

Local action Any impurities in the zinc, traces of iron for example, form miniature cells with the zinc on the surface of the plate. These miniature cells cause tiny currents to circulate continuously within the plate itself, and the zinc goes on dissolving even when the cell isn't in use. This effect is called local action.

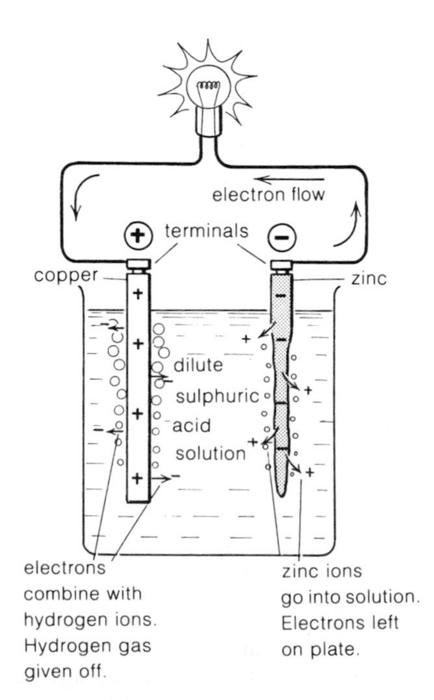

Figure 3 A simple cell. To enable the zinc to dissolve, and the acid to give off its hydrogen, electrons are forced to go round the external circuit

The dry cell

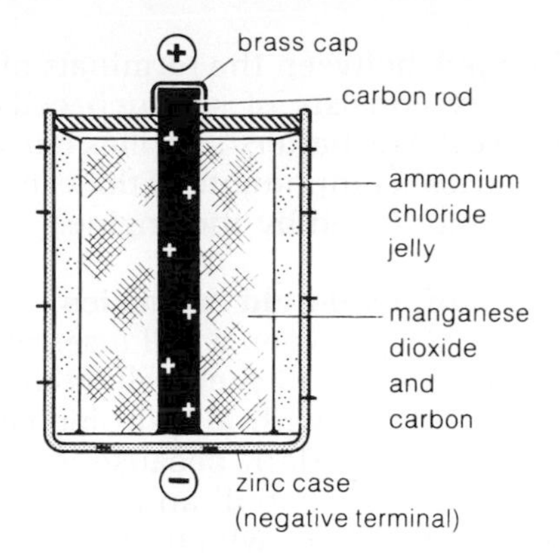

Figure 4 A dry cell. Here, the dissolving zinc is also the container for the cell. The hydrogen is removed chemically.

The cells you find in the handle of a torch are usually of a type known as dry cells. Instead of an acid, a dry cell uses ammonium chloride in 'dry' jelly form, so the cell doesn't have to be kept upright.

Figure 4 shows the structure of one type of dry cell. The positive terminal is on the central carbon rod, while the zinc case acts as the negative plate and terminal. The p.d. between the terminals is about 1.5 V. The manganese dioxide depolarizer is mixed with powdered carbon to improve its conducting ability, and packed around the carbon rod.

Dry cells will usually work after many months in storage, though they do deteriorate in time because of local action. They are more suitable for occasional rather than continuous use because the depolarizer is rather slow acting.

Batteries

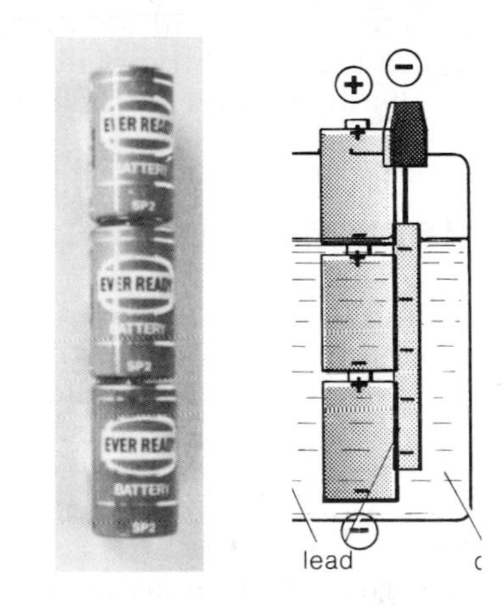

Figure 5 A battery of three cells.

Figure 5 shows several cells connected together to form a **battery**. A battery really means a collection of cells though the word is also used in everyday language to describe a single cell.

Cells connected as shown are said to be in **series**. Together, they store more energy than a single cell and will push a larger current (more electrons every second) along a conducting path.

Primary and secondary cells

The simple cell and the dry cell are both called **primary** cells. The chemical reactions which cause the electron flow are not readily reversible and the cells cannot easily be recharged. **Secondary** cells on the other hand must be supplied with charge before use by passing a current of electrons through them, but they can be recharged over and over again. Secondary cells are sometimes known as **storage cells** or **accumulators**.

The lead-acid cell

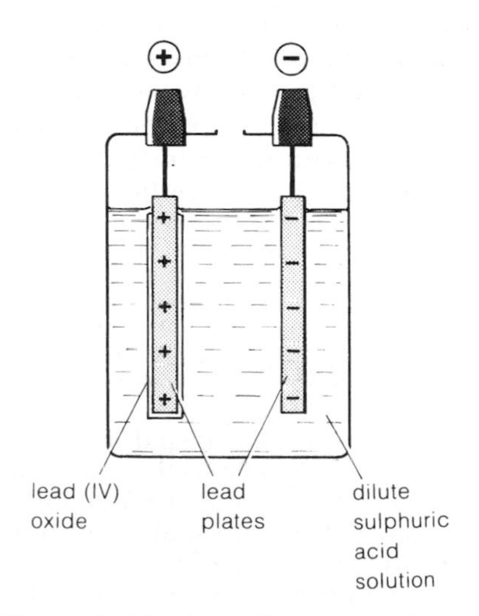

Figure 6 A lead-acid cell. In use, the sulphuric acid becomes dilute; both plates become coated with lead sulphate. Forcing a current through the cell reverses the chemical reaction and makes the cell ready for re-use

One of the most commonly used secondary cells is the lead-acid cell; a simple type is shown in figure 6. It can give out a much larger quantity of charge than a dry cell and is capable of supplying a greater current (see page 261). The cell consists of two sets of plates in a dilute solution of sulphuric acid. Both sets are made of lead but a layer of lead(IV) oxide is formed on the positive plates when the cell is first charged.

The cell discharges (gives out electrons) as the lead and lead dioxide on the plates are slowly converted to soft lead sulphate. The reaction dilutes the acid, making it less dense. When the cell is connected to a battery charger, electrons are pushed back through the cell. This reverses the chemical reactions; lead and lead(IV) oxide are built up on the plates again.

The p.d. between the terminals of a lead-acid cell is about 2 V. A car battery consists of six such cells arranged in series as shown in figure 7. The battery supplies current for the starter motor and other electrical components of the car, and is recharged by a generator or dynamo driven by the engine.

Care of lead-acid batteries The density of the acid solution becomes less when a cell loses charge so it is possible to check the state of charge of each cell by measuring the relative density of the acid solution with a hydrometer (as explained on page 109). The check is important because a 'flat' (fully discharged) battery soon becomes 'sulphated' and cannot then be recharged. A sulphated battery is one in which the soft lead sulphate has changed into a hard form that cannot be converted back into lead and lead(IV) oxide.

When a battery is fully charged, any further charging turns some of the water in the acid solution into hydrogen and oxygen gas that bubbles up round the plates. The battery must be kept topped up with distilled water to replace any water that is lost.

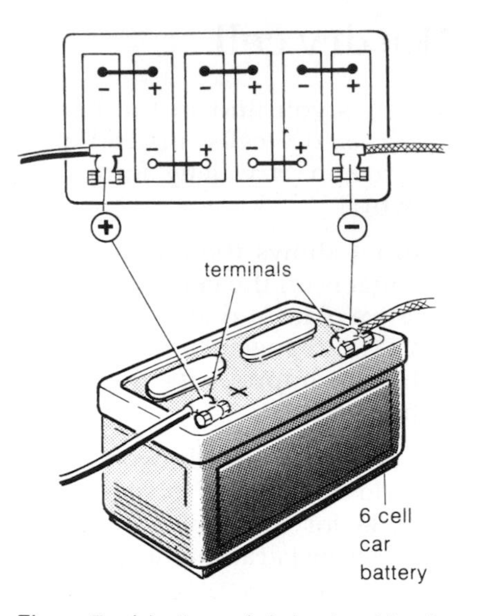

Figure 7 A battery of six lead-acid cells creates a p.d. of about 12 V

Alternative to petrol?

Britain has 40 000 battery-powered vehicles on its roads. This is far more than any other country, but still less than 2% of all British road vehicles. Most are milk floats – reliable and ideal for stop-start driving, but heavy, expensive, and very slow. Energy storage is the problem. It takes ten lead-acid car batteries to store as much usable energy as a gallon of petrol.

Lead-acid batteries are being improved all the time. But practical storage batteries will eventually be based on other materials. Sodium-sulphur batteries, now being developed, can store five times as much energy as lead-acid batteries of the same weight. And aluminium-air batteries are also showing great promise. They need no charging – just water, and a regular supply of new aluminium sheets. An electric car could travel about 15 000 kilometres every year on about 600 kilograms of aluminium. In the future, filling up with aluminium may become more common than filling up with petrol.

For stop-start driving at low speed, the battery powered vehicle is ideal. But its performance has so far proved unacceptable to most road users

Questions

1 Give the approximate p.d. across the terminals of a) a simple cell b) a dry cell c) a lead-acid cell.
2 What materials could be used for the positive and negative plates of a simple cell and in what liquid could they be placed?
3 What is polarization? How does it affect a simple cell and how can it be prevented? Give one other defect of a simple cell.
4 What materials act as the positive and negative plates of a dry cell? A dry cell contains manganese dioxide. What is its purpose?
5 Draw a diagram to show how several dry cells could be connected together to form a battery.
6 What is the difference between a primary cell and a secondary cell? Give one example of each.
7 What advantages does a lead-acid cell offer compared with a dry cell? What are its disadvantages?
8 How can you check the state of charge of a lead-acid cell? Why must the cell not be left 'flat' for any length of time? Why must a lead-acid cell be kept topped up with distilled water?

6.5

Current and voltage in a simple circuit

The potential difference between the terminals of a battery causes a current to flow along any conducting path that links them. Current and potential difference can be measured with meters, and measurements taken around a simple circuit reveal the basic rules that apply to each.

Figure 1 shows two light bulbs connected by copper wires to a battery made up of two cells. The conducting path through the bulbs, wire and battery is called a **circuit**. The diagram has been drawn using symbols selected from the chart in figure 2.

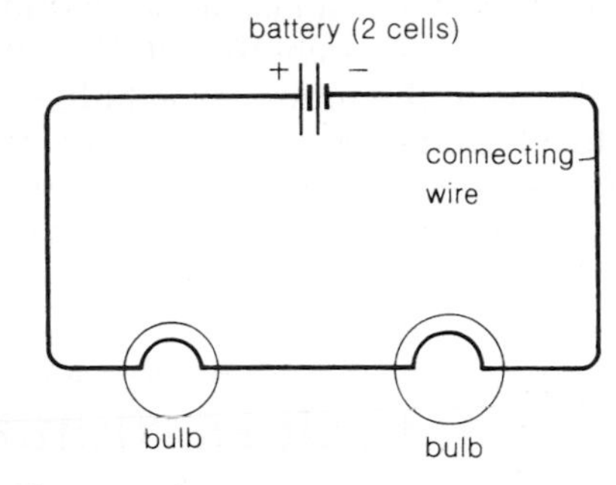

Figure 1 A simple circuit

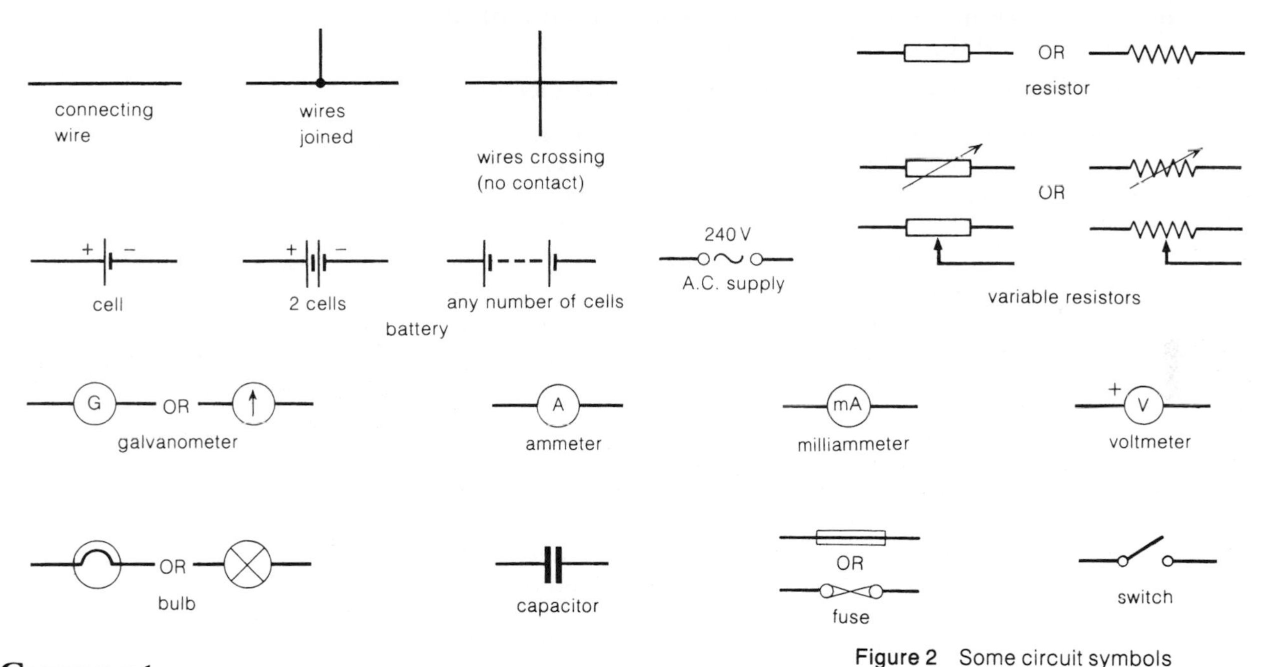

Figure 2 Some circuit symbols

Current

The current in a circuit is measured in **ampères** [A] and it indicates the rate at which charge is flowing. A current of 1 ampère is equivalent to a flow of about 6 million million million electrons every second, though this is *not* a definition of the ampère. The definition is given on page 294.

Smaller currents are often expressed using units of **milliampères** or **microampères**:

1 milliampère [mA] $= 10^{-3}$ A $\left[\frac{1}{1000}\text{A}\right]$

1 microampère [μA] $= 10^{-6}$ A $\left[\frac{1}{1\,000\,000}\text{A}\right]$

Typical current values are:

Current through a car headlight bulb: 4 A
Current through a small torch bulb: 0.2 A (200 mA)
Current through a pocket calculator: 0.005 A (5 mA)

Currents of around an ampère or so are measured using an instrument called an **ammeter** (see figure 3). The instrument is connected into a circuit so that all the current flows through it, but its design is such that it has little effect on the current itself – effectively the ammeter acts like another piece of connecting wire. Most ammeters are sensitive to current direction and can be damaged if connected the wrong way round. The terminal coloured red, or marked with the positive sign, should be connected to the side of the circuit which leads to the positive battery terminal.

Figure 3 An ammeter and its symbol

Current values round a simple circuit

In figure 4, three ammeters have been included in the circuit shown in the previous diagram. The ammeters measure the current at different points around the circuit but all three read the same:

The current is the same at all points round a simple circuit.

If this were not the case, electrons would be accumulating somewhere round the circuit or leaking away. As it is, electrons leave the negative terminal of a battery at exactly the same rate as they flow into the positive terminal.

Electrons move very slowly round most circuits, rarely travelling more than a few millimetres every second. The forces they exert on each other act almost instantaneously however. When electrons are pushed out from the battery terminal, electrons already in the bulbs and connecting wires immediately start to move.

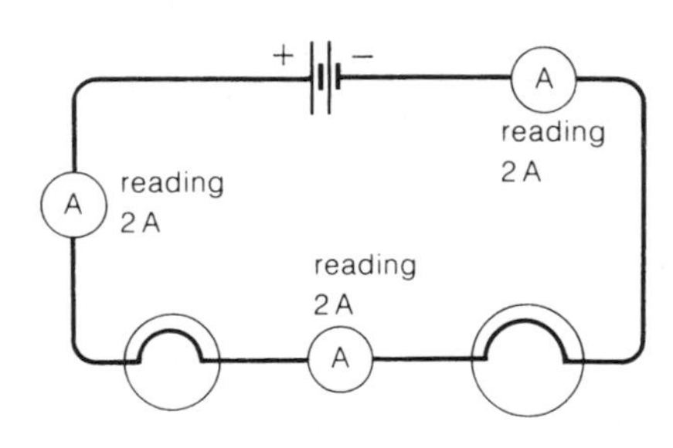

Figure 4 Using ammeters to check that the current at all points round a simple circuit is the same.

Current direction

When the first cells and batteries were made around the beginning of the nineteenth century, no one knew how charge moved round a circuit or in what direction. However, it became the convention to regard an electric current as a flow of positive charge from the positive terminal of a battery round to the negative. About a hundred years later, it was found that a current in a metal was actually a flow of negatively charged particles, called electrons, moving in the opposite direction, but it is still normal practice to use the **conventional** direction when marking currents on circuit diagrams.

Why not change the convention? It isn't really necessary as electrons do in effect cause a transfer of positive charge from the positive terminal of a battery round to the negative. Put another way, a flow of negative charge to the left is algebraically equivalent to a flow of positive charge to the right.

In this book, electron flow is indicated by arrows drawn to the side of the circuit, while the conventional current is shown by arrowheads drawn on the circuit itself. This is illustrated in figure 5.

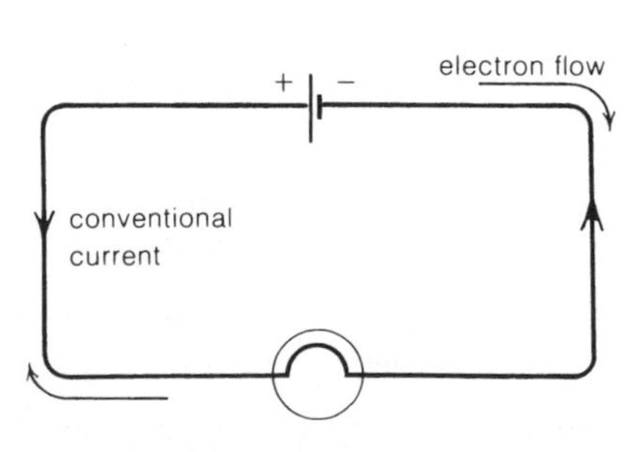

Figure 5 Electrons actually flow in the opposite direction to the conventional current. In this book, both directions are shown where it is useful to do so.

Charge and current

The total charge passing any point in a circuit depends on the current and the time for which it flows, these quantities being linked by the equation

charge = current × time

in symbols, $Q = It$

Charge is measured in coulombs [C] if current is measured in ampères [A] and time in seconds [s]. For example:

a charge of 6 coulombs passes if a current of 2 ampères flows for 3 seconds.

The definition of the coulomb also follows from the above equation:

1 coulomb of charge passes any point in a circuit when a steady current of 1 ampère flows for 1 second.

The link between charge and current provides a useful way of thinking of a current:

A current of 1 ampère means that charge is flowing at the rate of 1 coulomb every second (see figure 6)

A current of 2 ampères means that charge is flowing at the rate of 2 coulombs every second, and so on.

Note however that as far as definitions are concerned, it is the coulomb which is based on the ampère rather than the other way round.

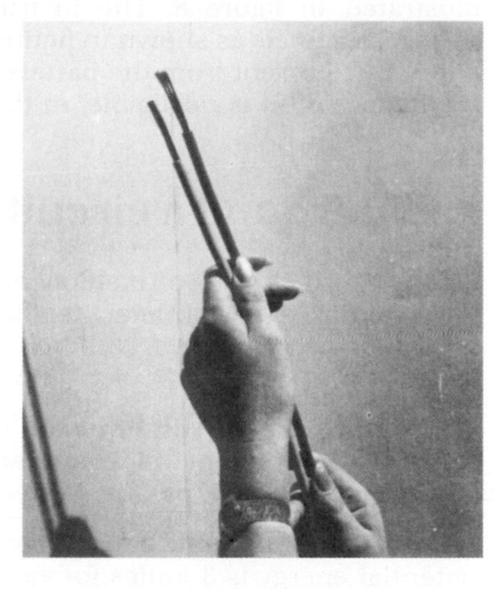

This cable is designed to carry a maximum current of 60 coulombs per second: 60 amperes

1 ampere → = 1 coulomb every second

Figure 6 The current in a circuit is a measure of the rate of flow of charge in it

Energy from a battery

In figure 7, the battery gives potential energy to the electrons it pushes out. It does so by forcing them closer together in the connecting wire – rather like the turns of a compressed coil-spring. The electrons lose all this potential energy as they flow round to the other terminal, the energy being given off as heat and light as the electrons pass through the bulbs.

Typical energy changes are shown on the diagram. The battery gives 4 joules of potential energy to each coulomb of charge it pushes out. 3 joules of potential energy are lost as the charge passes through the larger bulb, and the remaining 1 joule as it passes through the smaller bulb. The charge loses a negligible quantity of potential energy in passing through the copper connecting wires – the reasons for this are examined in unit 6.6.

P.D. across battery terminals

The p.d. or **voltage** across the terminals of the battery indicates the potential energy given to each coulomb of charge pushed out:

There is a p.d. of 1 volt across a battery if each coulomb of charge is given 1 joule of potential energy.

There is a p.d. of 4 volts across a battery if each coulomb is given 4 joules of potential energy, and so on.

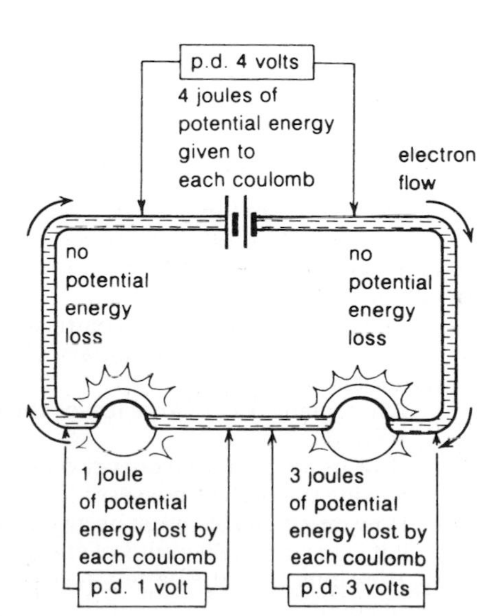

Figure 7 4 joules of energy are given to each coulomb of charge by the battery; this energy is converted to light and heat by the bulbs

The p.d. across the battery can be measured using a **voltmeter** illustrated in figure 8. The instrument is connected across the battery terminals as shown in figure 9. The voltmeter actually draws a very tiny current from the battery, but the effect on the current in the main circuit is negligible, in this case.

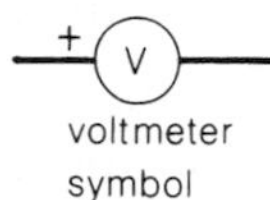

Figure 8 A voltmeter and its symbol

P.d.s. around a circuit

Figure 9 also shows voltmeters connected across the two bulbs in the circuit. The voltmeter reading in each case indicates the potential energy lost by a coulomb of charge as it passes through the bulb:

There is a p.d. of 1 volt between two points in a circuit if 1 joule of potential energy is changed into other forms when 1 coulomb of charge passes between the points.

There is a p.d. of 3 volts between two points in a circuit if the loss of potential energy is 3 joules for each coulomb, and so on.

The p.d. between the ends of any of the pieces of connecting wire is effectively zero because there is almost no loss of potential energy over these sections.

Figure 9 illustrates a principle which applies in all electrical circuits:

The sum of the p.d.s around a conducting path from one battery terminal to another is the same as the p.d. across the battery.

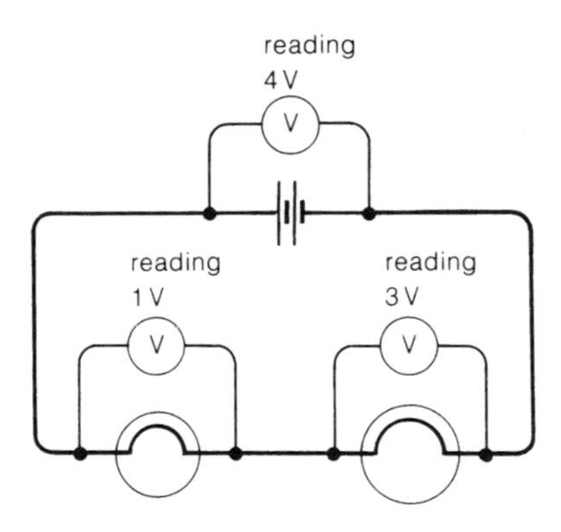

Figure 9 4 joules per coulomb are gained across the battery teminal; the *p.d.* across it is 4 V. The p.d.s across the bulbs are 3 V and 1 V respectively

Questions

1 In what units are the following measured:
a) current b) potential difference c) charge?

2 Express the following in milliampères
a) 0.75 A b) 0.02 A.

3 A current of 4 A flows from a battery when a light bulb is connected across its terminals. The p.d. across the terminals is 12 V.
a) What quantity of charge leaves the battery every second?
b) How much potential energy does each coulomb leaving the battery possess?
c) How much charge must be given out by the battery if it is to supply 60 joules of energy?
d) How long does the battery take to supply 60 joules of energy?

4 In 10 s, a charge of 25 C leaves a battery, and 200 J of energy are delivered to an outside circuit as a result.
a) What is the p.d. across the battery?
b) What current flows from the battery?

5 In figure 10, what type of meter is X? What type of meter is Y?
What will X and Y each read if the figures on the diagram show the readings on the other three meters?

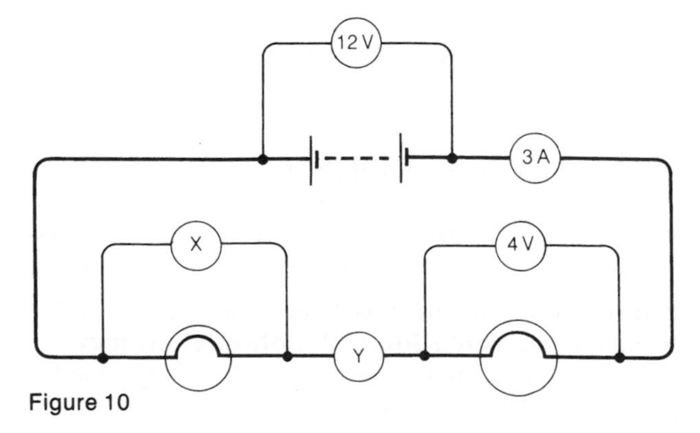

Figure 10

6.6

Ohm's law and resistance

Increase the p.d. across the ends of a conductor and a greater current flows. But the current depends on the conducting ability of the conductor as well as the p.d. across it. Some conductors offer more resistance to a current flow than others.

Ohm's law

In 1826, Georg Ohm carried out experiments with different metal wires to discover how the current through each depended on the p.d. applied across its ends. Typical results from a modern experiment are given in figure 1. The wire in this case is made of an alloy called nichrome; it is kept at a constant temperature throughout the experiment. If a graph of current against p.d. is plotted, all the usual features of a simple proportion are seen:

1 The graph is a straight line passing through the origin.
2 Doubling the p.d. doubles the current.
3 Dividing the p.d. by the current always gives the same value, 5 V/A in this case.

Experiments with other wires produce similar results. These can be summed up in a law now known as **Ohm's law** which states:

The current flowing through a metal conductor is directly proportional to the p.d. across its ends provided the temperature and other physical conditions remain constant.

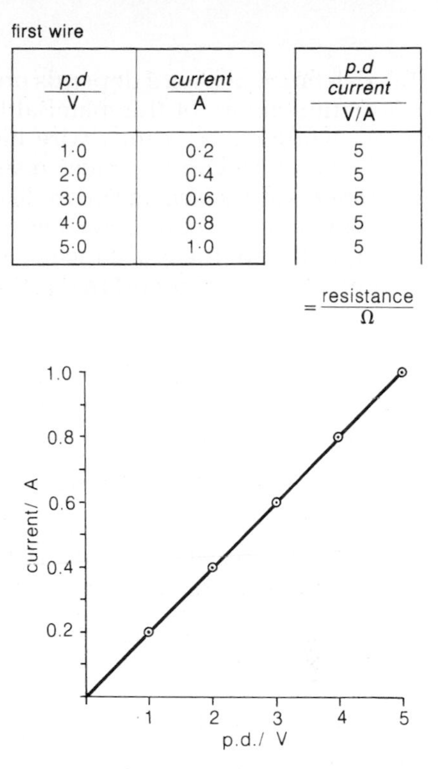

first wire

p.d / V	current / A	p.d / current / V/A
1·0	0·2	5
2·0	0·4	5
3·0	0·6	5
4·0	0·8	5
5·0	1·0	5

= resistance / Ω

Figure 1 The figures and graph show how the current increases in proportion to the p.d. applied across the ends of a wire

Resistance

Figure 2 gives readings from an experiment with a second length of wire. Once again, the current and p.d. are in direct proportion, but the current flowing for any given p.d. is only half what it was before. The second wire is a poorer conductor than the first wire, and is said to offer a greater **resistance** to the current flow.

Dividing p.d. by current gives 10 V/A for the second wire compared with 5 V/A for the first, so these values provide an indication of the amount of resistance offered by each wire. Resistance is in fact defined by the equation:

$$\textbf{resistance} = \frac{\textbf{p.d. across conductor}}{\textbf{current through conductor}}$$

With p.d. in volts [V] and current in ampères [A], resistance is measured in volts/ampère or **ohms** [Ω].

A conductor has a resistance of 1 Ω if a current of 1 A flows through it when a p.d. of 1 V is applied across its ends.

The first wire therefore has a resistance of 5 Ω while the second wire has a resistance of 10 Ω.

second wire

p.d / V	current / A	p.d / current / V/A
1·0	0·1	10
2·0	0·2	10
3·0	0·3	10
4·0	0·4	10
5·0	0·5	10

= resistance / Ω

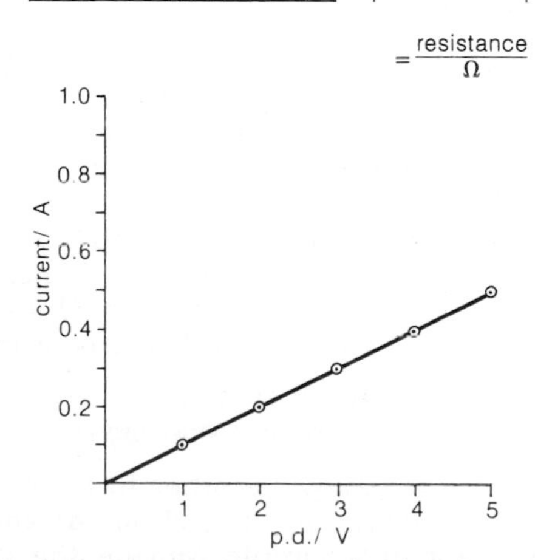

Figure 2 The current still increases in proportion to the p.d., but to a lesser extent – the *resistance* of the wire is greater

The resistance of a wire depends on its dimensions as well as on the conducting ability of the material from which it is made. All other factors being equal, a long wire has more resistance than a short one, and a thin wire has more resistance than a thick one. Typical resistance values are given in figure 3. The resistance equation applies whether a conductor obeys Ohm's law or not, but if Ohm's law isn't obeyed, the resistance of a conductor (i.e. the value of p.d./current) will vary depending on the current flowing.

	resistance
½ m nichrome	7Ω
1 m nichrome	14Ω
1 m nichrome	1Ω
1 m copper	0.02Ω

wire thicknesses as shown
wires at room temperature

Figure 3 Comparing resistances. A long, thin piece of nichrome wire has a higher resistance than a short thick piece; copper of the same dimensions will always have a much lower resistance

V, I, R equations

The resistance equation can be written using symbols:

$$R = \frac{V}{I}$$

where R is the resistance, V the p.d. and I the current. The equation can also be rearranged to give:

$$V = IR \text{ and } I = \frac{V}{R}$$

These are useful if the p.d. across a known resistance, or the current through it is to be calculated. For example:

A p.d. of 12 V is needed to make a current of 2 A flow through a wire of resistance 6 Ω (using $V = IR$).
A current of 4 A will flow if there is a p.d. of 20 V across a wire of resistance 5 Ω (using $I = V/R$).

Figure 4 shows a useful method of remembering the three versions of the resistance equation.

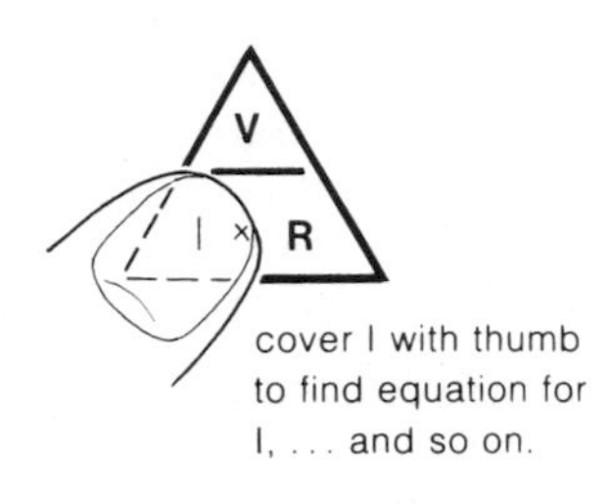

Figure 4 This triangle is useful for remembering V, I, and R equations

Resistors

Devices specially made to provide resistance are called **resistors**. Placed in a simple circuit, they each reduce the current flow. In the more complex circuits found in radios and TVs, they are used to keep currents and p.d.s at the levels needed for other circuit components to function properly.

A length of thin nichrome wire makes a simple resistor. Examples of commercially produced resistors are shown in figure 5. In some, the resistance is provided by a thin layer of carbon, while others contain a long thin alloy wire coiled to take up less space. The photograph also includes the small, rotary form of variable resistor, of the type used as a volume control in radios. Resistances range from a few ohms up to several million ohms. Higher resistance values are normally expressed in kilohms or megohms:

1 kilohm [kΩ] = 1000 Ω
1 megohm [MΩ] = 1 000 000 Ω

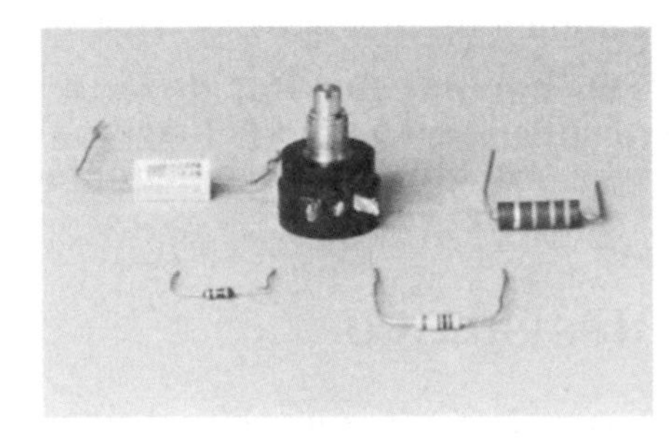

Figure 5 Various resistors

Figure 6a shows another form of variable resistor. Connected as shown, it is known as a **rheostat**, and is used for varying the current flowing in a circuit. Moving the position of the sliding contact changes the length, and therefore the resistance, of the thin coiled wire through which the current has to flow as it passes between terminals A and B.

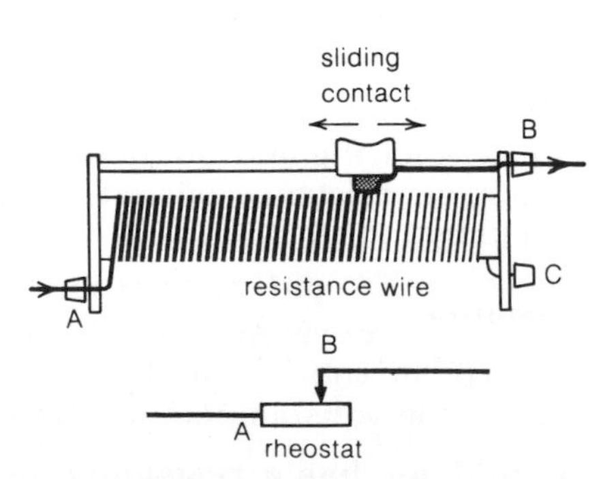

Figure 6a A variable resistor: moving the slide to the right increases the length of resistance wire through which the current has to pass.

The variable resistor has a third terminal, C, which enables it to be used as a **potential divider** or **potentiometer**. This is illustrated in figure 6b. Connected across the battery as shown, the potential divider can supply anything from zero to full battery p.d. depending on the position of the sliding contact.

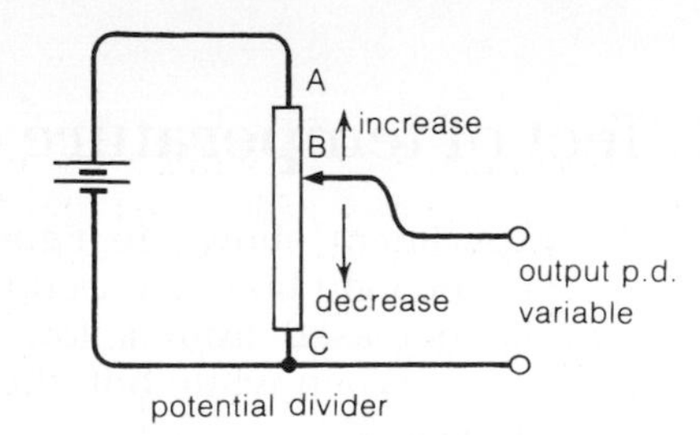

Figure 6b Using a variable resistor as a *source* of varying potential difference

Measuring resistance

Voltmeter-ammeter method Resistances up to about 50 Ω can be measured using the circuit shown in figure 7. The current through the resistor of unknown resistance R is set to any convenient value by adjusting the rheostat. R is then calculated using the readings given on the meters:

$$R = \frac{\text{voltmeter reading}}{\text{ammeter reading}}$$

For greater accuracy, a range of corresponding voltmeter and ammeter readings can be obtained and a graph of current against p.d. plotted as in figure 8. The value of x/y gives the unknown resistance R.

The circuit shown is not suitable for measuring high resistances. If the resistance is high, the current through the resistor is small, and the small current drawn by the voltmeter adds its effect to the reading on the other meter.

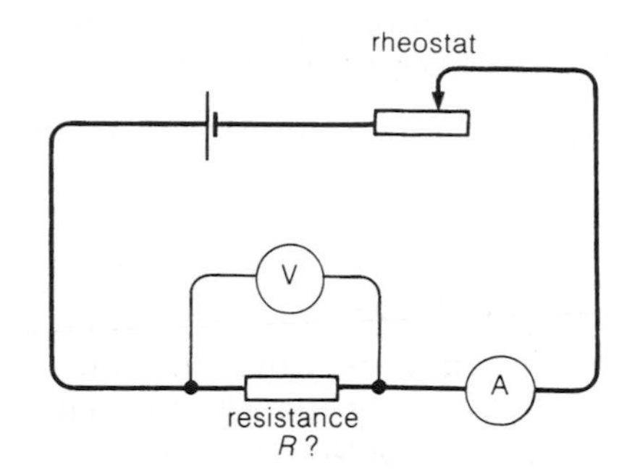

Figure 7 Measurements of the current through a resistor and the p.d. across it, can be used to find its resistance

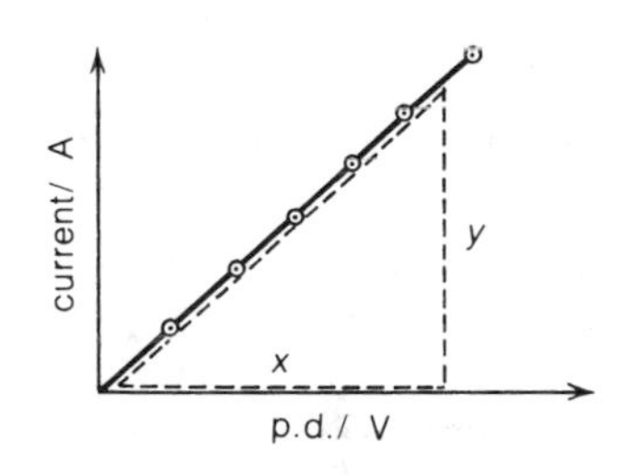

Figure 8 A graph of current against p.d. gives a more accurate method of finding the resistance of a wire

Substitution method Resistances of around 100 Ω and upwards can be measured using the circuit shown in figure 9. X is the resistor of unknown resistance, S is a resistance box containing standard resistors which can be arranged to provide different resistance values by turning the control knobs on the top. X and S are connected into the circuit alternately by moving the connector C between the two positions shown. The value of S is adjusted until the ammeter gives the same reading when the connector is in either position. X and S then have the same resistance.

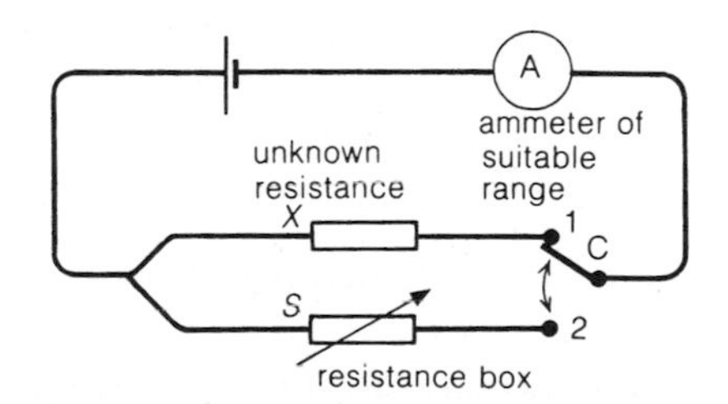

Figure 9 When the current through either circuit is the same, the variable resistance has been set to the same value as the unknown one

Resistance and heat

Any conductor which has resistance gives off heat when a current flows through it. Electrons forced through the conductor lose energy as they collide with its atoms, and the atoms vibrate more rapidly as a result.

Connecting wires used in circuits have as low a resistance as possible so that energy wasted as heat is kept to a minimum. On the other hand, wires of known resistance are used in making the elements of fires and kettles so that heat (thermal energy) is released at a specific rate when a given current is passed through as explained on page 264. The same principle is used in a light bulb where the current flowing through a fine tungsten wire or **filament** makes the filament glow white hot, as shown in figure 10.

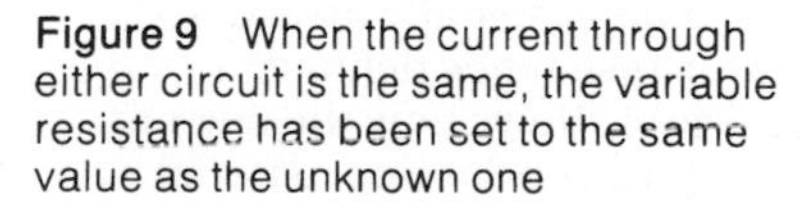

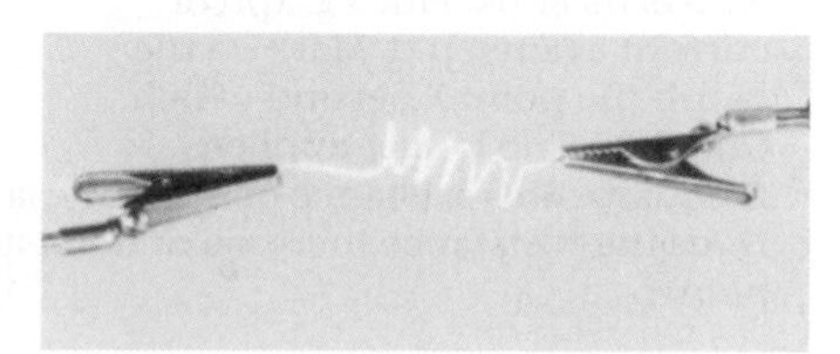

Figure 10 The current flowing through the resistance wire in the bulb is sufficient to make it glow white hot

Effect of temperature on resistance

In the experiment shown in figure 11, the current through the filament of the light bulb is varied by adjusting the rheostat. As the current is increased from a low value, the temperature of the filament rises. When white hot, the filament is at a temperature of more than 3000 °C.

Typical p.d. and current readings are given in figure 12a. The graph in figure 12b shows that the current is *not* proportional to the p.d. The calculations of p.d./current indicate that the resistance of the filament becomes greater as the temperature rises.

The resistance of most metals increases with temperature, though the increase is much less in some cases than in others. Constantan and manganin, both copper-based alloys, are often used in standard resistors because their resistances change very little unless they are heated strongly.

Semiconductors, such as silicon and germanium, are materials which conduct poorly in many cases when cold but become increasingly better conductors as they get warmer. In other words, their resistance decreases with temperature.

Figure 13 These thermistors have a resistance which decreases with temperature

The **thermistors** shown in figure 13 are made from semiconducting oxides of metals and in most cases, their resistance falls sharply as their temperature rises above room temperature. They are used in electronic circuits which are 'switched on' by a temperature change (see page 339).

Carbon isn't classed as a semiconductor, but it too shows a decrease in resistance when its temperature rises.

Figure 11 Using this circuit, the effect of temperature on the resistance of a wire can be found

p.d. / V	current / A	p.d./current / Ω
3·0	1·4	2·1
6·0	2·1	2·9
9·0	2·6	3·5
12·0	3·0	4·0

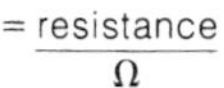

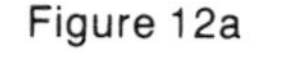

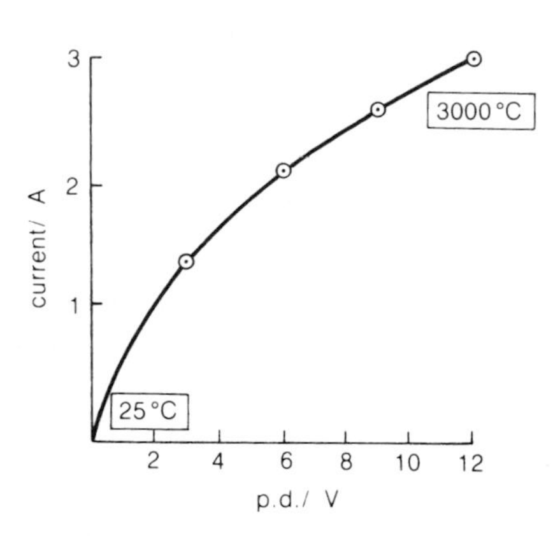

Figure 12b The figures and graph show that for increased p.d., the current does *not* increase in proportion: the increasing temperature of the wire increases its resistance

Questions

(Assume Ohm's law is obeyed unless the question states otherwise)

1 What is Ohm's law? In what unit is resistance usually measured? How is resistance defined?

2 Figure 14 shows how the current through a conducting material varies with the p.d. applied across its ends. Plot a graph of current against p.d. Mark on the graph the point X beyond which Ohm's law no longer applies. Calculate the resistance of the material up to point *X*. Does the resistance increase or decrease beyond this point?

p.d. / V	current / A
1·0	0·16
2·0	0·32
3·0	0·48
4·0	0·64
5·0	0·76
6·0	0·82

3 A resistor has a resistance of about 500 Ω. How could its resistance be measured experimentally?

4 What would be the effect on the resistance of a metal wire of a) increasing its length b) increasing its diameter c) increasing its temperature?

5 A p.d. of 15 V is needed to make a current of 2.5 A flow through a wire. What is the resistance of the wire? What p.d. is needed to make a current of 2.0 A flow throught the wire?

6 There is a p.d. of 6.0 V across the ends of a wire of resistance 12 Ω. What current flows? What p.d. would be needed to make a current of 1.5 A flow through it?

7 A current of 200 mA flows through a 4 kΩ resistor. What is the p.d. across the resistor?

6.7

Series and parallel circuits

There are two basic methods of joining resistors or other circuit components together. This unit examines these methods, the basic rules that apply in each case, and the techniques that can be used to solve circuit problems.

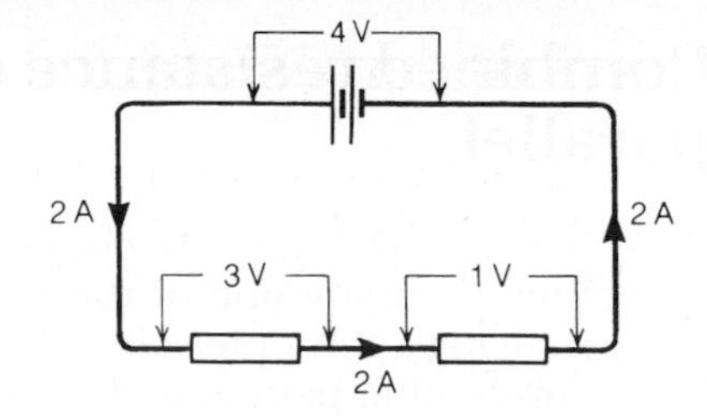

Figure 1 Resistors joined in *series*

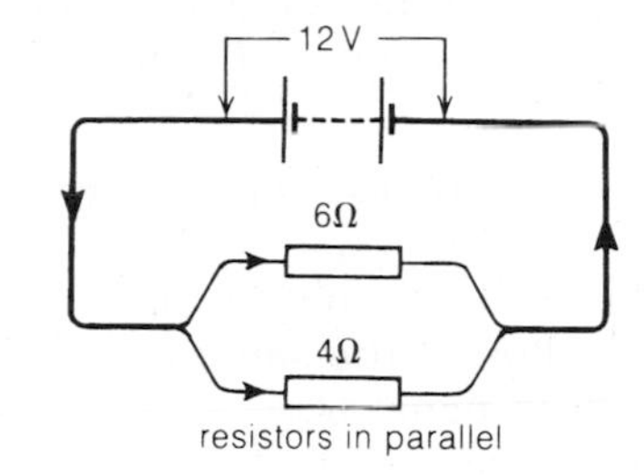

Figure 2 Resistors joined in *parallel*

Basic circuit rules

The two resistors in the circuit in figure 1 are joined in **series**. The basic rules which apply in this case were outlined in unit 6.5:

1 The current is the same at all points round the circuit.

2 The sum of the p.d.s across the resistors is the same as the p.d. across the battery.

The resistors in figure 2 are joined in **parallel**. Each resistor has the full battery p.d. across it, and the current through each can be calculated using $I = V/R$

$$\text{current through } 6\,\Omega \text{ resistor} = \frac{12\,\text{V}}{6\,\Omega} = 2\,\text{A}$$

$$\text{current through } 4\,\Omega \text{ resistor} = \frac{12\,\text{V}}{4\,\Omega} = 3\,\text{A}$$

The battery supplies both currents at the same time and, in the main circuit itself, both currents are carried by the same piece of wire. There is therefore a total current of 5 A in the main circuit.

The above example illustrates two rules which apply whenever resistors or other components are joined in parallel:

1 Each resistor in a parallel arrangement has the same p.d. across it.

2 Where a circuit divides into several parallel branches, the currents through different branches add up to equal the current in the main circuit (see figures 3a and 3b).

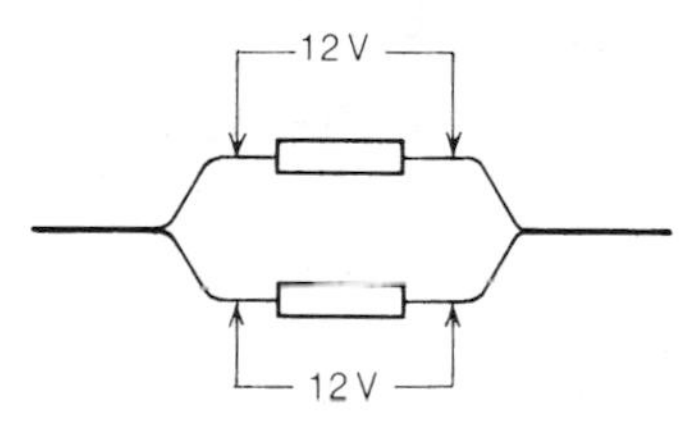

Figure 3a The p.d. across resistors joined in parallel is the same: they share the same source of p.d.

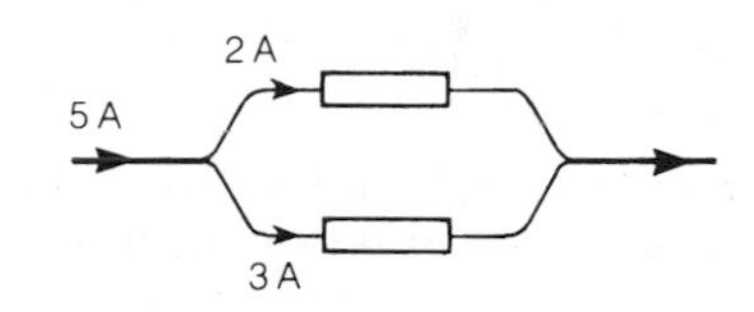

Figure 3b When resistors are joined in parallel, more current flows through the lower resistance. But the total current is the same as in the rest of the circuit

Combined resistance of resistors in series

If two or more resistors are connected in series, they give a *higher* resistance than any one of the resistors by itself. The effect is the same as joining several short lengths of resistance wire together to form a longer length. In figure 4 for example, the resistors of 2 Ω, 3 Ω, and 5 Ω, have a combined resistance of 10 Ω. In other words, a single 10 Ω resistor would cause the same current to flow in the circuit if it replaced the three resistors in series.

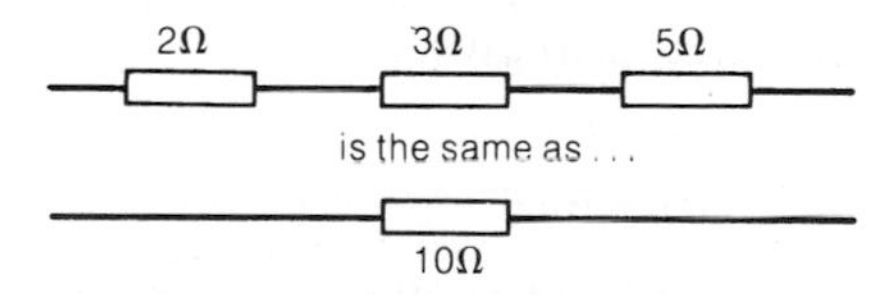

Figure 4 A 10 Ω resistor is equivalent to 2 Ω, 3 Ω and 5 Ω resistors in series

In general, if a number of resistors, R_1, R_2, R_3 etc have a combined resistance R when joined in series, then

$R = R_1 + R_2 + R_3$ and so on. This is illustrated in figure 5.

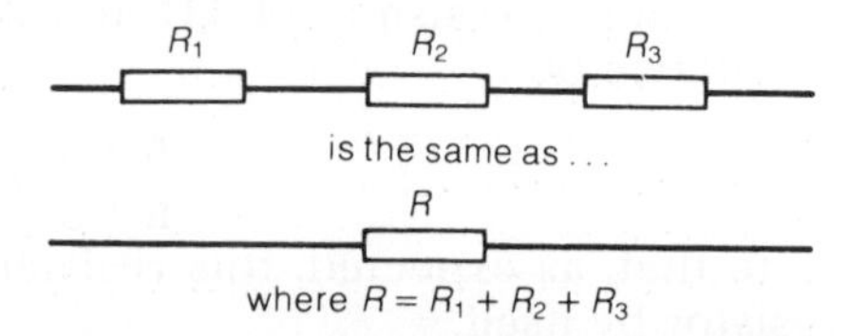

Figure 5 The general case from figure 4

Combined resistance of resistors in parallel

If two or more resistors are connected in parallel, they give a *lower* resistance than any one of the resistors by itself. The effect is the same as using a thick piece of resistance wire instead of a thin piece – the conducting path is wider than before.

Calculating the combined resistance of resistors in parallel isn't as simple as in the series case, but it can be done using the equation $V = IR$ and some of the basic circuit rules given at the start of the unit.

In the figure 6a, the two resistors R_1 and R_2 have a combined resistance R when joined in parallel. They could therefore be replaced by a single resistor R through which the same current I would flow if the same p.d. V were applied. This single resistor is shown in figure 6b.

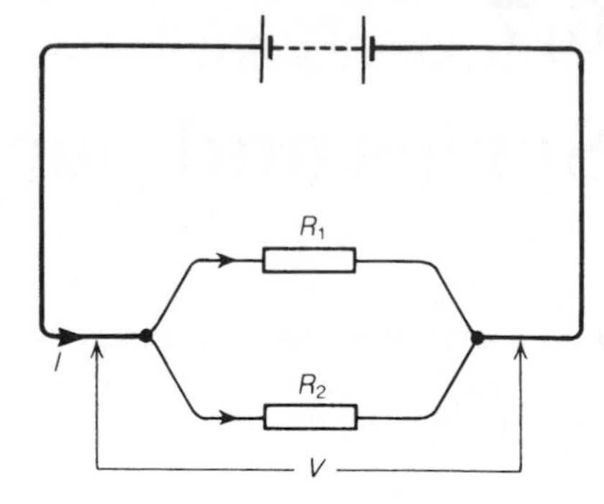

Figure 6a

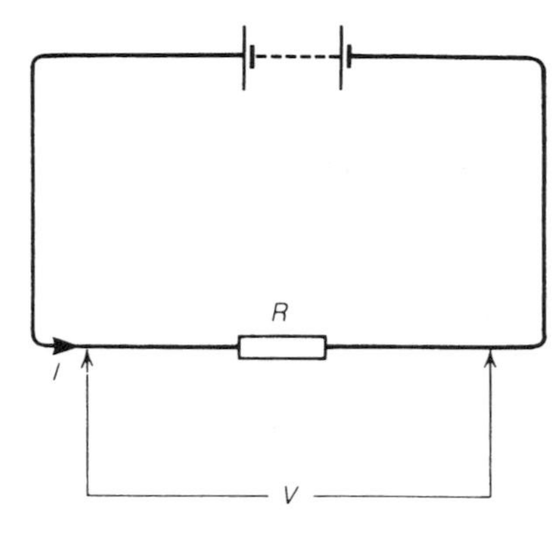

Figure 6b

Figures 6a, 6b It is possible to replace the parallel resistors R_1 and R_2 by a single equivalent resistor R. To find the relationship between R_1, R_2, and R, use the fact that the currents in the two circuits are the same.

In figure 6a: the current through $R_1 = \dfrac{V}{R_1}$

the current through $R_2 = \dfrac{V}{R_2}$

$$\therefore \text{ total current in the main circuit} = I = \frac{V}{R_1} + \frac{V}{R_2} = V\left(\frac{1}{R_1} + \frac{1}{R_2}\right)$$

In figure 6b: total current through $R = \dfrac{V}{R}$

This current is defined as being the same as the first one, so:

$\dfrac{V}{R} = V\left(\dfrac{1}{R_1} + \dfrac{1}{R_2}\right)$; dividing both sides by V gives $\boldsymbol{\dfrac{1}{R} = \dfrac{1}{R_1} + \dfrac{1}{R_2}}$

The same argument can be applied to any number of resistors in parallel. If resistors of R_1, R_2, R_3 etc have a combined resistance R when joined in parallel, then

$\boldsymbol{\dfrac{1}{R} = \dfrac{1}{R_1} + \dfrac{1}{R_2} + \dfrac{1}{R_3}}$ and so on; this is illustrated in figure 7.

Where two resistors only are joined in parallel, the equation can be rearranged to give:

$$R = \frac{R_1 R_2}{R_1 + R_2}$$

i.e. $\dfrac{\textbf{combined resistance of two resistors}}{} = \dfrac{\textbf{resistances multiplied}}{\textbf{resistances added}}$

For example, resistors of $3\,\Omega$ and $6\,\Omega$ in parallel have a combined resistance of:

$$\frac{6 \times 3}{6 + 3}\,\Omega, \text{ or } 2\,\Omega.$$

Note that, as expected, this resistance is *lower* than that of either resistor by itself.

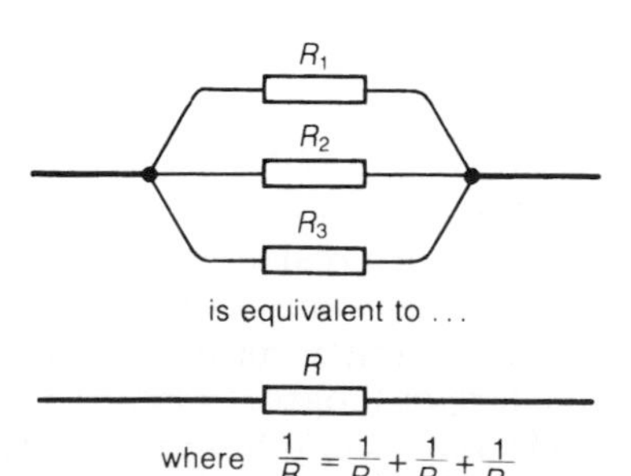

Figure 7 The resistors in parallel present a broader, "easier" conducting route. So their combined resistance is lower. Try arranging the resistors from figure 4 in parallel: see that the combined resistance is only about 1 Ω

Circuits and switches

Where several bulbs have to be powered by a single battery, as in a car lighting system for example, the usual practice is to connect the bulbs in parallel. This has two main advantages:

1 The full battery p.d. is applied across each bulb however many bulbs are connected into the system.

2 Each bulb can be switched on and off independently without the other bulbs being affected.

Figure 8a shows how two bulbs in parallel can each be controlled by a separate switch. If a switch is in the 'off' position in either case, contacts in the switch are separated and there is a break in the conducting path between the battery and the bulb. Breaking the conducting path in the branch containing one bulb however has no effect on the conducting path to the other bulb.

Figures 8b and 8c show alternative methods of drawing the same circuit.

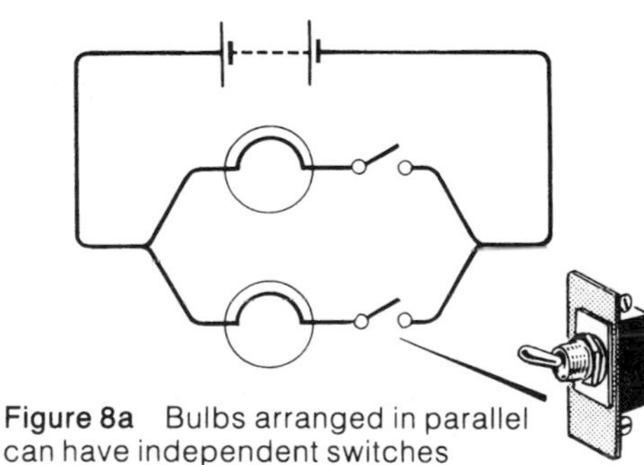

Figure 8a Bulbs arranged in parallel can have independent switches

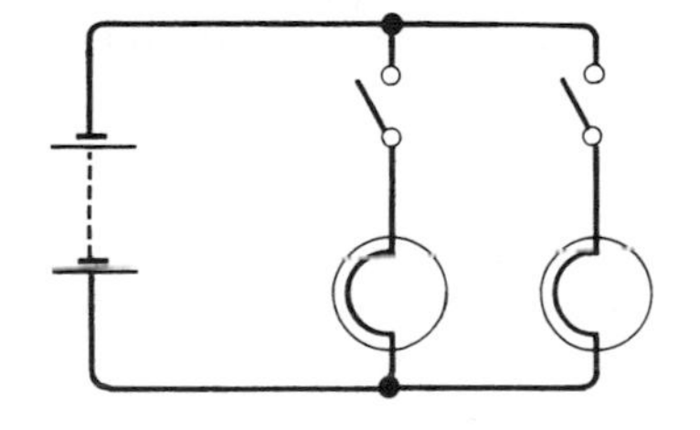

Figure 8b

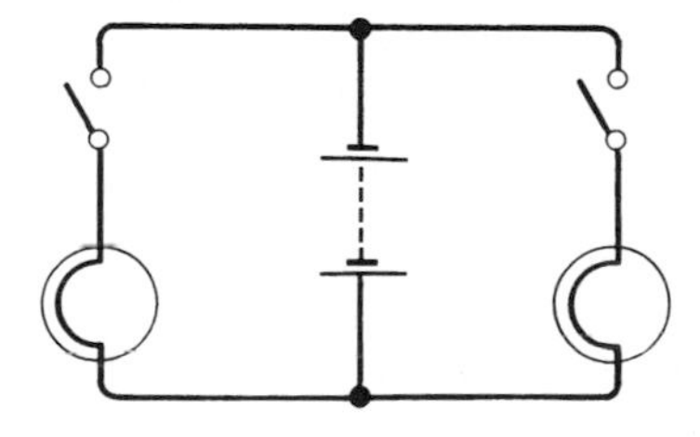

Figure 8c

Figures 8b, 8c these show two alternative ways of drawing exactly the same circuit as that shown in figure 8a

Circuit problems

Example 1 *In the circuit in figure 9, what is the p.d. across a) the* $3\,\Omega$ *resistor b) the* $6\,\Omega$ *resistor?*

a) The first stage is to calculate the total resistance in the circuit, and then use this information to find the current.

total resistance $= 3\,\Omega + 6\,\Omega = 9\,\Omega$

the current is now calculated using $I = V/R$

$$\text{current} = \frac{18\,\text{V}}{9\,\Omega} = 2\,\text{A}$$

Knowing that the $3\,\Omega$ resistor has a current of 2 A passing through it, the p.d. across the resistor can be calculated using $V = IR$

p.d. $= 2\,\text{A} \times 3\,\Omega = 6\,\text{V}$

the p.d. across the $3\,\Omega$ *resistor* $= 6\,\text{V}$

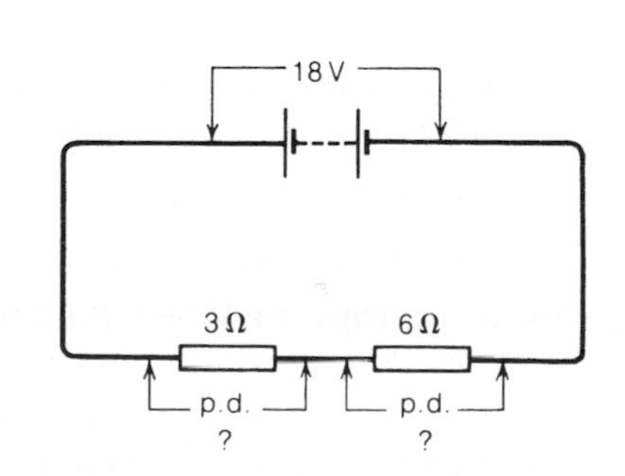

Figure 9

b) The p.d. across the $6\,\Omega$ resistor can be worked out in the same way. It can also be deduced from the fact that the p.d. across the two resistors must add up to 18 V, the p.d. across the battery. By either method,

the p.d. across the $6\,\Omega$ *resistor* $= 12\,\text{V}$

Example 2 *In the circuit in figure 10, a) what is the p.d. across the 3 Ω resistor? b) what current flows through the 3 Ω resistor?*

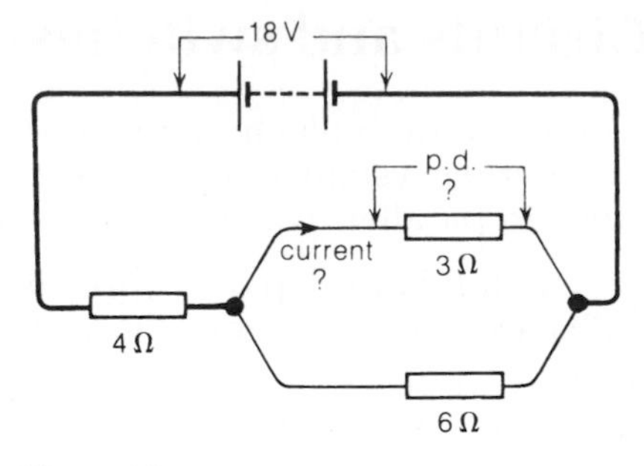

Figure 10

a) The first stage is to find the total resistance in the circuit:

The combined resistance of the parallel resistors is calculated using the equation:

$$\frac{1}{R} = \frac{1}{R_1} + \frac{1}{R_2}$$

$$\frac{1}{\text{combined resistance}} = \frac{1}{3\,\Omega} + \frac{1}{6\,\Omega} = \frac{1}{2\,\Omega}$$

∴ combined resistance of parallel resistors .= 2 Ω

This is in series with a resistance of 4 Ω, so the total resistance in the circuit = 2 Ω + 4 Ω = 6 Ω.

The current in the main circuit can now be calculated using $I = V/R$:

$$\text{current} = \frac{18\,\text{V}}{6\,\Omega} = 3\,\text{A}$$

Remembering that the parallel resistors are equivalent to a single 2 Ω resistor with a current of 3 A flowing through it, the p.d. across the resistors can be calculated using $V = IR$:

p.d. = 3 A × 2 Ω = 6 V

Alternatively, the p.d. across the 4 Ω resistor can be calculated (it works out at 12 V) and this value subtracted from the 18 V across the battery terminals. The remaining p.d. is across both the 3 Ω and 6 Ω resistors, so once again

the p.d. across the 3 Ω *resistor* = 6 V

b) Knowing that there is a p.d. of 6 V across the parallel resistors, the current through the 3 Ω resistor can be calculated using $I = V/R$:

$$\text{current} = \frac{6\,\text{V}}{3\,\Omega} = 2\,\text{A}$$

The current through the 3 Ω *resistor* = 2 A

Problem-solving techniques

Many circuit problems can be solved using the approach outlined in the examples above. Follow these suggestions:

1 Find the total resistance in the circuit

2 Use $I = V/R$ to find the current in the main circuit

3 Apply appropriate V, I, R equations to the different sections of the circuit.

A word of caution. When substituting V, I, or R values into equations, be sure that the values correspond. If for example you are calculating I for a resistor R, V must be the p.d. across that resistor alone and not any other p.d. that happens to feature on the circuit diagram. Equally, if R represents the combined resistance of several resistors in series, V must be the p.d. across *all* of them.

If at any stage of a problem you aren't sure what to do next, check to see if you can make use of the following:

1 The current in the main circuit is the same at all points.

2 Currents through parallel branches add up to equal the current in the main circuit.

3 Parallel resistors have the same p.d. across them.

4 P.d.s around *any* route between the battery terminals add up to the same value as the p.d. across the battery.

Sometimes, problems can be much simpler than they appear:

Example *In the circuit shown in figure 11, what is the current through the* 2 Ω *resistor when the switch is closed?*

Closing the switch 'short circuits' the 10 Ω resistor, and the combined resistance between A and B becomes zero. Effectively the circuit is then as shown in the diagram underneath. Calculating the current is a simple matter of using $I = V/R$:

$$\text{current} = \frac{6\,\text{V}}{2\,\Omega} = 3\,\text{A}$$

When the switch is closed, the current through the 2 Ω *resistor* = 3 A.

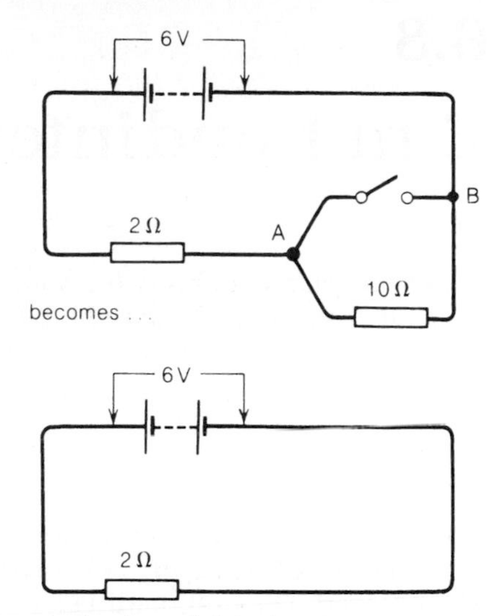

Figure 11

Questions

1 In figures 12a and 12b, what will be the readings on the meters X, Y and Z?

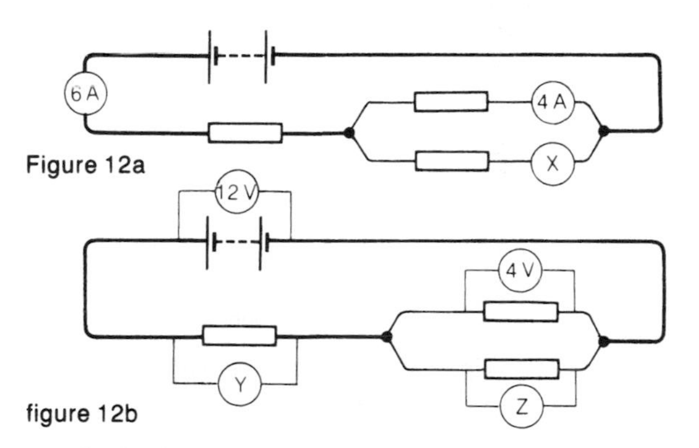

Figure 12a

figure 12b

2 Calculate the combined resistance in each case of the resistors in figures 13a, b, c, and d.

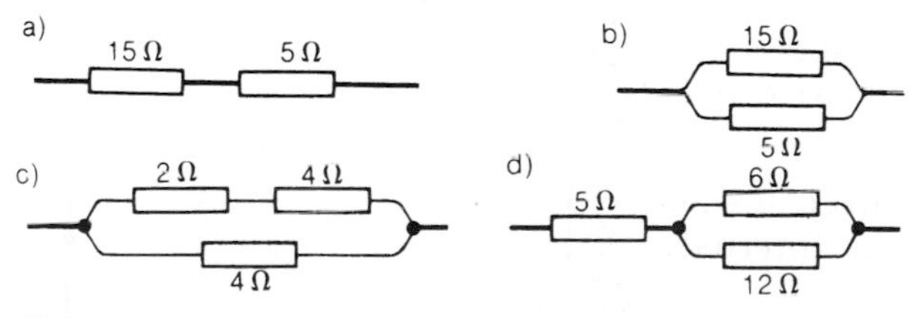

Figure 13

3 In figure 14, which resistor arrangement, A or B, has the lower combined resistance?

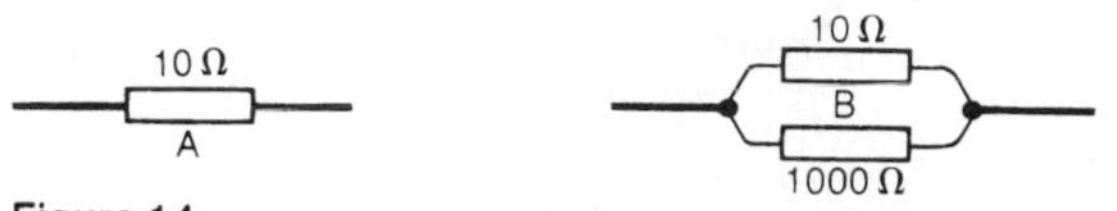

Figure 14

4 In figure 15, what is the p.d. across the 6 Ω resistor?

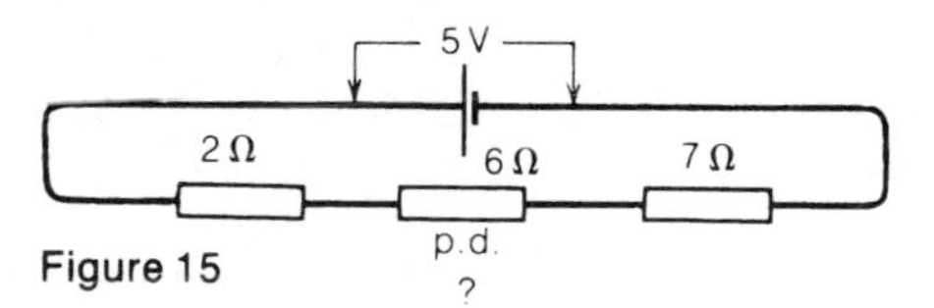

Figure 15

5 In figure 16, calculate a) the current in the main circuit b) the p.d across the 4 Ω resistor c) the current through the 12 Ω resistor.

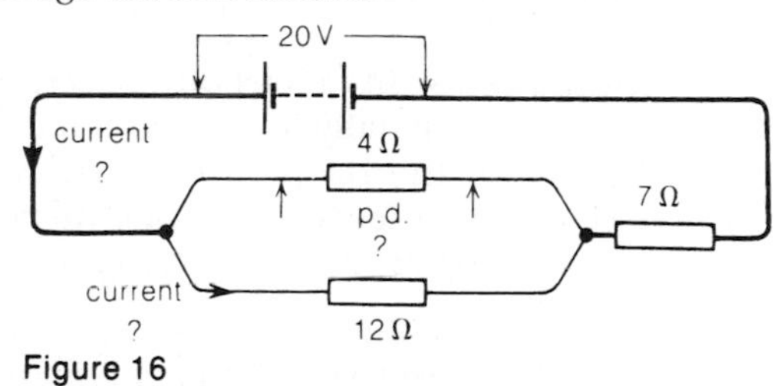

Figure 16

6 In figure 17, the switch is open. What will be the reading on the ammeter? What would the ammeter reading be if the switch were closed?

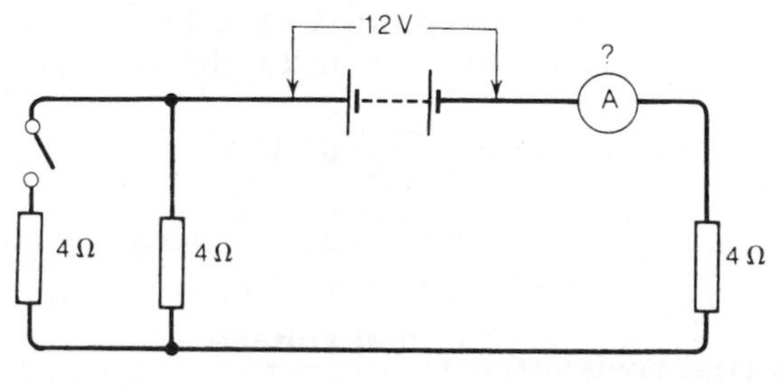

Figure 17

6.8

E.m.f. and internal resistance

A cell doesn't maintain the same p.d. under all conditions. If the current from a cell rises, the p.d. across its terminals drops. The cause of the drop in p.d. is resistance within the cell itself.

current	p.d. across terminals	'lost voltage'
0 A	1·5 V	0 V
0·5 A	1·2 V	0·3 V
1·0 A	0·9 V	0·6 V
1·5 A	0·6 V	0·9 V
2·0 A	0·3 V	1·2 V

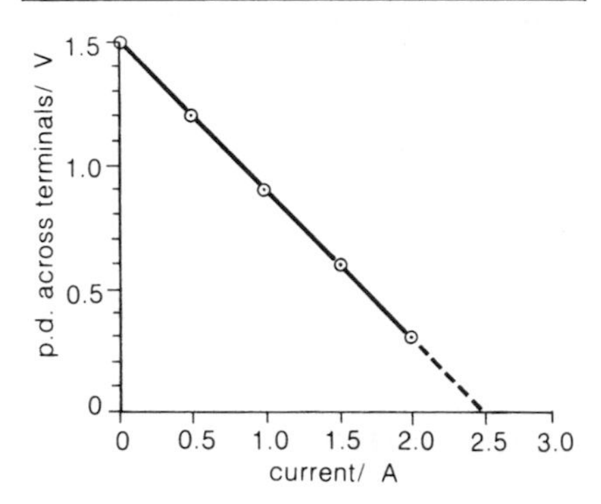

Figure 1 The table and graph show how the p.d. across a cell drops as the current flowing from it increases

P.D. and e.m.f.

The table and graph in figure 1 show how the p.d. across the terminals of a typical dry cell might vary as the current from the cell is increased. The first reading gives the p.d. when the cell is on **open circuit**, with no conducting path from one terminal round to the other (see figure 2a). The other readings are obtained using the circuit shown in figure 2b, and adjusting the rheostat to vary the current. Table and graph both show that the p.d. falls substantially from a maximum of 1.5V as the current is increased from zero.

The p.d. across the terminals of a cell when it is not supplying a current is called the electromotive force or e.m.f. of the cell.

E.m.f. is therefore measured in volts [V]; the dry cell described above has an e.m.f. of 1.5V. The e.m.f. of a cell can be measured experimentally by connecting a suitable voltmeter across the terminals of the cell when it is on open circuit. This method, though quick and convenient, isn't adequate for very accurate measurement of e.m.f. however, as the voltmeter itself draws a small current.

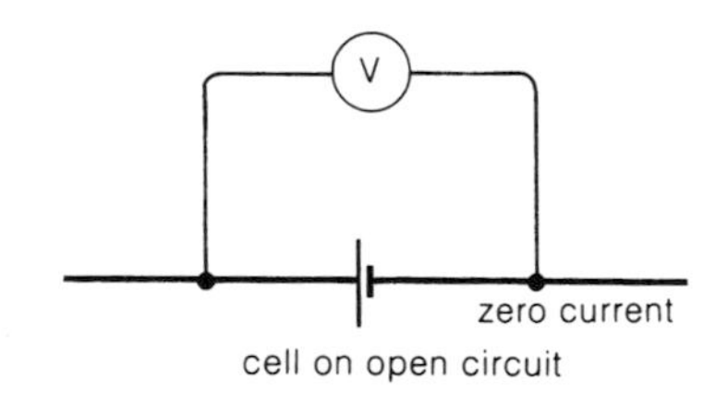

Figure 2a The maximum p.d. is provided when the cell is on "open circuit"

Internal resistance

Like other components in a circuit, a cell has resistance. This **internal resistance** is usually low – about 0.5 Ω or so for a dry cell – but it is the cause of the drop in p.d. which occurs when a cell supplies a current. Consider for example the performance of the dry cell described above when it is supplying a current of 1 A:

Chemical action within the cell causes an e.m.f. of 1.5V. The p.d. measured across the terminals of the cell is only 0.9 V, as shown in figure 3. The 'lost' 0.6 V represents the p.d. required to drive the current of 1 A through the cell itself.

The internal resistance of the cell can be calculated from these figures using the equation $R = V/I$. In this case, R is the internal resistance, V the 0.6 V 'lost' and I, the 1 A current flowing.

$$\text{internal resistance} = \frac{0.6\,\text{V}}{1.0\,\text{A}} = 0.6\,\Omega$$

This same value can be calculated using any corresponding figures worked out from the table or graph:

$$\textbf{internal resistance} = \frac{\textbf{'lost voltage'}}{\textbf{current}}$$

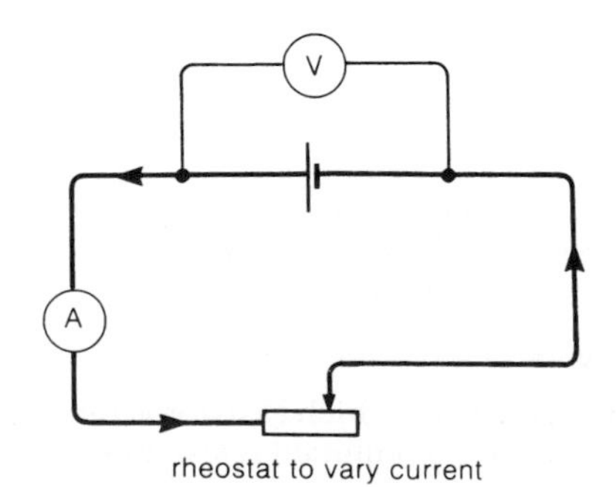

Figure 2b The rheostat can be used to vary the current taken

The maximum current which a cell can supply depends on its internal resistance. The dry cell described above supplies its maximum current if its terminals are directly connected by a piece of thick copper conducting wire (see figure 4). The internal resistance of the cell then provides the only resistance in the circuit and the current flowing is given by:

$$\text{current} = \frac{1.5\,\text{V}}{0.6\,\Omega} = 2.5\,\text{A} \qquad (\text{using } I = V/R)$$

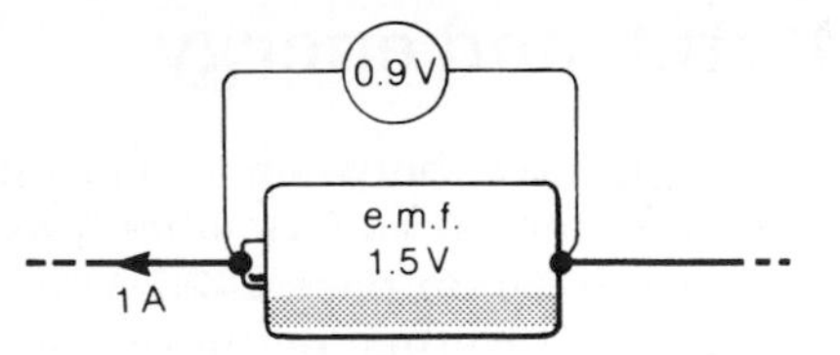

Figure 3 With a current of 1 A flowing, 0.6 V are needed to drive the current through the cell itself

Under these circumstances, the p.d. across the terminals of the cell is zero, all 1.5 volts having been 'lost' within the cell itself. The maximum current of 2.5 A can also be read off from the graph. If the internal resistance of the cell were lower, the maximum current which could be supplied would be greater.

'Short circuiting' the terminals of a cell isn't to be recommended. A 2 V lead-acid cell for example has an internal resistance of, typically, only about 0.02 Ω, and is therefore capable of supplying a current of around 100 A. A current of this size will damage the cell or melt the connecting wire.

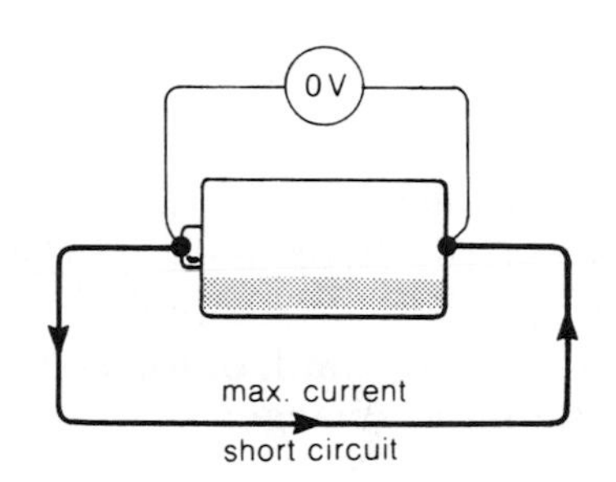

Figure 4 A maximum current flows when the cell is on short circuit. This only damages the cell however

Circuit problems

Where circuit calculations are involved, a cell with internal resistance can be regarded as a cell of fixed e.m.f. in series with a resistor. Figure 5 illustrates this in the case of the dry cell described in the paragraphs above.

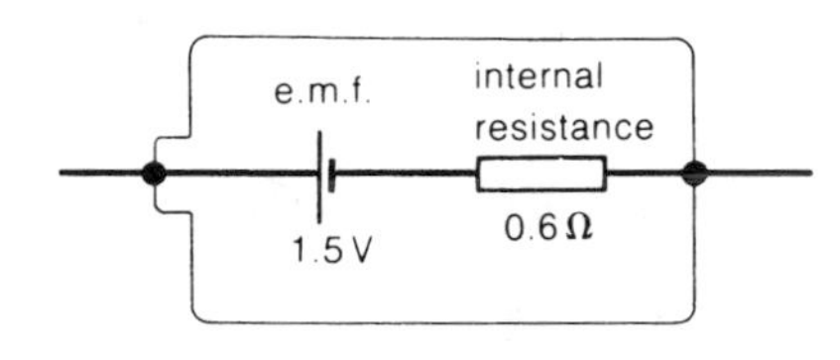

Figure 5

Example *A 2.4 Ω resistor is connected across the terminals of a dry cell of e.m.f. 1.5 V. If the internal resistance of the cell is 0.6 Ω, calculate a) the current flowing in the circuit b) the p.d. across the terminals of the cell.*

First stage is to draw a circuit diagram as in figure 6. A and B are the terminals of the cell.

a) The current can be calculated once the total resistance in the circuit has been found.

total resistance = 0.6 Ω + 2.4 Ω = 3.0 Ω

$$\therefore \text{current} = \frac{1.5\,\text{V}}{3.0\,\Omega} = 0.5\,\text{A} \qquad (\text{using } I = I/R)$$

the current in the circuit = 0.5 A

b) The p.d. across the terminals is also the p.d. across the 2.4 Ω resistor. Knowing that the current through this resistor is 0.5 A, the p.d. can be calculated using $V = IR$:

p.d. = 0.5 A × 2.4 Ω = 1.2 V

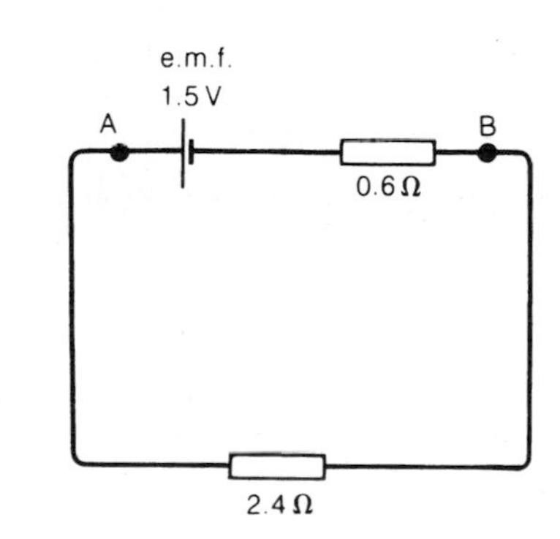

Figure 6

Alternatively the 'lost voltage' across the internal resistance can be calculated using $V = IR$,

'lost voltage' = 0.5 A × 0.6 Ω = 0.3 V

Subtracting this figure from the cell e.m.f. of 1.5 V also gives 1.2 V.

By either method, the p.d. across the terminals of the cell = 1.2 V

E.m.f. and energy

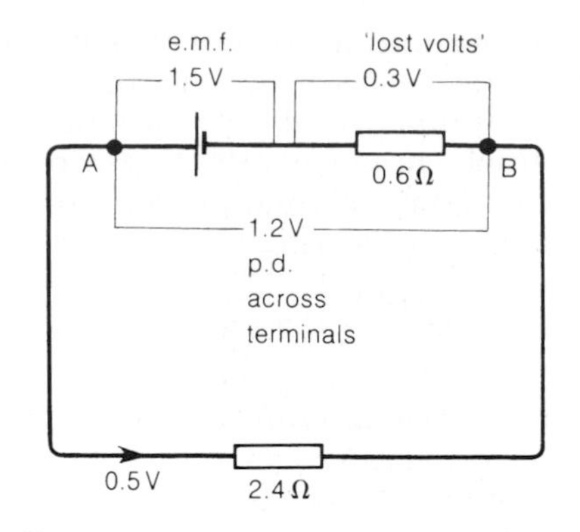

Figure 7 Each coulomb of charge gains energy by the action of the cell; loses it in the external circuit and within the cell itself

If there is a p.d. between two points in a circuit, then electrons gain or lose potential energy when passing between them (as explained in more detail on page 247). When this happens they become more 'squashed' together or more spread out. The p.d. is equal to the change of energy per unit of charge passing between the points. If energy is measured in joules, and the unit of charge is the coulomb, then p.d. is measured in volts.

Applying these ideas to charge passing round the circuit in figure 7:

Each coulomb gains 1.5 joules of potential energy on passing through the cell – as a result of chemical action within the cell; each coulomb loses 0.3 joules of potential energy within the cell itself because of the internal resistance – this energy is changed into thermal energy (heat); each coulomb loses the remaining 1.2 joules of potential energy in passing round the *external* circuit linking the cell terminals – the energy is changed into heat in the resistor.

Note that the e.m.f. of the cell indicates the total energy given to each unit of charge:

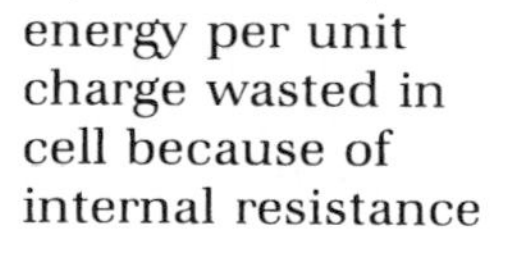

e.m.f.	=	**p.d. across terminals**	+	**'lost voltage'**
energy per unit charge gained in cell		energy per unit charge changed into other forms in external circuit		energy per unit charge wasted in cell because of internal resistance

Cells in series and parallel

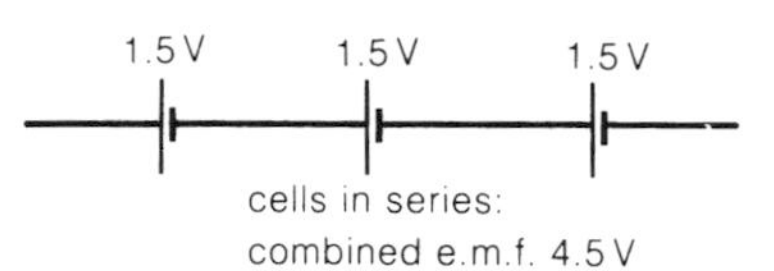

Figure 8 Cells arranged in series: this gives an increased p.d.

Groups of cells connected together are known as batteries. The battery in figure 8 is made up of three dry cells joined in series. The arrangement gives an increased e.m.f. because charge flowing round a circuit will pass through all three cells and gain potential energy from each of them. The combined e.m.f. in this case is 4.5 V, a figure calculated by adding up the individual e.m.f.s.

Figure 9 shows the three cells arranged in parallel. The combined e.m.f. in this case is only 1.5 V because charge flowing round a circuit can pass through one cell, but not all three. The arrangement does have some advantages however as each cell has to supply only a third of the current in the main circuit. This means that:

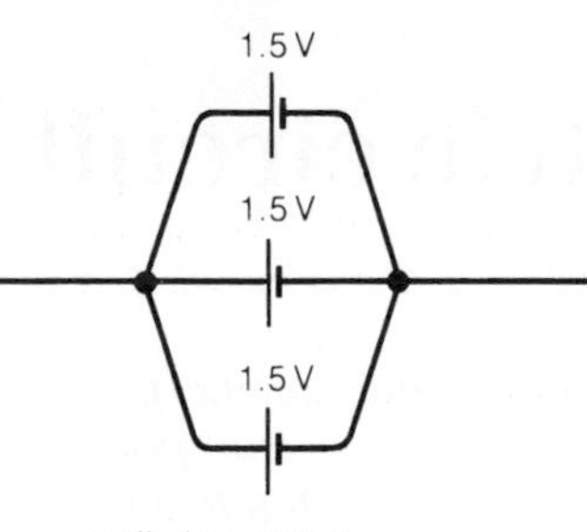

Figure 9 Cells arranged in parallel: the p.d. remains the same, but the ability to provide current is greater

1 The cells will last longer before going 'flat'.

2 The cells are capable of supplying a higher current than a single cell. Put another way, the internal resistances are in parallel, so the combined internal resistance is less than that of a single cell.

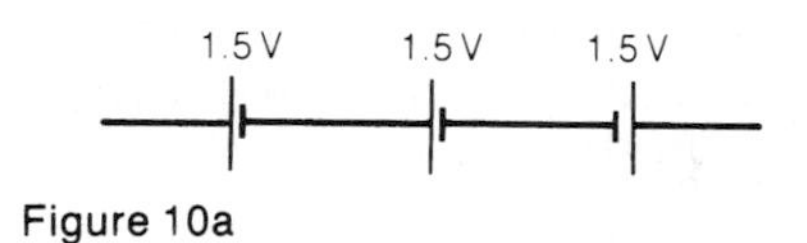

Figure 10a

Cells shouldn't be left connected in parallel when not in use. If one cell has a slightly greater e.m.f. than another, a small current will continue to flow round the conducting path that links them.

Example *Calculate the combined e.m.f.s of the groups of cells in figures 10a and 10b.*
a) Two of the cells are opposing each other, so their combined e.m.f. is zero. The third cell has an e.m.f. of 1.5 V, so the combined e.m.f. of all three is 1.5 V.
b) Trace your finger over any of the routes from point A to point B. Add up the e.m.f.s of the cells you pass over. Note in this case that a charge can pass through two cells but not all six, so the combined e.m.f. is 3 V.

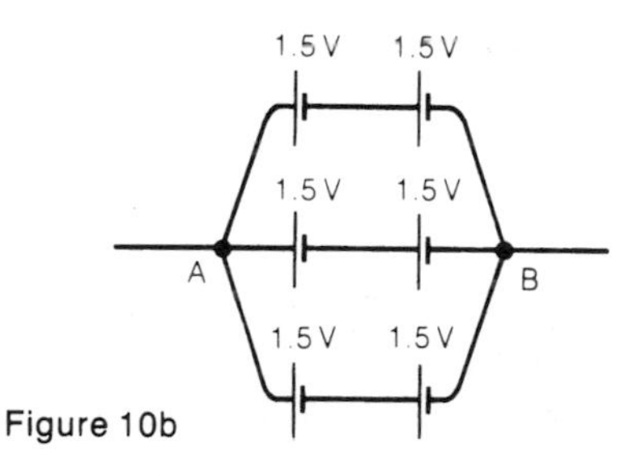

Figure 10b

Questions

1 When is the p.d. across the terminals of a cell equal to its e.m.f.?
2 How would you measure the e.m.f. of a lead-acid cell approximately?
3 A resistor is connected across the terminals of a cell of e.m.f. 2.0 V. If the p.d. across the cell is 1.5 V and the current flowing is 2 A, find
a) the internal resistance of the cell
b) the resistance of the resistor
c) the energy wasted in the cell when one coulomb of charge passes round the circuit.
4 A cell has an e.m.f. of 2.0 V and an internal resistance of 0.1 Ω. What is the p.d. across its terminals when it is supplying a current of 5 A? What is the maximum current the cell can supply?
5 What are the advantages of connecting several similar cells together in parallel? What problems can arise?
6 Calculate the combined e.m.f.s of the groups of cells in figures 11a, 11b, 11c and 11d.

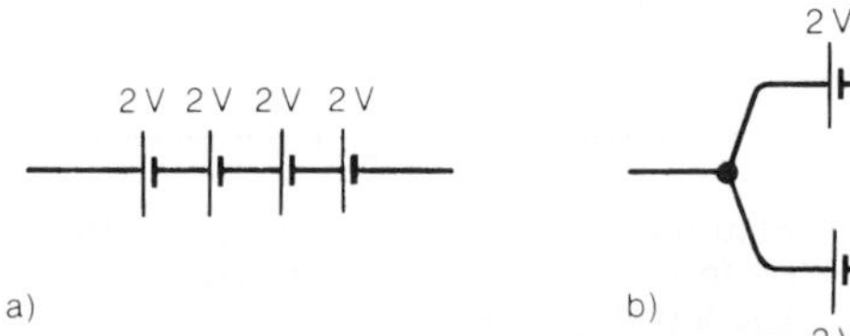

Figure 11a

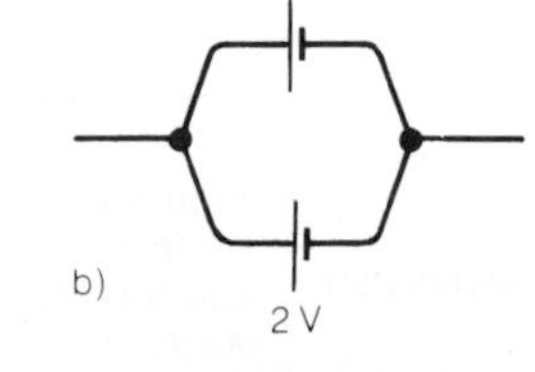

Figure 11b

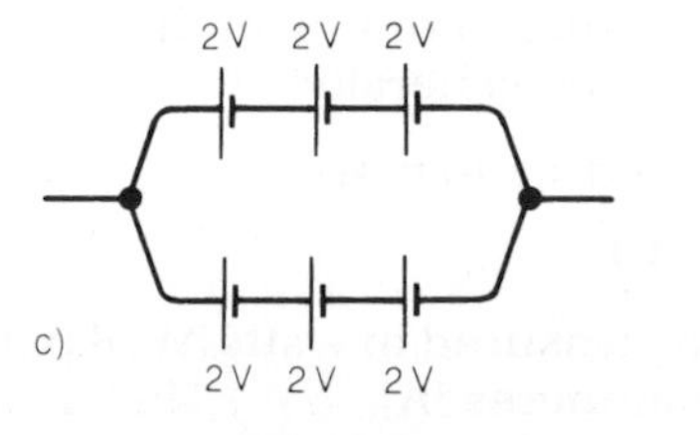

Figure 11c

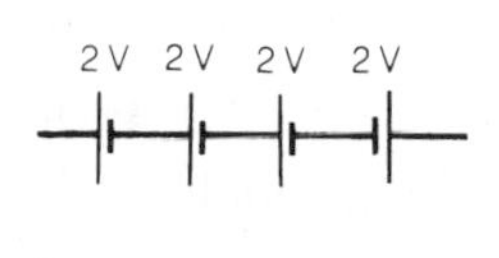

Figure 11d

6.9

Power in an electric circuit

When current flows through a resistor, all the potential energy lost by the charge is changed into heat. In many circuits, it is important to know the rate at which such energy changes are taking place.

Terms and definitions

The potential energy gained by charge in a battery is sometimes known as electrical potential energy – or just **electrical energy** for short – to distinguish it from other forms of potential energy. In the circuit in figure 1, chemical energy is changed into electrical energy in the battery, and electrical energy is changed into thermal energy (heat) in the resistor.

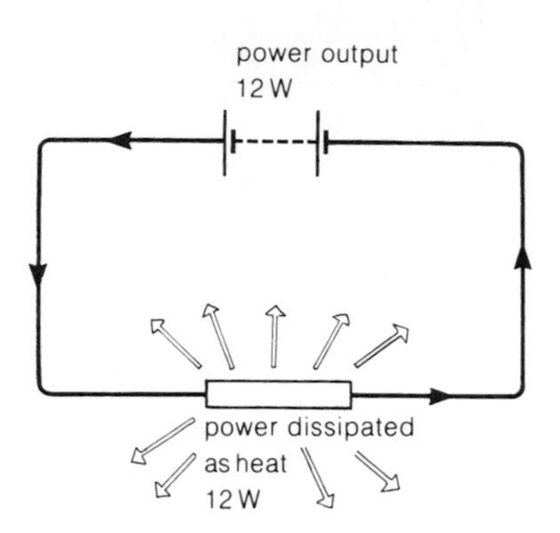

Figure 1 The power output from the battery is the same as the power given off as heat in the resistor. Power is measured in *joules per second*, or *watts*.

When energy changes from one form to another, the **power** indicates the rate at which the change is taking place:

$$\textbf{power} = \frac{\textbf{energy change}}{\textbf{time taken}}$$

If energy is measured in joules [J] and time in seconds [s], power is measured in joules/second or **watts** [W].

In the battery in figure 1, 12 joules of chemical energy are changed into electrical energy every second, so the **power output** of the battery is 12 watts. In the resistor, 12 joules of electrical energy are changed into thermal energy and lost from the circuit every second; the **power dissipated** (lost as heat) in the resistor is 12 watts.

Larger powers are measured in **kilowatts** [kW]; 1 kW = 1000 W.

The power output from this two-bar electric fire is 2000 watts

Power output of a battery

The power output of a battery depends on the p.d. across its terminals and on the current it supplies.

The battery in figure 2 has a p.d. of 6 V across its terminals. The current in the circuit is 2 A. Using the meanings of the volt and the ampère given in unit 6.5:

each coulomb of charge pushed out of the battery carries 6 joules of electrical energy, and 2 coulombs are pushed out every second. The battery therefore supplies 6 × 2, or 12, joules of electrical energy every second, i.e. its power output is 12 watts.

The power in this case is worked out by multiplying the p.d. by the current, a rule which applies generally:

$$\textbf{power} = \textbf{p.d} \times \textbf{current}$$

or in symbols, $P = VI$

Note again that power is measured in watts [W] if p.d. is measured in volts [V] and current in ampères [A].

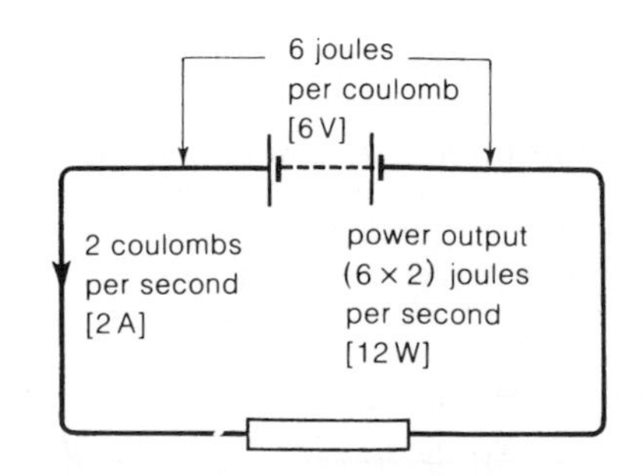

Figure 2 6 joules per coulomb from the battery at the rate of 2 coulombs per second: 12 joules per second, or *12 watts*.

Power dissipated in a resistor

The power dissipated as heat in a resistor can also be calculated using $P = VI$. The resistor shown in figure 3 has a p.d. of 6 V across it and a current of 2 A flowing through it,

power dissipated $= VI = 6\,\text{V} \times 2\,\text{A} = 12\,\text{W}$

The resistor therefore gives out 12 joules of thermal energy every second.

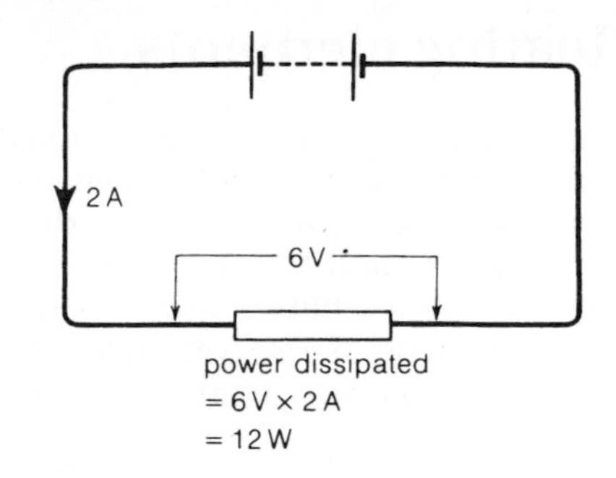

Figure 3 With a p.d. of 6 V across a resistor and 2 A flowing through it, the power dissipated by the resistor is 12 W.

Alternative equations for calculating power can be found by substituting $V = IR$ and $I = V/R$ in turn in the equation $P = VI$:

$$P = VI = IR \times I = I^2 R$$

$$P = VI = V \times \frac{V}{R} = \frac{V^2}{R}$$

Summarizing, three equations are available for calculating the power dissipated in a resistor:

$$\boldsymbol{P = VI} \qquad \boldsymbol{P = I^2 R} \qquad \boldsymbol{P = \frac{V^2}{R}}$$

Example *A p.d. of 12 V is applied across the ends of the 4 Ω resistor shown in figure 4. Calculate the power dissipated in the resistor (the thermal energy produced per second).*

In this case, $V = 12\,\text{V}$, $R = 4\,\Omega$, and P is to be found.

$$P = \frac{V^2}{R} = \left(\frac{12^2}{4}\right)\text{W} = 36\,\text{W}$$

The power dissipated in the resistor $= 36\,\text{W}$

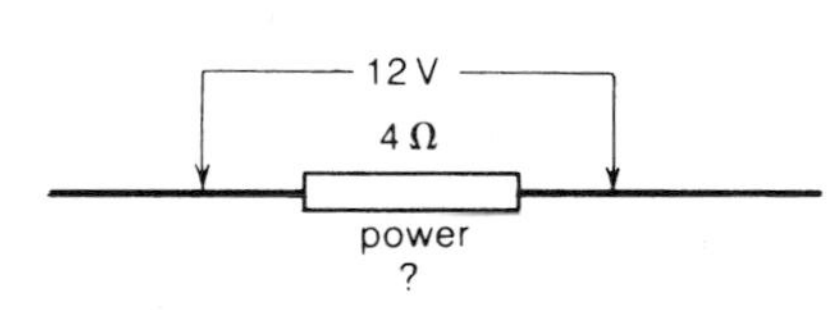

Figure 4

This isn't the only method of finding P. You could use $I = V/R$ to find the current through the resistor, and then either $P = VI$ or $P = I^2 R$ to calculate the power. All methods give the same answer: 36 W. Note that *any* of the power equations can be used in solving the problem, but one equation ($P = V^2/R$ in this case) provides a simpler route to the answer than the others.

Example *Calculate the power dissipated in the 5 Ω resistor in figure 5 when the current through the resistor is a) 2 A b) 4 A.*

The simplest method of solving both problems is to use $P = I^2 R$:

a) In this case, $P = (2^2 \times 5)\text{W} = 20\,\text{W}$

When the current is 2 A, *the power dissipated* $= 20\,\text{W}$

b) In this case, $P = (4^2 \times 5)\text{W} = 80\,\text{W}$

When the current is 4 A, *the power dissipated* $= 80\,\text{W}$

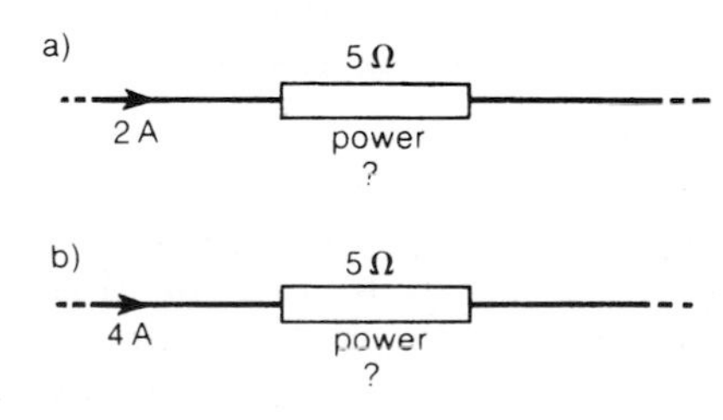

Figure 5

Comparing the answers to a) and b), note that a resistor gives out *four* times as much thermal energy every second if the current through it is doubled. You might be able to see why this is so without resorting to equations. For double the current, the p.d. must be twice what it was before. Twice as many coulombs therefore pass through the resistor every second, with each coulomb losing twice the previous quantity of potential energy.

Heating elements

Like all resistors, the elements used in electric fires, kettles, cookers and water heaters give off heat when a current is passed through. Most heating elements are made from lengths of nichrome resistance wire, coiled to take up less space, the particular advantage of nichrome wire being that it can be kept red hot in air without breaking. The X-ray photograph in figure 6 shows the coiled nichrome wire in a heating ring from a cooker.

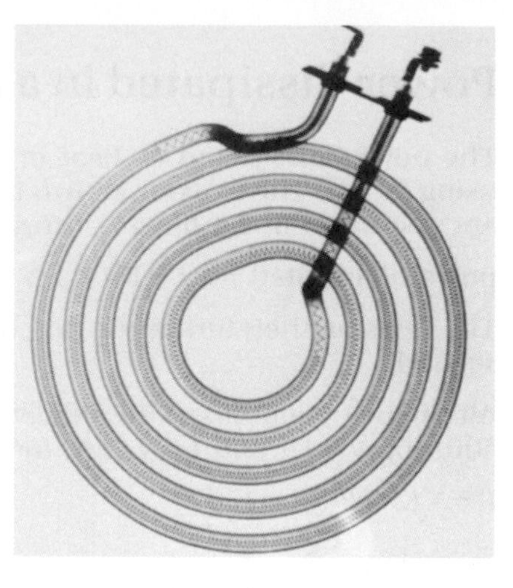

Figure 6 Nichrome wire is used inside heating elements: it does not melt or oxidise when red hot

Voltage and power ratings are often stamped on the side of heating elements. Figure 7 shows a small water heater rated at 12 V 48 W; this is designed to be used with a 12 volt supply, and will change 48 joules of electrical energy into thermal energy every second when connected to such a supply. The equation $P = V^2/R$ can be used to calculate the resistance which the heater element must provide in this case:

$P = 48\,W$, $V = 12\,V$, and R is to be found,

$$P = \frac{V^2}{R}$$

$$\therefore R = \frac{V^2}{P}$$

$$\text{rearranging, } R = \left(\frac{12^2}{48}\right)\Omega = 3\,\Omega$$

The heater element must provide a resistance of $3\,\Omega$.
Similar calculations show that, for any given supply voltage, the lower the resistance of a heating element, the greater is the power dissipated.

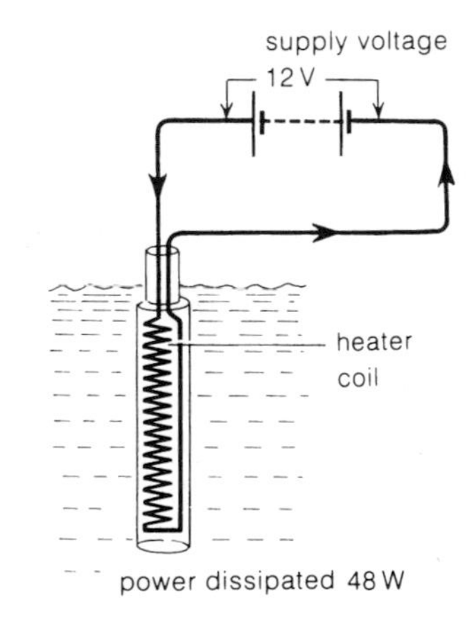

Figure 7

Filament light bulbs

The 'heating element' in a light bulb is a fine wire or **filament** of tungsten which becomes white hot when a suitable current is passed through. Tungsten is used because it has a high melting temperature (3400 °C) and can be kept white hot without melting. The filament would quickly burn up in air, but the bulb is filled with argon and/or nitrogen gases which do not react with the hot metal. Often, the filament is in the form of a 'coiled coil' (see figure 8) to reduce the cooling effect of the convection currents which circulate in the surrounding gases.

A 100 W light bulb gives out 100 joules of energy every second when connected to a suitable supply. Less than 10% of this power output is in the form of visible light however. A fluorescent light can provide the same level of illumination with only about a third of the power input.

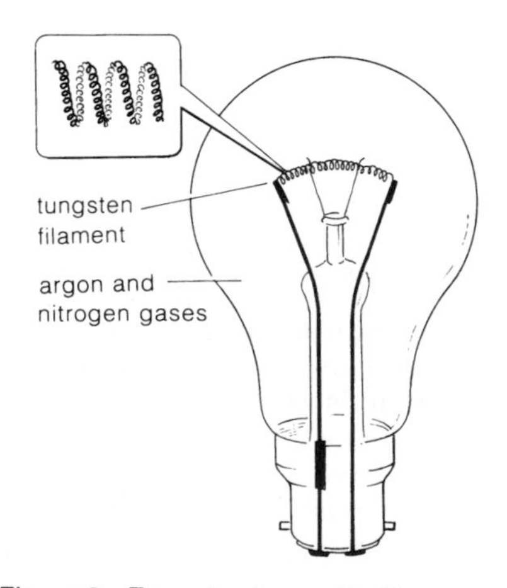

Figure 8 Tungsten is used in filament light bulbs: it does not melt even when white hot

Power losses in cables

In all calculations so far, the resistance of the connecting wires in a circuit has been taken as zero. In practice however, some circuits contain connecting wires in the form of long cables, and the effect of resistance can't be ignored. Consider for example the small water heater shown in figure 9. This is rated at 12 V 48 W, and is connected to a 12 V battery by wires with a total resistance of 1 Ω:

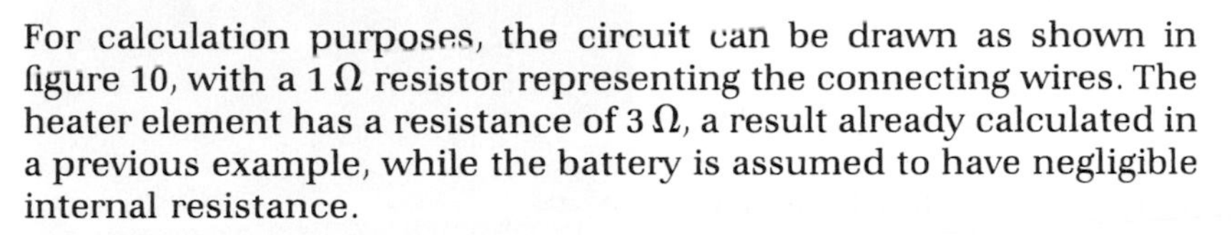

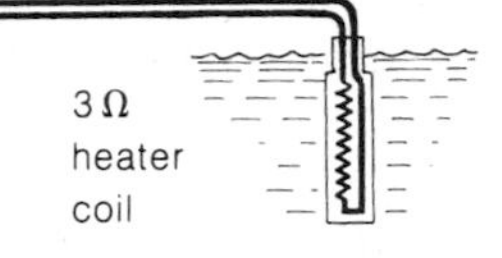

Figure 9 Long connecting wires can add substantially to the resistance of a circuit: electric lawn mowers suffer from this problem

For calculation purposes, the circuit can be drawn as shown in figure 10, with a 1 Ω resistor representing the connecting wires. The heater element has a resistance of 3 Ω, a result already calculated in a previous example, while the battery is assumed to have negligible internal resistance.

The total resistance in the circuit = 1 Ω + 3 Ω = 4 Ω

$\therefore$ the current in the circuit $= \dfrac{12\,\text{V}}{4\,\Omega} = 3\,\text{A}$ [using $I = V/R$]

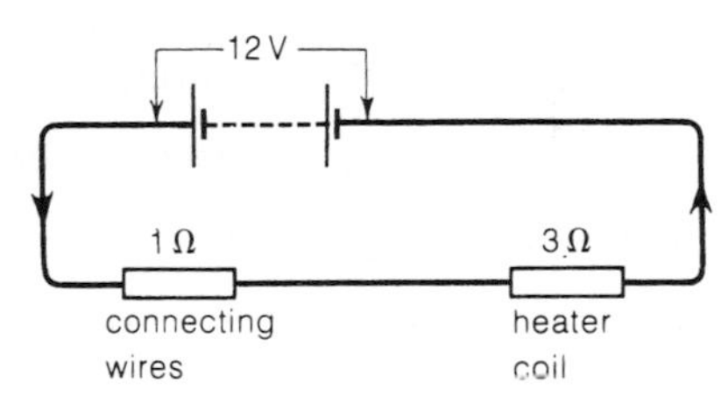

Figure 10 The circuit diagram for Figure 9

The equation $P = VI$ gives the power output of the battery:

power output of battery = 12 V × 3 A = 36 W

The equation $P = I^2R$ can be used to calculate the power dissipated in the heater and in the connecting wires:

power dissipated in heater = $(3^2 \times 3)$W = 27 W
power dissipated in connecting wires = $(3^2 \times 1)$W = 9 W

The results of these calculations are also shown in figure 11. Note that the heater is delivering power well below the rated value. There is a significant power loss (as heat) in the connecting wires.

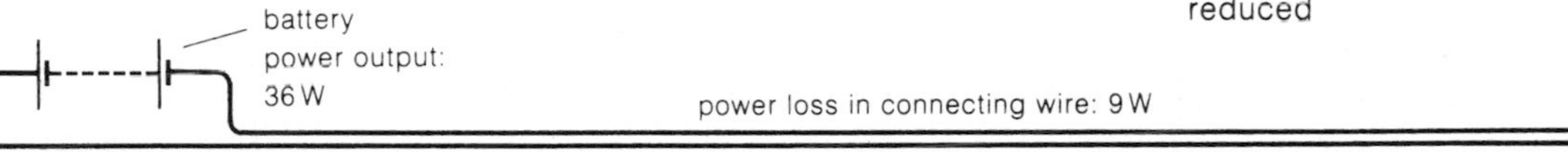

Figure 11 Because of the increased resistance, the current is lower and the power dissipated in the heater is much reduced

Questions

1 A water heater changes 200 joules of electrical energy into heat every second when connected to a 20 V supply.
a) What is the power of the heater?
b) What is the current through the heater?
c) What is the resistance of the heater?
d) If the supply voltage were to fall to 10 V, what current would then flow through the heater and what would the power output be?

2 A small water heater is rated at 12 V 60 W. Calculate the resistance of the heater element and the current through it.

3 Two resistors, of 3 Ω and 6 Ω, are to be connected to an 18 V battery. Calculate the total power dissipated in the resistors when they are connected a) in series b) in parallel.

4 In the circuit in figure 12, calculate
a) The current.
b) The power dissipated in the 2 Ω resistor.
c) The power dissipated in the 4 Ω resistor.
d) The power output of the battery.

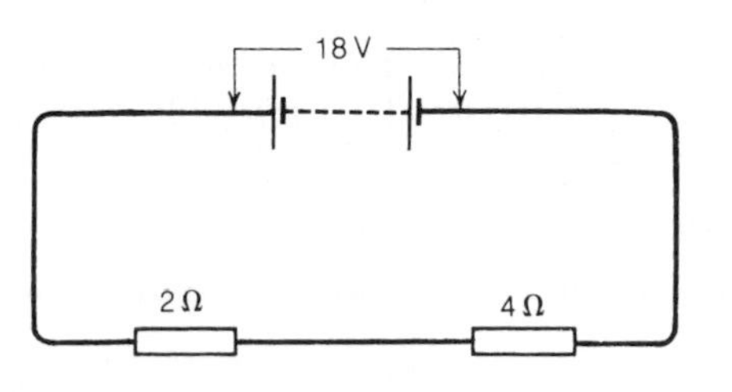

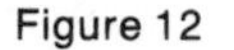

Figure 12

6.10

Electrical energy – calculations and costs

The electrical energy supplied or lost in a circuit can be calculated in many cases. Your local Electricity Board measures the electrical energy supplied to your home and charges you accordingly.

Chelsea bridge, London: you may like to estimate the number of bulbs, (each rated at 40 W), the energy used, and the bill for a night's illumination. But then again, you may not

Calculating energy

If the power output of a battery in a particular circuit is known, the electrical energy supplied by the battery in any given time can be calculated using the equation:

energy = power × time

in symbols, $E = Pt$

For example, a battery will supply 120 joules of electrical energy in 10 seconds if its power output is 12 watts [12 joules per second].

The equation can also be used to calculate the thermal energy released in any given time by a heating element or resistor. If the power P isn't known, alternative equations are available:

$$E = VIt \quad E = I^2Rt \quad E = \frac{V^2t}{R}$$

These are found by replacing P in the original equation by VI, I^2R, and V^2/R in turn.

Example *The small water heater in figure 1 is connected to a* 12 V *supply. If the heater element has a resistance of* 3 Ω, *how much thermal energy is given off in* 5 *minutes?*
If the energy [E] is to be in joules, the time [t] must be expressed in seconds; in this case, $t = 300$ s, $V = 12$ V, $R = 3\ \Omega$, and E is to be found.

$$E = \frac{V^2t}{R} = \left(\frac{12^2 \times 300}{3}\right)\text{J} = 14\,400\text{ J}$$

thermal energy given off by heater = 14 400 J

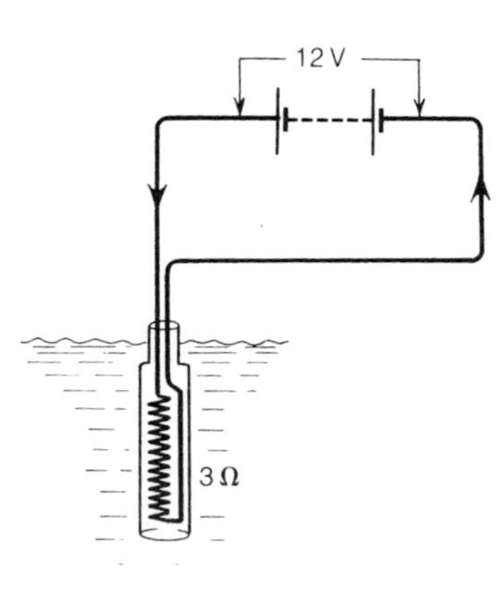

Figure 1

Energy transfer

Thermal energy lost from a heating element is gained by the surrounding material. Two possible effects of this gain in thermal energy are:

1 an increase in temperature;

2 a change of state – water changing into steam for example.

The equations used to calculate the thermal energy required for each change were given on pages 145 and 153.

For an increase in temperature,

thermal energy required = $m c(\theta_2 - \theta_1)$

where m is the mass, c is the specific heat capacity, and $(\theta_2 - \theta_1)$ the temperature increase. Energy is in joules [J], if mass is in kg, specific heat capacity in J/(kg K), and temperature in K or °C.

For a change of state, thermal energy required = mL

where m is the mass and L the specific latent heat. Energy is in joules [J] if mass is in kg and specific latent heat in J/kg.

Example *The beaker in figure 2 contains* 0.2 kg *of cold water. A current of* 4 A *flows through the heater when this is connected to a* 12 V *supply. If the heater is switched on for 210 seconds, what is the increase in temperature of the water? (Take the specific heat capacity of water to be* 4200 J/(kgK), *assume that no water evaporates, and that heat losses from the water are negligible.)*

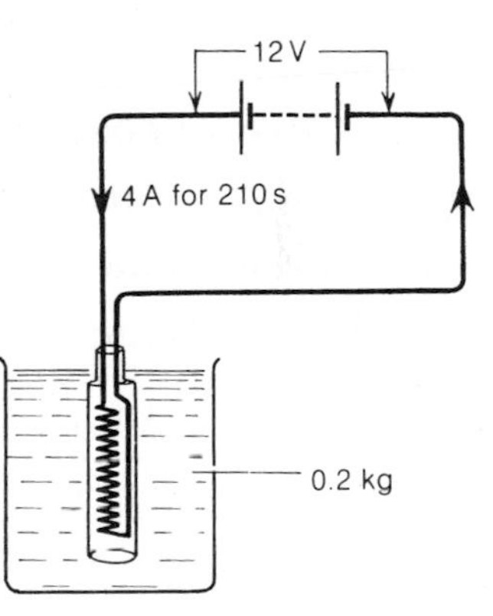

Figure 2

In choosing the appropriate equations, note that values of V and I are given in the problem, but not those of P or R. Note also that the temperature of the water rises, but there is no change of state.

$$\begin{array}{c}\text{thermal energy} \\ \text{given off by heater}\end{array} = \begin{array}{c}\text{thermal energy} \\ \text{gained by water}\end{array}$$

$$VIt = mc(\theta_2 - \theta_1)$$

If the unknown temperature increase is θ °C:

$$(12 \times 4 \times 210)\text{J} = (0.2 \times 4200 \times \theta)\text{J}$$

$$\text{which gives: } \theta = \frac{12 \times 4 \times 210}{0.2 \times 4200} = 12$$

the increase in temperature of the water = 12 °C

Example *A kettle, rated at* 2.1 kW, *is filled with* 1.5 kg *of water at* 20° C *(see figure 3). Calculate the time taken for the water in the kettle to reach boiling point. (Take the specific heat capacity of water to be* 4200 J/(kgK), *the boiling point of water to be* 100° C, *and assume that there are no heat losses from the kettle.)*

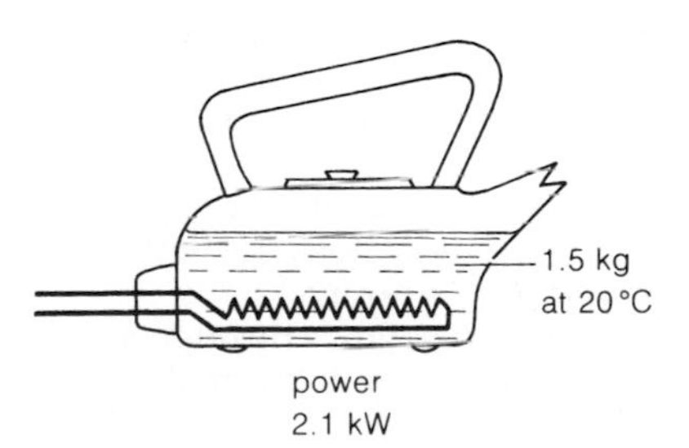

Figure 3

In this case, the temperature of the water rises by 80° C, but there is no change of state. As energy is to be measured in joules and time in seconds, the power of the kettle must be expressed in watts [2100 W].

$$\begin{array}{c}\text{thermal energy} \\ \text{given off by element}\end{array} = \begin{array}{c}\text{thermal energy} \\ \text{gained by water}\end{array}$$

$$Pt = mc(\theta_2 - \theta_1)$$

Call the unknown time t seconds:

$$(2100 \times t)\,\text{J} = (1.5 \times 4200 \times 80)\text{J}$$

$$\text{which gives: } t = \frac{1.5 \times 4200 \times 80}{2100} = 240$$

time taken = 240 seconds

Buying electrical energy

Electrical energy for kettles, lights, fires, and other household appliances is supplied by the local Electricity Board. Readings on the electricity meter (figure 4) give the total energy supplied in 'units'. The Board charges 5p or so for each unit.

Electricity Boards use the kilowatt hour [kWh] rather than the joule as their unit of energy measurement:

A kilowatt hour, or unit, is the energy supplied in 1 hour to an appliance whose power is 1 kW.

The kilowatt hour is a much larger unit of energy than the joule:

In 1 second, a 1 kW appliance is supplied with 1000 J of electrical energy;
in 1 hour (3600 s), a 1 kW appliance is supplied with 3600 × 1000 J of electrical energy, i.e. 3 600 000 J.

Therefore, **1 kWh = 3 600 000 J**

Energy in kWh is calculated using the equation:

energy = power × time

However, if energy is to be in kWh, power must be in kilowatts [kW] and time in hours [h]. For example, if a 3 kW fire is switched on for 4 hours, 12 kWh of electrical energy are supplied.

Figure 4 The electricity board's meter measures the energy supplied in kilowatt hours

Example *If electrical energy costs* 5p *per unit, what is the total cost of leaving* 4 *light bulbs, rated at* 100 W *each (see figure 5), switched on for* 8 *hours?*
The total power of the light bulbs is 400 W or 0.4 kW
Applying the equation given above:

energy = 0.4 kW × 8 h = 3.2 kWh

Each unit (i.e. each kWh) costs 5p; ∴ *total cost* = (3.2 × 5)p = 16p

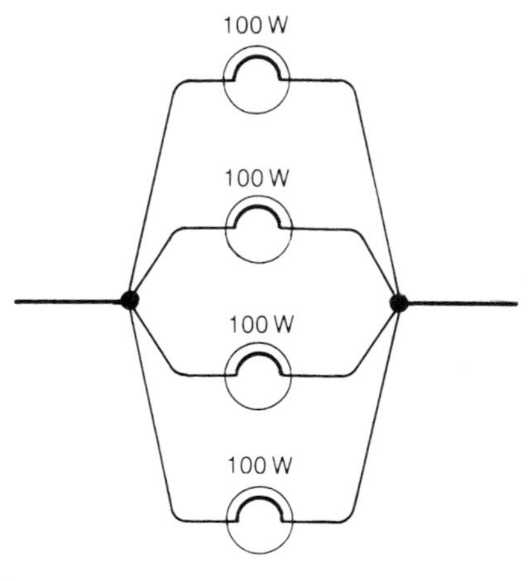

Figure 5

Questions

In questions 4, 5, 6, and 7, assume that heat losses to the surroundings are negligible.
specific heat capacity of water: 4200 J/(kg K);
specific latent heat of vaporization of water: 2 260 000 J/kg
boiling point of water: 100 °C

1 The power output of a battery is 60 W. How much electrical energy does the battery supply a) in 1 second b) in 1 hour?

2 A current of 5 A flows through a 4 Ω resistor for 20 seconds. How much thermal energy does the resistor give off?
What current must flow through the resistor if the same amount of thermal energy is to be given off in 5 seconds?

3 A heating coil is immersed in 0.2 kg of cold water. The coil is connected to a 12 V supply and a current of 5 A flows for 140 seconds. Calculate the temperature increase of the water.

4 A tank contains 40 kg of water at 10° C. If the water is heated by an element rated at 2.8 kW, how long is it before the water starts to boil?

5 Boiling water in a tank is heated by an element rated at 2.26 kW. What mass of water is changed into steam in 10 minutes?

6 A 3 kW fire is switched on for 6 hours. Calculate the thermal energy given off by the fire a) in kWh b) in joules.

7 If electrical energy costs 5p per unit, calculate the cost of
a) leaving a 3 kW fire switched on for 12 hours.
b) leaving a 120 W colour TV switched on for 8 hours.
c) heating a tankful of water from 20 °C to 60 °C, if there are 120 kg of water in the tank.

6.11
Mains electricity

When you plug into the mains, electrical energy comes from a power station rather than a battery, but the same basic circuit principles apply.

European Directive 1 Jan 1995

The UK's officially declared mains voltage value was set at 230 V to harmonize with Europe. However, as the actual value can vary, 240 V will be used in this book to make calculations easier.

A simple a.c. mains circuit

Figure 1 shows the circuit formed when a kettle is plugged into a mains socket. All the wires are insulated and contained in a single cable or 'flex' (see figure 2).

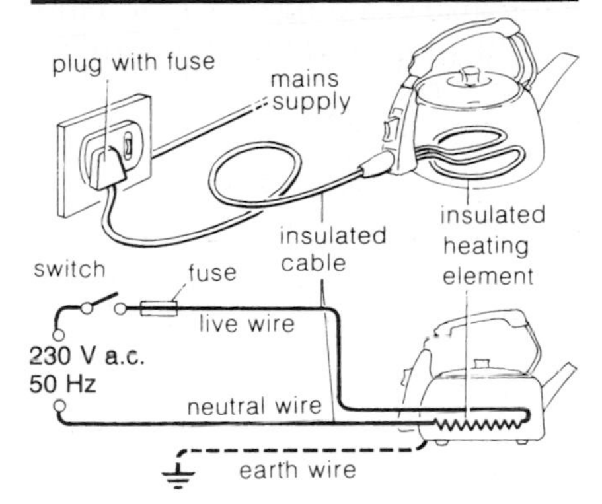

Figure 1 The circuit involved in boiling a kettle-full of water

Alternating current Unlike the one-way current from a battery, the current from a mains socket is pushed and pulled backwards and forwards through the circuit 50 times every second. The current is known as **alternating current** or **a.c.** Power stations supply a.c. because it is easier to generate than one-way **direct current (d.c.)** – see page 308.

In the UK, a.c. is supplied to the mains socket at a voltage of 230 V. The symbol for an a.c. supply is shown in figure 1.

The live wire The potential of this wire goes alternately negative and positive, making the current flow backwards and forwards through the circuit.

The neutral wire The Electricity Board earths the neutral wire by connecting it to a metal plate buried in the ground. Although current passes through the wire, it remains at zero potential. If you accidentally touch it, you don't get a shock.

The switch The switch is fitted in the live wire. The switch would work equally well in the neutral, but wire in the flex would then still be live when the switch was turned off. This would present a hazard if, for example, the cable were broken accidentally.

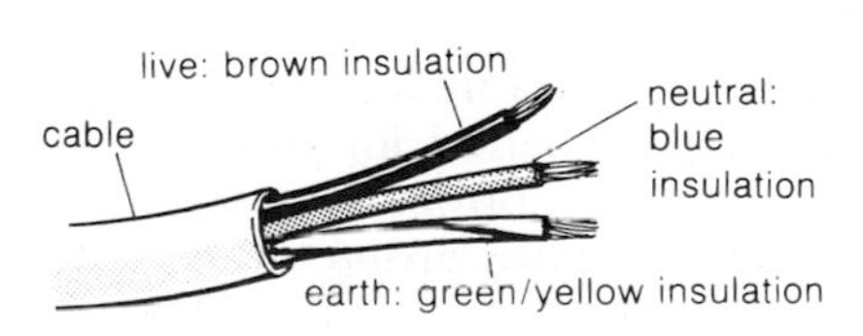

Figure 2 The three wires inside a flex are colour-coded

The fuse This is a short piece of thin wire which overheats and melts if current of more than a certain value flows through it. Like the switch, it is placed in the live wire, often in the form of a small cartridge (figure 3) inside the plug. If a fault develops in the kettle, and too high a current flows, the fuse 'blows' and breaks the circuit before the cable can overheat and catch fire.

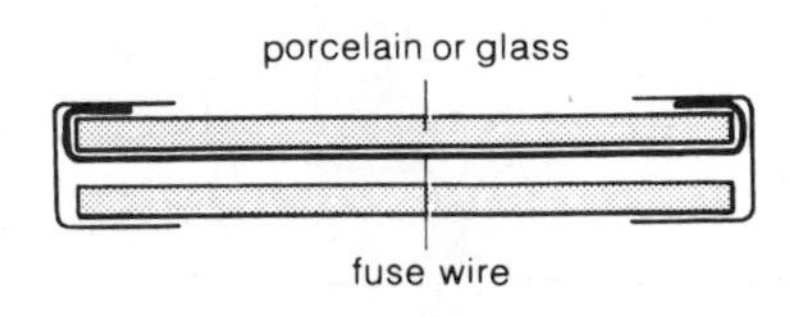

Figure 3 A cartridge fuse. If too high a current flows, the thin wire melts and breaks the circuit

The earth wire This is a safety wire which connects the metal body of the kettle to earth and prevents it becoming live if a fault develops. If for example the live wire were to work loose and touch the body of the kettle, a current would immediately flow to earth and blow the fuse. If there were no earth wire, the body of the kettle would remain live and a possibly lethal current would flow through anyone who happened to touch it.

In many houses, the earth wire is earthed by connecting it to the neutral wire in the supply cable.

Three pin plugs

Plugs provide a convenient and safe method of connecting different appliances into a mains circuit. In many countries, each appliance is sold with an unfused plug already moulded onto the end of the cable. Commonly used in the UK is the fused, square-pin plug shown in figure 4 below, which you can connect to the cable yourself. Figure 5 shows you how.

When wiring a plug, it is important to check that the three wires in the cable are connected to the correct terminals. The cable 'colour code' is shown in the diagram:

live: brown
neutral: blue
earth: yellow and green (or just green)

It is also important to check that a fuse of the correct value is fitted. If the fuse in a plug blows, you must switch off at the socket and pull out the plug; and fit a new fuse only when the fault has been traced and put right.

Fuse values

Fused plugs are normally fitted with either 3 A or 13 A fuses, though others are available. The fuse value must be greater than the current that normally flows through the appliance, but as close as possible to this value so that the fuse will blow before an overheating cable can cause a fire.

The correct fuse value can be worked out quite simply if the power of an appliance is known. Take for example a vacuum cleaner rated at 240 V 480 W:

The current through the vacuum cleaner is found using $P = VI$:

$$480\,\text{W} = 240\,\text{V} \times I$$

rearranging, $$I = \frac{480\,\text{W}}{240\,\text{V}} = 2.0\,\text{A}$$

So the plug should be fitted with a 3 A, rather than a 13 A, fuse.

Typical powers of household appliances are given in figure 6 at the top of the next page. Calculations like the one above show that 3 A fuses are suitable for appliances rated up to about 700 W, while 13 A fuses are needed for those rated between about 700 W and 3 kW. Cookers are connected permanently to a supply cable rather than to a plug, and are protected by a 30 A fuse in a 'fuse box'.

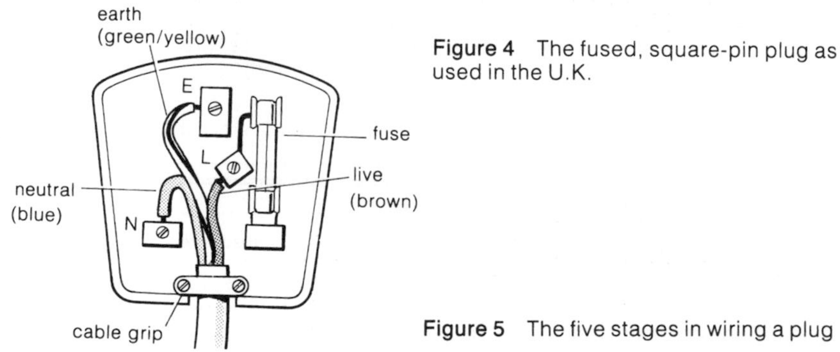

Figure 4 The fused, square-pin plug as used in the U.K.

Figure 5 The five stages in wiring a plug

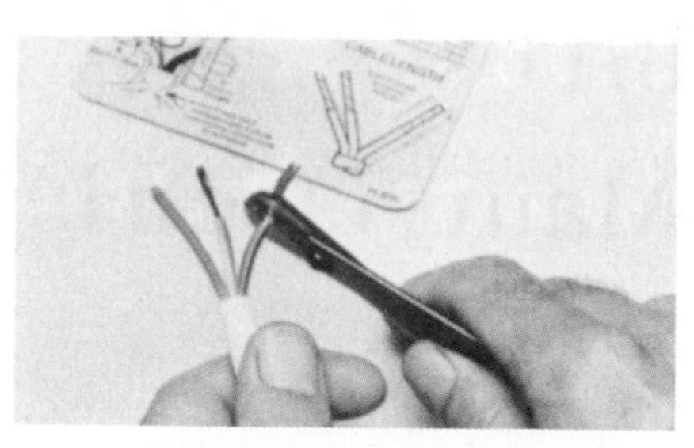

1. Remove the insulating plastic from the three wires

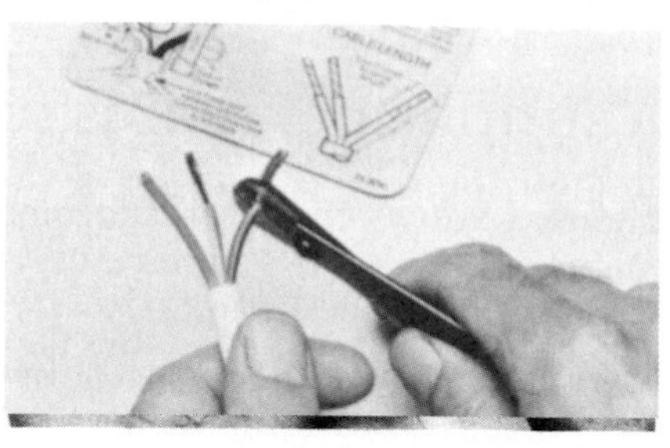

2. Wrap the wires *clockwise* round the studs, and ensure that the *outer* insulation is firmly gripped

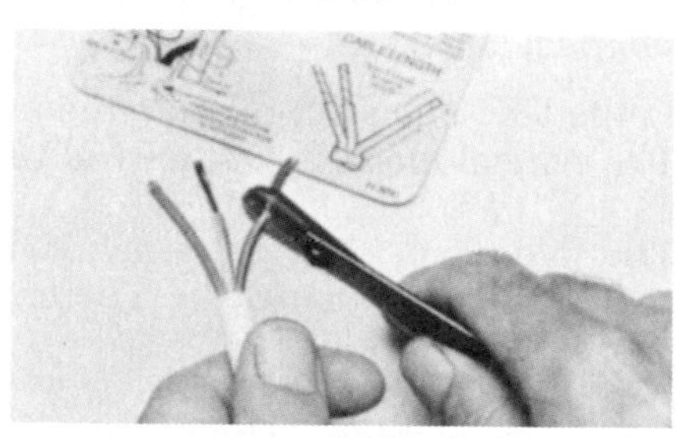

3. Tighten the studs

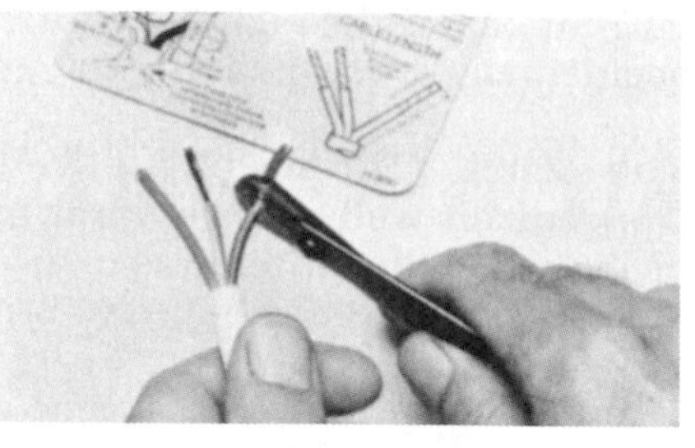

4. Replace the fuse: check it is of the right rating for the appliance

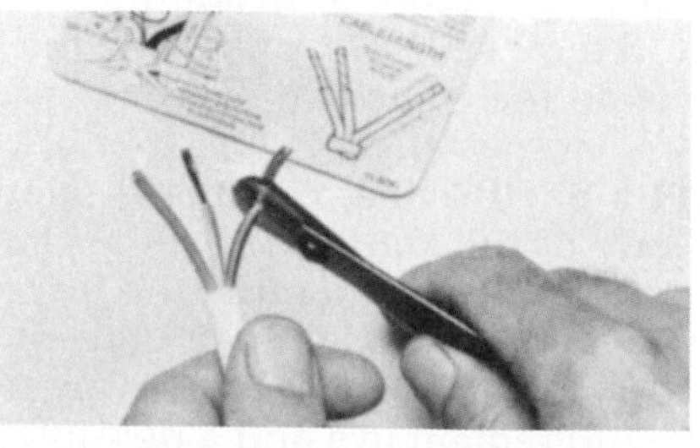

5. Check there are no loose strands anywhere in the plug before replacing the cover

Circuits around the house

The Electricity Board's cable into each house contains a live and a neutral wire. At the **consumer unit** or 'fuse box', these wires branch into several parallel circuits which carry current to the lights, the cooker, the immersion heater, and the various mains sockets, as shown in figure 7. The cable for each circuit contains an earth wire as well as a live and a neutral.

Each circuit passes through a fuse in the consumer unit:

upstairs lighting circuit: 5 A fuse
downstairs lighting circuit: 5 A fuse
circuit through cooker: 30 A fuse
circuit through immersion heater: 15 A fuse
together with a fuse (or fuses) for circuits through the mains sockets (see below).

Circuit breakers are now used in many consumer units instead of fuses. These are automatic switches which open when the current rises above the specified value, but can be closed again by pressing a button.

Typical powers [supply voltage: 240V]	
Cooker (max)	8000 W [8 kW]
Immersion heater	3000 W [3 kW]
Kettle	2400 W [2·4 kW]
Fire	2000 W [2 kW]
Iron	700 W
Drill	360 W
Colour TV	120 W
Table lamp	60 W

Figure 6 Typical powers for some household electrical appliances

Mains sockets

In many houses in Europe, the circuit through each mains socket is protected by its own fuse in the consumer unit. Unfused plugs are used in the sockets.

In many houses in Britain, the mains sockets are connected to a **ring main** as shown in figure 7. This is a cable which begins and ends at the consumer unit, with live, neutral and earth wires in the cable each forming a long loop or 'ring' around the house. An advantage of the system is that there are two conducting paths to each socket, so thinner cable can be used.

The ring main is protected by a single 30 A fuse in the consumer unit. In addition, each appliance is protected by the fuse in its plug.

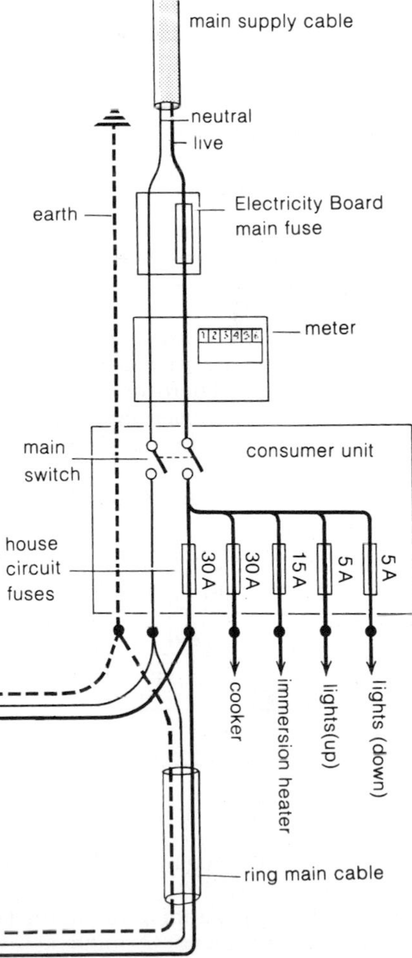

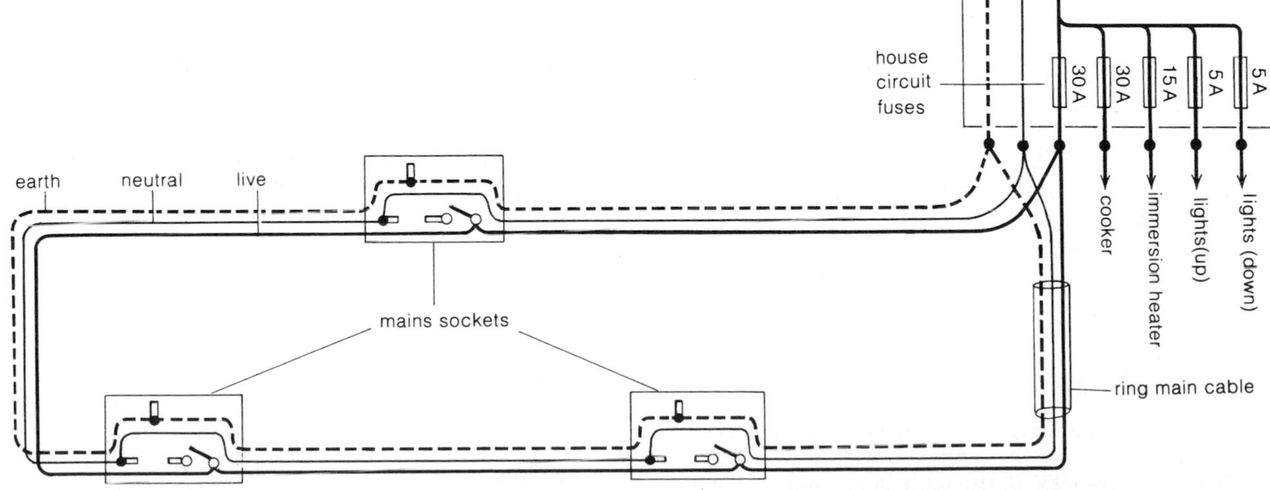

Figure 7 The consumer unit and ring main used in many households

Lights controlled by two-way switches

Lights on the upstairs landing can usually be controlled by either a downstairs or an upstairs switch. The switches used in this case aren't the simple 'on-off' type; technically they are known as **single-pole double-throw switches**, or two-way switches for short.

Figure 8 shows the circuit arrangement. A current flows through the bulb if the switches are either both up or both down. If the switches are in opposite positions however, the circuit is broken. Moving either switch reverses the effect of the other one.

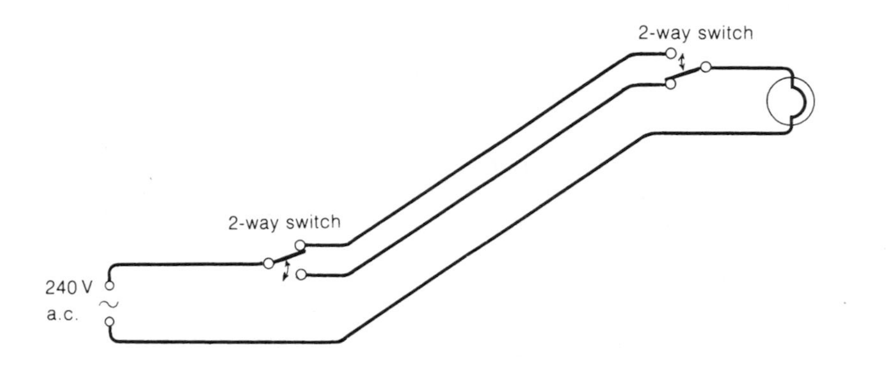

Figure 8 Flicking either 2-way switch will turn the light on if it is off, or off if it is on

Questions

1 What is the difference between an alternating currrent and a direct current? Give the voltage and the frequency of the a.c. mains in the UK.
2 What is the potential of the neutral wire in a mains supply cable?
3 Why are fuses and switches placed in the live rather than the neutral wire of a mains circuit?
4 Why is the metal body of an electric fire earthed?
5 Figure 9 shows a fused three pin plug. Copy and complete the diagram, and mark on the live, neutral and earth terminals. Indicate the colours used on the insulated wires leading to each terminal.

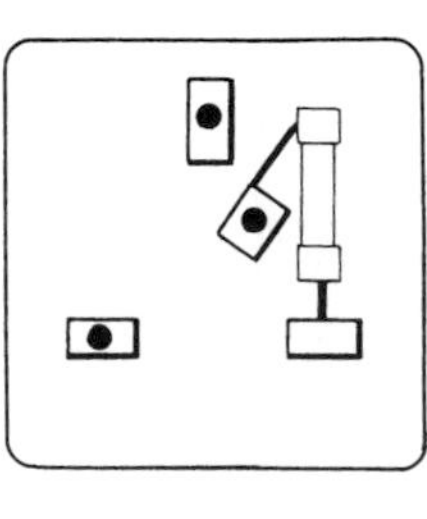

Figure 9

6 If the fuse in a plug blows, what steps should you take before replacing it?
7 Using the information given in the table on page 271, calculate the current through a) the kettle b) the drill c) the colour TV d) the table lamp. State in each case the current rating of the fuse which should be fitted in the plug.

8 If fuses of 250 mA, 500 mA, 1 A, 5 A, and 10 A were available, which would be most suitable for protecting an amplifier rated at 240 V 180 W?
9 Copy and complete figure 10 to show how the sockets can be connected to a ring main from the consumer unit.

Figure 10

10 Copy and complete figure 11 to show how the light bulb can be controlled by either of the switches.

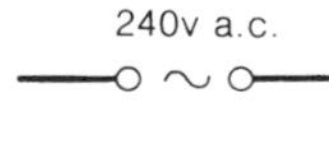

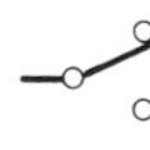

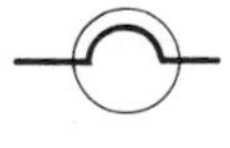

Figure 11

Electrical energy: part A

1 An aircraft flies just below a negatively-charged thunder cloud as shown in figure 1. Electrostatic charges are induced on the aircraft.

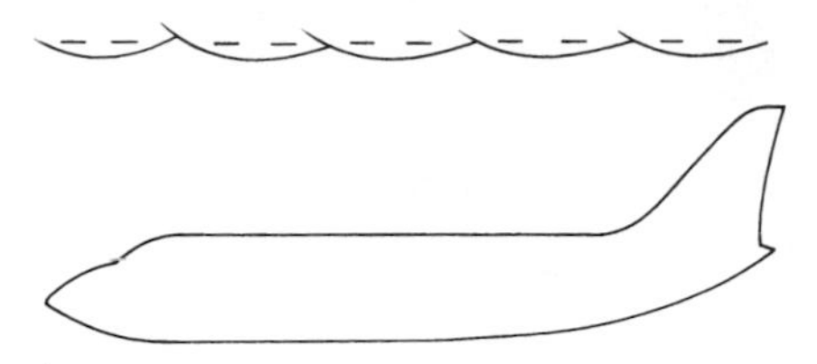

Figure 1

a) Copy the diagram. Draw on the positions and signs of the induced charges on the aircraft.
b) Explain, in terms of the movement of electrons, the distribution of the charges you have shown.
c) What would happen to the induced charges when the aircraft flies away from the cloud? (LEAG)

2 One millionth of a volt is the same as which one of these:
A a kilovolt
B a megavolt
C a microvolt
D a millivolt? (LEAG)

3 Which one of the circuits in figure 2 would be suitable for measuring the resistance of a lamp? (LEAG)

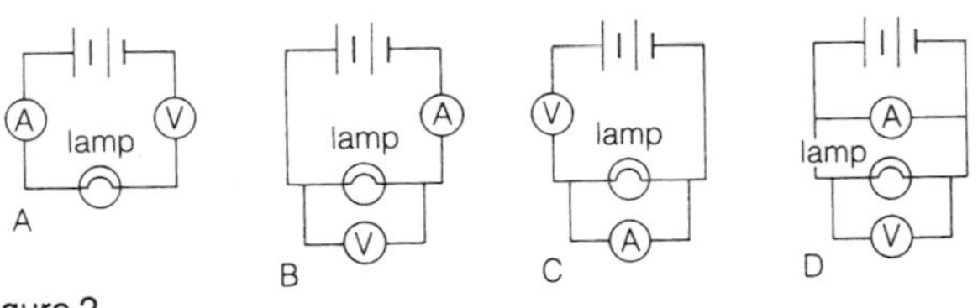

Figure 2

4 The graph in figure 3 shows how the current I changed with the voltage V applied to a sample of a material.

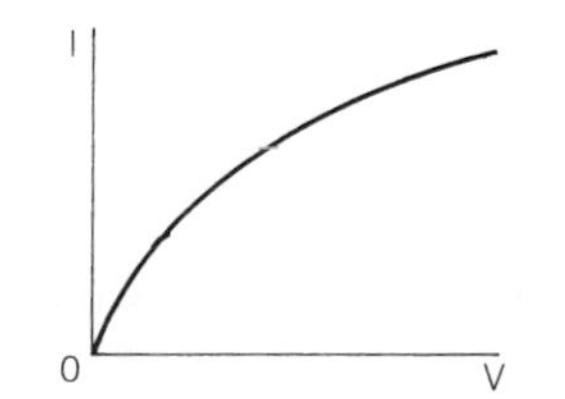

Figure 3

Does the shape of the graph show that
a) the resistance of the material decreases as the voltage increases
b) the resistance of the material is constant
c) the current decreases as the voltage increases
d) the resistance of the material increases as the voltage increases? (SEG)

5 The ammeters in the circuit in figure 4 have negligible resistance.

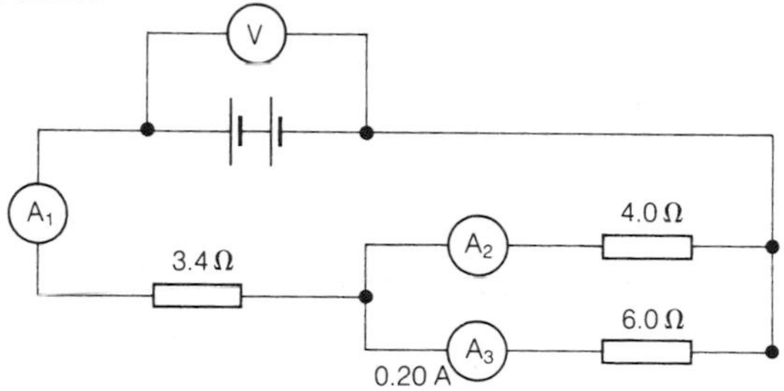

Figure 4

Using the values shown, calculate
a) the p.d. across the 6.0 resistor
b) the current through the ammeter A_2
c) the current through the ammeter A_1
d) the reading of the voltmeter across the cells. (MEG)

6 Figure 5 shows a lighting circuit taken from a motor car handbook. The filament lamps are all the same, and provide the side lights, the tail lights and the lamp which lights the number plate.
a) Why are the lamps wired in parallel?
A This way they require less current
B If one lamp fails the others remain lit
C This way they require less power
D If one lamp fails the others also fail
b) When the lamps are lit with normal brightness a current of 0.5 A is drawn by each lamp. What is the power of each lamp?
A 0.5 W
B 6 W
C 12 W
D 24 W
c) What is the resistance of each lamp?
A 0.5 Ω
B 6 Ω
C 12 Ω
D 24 Ω (LEAG)

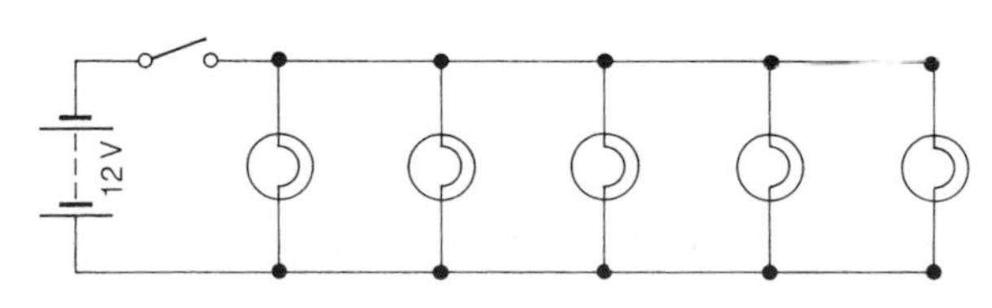

Figure 5

7 Figure 6 shows the inside of an electric plug wired for use on a 240 V 1 kW mains electric heater.

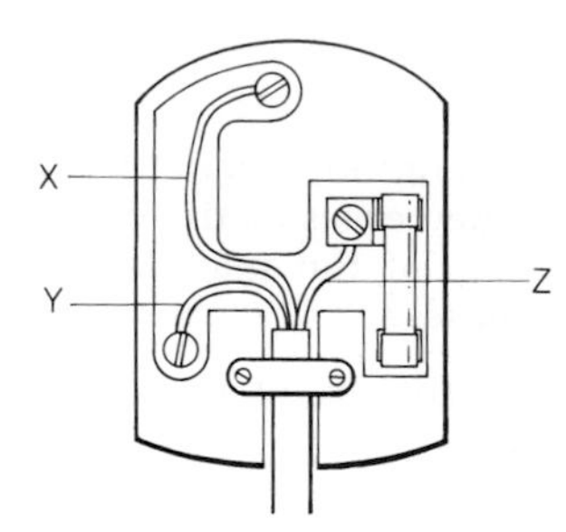

Figure 6

a) Copy and complete the table which describes wires X, Y and Z.

	Name of wire	Colour of insulation
X	earth	
Y		
Z		

b) To which part of the heater will the other end of wire X be connected?

c) Would a fuse rated at 3 A be suitable for use in this plug? Explain your answer. (MEG)

8 A new type of light bulb has recently been invented. It produces the same amount of light as an ordinary 1000 W bulb, but uses only 25 W of electrical power. It is expected to last for 5000 hours.

a) How many kilowatt-hours does a 100 W lamp use in 5000 hours?

b) How much will this cost? The electricity board charges 5 pence for 1 kilowatt-hour.

c) Both bulbs produce the same amount of light energy but the old type uses more electrical energy. Explain what happens to this extra energy. (SWEB/SEG)

9 Imagine you are trying to find out if a fuse has 'blown'. You are given the pieces of apparatus shown in figure 7.

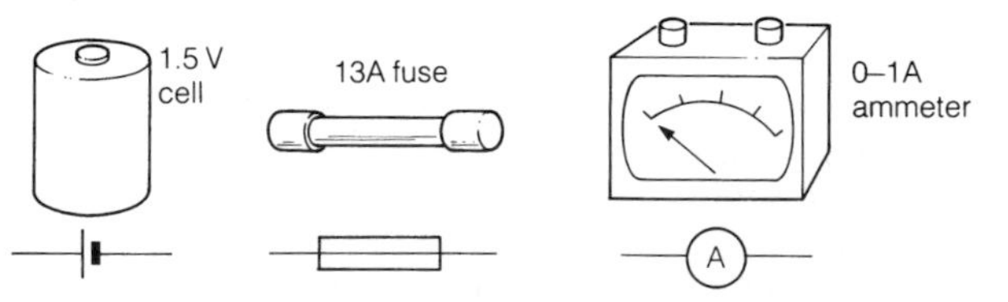

Figure 7

a) Draw a circuit diagram to show how you would connect these components.

b) Explain how you would tell whether or not the fuse had blown.

c) If the fuse was rated 100 mA you would need an extra component in your circuit.
   i) Why is this?
   ii) What extra component would you need?

(SWEB/SEG)

10 Figure 8 shows a 240 V electric hair drier with a plastic case.

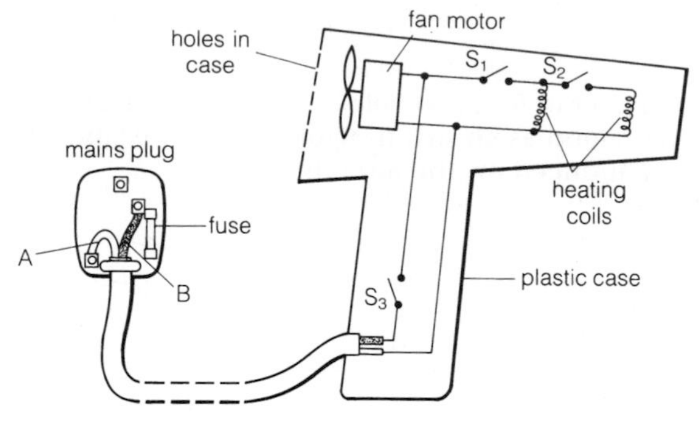

Figure 8

a) i) Write down the colours the cables in the plug should have.
   ii) One pin in the plug is not being used. Does this make the hair drier dangerous? Explain your answer.
   iii) Write down whether the heating coils are connected in series or in parallel.

b) Which switches need to be closed to switch on:
   i) the fan alone?
   ii) the fan and one heating coil?

c) When the hair drier is working at full power, the voltage is 240 V and the current in each coil is 2 A.
   i) What is the resistance of one heating coil?
   ii) The fan motor takes a current of 0.5 A. What is the total current from the supply when both coils are in use?
   iii) What size of fuse is needed in the plug? Choose your answer from: 1 A 3 A 7 A 10 A
   iv) Suppose you find a 13 A fuse in the hair drier! Explain why this might be dangerous.

(SWEB/SEG)

11 Figure 9 shows a long length of 2-core cable which is thought to be broken somewhere inside. This must be tested before it can be used with mains electricity.

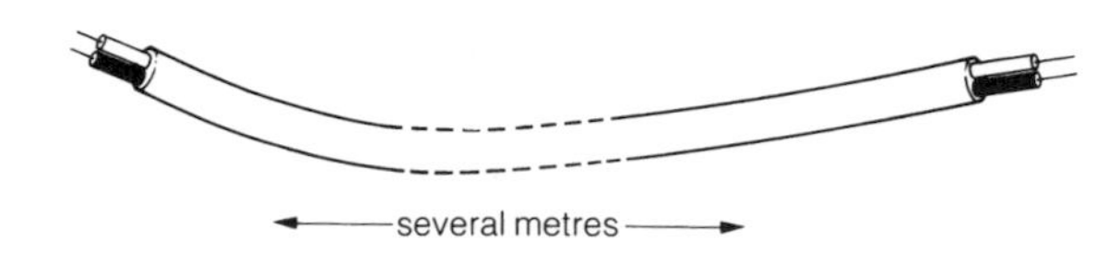

Figure 9

a) What equipment would you use to find out if this cable is broken?

b) Explain carefully how you would connect up this equipment.

c) How would you be able to tell whether or not the cable is broken? (SWEB/SEG)

Further questions

Electrical energy: part B

Some questions require a knowledge of concepts covered in earlier units

1 Draw a labelled diagram showing the essential structural features of a leaf electroscope. Describe in detail how the electroscope could be charged by the method of induction. Explain how it could then be used: a) to determine if another charged body had a charge of the same sign, b) to show that a damp cotton thread is a much poorer insulator than one which is perfectly dry. (L)

2 Two identical uncharged conducting spheres, each of which is on an insulating support, are placed as shown in figure 1. The spheres are in electrical contact.

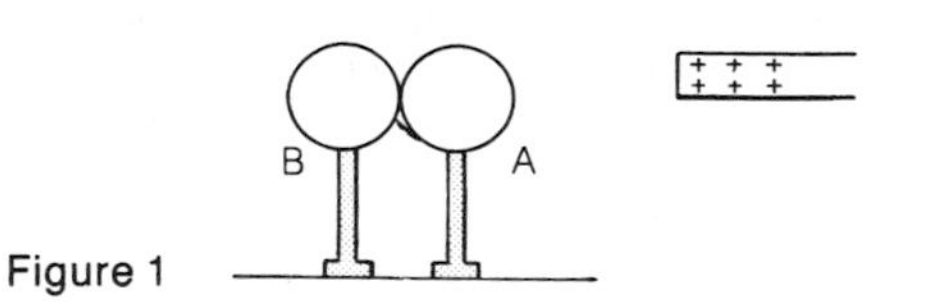

Figure 1

A positively charged rod is now moved into the position shown. The sphere B is then moved a distance away to the left. Finally the charged rod is removed. State what you know concerning the charges present on the two spheres A and B after this separation.
In what way, if any, would the final arrangement of the charges have been different if initially the charged rod had been brought much closer to the sphere A? (C)

3 a) In figure 2a, A and B are two conducting spheres on insulating stands. A is charged to a high positive potential and B, which is earthed, is brought near to it. Draw a diagram showing the resulting charge distribution on B.

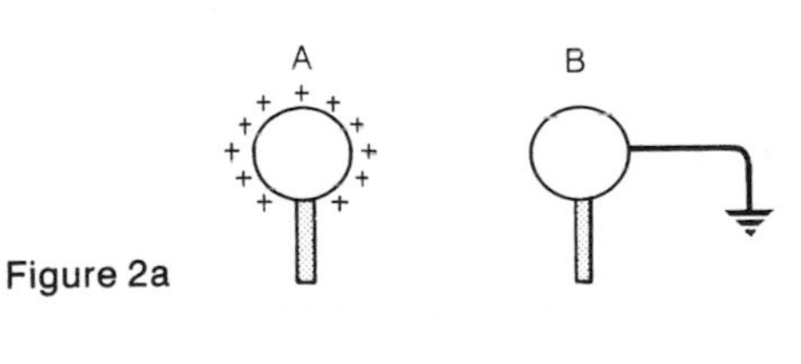

Figure 2a

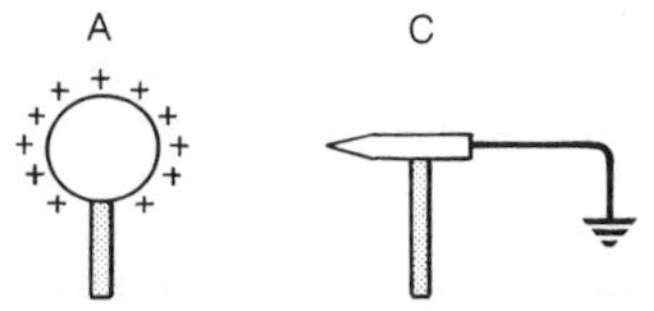

Figure 2b

b) B is replaced by the earthed metal needle C (see figure 2b), which is the same distance from A as was B. Draw a diagram to show the charge distribution on C, and explain why in this case A loses its charge much more quickly than it did in a).

c) Lightning conductors with pointed tops are put on high buildings to prevent them from being damaged by lightning but it is foolish to walk across an open space carrying an open umbrella in thundery conditions. Give the physical reasons for the above statement. (L)

4 What is meant by an electric field? Draw a diagram to show the electric field between two oppositely-charged parallel metal plates. (O)

5 Define capacitance. What is the potential difference between the plates of a 22 μF capacitor when the magnitude of the charge on each of its plates is 165 μC? (O)

6 A copper can A is much larger than a copper can B. Each is placed on the cap of an identical uncharged leaf electroscope. How would you give an identical charge to each can?
When you have done this, what would you observe about the deflection of each electroscope's leaf? Explain your answer and state what can be deduced about the capacitance of the two cans. (L)

7 a) Draw a labelled diagram of a parallel plate capacitor.
b) What is the relationship between capacitance, voltage and charge?
A 2 μF capacitor is charged to a potential of 200 V and then disconnected from the power supply.
c) What is 2 μF expressed in farads?
d) What is the size of the charge on each plate of the capacitor?
e) One plate of the capacitor carries a positive charge; the other plate is earthed. Explain why the earthed plate carries a negative charge. (O)

8 Two insulated metal spheres, at large distances from other conductors, A (capacitance 2×10^{-12} F) and B (capacitance 5×10^{-12} F) are each charged to a potential of +3 kV.
a) Calculate the charge on each sphere and state, if anything, what happens when the two are connected by a copper wire.
b) After discharging them, the two spheres are now each given a charge of $+10^{-9}$ C. Calculate the potential of each sphere and state what, if anything, happens when the two are connected by a copper wire. (O)

9 Figure 3 shows a circuit for measuring the resistance of a wire which is kept at constant temperature.
a) What is the purpose of i) the meter M_1; ii) the meter M_2; iii) the resistor R?

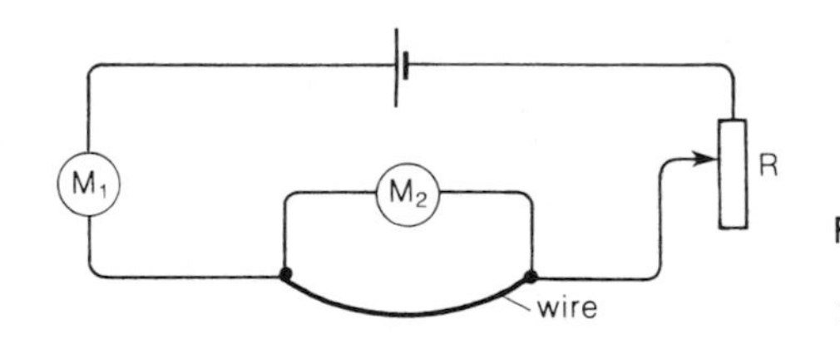

Figure 3

10 A 2Ω resistor is connected across a 12V battery. Calculate a) the current flowing b) the charge passing in 3 minutes.

11 How does the resistance of a household lamp bulb change when the current is switched off? (AEB)

12 The graph in figure 4 shows how the current I through a tungsten filament lamp varies with the voltage V across it.

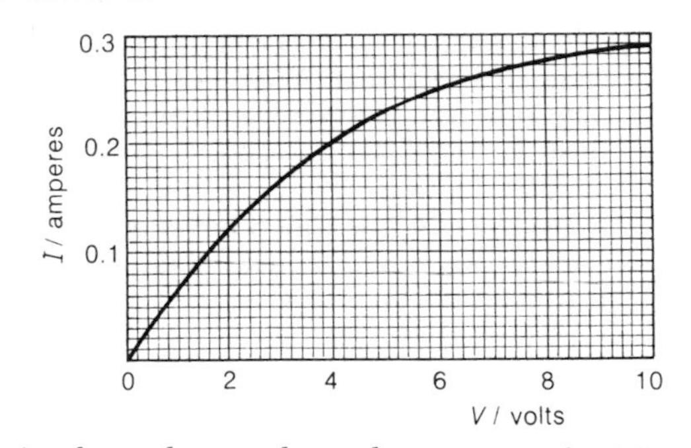

Figure 4

a) What is the voltage when the current is 0.2 ampères?
b) What is the resistance of the lamp when I is 0.2 ampères?
c) What is the increase in current when V is increased from 2V to 6V?
d) Does the resistance of the filament increase or decrease as V increases from 2V to 6V? What causes the change?
e) Draw a diagram of the circuit you would use to take readings to plot this graph. (O)

13 In figure 5, say which lamp or lamps is controlled by each of the switches. Answer this question by rewriting and completing the following sentences:
Switch X controls lamp(s) . . . .
Switch Y controls lamp(s) . . . .
Switch Z controls lamp(s) . . . . (SWEB)

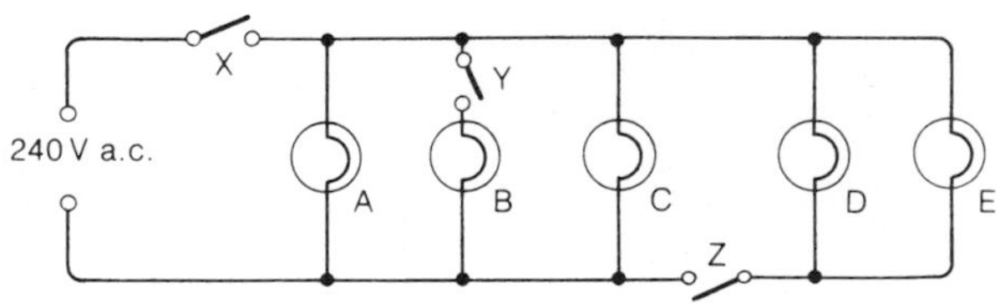

X Y 240 V a.c. A B C D E Z

Figure 5

14 In figure 6, the three resistors, X, Y, and Z, have equal resistance. If ammeter M reads 6A, what will be the reading on ammeter N?

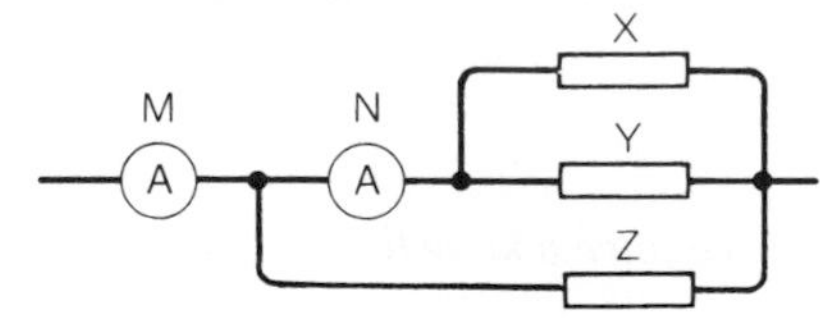

Figure 6

15 How you would connect three resistors, each of resistance 6 Ω so that the combination has resistance a) 9 Ω, b) 4 Ω.
In case a) the combination is connected across the terminals of a battery with a p.d. of 12V across its terminals.
Calculate c) the current through each resistor, d) the p.d. across each resistor. (SUJB)

16 A wire of resistance 18 Ω is cut into three equal lengths. If two of these are connected in parallel, what is their combined resistance?

17 a) Write down the formula for the effective resistance of two resistors in parallel. Calculate the effective resistance of a 12 Ω resistor and a 6 Ω resistor connected in parallel.
b) Suppose you are supplied with the following equipment:
a steady low voltage d.c. source of low internal resistance,
a 12 Ω resistor and a 6 Ω resistor,
a variable resistor of a maximum value of 10 Ω in 1 Ω stages,
connecting wire, a suitable ammeter.
Describe how, using only the above apparatus, you can check that the measured value of the resistance of the 6 Ω resistor and the 12 Ω resistor in parallel agrees with the calculated value. Draw a circuit diagram and state clearly the readings which you would take. (L)

18 A 1.5 V dry cell of internal resistance 2 Ω and two 4 Ω resistors are connected in two different ways (see figure 7).

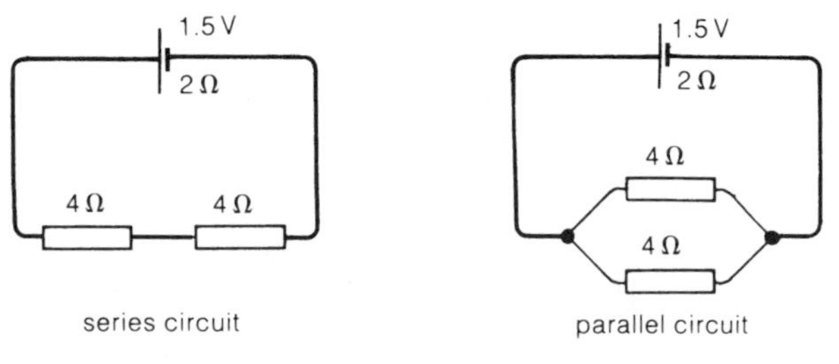

Figure 7

For the series circuit a) what is the current through the cell? b) what is the p.d. between the cell terminals?
For the parallel circuit c) what is the current through the cell? d) what is the current through each resistor? (O)

19 The circuit in figure 8 is to be used to find the internal resistance of the cell. With the key closed, the voltmeter reads 1.2 V. What value would this result give for the internal resistance of the cell? (AEB)

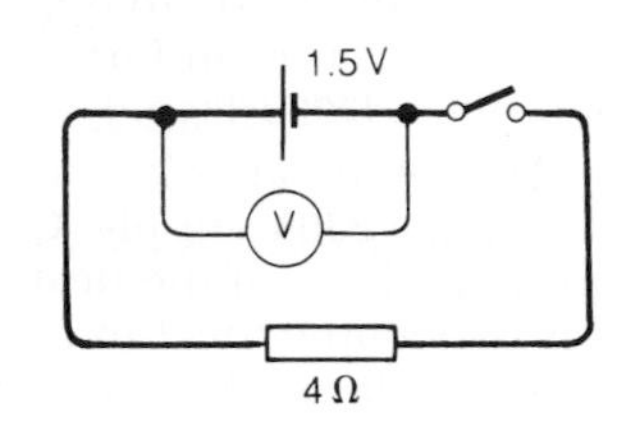

Figure 8

20 A cell of e.m.f. 1.5 V and internal resistance 3 Ω is connected in series with a cell of e.m.f. 2.0 V and internal resistance r so that the cells assist each other. When this arrangement is joined across the ends of a resistor of constant resistance 10 Ω, a current of 0.25 A is produced. Determine the value of r. (L)

21 Figure 9 shows a battery which has e.m.f. 4.5 V and internal resistance 0.75 Ω connected as shown to resistors of resistance 3.0 Ω and 1.0 Ω. Calculate

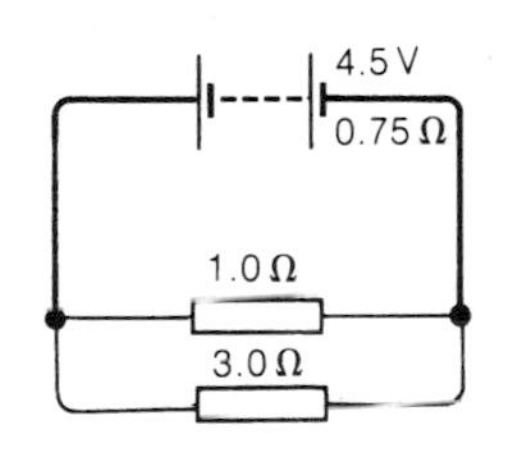

Figure 9

a) the combined resistance of the 3.0 Ω and 1.0 Ω resistors,
b) the current delivered by the battery,
c) the current passing through the 1.0 Ω resistor. (C)

22 You are provided with a suitable power supply, rheostat, a dual-purpose meter capable of indicating a maximum of 10 V and 5 A when used as a voltmeter and ammeter respectively, a bulb labelled 6 V 24 W, and a selection of connecting wires. Describe the procedure you would follow to determine the wattage of the bulb, when a potential difference of 5 V is applied to it. Include a circuit diagram in your description. (O&C)

23 A battery of e.m.f. 100 V and internal resistance 12 Ω is connected to a 180 Ω resistor by two leads each of resistance 4 Ω. Calculate
a) the current in the resistor.
b) the power dissipated in the resistor.
c) the potential difference between the ends of each lead,
d) the reading shown by the voltmeter connected across the terminals of the battery. (L)

24 Figure 10 shows two cells, each of e.m.f. 1.5 V and internal resistance 1.0 Ω, a 10 Ω resistor, a very high resistance voltmeter and a switch.
a) What will be the reading on the voltmeter when the switch is open, so that no current flows through the resistor?

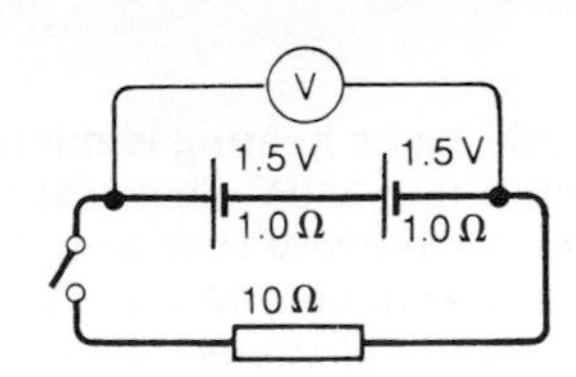

Figure 10

b) The switch is closed so that a current flows through the resistor.
i) What is the size of this current?
ii) What is the reading on the voltmeter?
iii) What is the total power generated by the cells?
iv) What is the power dissipated in the 10 Ω resistor? (O&C)

25 a) A 12 V 36 W heater is connected across a 12 V d.c. supply of negligible internal resistance. Calculate i) the current through the heater, ii) the resistance of the heater.
b) The supply is replaced by a battery which has an e.m.f. of 12 V and an internal resistance of 0.8 Ω. Calculate i) the current in the circuit assuming the resistance of the heater does not change, ii) the rate of heat production in the heater.
c) Explain why, when a current flows through the battery described in b), the potential difference across the heater is less than the e.m.f. of the battery.
Why does the potential difference between the terminals of the battery decrease when the current through the battery is increased? (C)

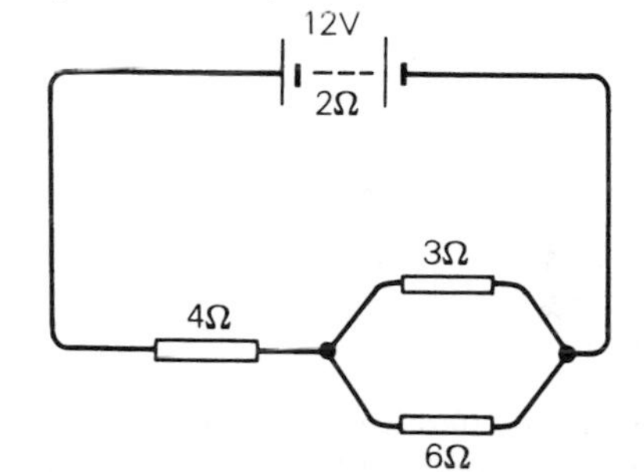

Figure 11

26 The 12 V battery in figure 11 has an internal resistance of 2 Ω. Calculate:
a) the combined resistance of the 3 Ω and 6 Ω resistors.
b) the total resistance in the circuit.
c) the current through the 4 Ω resistor and the p.d. across it.
d) the p.d. across the 3 Ω and 6 Ω resistors.
e) the currents through the 3 Ω and 6 Ω resistors.
f) the total power released by the battery.
g) the power dissipated in each of the three resistors.
h) the power dissipated in the battery because of its internal resistance. (C)

27 A cell of e.m.f. 1.5 V passes a current of 0.5 A through an external circuit for 120 s.
a) How much work is done by the cell in this time?
b) If the cell has an internal resistance of 0.4 Ω, how much energy is transformed within the cell? what is the form of the energy after this transformation?
c) How much energy has been transformed in the external circuit?
d) Deduce the resistance in the external circuit. (C)

28 An electric kettle has a heating element rated at 2 kW when connected to a 250 V electrical supply.

a) Calculate i) the current that would flow when the element was connected to a 250 V supply,

ii) the resistance of the element,

iii) the heat produced by the element in 1 minute.

b) If the cost of electricity were 5p a unit how much would it cost to use the kettle for 5 hours? (JMB)

29 A room is heated by a thermostatically controlled heater rated 3 kW, which, it is estimated, is switched on for 40% of an eight-hour period. The lighting is provided by two 150 W bulbs which are in use continuously. Calculate the cost of heating and lighting the room during the eight hours, the cost of electrical energy being 5p per kWh. (C)

30 How much energy is converted into heat in 1 minute in a 3 Ω heating coil carrying a current of 2 A?

31 Draw a circuit diagram showing how a 10 V d.c. electric motor may be operated at the correct p.d. using a 12 V battery. Include in your circuit meters to measure the current passing through the motor and to check the p.d. across it.
In an experiment to measure the efficiency of the above motor the current was 3 A and a mass of 6 kg was lifted vertically through 2 m in 5 s. How much energy was supplied to the motor in this time? How much potential energy did the mass gain? Calculate the efficiency of the motor. ($g = 10$ N/kg) (SUJB)

32 Figure 12 shows a network of resistors.

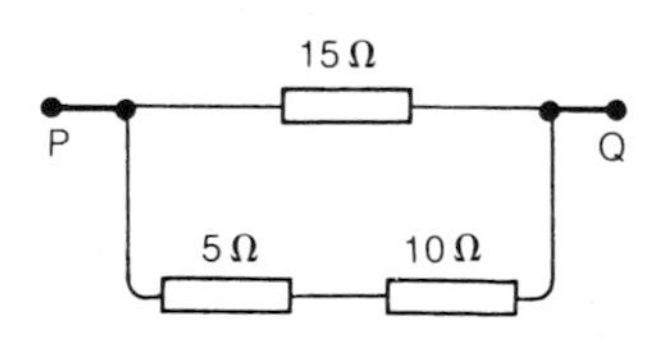

Figure 12

a) What is the effective resistance between P and Q?

b) i) If a potential difference of 30 V were applied across P and Q, what current would flow through the 10 Ω resistor?

ii) Calculate the energy dissipated by the 10 Ω resistor in 1 minute. (AEB)

33 a) Define *potential difference*.

b) A coil of resistance 20 Ω is joined in series with a coil X and a d.c. source of e.m.f. 15 V. If the potential difference across the 20 Ω coil is to be 10 V, calculate the resistance of X. (Neglect the resistance of the source and connecting leads.) Draw the circuit diagram; include an ammeter and a voltmeter to check this value for the resistance of X. What readings would you expect on these meters? Calculate the energy dissipated in the 20 Ω coil in 10 minutes if the current remains steady. If X and the 20 Ω coil had been joined in parallel, what would the current from the source have been? (SUJB)

34 The heating element in a kettle is rated at 240 V 2.4 kW. The kettle is filled with 1.5 kg of water at 20°C, and the element connected to a 240 V supply. Calculate:

a) the resistance of the element and the current through it.

b) the thermal energy released from the element every second.

c) the time taken for the kettle to boil, assuming no heat is lost to the surroundings. [Specific heat capacity of water = 4200 J/(kg K)]

35 The heating element of an electric iron takes a current of 4.0 A from a 250 V supply. Calculate:

a) the resistance, in ohms, of the heating element,

b) the rate of production of heat in the element.
If you had available 3 A and 10 A fuses, which should you use in the plug connecting the iron to the supply? (C)

36 A set of Christmas-tree lights, designed for use with 250 V mains, consists of twenty 12.5 V 2.5 W lamps in series.

a) What is the current, and what is the total resistance of the set when the lamps are alight?

b) What would you expect to happen if one of the lamps burns out?

c) Lamp failure is often taken care of by the manufacturers, who design each lamp so that it is short-circuited if the filament burns out. Supposing that two of the lamps are short-circuited at the same time, what is the p.d. across each of the remaining ones?

d) On the assumption that the resistance of each lamp alight is unchanged calculate the current now flowing.
Why would you really expect it to be appreciably less than the value you have calculated? (O)

37 Connecting plugs for domestic electrical appliances contain fuses. In which wire is the fuse placed and what colour lead is used for this wire? Why would a 30 A fuse not be recommended for a plug on a 60 W lamp on 240 V mains? (SUJB)

38 An electric filament lamp is marked '240 V 120 W'. Calculate the lamp resistance when in normal use. How many such lamps could be used in parallel across a source of 240 V which is protected by a 2 A fuse?

SECTION 7

MAGNETS AND CURRENTS

Magnetism puts coins on the spot

Clues to the origin of ancient coins may be revealed by looking at the coins' magnetic properties, thanks to a method developed in the physics laboratory at Durham University. Archaeologists need to know where coins came from to help them piece together the history of the cultures of the people who made them. But many coins, from sites in Central Asia for example, appear in local markets without any indication as to where they were found. Now, physicists at Durham have found that different series of certain types of ancient coins have characteristic magnetic properties.

The most interesting results have come from several series of coins from the Kushan Empire – a civilization which flourished during the first three or four centuries AD, and at its height stretched from what is now Southern Russia to Northern India. Physicists placed the coins in magnetic fields of different strengths, and found two distinct groups of coins. Coins of the late Kushan period showed only weak magnetization in moderate fields, and after the field was removed, scarcely any magnetization remained. Earlier coins could be magnetized 100 to 1000 times more strongly and the magnetization remained when the field was removed. Why the huge difference? It could be that different mines produced ores with different physical properties. Or perhaps the magnetic characteristics of the metal are affected by melt-

ing down coins already minted and reusing them. Whatever effect is responsible, the results open the way to using a simple magnetic measurement to establish many of a coin's characteristics. This could replace chemical and spectroscopic analysis of coins – expensive and time-consuming procedures often forbidden by museums.

A CENTURY AGO . . .

'There is no plea which will justify the use of high-tension and alternating currents, either in a scientific or a commercial sense. They are employed solely to reduce investment in copper wire and real estate.

'My personal desire would be to prohibit entirely the use of alternating currents. They are as unnecessary as they are dangerous . . . I can see therefore no justification for the introduction of a system which has no element of permanency and every element of danger to life and property.

'I have always consistently opposed high-tension and alternating systems of electric lighting, not only on account of danger, but because of their general unreliability and unsuitability for any general system of distribution.'

(Thomas Edison, inventor of the filament light bulb, 1889)

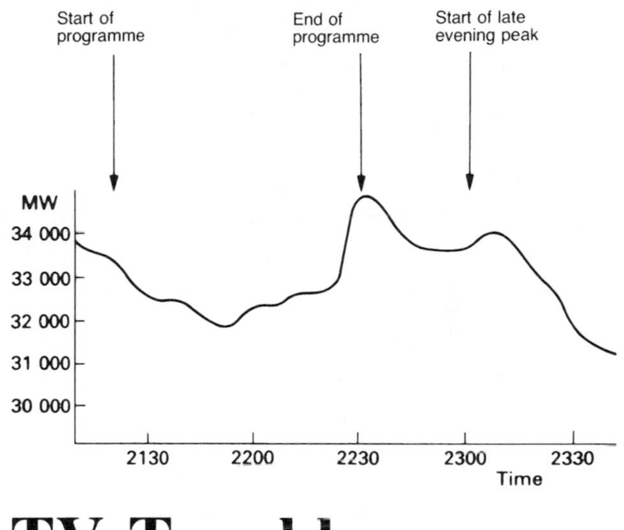

TV Trouble

Life would be much simpler for the Electricity Generating Board if the demand for electric power were constant. Unfortunately it isn't. More appliances are switched on in the daytime than at night, and more power is required in the winter than in the summer. Supply engineers have to anticipate these changes in demand, so that they are ready to connect more generators into the system when demand rises, and shut them off when it falls. If this were not done, the mains voltage would vary dramatically with possibly damaging consequences.

Some periods of rapid demand change are easy to predict – the early evening peak for example, when people arrive home from work and switch on their cookers and kettles. Others have to be found by a careful survey of the coming week's television programmes. The graph shows how the demand for power varied on the night a major TV programme was televised. Note the effect of the hot drinks and cooked suppers when the programme ended.

A 'light map' of the U.S.A. taken at night by a satellite 900 km up. How many cities can you locate?

7.1

Magnets

Magnetism is a mysterious force. Hang a magnet up and it always tries to point in the same direction. Put it near different materials, and it will attract some but not others. Whatever magnetism is, it seems to be present in every atom.

Figure 1 Iron filings are attracted towards the poles of a magnet

Magnetic poles

If a small bar magnet is dipped into iron filings, the filings cling in clumps around its ends, as shown in figure 1. The magnetic force pulling the filings seems to come from two points, known as the **poles** of the magnet. The straight line passing through these points is called the **magnetic axis** of the magnet.

A bar magnet, suspended at its centre by a length of nylon thread will swing round until its magnetic axis lies roughly north-south. This is illustrated in Figure 2. The reason why this happens is examined on page 286, but the effect is used to name the poles of the magnet:

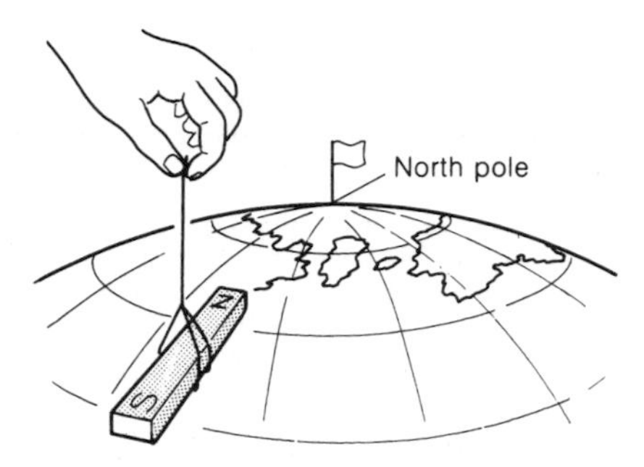

Figure 2 A freely suspended magnet will lie pointing approximately north-south: the north seeking end is called the *N pole*

The pole at the end pointing north is called a north-seeking pole, or N pole for short.

The pole at the end pointing south is called a south-seeking pole, or S pole for short.

If the N pole of a magnet is brought towards the N pole of a suspended magnet, the movement of this second magnet shows that there is a repulsion between the two poles, as shown in figure 3. Similar experiments show that there is also a repulsion between two S poles, but an attraction between a S and a N pole. These results can be summed up as follows:

Like poles repel each other;

unlike poles attract each other.

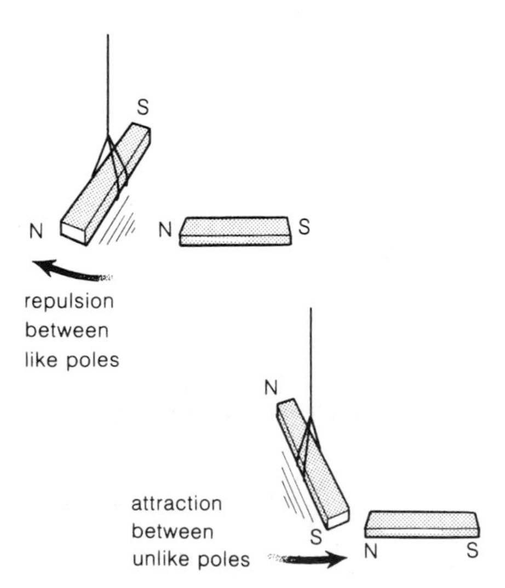

Figure 3 A freely suspended magnet can demonstrate repulsion – and attraction

Testing polarity

The repulsion between like poles can be used to test the **polarity** of a magnet, i.e. to find out which pole is which. For example, repulsion between the known N pole of a magnet and the unknown pole of a second magnet indicates that this is also a N pole.

Repulsion is the only sure means of finding out whether another object, placed near a magnet, is itself a magnet. If two magnets are placed near each other, unlike poles will of course attract, but attraction can also occur when unmagnetized objects are placed near a magnet.

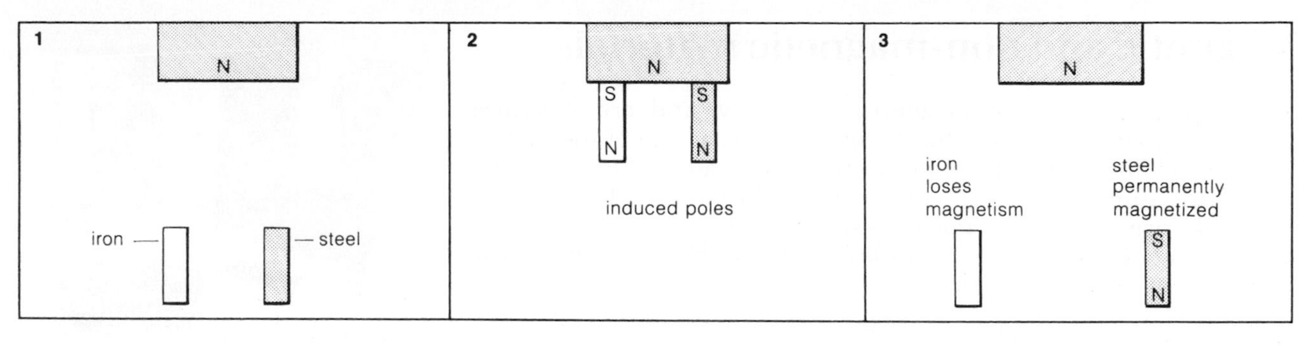

Figure 4 Both iron and steel can develop *induced poles*; but steel is much more likely to stay magnetized

Induced magnetism

Materials such as iron and steel are attracted to magnets because they themselves become magnetized in the presence of a magnet. This is illustrated in figure 4.

The magnet is said to **induce** magnetism in both metals, and a polarity test on each shows that the induced pole nearer the magnet is the opposite of the pole at that end of the magnet. It is the attraction between these unlike poles that holds each piece of metal firmly to the magnet.

If the steel is pulled well away from the magnet, it keeps some of the induced magnetism, and itself becomes a **permanent** magnet. Magnetism induced in the iron is only **temporary** however, and is virtually all lost when the iron is pulled well clear of the magnet.

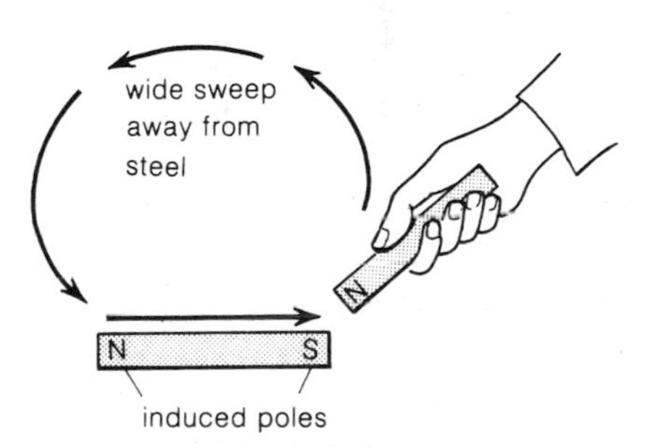

Figure 5 A piece of steel can be magnetized by stroking it with a magnet

Making magnets

Stroking method A piece of steel becomes permanently magnetized when placed near a magnet, but its magnetism is usually very weak. The steel can be magnetized more strongly by stroking it with a bar magnet as shown in figure 5. The magnet must be moved along the steel over and over again in one direction only, with wide sweeps between strokes. The pole produced at the end of the stroke is always the opposite of the stroking pole.

Electrical method The most effective method of magnetizing a piece of steel is to place it in a long coil or **solenoid** made up of several hundred turns of conducting wire, and then pass a large direct current through the coil for a second or so (see figure 6). This produces a very much stronger magnet than the stroking method. The principles involved are examined in more detail on page 289. The polarity of the magnet produced depends on the direction of the current through the coil. The polarity can be found using the **right-hand grip rule** illustrated in figure 7.

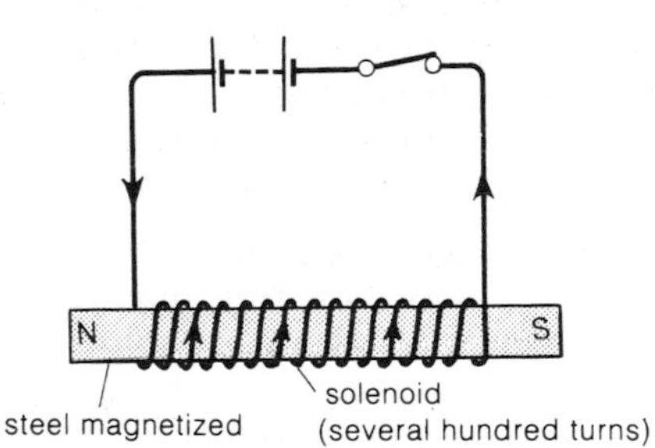

Figure 6 A piece of steel inside a *solenoid* also becomes magnetized. This is the most effective way of making a magnet.

Imagine your right-hand gripping the coil such that your fingers point the same way as the conventional current arrows. Your thumb then points towards the N pole.

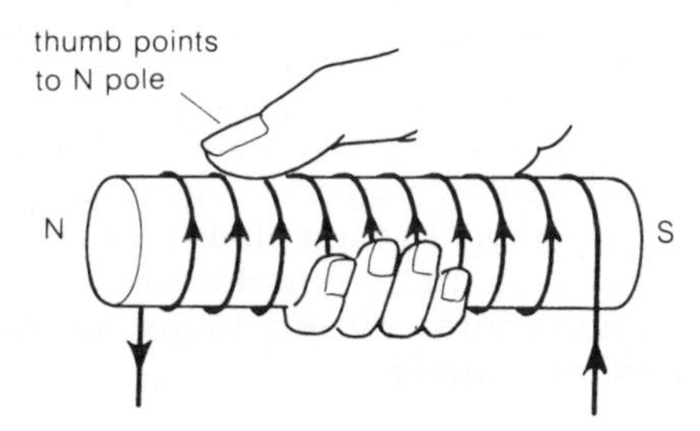

Figure 7 The *right hand grip rule* predicts which end will be the N pole

Magnetic and non-magnetic materials

Materials which can be magnetized strongly, and are therefore strongly attracted to magnets, all contain at least one of the metals iron, nickel and cobalt. Steel for example is an alloy of iron (mainly iron, with some other elements added). These strongly magnetic materials are known as **ferromagnetics**. They are classified as magnetically **hard** or **soft** depending on how well they retain their magnetism when magnetized.

Figure 8 "Hard" magnetic materials in use

Hard magnetic materials such as steel and alcomax (a steel-like alloy) are the most difficult to magnetize but do not readily lose their magnetism. They are used to make permanent magnets. Figure 8 shows a variety of permanent magnets – the iron oxide particles in cassette tape record sound as variations in the level of magnetism along the tape.

Soft magnetic materials, such as iron and mumetal (a nickel-based alloy) are relatively easy to magnetize but their magnetism is only temporary. They are used in electromagnets (see page 290) because in this case they remain magnetized only as long as a current is passing through a surrounding coil. Unlike permanent magnets, electromagnets can be switched on and off.

Materials such as brass, copper, aluminium, and non-metals, are commonly described as non-magnetic because they aren't attracted to small magnets and cannot apparently be magnetized. Experiments with very strong magnets however indicate that even these materials are influenced by magnetism to a slight extent.

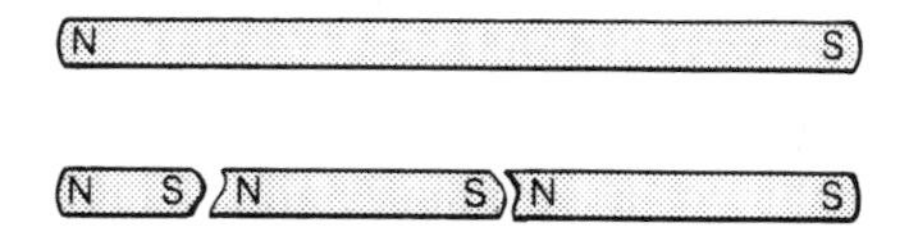

Figure 9 If a magnet is broken, *all* the pieces form complete magnets

Theory of magnetism

If a magnetized steel strip is broken into pieces as in figure 9, polarity tests show that each piece is itself a magnet. If the strip is broken into very much smaller pieces, these too are found to be magnets, and there is evidence to suggest that the smallest magnets of all lie within molecules themselves.

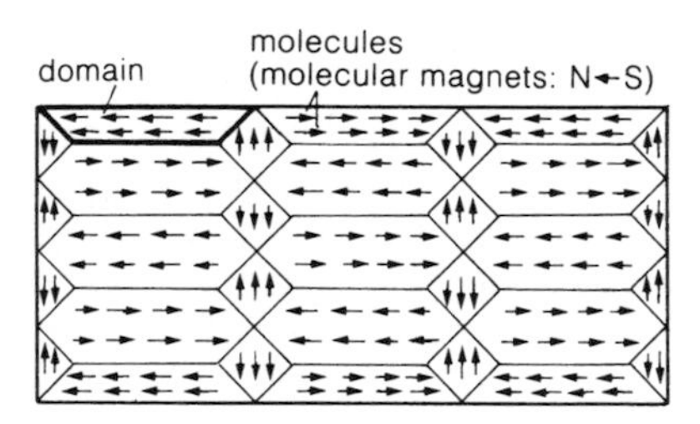

Figure 10a Unmagnetized ferromagnetic material

According to the generally-accepted theory of magnetism, each electron acts as a tiny magnet as it spins and moves around the nucleus of an atom. In some materials, the electron motions are such that the magnetic effects normally cancel out. In others, they do not cancel, and each molecule therefore behaves as a tiny magnet. **Ferromagnetic** materials are made up of 'molecular magnets' of this type.

In a ferromagnetic material, the molecular magnets line up with each other in groups called **domains**. Within any one domain, the magnetic axes of the atoms all lie in the same direction, but this direction varies from one domain to the next if the material is unmagnetized.

'free' poles

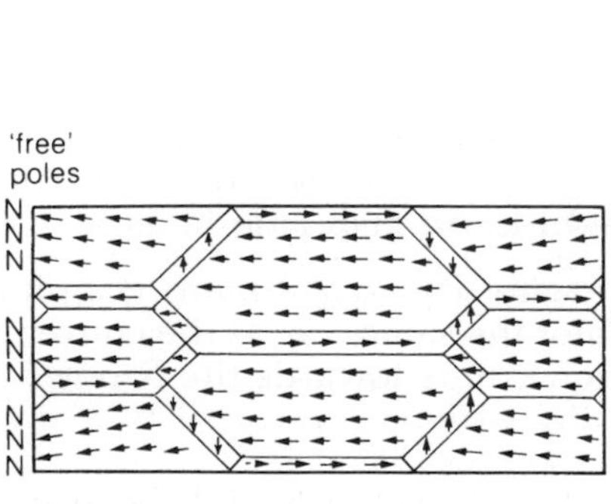

Figure 10b Partially magnetized ferromagnetic material

A simplified representation of the domains in an unmagnetized ferromagnetic material is given in figure 10a. In reality, domains vary considerably in their sizes and shapes, and a domain 0.1 mm across might contain 10^{17} molecules. Note that the magnetic axes of the domains form closed loops; overall, the magnetic effects of the domains cancel.

When a ferromagnetic material is magnetized, some domains grow at the expense of others, and some domain axes turn, so that more and more molecular magnets end up with their magnetic axes in the same direction. This is illustrated in figure 10b. Through most of the material, the poles of each molecule cancel out the effects of opposite poles near by, but uncancelled or 'free' poles are left at both ends. Repulsion between these 'free' poles causes the domain axes to fan out slightly. The 'free' poles around each end of the material together produce the effect of a single N or S pole just in from the end.

A material is said to be **magnetically saturated** when its molecular magnets all lie with their magnetic axes in the same direction. It isn't then possible to make the material into a stronger magnet.

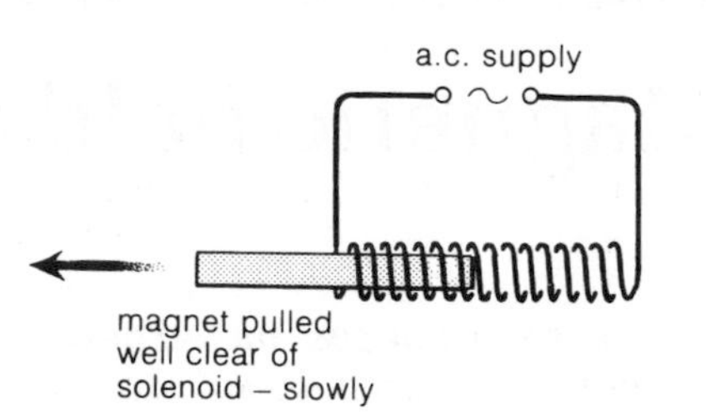

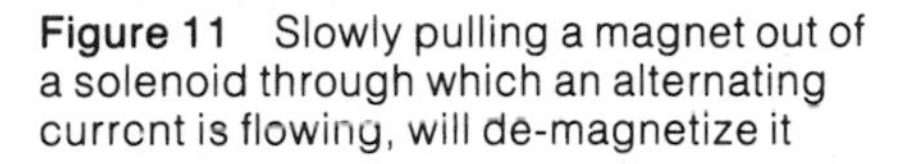
Figure 11 Slowly pulling a magnet out of a solenoid through which an alternating current is flowing, will de-magnetize it

Demagnetization

A magnet will become demagnetized if it is heated strongly because the vibrating molecular magnets eventually gain enough energy to overcome the forces holding them in a common direction. In some cases, the same effect can be produced by hammering the magnet or dropping it repeatedly.

In practice, the most effective method of demagnetizing a magnet without damaging it is to slowly withdraw it from a solenoid through which a large alternating current is passing (see figure 11). This takes the material through weakening cycles of magnetization in one direction and then the other, a process which disturbs the molecular magnets sufficiently to throw the domain axes out of line.

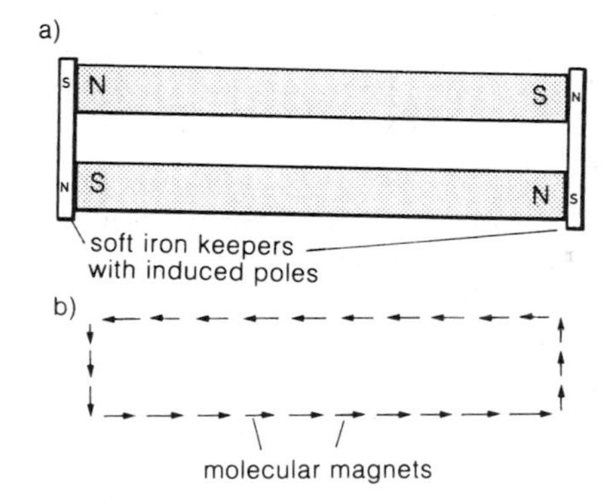

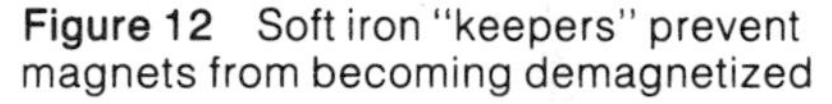
Figure 12 Soft iron "keepers" prevent magnets from becoming demagnetized

Self-demagnetization

A bar magnet becomes demagnetized over a period of time if stored by itself. This process of self-demagnetization starts at the ends of the magnet where the 'free' poles repel each other and gradually alter the alignment of the domain axes. The problem can be avoided by storing bar magnets in pairs with soft iron 'keepers' across their ends as shown in figure 12. The molecular magnets then lie in closed loops with no 'free' poles present to disturb the domains.

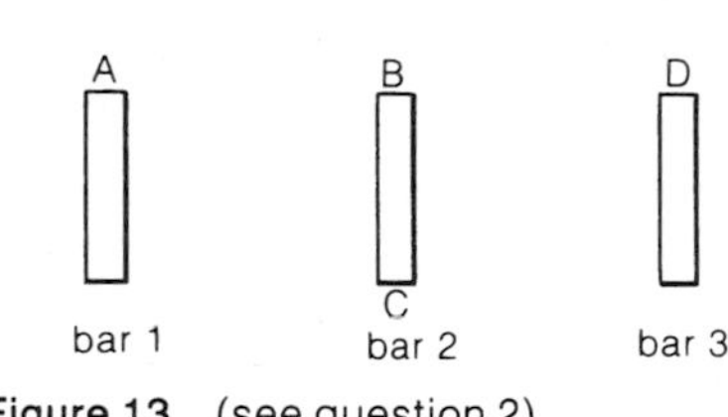

Figure 13 (see question 2)

Questions

1 What is meant by the 'N pole' of a magnet?
2 Figure 13 shows three metal bars. When the ends of different bars are brought together, it is found that A and B attract. A and C attract, but C and D repel. Use this information to deduce whether each of the bars is a magnet or not.
3 Name three ferromagnetic materials.
4 Give two ways in which a ferromagnetic material can be magnetized.
5 What is meant by a) a soft b) a hard ferromagnetic material? Give an example of each type. Which type would be used to make a permanent magnet?
6 What are domains? What is the difference between domains in a magnetized ferromagnetic material and those in one which is unmagnetized?
7 When is a ferromagnetic material magnetically saturated?
8 Give two ways in which a magnet can be demagnetized.
9 Why should bar magnets be stored in pairs with keepers across their ends?

7.2

Magnetic fields

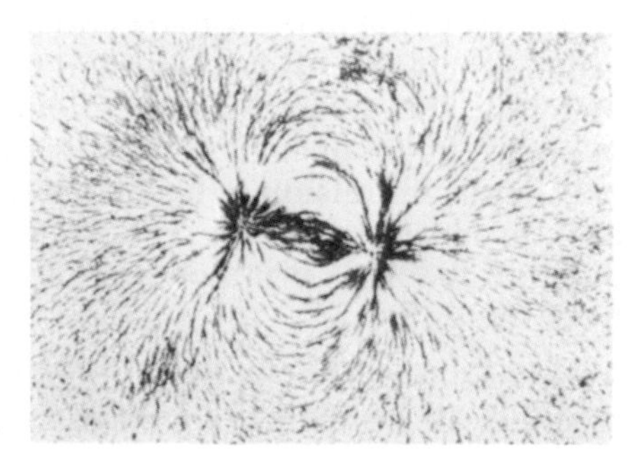

Figure 1 The field around a magnet, as revealed by iron filings

A magnet produces its strongest effects near its poles, but its magnetic influence is spread throughout the whole of the space around it. One source of magnetism influences just about every place on Earth.

Figure 1 shows the effect of sprinkling iron filings onto a piece of card covering a strong bar magnet. Like all other magnetic materials, the filings are affected by forces exerted by the poles of the magnet. A **magnetic field** is said to exist throughout the region in which these forces act.

Magnetic field direction

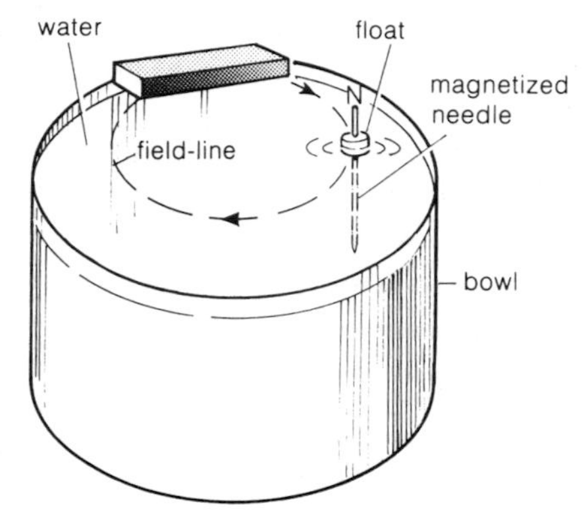

Figure 2 The 'free' N pole floats round from the N pole to S pole along the magnetic field line

By definition, **the direction of a magnetic field at any point is taken to be the direction of the force which would act on a 'free' N pole placed at that point.**

In practice, magnetic poles seem only to exist in pairs. Nonetheless, the N pole of the floating steel needle in the experiment in figure 2 effectively behaves as a 'free' N pole because the S pole of the needle is too far away from the magnet to be affected significantly by it. If the needle is floated just to one side of the N pole of the magnet, it is repelled by this pole, attracted by the opposite pole, and pulled in a curved path from one pole to the other as a result. The direction of its path at any point gives the direction of the resultant force acting and, therefore, the direction of the magnetic field.

The path taken by the moving N pole is called a **magnetic field line**. Any number of field lines can be traced out by releasing the floating needle from different points around the N pole of the magnet. In diagrams, arrowheads are marked on field lines to show that they run from the N pole of a magnet round to the S.

Figure 3 A plotting compass

Plotting magnetic field lines

Field lines around a magnet can be found more simply using a small plotting compass as shown in figure 3. This consists of a tiny magnet, called a needle, supported by a spindle through its centre so that it can turn freely. When the needle is in a magnetic field, its N pole is pulled one way, its S pole is pulled in the opposite direction, and the axis of the needle lines up with the field as a result.

Figure 4 shows how a plotting compass is used to plot a field line. The compass is placed near one end of the magnet, and pencilled dots are made on the paper to mark the positions of the ends of the needle. The compass is then moved so that its needle lines up with the previous dot made, and so on.

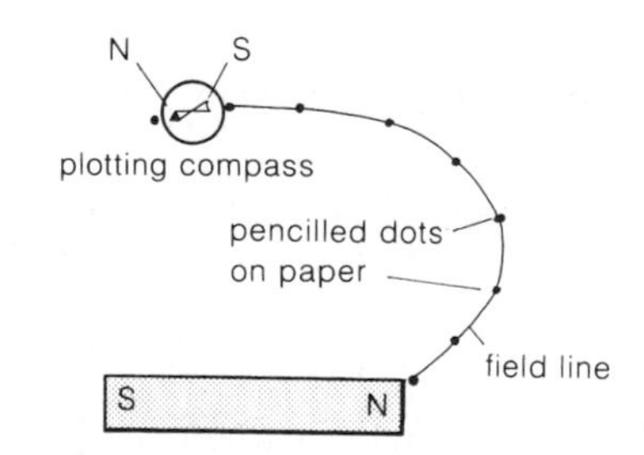

Figure 4 A plotting compass is used to trace field lines

By plotting different field lines, it is possible to build up a pattern showing the field around the magnet. When drawing such a pattern however, it is important to remember that the field extends above and below the paper as well as across its surface as shown in figure 5.

A plotting compass can be used in a weak or strong magnetic field. It isn't however very suitable for plotting highly curved field lines as the size of the needle affects the accuracy. In some cases, field patterns can be revealed by sprinkling iron filings onto a flat piece of card as in figure 1. When the filings become magnetized, each acts as a tiny plotting compass and lines up with the field. Though quick and convenient, this method isn't suitable for weak magnetic fields as the forces on the filings are not then strong enough to turn them.

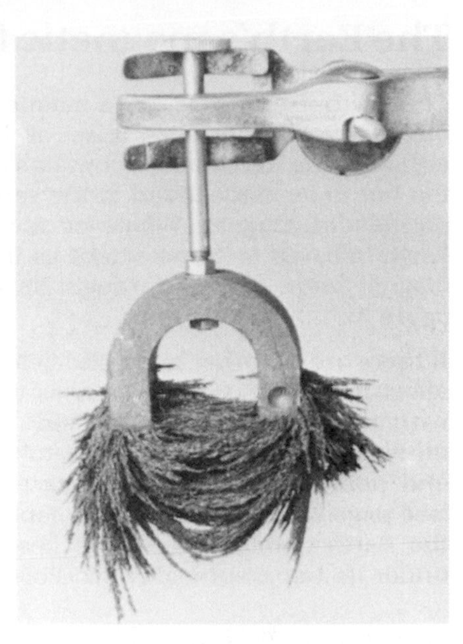

Figure 5 Magnetic fields are in three dimensions

The field around a bar magnet

Figure 6 shows the field pattern around a strong bar magnet. Any number of field lines could have been drawn, but only a selected few are normally included in such diagrams. The spacing of the lines is used to indicate the relative strength of the magnetic field. The closer the lines, the stronger the field, and the greater is the force which would act on any magnetic pole placed in that region.

Remember that the direction of a field line at any point gives:

1 the direction of the force which would act on a 'free' N pole

2 the direction in which the N pole end of a compass needle would point.

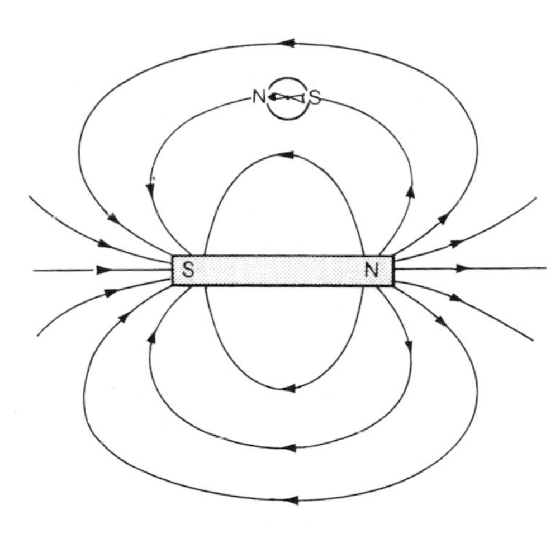

Figure 6 Typical field pattern around a bar magnet

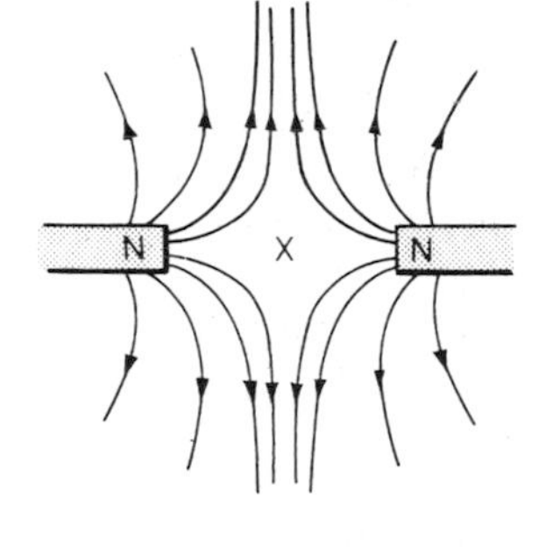

X = neutral point

Figure 7a Field pattern between two similar poles

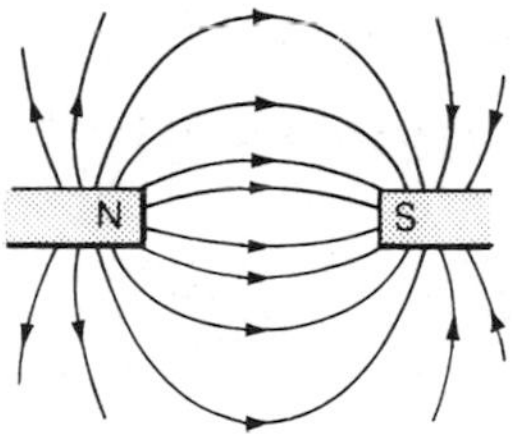

Figure 7b Field pattern between two different poles

Fields between magnets

If two magnets are placed near each other, their magnetic fields combine to produce a single magnetic field. Examples are shown in figure 7.

At point X in figure 7a, the field from one magnet exactly cancels out the field from the other. Put another way, the poles of the magnets would exert equal but opposite forces on a 'free' pole placed at this point. X is known as a **neutral point**.

The Earth's magnetic field

The Earth itself possesses a magnetic field. No one knows the exact cause of this field – although the Earth's core contains iron, it is far too hot to be magnetized in the same way as a permanent magnet. Whatever the cause, the Earth behaves to some extent as if a huge bar magnet were buried through its centre (see figure 8).

If there are no other magnetic fields around to affect it, a compass needle tries to align itself with the Earth's magnetic field. As with any other magnet which is free to turn, its N pole end points in the general direction of north (see page 280). From this, you can deduce that the Earth's magnetic S pole lies somewhere under its Geographical North Pole.

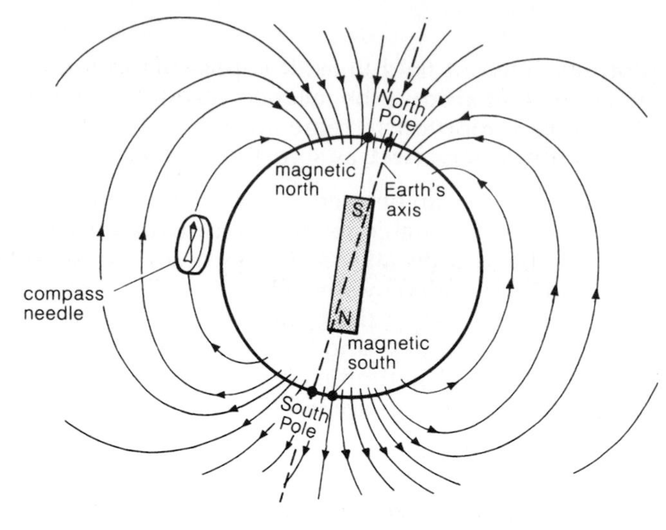

Figure 8 The Earth's magnetic field pattern

Angle of declination

At most points on the Earth's surface, a compass needle doesn't give the direction of true north because the Earth's magnetic poles are not quite in line with its north-south axis.

The angle between true north and the compass needle direction (magnetic north) is called the angle of declination.

This is shown in figure 9. Its value varies depending on where on the Earth's surface you happen to be. Navigators must make allowance for the local angle of declination when using a compass reading to set a course.

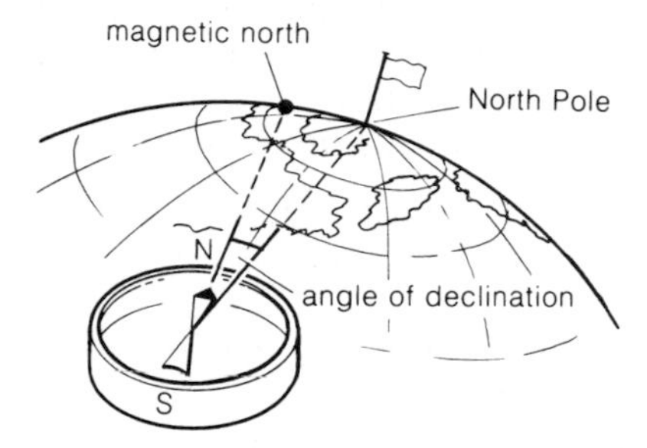

Figure 9 A free magnet does not point exactly towards the North pole

Angle of dip

Study the diagram of the Earth's magnetic field, and you will see that over much of the Earth's surface the magnetic field lines are inclined to the horizontal and cut the Earth's surface at an angle.

The angle between the Earth's magnetic field at any point and the horizontal is called the angle of dip.

This is shown in figure 10. With some local variations, the angle increases the closer you get to either the North or the South Pole. In the UK, the angle of dip is about 65°.

Angles of dip are measured using an instrument called a **dip circle** as shown in figure 11. Essentially this consists of a small magnetized needle mounted so that it is free to turn in a vertical plane only. Provided this plane lies along a magnetic north-south line, the needle aligns itself exactly with the Earth's magnetic field, and the angle of dip can be read off on the circular scale.

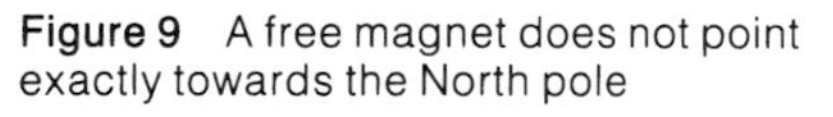

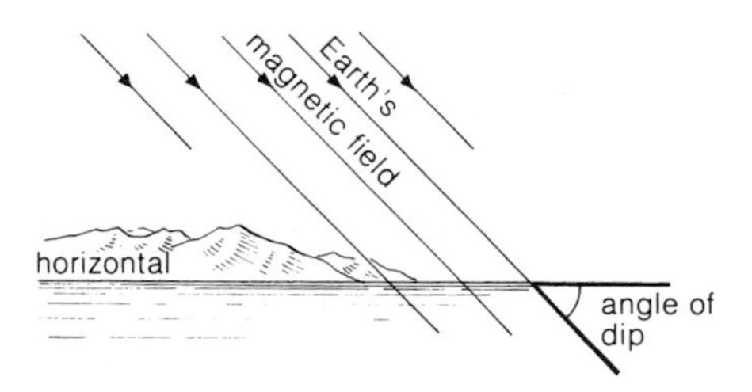

Figure 10 . . . neither does it lie horizontally

Figure 11 A dip circle can accurately indicate the angle of dip

When using a compass, you are not normally aware that the Earth's magnetic field is dipping towards the ground, because the spindle of the compass stops the needle tilting downwards. The magnetic field at a point is a vector quantity, and a horizontal compass needle is turned by the horizontal component of this field only (figure 12).

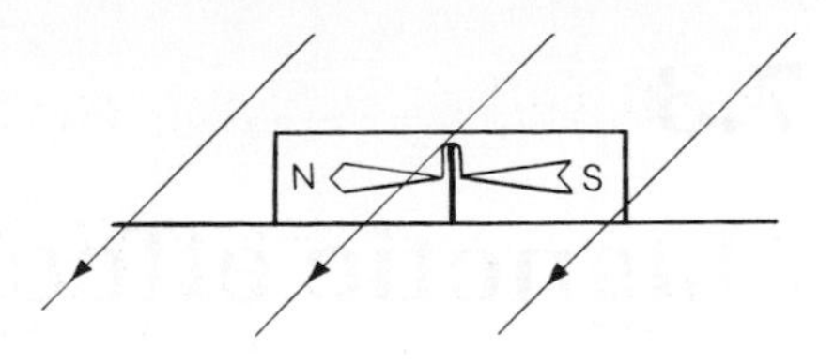

Figure 12 Only the horizontal component of the Earth's field turns the compass needle

Bar magnets in the Earth's magnetic field

In practice, two magnetic fields are present around most bar magnets – one due to the magnet itself, and one due to the Earth. Close to a small bar magnet, the field due to the magnet is usually so strong that the effect of the Earth's magnetic field can be ignored. Half a metre or so away from the magnet however, this isn't the case, and a small plotting compass is as strongly influenced by the Earth's magnetic field as it is by the field due to the magnet.

Figure 13 shows the combined magnetic fields produced in two cases – one with the N pole end of a bar magnet pointing north, and one with it pointing south. At each of the neutral points, X, the Earth's magnetic field exactly cancels the field due to the magnet.

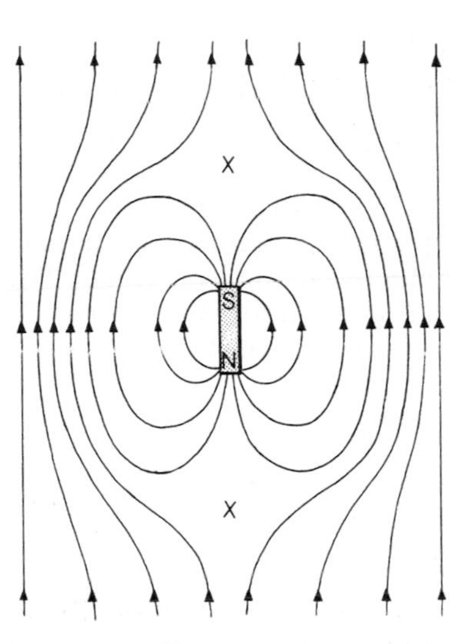

Figure 13a Field for a magnet lined up *against* the Earth's magnetic field

magnetic north

X N S X

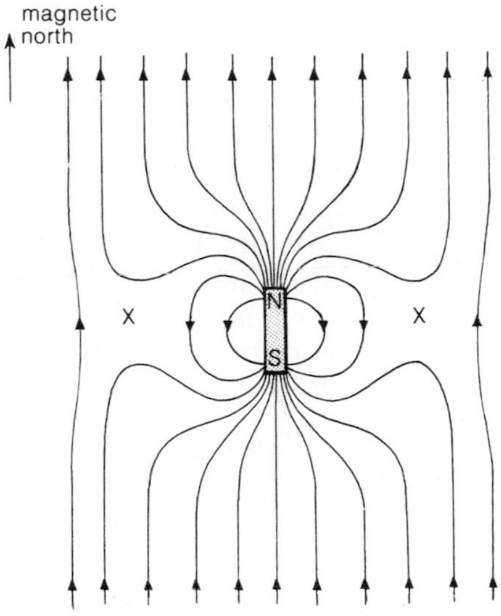

Figure 13b Field for a magnet lined up *with* the Earth's magnetic field

Questions

1 Copy and complete figure 14 to show a) the N and S poles of the magnet b) which way the N pole end of the compass needle would point c) which way a 'free' S pole at point O would move.

2 Give two ways in which it would be possible to find the field pattern round a strong bar magnet. Which of these methods is suitable for a weak magnetic field, and why?

3 Copy and complete figure 15 to show magnetic field lines and any neutral points. What is meant by a neutral point?

4 What is meant by the angle of dip? Where on the Earth's surface is the angle of dip a) about 90° b) about 0°?

5 In London, the angle of declination is about 7°. What does this mean?

6 Figure 16a represents the Earth's magnetic field in a particular region. Copy the diagram and add arrowheads to show the direction of the field lines. What does the spacing between the lines tell you about the field shown?
If the magnet in figure 16b were present in this field, show on a diagram approximately where you might expect to find neutral points.

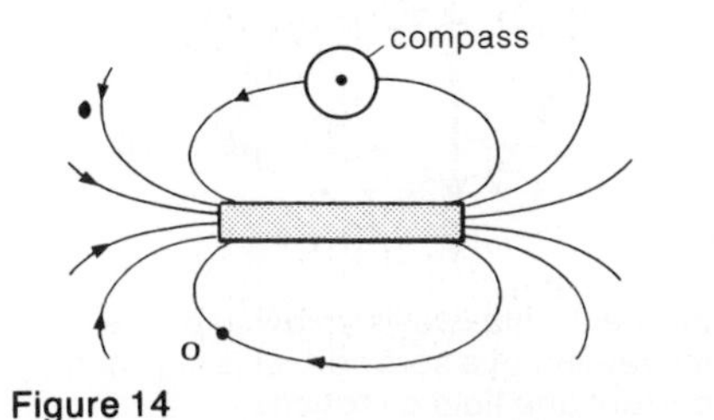

Figure 14

S

S

Figure 15

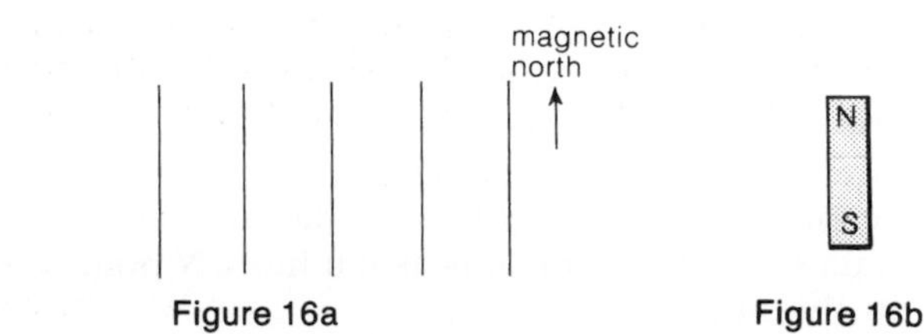

Figure 16a

Figure 16b

7.3

Magnetic effect of a current

Magnets aren't the only source of magnetism. Tests with iron filings and a plotting compass show that a wire has a magnetic field around it whenever a current is passed through.

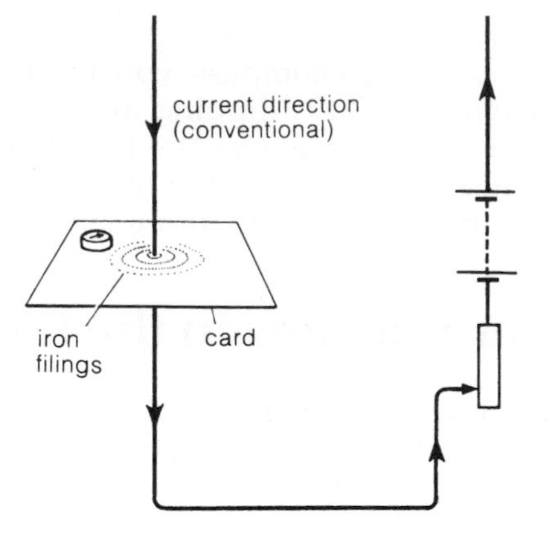

Figure 1 There is a magnetic field round a current-carrying wire

Field due to a current in a straight wire

If a current of a few amperes is passed through a wire as in figure 1, a weak magnetic field is produced. Increasing the current increases the strength of the field, and if the current is around 20 A, the field near the wire is strong enough to be detected using iron filings. The pattern of the filings on the card shows that the field lines are circles centred about the wire. The direction of the field at different points can be found using a plotting compass.

The field pattern close to the wire is illustrated in figure 2a. The field lines run clockwise about the wire, and the strength of the field decreases as you move further out. If the direction of the current is reversed, the field direction is also reversed, but the field pattern is otherwise unchanged. This is illustrated in figure 2b. Note the symbols used in the diagrams to represent the current direction in the wire – think of the tail of a dart going away from you, or the point of a dart coming towards you.

In practice, the Earth's field is also present around the wire in the above experiment. For simplicity, this effect is ignored in these diagrams.

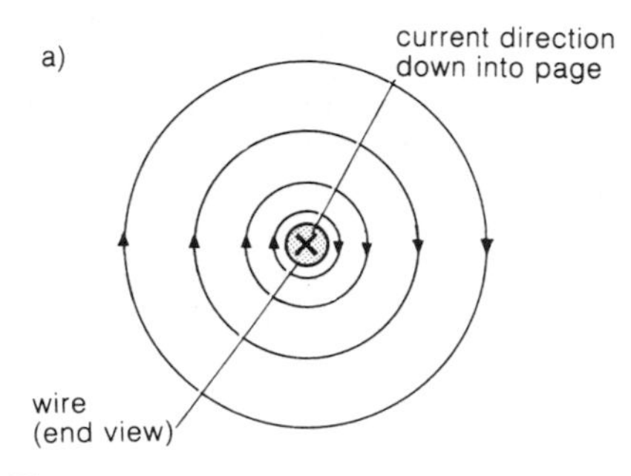

Figure 2a Field round a wire carrying current 'down into the page' (as in figure 1)

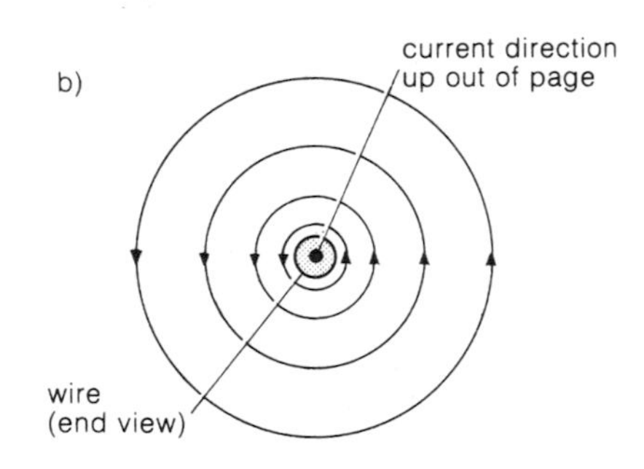

Figure 2b Field when the current is up out of page (opposite to figure 1)

Maxwell's screw rule

Figure 2 illustrates a rule, used to work out the direction of the magnetic field around a wire if the current direction is known:

Imagine a right-handed screw being turned so that it bores its way along the wire in the same direction as the current. The direction of turning gives the direction of the field as shown in figure 3.

This is known as **Maxwell's screw rule** – or Maxwell's corkscrew rule if you are more used to using a corkscrew than a screwdriver.

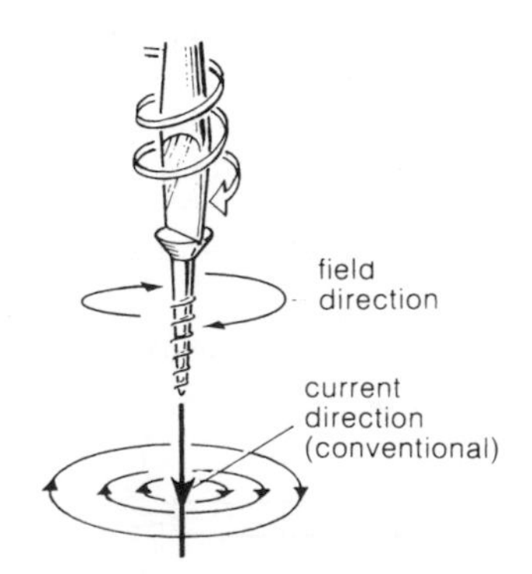

Figure 3 Maxwell's screw rule: The movement of a screwdriver is similar to current and field directions

Field due to a current in a coil

Figure 4 illustrates the field pattern produced by a current flowing in a circular coil. Note that Maxwell's screw rule applies to the field around any short section of the coil.

Outside the coil, the field lines run in loops from one face of the coil to the other. The field is similar to that produced by a short bar magnet, and the coil acts as if it has a N pole on one face and a S pole on the other.

Figure 4 Field pattern round a single loop of wire

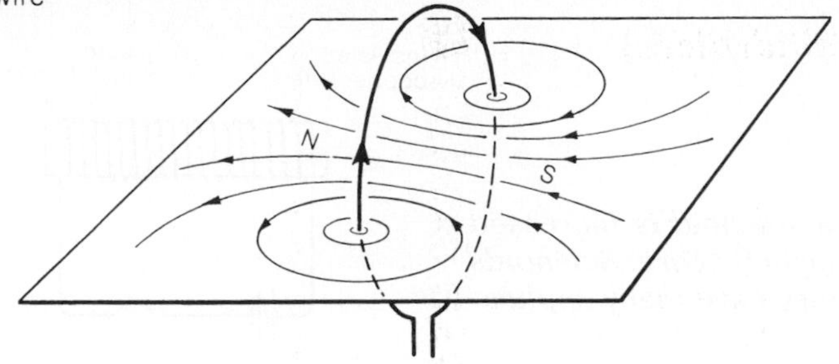

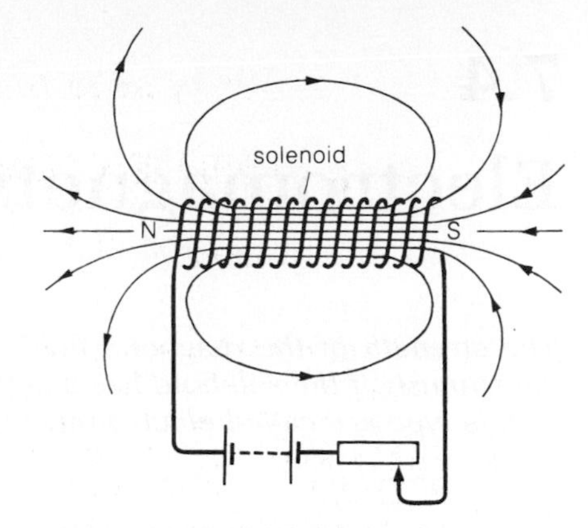

Figure 5 The field pattern round a solenoid is very similar to that round a single loop

Field due to a current in a solenoid

Figure 6 shows a long coil or **solenoid** made up of a number of turns of wire. When a current is passed through the solenoid, each turn acts as a single coil and produces a magnetic field as described above. Together, the turns give a combined field similar to the field around a long bar magnet, and the coil behaves as if it has a N pole at one end and a S pole at the other. The right-hand grip rule, described on page 281, can be used to work out the polarity. The rule is illustrated in figure 6.

Experiments show that, for a solenoid of any given length, the strength of the magnetic field can be increased by:

1 increasing the current

2 increasing the number of turns

The magnetic field is also very much stronger if a ferromagnetic material is present through the middle of the solenoid as explained in the next unit.

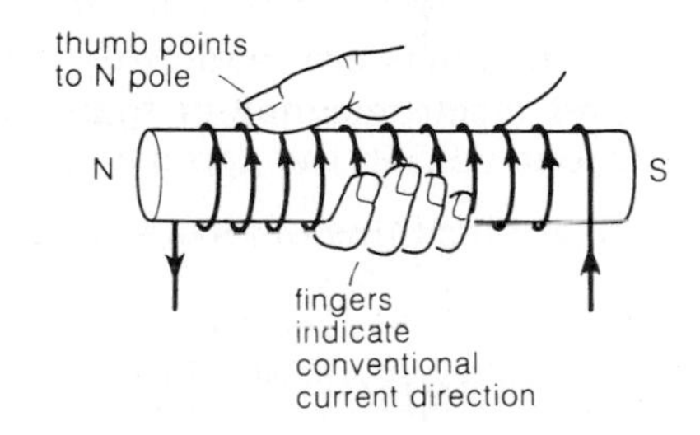

Figure 6 The 'right-hand grip rule'

Questions

In questions 1 and 2, neglect the Earth's magnetic field.

1 Figure 7 shows equal currents passing along two long, parallel, straight wires. Copy the diagram. Mark on it the magnetic field direction at P and at Q, and the positions of any neutral points.

2 Figure 8 shows a solenoid wound on a length of cardboard tubing. Copy the diagram. Draw field lines around the solenoid and show the field direction in each case. Mark the positions of the N pole and the S pole of the solenoid.
What would be the effect on the magnetic field at any point of a) increasing the current b) increasing the number of turns c) reversing the current direction d) placing a ferromagnetic material inside the solenoid?

3 Figure 9 shows a wire at right angles to the Earth's magnetic field. If a current is passed through the wire in the direction indicated, which of the points P, Q, R, or S, most closely corresponds to the position of a neutral point? How would the position of the neutral point be affected if the current through the wire were increased?

P Q

Figure 7

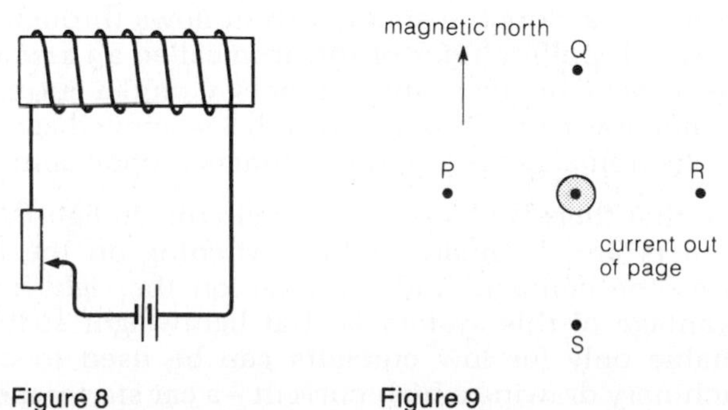

Figure 8 Figure 9

7.4

Electromagnets

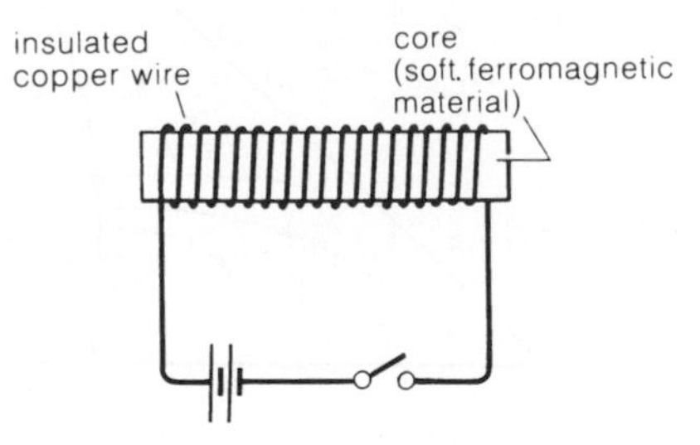

Figure 1 A simple electromagnet

The strength of the magnetic field around a solenoid is increased enormously if the solenoid has a soft ferromagnetic core. Solenoids of this type are called electromagnets, and they have many applications.

A simple electromagnet is shown in figure 1. The coil consists of several hundred turns of insulated copper wire, wound round a **core** made of a soft ferromagnetic material such as iron or mumetal as explained in more detail on page 282. A magnetic field is produced when a current is passed through the coil. This magnetizes the core, which in turn produces a magnetic field about a thousand times stronger than that from the coil alone. As the core is magnetically soft, its magnetism is only temporary, and is virtually all lost when the current through the coil is switched off.

Electromagnets are made in a variety of shapes and sizes. The U-shaped electromagnet in figure 2 produces a very concentrated magnetic field in the gap between its pole pieces.

Some applications of electromagnets are described below.

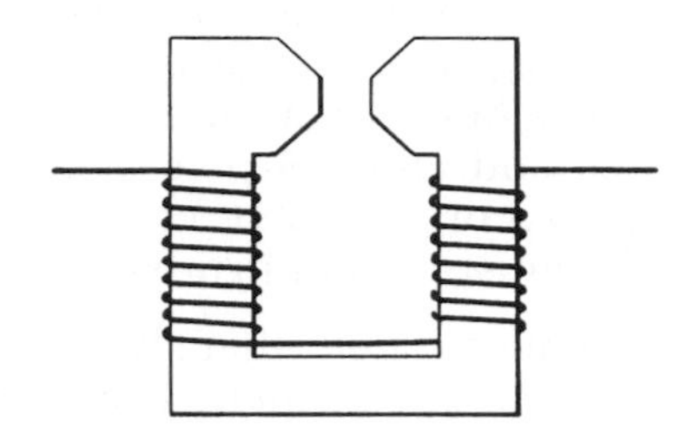

Figure 2 A U-shaped electromagnet

The magnetic relay

A magnetic relay is a switch operated by an electromagnet. The magnetic relay shown in figure 3a is being used to control the circuit connecting a battery to an electric motor.

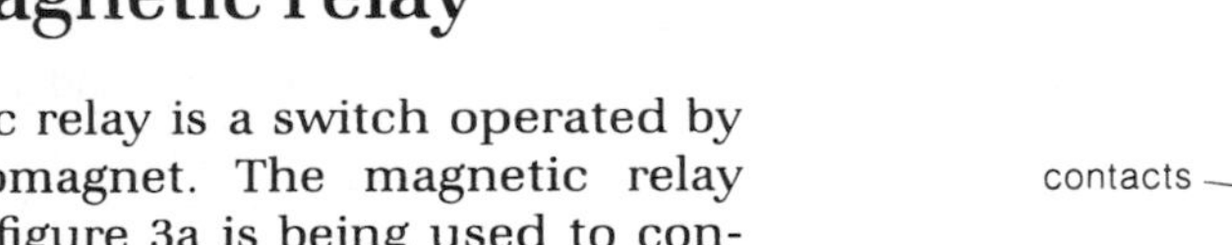

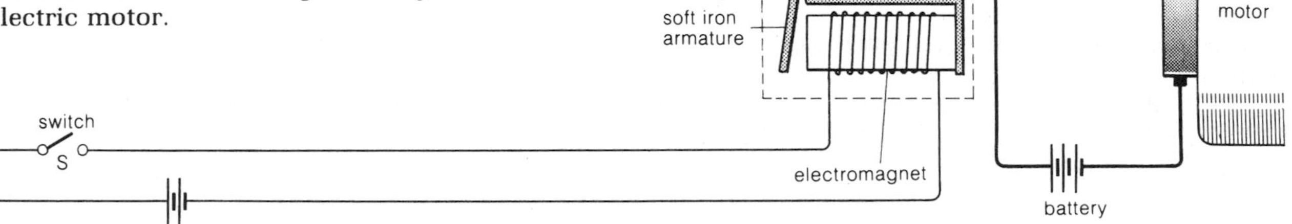

Figure 3a A *relay* uses a small current to operate a switch in another circuit (which may carry a much greater current)

The motor is switched on and off by closing and opening the switch S. When S is closed, a small current flows through the electromagnet, which pulls a block of soft-iron called an **armature** towards it. This movement closes the contacts, C, and a current flows through the motor as a result. When switch S is opened again, the current to the electromagnet is cut, the contacts C open, and the motor stops.

Note that there are two separate circuits in figure 3a, and the relay contacts are 'normally open'. Switching on the left-hand circuit closes the contacts and switches on the right-hand circuit. One advantage of this system is that lightweight switches and cables suitable only for low currents can be used to control electrical machinery drawing a high current – a car starter motor for example.

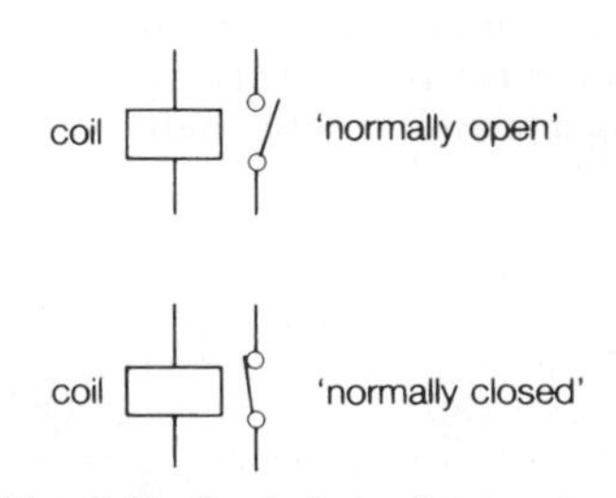

Figure 3b Symbols for the two types of relay

'Normally closed' relays operate in the opposite way to that just described. Switching on the first circuit *opens* the relay contacts and switches the second circuit off. Symbols for the two types of relay are shown in figure 3b.

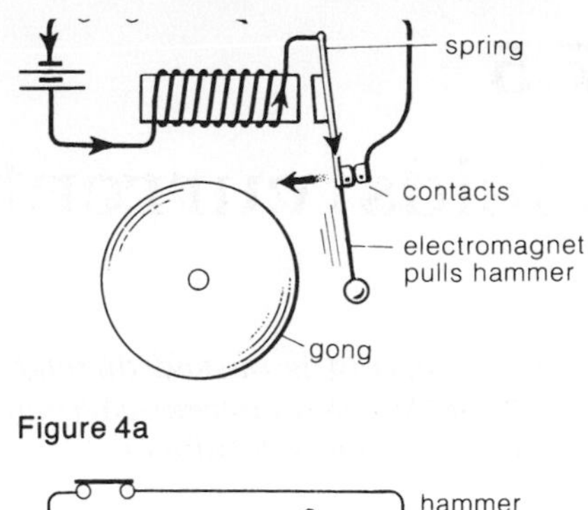

Figure 4a

The electric bell

An electric bell contains an electromagnet that switches itself off and on very rapidly, moving the bell hammer as it does so. The action is shown in figure 4.

When the bell switch is pressed, current flows through the electromagnet, and the hammer is pulled across to strike the gong, as shown in figure 4a. The movement pulls the contacts apart, which switches off the current through the electromagnet, as shown in figure 4b. The hammer springs back, the contacts close again, and the process repeats itself until the bell switch is released.

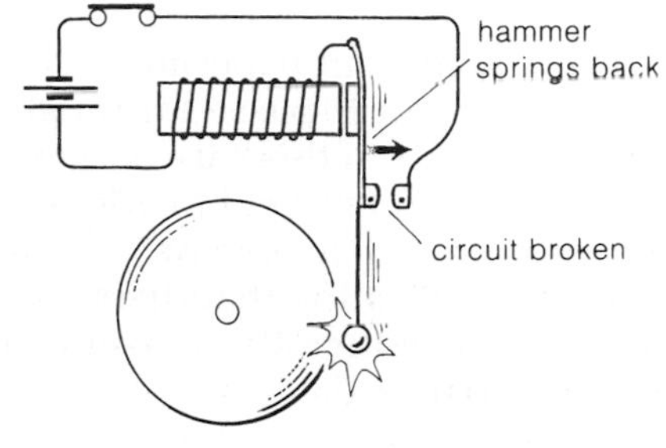

Figure 4b

Figures 4a, 4b The electric bell: the electromagnet pulls the hammer towards the gong, but in so doing switches itself off; so the hammer springs back only for the current to flow once again

The telephone

An electromagnet is used in the earpiece of a telephone. A simple telephone system is shown in figure 5.

When you speak into the mouthpiece of the telephone, each sound compression pushes the thin aluminium plate or **diaphragm** inwards and squashes together the carbon granules behind it. This causes a surge of current in the main circuit because the carbon granules have a lower resistance when they are pushed more closely together. As you speak, the sound compressions from your mouth are 'changed' into a whole series of current surges in the circuit.

In the earpiece, the current surges are changed back into sound compressions. The thin iron diaphragm in the earpiece is pulled strongly by a small permanent magnet and an electromagnet, the pull on it increasing every time there is a surge in the current through the electromagnet. A whole series of surges sets the diaphragm vibrating. As it vibrates, it sends out sound compressions which are a copy of the sound compressions entering the mouthpiece.

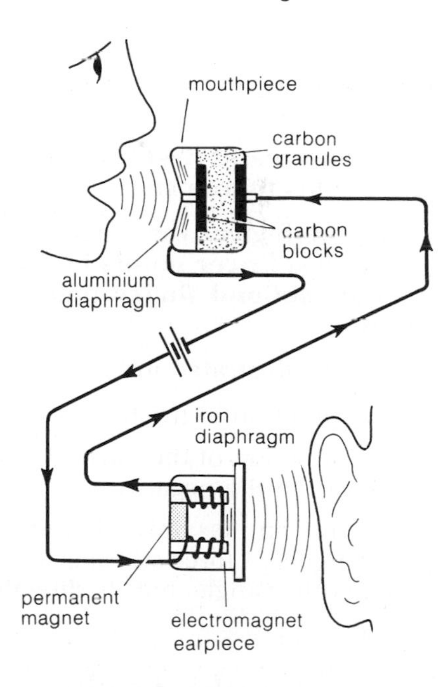

Figure 5 The telephone: Speaking into the mouthpiece varies its resistance; the varying current produced makes the earpiece vibrate similarly

Questions

1 What material would you use for a) the coil b) the core of an electromagnet? What is the purpose of the core? Why should the core be made of a soft rather than a hard magnetic material?
2 Give one advantage of using a magnetic relay to switch electrical machinery on and off.
3 What makes the hammer of an electric bell strike the gong? Why does the hammer move back again after the gong has been struck?
4 In a telephone mouthpiece, what do the sound compressions do to the carbon granules? What effect does this have? What makes the diaphragm in the earpiece vibrate?

7.5

Fields, currents and forces

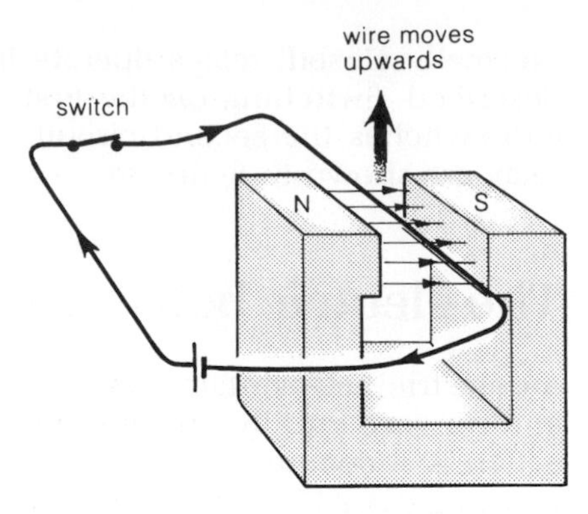

Figure 1 A force acts on a current-carrying conductor in a magnetic field

When a current is passed through a conductor, a magnetic field is produced. When a current-carrying conductor is placed in a magnetic field, a force is produced.

Figure 1 shows an experiment to demonstrate the force produced when a current is passed through a conductor in a magnetic field. The conductor in this case is a length of stiff copper wire, and it is at right angles to the field provided by a U-shaped magnet. When the switch is closed, a current flows through the wire. The wire moves upwards, indicating that there is an upward force acting on it. If the direction of either the current or the field is reversed, the wire moves downwards.

Note in each case that the force on the wire acts at right angles to both the current direction and the field direction. The wire is moved *across* the field; it is *not* attracted to either pole of the magnet.

Fleming's left-hand rule

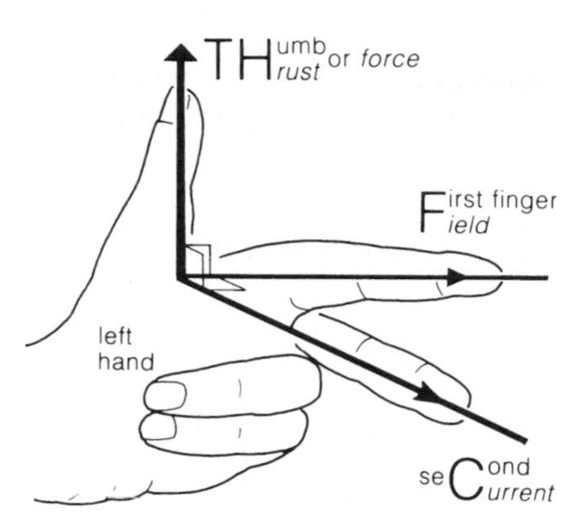

Figure 2 Fleming's left-hand rule indicates the link between force, field and current directions

Knowing the directions of the current and the field in the above experiment, the direction of the force can be found using **Fleming's left-hand rule**:

If the thumb and first two fingers of the left hand are held at right angles to one another,

the THumb gives the direction of the THrust (the force)
if the First finger points in the same direction as the Field
and the seCond finger points in the same direction as the Current.

The rule is illustrated in figure 2. In applying the rule, remember:

1 The direction of the field is from the N pole to the S pole.

2 The direction of the current is from the positive [+] terminal of the battery to the negative [−], i.e. *conventional* current direction.

3 The rule applies only where the current and field directions are at right angles. A force still acts if the current and field directions are at some other angle, but its direction is more difficult to predict.

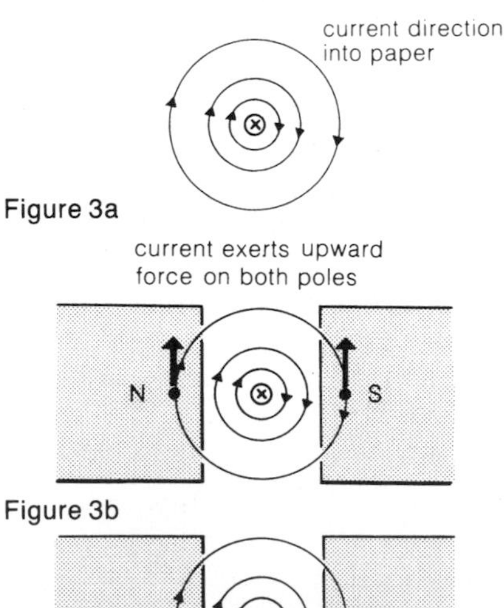

Why the rule works

You can see why Fleming's left-hand rule works if you consider the forces which the current in the wire will exert on magnetic poles placed either side of it.

Figure 3a

current exerts upward force on both poles

Figure 3b

poles exert downward force on wire
(diagram does not show field produced by magnet)

Figure 3c

Figures 3a, 3b, 3c Explaining Fleming's left-hand rule. (Diagrams do not show field produced by magnet)

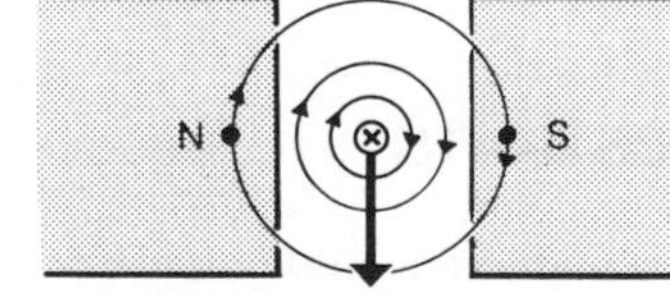

Figure 3a shows an end view of the magnetic field around a wire carrying a current. The direction of the field is given by Maxwell's screw rule described on page 288. The field direction at any point indicates the direction of the force which would act on a N pole placed at that point. The force on a S pole would be in the opposite direction to the field.

In figure 3b, a U-shaped magnet has been placed with its poles either side of the wire. Study the field directions shown, and you will see that the current in the wire exerts an upward force on each pole. It follows from Newton's third law that the poles exert a downward force on the wire. This is illustrated in figure 3c; check for yourself that the direction of the force is the same as that given by Fleming's left-hand rule.

Figure 4a A moving-coil loudspeaker

Magnitude of the force on a current-carrying conductor

Experiments with current-carrying conductors placed at right angles to a magnetic field show that the force acting depends on:

1 the strength of the magnetic field
2 the size of the current
3 the length of the conductor

An increase in any of these three produces an increase in the magnitude (size) of the force acting.

Applications

Several devices make use of the fact that a force acts on a current-carrying conductor when it is in a magnetic field. These include the d.c. electric motor (see unit 7.6), moving coil ammeters and voltmeters (see unit 7.7), and the moving-coil loudspeaker.

The moving-coil loudspeaker Most loudspeakers are of the moving-coil type shown in figures 4a and 4b. It has three main parts:

1 A cylindrical permanent magnet which produces a strong **radial** ('spoke-like') magnetic field.
2 A coil which is free to move short distances backwards and forwards in the magnetic field.
3 A stiff paper cone attached to the coil.

The wire in the coil lies at right angles to the magnetic field. If a current is passed through the coil, a backward or forward force acts on it; this follows from Fleming's left-hand rule.

If an alternating current is passed through the coil, the coil is pushed alternately backwards and forwards. This makes the paper cone vibrate, and sound waves are given out as a result. The nature of the sound produced depends on the frequency and amplitude of the alternating current flowing through the coil. The current could be supplied by a signal generator. Alternatively, it could be supplied by an amplifier connected to, say, the pick-up on a guitar.

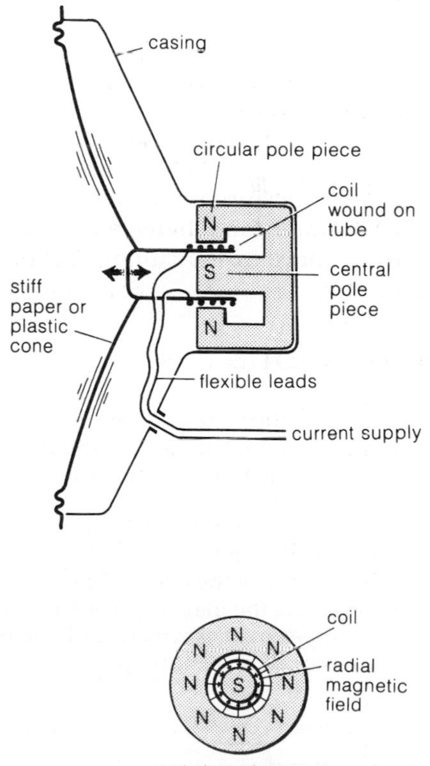

Figure 4b The varying current supply in the coil creates a varying force on it: the force makes the cone move

Force between current-carrying conductors

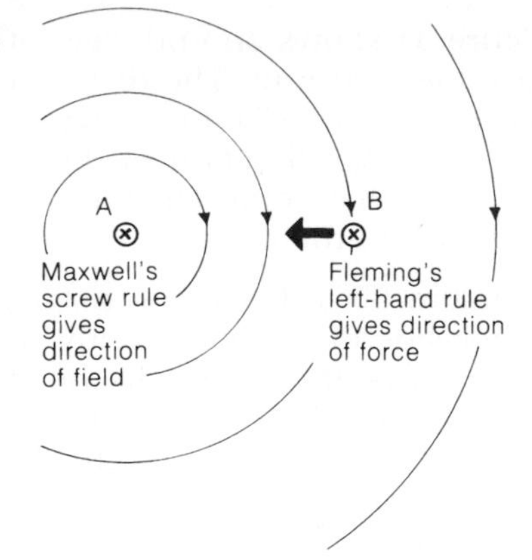

Figure 5 Force between two parallel wires (current going into the page); only the field due to wire A is shown

If currents flow in the same direction through two parallel wires as in figure 5, there is a weak attraction between them. Each wire experiences a force because each carries a current and each is in the magnetic field produced by the current in the other wire.

Figure 5 also shows how to work out the direction of the force on one of the wires, in this case wire B. Wire A produces a magnetic field whose direction is given by Maxwell's screw rule. Knowing this direction, and the direction of the current through B, the direction of the force on B can be found using Fleming's left-hand rule. The force is to the left. A similar argument shows that the force on wire A is to the right.

If the currents flow in opposite directions, they repel each other.

Defining and measuring the ampère

The ampère is defined in terms of the force between two current-carrying conductors:

The ampère is the current which, flowing through two infinitely long, parallel, straight, thin wires placed one metre apart in a vacuum, produces a force of 2×10^{-7} newton on each metre length of wire.

This is illustrated in figure 6.

In theory, the value of any unknown current can be found by passing the current through two long, parallel wires and measuring the force between them. In practice, this isn't very satisfactory, because the force is far too small to measure accurately. Accurate current measurements can however be made by measuring the rather larger force between two current-carrying coils. This is done with a complex instrument called a **current balance**.

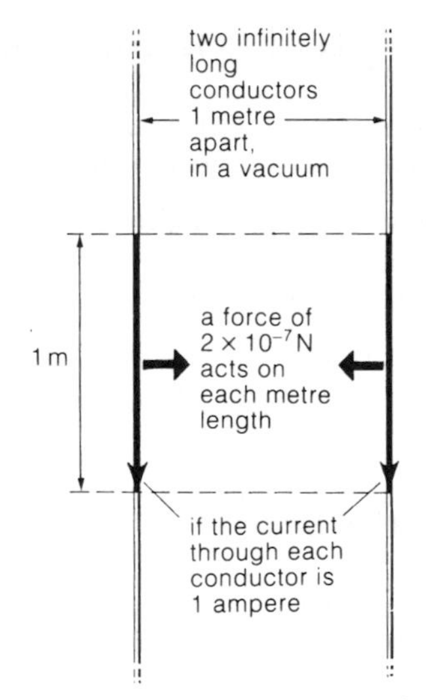

Figure 6 The definition of the ampère

Questions

1 Figure 7 shows the end view of a conductor in a magnetic field. The current direction is into the paper. Copy the diagram, showing the direction of the force acting on the wire.

2 If the wire AB in figure 8 is allowed to move, in which direction will it do so?
What would be the effect of a) reversing the current b) reversing the magnetic field c) reversing both?
Give three changes which could be made to increase the force on the conductor.

3 Why does the cone of a moving coil loudspeaker vibrate when an alternating current is passed through the loudspeaker coil?

4 Figure 9 shows two parallel wires. What would be the effect of passing a) upward currents through both wires b) an upward current through wire A and a downward current through wire B?

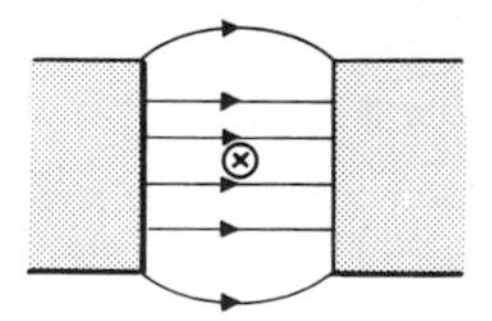

Figure 7

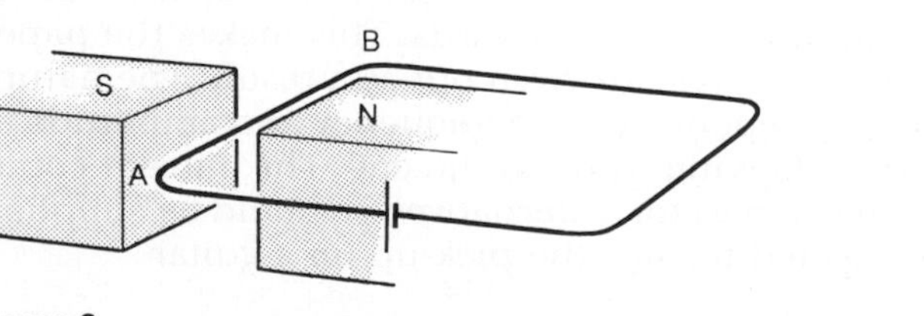

Figure 8

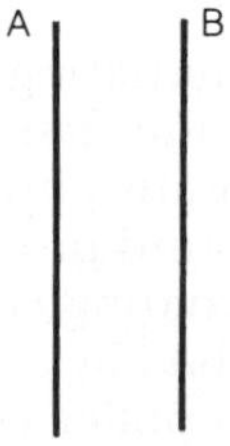

A B

Figure 9

7.6
Electric motors

A turning effect can be produced by passing a current through a coil which is in a magnetic field. This is the principle behind a common type of electric motor.

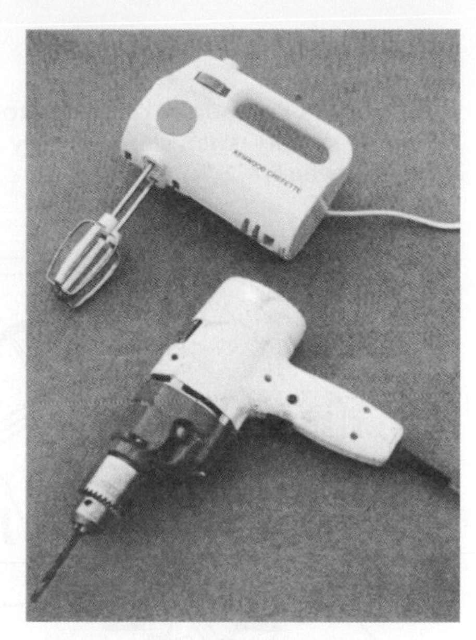

Familiar electric motors

Turning effect on a coil in a magnetic field

In figure 1a, the coil between the poles of the magnet consists of a single turn of wire. The current through the left-hand side of the coil flows in the opposite direction to the current through the right-hand side. From Fleming's left hand rule, it follows that an upward force is acting on one side of the coil, and a downward force on the other. Together, these forces form a *couple* explained on page 56 whose turning effect makes the coil move in a clockwise direction.

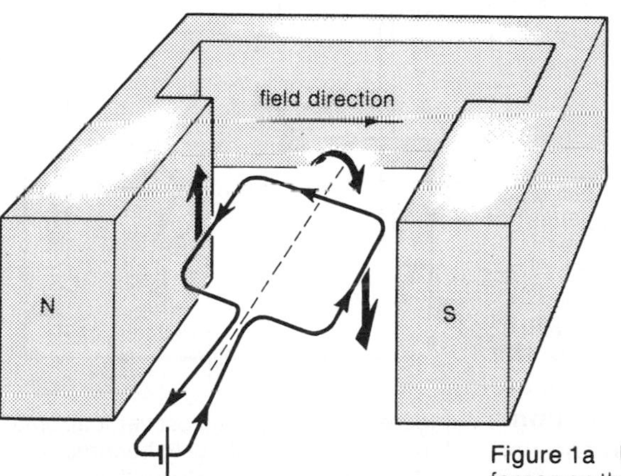

Figure 1a Upward and downward forces on the coil tend to make it rotate

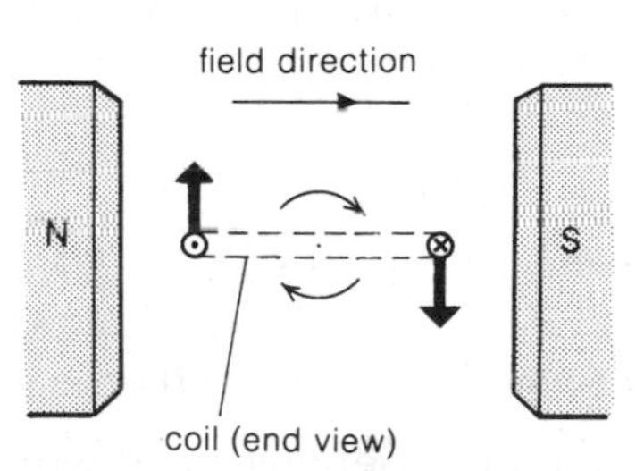

Figure 1b There is a maximum turning effect when the face of the coil is in line with the field

The turning effect on a coil is greatest when the coil lies with its face parallel to the magnetic field, as in figure 1b. In this position, the forces are furthest apart, so the couple has its highest value.

As the coil turns, the upward and downward forces move closer together, and the couple is reduced. When the face of the coil is at right angles to the field, as in figure 1c, the forces have no turning effect because both act in the same line. The couple is zero.

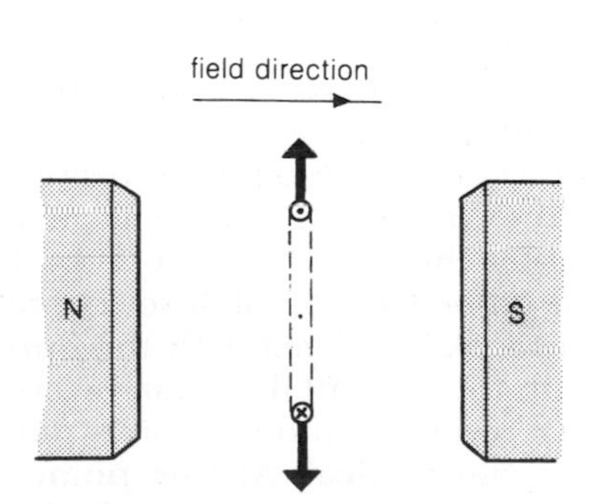

Figure 1c There is no turning effect when the face of the coil is at right angles to the field

The turning effect on a coil lying at any given angle to a magnetic field can be increased by:

1 increasing the current
2 increasing the number of turns in the coil
3 increasing the strength of the magnetic field
4 increasing the area of the coil – lengthening the coil increases the force on each side; widening the coil increases the turning effect of each force.

Forces also act on the near and far ends of the coil in each case. However, the turning effects of these forces cancel out.

A simple d.c. motor

D.C. motors are motors designed to operate from a direct ('one way') current supply such as a battery. A simple type of d.c. motor is shown in figure 2.

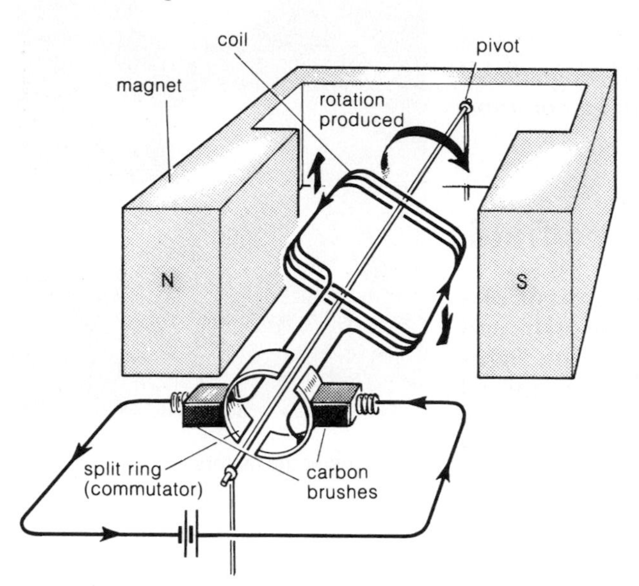

Figure 2 A simple d.c. motor. The commutator reverses the current direction every half-turn of the coil. So the current is always flowing "into the paper" on the right, and "out of the paper" on the left. The turning effect on the coil is thus always in the same direction

The motor contains a coil which is free to rotate between the poles of a U-shaped permanent magnet. The coil is connected to the battery leads by two carbon contacts known as **brushes**. These are pushed against the two halves of the copper split ring or **commutator** by two small springs. The commutator is fixed to the coil and turns round with it.

With a current passing through the coil as in figure 2, the coil turns in a clockwise direction. The forces acting are shown in figure 3a. If these forces were to continue to act in the directions shown, the coil would finally come to rest in a vertical position. However, as the turning coil overshoots the vertical, the commutator changes the direction of the current through it. You can see how by studying figures 3b and 3c, which show the coil just before and just after it passes through the vertical. As the coil passes from one position to the other, each brush loses contact with one half of the commutator and makes contact with the opposite half.

With the current direction reversed, the forces acting on the coil are also reversed, and the coil is pulled round another half turn until it is again vertical. At this point, the current and forces are again reversed and the coil is pulled round another half turn, and so on. In this way, the coil continues to rotate half a turn at a time in a clockwise direction. Note that, in this case, an *upward* force always acts on the half of the coil which is on the *left*, while a *downward* force always acts on the half of the coil which is on the *right* – hence the continuous clockwise turning action.

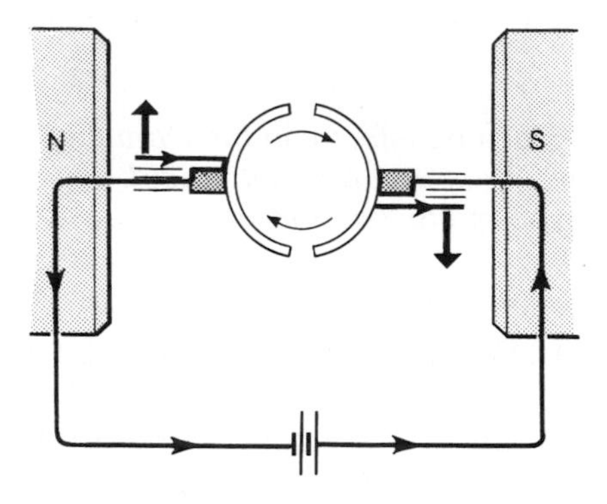

Figure 3a The coil horizontal: maximum turning effect

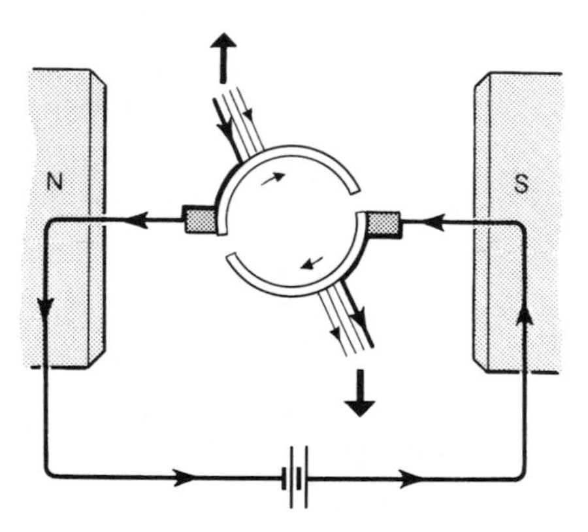

Figure 3b The coil almost vertical: small turning effect and about to become zero; but it moves round due to inertia

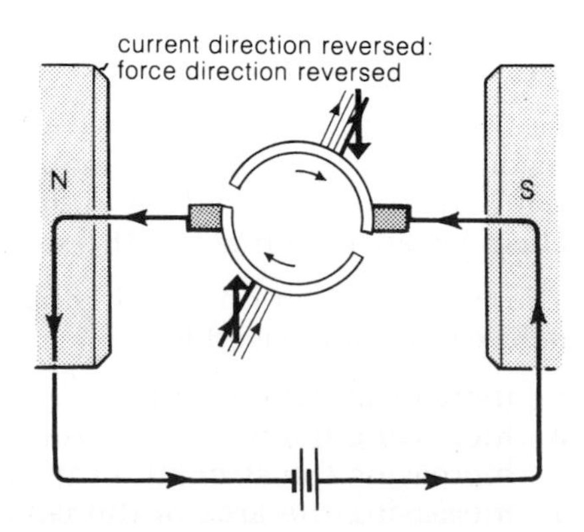

Figure 3c The coil just past vertical: the current direction has reversed, and so has the force direction. As it moves on towards the horizontal, the turning effect increases

Practical motors

The simple motor just described produces a low turning effect and is jerky in its action, particularly at low speeds. Practical motors give a much improved performance for the following reasons:

1 several coils may be used, each set at a different angle, and each connected to its own pair of commutator pieces or segments. This gives a greater turning effect and smoother running.

2 each coil consists of hundreds of turns of wire, and is wound on a soft-iron core called an **armature**. The armature becomes magnetized and increases the strength of the magnetic field.

3 The pole pieces are curved as shown in figure 4. With the armature, they produce a near-radial magnetic field. This means that each coil has its face parallel to the field for much of its rotation, so the couple acting is almost constant.

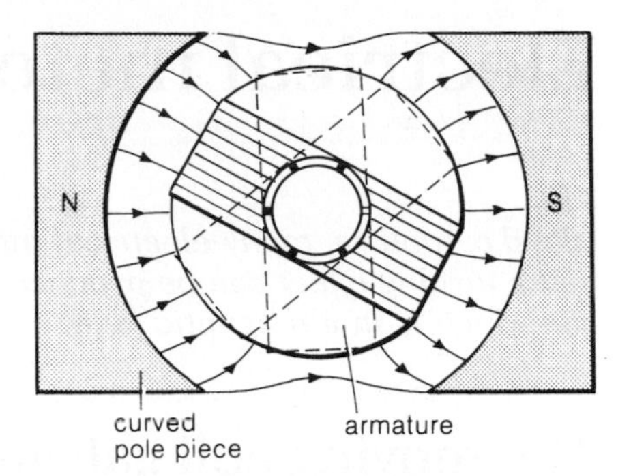

Figure 4 With curved pole-pieces, the field is almost radial: the face of the coil remains in line with the field

In some motors, the field is provided by an electromagnet rather than a permanent magnet (see figure 5). One advantage of this is that the motor will operate from an a.c. supply (see page 269). Although the current from the supply is constantly changing direction, the magnetic field changes direction to match it. As a result, the turning effect is always in the same direction, so the motor rotates in the normal way. Another advantage is compactness; an electromagnet can be stronger than a permanent magnet of the same size.

Several of the features mentioned above can be seen in the electric motor shown in figure 5.

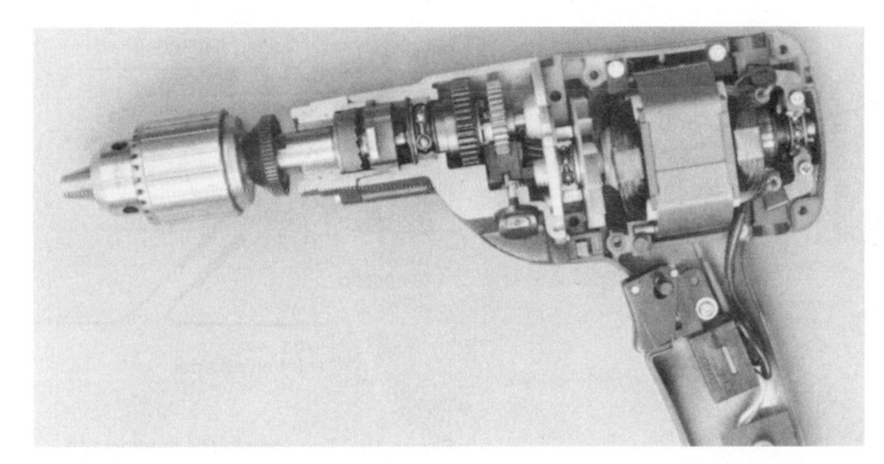

Figure 5 An electric drill: note the many commutator segments, and the electromagnetic pole pieces

Questions

1 Figure 6 shows an end view of a coil carrying a current in a magnetic field. Copy and complete the diagram to show the forces acting on the two sides of the coil. Redraw the diagram to show the position of the coil when the turning effect is a) at its greatest b) zero. Give three ways in which the turning effect on a coil in position a) could be increased.

2 In a d.c. motor, what is the function of the commutator? Give one advantage of using an electromagnet in a motor, rather than a permanent magnet.

Figure 6

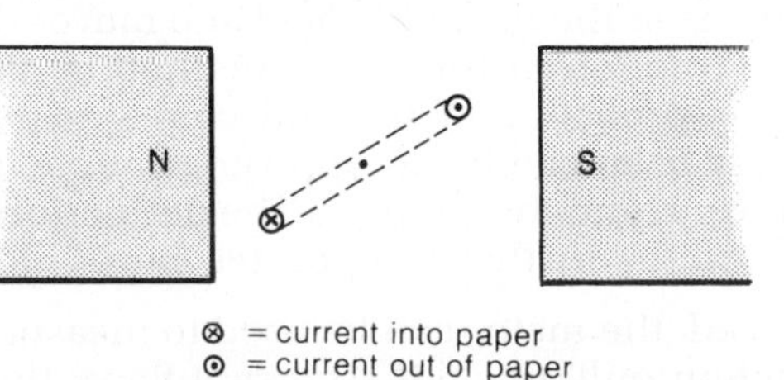

3 Give two ways in which the design of the motor in figure 2 could be improved.

7.7

Electrical meters

Like d.c. motors, many electrical meters are based on the principle that a turning effect can be produced by passing a current through a coil which is in a magnetic field.

The moving coil galvanometer

A **galvanometer** is any instrument that can detect small currents. If a galvanometer is calibrated so that the scale accurately indicates the size of the current in milliampère [mA], it is then called a **milliammeter**. Most milliammeters are of the 'moving coil' type; one is shown in figures 1a and 1b:

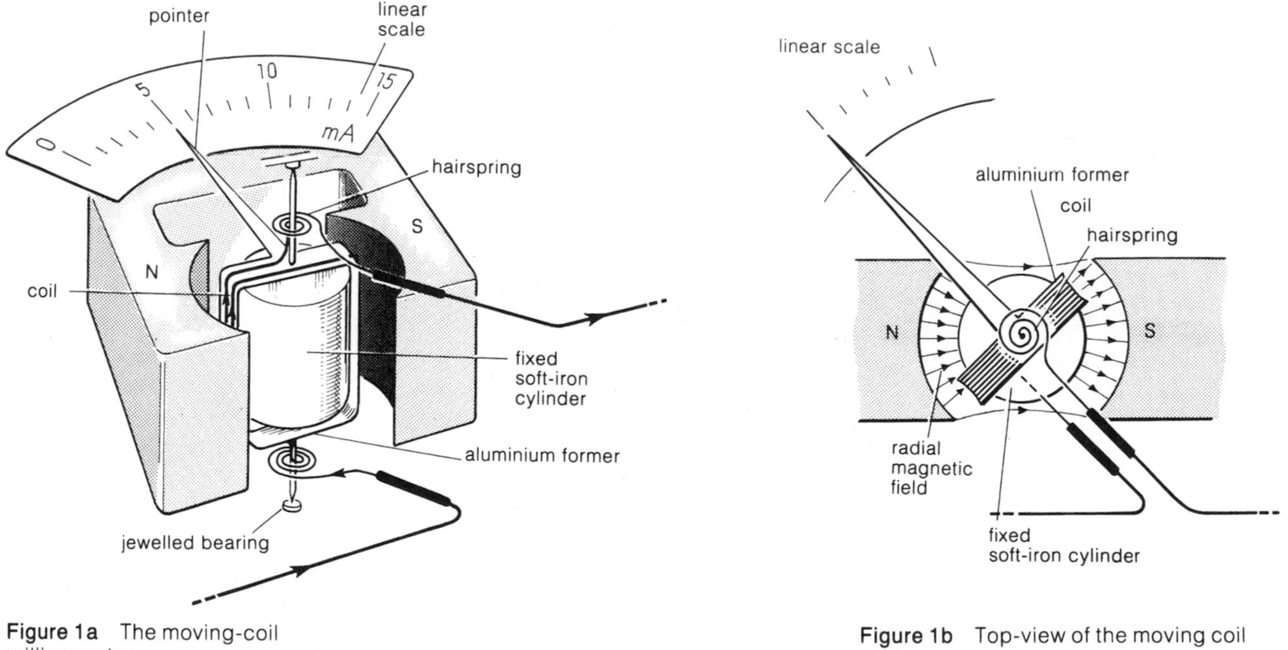

Figure 1a The moving-coil milliammeter

Figure 1b Top-view of the moving coil milliammeter

How the meter works When a current is passed through the coil, the coil turns in the magnetic field and moves the pointer across the scale. The rotation of the coil is resisted by the two spiral springs (called hairsprings), and the coil comes to rest when the couple turning it is balanced by an opposing couple from the springs. The higher the current, the greater is the **deflection** of the coil (the angle turned), and the further the pointer moves along the scale.

As described, the meter can be used to measure currents flowing in one direction only (d.c.). If a current flows through the coil in the opposite direction, it starts to move the pointer backwards. The movement is resisted by an end-stop.

Design features In figure 1, note the following:

1 The hairsprings spiral in opposite directions. When the coil is deflected, one spring uncoils while the other coils more tightly. They are arranged in this way so that the effect of expansion cancels out.

2 The coil is wound on a light aluminium frame or **former**. Like any closed loop of conducting material (see page 305), this resists motion through a magnetic field, and stops the coil travelling past its final position whenever it is deflected. The motion is said to be **damped**. With no damping effect, coil, springs and pointer would overshoot their final position and oscillate backwards and forwards for several seconds before coming to rest.

3 The meter has a **linear** scale, with numbers evenly spaced. In other words, the deflection of the pointer is directly proportional to the current. Two features make this possible:

first, the springs obey Hooke's law (explained on page 63) and produce an opposing couple which is directly proportional to their deflection.

Second, the coil turns in a radial magnetic field of constant strength. This is achieved by giving the magnet curved pole pieces with a cylinder of soft iron fixed between them.

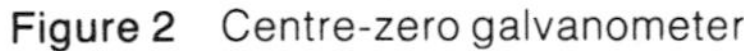
Figure 2 Centre-zero galvanometer

Full-scale deflection The current needed to move the pointer to the end of the scale is called the **full-scale deflection current**. The meter in figure 1 has a full-scale deflection current of 15 mA.

Sensitivity One meter is said to be more **sensitive** than another if it gives a greater pointer deflection for any given current. The sensitivity of the meter in figure 1 could be increased by

1 increasing the number of turns in the coil
2 increasing the area of the coil
3 using a stronger magnet
4 using weaker springs, i.e. springs which must be turned further to produce any given opposing couple.

Centre-zero galvanometers Some galvanometers have a zero in the centre of the scale, as shown in figure 2. The pointer deflects to the right or to the left depending on the current direction.

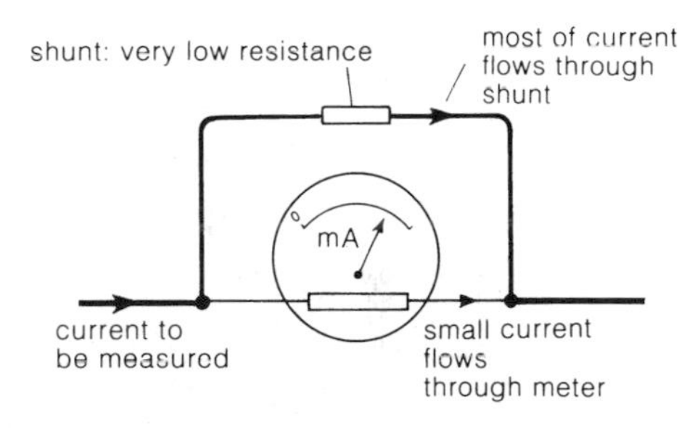

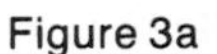
Figure 3a

Converting a milliammeter into an ammeter

A milliammeter can be adapted to measure much higher currents by connecting a resistor called a **shunt** across its terminals as shown in figure 3. Provided the shunt has a much lower resistance than the meter, nearly all of the current entering the combination bypasses the meter and flows through the shunt instead. A much larger total current is therefore needed to produce any given deflection on the meter. If the meter is given a new scale marked in amperes (A), the meter and shunt together form an ammeter.

Knowing the resistance of a milliammeter and its full-scale deflection current, it is possible to calculate the shunt resistance required to produce an ammeter measuring up to any given current value.

Figure 3b

Figures 3a, 3b An ammeter is formed by connecting a low resistance *shunt* across a milliammeter

Example *A milliammeter has a resistance of* 5 Ω *and a full-scale deflection current of* 10 mA. *The meter is to be adapted to measure currents up to* 1 A. *Calculate the shunt resistance required.*

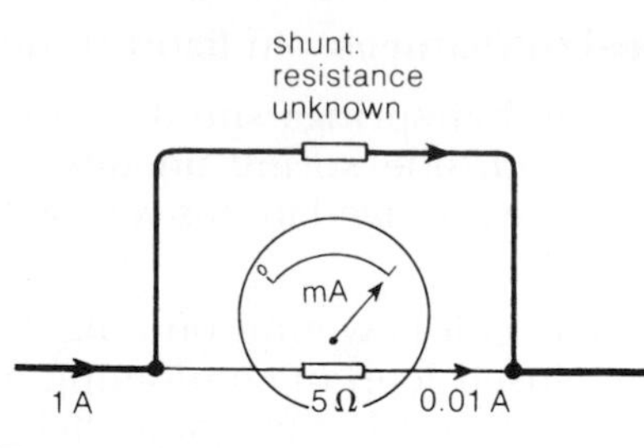

Figure 4

Figure 4 shows the meter with a suitable shunt connected, and the pointer fully deflected. A total current of 1 A is flowing through the meter and the shunt, but the current through the meter itself is only 10 mA [0.01 A].

The problem is solved in three stages:

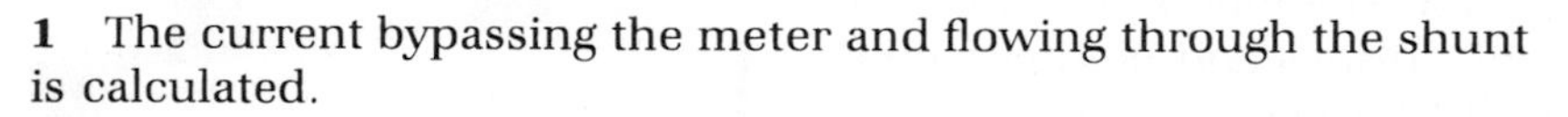

1 The current bypassing the meter and flowing through the shunt is calculated.

2 The p.d. across the meter, and therefore across the shunt, is calculated using $V = IR$.

3 Knowing the p.d. across the shunt and the current through it, the shunt resistance is calculated using $R = V/I$

Working through these stages:

1 $$\text{Current through shunt} = \frac{\text{total current through}}{\text{meter and shunt}} - \frac{\text{current through}}{\text{meter}}$$
$$= 1\,\text{A} - 0.01\,\text{A}$$
$$= 0.99\,\text{A}$$

2 $$\frac{\text{P.d.}}{\text{across meter}} = \frac{\text{current}}{\text{through meter}} \times \frac{\text{resistance}}{\text{of meter}} \quad [\text{using } V = IR]$$
$$= 0.01\,\text{A} \times 5\,\Omega$$
$= 0.05$ V, which is also the p.d. across the shunt.

3 $$\frac{\text{Resistance of}}{\text{shunt}} = \frac{\text{p.d. across shunt}}{\text{current through shunt}} \quad [\text{using } R = V/I]$$
$$= \frac{0.05\,\text{V}}{0.99\,\text{A}}$$
$$= 0.051\,\Omega$$

A shunt of resistance 0.051 Ω *is required.*

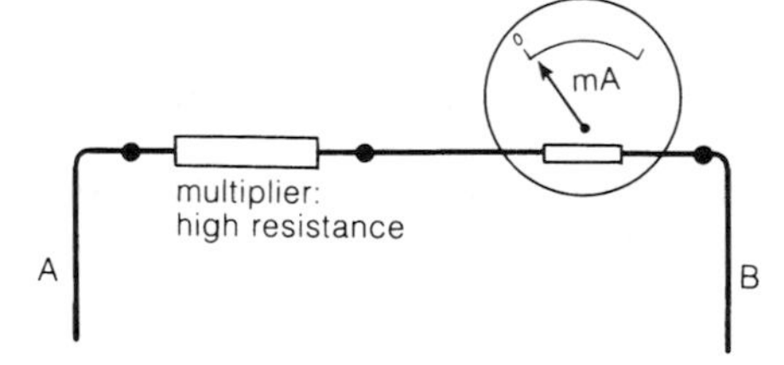

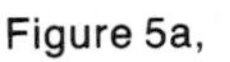

Figure 5a,

As the above answer suggests, shunts normally have a very low resistance. Most are made from short lengths of thick manganin resistance wire or strip.

The resistance of the shunt is very low, so the combined resistance of the meter and shunt is also very low. In practice, this is an advantage because the current through a circuit should, ideally, be the same whether an ammeter is included in the circuit or not.

Converting a milliammeter into a voltmeter

A milliammeter can be adapted for use as a voltmeter by placing a high resistance in series with it as shown in figure 5. The resistor added is sometimes called a **multiplier**.

Figure 5b

Figures 5a, 5b A voltmeter is formed by connecting a high resistance *multiplier* to a milliammeter

The moving coil voltmeter shown on page 248 is essentially a milliammeter with a multiplier added. Here however, the multiplier is housed inside the meter case, rather than added as an extra.

The effect of the milliammeter and multiplier in figure 5a is as follows:

If the two leads A and B are connected to, say, the terminals of a battery as shown in figure 6, the p.d. between A and B makes a small current flow through the multiplier and meter. The pointer of the meter is deflected as a result. The higher the p.d., the greater the current through the meter, and the greater the deflection of the pointer. If the meter is given a new scale marked in volts, the meter and multiplier together form a voltmeter.

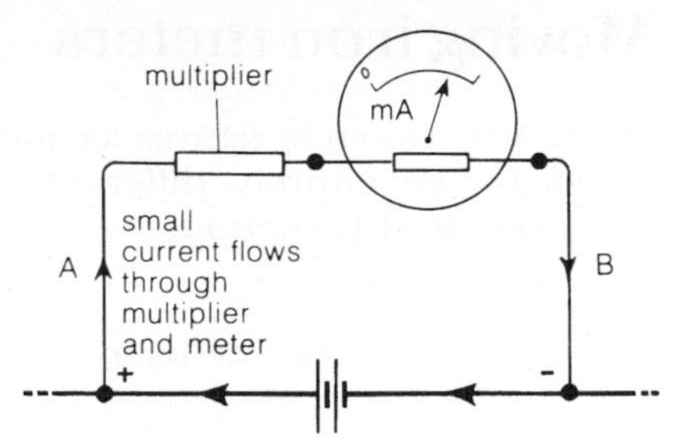

Figure 6 The high resistance multiplier only permits a small current to flow through the meter

Knowing the resistance of a milliammeter and its full-scale deflection current, it is possible to calculate the multiplier resistance required to produce a voltmeter measuring up to any given p.d.

Example *A milliammeter has a resistance of* 5 Ω *and a full-scale deflection current of* 10 mA. *The meter is to be adapted to measure p.d.s up to* 10 V. *Calculate the multiplier resistance required.*

Figure 7 shows the meter with a suitable multiplier connected in series. A p.d. of 10 V is being applied across the combination. The pointer is fully deflected, so there is a current of 10 mA [0.01 A] flowing through the meter and multiplier.

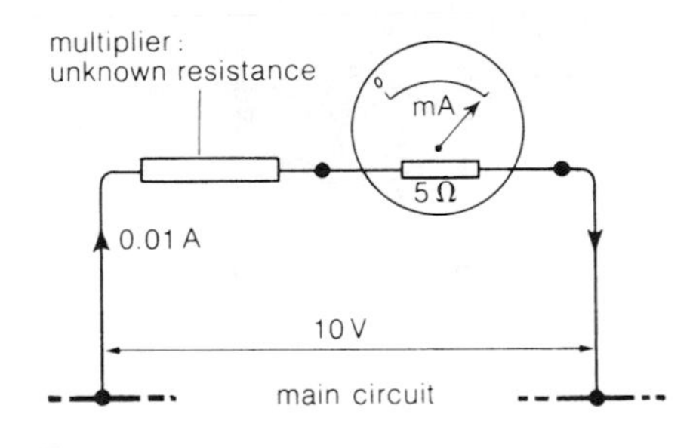

Figure 7

The problem is solved in two stages:

1 Knowing the p.d. across the meter and multiplier, and the current flowing through them, the total resistance of the combination is calculated using:

$$R = \frac{V}{I}$$

2 Knowing the resistance of the combination, and the resistance of the meter alone, the resistance of the multiplier can be found.

Working through these stages,

1 resistance of combination $= \dfrac{\text{p.d. across combination}}{\text{current}}$ [using $R = V/I$]

$$= \frac{10\,\text{V}}{0.01\,\text{A}}$$

$$= 1000\,\Omega$$

2 resistance of multiplier $=$ resistance of combination $-$ resistance of meter

$$= 1000\,\Omega - 5\,\Omega$$

$$= 995\,\Omega$$

A multiplier with a resistance of 995 Ω *is required.*

Note that the multiplier in this case converts a meter measuring currents up to 0.01 A into one measuring p.d.s up to 10 V. The numbers on the scale of the meter have been 'multiplied' by 1000.

Note also that the meter and multiplier form a voltmeter of total resistance 1000 Ω. As a voltmeter is connected *across* different components in a circuit, it is desirable that it should have as high a resistance as possible. The greater its resistance, the less current it draws from the main circuit.

Ideally, the presence of a voltmeter should have no effect at all on the current in a circuit.

Moving iron meters

The meter shown in figures 8a and 8b is a **moving-iron** ammeter. It works on an entirely different principle from the other meters described in this section:

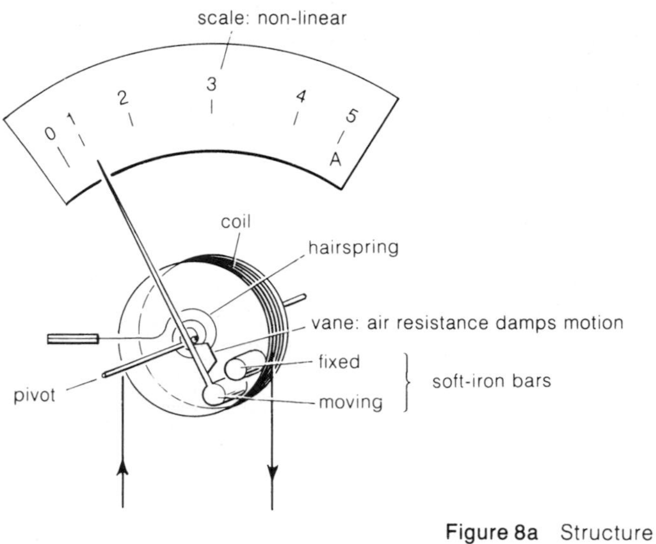

Figure 8a Structure of a moving iron ammeter

Figure 8b Moving iron ammeter

The meter consists of a fixed coil, inside which are two soft-iron bars, as shown in figure 8a. One of the bars is fixed, the other is attached to a pivoted pointer whose movement is opposed by a hairspring. When a current is passed through the coil, the iron bars become temporarily magnetized in the same direction and repel each other. The higher the current, the more the bars are pushed apart, and the further the pointer moves along the scale.

Moving-iron meters are not sensitive enough to measure small currents accurately, and they do not have a linear scale. They are however cheaper than moving-coil instruments, and more robust. Their other big advantage is that the iron bars repel whatever the direction of the current, so the meters can be used to measure alternating currents (a.c.).

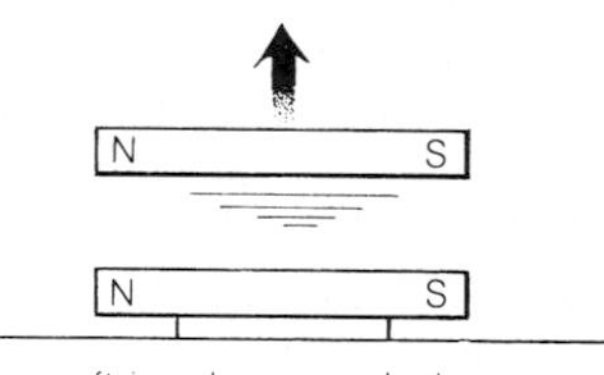

Figure 8c Soft-iron bars inside the meter repel one another as they become magnetized the same way

Questions

1 Explain why, in a moving-coil meter,
   a) two hairsprings are used
   b) the magnet has curved pole pieces
   c) the coil is wound on an aluminium former.

2 Give two ways in which the sensitivity of a moving-coil galvanometer could be increased.

3 Draw diagrams to illustrate how a milliammeter could be adapted to measure a) a current of several amperes b) a p.d. of several volts (values of components are not required).

4 What advantages does a moving-iron meter have over a moving-coil meter? What are its disadvantages?

5 A moving-coil meter has a resistance of 5 Ω and a full-scale deflection current of 100 mA. Calculate
   a) the shunt resistance required to convert the meter into an ammeter reading up to 1 A
   b) the multiplier resistance required to convert the meter into a voltmeter reading up to 5 V
   c) the maximum p.d. which could be measured with the meter if a multiplier of resistance 95 Ω is connected to it
   d) the maximum current which could be measured by the meter if a shunt of resistance $\frac{5}{19}$ Ω is connected across it.

7.8

Electromagnetic induction

By 1821, scientists had discovered that motion could be produced by passing a current through a conductor in a magnetic field. Ten years later, Michael Faraday found that the reverse was possible: a current could be produced by moving a conductor through a magnetic field.

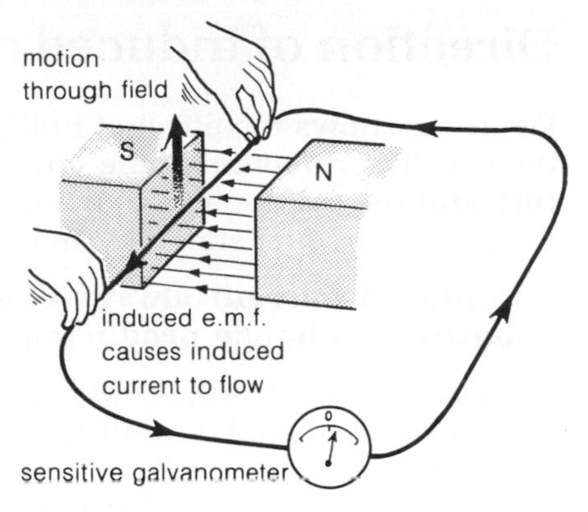

Figure 1a

Induced e.m.f. and current: Faraday's law

When a wire is moved across a magnetic field as in figure 1a, a small e.m.f. is produced in the wire. The effect is called **electromagnetic induction**; an e.m.f. has been **induced**. If the wire forms part of a circuit as in the diagram, the induced e.m.f. causes the current to flow. The current can be detected using a sensitive galvanometer.

The induced e.m.f. is present only while the wire is moving, and cutting through magnetic field lines. There is no induced e.m.f. if the wire is held still, nor if it is moved parallel to the field (figure 1b).

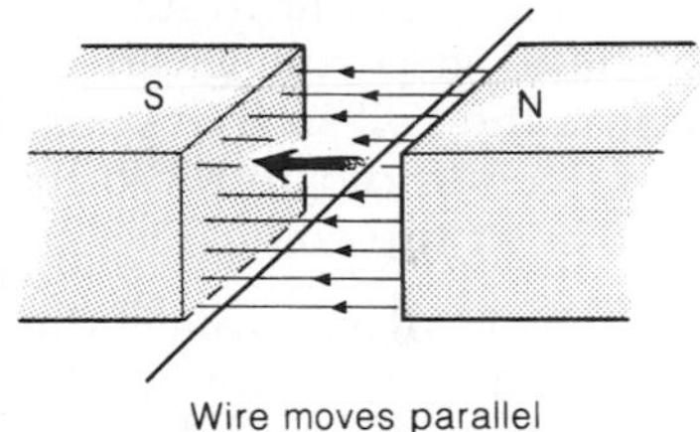

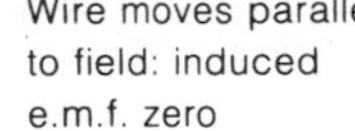

Wire moves parallel to field: induced e.m.f. zero

Figure 1b

The size of the induced e.m.f. can be increased by:

1 moving the wire at a higher speed
2 using a stronger magnet
3 increasing the length of wire in the magnetic field – for example, by looping the wire through the field several times as in figure 1c.

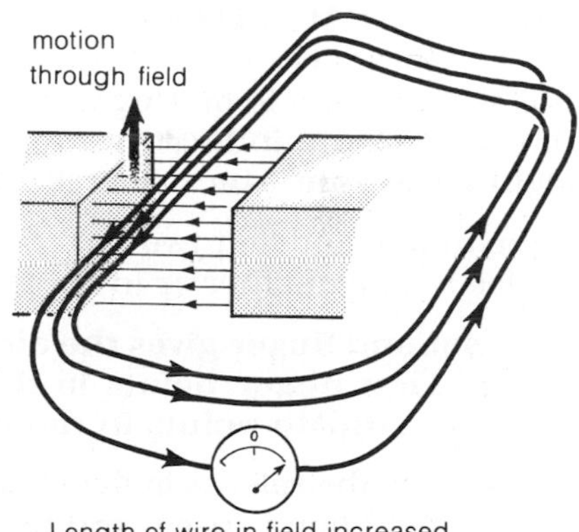

Length of wire in field increased
Induced e.m.f. and current increased

Figure 1c

The above results are summed up by **Faraday's law of electromagnetic induction**. This is sometimes stated as follows:

The e.m.f. induced in a conductor is directly proportional to the rate at which the conductor cuts through the magnetic field lines.

In applying this law, remember that magnetic field lines are used to represent the strength as well as the direction of a field. The stronger the field, the closer together the lines (see page 285).

Induced e.m.f. and current in a coil

If a bar magnet is moved into a coil as in figure 2a, an e.m.f. is induced in the coil. In this case it is the magnet which is being moved, but the motion still causes field lines to be cut by the wire.

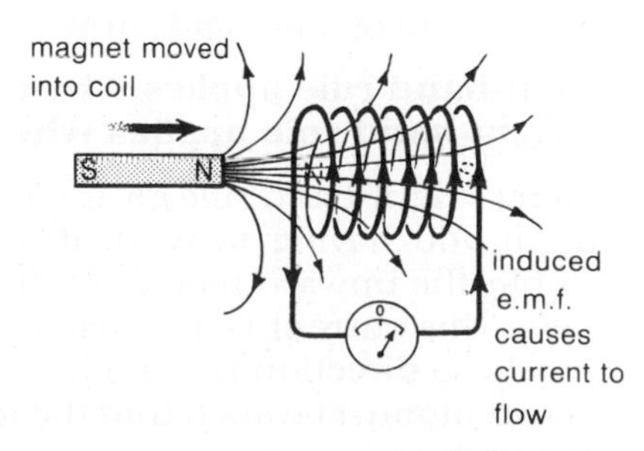

Figure 2a Moving the magnet rather than the wire still produces the same effect

The coil in figure 2a forms part of a complete circuit, so a current is also induced when the magnet is moved. The size of the induced e.m.f., and therefore of the induced current, can be increased by:

1 moving the magnet at a higher speed
2 using a stronger magnet
3 increasing the number of turns in the coil, as this increases the length of wire cutting through the magnetic field.

No e.m.f. or current is induced if the magnet is held still because no magnetic field lines are then being cut.

Direction of induced current: Lenz's law

Figure 2b shows the effect of pulling the magnet in figure 2a out of the coil. The direction of the induced current has now reversed. In this, and the previous case, the direction of the induced current is given by a law first stated by Heinrich Lenz in 1834:

An induced current always flows in a direction such that it opposes the change producing it.

In figure 2a for example, the motion of the magnet is opposed because the induced current turns the coil into a weak electromagnet with its N pole repelling the approaching N pole of the magnet. In figure 2b, the motion of the magnet is again opposed. This time, the coil attracts the N pole of the magnet as it is pulled away. Knowing which ends of the coil have become N and S in each case, the direction of the induced current can be worked out using the right-hand grip rule given on page 281. Imagine your right hand gripping the coil so that your thumb is pointing towards its N pole; your fingers indicate the direction of the (conventional) current in the turns of the coil.

Lenz's law is an example of the law of conservation of energy. A force must oppose any motion that produces a current because work must be done if electrons are to gain energy.

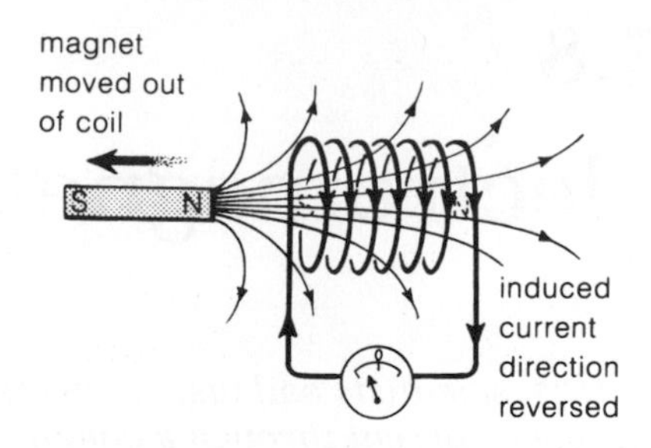

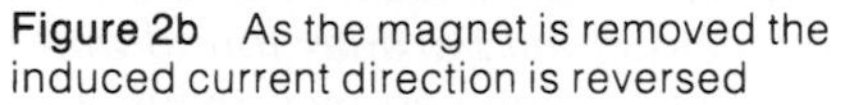
Figure 2b As the magnet is removed the induced current direction is reversed

Direction of induced current: Fleming's right-hand rule

If a straight wire is moving at right angles to a magnetic field, the direction of the induced current can be found using **Fleming's right-hand rule**. The rule is illustrated in figure 3a.

If the thumb and first two fingers of the right hand are held at right angles to one another as in the diagram:

the seCond finger gives the direction of the induced Current if the First finger points in the same direction as the Field and the thuMb points in the direction of the Motion.

In applying the rule, as in figure 3b, remember that the *conventional* current direction is used, and that the field direction is from the N pole of the magnet to the S pole.

It is important not to confuse Fleming's right-hand rule with the left-hand rule given on page 292. Both apply to conductors at right angles to a magnetic field. However,

the left-hand rule applies when a *current* causes *motion*
the right-hand rule applies when *motion* causes a *current*.

Fleming's right-hand rule gives you the result of applying Lenz's law without you having to work through every stage. In figure 4 for example, the upward motion of the wire causes an induced current to flow. This current is in a magnetic field, so there is a force acting on it whose direction is given by the left-hand rule. The direction of the current must be such that the force opposes the upward motion of the wire.

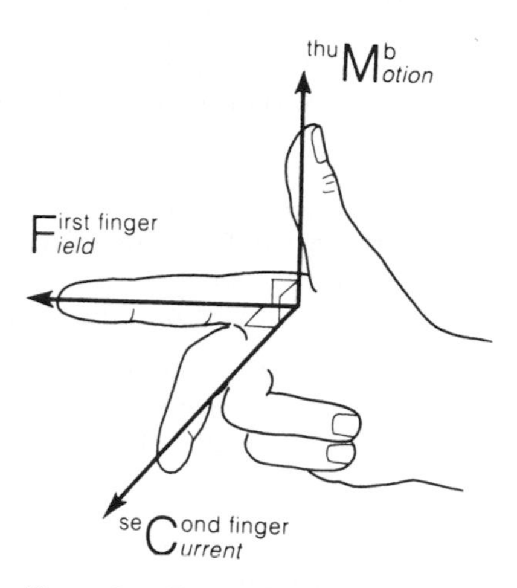

Figure 3a Fleming's right-hand rule gives the direction of an induced current.

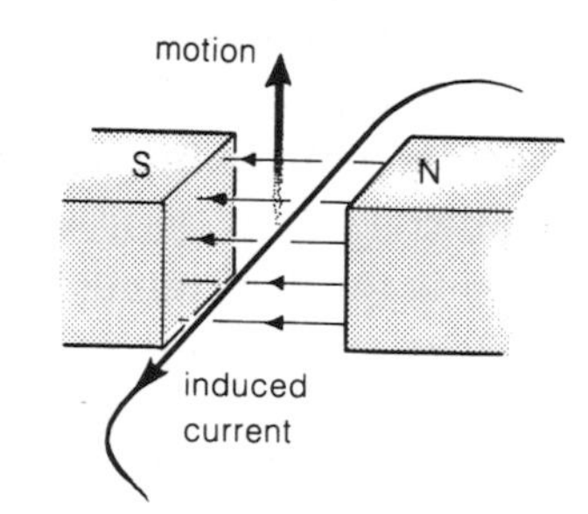

Figure 3b Applying Fleming's right-hand rule

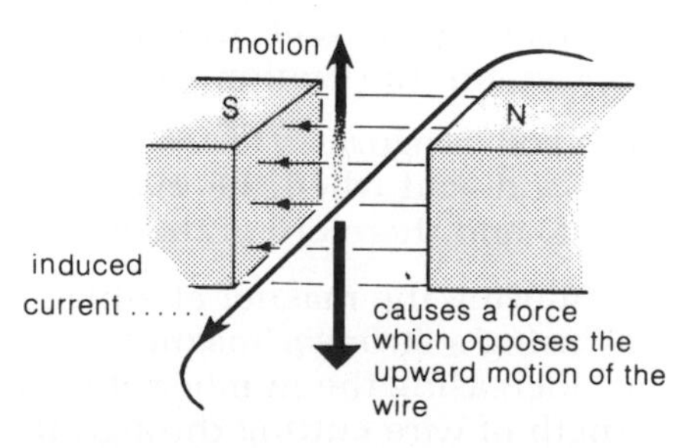

Figure 4 The direction of any induced current will always be such as to oppose the motion creating it

Eddy currents

If the aluminium disc in figure 5a is spun round fast, it may take quite a long time for frictional forces to bring it to rest. If however the disc is spun between the poles of a strong magnet as in figure 5b, it stops moving almost immediately. Being made of aluminium, the disc conducts well, and currents are induced in it as it moves through the magnetic field. These currents are known as **eddy currents**, and they have a magnetic effect which, by Lenz's law opposes the motion of the disc.

The coil of the moving-coil galvanometer is wound on an aluminium frame, as described on page 299. The damping effect on the coil's motion is caused by eddy currents which are induced in the frame as it turns through a magnetic field.

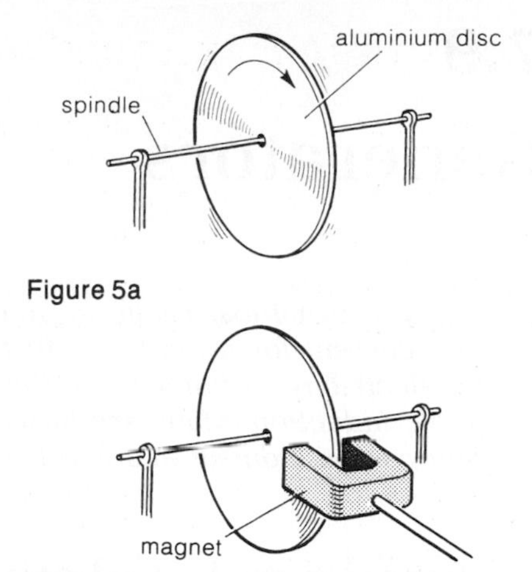

Figure 5a

Figure 5b

Figures 5a, 5b Induced eddy currents rapidly bring a spinning aluminium disc to a halt

Moving-coil microphone

When you speak into a moving-coil microphone as shown in figure 6, sound waves set the diaphragm vibrating. This moves a small coil backwards and forwards through the magnetic field from a cylindrical magnet, and a small alternating current is induced in the coil as a result. When amplified (made larger), the current can be used to drive a loudspeaker.

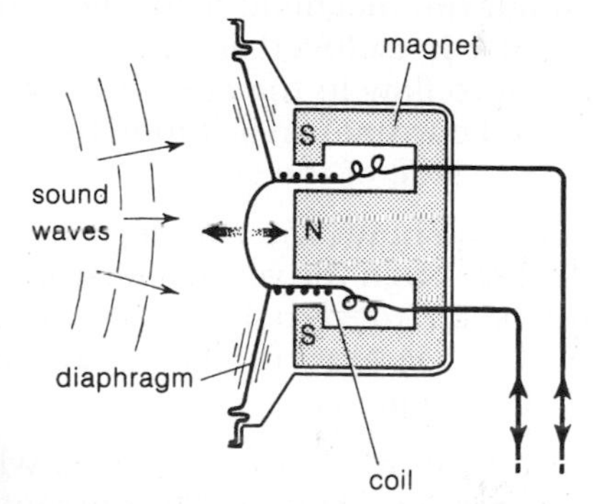

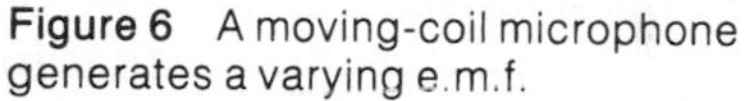

Figure 6 A moving-coil microphone generates a varying e.m.f.

An early moving-coil microphone in use at the B.B.C.

Questions

1 The wire in figure 7 is being moved downwards through the magnetic field. Copy the diagram and mark on the direction of the induced current.
What would be the effect of
a) moving the wire at a higher speed?
b) moving the wire upwards rather than downwards?
c) using a stronger magnet?
d) holding the wire still in the magnetic field?
e) moving the wire parallel to the magnetic field lines?

2 Figure 8 shows a bar magnet being pushed into a coil. Copy the diagram, work out which end of the coil becomes N and which S, then mark on the direction of the induced current.
Give three ways in which the size of the induced current could be increased.
Give three ways in which the direction of the induced current could be reversed.

3 Aluminium is usually regarded as a non-magnetic material. Yet a freely-spinning aluminium disc quickly stops moving if a magnet is brought close to it. Why?

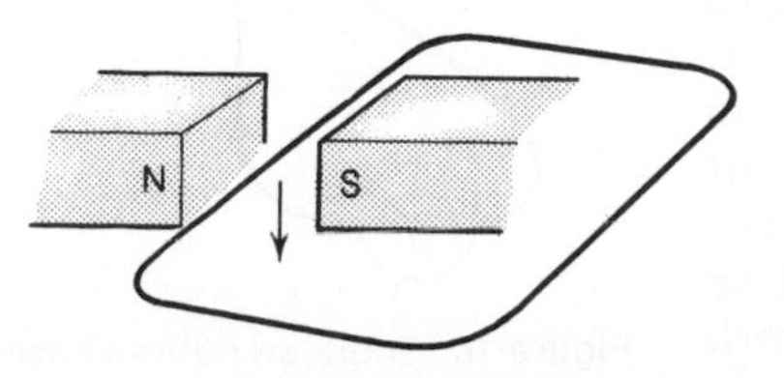

Figure 7

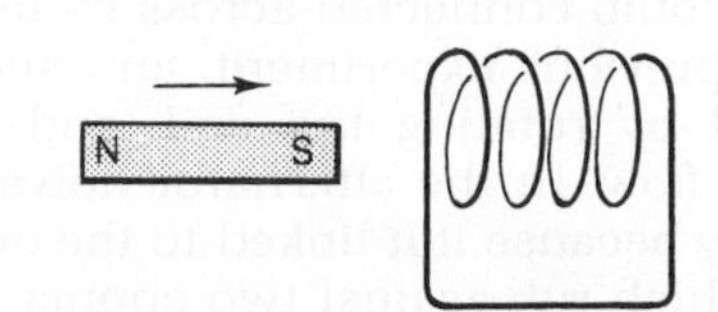

Figure 8

7.9

Generators

Generators (dynamos) range in size from the small dynamos that provide current for cycle lights to the huge alternators that supply mains electricity to homes and factories. All make use of electromagnetic induction. Many are based on the simple principle that a current can be induced in a coil by rotating it in a magnetic field.

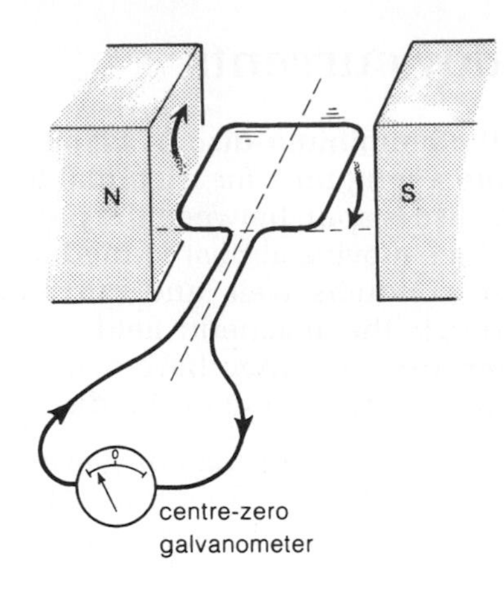

Figure 1a Induced current flows

Induced e.m.f. and current in a rotating coil

When the coil in figure 1a is rotated, one side moves upwards through the magnetic field, the other side moves downwards, and an e.m.f. is induced in the coil as a result. This e.m.f. causes a current to flow in the coil and outside circuit. The direction of this induced current can be found by applying Fleming's right-hand rule to either side of the coil.

Figure 1b shows the rotating coil one quarter of a turn later as it passes through the vertical position. The induced e.m.f. and current have fallen to zero because the two sides of the coil are now travelling parallel to the magnetic field and no field lines are therefore being cut.

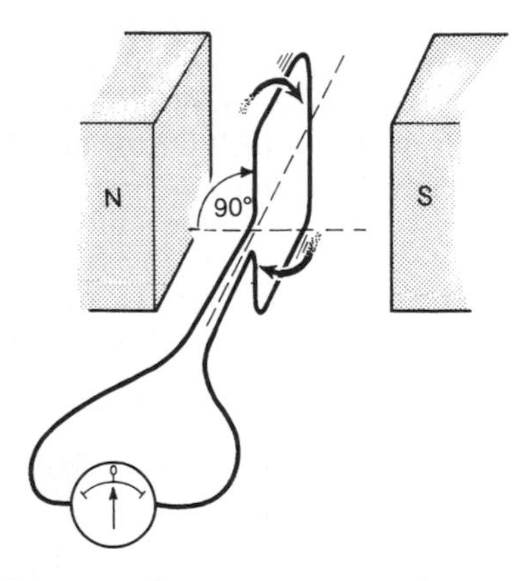

Figure 1b No induced current flows

Figure 1c shows the rotating coil when it has completed half a turn and is again horizontal. The current in the outside circuit now flows in the opposite direction to that in figure 1a, because the side of the coil which was moving upwards is now moving downwards, and vice versa.

A quarter turn later, the coil is vertical and the induced e.m.f. and current zero. As the coil completes one revolution, the situation is – apart from the twist in the wire – the same as in figure 1a.

If it were possible to go on turning the coil in this way, a current would continue to flow alternately forwards and backwards in the circuit. In other words, the rotating coil would provide a continuous supply of **alternating current (a.c.)**.

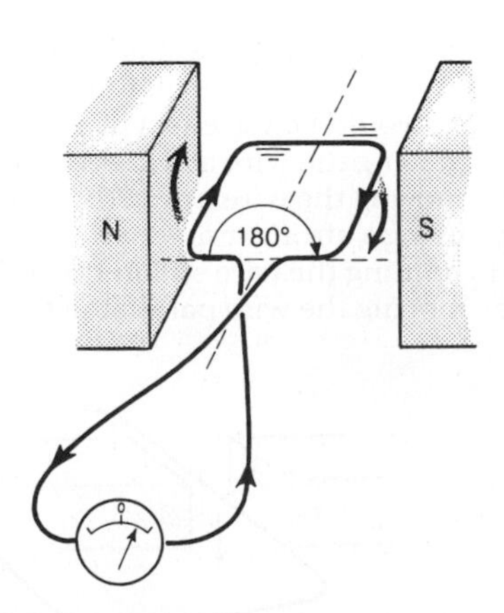

Figure 1c Induced current flows in the opposite direction

A simple a.c. generator

A.C. generators are known as alternators. A simple alternator is shown in figure 2; this one is being used to drive a current through a small light bulb connected across its terminals.

As in the previous experiment, an alternating e.m.f. is induced or **generated** by rotating the coil, and this causes an alternating current to flow. In the alternator however, the coil can be rotated indefinitely because it it linked to the outside circuit by two carbon brushes which rub against two copper **slip rings** fixed to the ends of the coil.

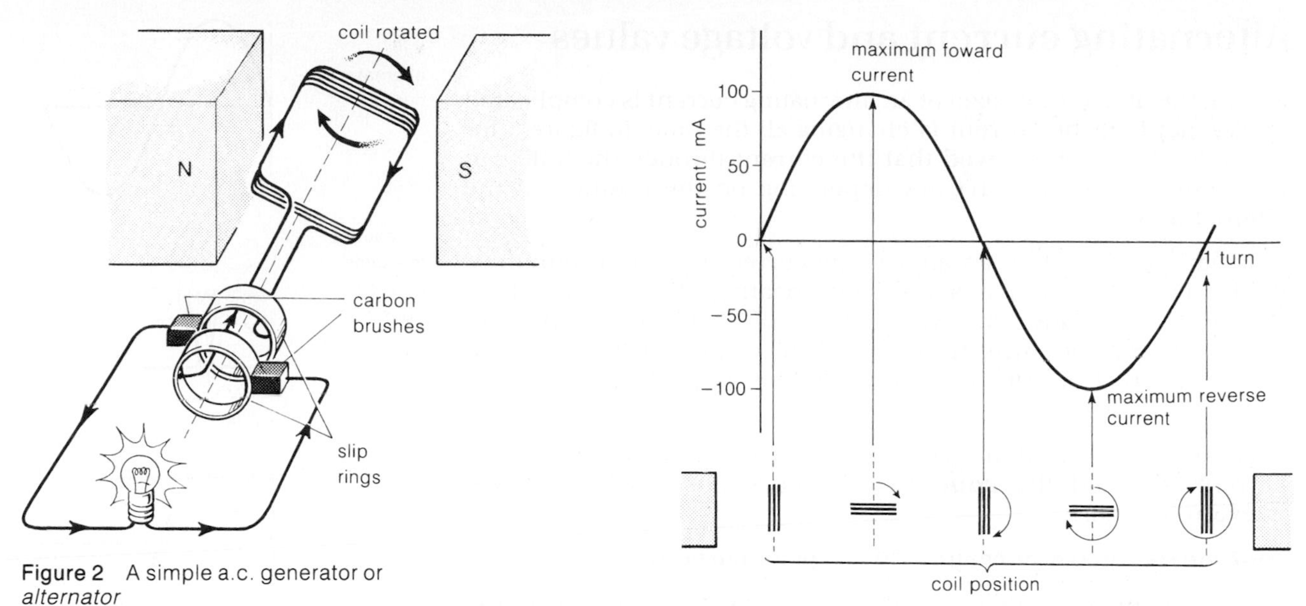

Figure 2 A simple a.c. generator or *alternator*

Figure 3 Alternator current output

The alternating current produced is illustrated by the graph in figure 3. This shows how the current through the bulb varies during one complete rotation of the coil, starting with the coil vertical. The current is taken as positive in one direction and negative in the other. It is greatest when the coil is horizontal and cutting through magnetic field lines most rapidly. The current is zero and on the point of changing direction when the coil is vertical.

The induced e.m.f., and therefore the current is increased by:

1 using a coil with more turns

2 using a stronger magnet

3 winding the coil on a soft-iron armature

4 rotating the coil at a higher speed.

Increasing the speed of rotation also increases the frequency of the alternating current generated.

Practical alternators

Unlike the simple alternator described on the previous page, practical alternators usually have a fixed set of coils arranged around a rotating electromagnet. Slip rings and brushes are still used, but these carry only the small fraction of the generated current that is used to power the electromagnet. Current supplied by the alternator does not therefore have to flow through moving contacts of any kind.

Most cars are now fitted with alternators like the one shown in figure 4. The alternator is turned by the engine. The a.c. it generates is turned into 'one way' direct current (d.c.) by devices called diodes (explained further in unit 8.4) and used to charge the battery.

In power stations, very much larger alternators are used to generate current for the a.c. mains.

Figure 4 This car alternator (shown bottom right) uses rotating electromagnets and fixed coils

Alternating current and voltage values

Stating the magnitude (size) of an alternating current is complicated by the fact that the current is changing all the time. In figure 3 for example, the graph showed that the current through the bulb can be anything from 0 to 100 mA depending on the position of the alternator coil.

Calculations show that an alternating current with a maximum or **peak** value of 100 mA has the same heating effect on the bulb filament as a steady 'one way' current of about 70 mA. In other words, in terms of power transfer, an alternating current is equivalent to a direct current of around 70% of its peak value. This is illustrated in figure 5.

For reasons which will not be examined here, this equivalent direct current is called the **root mean square current** – or **r.m.s. current** for short:

root mean square current = 70 % × peak current

A similar principle applies to alternating e.m.f.s and p.d.s – or **a.c. voltages** as they are commonly known:

root mean square voltage = 70% × peak voltage

Where a.c. voltage and current values are given, these are normally root mean square values unless otherwise stated. For example, the 240 V a.c. mains voltage is a root mean square value. Peak mains voltage is about 340 V.

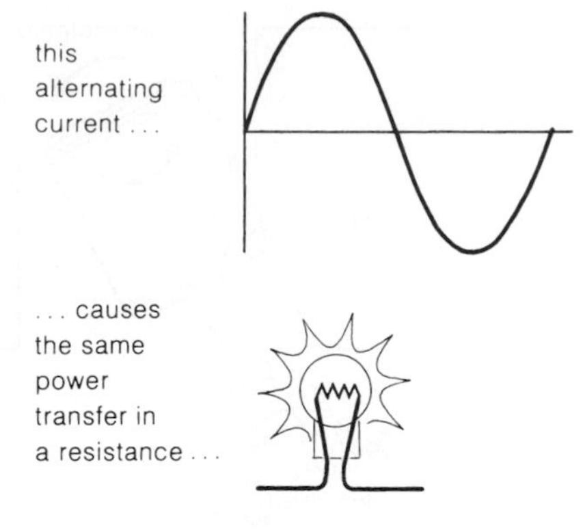

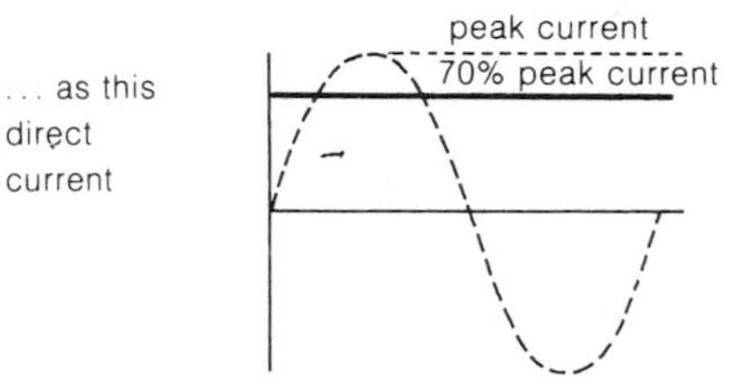

Figure 5 An alternating current is equivalent to a direct current of about 70% of its peak value

d.c. generators

A simple d.c. generator is shown in figure 6. Like the simple alternator, it contains a coil which is rotated in a magnetic field. Instead of two slip rings however, the generator has a single copper split-ring or commutator like that found on a simple d.c. motor. This reverses the connections with the outside circuit every time the coil passes through the vertical, so that the current in the outside circuit always flows in the same direction. This is illustrated by the graph in figure 7.

The action of a commutator is discussed in detail on page 296.

A d.c. generator, made in 1878

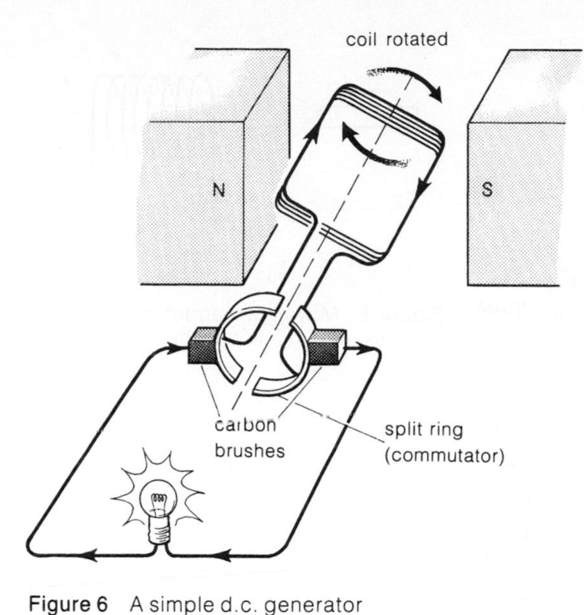

Figure 6 A simple d.c. generator

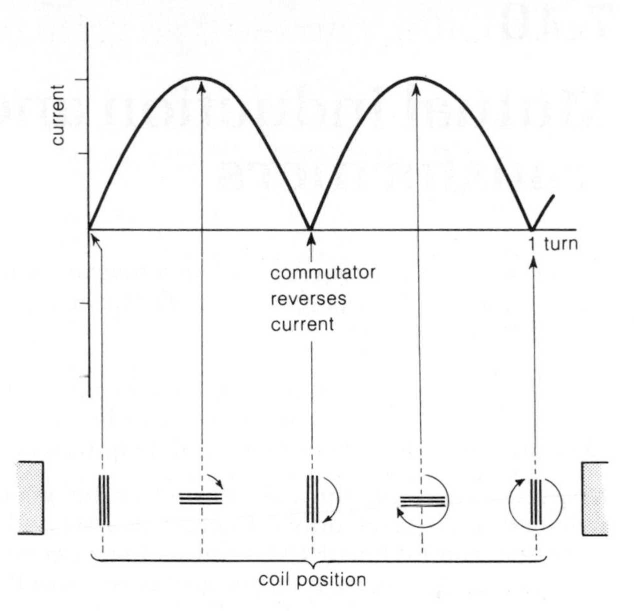

Figure 7 d.c. generator current output

The d.c. generator described above is in reality just a simple d.c. motor worked in reverse. Instead of a current being passed through the coil to produce rotation, the coil is rotated to produce a current.

Practical d.c. generators are similar in design to practical d.c. motors. Most have a soft-iron armature with several coils wound on it. Each coil is set at a different angle, and each is connected to its own pair of commutator segments. This gives a more or less constant d.c. output. In many cases, the coils are rotated between the curved pole-pieces of an electromagnet rather than a permanent magnet.

Questions

1 Figure 8 shows an end view of the coil in a simple alternator. The coil is being rotated in the direction indicated. The coil is connected through brushes and slip rings to an outside circuit.

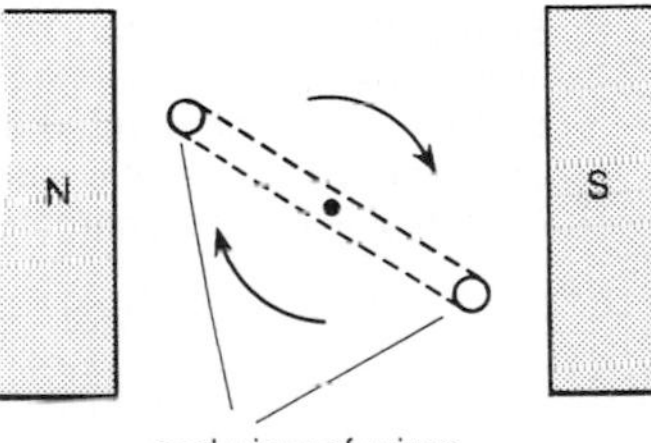

Figure 8

Copy the diagram and indicate the direction of the induced current.

Give three ways in which the current generated by the alternator could be increased.

The current varies as the coil rotates. Redraw the diagram to show the position of the coil when the current is a) a maximum b) zero.

2 What is the major difference between the simple alternator in figure 2 and most practical alternators?

3 In what way is the design of a simple d.c. generator different from that of a simple alternator?

4 The alternating current passing through a resistor has an r.m.s. value of 7A. The r.m.s. voltage across the reistor is 14V. Give approximate values for the peak voltage and current. Calculate the power dissipated in the resistor.

7.10

Mutual induction and transformers

It isn't necessary to move a coil or a magnet in order to induce an e.m.f. in the coil. Changing the current through another coil can have exactly the same effect.

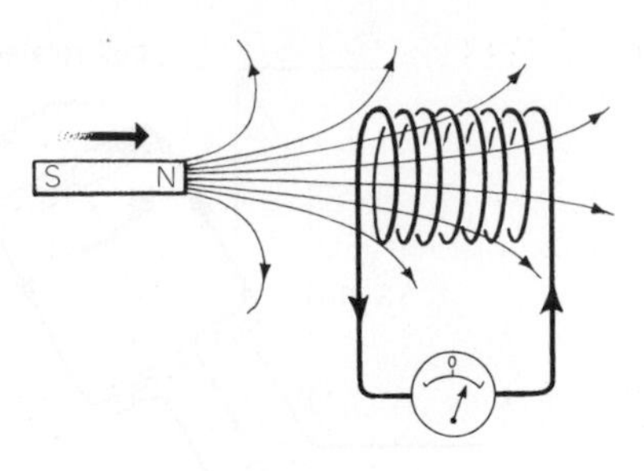

Figure 1 Moving a magnet into a coil induces an e.m.f. in the coil

An e.m.f. can be induced in a coil by moving a magnet in or out of it as in figure 1. But the same effect can be achieved by placing an electromagnet close to the coil and simply switching it on or off.

When the current through the electromagnet in figure 2a is switched on, it takes only a fraction of a second for it to reach its maximum value. During this time, as the magnetic field rises to full strength, field lines rush closer and closer together and are cut by the coil as they sweep past it. This induces an e.m.f. in the coil, and a short burst of induced current flows through the galvanometer. The effect is the same as if a magnet had been pushed rapidly into the coil.

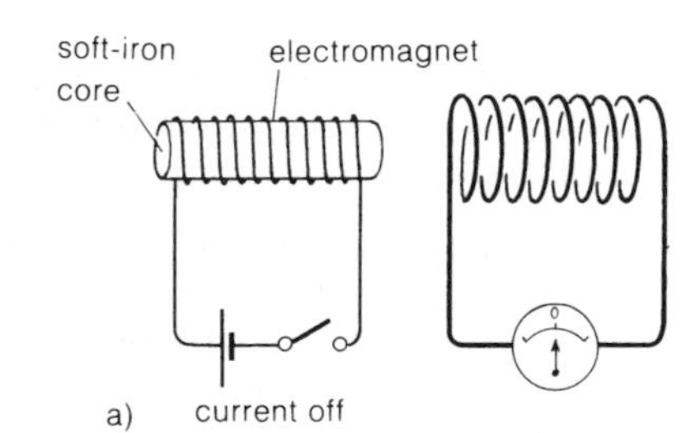

Figure 2a With no current in the electromagnet, no e.m.f. is generated

When the current is steady and the magnetic field unchanging as in figure 2b, no e.m.f. is induced in the coil. If the switch is now opened, the current through the electromagnet falls to zero, and field lines rush outwards as the magnetic field rapidly weakens. An e.m.f. is again induced in the coil, but this time the burst of induced current is in the opposite direction. The effect is the same as if a magnet had been pulled rapidly out of the coil.

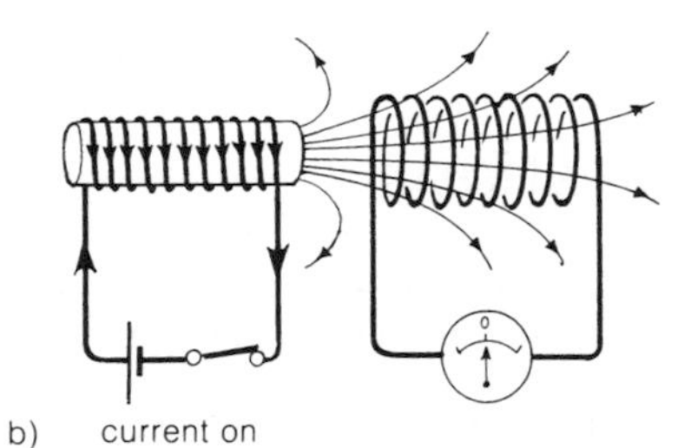

Figure 2b In the instant that the current is switched on and the magnetic field builds up, a large e.m.f. is induced; but once the magnetic field is steady, no current flows in the solenoid

The experiment still works if the soft-iron core is removed from the electromagnet, though the induced e.m.f. is less than before because the magnetic field at any instant is much weaker. On the other hand, the induced e.m.f. is greatly increased if both coils are wound on a soft-iron core forming a complete loop as in figure 3. The **primary** coil acts as the electromagnet, and the core guides the magnetic field so that field lines pass through the **secondary** coil.

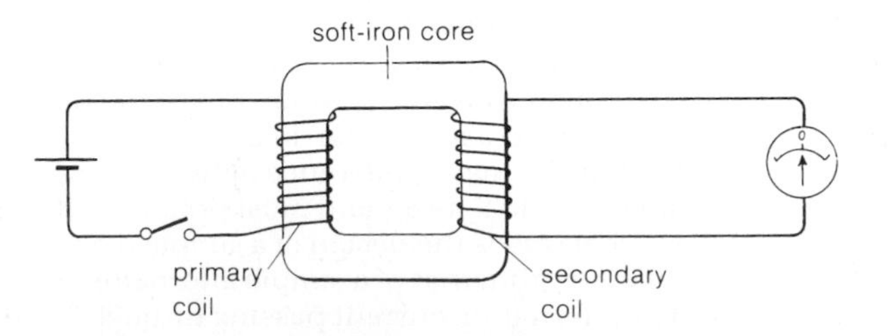

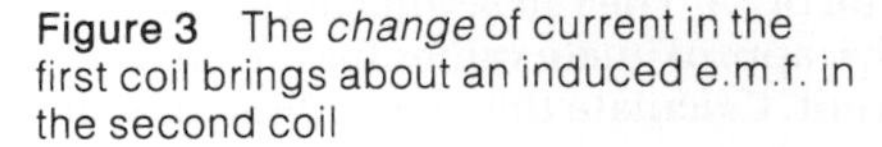

Figure 3 The *change* of current in the first coil brings about an induced e.m.f. in the second coil

In each of the above experiments, it is a *change* in the current through one coil which brings about an induced e.m.f. in the other. The effect is called **mutual induction**. Mutual induction has many applications, including the heart pacemaker shown in figure 4, the car ignition system described at the end of the section, and the transformer described in the next section.

a.c. mutual induction: the transformer

If the current through the primary in figure 3 is switched on and off repeatedly, induced current flows alternately backwards and forwards through the secondary. Much the same effect can be produced more simply by passing alternating current (a.c.) through the primary. Unlike the d.c. used in the previous experiments, a.c. is changing all the time.

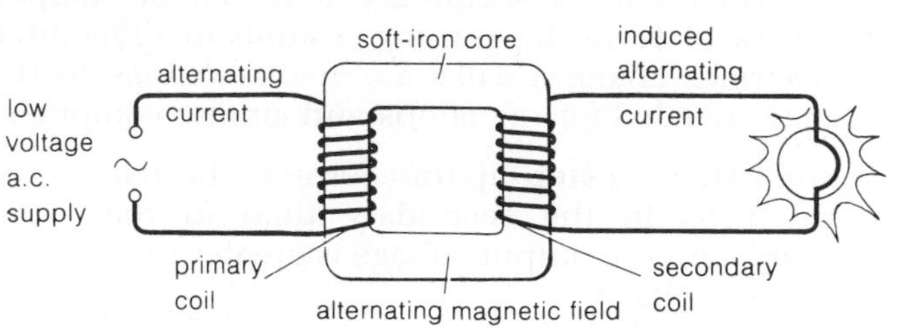

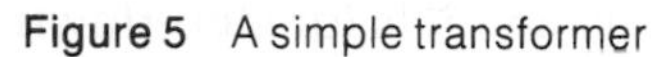
Figure 5 A simple transformer

The primary in figure 5 is connected to a low voltage a.c. supply. The secondary is connected to a small bulb. As current flows backwards and forwards through the primary, it sets up an alternating magnetic field in the core. This induces an alternating e.m.f. and current of the same frequency in the secondary, so the bulb lights up.

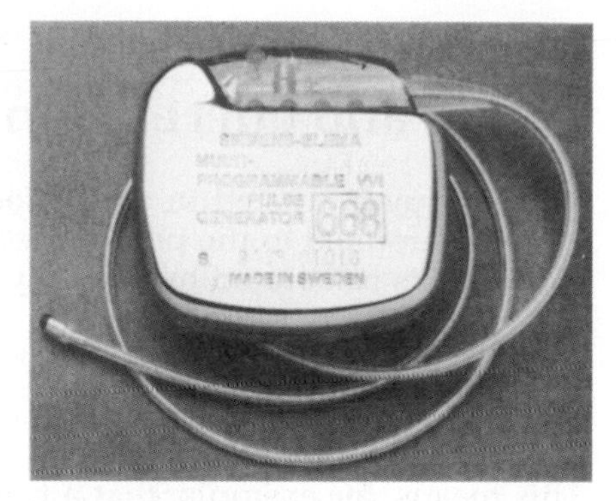

Figure 4 The heart pacemaker: pulses of current through a primary circuit in the pacemaker unit induce pulses of current in a secondary circuit in the patient's chest. These are used to trigger heart beats

The primary and secondary coils and core together form a **transformer**. When a transformer is connected to an a.c. supply, electrical energy is continuously transferred from the primary to the secondary circuit.

Step-down and step-up transformers

By choosing suitable numbers of turns for the coils of a transformer, it is possible to make the alternating e.m.f. (a.c. voltage) in the secondary different from the voltage across the primary. In other words, a transformer can be used to change or **transform** an a.c. voltage from one value to another.

A simple equation links the a.c. voltages across the primary and secondary coils and the number of turns on each:

$$\frac{\textbf{secondary voltage}}{\textbf{primary voltage}} = \frac{\textbf{number of secondary turns}}{\textbf{number of primary turns}}$$

In symbols, $$\frac{V_2}{V_1} = \frac{N_2}{N_1}$$

Strictly speaking, the equation applies only where the coils have negligible resistance and all field lines produced by the primary pass through the secondary. However, the equation is a good enough approximation for most practical purposes.

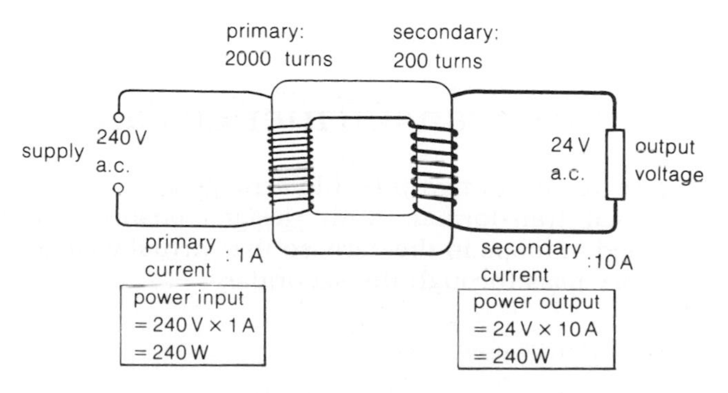

Figure 6 A step-down transformer produces a reduced voltage in the secondary

Figure 6 illustrates how the equation applies in one particular case. Here, the secondary has only one tenth of the turns of the primary, so the secondary voltage is only one tenth of the primary voltage:

$$\frac{24\,\text{V}}{240\,\text{V}} = \frac{200\text{ turns}}{2000\text{ turns}}$$

The transformer in this case is called a **step-down** transformer. It has fewer turns in the secondary than in the primary, and the secondary or **output** voltage is less than the primary or **input** voltage. Most laboratory power supplies contain a step-down transformer with a secondary that can be 'tapped' at different numbers of turns by turning a knob to different positions. The transformer changes 240 V a.c. mains voltage to the much lower voltages needed for ray lamps and similar equipment.

Figure 7 shows a **step-up** transformer. This has more turns in the secondary than in the primary, and the output voltage is greater than the input voltage.

The circuit symbols for step-down and step-up transformers are given in figure 8.

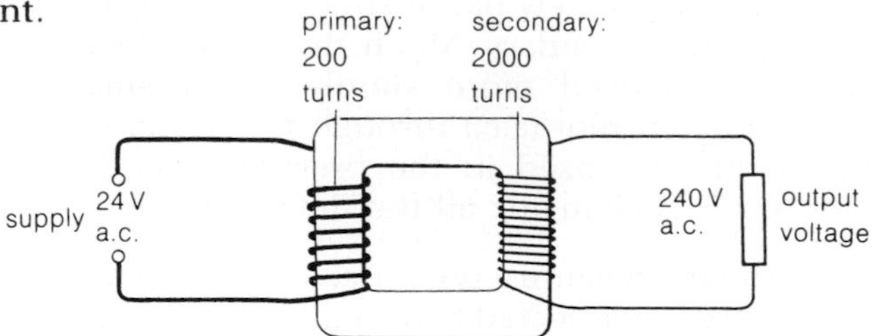

Figure 7 A step-up transformer produces an increased voltage in the secondary

Power and currents in a transformer

If no energy is wasted in a transformer, all the power (energy per second) supplied to the primary will be delivered by the secondary. As power is calculated by multiplying voltage by current,

$$\text{primary voltage} \times \text{primary current} = \text{secondary voltage} \times \text{secondary current}$$

In symbols: $V_1 I_1 = V_2 I_2$

This means, for example, that a transformer which *reduces* voltage will *increase* current in the same proportion so that the power remains the same. An example of this was given in figure 6. Here, the voltage is reduced from 240 V to 24 V, the current is increased from 1 A to 10 A, and the power output, (24 × 10)W is the same as the power input, (240 × 1)W.

Transformers therefore transform current as well as voltage. Use is made of this fact in the soldering 'gun' in figure 9. The gun contains a step-down transformer with only a few turns in the secondary. The primary draws a small current from the mains, but a very large current flows in the secondary and the tiny heating element connected to it.

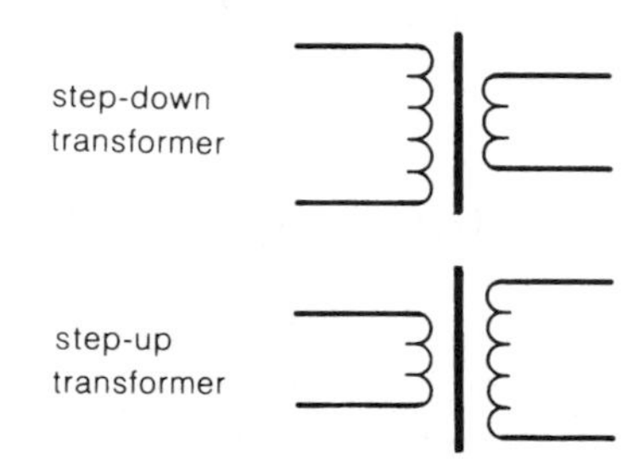

Figure 8 Circuit symbols

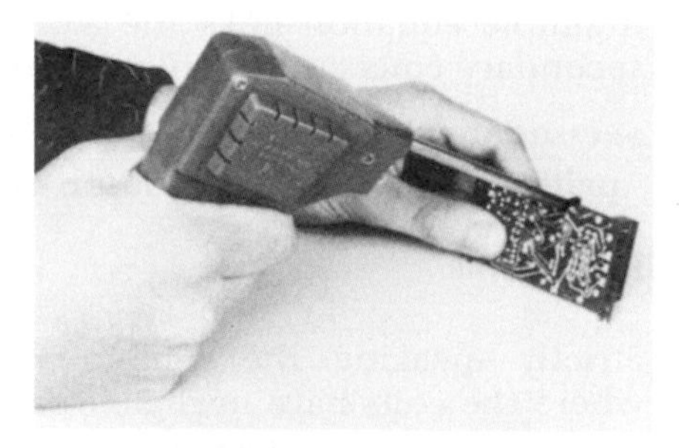

Figure 9 This soldering "gun" contains a step-down transformer

Practical transformer design

Figure 10 shows two possible arrangements of coils and core in a practical transformer. Both designs ensure that field lines are trapped in loops in the core, so that virtually all field lines from the primary pass through the secondary.

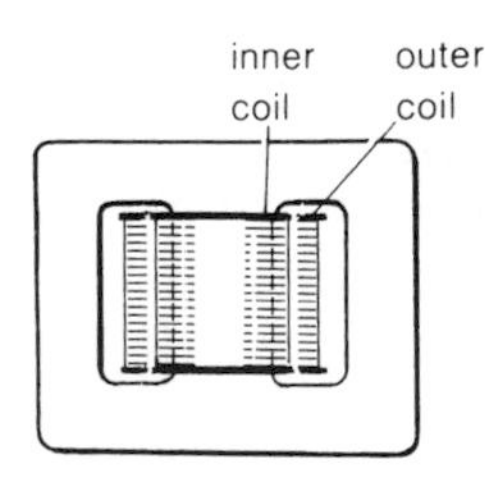

Figure 10 Practical transformer design

No practical transformer is 100% efficient, and some of the electrical energy supplied to the primary is lost as thermal energy (heat). Energy losses are caused by several factors:

1 The resistance of the coils. As the coils have resistance, they give off heat when a current flows through. Coil resistance is kept as low as possible by making the coils from thick copper wire.

2 Magnetization and demagnetization of the core. Work has to be done to alter sizes and direction of domains (see page 282) and heat is released in the process. These energy losses are reduced by making the core from a soft magnetic material such as iron which magnetizes and demagnetizes easily.

3 Eddy currents in the core. These occur because the core is itself a conductor in a changing magnetic field. They are reduced by making the core from thin sheets of soft iron which are insulated from each other so that their total resistance is high. A core made in this way is said to be **laminated** (layered).

Well-designed large transformers can have efficiencies as high as 99%. In other words, the power output is 99% of the power input, and only 1% of the energy supplied every second is wasted as heat.

Michael Faraday

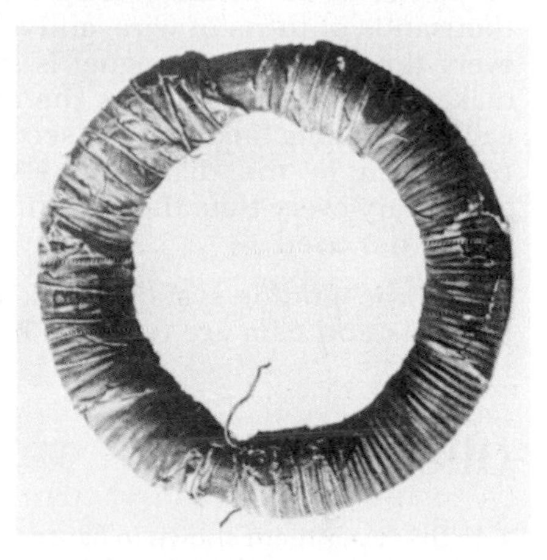

The first transformer – made by Faraday in 1831

Transformer calculations

Example *The primary of the transformer in figure 11 is connected to an a.c. supply, and the secondary to a bulb rated at 12 V 36 W. There are 4000 turns on the primary and 200 on the secondary. If the bulb is delivering its rated power, find*
a) the supply voltage b) the current through the primary c) the power taken from the supply.
Assume that the efficiency of the transformer is 100%.

Part a) is solved by applying the transformer equation:

$$\frac{V_2}{V_1} = \frac{N_2}{N_1}$$

$$\frac{12\,\text{V}}{\text{supply voltage}} = \frac{200}{4000}$$

rearranging: $\text{supply voltage} = \dfrac{4000}{200} \times 12\,\text{V} = 240\,\text{V}$

the supply voltage is 240 V

Part b) is solved using the power equation: $V_1 I_1 = V_2 I_2$,

where $V_2 I_2$ is already known to be 36 W.

$\therefore$ 240 V × primary current = 36 W

rearranging, $\text{primary current} = \dfrac{36\,\text{W}}{240\,\text{V}} = 0.15\,\text{A}$

the current through the primary is 0.15 A

Part c) doesn't need any calculation. If the tranformer is 100% efficient, the power taken from the supply is the same as the power delivered to the bulb:

the power taken from the supply is 36 W

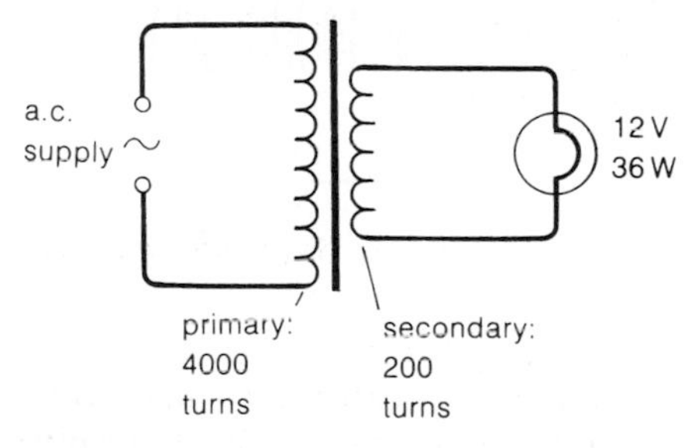

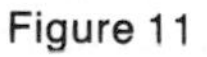

Figure 11

Coil ignition

Mutual induction is used in many cars to produce high voltage sparks in the plugs. A simplified version of one type of coil ignition system is shown in figure 12.

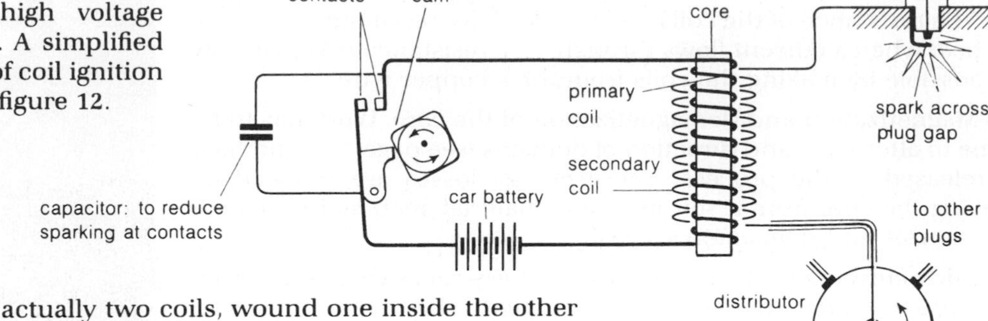

Figure 12 A sudden drop of current in the primary causes a sufficiently high e.m.f. in the secondary to make a strong spark

The **ignition coil** is actually two coils, wound one inside the other around a soft-iron core. The primary coil and core act as an electromagnet which is switched off every time one of the cams (bumps) on the spinning distributor shaft pushes apart the contacts or 'points' in the primary coil circuit. The secondary coil consists of thousands of turns of wire, and a very high voltage is induced in it every time the electromagnet is switched off. The induced voltage makes a spark jump across the air gap in one of the spark plugs, each plug being linked to the secondary coil in turn by the spinning rotor arm in the distributor. A voltage is also induced in the secondary every time the current in the primary is switched on, but this is not used.

Electronic ignition systems work along similar lines, except that the contacts and cam are replaced by a transistorized switch.

Questions

1 In the experiment shown in figure 13, what happens when a) the switch is closed b) the switch is left in the closed position c) the switch is then opened?

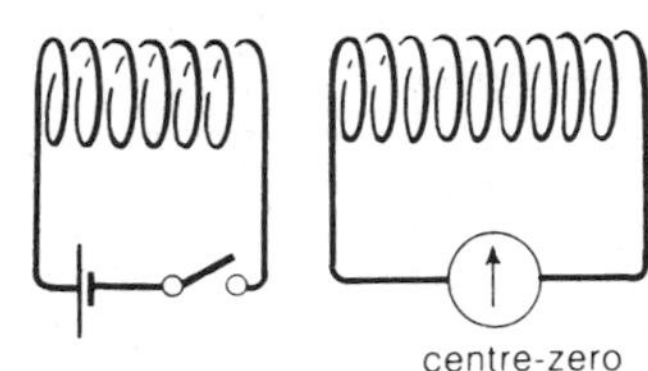

Figure 13

What would be the effect of placing a soft-iron core through the coils and then repeating the experiment? What would be the effect of replacing the battery and switch by an alternating current supply?

2 How does a step-up transformer differ from a step-down transformer?

3 Explain why the core of a transformer is a) laminated b) made from a soft magnetic material such as iron.

4 A transformer has a primary of 6000 turns and a secondary of 150 turns. The primary is connected to an a.c. supply of 240 V.
A 3 Ω resistor is connected across the secondary.
Draw a circuit diagram to show the above arrangement.
Assuming that the transformer has an efficiency of 100%, calculate
a) the voltage across the resistor
b) the current through the resistor
c) the current in the primary circuit
d) the power taken from the supply
e) the power dissipated (wasted as heat) in the resistor.

5 A 6 V 12 W bulb operates from a transformer connected to a 240 V a.c. supply. If there are 8000 turns on the primary, how many turns are there on the secondary? What is the current in the primary? (Assume that the transformer is 100% efficient)

7.11

Generating and transmitting power for the mains

Mains power is generated in huge alternators sited in power stations, and is sent across the country mainly through overhead cables. There are many transformers between a power station and the mains sockets in your house, and they play a vital part in reducing power losses in the system.

Power generation and the Grid

The photos on this page show parts of the system by which electric power is generated and transmitted. Inside the power station there are several alternators, each of which generates at a voltage of 25 000V and supplies a current of up to 20 000A. Each alternator is driven by huge turbines which are spun round by the action of high pressure steam. The steam comes from water heated in a boiler by burning coal, gas or oil, or by using the heat from a nuclear reactor.

Current generated by each alternator flows through a huge step-up transformer which transfers power to the overhead cables in the form of a much smaller current at greatly increased voltage (275 000V or 400 000V). The cables, which may be hundreds of kilometres long, have a small but significant resistance. Reducing the current which they have to carry greatly reduces power losses due to heating effects.

Above A power station alternator
Below A step-up transformer
Below, left 400 000 V transmission cables

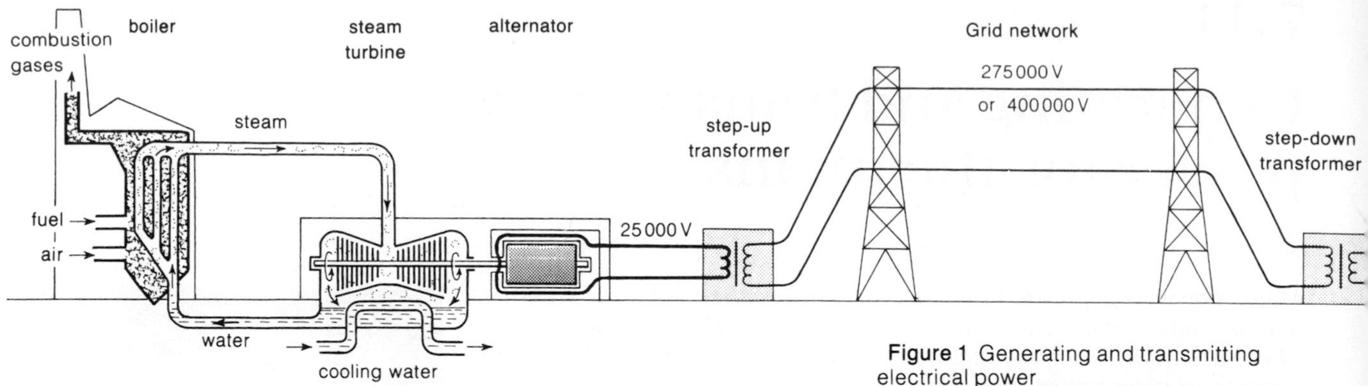

Figure 1 Generating and transmitting electrical power

The overhead cables feed power to a nationwide supply network called the **Grid**. Using the Grid, power stations in areas where the demand for power is low can be used to supply any area where the demand is high. Power from the Grid is distributed by a series of substations. Transformers in these substations reduce the voltage in stages to 230 V for home consumers, though industry usually takes its power at a higher voltage.

A substation transformer

Advantages of high voltage transmission

The advantage of transmitting (sending) power through a cable at the highest possible voltage can be demonstrated in two ways:

By calculation Power (energy per second) is calculated by multiplying voltage by current. If 2000 W of power is to be transmitted through a cable of resistance 2 Ω, it could be carried for example, by a current of 10 A at 200 V, or a current of 1 A at 2000 V. This is illustrated in figure 2.

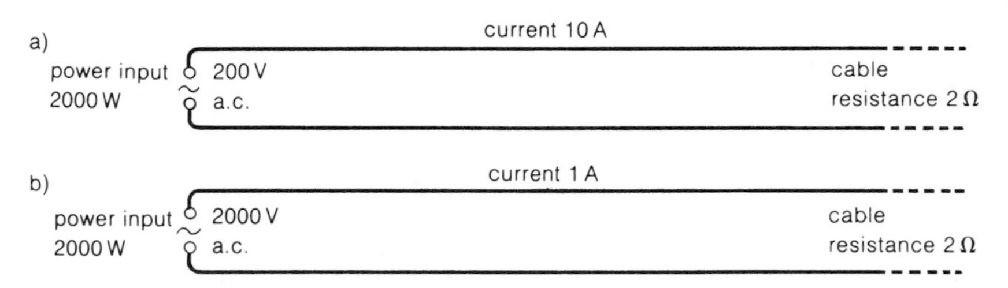

Figure 2 The same power input can be transmitted at two different voltages; but the power wasted as heat is much less in the higher voltage cable

The power dissipated (wasted) as heat in the cable can be calculated in each case using the equation given on page 263:

$$\text{power dissipated} = (\text{current})^2 \times \text{resistance}$$

In symbols: $P = I^2 R$

In case a), power dissipated $= (10^2 \times 2)\,\text{W} = 200\,\text{W}$

which is 10% of the power fed into the cable.

In case b), power dissipated $= (1^2 \times 2)\,\text{W} = 2\,\text{W}$

which is only 0.1% of the power fed into the cable.

The calculations show that transmitting power at an increased voltage gives a greatly reduced power loss in the cable.

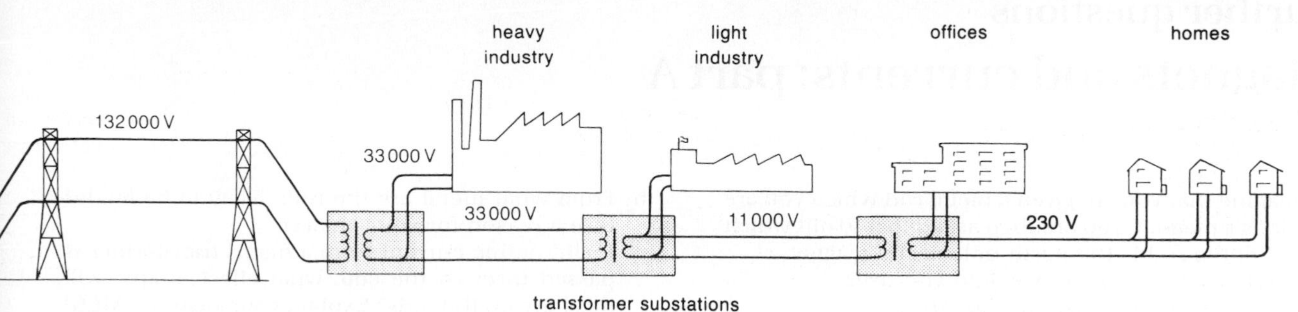

By experiment In figure 3a, a bulb rated 12 V 24 W is connected to an a.c. supply by two lengths of resistance wire. These represent very long transmission cables. Because of power losses in the wires, the bulb does not receive its rated power and glows dimly. In figure 3b, small transformers have been used to step up the voltage at the supply end of the wires and to step it down again at the bulb end. This time, power losses are much less because of the reduced current in the wires. The bulb now glows brightly.

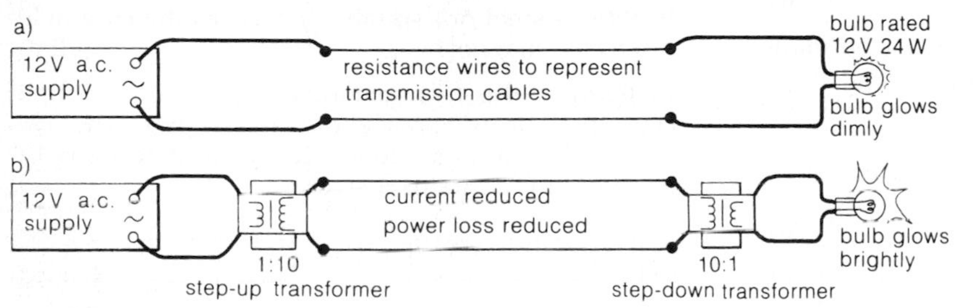

Figure 3a

Figure 3b

Figures 3a, 3b This experiment demonstrates the effect of power loss at low voltages

Advantage of a.c. for power transmission A.C., rather than d.c., is normally used for long distance power transmission because transformers make it possible to change voltage with very little loss of power. D.C. voltages can be changed, but the process is relatively difficult and expensive if high powers are involved. Transformers of course will not work with d.c.

Questions

1 What drives the alternators in a power station?
2 What is the main advantage of the Grid network?
3 What are substations for?
4 Why is power transmitted at very high voltage?
5 What is the advantage of transmitting a.c. rather than d.c.?
6 20 kW of power is fed to a transmission cable of resistance 5 Ω. Calculate the power wasted in the cable if power is transmitted a) at 200 V b) at 200 000 V.

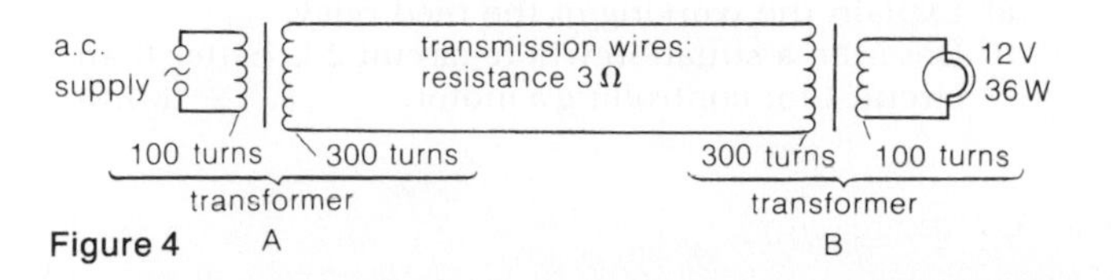

Figure 4

7 The bulb in figure 4 is rated at 12 V 36 W. If each transformer delivers the same power as it receives, and the bulb is receiving its rated power, calculate
a) the current through the bulb
b) the current through the transmission wires (use the answer to (a) and the turns information given beside transformer B)
c) the current through the supply (use the answer to b) and the turns information given by transformer A)
d) the power lost in the transmission wires (use the answer to b) and the given resistance of the wires)
e) the power output of the supply (remember that the supply has to provide power for the bulb and for heating the transmission wires)
f) the supply voltage (use the answers to c) and e))

Further questions

Magnets and currents: part A

1 Imagine that you are given a metal rod which you are told is a magnet. You are then asked to find out which end of the rod is the north pole of the magnet.
   a) What other apparatus would you use?
   b) How would you use this apparatus?
   c) Explain how you would be able to tell which end is the north pole.
   d) Give the names of two different substances which could be used to make a magnet. (SWEB/SEG)

2 Which of the following materials could be used to make the needle of a pocket navigating compass?
   A Magnesium.
   B Soft iron.
   C Aluminium.
   D Steel.
   E Brass. (LEAG)

3 Which of the following components must be made from a material which retains magnetism?
   A The commutator for a d.c. motor.
   B The magnet in a moving-coil meter.
   C The core for a transformer.
   D The core of an electro-magnet.
   E The slip-rings of an a.c. generator. (LEAG)

4 Figure 1 shows a thick copper rod passing through a hole in a drawing board. The ends of the rod are connected to a power supply and a large electric current passes from A to B. Iron filings are then sprinkled onto the drawing board near to the hole.

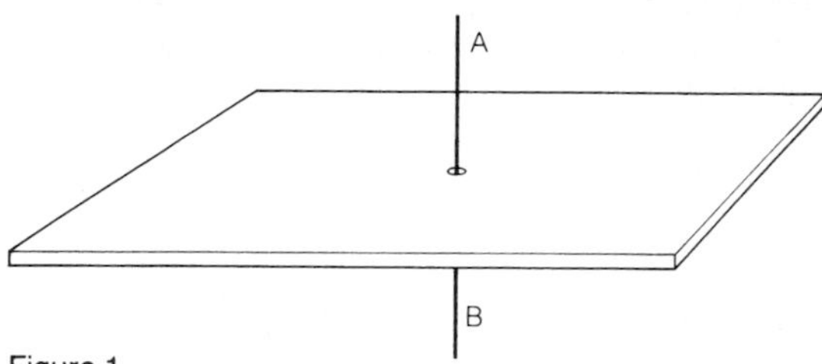

Figure 1

   a) Draw the pattern which the iron filings make.
   b) What would happen if copper filings were used instead of iron filings? Give a reason for your answer. (SWEB/SEG)

5 Two metal rods are placed in a long coil as shown in figure 2. When a direct current flows through the coil, the rods move apart. When the current is switched off, the rods return to their original positions.

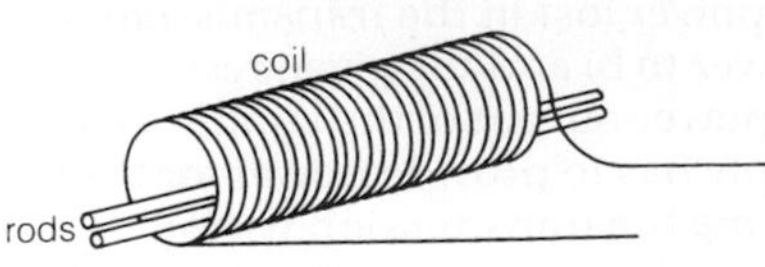

Figure 2

   a) Why did the rods move apart?
   b) From what metal are the rods likely to be made? Give a reason for your answer.
   c) If alternating current from a mains transformer is passed through the coil, what effect, if any, will this have on the rods? Explain your answer. (MEG)

6 A coil of wire is wound on an iron rod as shown in figure 3 to form an electromagnet.

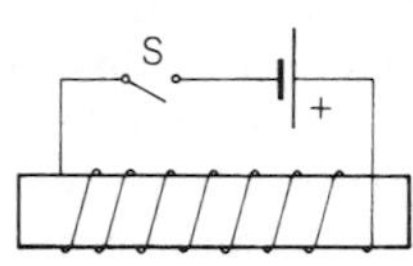

Figure 3

   a) Copy the diagram. Mark on any magnetic poles formed when the switch is closed.
   b) Why is steel not suitable for use as the core of an electro-magnet? (LEAG)

7 In figure 4 the two rectangles represent two light cylindrical iron cores about 1 cm apart. The two electrical circuits are identical except that the right-hand circuit contains a switch.

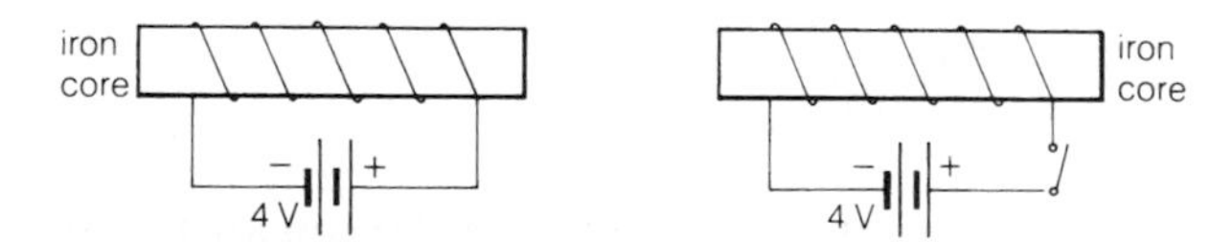

Figure 4

After the switch is closed, will the two iron cores:
   A attract each other all the time
   B repel each other all the time
   C have no force of attraction or repulsion between them
   D attract each other for just a brief moment? (LEAG)

8 The circuits in figure 5 show two ways of switching an electric motor on and off. Circuit 1 uses a simple switch. Circuit 2 uses a reed relay as a switch.

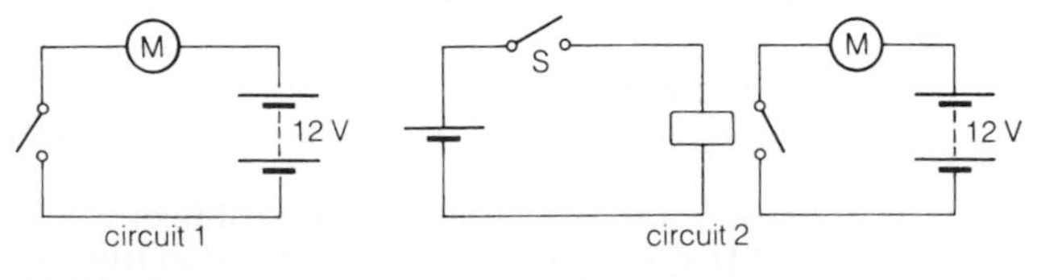

Figure 5

   a) Explain the working of the reed relay.
   b) Describe a situation where circuit 2 is better than circuit 1 for controlling a motor. (MEG)

9 A horizontal wire carries a current as shown in figure 6 between magnetic poles N and S.

Figure 6

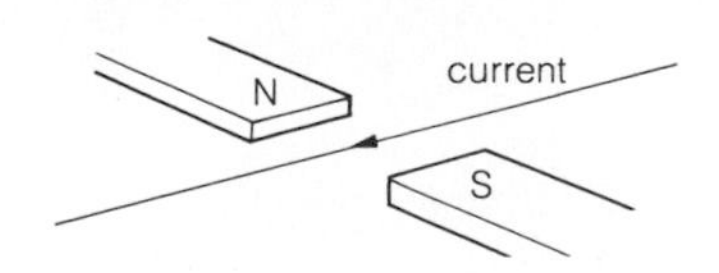

Is the direction of the force on the wire due to the magnet:

A from N to S
B from S to N
C in the direction of the current
D vertically upwards? (SEG)

10 Figure 7 illustrates an arrangement for measuring the energy input and the energy output of an electric motor. It is found that the 6 kg mass, initially at rest, can be raised 3 m in 5 s and that the readings of the ammeter and voltmeter are 2 A and 24 V respectively. (The gravitational field strength is 10 N/kg.)

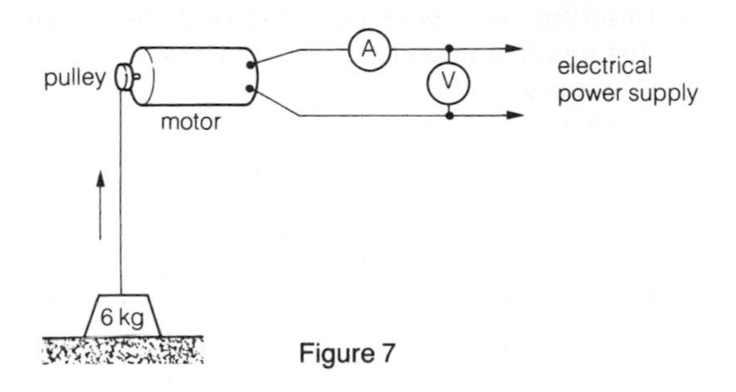

Figure 7

a) How many coulombs of electricity have passed through the motor?
b) What is the energy input from the supply?
c) How much energy was used in lifting the 6 kg mass?
d) Assuming that the 6 kg mass moves with a steady speed, what is i) its speed, ii) its kinetic energy?
e) It is found that the gain in energy of the mass is never equal to the energy put in by the electrical supply. Account for this difference in energy.

(MEG)

11 When the switch S is closed in figure 8, the pointer of the meter moves to the right.

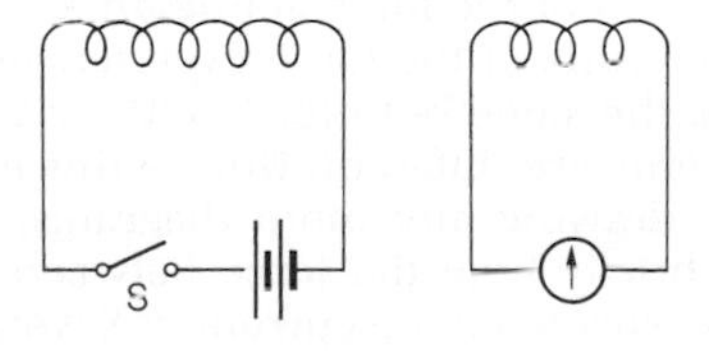

Figure 8

If S is kept closed, will the pointer

A return to zero
B stay over on the right
C move to the left and stay there
D move to and fro until S is opened? (SEG)

12 Study figure 9 which shows part of a motorcycle's electrical system, then answer the questions which follow:

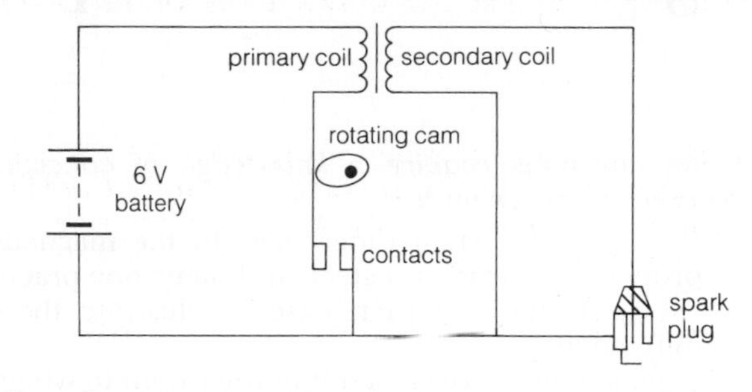

Figure 9

a) When the contacts are open as shown:
   i) is there a current in the primary coil?
   ii) What can you say about the magnetic field around the primary coil?
b) What happens to the magnetic field
   i) as the contacts touch?
   ii) as the contacts open again after being closed?
c) i) Why does a voltage appear across the secondary coil while the contacts are opening?
   ii) Why does the secondary coil have a large number of turns?
d) Part of the circuit can be connected through the metal of the motorcycle engine. Suggest a reason for this.
e) The wire taking the current to the top of the spark plug has thick PVC around it. Suggest a reason for this. (SWEB/SEG)

13 This question is about supplying a consumer with electrical power from the National Grid system.
In figure 10, the voltage across the power lines supplying alternating current to an isolated house is 12 000 V. The device D changes the voltage of the supply to 240 V.

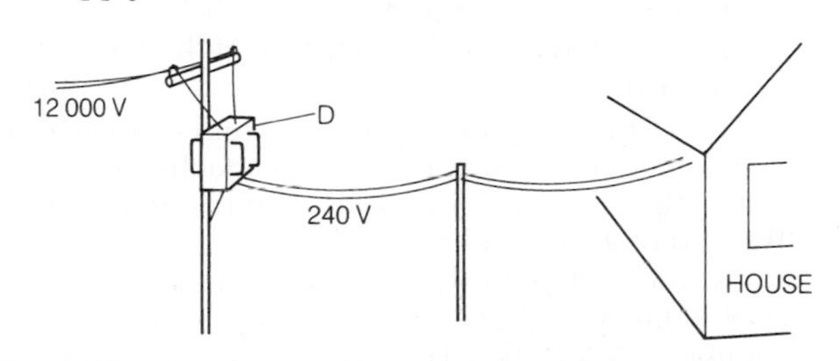

Figure 10

a) What do we call the device D?
b) Why is the supply not transmitted all the way at 240 V?
c) Why cannot the 12 000 V supply be used, unchanged, in the house? Give TWO reasons.
d) Why is alternating current used? (SEG)

Further questions

Magnets and currents: part B

Some questions require a knowledge of concepts covered in earlier units

1 a) Give two important differences in the magnetic properties of iron and steel, and name one practical application in each case to illustrate these differences.

b) Explain why there is a limit to the extent to which an iron or steel rod can be magnetised, and why a bar magnet loses its magnetic properties when it is strongly heated. (O)

2 Describe carefully how a solenoid may be used to demagnetise a bar magnet. (SUJB)

3 Draw diagrams showing the magnetic field produced by a) two bar magnets with North-poles facing each other, b) a straight wire carrying a direct current, and c) a solenoid carrying a direct current. (WJEC)

4 Figure 1 shows two vertical wires A and B, viewed from above, which are equidistant from the point P. Ignoring the effect of the Earth's magnetic field, draw separate diagrams to show the direction in which a small compass needle placed at P will set:

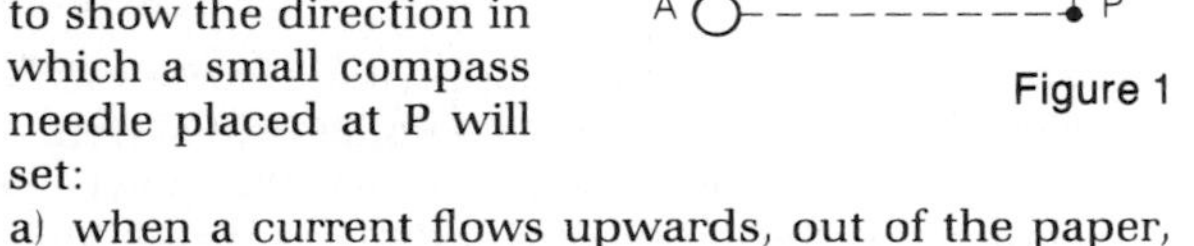

Figure 1

a) when a current flows upwards, out of the paper, through wire A and there is no current in B;

b) when a current flows upwards through wire B and there is no current in A;

c) when each wire carries a current of the same value upwards. (O&C)

5 Figure 2 shows an incomplete diagram of an electric bell and circuit. The armature A is attached to a fixed pillar P by a strip of springy steel and is situated near the end of a solenoid S.

Copy and complete the diagram to show the additions required to enable the bell to ring continuously when the switch B is closed. Label your additions to the diagram.

State the materials which are used for the armature A and for the core C of the solenoid S. (C)

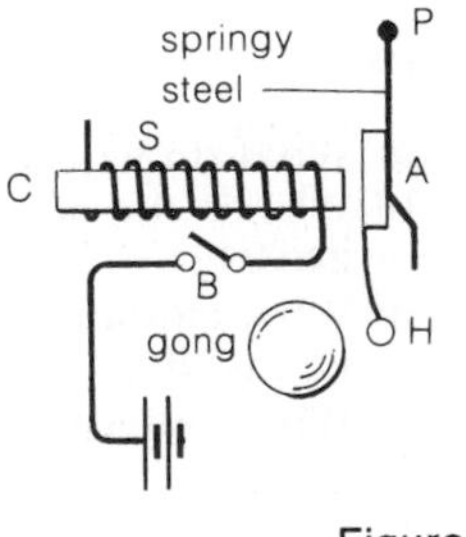

Figure 2

6 A coil of insulated wire is wound around a U-shaped core AB and connected to a battery, to make an electromagnet, as shown in figure 3.

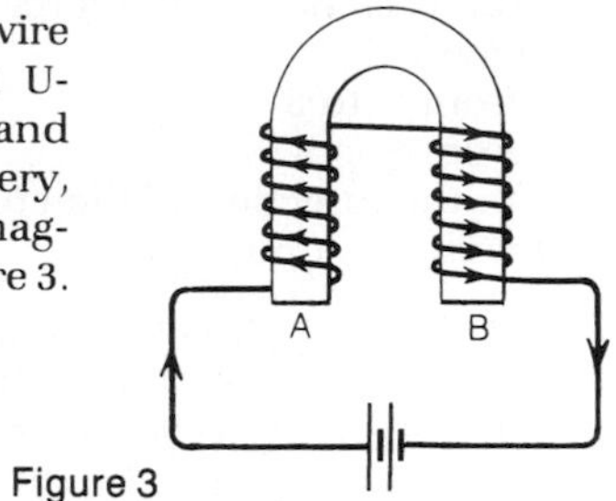

Figure 3

a) State the polarity of end A and show in a diagram how you deduced this.

b) Describe two ways of increasing the strength of this electromagnet.

c) Why must the wire be insulated?

d) What material would be used for the core and what are its properties that make it suitable? (O)

7 Figure 4 represents in perspective part of two long parallel straight wires, each carrying steady currents, 5 A in X and 3 A in Y, with the directions as shown. A very simple sketch of the magnetic field due to the current in X alone is also shown.

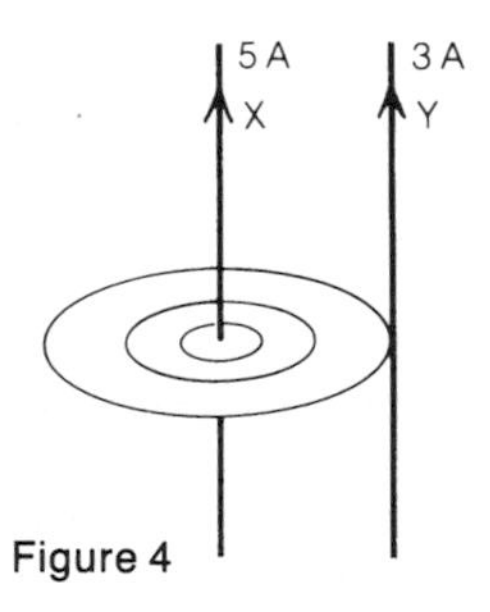

Figure 4

a) Copy the diagram, and add to it arrows indicating the direction of the magnetic field at each field line, and also a clearly labelled arrow showing the direction of the force acting on Y.

b) Draw a second diagram, showing the directions of the magnetic field at each field line, for the magnetic field due to the current in Y alone; and show on this diagram by a clearly labelled arrow the direction of the force acting on X.

c) The magnitude of the force experienced per unit length is the same both for X and for Y, although the currents are different. How is this explained?

d) Without drawing any more diagrams, state and explain briefly how the force between the wires would be affected if the current in X were changed to 3 A and the current in Y were changed to 4 A, the directions being unaltered.

e) Without drawing any more diagrams, state and explain briefly how the force between the wires would be affected if the direction of one of the currents only were reversed, the magnitudes of the currents being unaltered. (O)

8 Draw a labelled diagram showing the chief features of a moving-coil galvanometer. What features of the design i) give a sensitive meter, ii) give a uniform scale, iii) cause the pointer to take up its reading without swinging?

9 A moving-coil galvanometer of resistance 10 Ω gives full-scale deflection for a current of 2 mA.
a) What is the p.d. across it?
b) Show that with a resistance of 990 Ω in series it can be used as a voltmeter measuring up to 2 V. (O)

10 A milliammeter, of resistance 40 Ω, gives its full scale deflection for a current of 0.1 A. Calculate the value of the resistor, placed in parallel with the meter, which will allow a current up to 10 A to flow through the combination. (O)

11 A moving coil meter has resistance 5 Ω and gives full scale deflection for a current of 10 mA. Show clearly how an appropriate resistor may be used to convert this meter into a) an ammeter reading to 1.0 A, b) a voltmeter reading to 20 V. Calculate the resistance value in each case. (SUJB)

12 A galvanometer has a resistance of 40 Ω and gives a full scale reading when carrying a current of 2.0 mA. Calculate the value of the resistor which must be added to the galvanometer in order that it shall give a full scale reading when a potential difference of 10 V is applied. Draw a diagram showing how the resistor is connected to the galvanometer. Draw a second diagram showing your resistor and galvanometer connected to measure the potential difference across a lamp which is connected to a cell. (L)

13 Describe, with a clear diagram, the structure and action of an ammeter for measuring a.c. Would your meter be suitable for measuring d.c. also? Give a reason. (SUJB)

14 A stiff wire AB is held between the pole pieces of a permanent magnet and connected to a centre-zero galvanometer with flexible wire, as shown in figure 5.

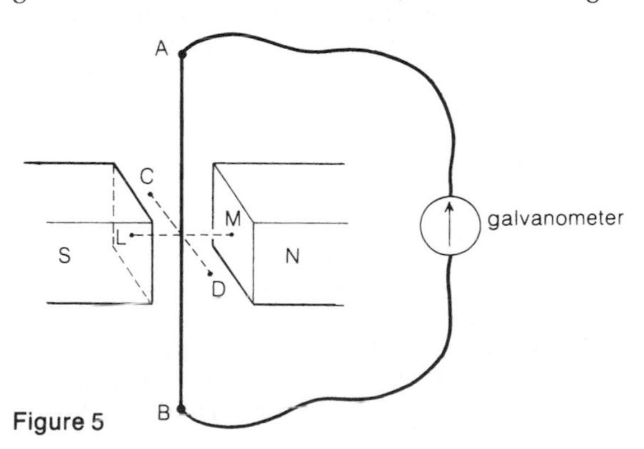

Figure 5

The wire AB is kept vertical and moved slowly back and forth along the line CD. Describe and explain the behaviour of the galvanometer needle
a) as the wire AB is moved towards C;
b) when the wire AB is momentarily at rest;
c) as the wire AB is moved towards D.
In which direction does current flow when the wire AB is moving towards C? Describe and explain the behaviour of the galvanometer-needle if the wire AB is kept vertical and moved slowly to and fro along the line LM. (O&C)

15 The leads from an air cored solenoid are connected across a sensitive centre zero milliammeter. What would you expect to observe if one pole of a strong bar magnet is moved into the end of the solenoid and then withdrawn? State how the rate at which the magnet is moved may affect the observations. (SUJB)

16 In figure 6 the N pole of the permanent magnet was thrust into the centre of the aluminium ring. The ring moved to the right, indicating that it behaves like a magnet.

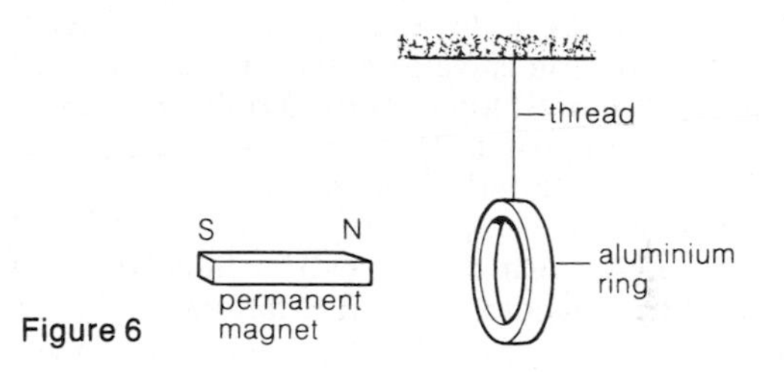

Figure 6

a) State the polarity of the face of the ring nearest the permanent magnet.
b) Draw a diagram to show the direction in which the induced current flows in the ring.
Explain why this current is produced. (AEB)

17 Figure 7, which is incomplete, illustrates the principle of a simple a.c. generator. Re-draw the diagram showing in addition slip rings, brushes and load resistor.

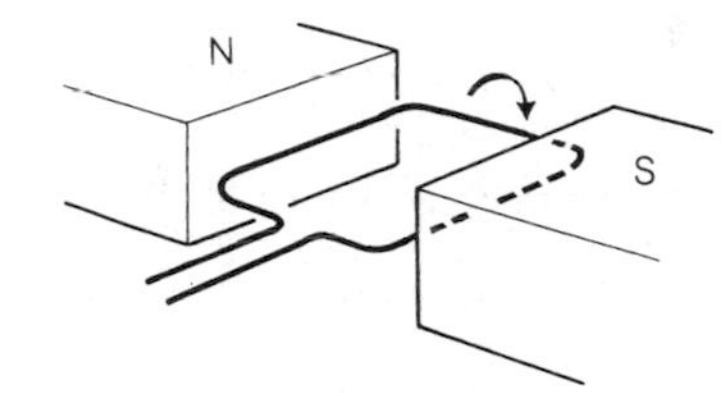

Figure 7

Considering the coil to be rotating in a clockwise direction as indicated:
a) show the direction of the induced current in the coil;
b) sketch a graph to show the waveform generated;
c) state at what positions of the coil the e.m.f. is a maximum and a minimum. Explain your answers;
d) give two ways in which the e.m.f. alters if the coil is rotated faster between the pole pieces;
e) explain how the generator can be modified to produce a d.c. current. (WJEC)

18 A single coil of wire is held between the poles of a magnet (see figure 8). The ends of the coil are connected to a sensitive centre-zero galvanometer. When the coil is lifted out of the field of the magnet the galvanometer deflects momentarily to the left.

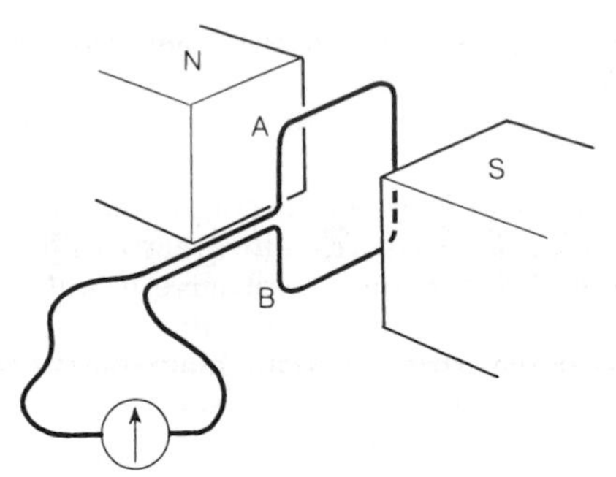

Figure 8

a) Explain why the galvanometer deflects.
b) Explain how the deflection could be made larger with the same apparatus.
For each of the following cases, describe what is observed on the galvanometer, giving reasons.
c) The coil is moved towards the S pole of the magnet without being rotated.
d) The coil is rotated through 90° in a clockwise direction.
e) In this new position the coil is lifted out of the field of the magnet without rotation. (O)

19 Draw a labelled diagram to show the essential structure of a transformer suitable for operating a 12 V 36 W lamp from 240 V mains. Explain the action of the transformer and calculate the current you would expect to be drawn from the mains.
Explain why it would be dangerous to connect this transformer the other way round with the 'output' terminals to the mains. (SUJB)

20 Figure 9 shows a simple demonstration transformer, intended to convert a 24 V 50 Hz a.c. supply to give 240 V 50 Hz.

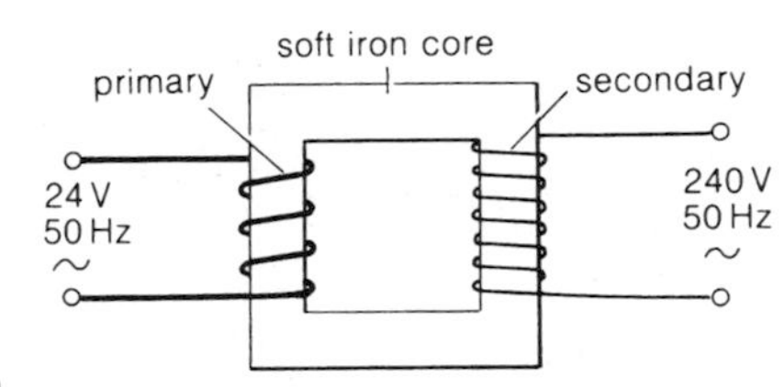

Figure 9

a) Why cannot such a transformer be used for steady direct current?
b) What is the purpose of the 'soft iron' core?
c) Why is the core usually made of sheets of a magnetic material (which has a high resistivity) that are insulated from one another?
d) If the primary has 50 turns, how many turns should there be on the secondary? Explain briefly.
e) If the primary current is 25 A, what is the greatest possible value (in theory) of the secondary current, and why will the value of the secondary current always be less than this? (O)

21 a) It is required to run a 6 V 24 W lamp from a 240 V a.c. mains using a transformer as shown in figure 10a.
i) Calculate the current that would be taken by the lamp when operating normally.

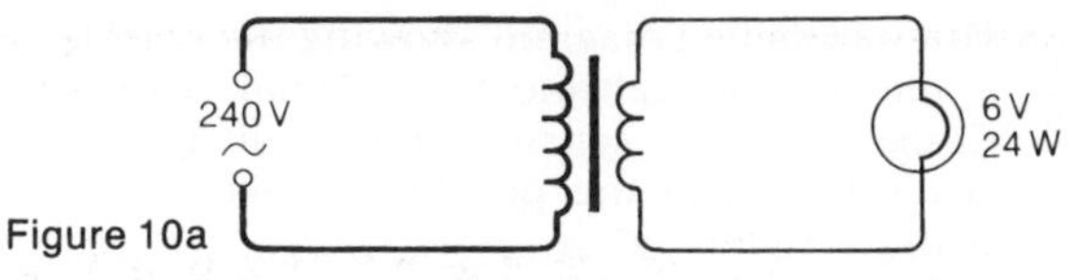

Figure 10a

ii) Calculate the turns ratio of the transformer you would use.
iii) Calculate the current taken by the primary coil of the transformer, assuming it to be 100% efficient.
iv) Why, in practice, is the efficiency of the transformer less than 100%?

b) Alternatively the 6 V, 24 W lamp can be operated normally from a 240 V d.c. supply using a suitable fixed resistor, R, as shown in figure 10b.

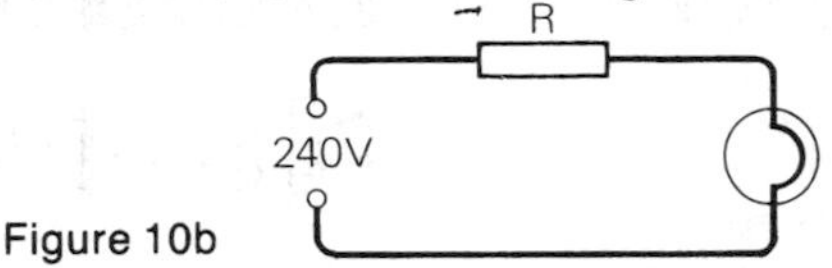

Figure 10b

i) What is the resistance of the lamp?
ii) What is the p.d. across the resistor?
iii) What is the resistance of the resistor?
iv) How much energy is dissipated in the resistor in 1 s?

c) Why may the method used to light the lamp described in a) be preferable to that described in b)? (L)

22 A power station generator produces an e.m.f. of 33 000 V at a frequency 50 Hz. The domestic supply is approximately 250 V, 50 Hz. Explain how the output of the power station can be modified for use in the home. (L)

23 A consumer receives a power of 30 kW at 600 V at his end of the transmission lines. If the resistance of the transmission lines is 0.2 Ω, calculate:
i) the current flowing in the lines;
ii) the voltage drop in the lines;
iii) the power wasted in the lines.
What would be the corresponding values for the current flowing, the voltage drop and the power wasted if the 30 kW of power were received at 60 kV instead of 600 V? (O)

24 a) Describe a simple experiment which illustrates electromagnetic induction.
b) Explain briefly how electromagnetic induction is applied i) in the generation of alternating current and ii) in the transformers used to step up or step down alternating voltages.
c) With all the domestic appliances in use, a house takes 20 kW at 240 V. Supposing that this came direct from a generator through cables of total resistance 0.5 Ω, what would be the 'lost' volts in the cables and the power wasted as heat in the cables?
d) Supposing that the same power of 20 kW were transmitted at 120 000 volts from a generator through cables of total resistance 0.5 Ω, and then the voltage were transformed down to 240 V using perfectly efficient transformers, what would be the 'lost' volts in the cables, and the power wasted as heat in the cables? (O)

SECTION 8

ELECTRONS AND ATOMS

Life with Nuclear Power?

Taken from 'Life with nuclear energy' issued by the Nuclear Power Information Group.

Three barriers – intrinsic safety, engineered protection and regulation – provide extremely high standards of safety for nuclear plants. If plant failures and human errors were to result in severe damage to a reactor, the *most likely* consequence is that nobody at the plant or outside would be harmed. In the *worst-imaginable* case, with a big release of radioactivity, an unfavourable wind and failure of emergency plans, a nuclear accident would conceivably cause several thousand deaths. But so too could other types of worst-imaginable accident, such as an aircraft crash on a crowded football stadium. The chance of *worst-imaginable* accidents is extremely small – much smaller for nuclear plants than other industrial activities.

In all industrial activities, and for that matter all human activities, accidents do happen from combinations of equipment failures and human errors. The nuclear industry is no exception. Many protection systems are provided to stop accidents happening and to minimize the consequences if they do.

The most serious accidents in the West have been in 1957, at a military plutonium production reactor at Windscale in Britain, and in 1979, at the Three Mile Island reactor in the US. Both involved very severe damage through overheating of the core of the reactors. At Windscale the fuel caught fire, but there were still enough protective systems to ensure that the release of radioactivity to the environment was very small in both cases. Nobody suffered harmful effects from radiation.

Death from Nuclear Disaster?

The damaged reactor buildings at Chernobyl in the Soviet Union; the arrow points to the source of the explosion

The accident at the Russian nuclear power station at Chernobyl happened swiftly and unexpectedly. It came in the early hours of April 26 1986, when number four reactor was being shut down for routine maintenance. Suddenly, and without warning, a violent explosion blew the top off the reactor vault. Two workers were killed outright, hundreds more received massive radiation overdoses, and a huge cloud of radioactive gas and dust was blasted high into the atmosphere. Days later, the cloud had spread all over Northern Europe. Radiation levels had weakened, but radioactive rain was falling.

In Britain and America, the nuclear authorities were quick to point out that a similar disaster couldn't happen in the West: reactor designs were inherently safer, and safety standards were higher. But not everyone was convinced. Many remembered the accident at Three Mile Island seven years earlier, when a nuclear reactor came close to a massive meltdown.

It began at 4 am on Wednesday 28 March 1979, with the wail of sirens in Unit 2 control room. A pump in the secondary cooling system had failed, turning off the plant's turbine. Three Mile Island no longer generated electricity, but inside the containment building the reactor generated heat at full capacity. Within minutes, the core temperature had soared to 1500°C – hot enough to rupture the zirconium-clad fuel rods – and radioactive water had gushed through a faulty pressure relief valve to flood the reactor basement.

By 5.40 am the core temperature had soared to 2000°C – high enough theoretically to trigger core collapse and a massive meltdown.

At about 6 am, relief valves in the waste tanks opened, venting radioactive steam into the atmosphere and creating a hot plume that drifted over Harrisburg, Pennsylvania.

At 6.50 am, the radiation alarm sounded, and an on-site emergency was declared. This was followed at 11 am by a general emergency – the first ever at a commercial nuclear facility.

Outside the plant, the vice-president of the company which operates the Three Mile Island facility, told the gathering swarm of reporters that the situation was under control. 'When we say general emergency,' he explained, 'it does not mean that an emergency exists. There was nothing that was catastrophic or unplanned for.'

Throughout the day, the company continued to utter reassuring but conflicting statements. By Thursday morning, it was not clear what was happening in the plant, but the company's assurances had worn thin. The extent of the crisis was revealed on Friday. At a 9 am news conference, the Emergency Management Agency reported a large release of radioactive gas into the atmosphere a few hours earlier, and that 40 000 gallons of low-level waste water had been jettisoned into the Susquehanna River. Governor Dick Thornburgh ordered local schools to be closed and advised pregnant women and pre-school children within a five mile radius of the plant to be evacuated. Civil defence staff were instructed to draw up evacuation plans for the 165 000 residents within a ten mile radius. A huge and potentially explosive gas bubble had formed above the core – which took an extremely worrying week to reduce in size. A spokesman then stated: 'There are no more problems.' But the problems at Three Mile Island had just begun . . .

8.1
Electron beams

Given enough energy, electrons can escape from a wire and travel through a vacuum. Experiments with beams of 'escaped' electrons give a great deal of information about the nature of electrons themselves.

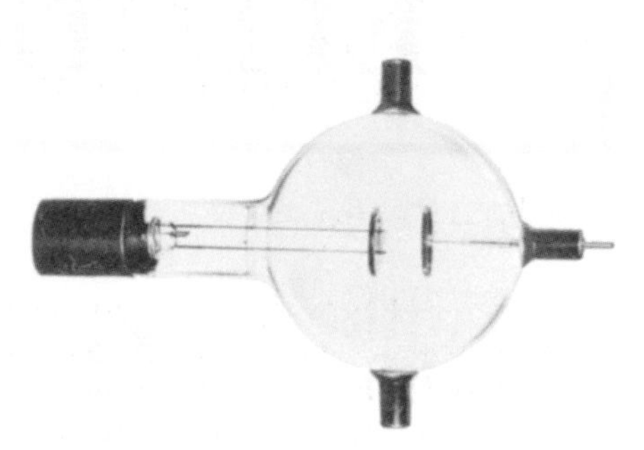

Figure 1 Equipment to demonstrate thermionic emission

Thermionic emission

If a tungsten filament is heated to around 2500 K, it gives off electrons. This happens because some of the electrons in the white hot metal have enough energy at this temperature to break free from the metal surface and escape into the space surrounding it. This is called **thermionic emission**; in some ways it is like evaporation, except that electrons rather than whole molecules are emitted.

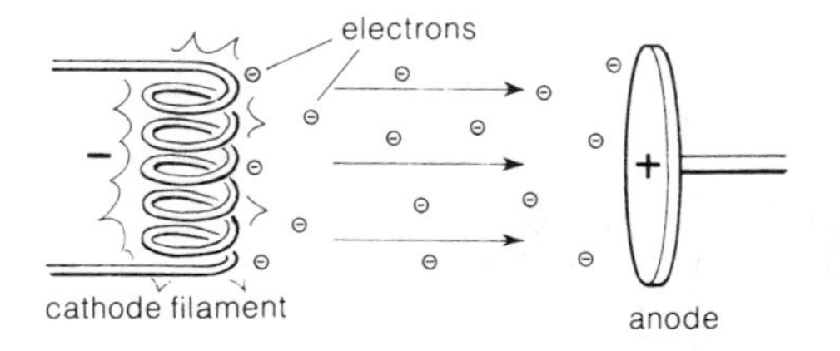

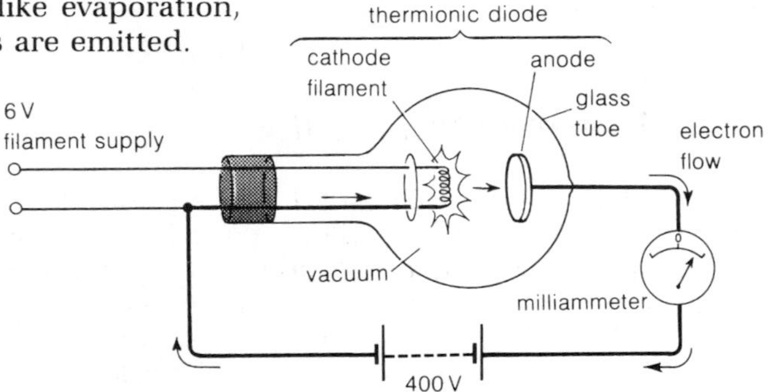

Figure 2 The hot cathode emits electrons; the high p.d. between it and the anode causes the electrons to travel across the gap

Figure 2 shows an experiment to demonstrate thermionic emission. The glass tube contains two pieces of metal called **electrodes**. One of these, a metal plate called the **anode**, is connected to the positive terminal of a 400 V d.c. supply. The other, a small tungsten filament called the **cathode**, is connected to the negative terminal. The filament is heated by passing a small current through it from a 6 V supply. The air has been removed from the tube so that the electrons are free to move without colliding with air molecules.

Despite the gap between the electrodes, a small current flows round the main circuit when the filament is white hot. The hot filament emits electrons which, being negatively charged, are pushed away from the cathode and pulled across to the anode, as shown in figure 2.

If the current through the filament is switched off, thermionic emission ceases and no current flows round the main circuit. If the filament is heated with the anode and cathode connections reversed as in figure 3, there is again no current in the main circuit. This time, the electrons are repelled by the anode, so none can reach it.

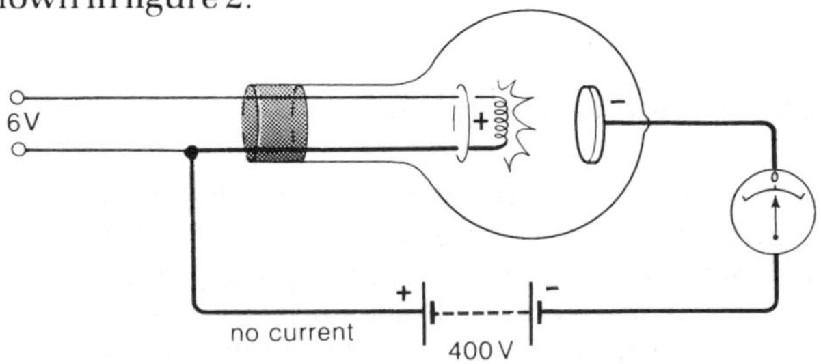

Figure 3 If the high p.d. is connected the other way round, no current flows

The tube and electrodes in figure 1 together make up a **thermionic diode**. An important feature of a diode is that it allows electrons to flow in one direction only. This is examined in more detail on page 336.

Most metals emit electrons above about 2500 K, though tungsten is one of the few that can be heated to this temperature without melting. With some metal oxides, thermionic emission takes place at lower temperatures. Figure 4 shows a diode in which the cathode is a nickel cylinder coated with a mixture of barium and strontium oxides. The coating emits electrons when the cylinder is heated to about 1100 K by the filament inside it.

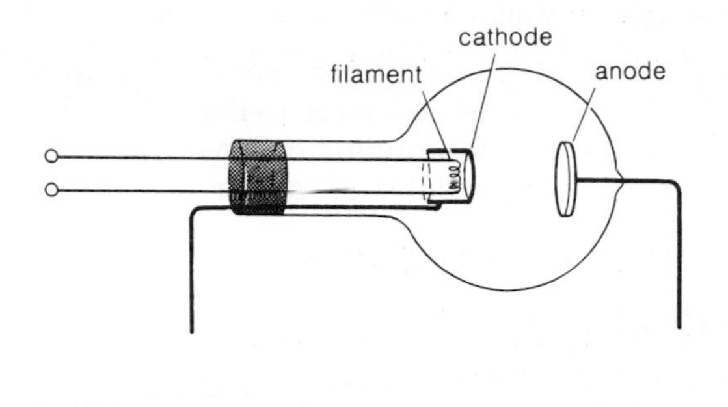

Figure 4 A thermionic diode: the separate cathode, which has been coated with a metal oxide, is heated by the filament

Cathode rays

Beams of fast-moving electrons are known as **cathode rays**. The name dates from the late 1800's when 'radiation' travelling from a cathode to an anode had been discovered but its exact nature was unknown. The experiments described below demonstrate some of the basic properties of cathode rays. In their original form, these experiments led to the discovery of the electron.

The Maltese cross tube

The tube illustrated in figure 5 has the same basic parts as a thermionic diode. Here however, the anode has a hole in it, and the end of the glass tube is coated with a fluorescent material which forms a screen. Between the anode and the screen is a metal obstacle shaped like a Maltese cross.

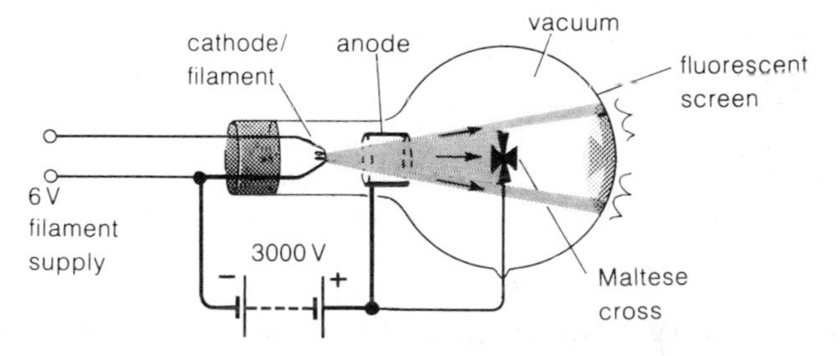

Figure 5 In a 'Maltese Cross' tube, a metal cross lies in the path of a beam of electrons

Electrons leaving the cathode are accelerated towards the anode, but most pass through the hole and travel on towards the screen. Some of these electrons strike the screen, making it fluoresce (glow) with a green light. The rest are stopped by the metal cross, causing a cross-shaped 'shadow' to appear on the screen where no fluorescence occurs. This cathode ray shadow is exactly the same shape as the shadow formed by light rays from the bright filament, so electrons reaching the screen must have travelled in straight lines.

Julius Plücker first used a Maltese cross tube in 1859 to show that cathode rays could cause fluorescence and travel in straight lines.

The deflection tube

In the tube shown in figure 6, a narrow beam of electrons leaves the hole in the anode and passes between two horizontal metal plates. The fluorescent screen between the plates shows the path of the beam. If a high p.d. is applied across the plates as shown in the diagram, the beam is deflected by the electric field. In other words, the moving electrons alter course as they are attracted by the positive plate and repelled by the negative plate.

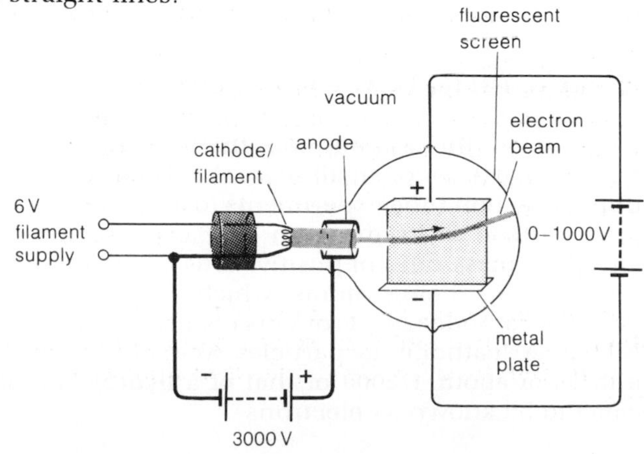

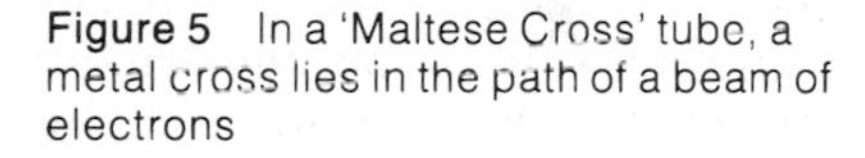

Figure 6 In the "deflection tube" the electron beam is deflected as it passes through an electric field

In figure 7a, the horizontal plates are not in use. Instead, the electron beam passes through a horizontal magnetic field produced by two current-carrying coils placed on either side of the tube. Electrons moving in the direction shown are equivalent to a *conventional* current (see page 246) in the opposite direction, and experience a force whose direction is given by Fleming's left-hand rule. The beam in this case is deflected upwards – figure 7b shows how the rule applies.

J.J. Thomson carried out similar experiments in 1897. Observing the ways in which cathode rays were deflected by electric and magnetic fields, he came to the conclusion that the rays were fast-moving negatively charged particles of matter.

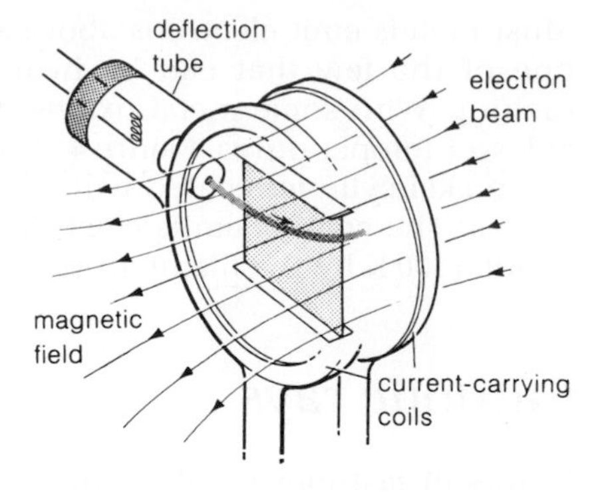

Figure 7a A horizontal magnetic field will produce a similar deflection

The Perrin tube

Figure 8 shows the modern version of the apparatus first used by the French scientist Jean Perrin to demonstrate that cathode rays carried a negative charge.

The metal container on one side of the tube is connected to a gold leaf electroscope. Using a charged rod (as explained on page 230), a negative charge is induced in the electroscope, so that the leaf shows a small deflection.

With a current through the filament, and a high p.d. between the cathode and the anode, a narrow beam of electrons passes through the hole in the anode and strikes the fluorescent screen. The bright dot on the screen shows the position of the beam. Using a small magnet, the beam is deflected upwards so that it strikes the metal container. This makes the leaf of the electroscope rise further, indicating that the container and the electroscope are gaining negative charge as they collect electrons from the beam.

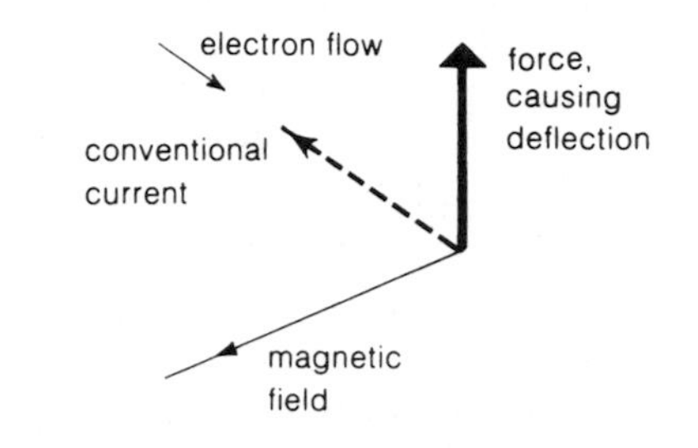

Figure 7b Fleming's left-hand rule predicts the direction of the force on the electrons and thus the deflection of the beam. When applying this rule, remember that the conventional current direction is *opposite* to the electron flow

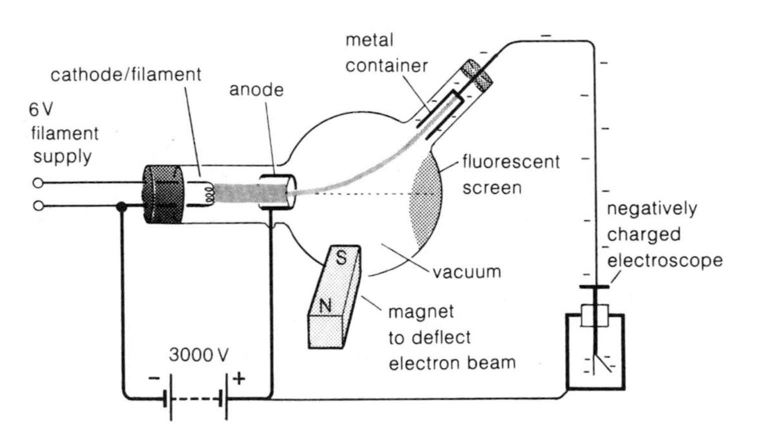

Figure 8 The Perrin tube: passing the electron beam into a metal container linked to a negatively-charged electroscope demonstrates that the electrons are negatively charged

Measuring the electron

Working on the assumption that cathode rays were negatively charged particles of small mass, J.J. Thomson used measurements from deflection tube experiments to calculate the ratio of the charge on a particle to its mass. In later experiments, he measured the charge-to-mass ratio of positive ions of several different gases – the positive ions being atoms which had lost negative charge by emitting cathode rays. From his results, he concluded that negatively charged 'cathode ray particles' were present in all atoms, and had a mass of about 1/2000 of that of a hydrogen atom. The particles came to be known as electrons.

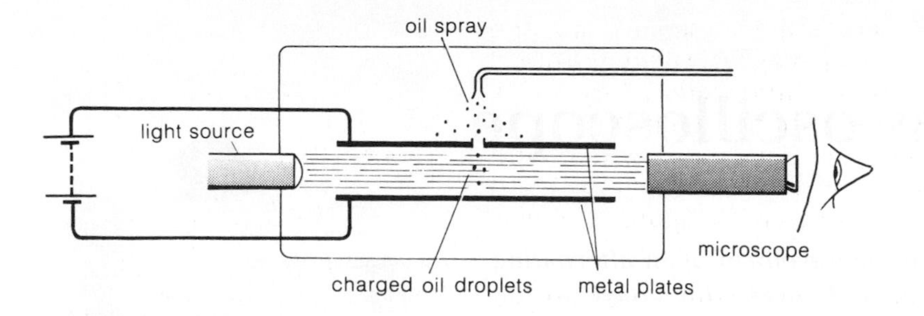

Figure 9 Millikan's oil drop experiment. He found that the charge on the droplets was always a multiple of 1.6×10^{-19}C, this being the charge on one electron

The first accurate measurement of the charge on an electron was made by Millikan in 1911. His experiment is shown in simplified form in figure 9. Millikan measured the terminal velocities of tiny charged oil droplets as they fell through the air between two horizontal metal plates. As the plates had been connected to the terminals of a high voltage d.c. supply, they exerted a vertical force on each charged oil droplet. The terminal velocity of each droplet therefore varied according to the charge it carried.

From terminal velocity and other measurements, Millikan was able to calculate the charges on a hundred or so different droplets. In each case, he found that the charge was always a multiple of a basic charge of 1.6×10^{-19} C. As the oil droplets had become charged by gaining or losing a particular number of electrons, he concluded that each electron carried a charge of 1.6×10^{-19} C.

With the charge on an electron known, as well as its charge-to-mass ratio, the mass of the electron could be calculated. It was found to be 9.1×10^{-31} kg.

Questions

1 Figure 10 shows a thermionic diode
  a) What happens when the filament becomes white hot, and what is the effect called?
  b) Why is all the air removed from the tube?
  c) Copy the diagram and label the anode and the cathode. On your diagram, show how a battery must be connected if a current is to flow through the diode. Why would no current flow if the battery connections were reversed?
  d) In the diode shown, the filament also acts as one of the electrodes. Draw a diagram to show an alternative arrangement.

2 What are cathode rays?

3 Cathode rays are deflected by a magnetic field. Give three other properties of cathode rays.

4 Which way will the electron beam in figure 11 be deflected when the switch is closed?

5 Which way will the electron beam in figure 12 be deflected if magnetic poles are moved into the positions shown?

6 What experimental evidence is there to suggest that electrons carry a negative charge?

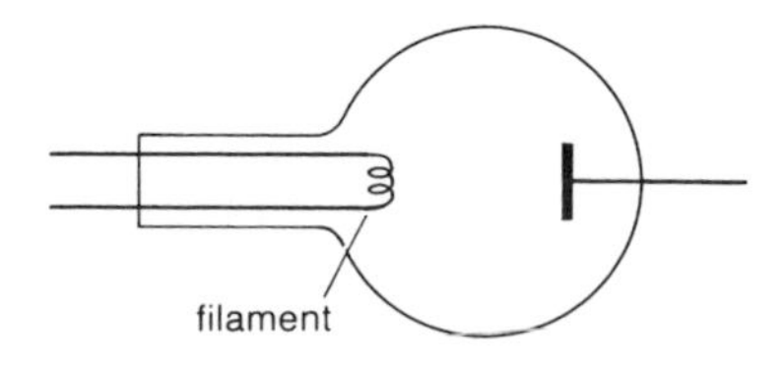

Figure 10

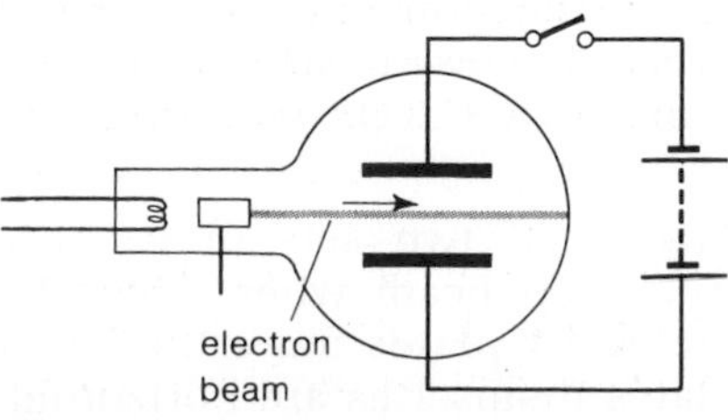

Figure 11

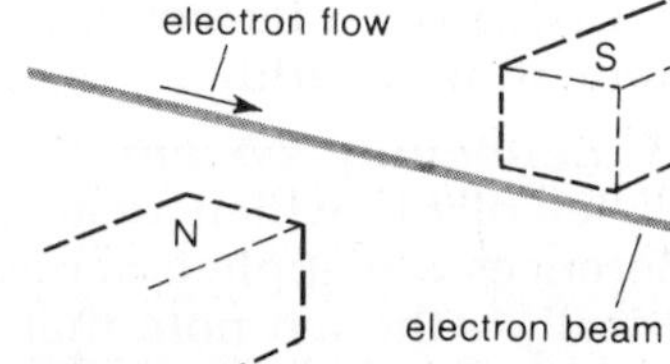

Figure 12

8.2

The cathode ray oscilloscope

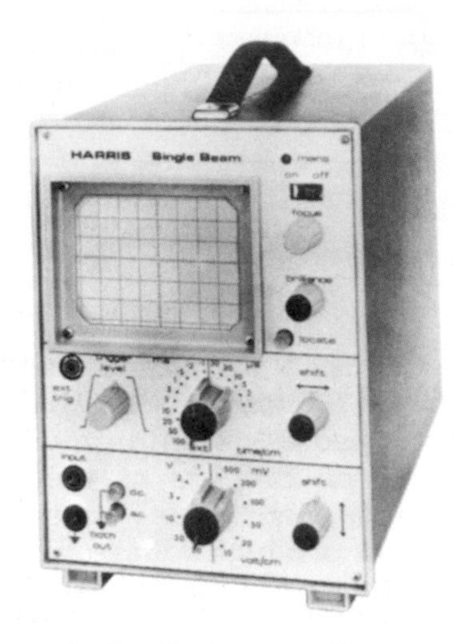

Figure 1 A cathode ray oscilloscope

A simple oscilloscope can show you the waveform of an alternating voltage; a t.v. set is a complex one. In both cases, the image on the screen is being traced out by a moving beam of electrons.

A cathode ray oscilloscope, or C.R.O. for short, is shown in figure 1. The main feature of the C.R.O. is a **cathode ray tube**, one end of which forms the screen. Controls on the front of the C.R.O. are used to make adjustments to the complex electronic circuits which control the movement and the intensity of a narrow beam of electrons inside the tube. The structure is shown in figure 2. The tube has three main parts:

1 A fluorescent screen At the wide end of the tube, the inside surface of the glass is coated with zinc sulphide. This fluoresces when the electron beam strikes it, and a bright spot is seen on the outside surface of the screen as a result.

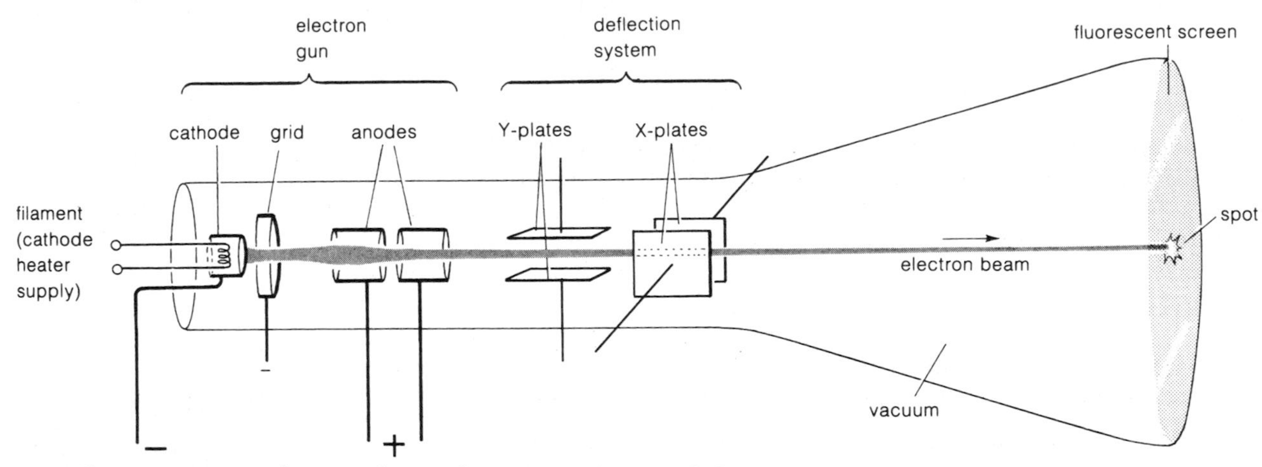

Figure 2 The structure of a cathode ray oscilloscope: an electron beam is deflected horizontally by the X-plates and vertically by the Y-plates

2 An electron gun This produces the narrow beam of electrons. Main parts of the gun are a filament, a cathode, and two anodes in the form of open-ended cylinders. The two anodes are at different positive potentials, and their effect is not only to accelerate the electrons, but also to make them converge to a sharp focus on the screen. The **focus control** on the front of the C.R.O. varies the converging effect of the anodes by altering the p.d. between them.

Between the cathode and the anodes is a ring-shaped electrode called the **grid**. If this is at a more negative potential than the cathode, it repels electrons and reduces the number emitted by the gun. This makes the spot on the screen less bright. When you turn the **brilliance control** on the front of a C.R.O., you vary the potential of the grid.

3 A deflection system This consists of two sets of roughly parallel plates which deflect the electron beam when potential differences are applied across them. The **Y-plates** move the beam vertically – though note that the plates themselves are horizontal. The **X-plates** move the beam horizontally.

Vertical deflection

Figure 3a shows the screen of a C.R.O. when the p.d.s (voltages) across both sets of deflection plates are zero.

Figure 3b shows the screen when a steady d.c. voltage is being applied to the Y-plates. The top plate is positive and therefore attracts the electron beam, so the deflection of the spot is upwards. The deflection is directly proportional to the applied voltage.

Figure 3c shows the effect of applying an a.c. voltage of frequency 50 Hz to the Y-plates. The spot oscillates up and down fifty times a second, but so rapidly that it is seen as a continuous line.

A voltage is applied to the Y-plates by connecting an a.c. or d.c. source to the input terminals on the front of the C.R.O. In most cases however, the connection to the plates is not direct. An amplifier is used to amplify (magnify) the voltage to a level which will cause a suitable deflection of the spot. You can vary the degree of amplification by turning the **gain control** to different positions. In figure 3d for example, the a.c. voltage across the input of the C.R.O. is the same as in figure 3c, but the gain has been increased.

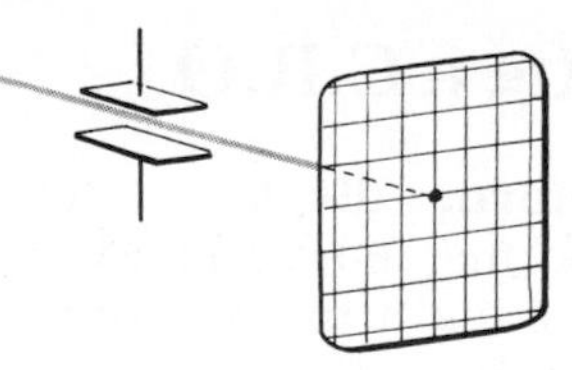

Figure 3a Zero voltage across both plates

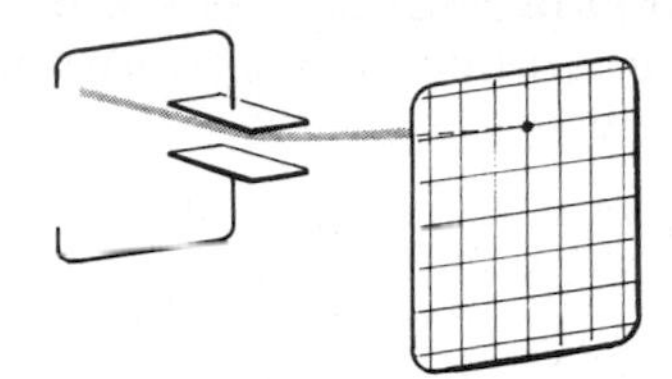

Figure 3b d.c. voltage across Y-plates

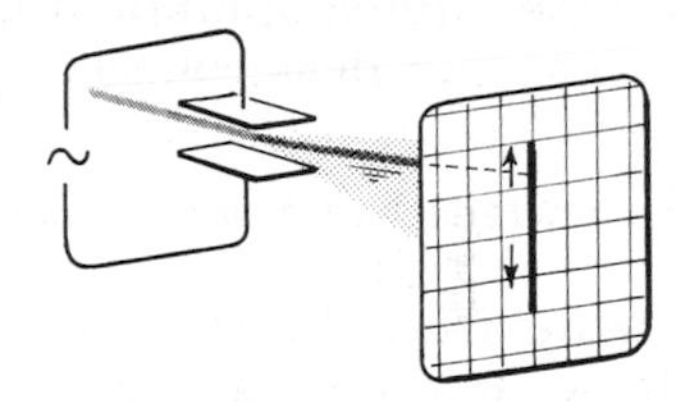

Figure 3c a.c. voltage across Y-plates

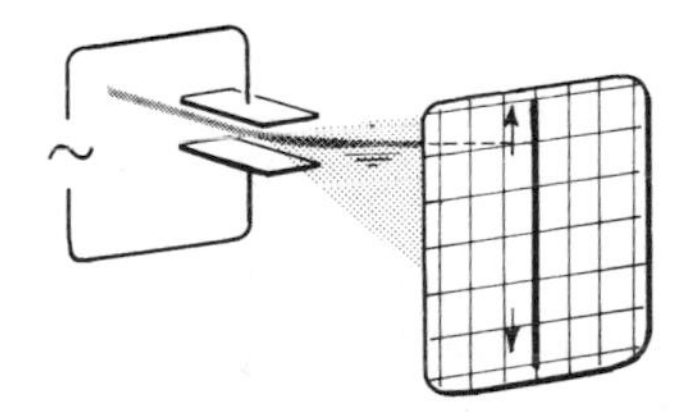

Figure 3d a.c. voltage across Y-plates with the *gain* increased

Horizontal deflection: time base circuit

In figure 3c, the spot is oscillating up and down. If at the same time it were to be moved rapidly across the screen at a steady speed, it would trace out a wavy line as in figure 4. The line is the **waveform** of the a.c. voltage – in effect, it is a graph of voltage against time. You can produce a similar waveform if you move a pencil point rapidly up and down on a piece of paper, pulling the pencil across the paper at the same time. This is also illustrated in the diagram.

The **time base** circuit in a C.R.O. automatically applies a changing voltage to the X-plates which makes the spot move from left to right across the screen at a steady speed. Figure 5 shows how this applied voltage varies with time. When the spot completes one sweep of the screen, it flies back to its starting point, and the process is repeated.

The spot traces out a new waveform every time it crosses the screen. By turning the **time base control**, the speed of sweep can be adjusted so that each new waveform is 'drawn' in the same position as the one before and the image on the screen is stationary. In some cases the waveform may be re-drawn on the screen thousands of times a second, so you only see a continuous image on the screen.

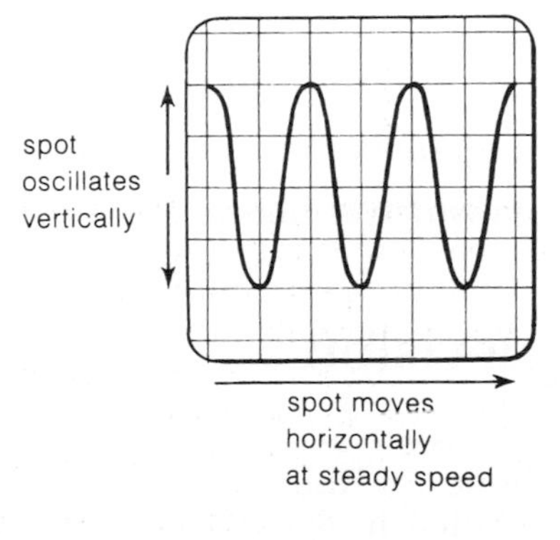

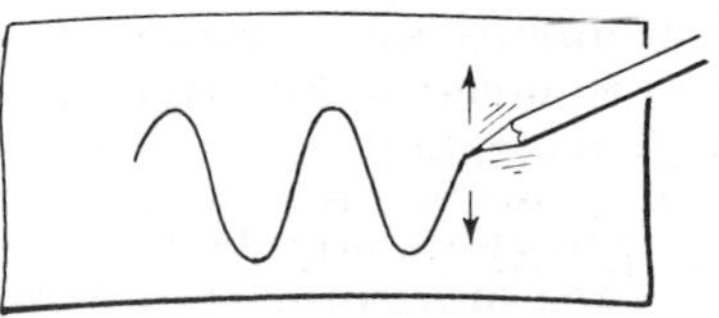

Figure 4 The a.c. voltage of figure 3c, with the spot moved steadily across the screen. The effect is similar to moving a pencil up and down while pulling it sideways across a piece of paper.

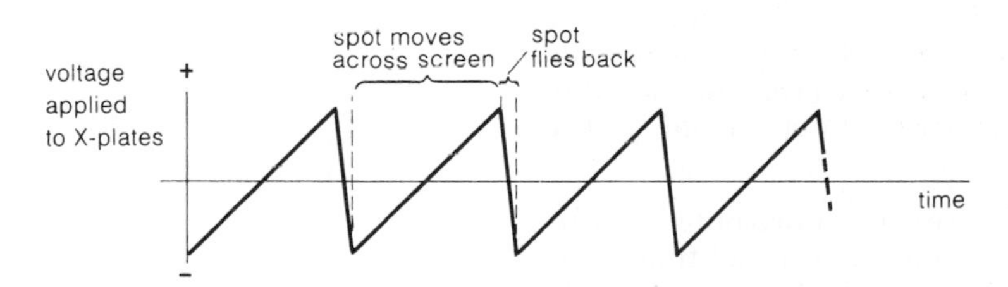

Figure 5 A time base: a steadily changing voltage moves the spot steadily across the screen: a rapid change makes it fly back to its starting point

Using a C.R.O.

Waveform display C.R.O.s are widely used for displaying waveforms. For example, sound waveforms can be studied by connecting a microphone to the input terminals of a C.R.O. (see page 215).

Measuring voltage As the deflection of the spot depends on the voltage applied to the input terminals, a C.R.O. can be used as a voltmeter.

In figure 6, the screen of a C.R.O. is displaying the waveform of an a.c. voltage. The gain control of the instrument is set at '2 volts/cm', which means that the spot is deflected 1 cm vertically for every 2 volts across the input terminals.

In this case, the amplitude of the waveform = 1.5 cm

$\therefore$ peak voltage = 1.5 cm × 2 V/cm

= 3.0 V

The a.c. voltage has a peak value of 3.0 V.

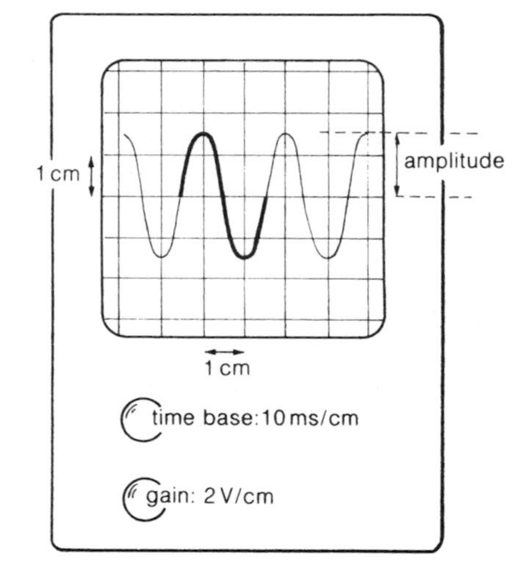

Figure 6 With the above settings, it takes the spot 10 milliseconds to move 1 cm to the right; the peak voltage is 3 V

Measuring frequency The frequency of an a.c. voltage can be found by making horizontal distance measurements on its waveform on a C.R.O. screen.

In figure 6, the time base control of the C.R.O. is set at '10 ms/cm', which means that the spot takes 10 milliseconds to move 1 cm horizontally across the screen.

On the screen, one complete wave occupies a distance of 2.0 cm,

$\therefore$ the time taken for one complete wave to be traced out

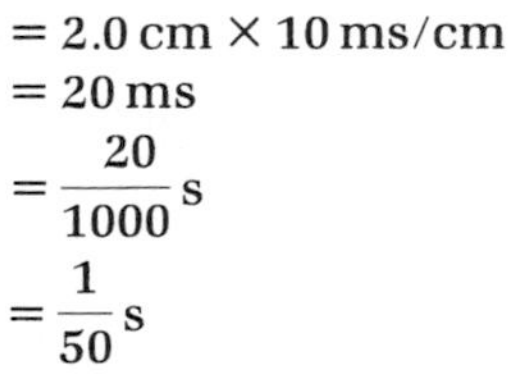

$$= 2.0\ \text{cm} \times 10\ \text{ms/cm} = 20\ \text{ms} = \frac{20}{1000}\ \text{s} = \frac{1}{50}\ \text{s}$$

$\therefore$ the number of complete waves traced out every second = 50

From which it follows that *the frequency of the a.c. voltage is* 50 Hz.

Television

A television set is essentially a sophisticated type of C.R.O. A cathode ray tube from a monochrome ('black and white') set is illustrated in figure 7. Like the tube in a C.R.O., this tube has an electron gun and a fluorescent screen. Here however, the electron beam is deflected by magnetic coils rather than by deflection plates inside the tube.

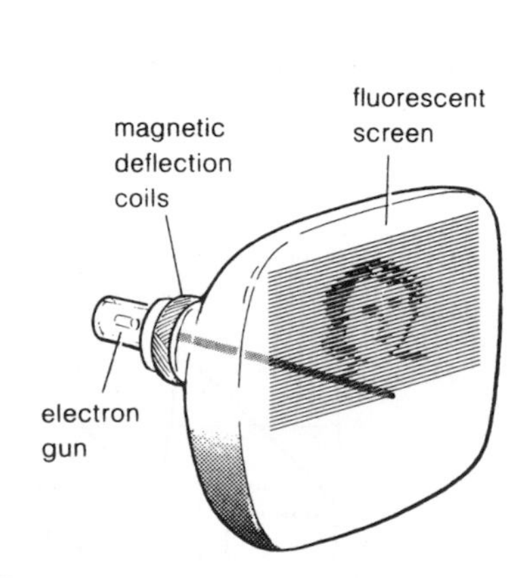

Figure 7 A cathode ray tube is the basis of the "black and white" t.v. set

Two time base circuits are used in a TV set. The horizontal or **line** time base moves the spot across the screen; the vertical time base moves it down the screen at a much slower rate. This produces a series of parallel lines which make up the picture area. The blacks, whites and greys of the picture are produced by varying the brightness of the spot as it travels over the screen. The moving spot 'draws' 25 separate still pictures every second, though to the eye, these appear as a continuously moving picture.

Needless to say, producing a colour picture is a much more complicated process. Most colour tubes have three electron guns and a screen which is coated with many thousands of phosphor strips as illustrated in figure 8.

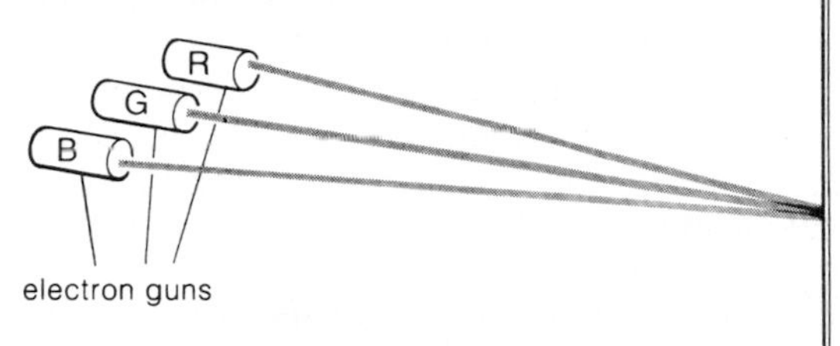

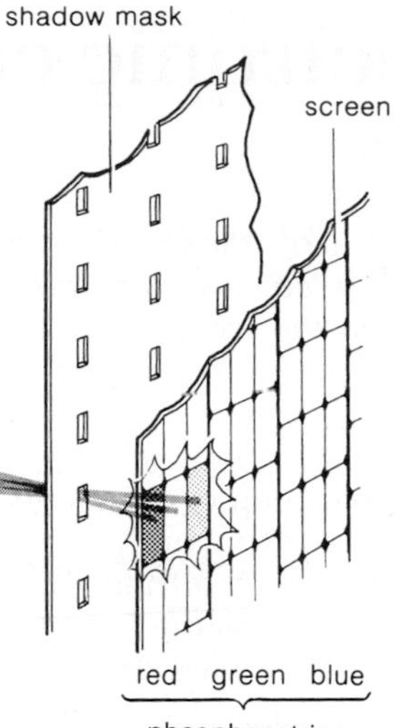

Figure 8 In a colour t.v. set three electron beams are used, each of which is controlled independently

These fluoresce red or green or blue when an electron beam strikes them. A **shadow mask** just behind the screen ensures that each electron gun can cause fluorescence in strips of one 'colour' only. In effect, the colour tube produces separate red, green and blue pictures on top of each other, and the eye sees these from a distance as a single full colour picture (see page 192).

Questions

1 In the tube of a cathode ray oscilloscope, a) what are the two functions of the anodes? b) what is the function of the grid?

2 Figure 9 represents the screen and deflection plates of a C.R.O.
  a) Copy figure 9a, and label the X-plates and the Y-plates.
  b) Copy figures 9b and 9c. In each case, indicate which plate or plates must be + and which − to produce the spot deflection shown.
  c) How is the line in figure 9d produced?
  d) Give two ways in which the length of the line in figure 9d could be increased.

3 Figure 10 shows how the time base voltage in a C.R.O. varies with time.
  a) To which plates is the time base voltage applied?
  b) What happens to the spot between points A and B on the graph?
  c) What happens to the spot between points B and C on the graph?

4 Figure 11 shows the waveform produced on the screen of a C.R.O. when an a.c. supply is connected to the input terminals.
If the gain control is set to 50 mV/cm and the time base control to 10 ms/cm, calculate the peak voltage of the a.c. supply and its frequency.

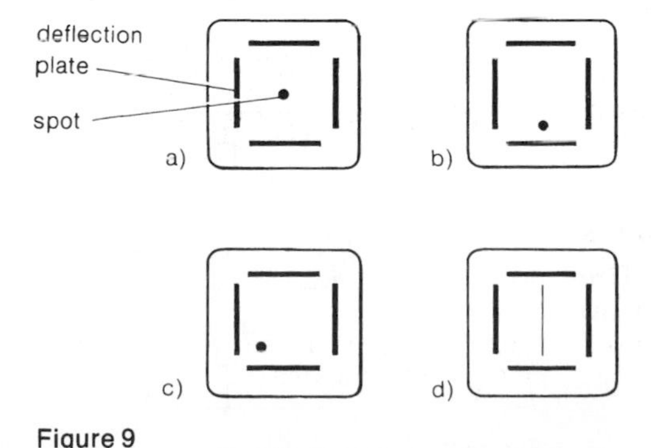

Figure 9

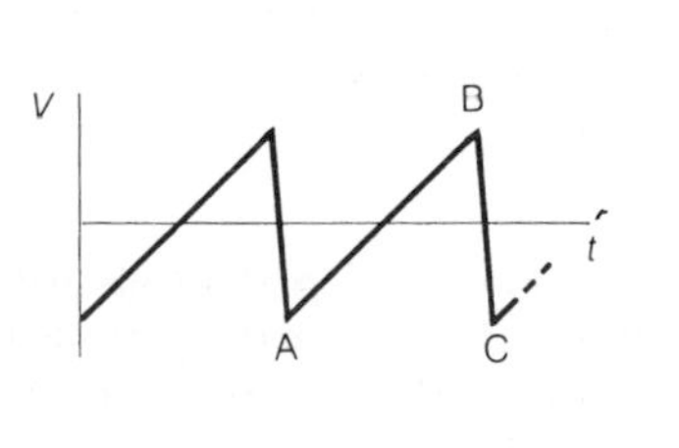

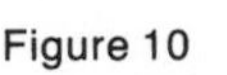

Figure 10

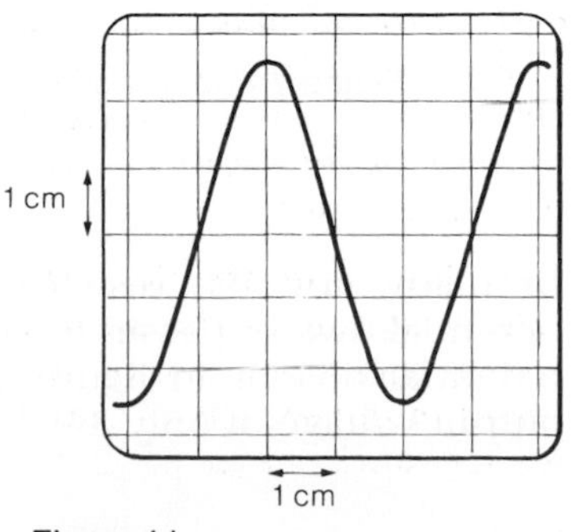

Figure 11

8.3

Electronic components

Circuits like those in oscilloscopes, TVs and radios are called electronic circuits. The components below all have important parts to play in the electronic circuits on the next few pages.

Electronic circuits are designed on a large scale. The design may be shrunk photographically before manufacture.

Resistors keep currents and voltages at the levels needed for other components to work properly. The resistance in ohms is marked on the side using either the **colour code** or the **resistance code** (figure 1). The value is only approximate. When resistors are made, the resistance can change slightly from one to the next.

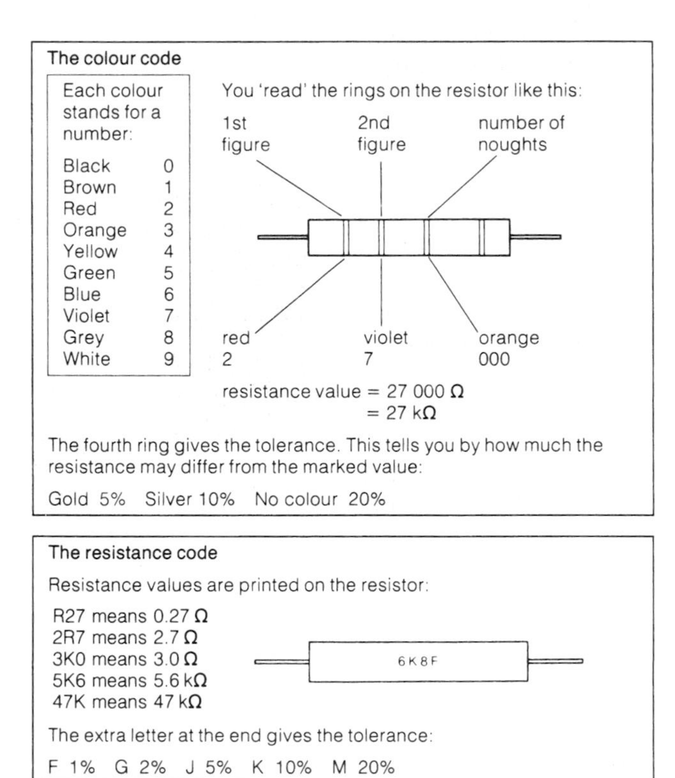

Figure 1 Codes for resistance

In many circuits, resistors are used as **potential dividers**. A potential divider passes on just part of any voltage put across it. The potential divider in figure 2 has an input voltage of 12 V, but an output voltage of only 3 V.

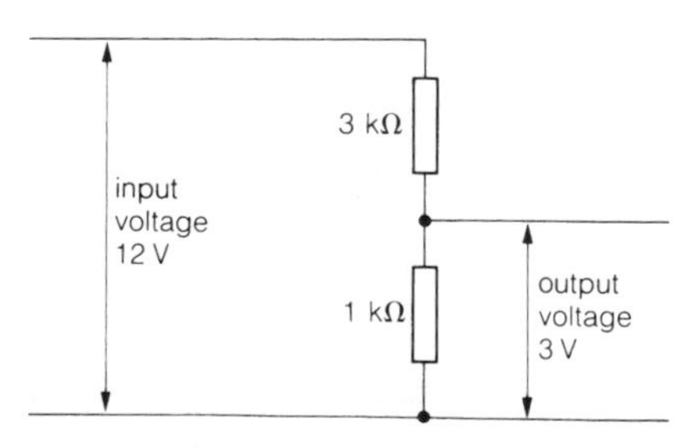

Figure 2 Potential divider

Thermistors, as shown in figure 3, are very sensitive to temperature change. With most types, the resistance falls sharply when the temperature rises.

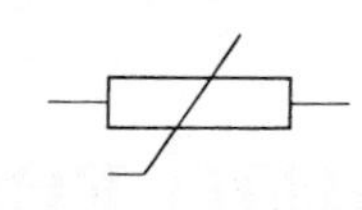

Figure 3 Thermistor and symbol

Light dependent resistors [**LDRs**] or **photoresistors** as shown in figure 4 are sensitive to light. Their resistance falls when the light intensity rises.

Capacitors (figure 5) store small amounts of charge, usually only for a short time. Their capacitance value is normally given in microfarads [μF] or picofarads [pF]:

1 μF = 1 000 000 pF

Electrolytic capacitors have the largest capacitance values. But their dielectrics are damaged if they are connected the wrong way round. One terminal is marked +, so that you know which way to make the connection.

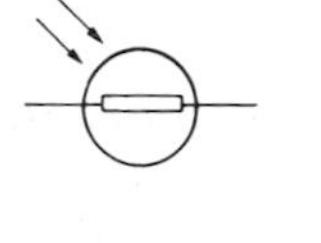

Figure 4 LDR, and symbol

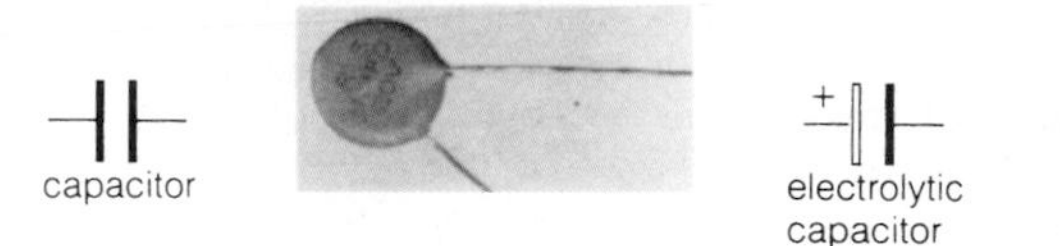

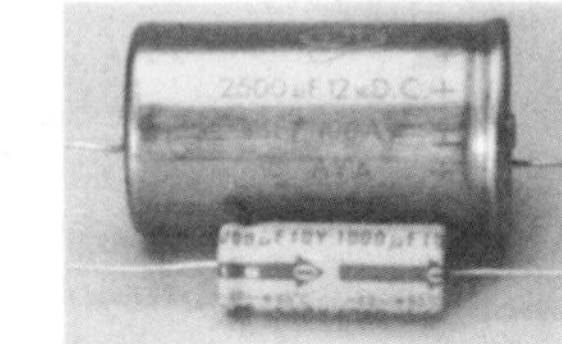

Figure 5 Capacitors, and symbols

Light emitting diodes, or **LEDs**, as shown in figure 6 are used as indicator lights on electronic equipment. They glow red, green or yellow when a current flows through. LEDs can also be used in digital displays. Seven of them, lighting up in different combinations, can display any number from 0 to 9.

LEDs won't light unless they are connected the right way round. The arrowhead in the symbol has to match the conventional current direction in the circuit. They are easily damaged by current, and a series resistor is usually needed to limit the current flowing through.

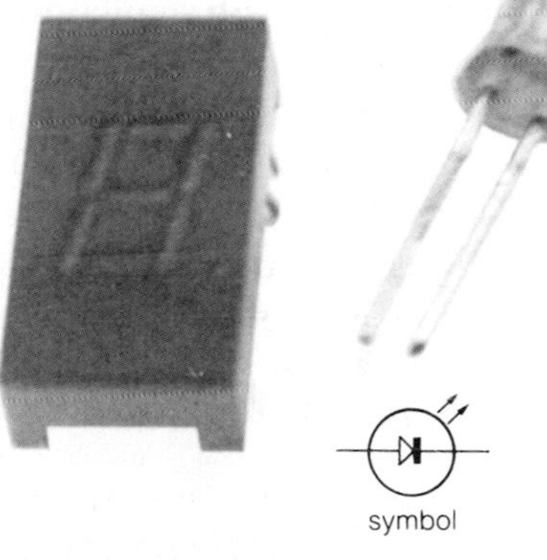

Figure 6 Light emitting diodes

Questions

1 What is the resistance of each of the resistors in figure 7? What is the tolerance of each? Explain what is meant by the tolerance.

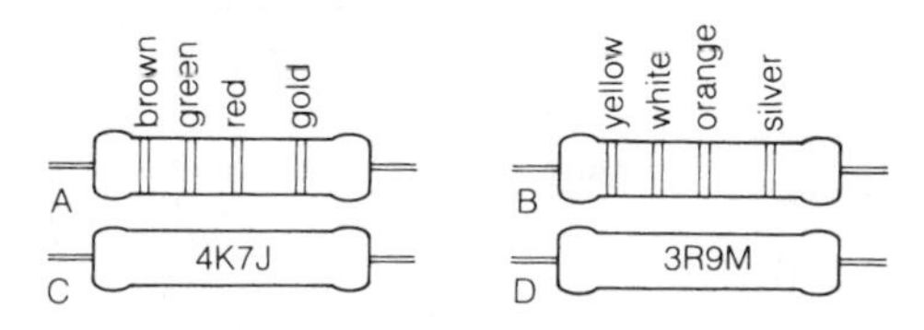

Figure 7

2 *electrolytic capacitor* *thermistor* *LDR* *potential divider* *resistor*

Which of these
a) passes on a proportion of a voltage
b) stores small amounts of charge
c) is sensitive to light
d) has three terminals
e) has a resistance which falls when the temperature rises
f) has + and − terminals which must be connected the right way round?

3 You have an 8 V supply, and a selection of resistors of the following values: 1 kΩ 2 kΩ 3 kΩ 4 kΩ
Draw diagrams to show how you would use them to make a potential divider with an output voltage of
a) 2 V b) 4 V c) 6 V d) 1 V

8.4

Diodes and rectification

Diodes are electrical one-way valves. Most are made from silicon and are the simplest of all silicon chips.

A diode is any two-terminal device which allows current to pass through it in one direction but not the other. At one time, diodes were commonly of the thermionic type described on page 308. Nowadays, most are of the 'solid state' type shown in figure 1. These usually consist of a specially treated crystal or **chip** of silicon, sealed in a case with terminals attached.

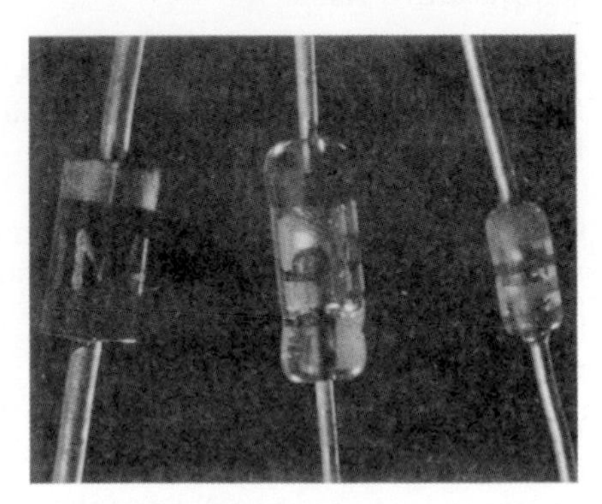

Figure 1 Solid state diodes

Semiconductors: *n*- and *p*-type

Silicon belongs to a group of materials known as **semiconductors**. Pure semiconductors act as insulators when very cold, and conduct only poorly at normal room temperatures. However, they can be given greatly improved conducting abilities by 'doping' them with small amounts of certain other substances. They then become either ***n*-type** or ***p*-type** semiconductors depending on the substance added.

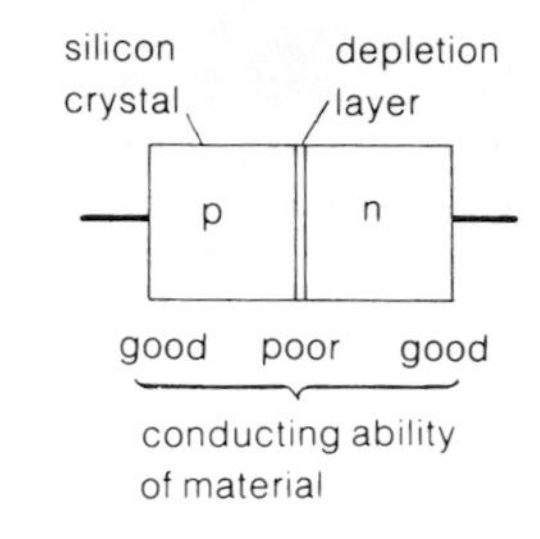

Figure 2a *p-n* junction diode

Silicone doped with phosphorus is an example of an ***n***-type material. The presence of the phosphorus atoms increases the number of ***n***egative electrons which are free to move through the material.

Silicon doped with boron is an example of a ***p***-type material. The boron atoms create gaps or ***p***ositive holes in the electron structure and electrons are then able to move through the material by jumping from one hole to another.

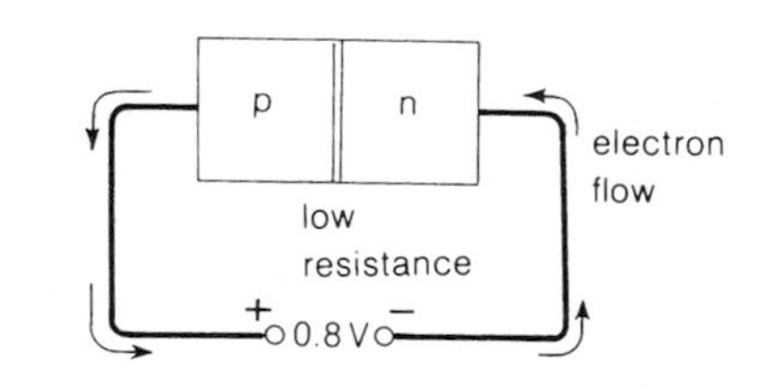

Figure 2b the diode 'forward biased'

The *p-n* junction diode

A diode can be produced by doping a crystal of pure silicon so that a junction is formed between *p*-type and *n*-type regions. This is illustrated in figure 2a. *p*-type material meets *n*-type material across a narrow layer called the **depletion layer**. Material in this layer conducts very poorly.

If a p.d. is applied across the crystal as in figure 2b, a current flows. Despite the poor conducting ability of the material in the depletion layer, the layer is so narrow that its resistance is low. The crystal in this case is said to be **forward biased**. Note that:

1 the *p*-type material has been made ***p***ositive
2 the p.d. is assisting the natural tendency of electrons in the *n*-type region to move across and fill holes in the *p*-type.

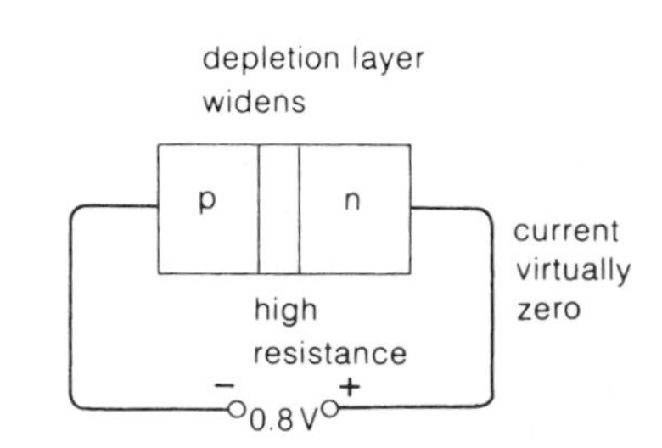

Figure 2c the diode 'reverse biased'

If the p.d. is applied in the opposite direction as in figure 2c, the current through the crystal is virtually zero. This is because electrons and holes are pulled away from the depletion layer, making it much wider and greatly increasing its resistance. The crystal in this case is said to be **reverse biased**.

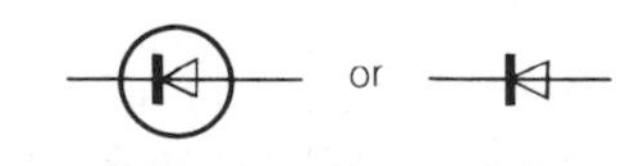

Figure 2d diode symbols

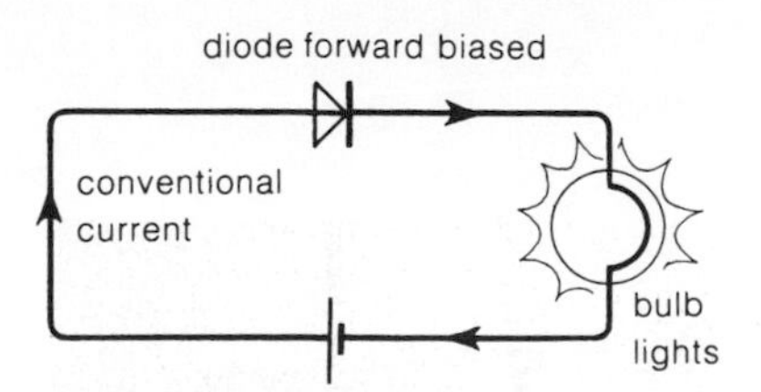

Figure 3a With the diode forward biased, a current flows

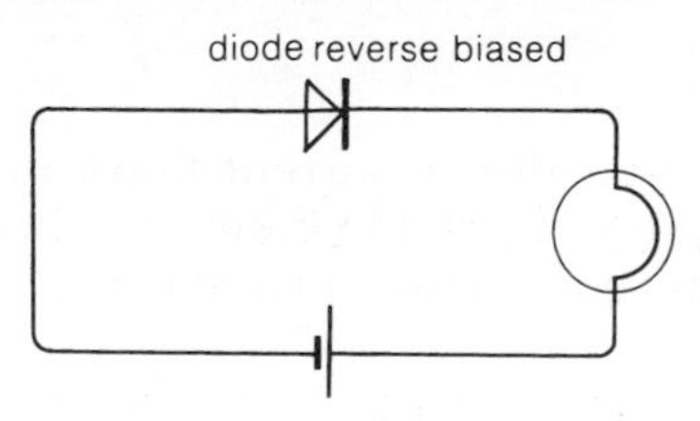

Figure 3b With the diode reverse biased, virtually no current flows

The action of a *p-n* junction diode can be demonstrated using the circuits in figure 3. The diode in each case is drawn symbolically, the arrowhead showing the direction in which conventional current can flow. The bulb only lights up if the diode is forward biased.

Junction diode: variation of *I* with *V*

The circuit shown in figure 4 can be used to find out how the current *I* through a junction diode varies with the p.d. *V* across it. *V* is set at different values by adjusting the rheostat. The corresponding value of *I* is measured in each case.

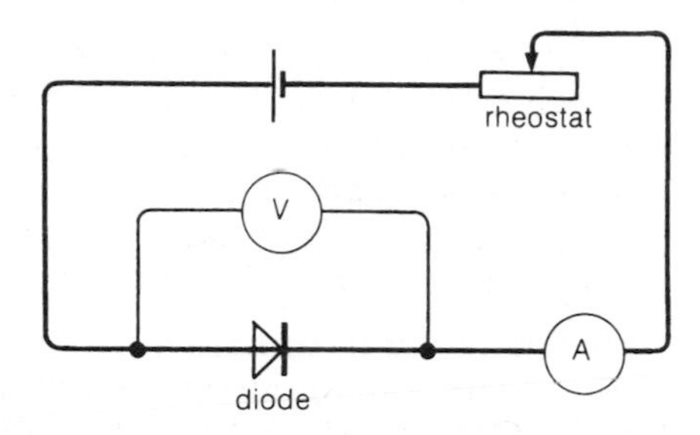

Figure 4 This circuit can be used to study the properties of a diode

Typical results are shown in graphical form in figure 5. This is the **characteristic curve** of the diode. Points to the right of the current axis are obtained with the diode forward biased as shown. Points to the left of the current axis are obtained with the battery connections reversed. Note that the diode does *not* obey Ohm's law.

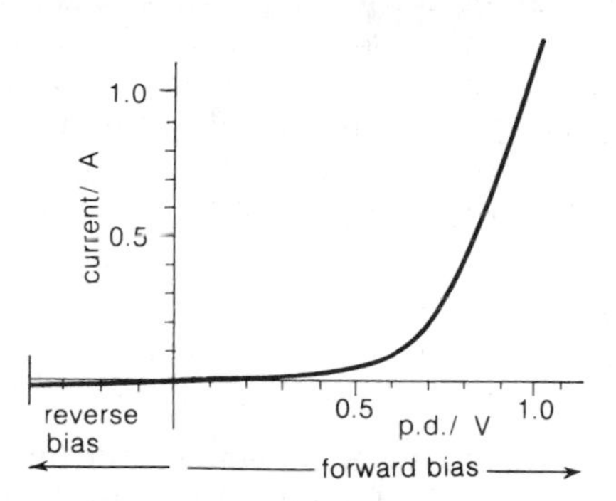

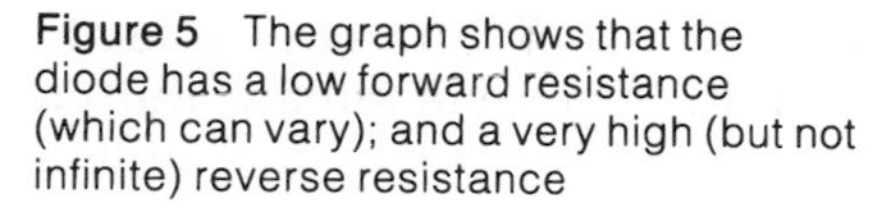
Figure 5 The graph shows that the diode has a low forward resistance (which can vary); and a very high (but not infinite) reverse resistance

An ideal diode would have zero resistance in the forward direction and infinite resistance in the reverse direction. The graph shows that this isn't the case in practice. This particular diode has a forward resistance of around an ohm. When reverse biased, it allows a small current (less than a microampère) to flow through.

Half-wave rectification

Diodes are also known as **rectifiers**. They can be used to change a.c. into d.c., a process called **rectification**.

A simple rectification circuit is shown in figure 6. The resistor represents the piece of equipment or the **load** which is to be supplied with direct current. Without the diode, the current would flow alternately backwards and forwards through the load. With the diode present, the current can flow in the forward direction only.

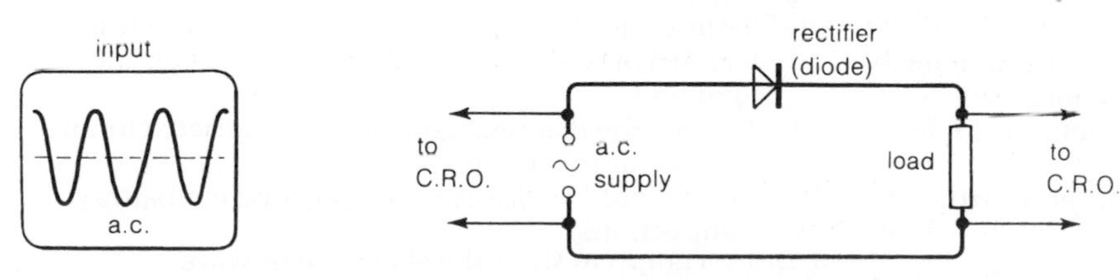

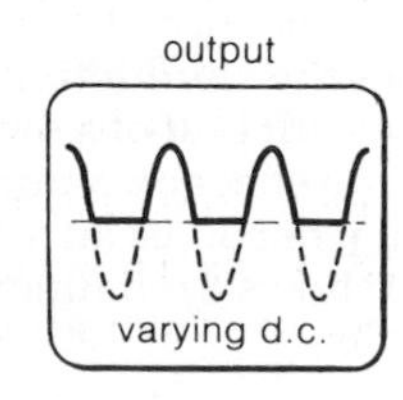

Figure 6 Half-wave rectification: the negative part of the a.c. current is prevented from passing

The effect of the diode can be seen by connecting the input terminals of a C.R.O. first across the a.c. supply and then across the load. The final waveform on the screen is the positive half only of the original a.c. waveform – hence the term 'half-wave' rectification.

Smoothing

In the circuit in figure 7, the 'one-way' direct current flows in a series of surges with brief periods of zero current in between. These surges can be partly smoothed out by connecting a large capacitor across the load as shown in figure 8.

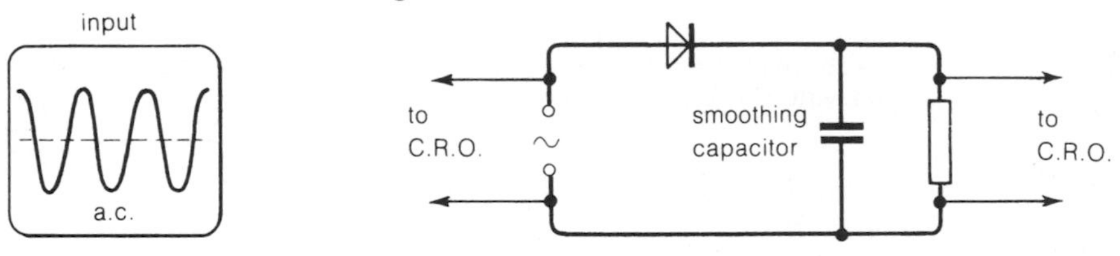

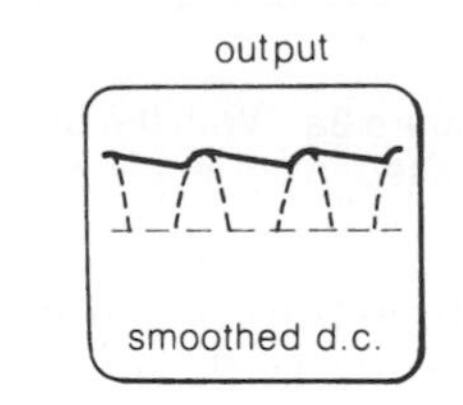

Figure 7 Half-wave rectification with smoothing

The capacitor charges up when current flows from the diode, then discharges through the load when the current from the diode is zero. Smoothed in this way, the current through the load is similar to the steady direct current which would flow from a battery.

Full-wave rectification

Figure 8 shows a circuit to produce **full-wave rectification**. Using an ingenious arrangement of diodes, called a **bridge rectifier**, this reverses the negative half of each a.c. cycle, instead of just blocking it. Follow the current arrows to see how the circuit works.

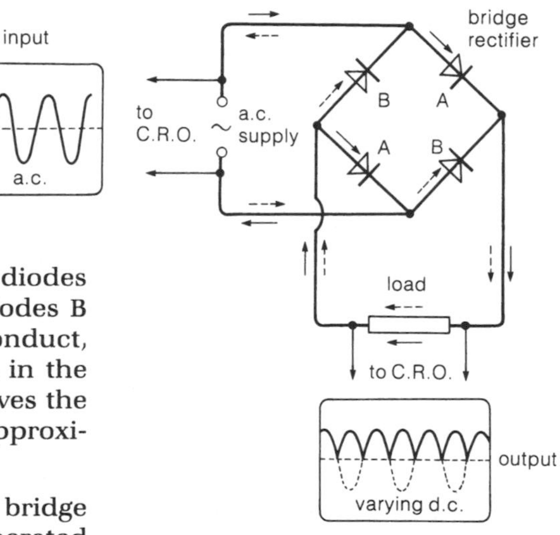

Figure 8 Full-wave rectification. The negative part of each a.c. cycle is reversed.

During the forward half of each a.c. cycle (solid arrows), the diodes marked A are forward biased and therefore conduct, but diodes B block. During the reverse half (broken arrows), diodes B conduct, but diodes A block. The result is that current always flows in the same direction through the load, no matter which way it leaves the supply. When smoothed, the current is an even better approximation to steady d.c. than in figure 7.

Combine a transformer (to reduce mains voltage) with a bridge rectifier and a smoothing capacitor, and you have a mains-operated d.c. power supply – as used in radios, instead of batteries.

Questions

1 Name a semiconductor material.
How does the conducting ability of a semiconductor compare with that of a metallic conductor?
What is the effect on a pure semiconductor of doping it so that it becomes an *n*-type material?

2 Figure 9 shows a *p*-*n* junction diode in series with a small bulb.

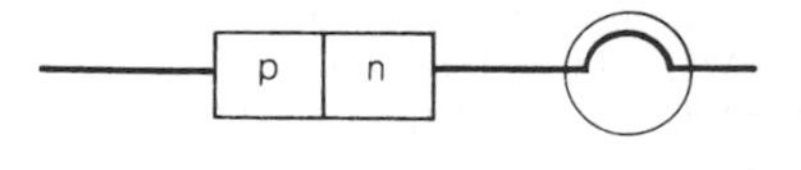

Figure 9

a) Copy and complete the diagram to show how a battery should be connected so that the diode is forward biased. What would be the effect on the light bulb?
b) Redraw the diagram using the appropriate circuit symbol for the diode.
c) What would be the effect of reversing the battery connections?

3 Draw a graph to show the effect of half-wave rectification on an alternating current.
Draw a circuit which could be used to produce half-wave rectification.
What would be the effect of connecting a large capacitor across the load?

8.5

Transistors

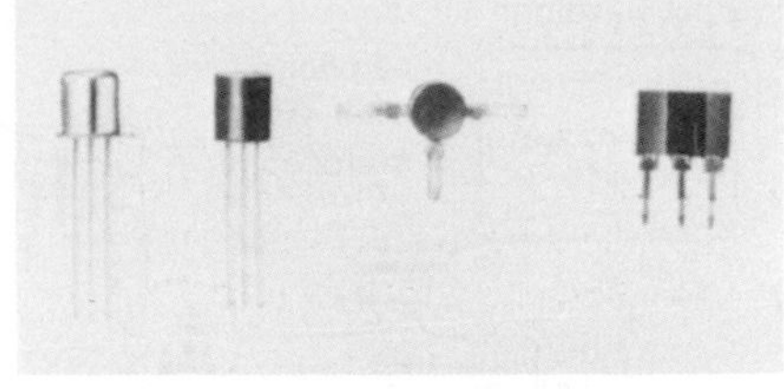

Figure 1 A variety of transistors

The first transistor was made in 1948. Since then, the science of electronics has been revolutionized by this tiny semiconductor device which can amplify and act as a high-speed switch.

Examples of transistors, complete with outer cases, are shown in figure 1. Like a junction diode, a transistor consists of a specially treated semiconductor crystal, but it has three terminals instead of two. It is important because it can link two circuits in such a way that the current through one controls the current through the other.

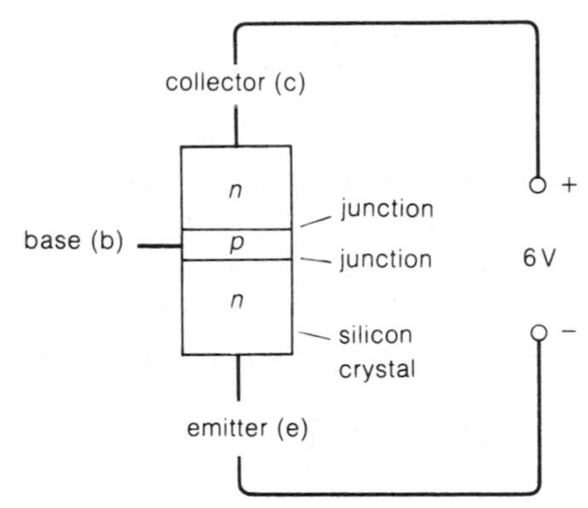

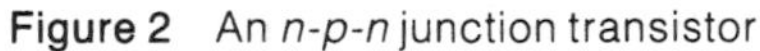

Figure 2 An *n-p-n* junction transistor

The n-p-n junction transistor

Figure 2 shows in simplified form the structure of one type of transistor, the n-p-n junction transistor. Basically, this consists of a semiconductor chip, usually silicon, doped so that two p-n junction diodes are formed back to back. The n-type regions are the **emitter** and **collector** of the transistor; the thin p-type region is the **base**.

If the battery in figure 2 is connected across the transistor as shown, the upper junction becomes reverse biased, so no current flows.

Figure 3 shows what happens if a small p.d. is applied across the lower junction so that it is forward biased. Not only does a current flow through the base, but this current greatly reduces the 'current blocking' effect of the upper junction. The junction starts to conduct and current flows between emitter and collector. Typically, a small current in the base or **input** circuit can cause a current fifty to a hundred times larger to flow in the collector or **output** circuit.

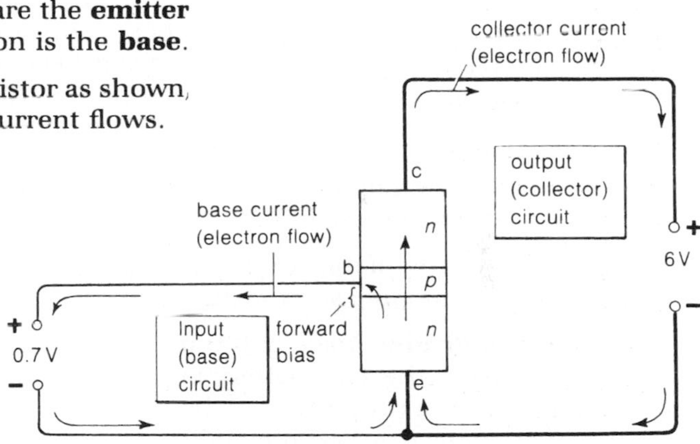

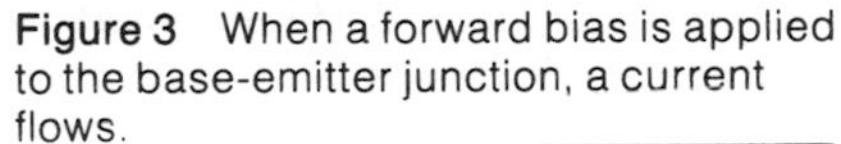

Figure 3 When a forward bias is applied to the base-emitter junction, a current flows.

In figure 4, figure 3 has been redrawn using the circuit symbol for an n-p-n junction transistor. The arrow in the symbol indicates the conventional current direction, though the general movement of electrons is from emitter to base to collector. With the output and input circuits joined as shown, the transmitter is said to be in **common-emitter** mode.

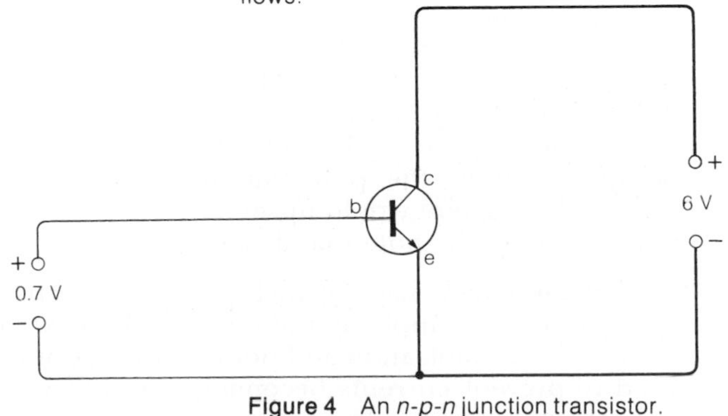

Figure 4 An *n-p-n* junction transistor. Note that the emitter has a common connection with the input and output circuits

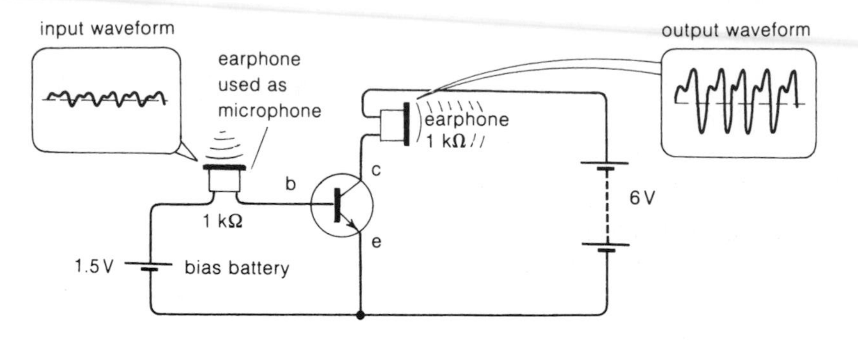

Figure 5 A simple transistor amplifier. The small variations in the base current affect the resistance of the transistor so that similar, but much greater current variations occur in the collector circuit

The transistor as an amplifier

A transistor can be used to amplify ('magnify') current changes because a small change in base current produces a large change in collector current.

A simple transistor amplifier circuit is shown in figure 5. As in the previous circuit, the base-emitter junction is forward biased. In this case however, the base current is varying because of the small alternating voltage produced by the microphone. The small changes in base current cause much larger changes in collector current. The collector circuit includes an earphone through which you would hear an amplified version of the original sound. Typical input and output waveforms are illustrated in the diagram.

A simple transistor amplifier in use

An improved version of the circuit, using one battery only, is shown in figure 6. Note the following features:

The potential divider (page 251 and 332) This replaces the 1.5 V battery in figure 5. It applies a proportion of the main battery voltage across the emitter-base junction, and keeps the junction forward biased.

The input capacitor This passes on current changes from the microphone (see pages 241 and 333) but blocks the steady current which might otherwise flow through the microphone from the potential divider. Such a current would upset the biasing effect of the potential divider.

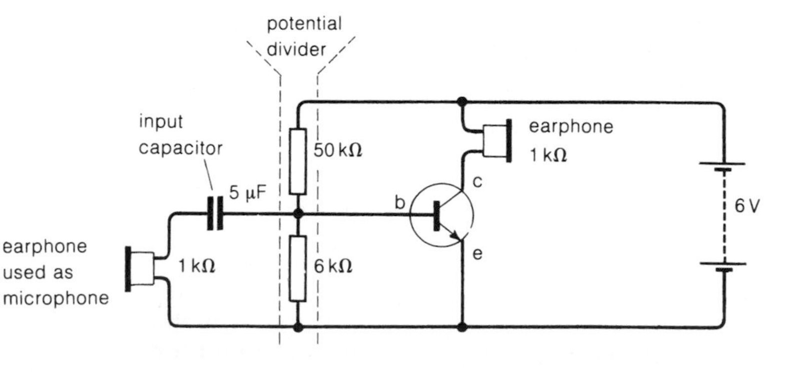

Figure 6 An improved transistor amplifier. The potential divider replaces the smaller battery in figure 5

Practical amplifiers usually include several transistors, with the output from one supplying the input of the next, and so on. This gives greater amplification and power output. Components are also added to prevent currents becoming unstable and the transistors overheating.

The transistor as a switch

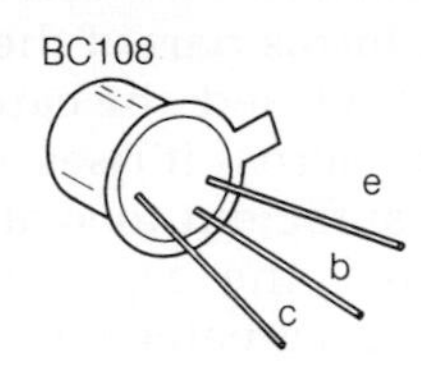

In a transistor, no current can flow in the collector circuit unless a current flows in the base circuit. This means that a transistor can be used as a switch, which is turned on or off by changes in the base current. The switching circuits described below all work in this way. Like the amplifier, each uses a potential divider to drive current round the base circuit. But each has a **photoresistor**, **thermistor**, or other energy-sensitive device in the potential divider which can alter the base current sufficiently to switch the transistor on or off.

Light-operated switch (figure 7) This circuit makes the bulb light up in the light and go off in the dark.

The potential divider contains a light-dependent resistor, or photoresistor. In darkness, the photoresistor has a resistance of 1 MΩ. This means that only a small fraction of the full 6 volts is dropped across the resistor, R, and the base current is too low to switch the transistor on. In bright light however, the resistance of the photoresistor falls to only a few hundred ohms. This gives R a much greater share of the available voltage. The base current rises, the transistor is switched on, and the bulb lights.

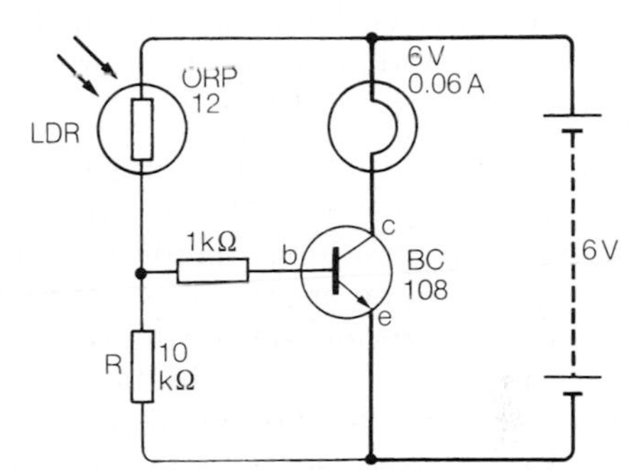

Figure 7 Light-operated switch

Swopping the positions of the photoresistor and R, produces a circuit which keeps the bulb off in the light and on in the dark. A circuit of this type could be used to switch on a car's parking lights at dusk, or to trigger a burglar alarm when a light beam is cut.

Heat-operated switch (figure 8) This circuit is similar to figure 7, except that the photoresistor has been replaced by a thermistor. If heat is applied to the thermistor, its resistance drops, the transistor is switched on as before, and the bulb lights.

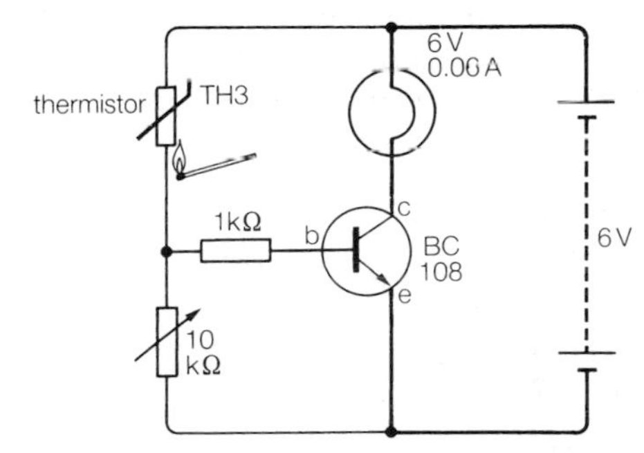

Figure 8 Heat-operated switch

Sound-operated switch (figure 9) As in the amplifier, the potential divider includes a microphone. If, say, you shout into the microphone, enough base current is produced to switch on the transistor. This sends current to the **thyristor**, or **silicon controlled rectifier**. This is a special type of diode which won't conduct until a current enters its gate, but stays on, or **latched**, until its circuit is broken. As the bulb is in the thyristor circuit, it remains lit, even when the sound has died away.

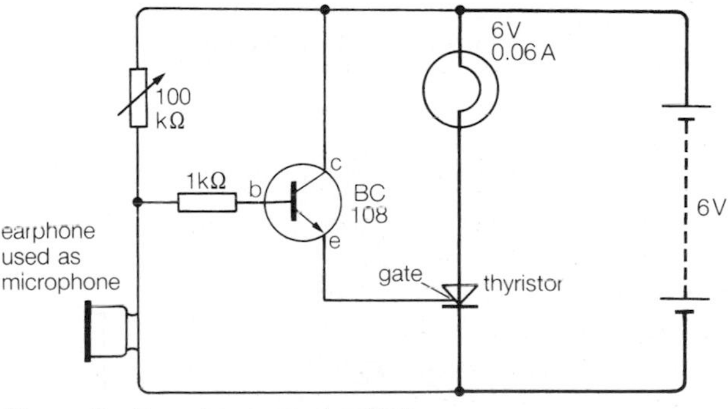

Figure 9 Sound-operated switch

Time-operated switch (figure 10) Here, a capacitor forms part of the potential divider. When switch S_1 is closed, the capacitor slowly charges. As the voltage across it rises, so does the base current, until it is sufficient to switch on the transistor. If R is increased, the capacitor charges more slowly, and there is a greater time delay between closing S_1 and transistor switch on. To reset the circuit, the capacitor is discharged by closing S_2.

As an improvement on previous circuits, the bulb is not placed in the collector circuit. Instead, it is operated by a magnetic relay of the 'normally open' type (see page 290 and figure 11).

The diode is required to protect the transistor when the relay operates. When the transistor switches 'off', the changing magnetic field in the relay coil could induce enough current to damage the transistor. The diode stops this happening. It short-circuits the coil, and conducts the currently harmlessly away. While the transistor is 'on', the diode is reverse biased, so it doesn't conduct.

Magnetic relays could be fitted to any of the circuits on the previous page. This would enable them to switch on equipment drawing much more power – such as alarm bells, or mains-operated appliances.

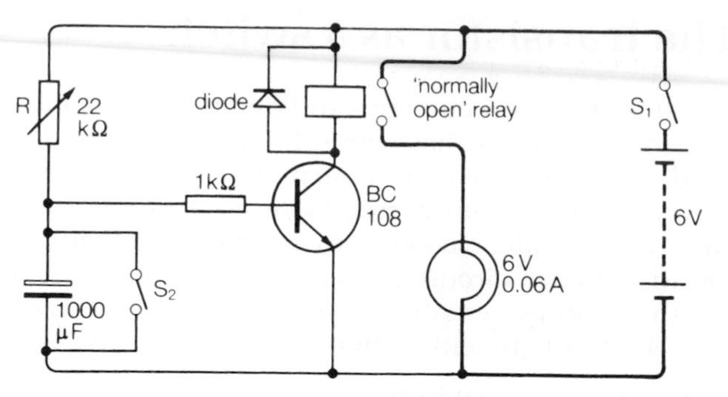

Figure 10 Time operated switch.

Figure 11 A magnetic relay

Questions

1 a) Redraw figure 12 using the appropriate symbol for the n-p-n junction transistor. Mark on your diagram the emitter, base, and collector of the transistor.

b) What does the arrow in the transistor symbol indicate?

c) By moving C, what is the maximum voltage which can be applied across terminals A and B of the transistor? What is the minimum voltage?

d) If C is moved upwards from the position shown, what is the effect on the base current? What is the effect on the collector current?

2 In figure 6, why is the input capacitor necessary?

3 Draw a circuit diagram to show how a photoresistor and transistor could be used to switch on a small bulb when darkness falls.

4 Redraw figure 10, replacing the 'normally open' relay by a 'normally closed' one (see page 290). Describe what then happens to the bulb from the instant switch S_1 is closed onwards. What is the effect of decreasing the resistance of R? Suggest a practical use for the circuit you have drawn.

5 Draw a circuit diagram to show how a thermistor and transistor could be used in a fire alarm (base your design on figure 8, but, instead of the bulb, include an electric bell operated by a magnetic relay).

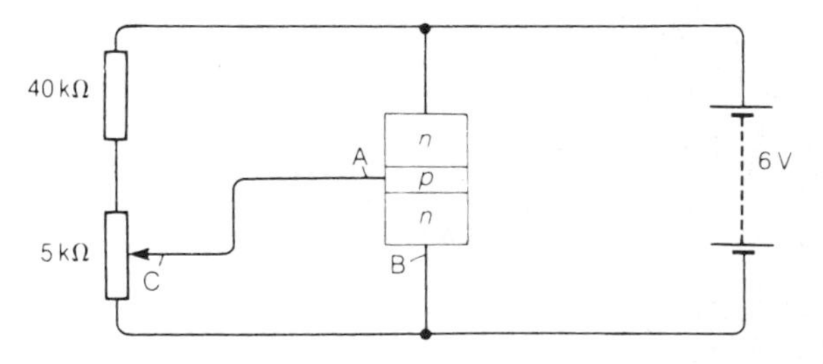

Figure 12

8.6
Microchips

Transistor circuits can be formed on a single, tiny piece of silicon. The result is a microchip.

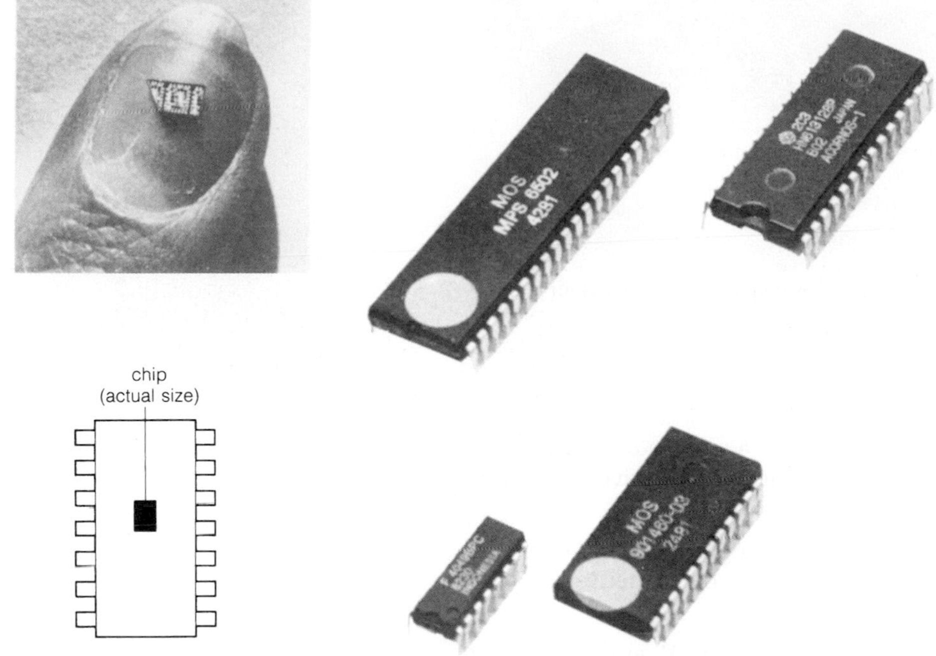

Figure 1 Integrated circuit

Figure 1 shows a selection of **integrated circuits** [**ICs**] or **microchips**. In each, transistors, resistors, and diodes have been formed on one wafer-thin chip of silicon. The result is a complete amplifier or switching system only a few millimetres square.

The development of complex circuits using microchips has helped to reduce powerful computers in size, so that they can be used at the desk for a wide range of tasks

Amplifier ICs

Figure 2 Amplifier ICs have many applications: above is a cassette player and a voltmeter

Both the items in figure 2 contain amplifier ICs:

In the cassette player, the amplifier changes a small a.c. voltage from the pick-up into an a.c. voltage large enough to drive the loud-speaker.

In the electronic voltmeter, an amplifier changes a small a.c. or d.c. input voltage into a voltage large enough to be measured by the meter.

Gain If an amplifier has an input voltage of 0.5 V, and an output voltage of 5.0 V, its voltage gain is 10:

$$\textbf{voltage gain} = \frac{\textbf{output voltage}}{\textbf{input voltage}}$$

Operational amplifiers

Operational amplifiers, or **op-amps**, are used in measuring instruments and certain types of computer. Figure 3 shows an op-amp in symbol form. It has two inputs and one output. It needs a three-terminal power supply, though these connections aren't always shown in circuit diagrams.

A commonly used op-amp is the Type 741 shown in figure 4 below.

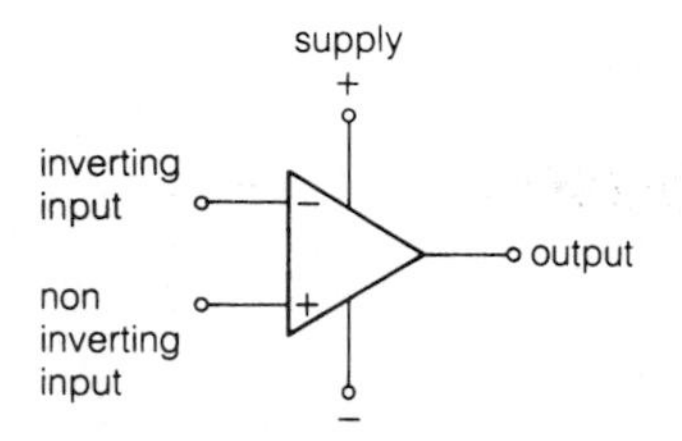

Figure 3 Operational amplifier

Figure 4

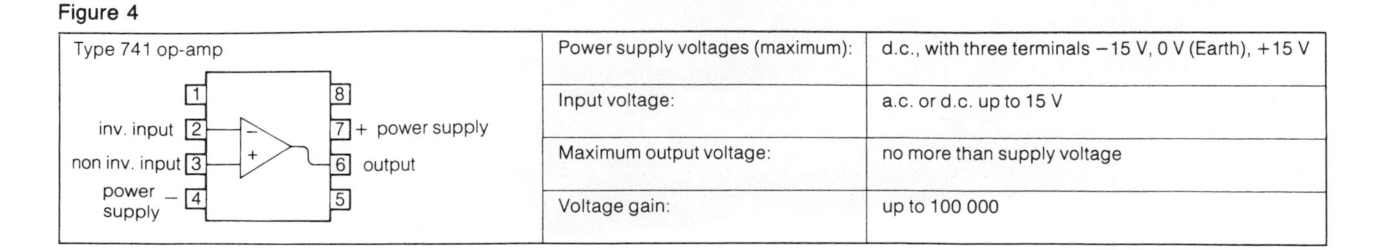

Power supply voltages (maximum):	d.c., with three terminals −15 V, 0 V (Earth), +15 V
Input voltage:	a.c. or d.c. up to 15 V
Maximum output voltage:	no more than supply voltage
Voltage gain:	up to 100 000

The op-amp as an inverting amplifier

The op-amp in figure 5 is being used as an **inverting amplifier**. When the input is −, the output is +, and vice versa. In other words, the amplifier gives an inverted output.

The voltage to be amplified is fed to the inverting input of the op-amp. The non-inverting input isn't needed. So it is connected to the Earth (0 V) terminal on the power supply.

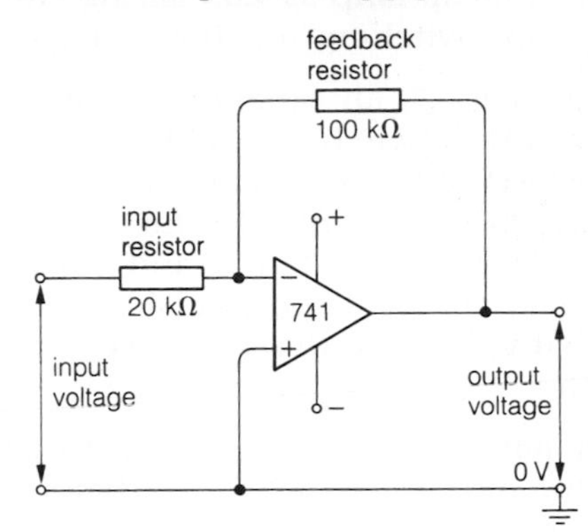

Figure 5 An inverting amplifier

The gain of an op-amp is far too high for most purposes. So a **feedback resistor** is used to reduce it. The resistor 'feeds back' some of the inverted output voltage so that it partly cancels out the input voltage. This in turn means a lower output voltage. Not only is the gain cut. There are other advantages as well:

1 the op-amp can handle higher a.c. frequencies without any loss of gain;

2 the gain can be set to any particular value by choosing suitable input and feedback resistances.

There is an equation for calculating the gain:

$$\textbf{voltage gain} = \frac{\textbf{resistance (feedback)}}{\textbf{resistance (input)}}$$

For example, in figure 5:
resistance (feedback) = 100 kΩ
resistance (input) = 20 kΩ
so, voltage gain $= \frac{100\ \text{k}\Omega}{20\ \text{k}\Omega} = 5$

Note that:

the gain would be *higher* if
the resistance of the feedback resistor was *higher*,
or
the resistance of the input resistor was *lower*

the equation isn't accurate if the output voltage is close to the supply voltage. The output voltage can never be more than the supply voltage.

To measure the gain of the amplifier:
connect a signal generator across the input so that it gives about 0.5 V a.c. Then measure the input and output voltages accurately with an oscilloscope or electronic voltmeter.

The op-amp as a switch

Like a transistor, an op-amp can be used as a switch as shown in figure 6. It has such a high gain, that it takes an input voltage of less than 0.5 mV to switch the op-amp 'on' so that its output voltage is nearly the full supply voltage (12 V in this case).

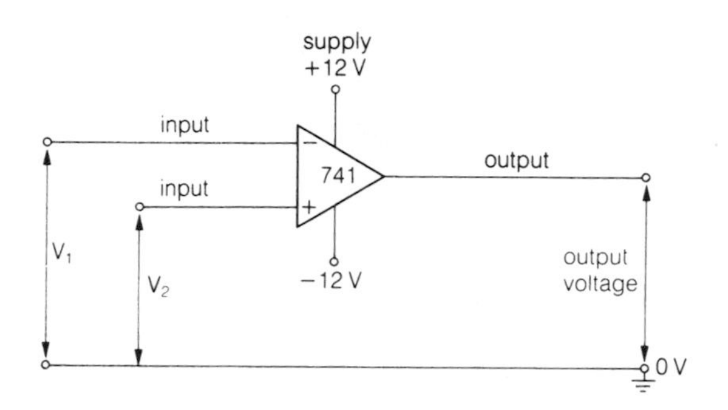

Figure 6 When switched 'on', this op-amp gives an output voltage of 12 V

However, unlike a transistor, an op-amp has *two* inputs. When it is switched 'on', the output can be either + or −:

If $V_1 > V_2$, the output is −
If $V_2 > V_1$, the output is +

So the op-amp is *comparing* the input voltages before switching 'on'. It is acting as a **comparator**.

In figure 7, an op-amp is being used in a light-operated switch. The LED will only light up if the output is +. The circuit has two potential dividers, one for each input. In the dark, the LDR (ORP 12) has a high resistance, and resistor R only has a small share of the available voltage. This means that $V_1 > V_2$, so the output is −, and the LED stays off.

If light shines on the LDR, its resistance falls. Now, the voltage across R rises. When $V_2 > V_1$, the output goes +, and the LED comes on.

Questions

1

amplifier	input voltage	output voltage
A	0.5 V	4.0 V
B	0.1 V	3.0 V
C	1.0 V	8.0 V
D	2.0 V	10.0 V

a) Which amplifier gives the greatest voltage gain?
b) Which amplifiers give the same voltage gain?

2 In figure 7,

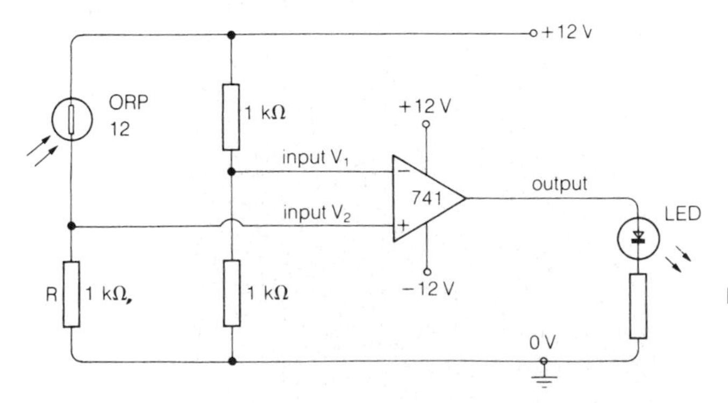

Figure 7 Light-operated switch

a) What would be the output voltage if $V_1 > V_2$?
b) What would be the output voltage if $V_2 > V_1$?
c) Why does the output have to be + for the LED to light?
d) Why is there a resistor in series with the LED?
e) What would be the effect of reversing the LED?
f) What would be the effect of swopping the positions of the LDR and the resistor R?

3 In figure 8,
a) Which terminal of the operational amplifier is the inverting input?

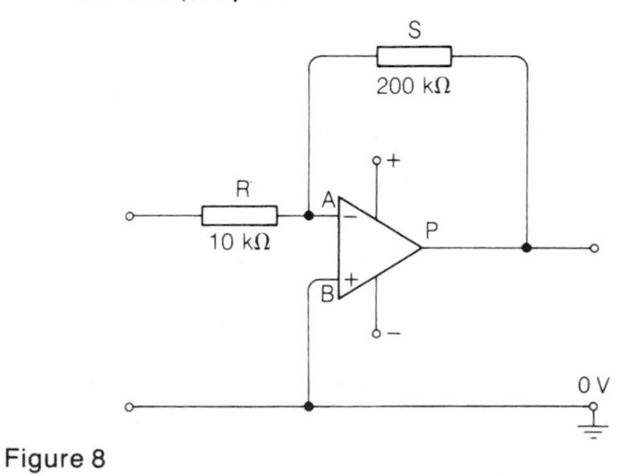

Figure 8

b) The output voltage is inverted. What does this mean?
c) Which resistor is the feedback resistor?
d) What is the purpose of the feedback resistor?
e) What is the voltage gain of the amplifier?

8.7
Logic gates

ICs are used in the most advanced switching systems. They can control anything from a security lock to a video recorder.

Figure 1 A logic chip

The door of a safe can be controlled by groups of switches called **logic gates**. If the buttons are pressed in the right order, the door opens. But if they are pressed in the wrong order, the alarm goes off.

Logic gates can be made in integrated circuit form as shown in figure 1. Each 'package' can hold up to four separate gates. Twelve of the 'pins' are used for making connections to the gates. Two others are for the power supply.

The gates needed to protect a safe are quite complicated. The gate used in figure 2 is much simpler – just two switches in a box:

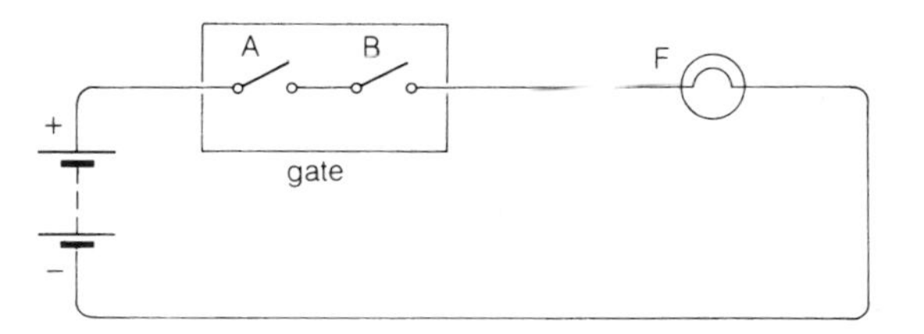

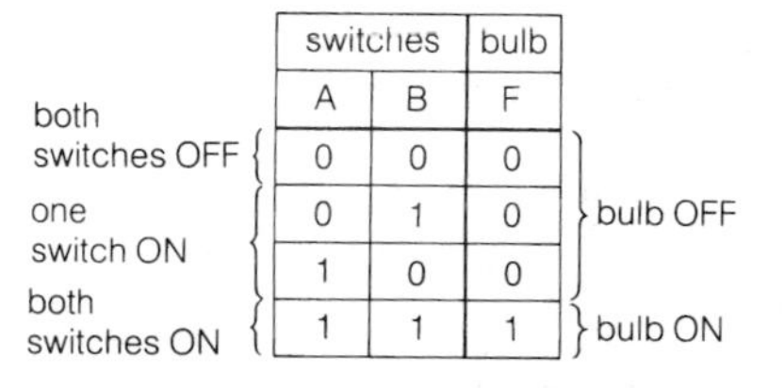

	switches		bulb	
	A	B	F	
both switches OFF	0	0	0	bulb OFF
one switch ON	0	1	0	
	1	0	0	
both switches ON	1	1	1	bulb ON

Figure 2 Simple logic gate with truth table

if both switches are closed the bulb comes ON. but if either switch is open, the bulb stays OFF.

The **truth table** shows the result of every possible switch setting. The table uses two logic numbers; 0 and 1: **Logic 0 stands for OFF. Logic 1 stands for ON.**

Most logic gates don't contain ordinary 'lever' switches. They use transistor switches instead. They have one or more input lines, and one output line.
A line is ON if there is a positive voltage on it (usually +5 V).
A line is OFF if the voltage on it is zero (0 V).
A separate d.c. supply is needed to power the gate, usually a 5 V supply.

Figure 3 A burglar alarm: when the door opens, the alarm will sound unless the correct 3-digit code is entered into the logic circuits

Types of gate

There are several types of logic gates. The main ones are shown in symbol form in figure 4, along with their truth tables. Unlike simple switches, some of the gates have more than one input. They also have power supply connections, but these aren't shown.

Figure 4

Logic gates

Logic 0: Off   Logic 1: ON

NOT gate

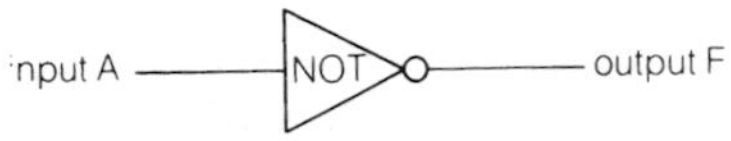

input A — NOT — output F

The output is ON if the input is NOT ON. This type of gate is called an inverter: it changes OFF to ON, and ON to OFF.

Input	Output
A	F
0	1
1	0

AND gate

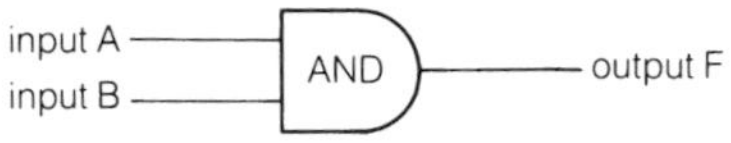

input A, input B — AND — output F

The output is ON if both input A AND input B are ON.

Inputs		Output
A	B	F
0	0	0
0	1	0
1	0	0
1	1	1

NAND gate

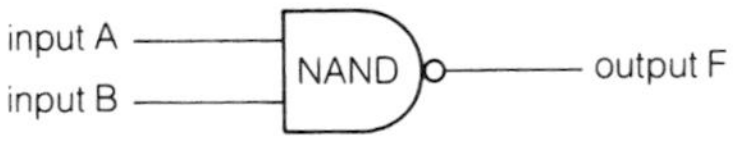

input A, input B — NAND — output F

A NAND gate is equivalent to an AND gate followed by a NOT gate. Its output is ON if input A AND input B are NOT both ON. Its output is that of an AND gate inverted.

Inputs		Output
A	B	F
0	0	1
0	1	1
1	0	1
1	1	0

OR gate

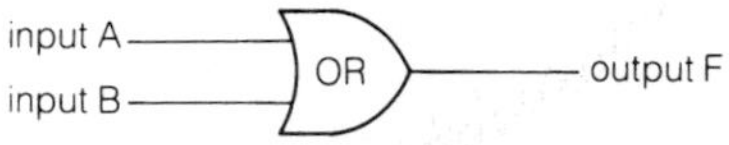

input A, input B — OR — output F

The output is ON if either input A OR input B is ON.

Inputs		Output
A	B	F
0	0	0
0	1	1
1	0	1
1	1	1

NOR gate

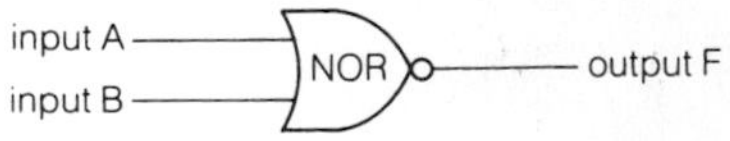

input A, input B — NOR — output F

A NOR gate is equivalent to an OR gate followed by a NOT gate. Its output is ON if neither input A NOR input B is ON. Its output is that of an OR gate inverted.

Inputs		Output
A	B	F
0	0	1
0	1	0
1	0	0
1	1	0

Security with logic

The manager of a camera shop wants an electric lock on her store room door. She needs a system without combinations or numbers to remember, so that any member of staff can unlock the door quickly. She decides on a two-switch system. If a hidden switch is turned on first, a main switch will open the door lock. But if the hidden switch is left off, the main switch will turn on an alarm instead.

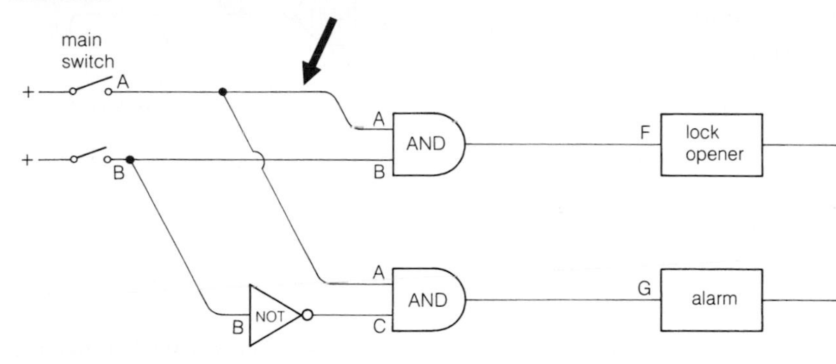

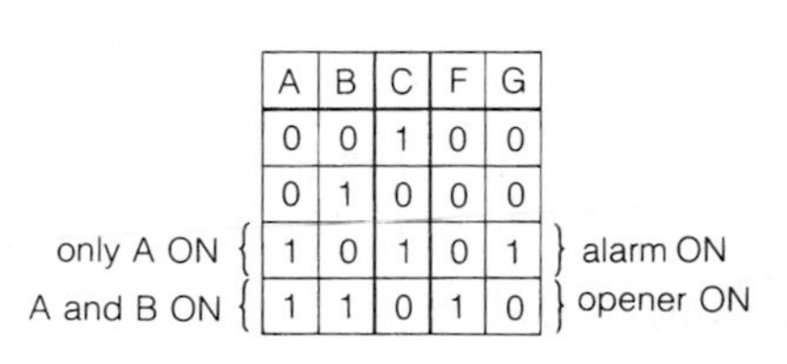

	A	B	C	F	G	
	0	0	1	0	0	
	0	1	0	0	0	
only A ON	1	0	1	0	1	alarm ON
A and B ON	1	1	0	1	0	opener ON

Figure 5 Security system with truth table

Figure 5 shows the system of logic gates that she decides to use, together with its truth table. For simplicity, complete circuits aren't shown. Nor are the connections to the power supply.

A bistable

The arrangement of gates in Figure 6 is called a **bistable**. Two NOR gates have been 'cross-coupled' so that the output from one acts as one of the inputs of the other.

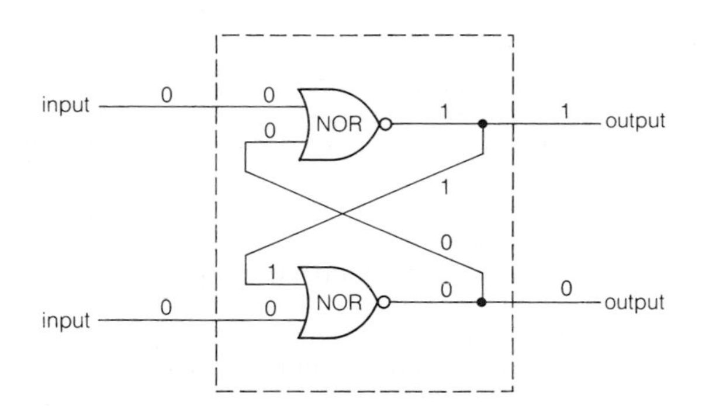

Figure 6 A bistable

The bistable has *two* stable states:
if both inputs are OFF, either the top output is ON and the bottom output is OFF, as shown,
or vice versa.

If the top input is made positive (ON), the bistable 'flips' to the other state. The top output goes OFF, and the bottom output comes ON. The input signal doesn't have to be maintained. One brief positive 'pulse' is all that is needed to make the change. Unlike a simple gate, a bistable 'remembers' that an input has been made.

A latch

A bistable with just one of its outputs in use is called a **latch**. By applying pulses to its inputs, the output can be made to go ON and OFF alternately.

In figure 7, a latch is being used to control an electric motor, via a relay. Normally, the input voltages are zero (logic 0). But pressing a switch gives an input voltage of +5 V (logic 1). The motor starts when the left-hand switch is pressed. And stops when the right-hand switch is pressed. The 'touch' buttons used to start and stop a video recorder are connected in this way.

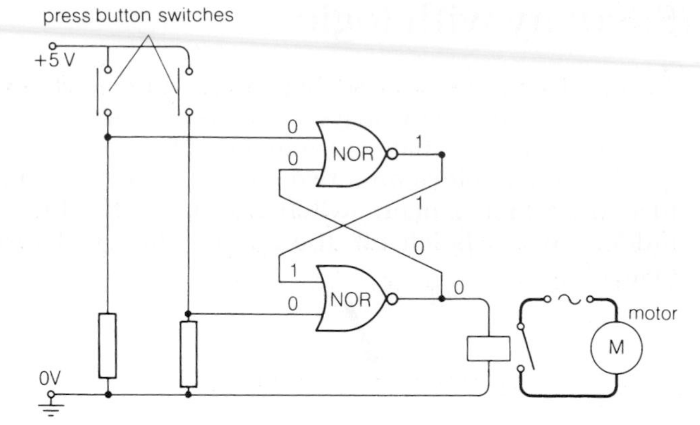

Figure 7 A bistable used as a latch. One button starts the motor. The other stops it.

Questions

1 AND OR NAND NOR NOT

In which logic gates

a) is the output ON when the input is OFF.
b) is the output ON when both inputs are ON
c) is the output ON when both inputs are OFF?

2 Figure 8 shows the fire safety system installed in a furniture showroom. Smoke turns the smoke detector ON. Heat turns the heat detector ON. Logic gates are used to turn the alarm and the fire extinguisher ON.

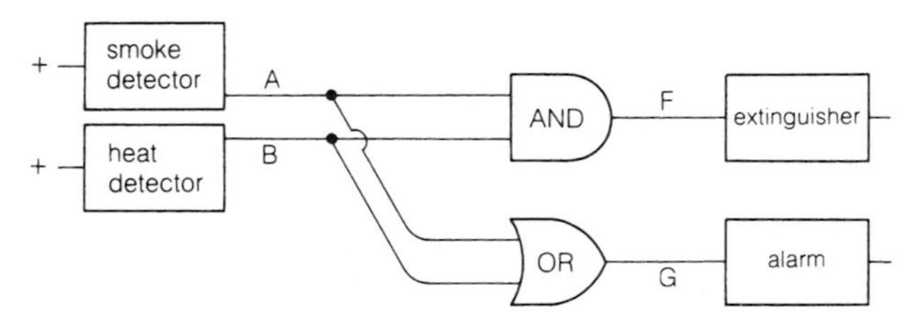

Figure 8

a) Write down the truth table for the system, using column headings A, B, F, and G.
b) What happens if smoke is detected but no heat?
c) What must happen to turn both the alarm and the extinguisher ON?

3 For each of the gate systems in figure 9, write down the truth table, using column headings A, B, C, and F. Use the table to work out whether the output is ON or OFF when

a) both inputs are ON b) both inputs are OFF
c) only one input is ON.

What type of gate is produced by each system?

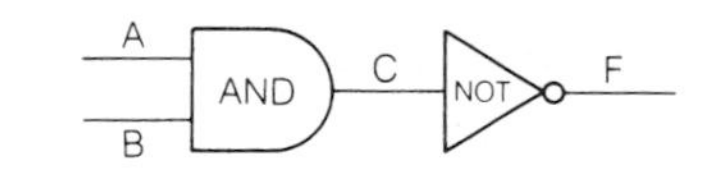

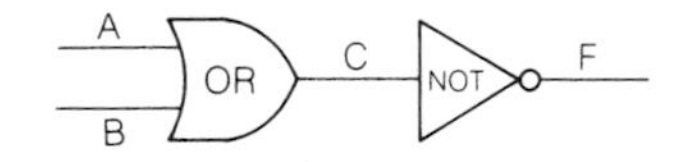

Figure 9

4 When the store room lock in figure 5 on page 347 was installed, a mistake was made. The NOT gate was fitted in the top input line (see arrow) instead of the bottom one.

a) What happened when the manager turned the hidden switch ON?
b) What happened when she then turned the main switch ON?

(Redraw the diagram and construct a truth table to help you to work out the answers).

8.8 Photoelectrons and X-rays

A photocell and an X-ray tube use opposite effects. In one, electromagnetic radiation causes emission of electrons; in the other, electrons give rise to electromagnetic radiation.

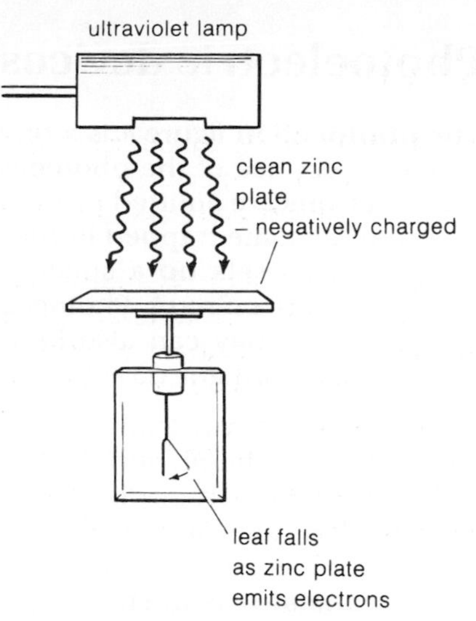

Figure 1 Ultra violet light makes a negatively charged zinc plate emit electrons. Red light has no effect

The photoelectric effect

Some materials emit electrons when electromagnetic radiation such as light strikes them. This is known as the **photoelectric effect** and it can be demonstrated using the apparatus shown in figure 1. Here, a zinc plate, freshly rubbed with emery paper to expose clean metal, has been given a negative charge and placed on an electroscope. When the ultraviolet lamp is switched on, the leaf falls, indicating that negative charge in the form of electrons is being emitted from the plate. This and similar experiments show that, for any given material:

1 only radiation beneath a certain wavelength causes electrons to be emitted. Most metals require ultraviolet wavelengths or shorter, though some metals and semiconductors emit electrons in visible light.

2 beneath this wavelength, even the weakest of lamps will make the material emit some electrons. But above this wavelength, none are emitted no matter how bright the lamp.

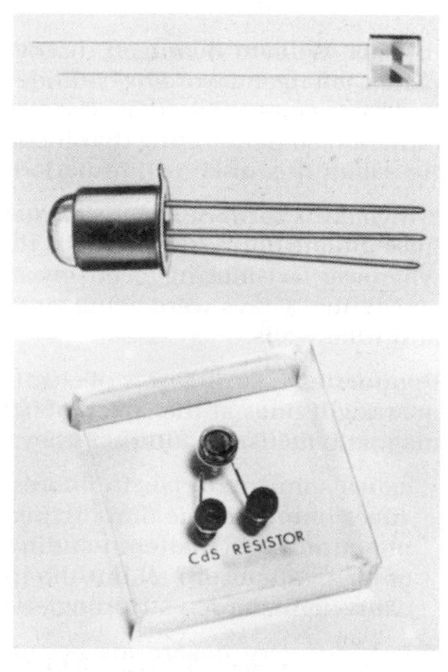

Figure 2 A photodiode, a phototransistor and a LDR

The quantum theory and photons

When first discovered, the above results proved very difficult to explain. However, in 1905, Einstein offered a solution by applying the **quantum theory** previously developed by Max Planck. According to this theory, light is emitted in 'packets' or **quanta** of energy known as **photons**. Each photon is in effect a 'particle' of light energy. The shorter the wavelength of the light, the higher is the energy of the photon. A photon of blue light for example has about twice the energy of a 'red' photon, and an 'ultraviolet' photon has higher energy again.

Einstein suggested that, when a photon strikes the surface of a metal, its energy may be absorbed by an electron. The electron needs a certain minimum energy to escape from the metal. Any surplus energy is transferred to the electron as kinetic energy (see figure 3). Expressed as an equation:

energy of photon = **energy required to remove electron from metal** + **kinetic energy of escaping electron**

In the case of zinc, each photon in an ultraviolet beam has more than enough energy to free an electron from the surface of the metal. On the other hand, a 'red' photon does not.

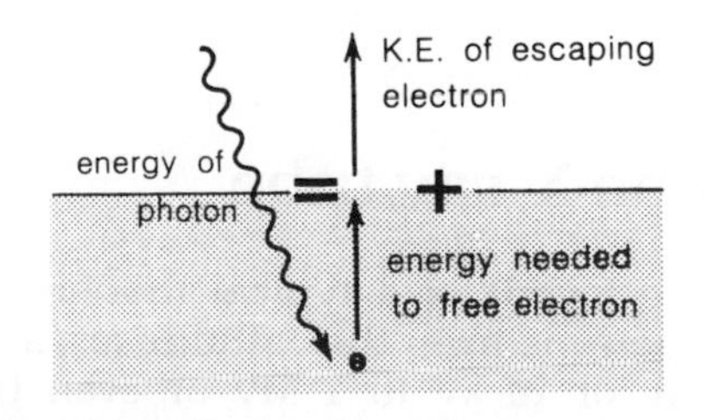

Figure 3 Energy from the photon of light is used to free the electron and to give it kinetic energy

Photoelectric devices

The **photocell** in figure 4 is a tiny light-powered electric cell which makes direct use of the photoelectric effect. When light strikes the cell, electrons are emitted from the selenium. They cross the barrier layer and become trapped in the transparently thin metal layer just above it. This sets up a small p.d. which can be used to drive a current round a circuit. Photocells of this type are found in camera light meters. They can also be used as **solar cells** for converting energy from sunlight directly into electrical energy.

The devices in figure 2 all make use of the fact that light can 'loosen' electrons in a semiconductor so that the conducting ability is improved. The **LDR** or photoresistor is described on page 339. The **photodiode**, acts a normal diode in the dark, but its 'current blocking' effect reduces as the light (or infrared) intensity rises. The **phototransistor** is in effect a photodiode which also amplifies.

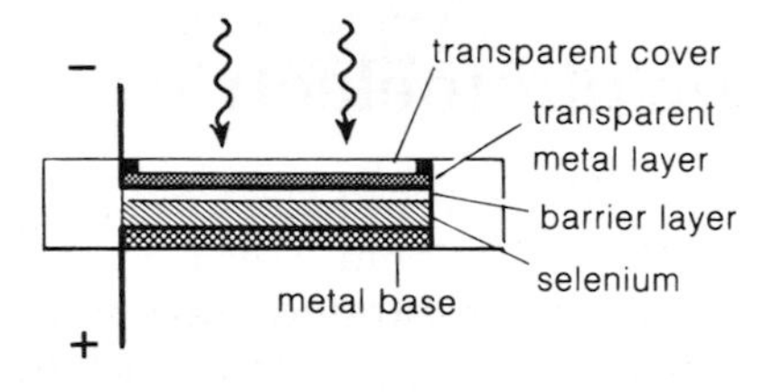

Figure 4 In a photocell, the electrons given off are "caught" and used in a circuit

X-rays

In 1895, William Roentgen discovered that wrapped photographic plates, left near a working cathode ray tube, became fogged as if they had been exposed to light. The tube apparently emitted an invisible radiation so penetrating that it could travel through solid materials. He called this unknown radiation **X-rays**.

X-rays are a form of electromagnetic radiation (see unit 5.11). They have much shorter wavelengths than visible light, and are produced whenever fast-moving electrons are stopped very rapidly. In Roentgen's tube, X-rays were being emitted as electrons struck the anode and tube walls.

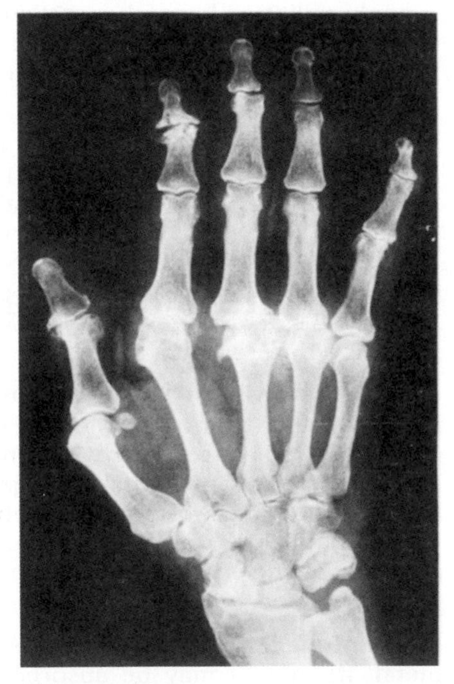

Figure 5 X-ray photograph of a hand

Properties Like all forms of electromagnetic radiation, X-rays travel in straight lines at the speed of light and are not bent by electric or magnetic fields. In addition, they:

cause some materials to fluoresce;
affect photographic film (figure 5);
penetrate solid matter including metals, though a few millimetres of lead will absorb all but the most penetrating;
cause ionization – stripping electrons from atoms in their path.

X-rays are very hazardous because their ionizing effect damages living cells. They have the same properties and present the same dangers as gamma rays discussed on page 353.

The X-ray tube

The layout of a typical modern X-ray tube is shown in figure 6. Electrons from the cathode are accelerated towards the anode by a p.d. of 10 kV to 1 MV or even higher. X-rays are emitted as the electrons strike the target on the face of the anode. Only about 0.2% of the energy of the electron beam is converted into X-ray photons. The rest is wasted as heat.

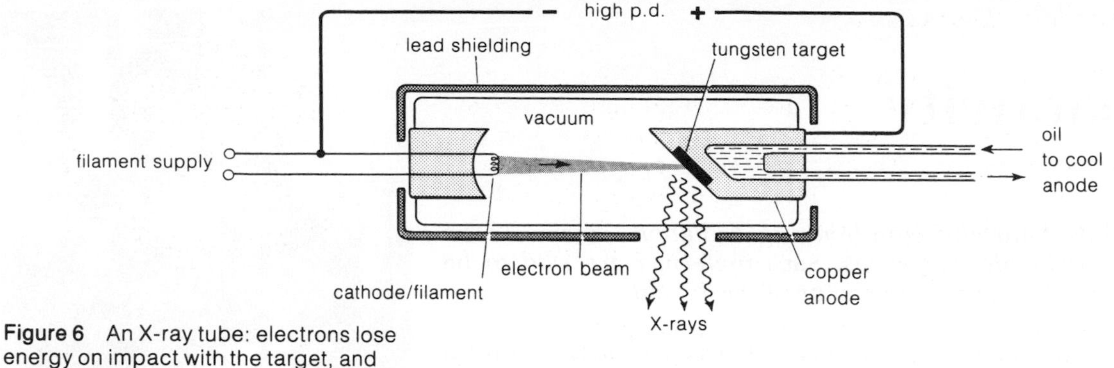

Figure 6 An X-ray tube: electrons lose energy on impact with the target, and X-rays are given off

The target is made of tungsten to withstand the high temperatures involved, and the anode is oil-cooled.

The filament current controls the intensity of the X-rays The higher the current, the more electrons given off every second, and the more X-ray photons emitted every second.

The accelerating p.d. controls the penetrating power of the X-rays. The higher the p.d., the higher the energies of the photons produced and the shorter their wavelengths. Very short wavelengths are the most penetrating and are known as **hard** X-rays. Longer wavelengths are the least penetrating and are known as **soft** X-rays.

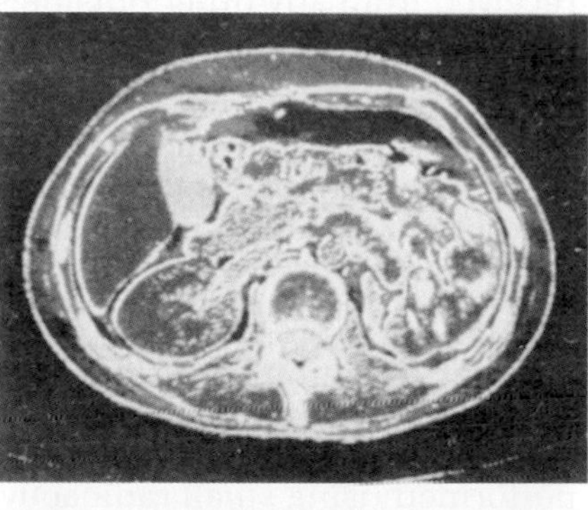

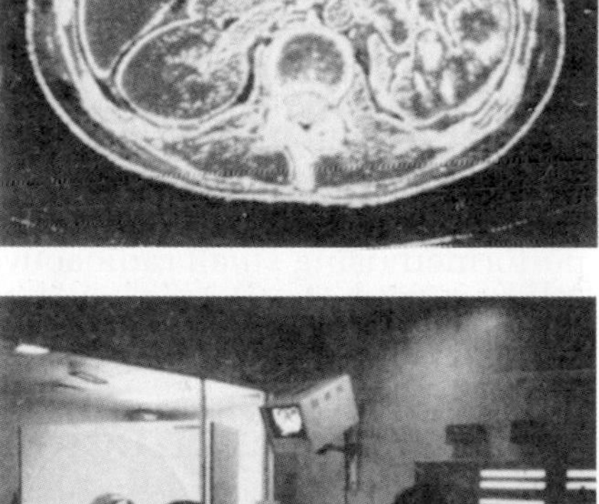

Figure 7 A body scanner can produce a detailed cross-section through the human body

Using X-rays

Soft X-rays penetrate flesh more readily than they do bone, which makes X-ray 'shadow' photographs like the one in figure 5 possible. The X-ray body scanner in figure 7 gives, with the aid of a computer, a more detailed image of any section of the human body. When medical X-ray photographs are taken, radiation doses are kept very low to reduce the risk to the patient. Concentrated beams of hard X-rays are however sometimes used to destroy cancerous cells.

In industry, X-ray photographs are used to check for flaws in welded metal joints. In research work, interference patterns produced when X-rays are scattered by atoms give important information about the arrangement of atoms in different materials.

Questions

1 A charged zinc plate rests on the cap of an electroscope. Ultraviolet radiation is falling on the plate. What happens to the leaf of the electroscope if:
  a) the charge on the plate is positive?
  b) the charge on the plate is negative?
  c) the charge on the plate is negative, but red light is used instead of ultraviolet?

2 Which has most energy, a 'red' photon, a 'blue' photon, or an 'ultraviolet' photon? Which has least energy?

3 Photons in a light beam have just enough energy to enable an electron to escape from a metal. What would be the effect of using light of a) shorter b) longer wavelength?

4 Give one application of the photoelectric effect.

5 When are X-rays produced?

6 When electrons in an X-ray tube strike the target, what happens to most of their energy?

7 If the accelerating p.d. in an X-ray tube is increased, what is the effect on the wavelengths of the X-rays produced?

8 How could a) the intensity b) the penetrating power of the radiation from an X-ray tube be increased?

9 What is the difference between hard and soft X-rays?

10 Why are X-rays hazardous?

11 Give three properties of X-rays.

8.9
Radioactivity

Some materials naturally emit ionizing radiation which can be penetrating and highly dangerous. Such materials are said to be radioactive, and their radiation can be of three kinds.

In 1896, Henri Becquerel discovered that uranium salts continuously emitted radiation which could ionize air and affect a wrapped photographic plate. The radiation seemed to be a property of the uranium itself, and was not apparently due to light, heat, or energy received from any other outside source. The effect is now known as **radioactivity**; uranium is a naturally **radioactive** material.

In 1898, Marie Curie (shown in her laboratory in figure 1) extracted from an ore called pitchblende two substances which were much more strongly radioactive than uranium. These were the elements polonium and radium. Since then, many other radioactive materials have been identified.

Figure 1 Madame Curie in her laboratory

The radiation emitted by a radioactive material is sometimes called **nuclear** radiation because it is now known to come from the nuclei of atoms in the material (see units 8.11 and 8.12). In laboratory work, experiments to study the properties of nuclear radiation are performed using small radioactive sources in which the radioactive material lies behind wire gauze.

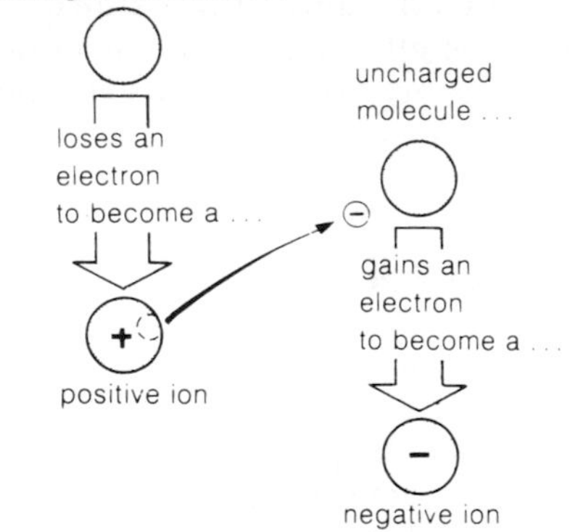

Figure 2 A positive and a negative ion are formed when an electron is transferred from one molecule to another

The ionizing effect of nuclear radiation

If electrons become detached from molecules in a gas, ions are produced and the gas is **ionized** (see page 252). This is illustrated in figure 2. The molecules with electrons missing are the positive ions; detached electrons join on to other molecules to form negative ions. As ions in a gas are free to move, an ionized gas can conduct.

Figure 3 shows an experiment to demonstrate the ionizing effect of the radiation from a laboratory radium source. The source is brought near the cap of a charged leaf electroscope, and the air around the cap becomes ionized. Ions with a charge opposite to that on the electroscope are attracted to the cap where they neutralize some of the charge on it. The leaf falls as a result. In effect, the ionized air conducts charge away from the electroscope.

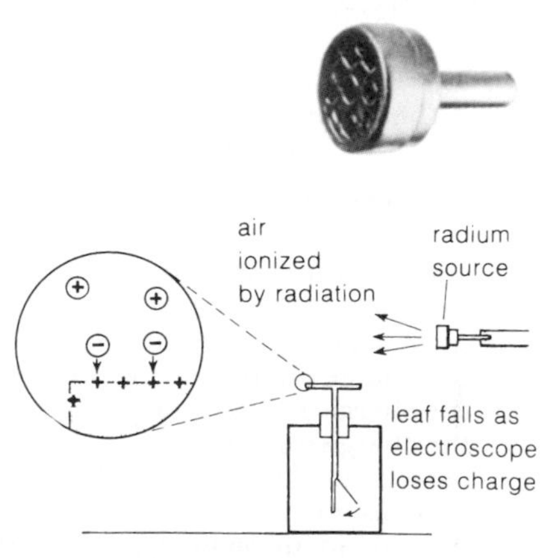

Figure 3 The radium source (also shown on the inset photograph) causes positive and negative ions to form in the air; one of the sets of ions neutralizes the charge on the electroscope

Alpha-, beta- and gamma-radiation

Around 1900, various experiments were carried out to measure the penetrating and ionizing power of the newly-discovered radiation and to study its behaviour in electric and magnetic fields. The results suggested that the radiation could be of three types; these were named **alpha**, **beta**, and **gamma** rays. Alpha and beta 'rays' seemed to be tiny charged particles shot out at random, while gamma-rays showed all the features of very short wavelength X-rays.

More is now known of the nature and properties of the three types of nuclear radiation:

Alpha- (α-) particles An alpha-particle is the nucleus of a helium atom (described further on page 359) and is made up of two protons and two neutrons. Because of the protons present, it carries a positive charge equal in magnitude to the charge on two electrons.

Alpha particles emitted from any one material all travel at approximately the same speed – up to about one tenth of the speed of light. They tend to attract electrons away from nearby atoms, and are the most strongly ionizing of the nuclear radiations. They are however the least penetrating. Their range in air is only a few centimetres, and they can be stopped by a thick sheet of paper. Collisions with atoms finally bring them to rest.

Figure 4a Tracks caused by α-particles

Beta- (β-) particles These are electrons emitted with varying speeds which can be as high as nine-tenths of the speed of light. Like all electrons, they carry a negative charge. They are much lighter than alpha particles (less than one seven thousandth of their mass) and much less ionizing. They are however more penetrating. The fastest beta-particles have a range of a metre or so in air, though they can be stopped by a few millimetres of Perspex or aluminium. If you can remember the basic structure of the atom given on page 230, it may seem odd that a nucleus can emit electrons. An explanation is given in unit 8.11.

Figure 4b Tracks caused by β-particles

Gamma- (γ-) rays These are electromagnetic waves of very short wavelength; in other words, high energy photons. They result from energy changes inside atomic nuclei, as distinct from X-rays which are caused by energy changes outside. A gamma-photon is often emitted at the same time as an alpha or beta particle.

Gamma-rays are the least ionizing of the nuclear radiations, but the most penetrating. Their intensity is greatly reduced by several centimetres of lead, but they are never completely absorbed.

The nature and properties of alpha-, beta- and gamma-radiation are summarized in figure 5. Radioactive sources like that shown in figure 3 may give out one, two, or all three types of radiation depending on the materials they contain.

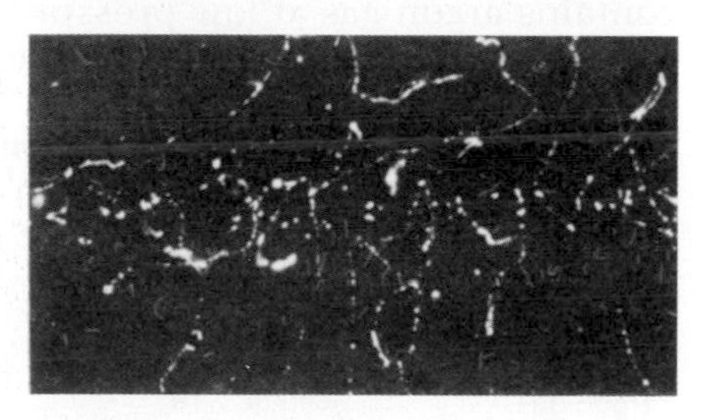

Figure 4c Electron tracks caused by γ-rays

Figure 5 Characteristics of the three types of nuclear radiation

	nature and charge	approximate mass	ionizing effect	absorbed by	deflection in electric or magnetic field
alpha- (α-) particle	2 protons 2 neutrons helium nucleus	4 proton masses	strong	sheet of writing paper	very small
beta- (β-) particle ⊖	electron	$\frac{1}{1800}$ proton mass	weak	about 5mm of aluminium	large
gamma- (γ-) ray	no charge high energy photon	—	very weak	Never fully absorbed: 25mm of lead reduces intensity to half	zero

The effect of electric and magnetic fields

Like any stream of moving charged particles, alpha- and beta-rays are deflected by electric and magnetic fields. Beta-particles show much greater deflections than alpha-particles because of their lower mass.

Figure 6 illustrates the effect of a magnetic field on the three types of radiation. Alpha- and beta-particles are deflected in directions given by Fleming's left-hand rule (remember that negative charge travelling to the right counts as a conventional current to the left). Being electromagnetic waves, gamma-rays are not deflected by a magnetic field.

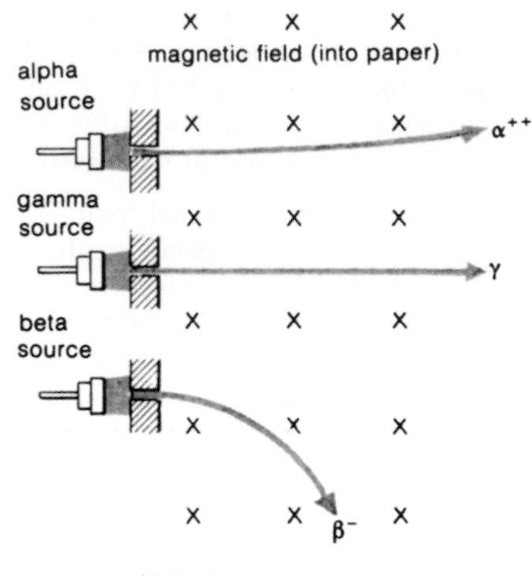

Figure 6 Response of α, β and γ radiation to a magnetic field

Detecting nuclear radiation

Most methods of detecting alpha-, beta- and gamma-rays are based on the fact that these radiations have an ionizing effect.

Leaf electroscope Alpha radiation can be detected using a charged electroscope as in figure 3. The more intense the radiation, the faster the leaf falls. The method isn't suitable for detecting beta- and gamma-radiation as these cause insufficient ionization of the air.

Geiger-Müller tube This is illustrated in figure 7. The tube contains argon gas at low pressure. A p.d. of about 400 V is applied between the central wire electrode and the aluminium wall of the tube. The end of the tube is sealed by a mica 'window' thin enough to allow alpha particles to pass into the tube as well as beta and gamma radiation.

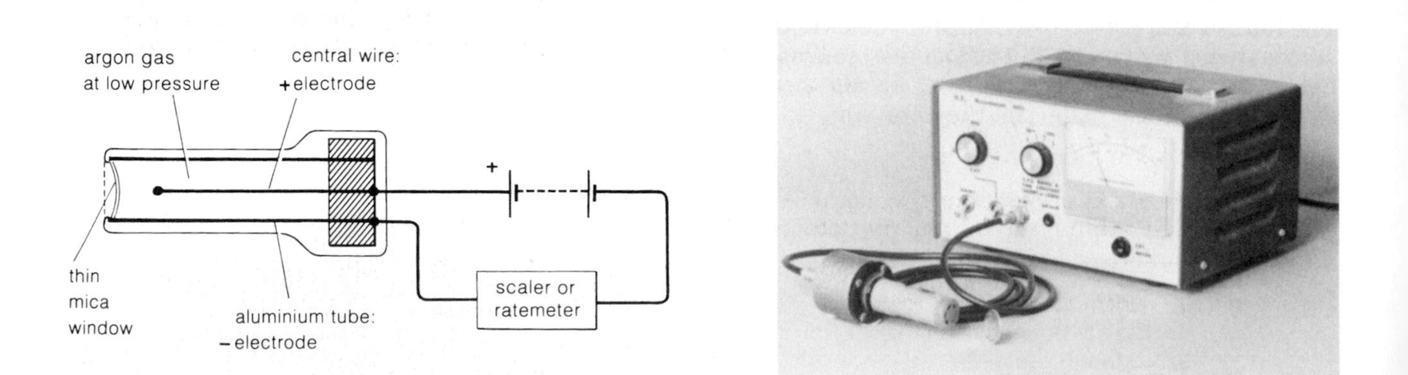

Figure 7 The Geiger-Müller tube

When a charged particle or gamma-radiation enters the tube, the argon gas becomes ionized. This triggers a whole avalanche of ions between the electrodes. For a brief moment, the gas conducts and a pulse of current flows in the circuit. The circuit includes either a **scaler** or a **ratemeter**. A scaler counts the pulses and shows the total on a display. A ratemeter indicates the number of pulses or counts per second. The complete apparatus is often called a **Geiger counter**.

Figure 8 shows how a Geiger-Müller tube and ratemeter can be used to measure the range of alpha-particles in air. The alpha-source is moved away from the window of the tube. As the window passes beyond the maximum range of the alpha-particles, the reading on the ratemeter drops.

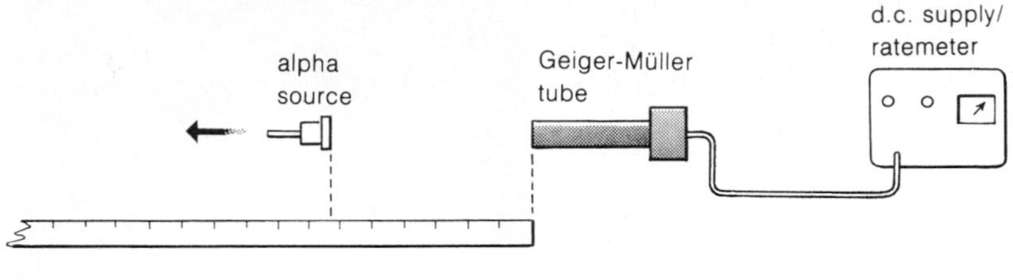

Figure 8 Using a Geiger-Müller tube to measure the range of α-particles

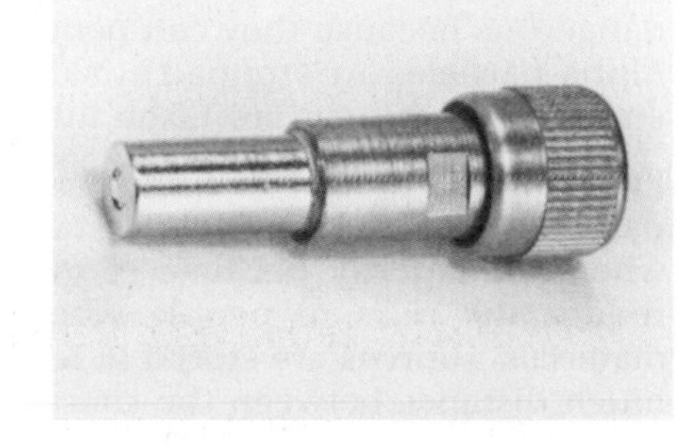

Figure 9 A solid-state detector

Solid-state detector In the solid-state detector shown in figure 9, incoming radiation ionizes atoms in a reverse-biased *p-n* junction diode. This makes the diode conduct, and a pulse of current flows in the circuit as a result. As with a Geiger-Müller tube, current pulses are detected using a scaler or a ratemeter.

Cloud chamber Figure 10 shows a simple form of cloud chamber, a device which enables the tracks of charged particles to be seen. In this case, the chamber contains a weak alpha-source.

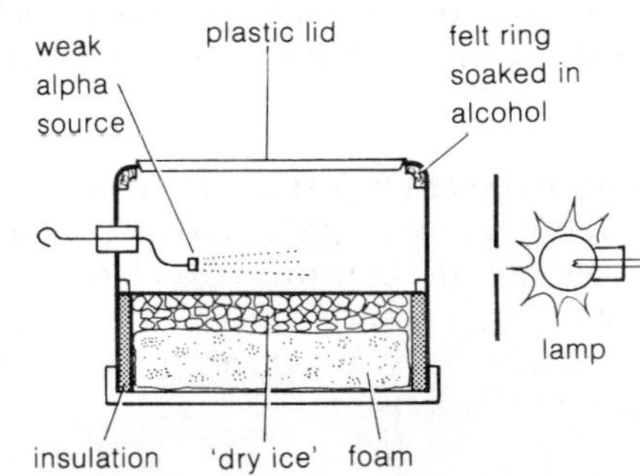

Figure 10 A cloud chamber. The cold alcohol vapour condenses easily on ionized air; this reveals the paths of α-particles

The felt ring round the top of the chamber is soaked in alcohol. The bottom of the chamber is cooled by 'dry ice' (solid carbon dioxide) to around −80 °C. As the alcohol vapour spreads downward through the chamber, it is cooled beyond the point at which it would normally condense. Each time an alpha-particle is shot out from the source, it ionizes the air in its path and alcohol readily condenses around these ions. A narrow cloud made up of millions of tiny alcohol droplets forms along the track of the alpha-particle. This cloud looks rather like the trail left by a high-flying aircraft, and is visible because it reflects light from the lamp.

The photographs in figure 4 were all cloud chamber tracks. Look back and note the following points:

Alpha-particle tracks are thick and straight, with the occasional deflection if an alpha particle collides with an air molecule.

Beta-particle tracks are thin and crooked. The particles cause much less ionization and, being light, are continually being pushed off course by air molecules nearby.

Gamma-photons don't produce tracks as such. The tracks seen are those caused by electrons which have absorbed energy from photons and have escaped from atoms.

Nowadays, high-speed particles from the atom are often studied using a **bubble chamber** as shown in figure 11. In this, the tracks of the particles are seen as a trail of bubbles in a liquid.

Figure 11 A photograph of the massive *bubble chamber* assembled at CERN, Geneva

Dangers and uses of radioactive materials

Like all ionizing radiations, nuclear radiations are potentially very dangerous because they damage or destroy living cells. High levels of radiation can kill. Marie Curie died as a result of radiation doses from the radium she purified. Lower levels may lead to cancer or eye cataracts later in life, or cause cell mutations (changes) which may be passed on to future generations. Gamma-rays are the most dangerous because they can penetrate to tissues deep in the body. Alpha-particles are stopped by skin, but become hazardous if alpha-emitting substances are taken into the body.

Everyone is exposed to a small amount of **background** radiation, partly due to radioactive substances present in rocks and elsewhere, and partly because of **cosmic radiation** from space. To reduce the risks to people who work with strongly radioactive materials, sources are stored in lead or concrete containers with as much distance between the source and the user as possible.

Safe disposal of radioactive waste is a particular problem because it often contains substances which are readily absorbed by the body if food or water become contaminated. The body's chemical system cannot for example tell the difference between calcium, which is good for you, and radioactive strontium which isn't. On the other hand, radioactivity can be extremely useful. Two applications are shown in figures 12 and 13. Others are discussed later.

Figure 12 A strontium-90 source in use to control the thickness of tyre cord. Some beta radiation penetrates the tyre cord: if the level rises, the cord is too thin. The machine making the cord is adjusted accordingly

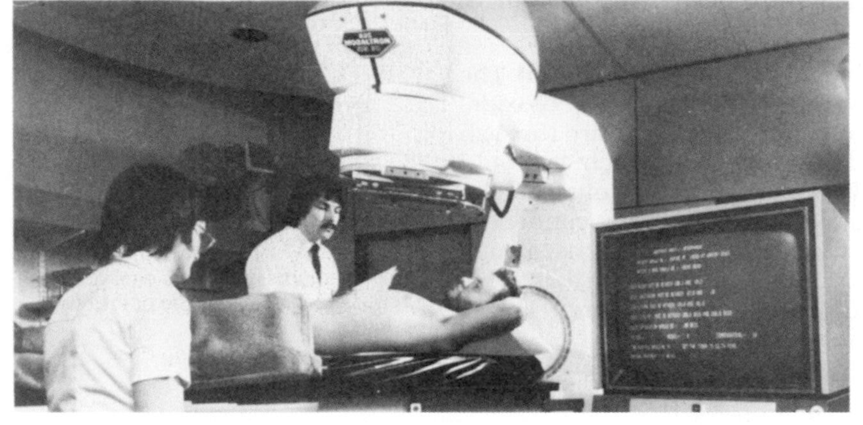

Figure 13 A computer-controlled, three-dimensional beam of radiation is used to kill cancerous cells, at the same time keeping side effects to a minimum

Questions

1 Name three types of nuclear radiation. State which type of radiation
a) is a form of electromagnetic radiation, b) carries a positive charge, c) carries a negative charge, d) is made up of helium nuclei, e) is made up of high-speed electrons, f) has the same characteristics as X-rays, g) is the most highly ionizing, h) is the most penetrating, i) can be stopped by skin or paper, j) is never completely absorbed

2 Figure 14 shows the effect of a magnetic field on three types of nuclear radiation. Which types of radiation are A, B, and C?

3 Which type of nuclear radiation could be detected using a leaf electroscope? Describe how. Why is this method not suitable for detecting other forms of nuclear radiation?

4 How is the range of alpha particles in air measured?

5 How do the tracks made by alpha particles in a cloud chamber differ from those made by beta particles?

6 Which type of nuclear radiation is potentially the most dangerous? Why?

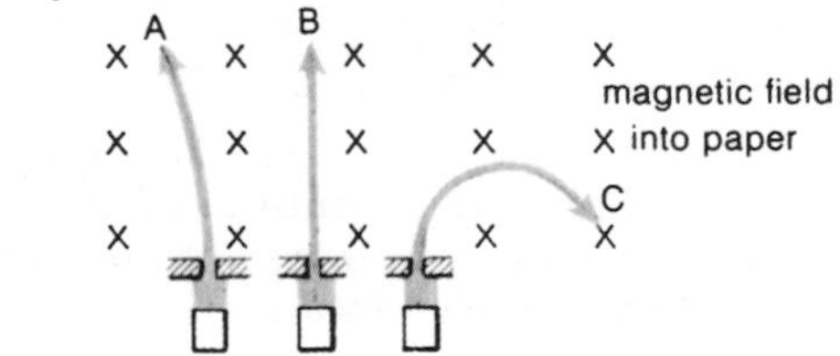

Figure 14

8.10
The structure of the atom

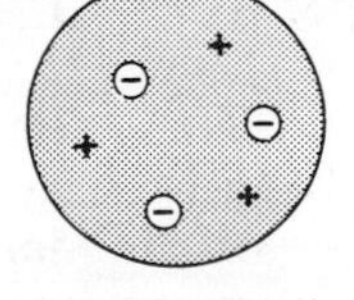

Figure 1 The "Plum pudding" model of the atom

Atoms are far too small to be seen, but a great deal is now known about their structure. Much of the early information was gained by bombarding atoms with alpha particles.

At different times, scientists have proposed various descriptions or **models** of the atom to match the experimental evidence available. Following Thomson's discovery of the electron in 1897, one of the first atomic models proposed was the 'plum pudding' model illustrated in figure 1. The atom was assumed to be a sphere of positive charge with negatively charged electrons spread through it rather like currants in a pudding. In 1911, Geiger and Marsden performed a series of experiments under the direction of Ernest (later Lord) Rutherford which led to a new model of the atom.

Ernest Rutherford as a young man

Rutherford's scattering experiments

Figure 2 shows in simplified form the apparatus used by Geiger and Marsden. A thin gold foil was bombarded with alpha particles from a radium source. Alpha particles leaving the foil were detected by observing the flashes of light or **scintillations** which they caused on a glass screen coated with zinc sulphide.

Geiger and Marsden found that most of the alpha particles passed through the foil virtually undeflected. To their surprise however, they discovered that a few (less than 1 in 8000) were bouncing off the foil with deflections of 90° or more. To Rutherford, these results suggested that the positive charge of an atom and most of the mass were concentrated in a tiny region which he called the **nucleus**. The rest of the atom was largely empty space. The alpha particles which had rebounded from the foil were those very few which had come close enough to nuclei to be repelled strongly by their highly concentrated charges. This is illustrated in figure 3.

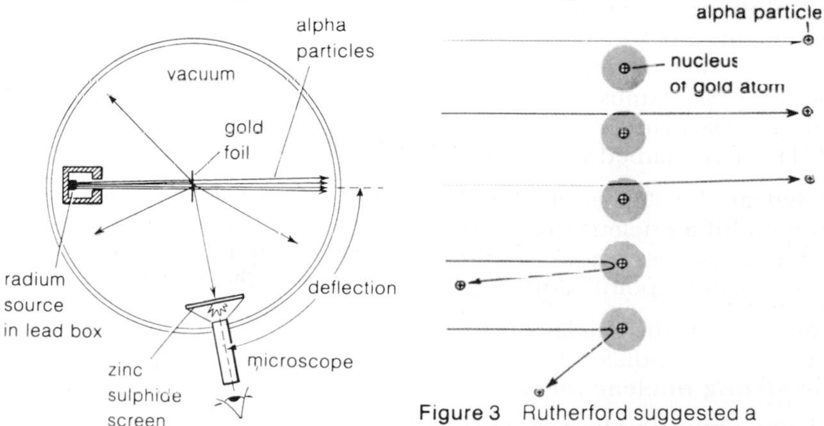

Figure 2 Scattering experiments performed by Geiger and Marsden

Figure 3 Rutherford suggested a reason for the occasional scattering: only the tiny nucleus affects the path of the α-particles

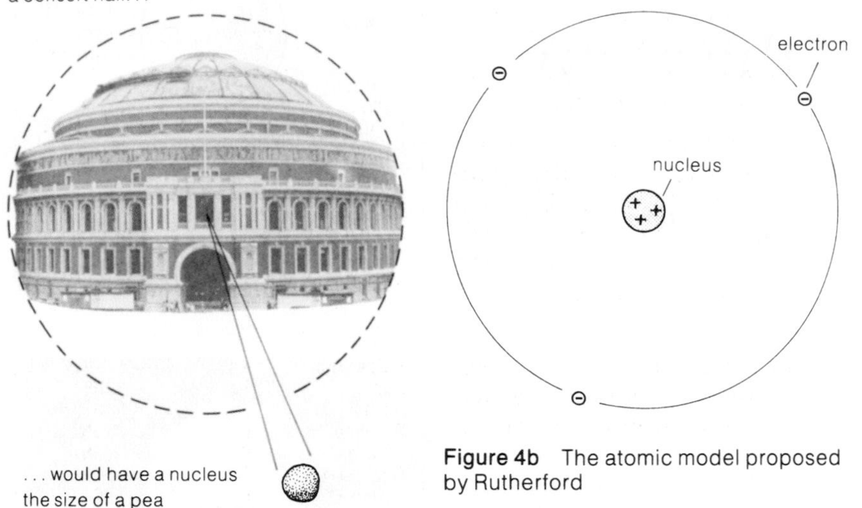

Figure 4a Comparing the sizes of an atom and its nucleus

Figure 4b The atomic model proposed by Rutherford

The nucleus is very small indeed. If an atom were the size of a concert hall, its nucleus would be no larger than a pea, as shown in figure 4a. Figure 4b shows the detailed model of the atom proposed by Rutherford. Negatively charged electrons orbit a tiny, positively charged nucleus containing over 99.9% of the mass of the atom. The electrons are held in orbit by the attraction between opposite charges.

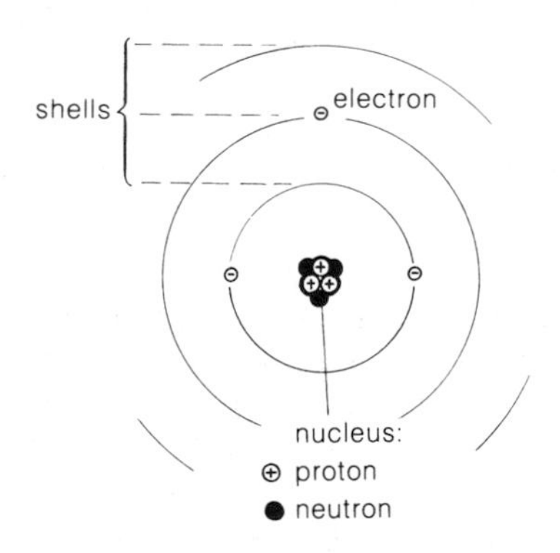

Figure 5a The Rutherford-Bohr atomic model

The Rutherford-Bohr model of the atom

In the years following the scattering experiments, various improvements were made to Rutherford's nuclear model of the atom. In 1913, Niels Bohr applied the quantum theory to electron orbits and was able to explain how atoms emitted light. In 1919, Rutherford bombarded nitrogen gas with very fast alpha particles and found that positively charged particles were being knocked out of the nitrogen nuclei. The particles, which were about 1800 times heavier than electrons, were called **protons**. In 1932, James Chadwick discovered that the nucleus also contained particles of about the same mass as protons, but uncharged. These he named **neutrons**.

Figure 5a illustrates what is now known as the Rutherford-Bohr model of the atom. Fast-moving electrons orbit a nucleus made up of protons and neutrons. The numbers of particles vary from one type of atom to another, but the following general points apply.

1 Protons and neutrons are bound together in the nucleus by a force strong enough to overcome the very high repulsion between the protons. This force is known as the **strong nuclear force**.

2 Electrons and protons carry equal but opposite charges. In a neutral atom, the number of electrons is the same as the number of protons.

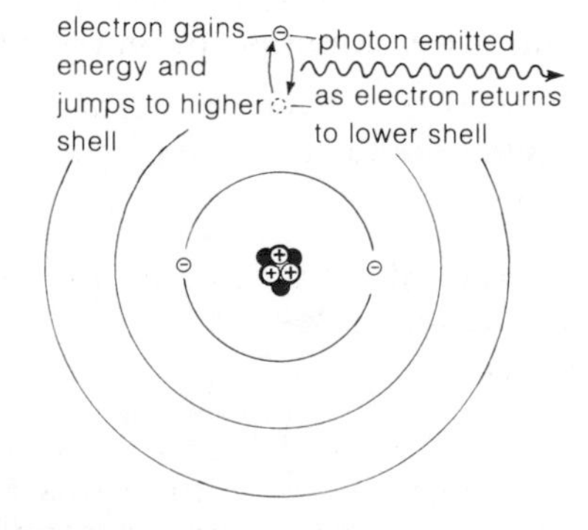

Figure 5b This model can explain the way in which atoms emit light

3 Electrons orbit the nucleus at certain fixed levels only. These levels are called **shells**. Electrons normally occupy shells as near the nucleus as possible, but there is a limit to the number of electrons which each shell can hold – no more than 2 electrons in the first shell, 8 in the second, 18 in the third for example.

4 If an electron gains energy in some way, it may jump to a vacant space in a higher shell as shown in figure 5b. The atom is then said to be **excited**. The electron quickly returns to the lower shell, losing energy as it does so and emitting a photon in the process. This could be an infrared, visible, ultraviolet or X-ray photon depending on the energy released. The larger the energy loss, the shorter the wavelength of the electromagnetic radiation emitted.

In advanced work, the Rutherford-Bohr model of the atom has been replaced by a mathematical model based on the quantum theory. For convenience however, atoms are still commonly drawn according to the Rutherford-Bohr model.

Elements and atoms

All naturally-occurring materials are made up from about 90 basic substances called **elements**, an atom being the smallest 'piece' of any one element. Some of these elements are listed in figure 6, together with the chemical symbols for their atoms.

Atoms of different elements have different numbers of protons in their nuclei, and therefore different numbers of electrons in orbit. This is also shown in figure 6. The lightest atom, hydrogen, has 1 proton (and therefore 1 electron), while the heaviest naturally-occurring atom, uranium, has 92. Atoms of three of the elements are illustrated in figure 7, using the Rutherford-Bohr model.

The number of protons in the nucleus of an atom is called its atomic number or proton number. Counting the charge on a proton as +1 unit, the atomic number gives the number of units of charge in the nucleus. In a neutral atom it also gives the number of electrons. When atoms join, electrons provide the bonds which link them, so it is the atomic number which decides the particular chemical properties of an atom, i.e. the ways in which it will combine with other atoms.

Protons and neutrons are known as **nucleons** because they are found in the nucleus. **The total number of protons and neutrons in an atom is called its mass number or nucleon number.**

If an element with a chemical symbol X has a mass number A and an atomic number Z, its nucleus is represented by the symbol:

$${}^{A}_{Z}X$$

For example, the nuclei of the atoms in figure 7 would be represented as follows:

$${}^{1}_{1}H \quad {}^{4}_{2}He \quad {}^{9}_{4}Be$$

In each case, the difference between the top and bottom numbers gives the number of neutrons in the nucleus. For example, the hydrogen atom has no neutrons, while the beryllium atom has 5.

Element	Symbol for atom	Atomic number Z (number of protons or electrons)
Hydrogen	H	1
Helium	He	2
Lithium	Li	3
Beryllium	Be	4
Boron	B	5
Carbon	C	6
Nitrogen	N	7
Oxygen	O	8
Sodium	Na	11
Magnesium	Mg	12
Aluminium	Al	13
Silicon	Si	14
Chlorine	Cl	17
Potassium	K	19
Iron	Fe	26
Cobalt	Co	27
Bromine	Br	35
Krypton	Kr	36
Strontium	Sr	38
Iodine	I	53
Xenon	Xe	54
Caesium	Cs	55
Barium	Ba	56
Lanthanum	La	57
Gold	Au	79
Mercury	Hg	80
Lead	Pb	82
Polonium	Po	84
Radon	Rn	86
Radium	Ra	88
Thorium	Th	90
Uranium	U	92

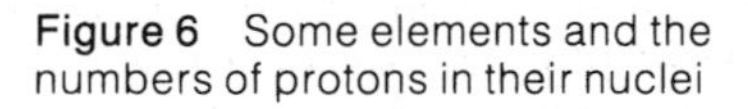

Figure 6 Some elements and the numbers of protons in their nuclei

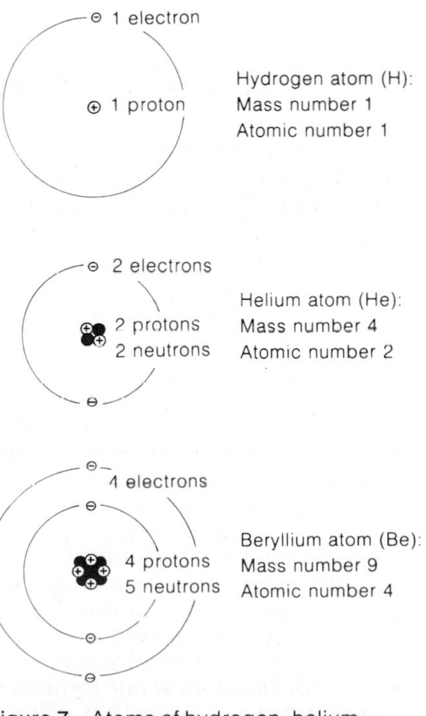

Figure 7 Atoms of hydrogen, helium, and beryllium

Nuclides and isotopes

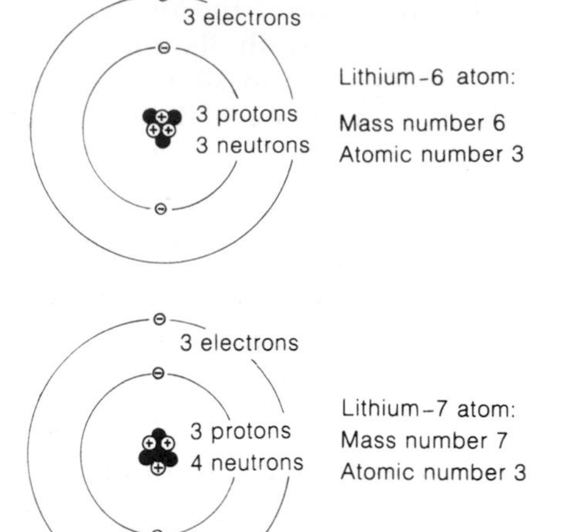

Figure 8 Atoms of lithium-6 and lithium-7

All atoms of a particular element have the same number of protons in their nuclei, but not necessarily the same number of neutrons. Lithium for example exists naturally in two forms, atoms of which are illustrated in figure 8. One form has 3 neutrons in the nucleus, giving a mass number of 6; the more common form has 4 neutrons in the nucleus, giving a mass number of 7. The nuclei are represented by the symbols $^{6}_{3}Li$ and $^{7}_{3}Li$.

Each form of an element is called a **nuclide** and is normally referred to by its mass number, e.g. lithium-6, lithium-7. Most elements occur naturally as a mixture of nuclides.

Nuclides with the same atomic number but different mass numbers are called isotopes.
Isotopes are therefore nuclides of the same element which have different numbers of neutrons in their nuclei.

To illustrate the various terms:

Lithium is an element.
Lithium-6 is a nuclide; Lithium-7 is a nuclide.
Lithium-6 and lithium-7 are isotopes.

Fundamental particles

Figure 9 shows the first stage of the giant accelerator at Geneva in Switzerland, where beams of electrons, protons and other particles are shot at each other at speeds approaching that of light. The impacts reveal useful information about the structure of the nucleus and its particles. They also give rise to short-lived particles which don't exist in 'ordinary' matter.

Matter seems to be made up of two groups of particles. Particles like electrons which do not feel the strong nuclear force and are not bound in the nucleus are called **leptons**. Those, such as neutrons and protons, which do feel the strong nuclear force are known as **hadrons**. Physicists now believe that hadrons are built up from more fundamental particles called **quarks** – but whether these consist of even more basic particles, no one yet knows.

Figure 9 This device (similar to a massive van de Graaff generator) provides a voltage of 750 kV which is used to accelerate protons. They then enter a massive accelerator, 2.2 km in diameter, after which they are collided with other sub-nuclear particles

Questions

1 What evidence is there to suggest that most of the mass of an atom is concentrated in a small, positively charged nucleus?
2 Which particle in an atom carries a) positive charge b) negative charge c) no charge? Which of these particles is/are present in the nucleus?
3 In the Rutherford-Bohr model of the atom a) what are electron shells? b) how does the atom emit light?
4 Explain the terms atomic number and mass number. What are their alternative names?
5 How does an atom of one element differ from an atom of another?
6 What are isotopes? Give an example.
7 The nucleus of a nuclide of chlorine is represented by the symbol $^{35}_{17}Cl$. How many a) protons b) electrons c) neutrons does this chlorine atom have? Write down the symbol for the nucleus of the chlorine nuclide containing two more neutrons.
8 Using information given in figure 6, write down the symbol for the nucleus of a) carbon-12 b) carbon-14 c) aluminium-27. How many neutrons are present in each nuclide?
9 $^{56}_{26}X$ represents a nucleus of an unknown element X. Use the information given in figure 6 on page 359 to identify the element.

8.11

Radioactive decay

Alpha and beta particles are emitted by unstable nuclei which break up. The process is called radioactive decay, and it occurs when atoms of one element change into atoms of another.

Not all combinations of protons and neutrons are stable. Radioactive nuclides such as radium-226, iodine-131, and carbon-14, have unstable nuclei which disintegrate or **decay** in time, each emitting an alpha or beta particle as it does so. The emission of the particle makes the nucleus more stable, but alters the numbers of protons and neutrons it contains. When a nucleus decays, it therefore becomes a nucleus of a different element. The nucleus which decays is called the **parent** nucleus; the nucleus which results is called the **daughter** nucleus. Daughter nucleus and emitted particles are together known as the **decay products**.

Alpha decay

An alpha particle is a nucleus of helium-4. It consists of two protons and two neutrons and can be represented as $^{4}_{2}He$ or $^{4}_{2}\alpha$.

A typical alpha decay process is illustrated in figure 1. A nucleus of radium-226 (atomic number 88) emits an alpha particle, which leaves the nucleus with two protons and two neutrons less than before. The mass number of the nucleus therefore drops to 222 and the atomic number to 86. According to the table on page 333, radon has an atomic number of 86, so it is a nucleus of radon-222 which forms.

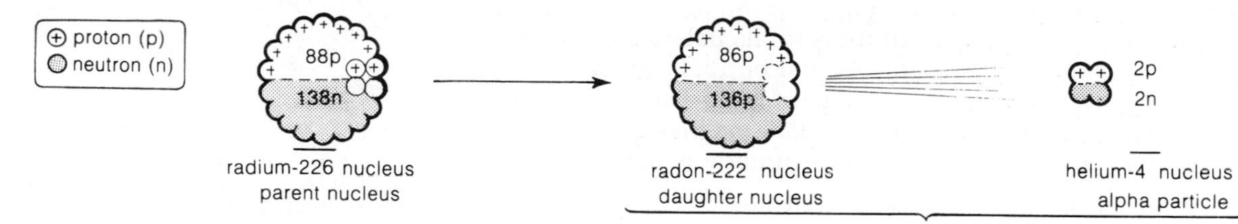

Figure 1 α-decay of radium-226: a helium-4 nucleus is given off

The decay process can be written in the form of an equation:

$$^{226}_{88}Ra \rightarrow ^{222}_{86}Rn + ^{4}_{2}\alpha$$

Note that the top numbers balance on both sides of the equation: $226 = 222 + 4$. This expresses the fact that the total number of nucleons (226) stays the same during the decay process.

Note that the bottom numbers also balance: $88 = 86 + 2$. This expresses the fact that the total charge (+88 units) stays the same during the decay process.

The process of alpha decay is summarized by the following rule:

When a nuclide decays by alpha emission, it becomes a nuclide with an atomic number two less than before, and a mass number four less.

Beta decay

A beta particle is an electron. It has negligible mass compared with a proton or neutron, and carries a charge of −1 unit. In nuclear equations, it is written as ${}^{0}_{-1}e$ or ${}^{0}_{-1}\beta$.

A typical beta decay process is illustrated in figure 2. Inside a nucleus of iodine-131 (atomic number 53), a neutron ceases to exist and a proton and electron are created in its place (together with a tiny, uncharged and almost undetectable relative of the electron called an antineutrino). The electron and antineutrino are shot out of the nucleus at high speed. As a proton has now replacd a neutron in the nucleus, the mass number is unchanged at 131, but the atomic number rises to 54. A nucleus of xenon-131 has been formed.

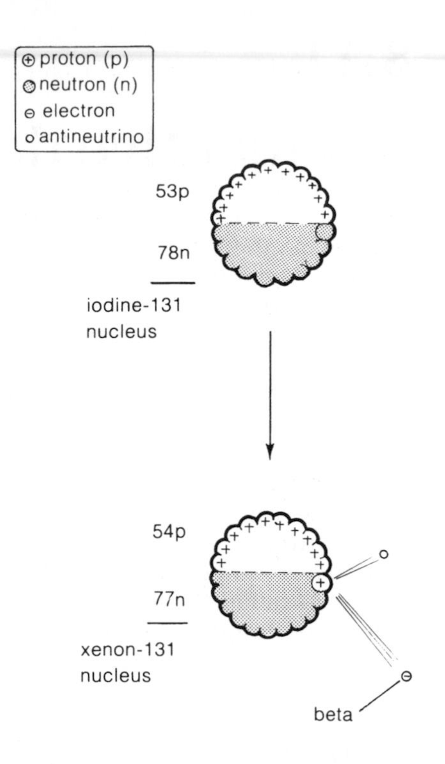

Figure 2 β-decay of iodine-131: a neutron breaks up to give a proton, electron, and (almost undetectable) antineutrino

Ignoring the antineutrino, the decay process can be written:

$${}^{131}_{53}I \rightarrow {}^{131}_{54}Xe + {}^{0}_{-1}\beta$$

Again, note that the top and bottom numbers balance on both sides of the equation; $131 = 131 + 0$ and $53 = 54 - 1$. During the decay process, there is no change in the number of nucleons or in the total charge.

The process of beta decay is summarized by the following rule:

When a nuclide decays by beta emission, it becomes a nuclide with an atomic number one greater than before, but with the same mass number.

Example *The nuclide sodium-24 decays by beta emission. Name the daughter nuclide produced and write down the equation for the decay process.*

The table on page 333 gives the atomic number of sodium (Na) as 11. Calling the unknown daughter nuclide ${}^{A}_{Z}X$, the decay process can be written:

$${}^{24}_{11}Na \rightarrow {}^{A}_{Z}X + {}^{0}_{-1}\beta$$

To make the top and bottom numbers totals balance on both sides of the equation, A must be 24 and Z must be 12. In other words, the unknown nuclide X has a mass number of 24 and an atomic number of 12. According to the table on page 359, magnesium (Mg) has an atomic number of 12, so the unknown nuclide is magnesium-24.

The decay process can therefore be written:

$${}^{24}_{11}Na \rightarrow {}^{24}_{12}Mg + {}^{0}_{-1}\beta$$

Gamma emission

With some nuclides, the emission of an alpha or beta particle from a nucleus leaves the protons and neutrons in an 'excited' arrangement with more energy than normal. As the protons and neutrons rearrange to become more stable, they lose energy which is emitted as a photon of gamma radiation.

Gamma emission causes no change in mass number or atomic number.

Rate of decay and half-life

During radioactive decay, nuclei disintegrate at random at a rate which is unaffected by chemical change or temperature. In any sample of iodine-131 for example, on average, 1 nucleus disintegrates every second for every 1 000 000 nuclei present. Other nuclides decay at different rates.

The decay of a sample of iodine-131 is illustrated in figure 3.

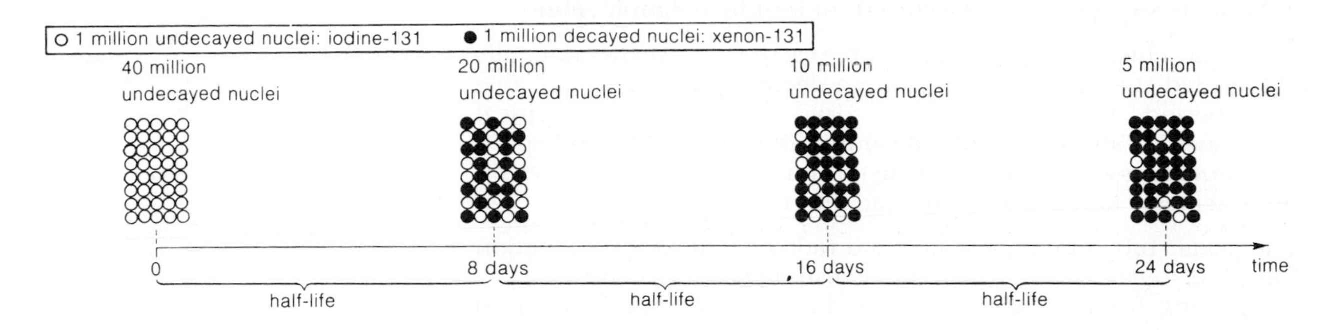

Figure 3 The length of time it takes for half of any quantity of iodine-131 to decay, is called its *half life*.

To begin with, there are 40 million undecayed nuclei. 8 days later, half of these have disintegrated to form nuclei of xenon-131 and beta particles. With the number of undecayed nuclei now halved, the number of distintegrations over the next 8 days is also halved. The decay process continues in this way, with the number of undecayed nuclei halving every 8 days. Many months later, all the nuclei are likely to have disintegrated.

The half-life of a radioactive nuclide is the time taken for half the nuclei present in any given sample to decay.

Iodine-131 therefore has a half-life of 8 days. Half-lives of some other nuclides are given in figure 4. In an old universe, it might seem strange that any nuclides with short half-lives are left at all. However, some are radioactive daughters of long-lived parents, while others are produced artificially (see later in the section).

Nuclide	Half-life
Uranium-235	7×10^8 years
Radium-226	1620 years
Sodium-24	15 h
Radon-220	52 s
Polonium-212	3×10^{-7} s

Figure 4 Half-lives of different nuclides

'Waste' nuclides in the used fuel-rods from nuclear power stations have half-lives of many thousands of years

The average number of disintegrations per second in a radioactive sample is called the **activity**. The decay curve in figure 5 shows how the activity of the iodine-131 sample in figure 3 varies with time. As the activity of the sample is always proportional to the number of undecayed nuclei, it too halves every 8 days. Take any point on the graph, then move to a point at half the height on the activity scale, and you will have moved 8 (days) along the time scale. From this, it follows that half-life can be defined in another way:

The half-life of a radioactive nuclide is the time taken for the activity of any given sample to fall to half its original value.

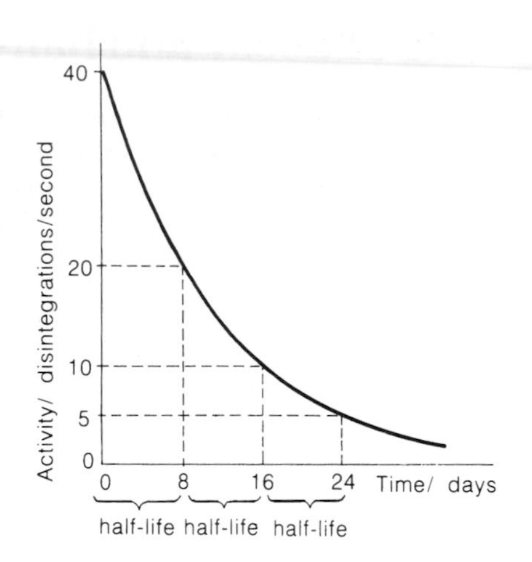

Figure 5 The decay of iodine-131: its half-life is 8 days

The decay curve of a radioactive source can be found experimentally – provided the activity of the source decreases over a practicable time period (minutes, hours or days for example). A typical experiment is shown in figure 6; source and geiger tube are kept in fixed positions and the **count-rate** (in counts per second) shown on the ratemeter noted at regular time intervals.

The geiger tube detects background radiation as well as radiation from the source, so each ratemeter reading has to be corrected by subtracting from it the count-rate due to background radiation alone. The corrected count-rate is then plotted against time to give the decay curve. The half-life of the source can be found from the curve as indicated above.

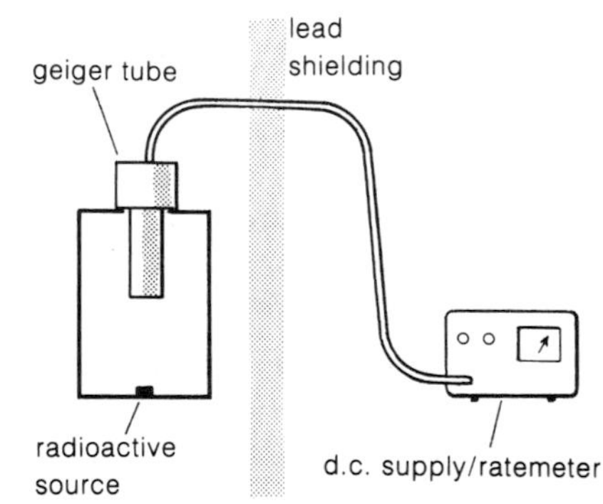

Figure 6 Apparatus for finding the half-life of a radioactive material

Artificial nuclides

Many radioactive nuclides are produced artificially by a reverse of the decay process – stable nuclei absorb nuclear particles or gamma photons which strike them, and become unstable as a result. For example, radioactive cobalt-60 is produced when cobalt-59, the stable and only naturally occurring nuclide of cobalt, is bombarded with neutrons from the core of a nuclear reactor. Cobalt-60 is a strong gamma emitter, and is used in the treatment of cancer (see figure 7). It is also used as a radiation source for X-ray type photographs of metals.

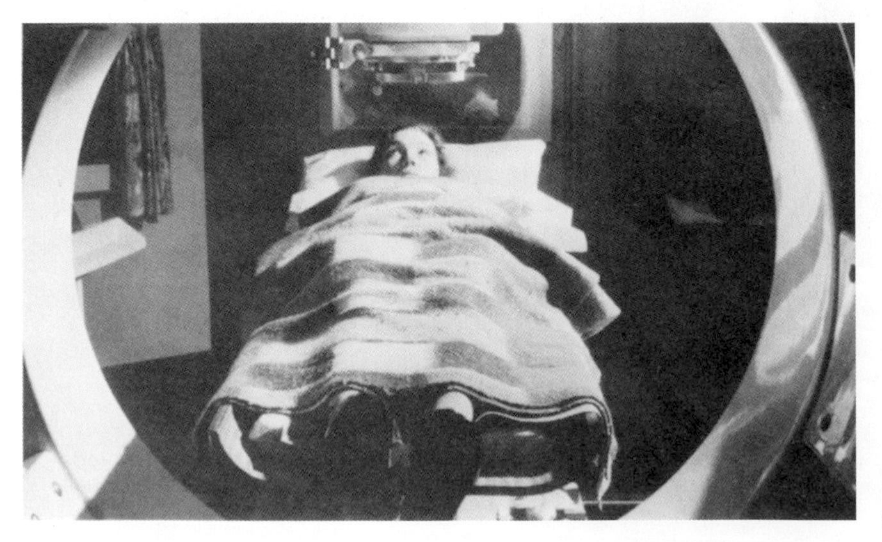

Figure 7 Using cobalt-60 for the treatment of cancer

Carbon dating

Carbon-14 is a rare radioactive nuclide of carbon produced when cosmic radiation from space bombards nitrogen-14 in the Earth's upper atmosphere. It has a half-life of 5600 years. Plants and animals absorb and give out carbon-14 while they are alive, so the percentage of carbon-14 in their tissues stays constant. When they die, the percentage of 'trapped' carbon-14 reduces as the nuclide decays by beta emission. By measuring the activity of the carbon-14 present in dead plant or animal tissues, it is possible to estimate their age from a decay curve. This is known as **carbon dating**, and it is one of the many methods used by archaeologists to determine the age of ancient remains (see figure 8).

Figure 8 Carbon-dating techniques are being applied to the 'Turin Shroud' – a piece of cloth reputed to be the shroud in which Christ was placed after his death. The scientific tests suggest that the cloth was probably made in the 14th century, though no one has discovered what caused the image.

$\frac{\text{Time}}{\text{minutes}}$	$\frac{\text{Activity}}{\text{average disintegrations per second}}$
0	75·8
10	57·4
20	43·5
30	33·0
40	25·0
50	19·0
60	14·4
70	10·9
80	8·3

Figure 9 (see question 7)

Questions

Where necessary, use information given in the table of atomic numbers on page 359.

1 Explain the following terms: parent nucleus; daughter nucleus; decay products.
2 What symbols are used in nuclear equations to represent a) an alpha particle b) a beta particle?
3 The following equation represents the decay of thorium-232:

$${}^{232}_{90}\text{Th} \rightarrow {}^{A}_{Z}\text{X} + {}^{4}_{2}\alpha$$

What are the values of A and Z? Name the nuclide X. Name two quantities which do not change during the decay process.
4 Polonium-210 decays by alpha emission. What nuclide is formed as a result? Write down an equation for the decay process.
5 Carbon-14 decays by beta emission. What nuclide is formed as a result? Write down an equation for the decay process.
6 Give two ways in which the half-life of a radioactive nuclide can be defined.
7 The table in figure 9 shows how the activity of a sample of iodine-128 varies with time. Plot a graph of activity against time.
From your graph, find the time taken for the activity to drop from a) 60 to 30 disintegrations/second b) 40 to 20 disintegrations/second.
What is the half-life of iodine-128?
How long would the activity of a sample of iodine-128 take to drop from 300 to 75 disintegrations/second?

8.12

Nuclear energy

Natural radioactive decay releases large quantities of energy, but at a slow rate. In a nuclear reactor, the decay process is speeded up considerably.

The protons and neutrons in a nucleus possess nuclear potential energy – or **nuclear energy** for short. When a radioactive material decays, protons and neutrons are pulled into arrangements in which they are more tightly bound to one another, and nuclear energy is lost. Much of this energy is turned into heat (thermal energy) as fast-moving decay products collide with other atoms and make them move faster. For example, radioactive decay is responsible for the high temperature of the rocks beneath the Earth's surface.

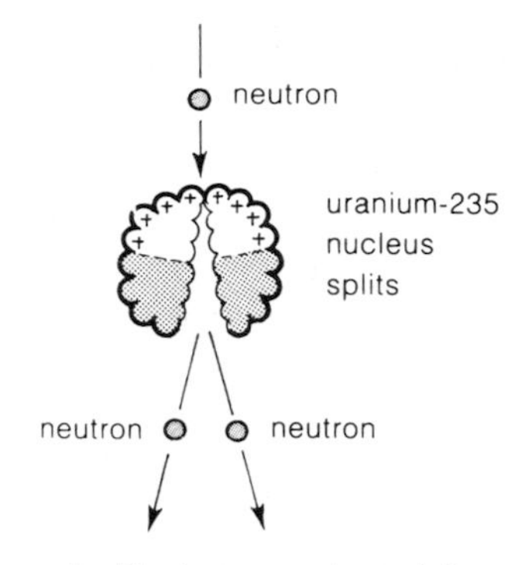

Figure 1 Fission: a neutron strikes a nucleus of uranium-235. The nucleus splits, and gives off even more neutrons, as well as energy

During radioactive decay, the energy released per atom is around a million times greater than that given off during a chemical change such as burning. However, rates of decay are usually very slow. Much faster decay can be produced in some cases by bombarding nuclei with other particles. Neutrons make the most effective missiles; being uncharged, they are not repelled by nuclei, and can penetrate and disrupt them relatively easily.

Any process in which a particle penetrates a nucleus and changes it in some way is called a **nuclear reaction**.

Fission

Natural uranium is a dense radioactive metal consisting mainly of two isotopes; uranium-238 (over 99%) and uranium-235 (less than 1%). Uranium-235 has properties which enable energy to be obtained from it very rapidly.

Figure 1 shows what can happen if a neutron (symbol $^{1}_{0}n$) strikes and penetrates a nucleus of uranium-235. The nucleus becomes highly unstable and splits into two lighter nuclei of roughly equal mass, shooting out two or three neutrons as it does so. This splitting process is called **fission**, and the fragments are thrown apart with far more energy than would be released by natural decay. The nucleus can split in several ways, but a common reaction is:

$$^{235}_{92}U + ^{1}_{0}n \rightarrow ^{144}_{56}Ba + ^{90}_{36}Kr + ^{1}_{0}n + ^{1}_{0}n \text{ (+energy)}$$

The barium and krypton nuclei are themselves radioactive.

Just occasionally, fission occurs **spontaneously** – a uranium-235 nucleus splits without first being struck by a neutron. If neutrons from the fission of one nucleus go on to split other nuclei, a **chain reaction** may result as illustrated in figure 2. The number of nuclei undergoing fission multiplies rapidly, and an enormous quantity of energy is released in a short space of time.

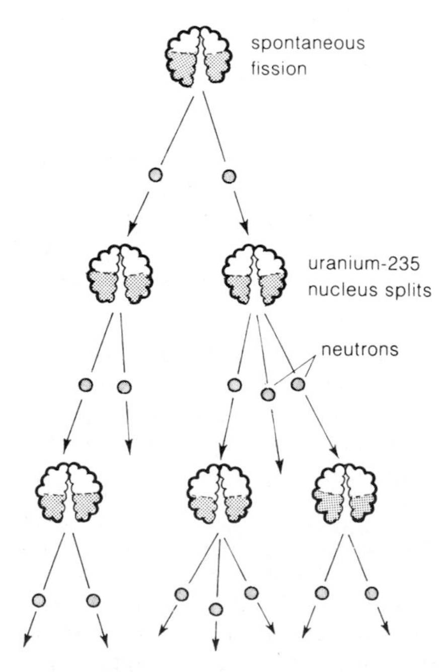

Figure 2 *A chain reaction* develops if the neutrons go on to strike even more uranium nuclei; a vast amount of energy is given off

The uranium-235 present has to be above a certain **critical mass** for a chain reaction to occur, otherwise too many neutrons escape without causing further fission.

In the first 'atom' bomb, an uncontrolled chain reaction was started by bringing two lumps of uranium-235 together so that the critical mass was exceeded.

Nuclear reactors

In a nuclear reactor, a controlled chain reaction takes place, and thermal energy is released at a steady rate.

Figure 3 shows the basic layout of an advanced gas-cooled reactor (AGR) as used in some nuclear power stations. Heat given off by the fission of uranium-235, is carried away by carbon dioxide gas at high pressure, and used to make steam as in a conventional power station (as described on page 315.

The reactor consists of a pressure cylinder containing:

Nuclear fuel elements These are made from uranium dioxide, the natural uranium present being enriched with extra uranium-235. As much electrical energy can be produced with 1 kg of enriched uranium fuel as with 55 tonnes of coal.

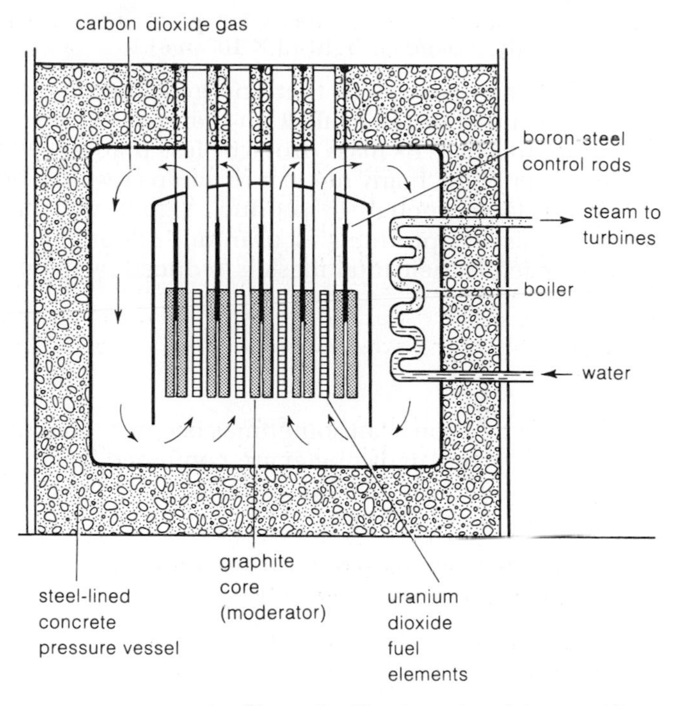

Figure 3 Structure of an Advanced Gas-cooled Reactor

A graphite core Fission neutrons have to be slowed down for a chain reaction to be maintained. This is because slow neutrons are more effective in causing fission, and, unlike fast neutrons, are not absorbed by uranium-238. Graphite blocks are used to slow down the neutrons; the graphite acts as a **moderator**.

Control rods Once fission has started, its rate is controlled by raising or lowering boron-steel control rods. The boron absorbs neutrons. If the rods are raised, more neutrons are able to cause fission, and the core temperature rises. The reactor can be shut down by keeping the rods in the lowered position.

Spent fuel elements are removed from the reactor core and sent to a reprocessing plant. Here, unused uranium is separated from the radioactive waste products, together with small quantities of plutonium-239. Plutonium-239 is produced in the reactor as uranium-238 is bombarded with neutrons. It has similar properties to uranium-235 and can be used as a fuel in a fast-breeder reactor. Its only other major use is in the production of nuclear weapons, and it is the most hazardous substance known.

The core of the Dounreay prototype reactor being loaded with (dummy) fuel rods

Energy and mass

In 1905, Albert Einstein, shown in figure 4, proposed the theory that energy itself possesses mass. If an object gains energy, its mass increases; if it loses energy, its mass decreases. The change of mass m is linked to the energy change E by the equation

$$\boldsymbol{E = mc^2}$$

where c is the speed of light $[3 \times 10^8\ \text{m/s}]$.

The value of c^2 is so high that energy gained or lost by everyday objects has no detectable effect on their mass. For example, if a fast car comes to a halt, its mass drops by less than 0.000 000 000 000 01%! On the other hand, many nuclear reactions result in energy changes which are large enough to produce significant changes in mass. When the fission products of uranium-235 are slowed down in a nuclear reactor, their total mass is reduced by about 0.1%.

Figure 4 Einstein, photographed in 1944

Fusion

Protons and neutrons can sometimes become more tightly bound to one another if two light nuclei are combined or 'fused' together to form a single nucleus. This process is called **fusion**. Like fission, it results in a release of energy.

Figure 5 illustrates the fusion of nuclei of hydrogen-2 (deuterium) and hydrogen-3 (tritium) to form a nucleus of helium-4 and a neutron. The reaction can be written:

$${}^{2}_{1}\text{H} + {}^{3}_{1}\text{H} \rightarrow {}^{4}_{2}\text{He} + {}^{1}_{0}\text{n} \quad (+\text{energy})$$

This reaction produces far more energy for a given mass of material than any fission reaction.

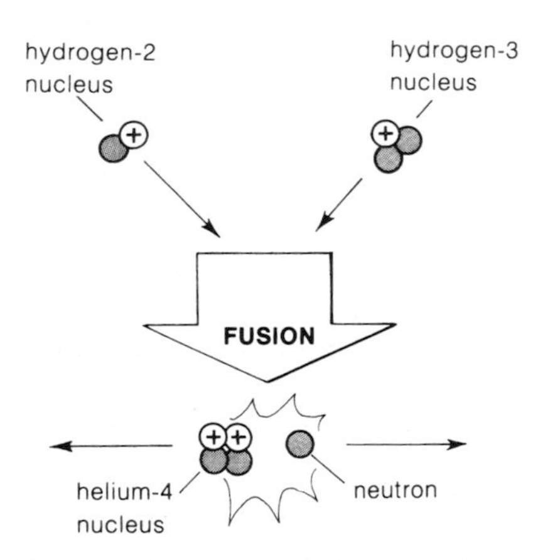

Figure 5 Fusion: energy is given off when two small nuclei combine

Nuclei have to collide at very high speeds for fusion to occur, otherwise the repulsion between their charges prevents them coming close enough to combine. In practice, this means raising the temperature of hydrogen gas to a staggering 100 million K! Fusion reactions are sometimes known as **thermonuclear** reactions because thermal energy has to be supplied before energy can be released.

The Sun is powered by fusion. In the Sun's core, vast quantities of energy are released as thermonuclear reactions convert hydrogen into helium. Man's experiments with uncontrolled fusion have resulted in the 'hydrogen' bomb, in which an uncontrolled fission chain reaction supplies the heat needed for thermonuclear reactions to start.

The technical problems of achieving controlled fusion are however very much greater than with fission. No ordinary container can hold hydrogen at the temperature necessary for fusion to start, and resist its expansion so that fusion reactions can be maintained. Research into such problems is being carried out using devices like the **tokamak** in figure 6. This uses a doughnut-shaped magnetic field to trap the hot, ionized gas.

If thermonuclear reactors ever become a practical proposition, they could solve many of the world's energy problems. The nuclides of hydrogen used as fuel are found in sea-water, and the waste product, helium, is inert and non-radioactive.

Figure 7 A solar power cut: the Sun eclipsed by the Moon

Figure 6 The 'tokamak' is one of the machines being built which attempts to harness fusion power.

Questions

1 Radioactive decay releases heat (thermal energy). Where does this energy come from?

2 Which of the following processes releases the most energy per kilogram of material – natural radioactive decay, fission, fusion, chemical reactions? Which releases the least?

3 How many neutrons are released by the following nuclear reaction?

$$^{235}_{92}U + ^{1}_{0}n \rightarrow ^{148}_{57}La + ^{85}_{Z}X + \ldots\ldots\ldots\ldots$$

What is the value of Z? Use the table on page 359 to identify the unknown nuclide X.

4 Explain what is meant by the terms a) fission b) spontaneous fission c) chain reaction.

5 In an advanced gas-cooled reactor
   a) how is heat removed from the reactor core, and what use is made of the heat?
   b) what material is used as a moderator? What is the purpose of the moderator, and why is it necessary?
   c) how do the control rods control the rate of fission?

6 What is fusion? Why are very high temperatures required for fusion to occur?

Further questions

Electrons and atoms: part A

1 The trace on an oscilloscope is shown in figure 1(a). A student then alters one of the controls and obtains the trace in figure 1(b).

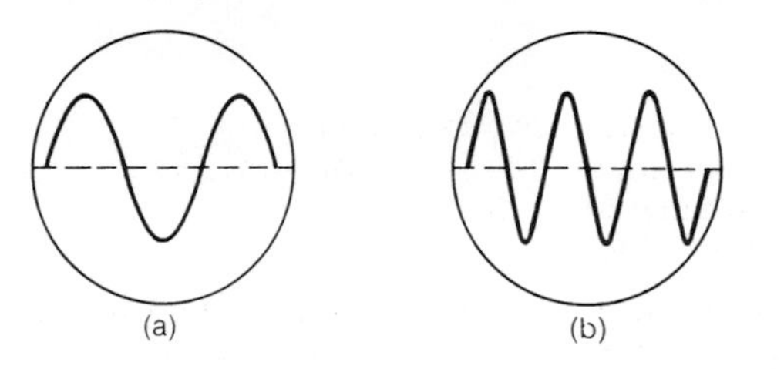

Figure 1

Which control is altered:
A The Y gain
B The Y shift
C The X shift
D The time-base frequency? (SEG)

2 From the following list of components, copy and write down the one which best fits each of the sentences given below. Each may be used once, more than once or not at all.

diode: npn transistor: LED: LDR: thermistor: relay

a) The .... normally has three terminals called the emitter, base and collector.
b) The resistance of a .... decreases a lot as the temperature increases.
c) The resistance of a .... decreases a lot as the light intensity increases.
d) A .... is an electromagnetic switch which allows a larger current to be switched on by a smaller current. (NEA)

3 Figure 2 shows a simple d.c. power supply circuit.

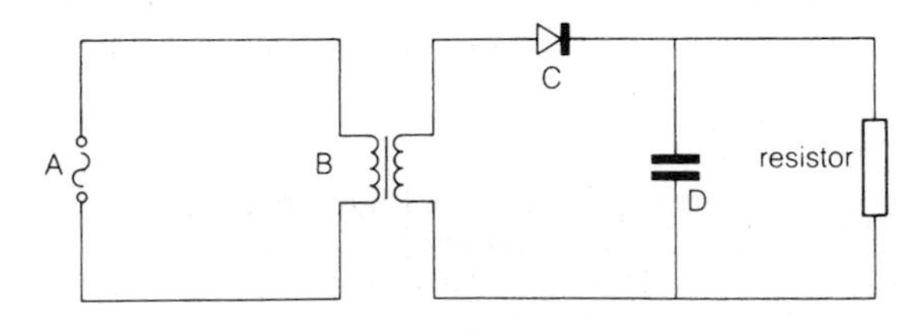

Figure 2

a) What do each of the circuit symbols A, B, C, and D stand for?
b) What is the purpose of the component labelled B?
c) What is the purpose of the component labelled C?
d) What is the purpose of the component labelled D? (SWEB/SEG)

4 The circuit diagram in figure 3 shows how a transistor may be used to make a switch which makes a light come on when it gets dark.

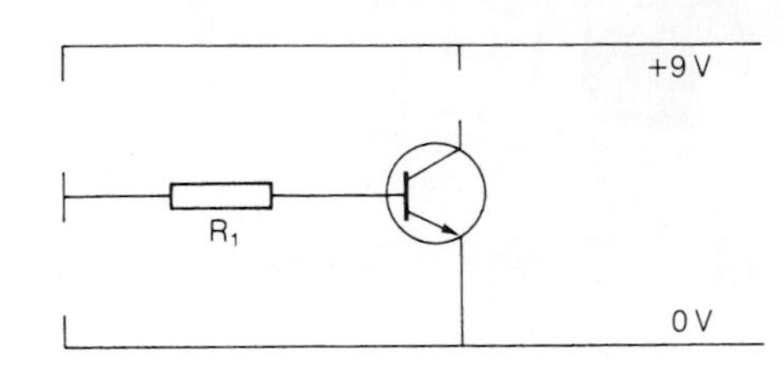

Figure 3

a) Copy and complete the circuit by adding in the correct spaces and labelling a light sensitive resistor, a variable resistor, and an electromagnetic relay.
b) Explain how the circuit will operate the relay and hence switch a bulb on as it gets dark.
c) Why is a variable resistor preferred to a fixed resistor? (LEAG)

5 Figure 4 shows a relay being driven by a transistor. The diagram also shows a 12 V power supply and electric immersion heater connected to the relay contacts.

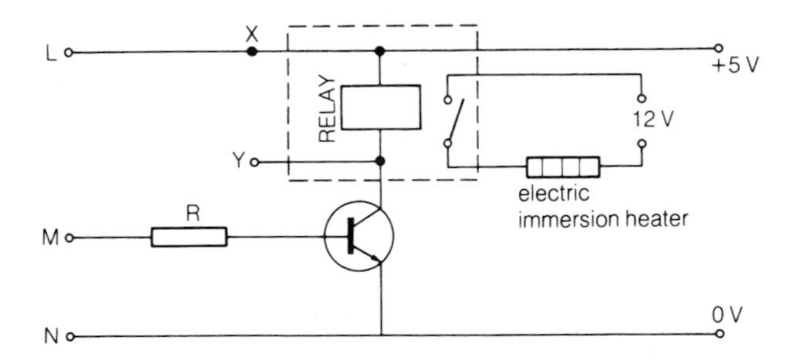

Figure 4

a) A component is usually placed between X and Y to protect the transistor.
   i) What is the name of this component?
   ii) Copy the diagram and draw in the component between X and Y.
b) The relay is of the 'normally-open' type, i.e. the contacts are open when the relay coil has no current flowing through it.
What is the state of the contacts when the voltage between M and N is (i) 0 V, (ii) + 5 V?
c) i) The circuit in Figure 4 can be made to operate as a thermostatically controlled heating circuit by the addition of a variable resistor and another component at the input to the transistor. On your diagram draw in these two components, correctly connected, so that the circuit will work.
   ii) Which component can adjust the temperature at which the relay operates?

6 The output of a logic circuit can be displayed using an LED and associated series resistor, as in the circuit shown in figure 5.

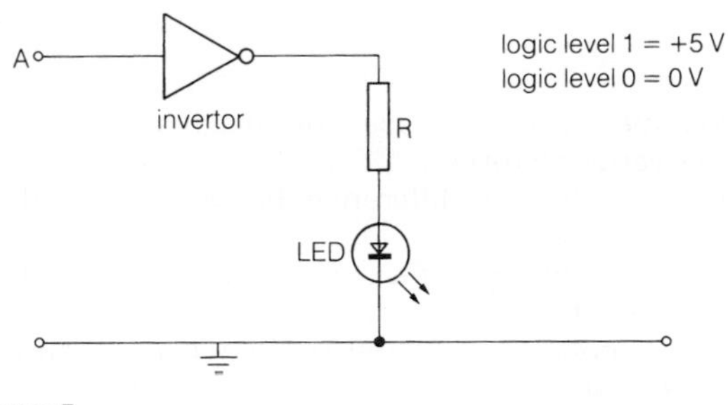

Figure 5

a) When the LED is lit, what is the logic state at A?
b) The LED has a potential difference drop of 2 V across it and a current of 10 mA flowing through it when it is lit. What is the potential difference across resistor R when the LED is lit?
c) Calculate the resistance of R,
d) Why is the resistor R needed? (NEA)

7 Part of an alarm system is shown in figure 6. The system requires an alarm siren to sound when either of two switches A and B are operated but off when the switch S is set to logic 1 by the security guard.

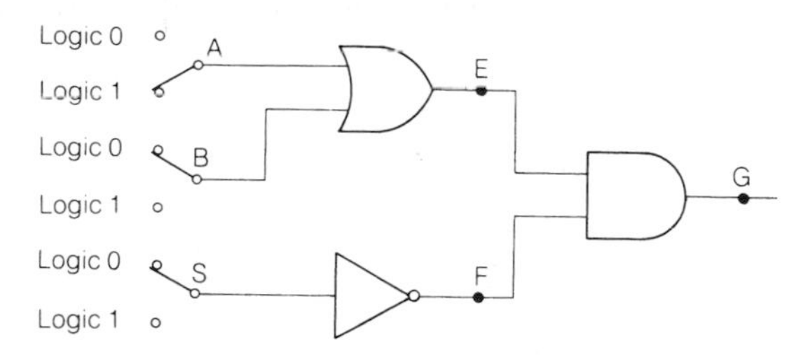

Figure 6

a) Copy and complete the truth table below for the logic circuit.

S	B	A	F	G
0	0	0		
0	0	1		
0	1	0		
0	1	1		
1	0	0		
1	0	1		
1	1	0		
1	1	1		

Figure 7

b) The output from the gate G is sufficient to operate the alarm siren. Draw a circuit diagram for a suitable driver circuit which would be capable of switching a mains operated siren and which could be connected directly to G (the mains connection and the siren may be omitted). (NEA)

8 Figure 8 shows a basic inverting amplifier using a 741 operational amplifier.

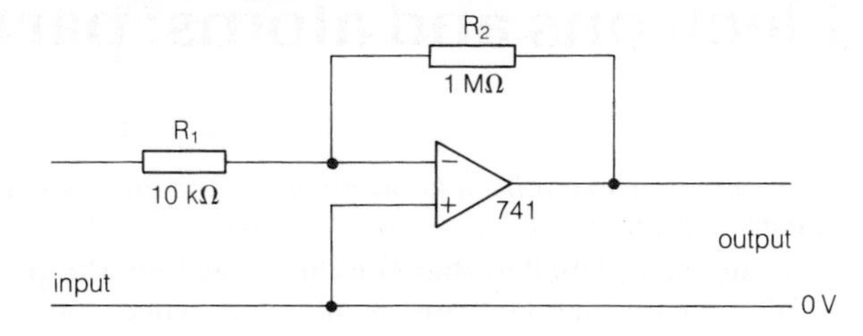

Figure 8

a) Calculate the voltage gain.
b) Assuming that $R_1 = 10\,k\Omega$, what value would be suitable for R_2 in order to obtain a voltage gain of 50? (LEAG)

9 This question is about testing the thickness of paper using a radioactive source which emits beta particles. The source is put on one side of the paper and the Geiger counter on the other. The paper rolls from the papermaking plant onto the roller as shown in figure 9.

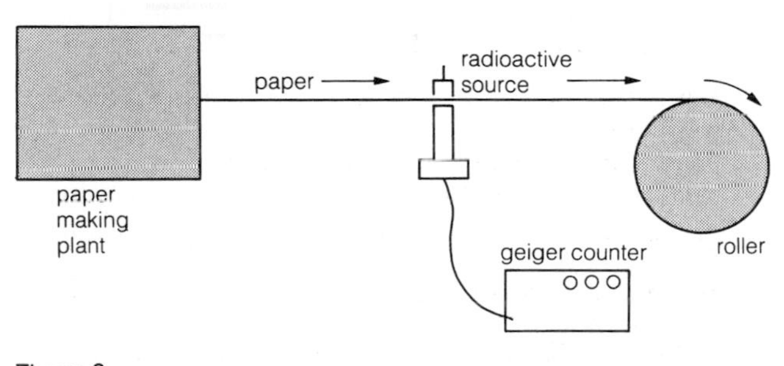

Figure 9

a) What are beta particles?
b) Why are beta particles more suitable than alpha particles or gamma rays for this job?
c) Write down one precaution you would take when handling this radioactive source.
d) Chris is worried that the paper ends up radioactive. What would you say in answer to this?

The table shows the reading on the counter during 70s:

Table of results

times in seconds	10	20	30	40	50	60	70
total count since the start	50	100	150	195	235	275	315
count in 10 seconds	50	50	50				

e) Copy and complete the table to show the count in each 10 second time period.
f) Look at the table of results. What happened to the thickness of the paper? Why do you say this?
g) i) At what time did the paper begin to change thickness?
ii) The paper was moving at 3 m/s. What length of paper passed the source before it changed thickness? (SWEB/SEG)

Further questions

Electrons and atoms: part B

Some questions require a knowledge of concepts covered in earlier units

1 Draw clear, labelled diagrams to show how the path of a beam of electrons is affected when passing between
   a) the poles of a strong magnet,
   b) two flat metal plates across which a source of high p.d. is connected. Your diagrams should show clearly the electron paths before entering each field as well as after leaving the field. (SUJB)

2 Figure 1 shows a simplified version of a tube designed to illustrate the effect of an electrostatic field on a stream of electrons. Explain carefully:

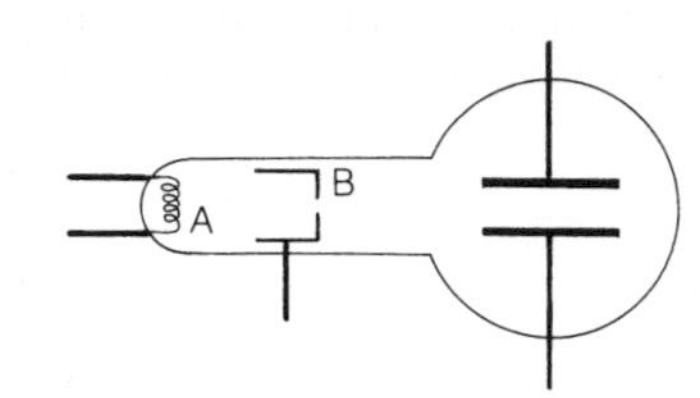

Figure 1

   a) How A is made to emit electrons, naming a suitable substance for A and giving a reason for your choice.
   b) The purpose of B and how this purpose is achieved.
   c) How you would establish the nature of the charge on the electrons.
   d) Why the air must be removed from the tube.
   e) How the path of the electron beam is made visible. (L)

3 Electrons are accelerated through a potential difference of 5 kV and 3×10^{13} electrons strike a metal block every second. If all their kinetic energy is converted into heat, what is the rate of production of heat? (*Charge on the electron* $= 1.6 \times 10^{-19}$ C.) (C)

4 Figure 2 represents the heated cathode C, the accelerating anode A, one pair of deflector plates P_1, P_2, and the screen of a simple cathode ray tube. The whole is in a vacuum. The broken line represents the path of a narrow beam of electrons passing from C to the screen.

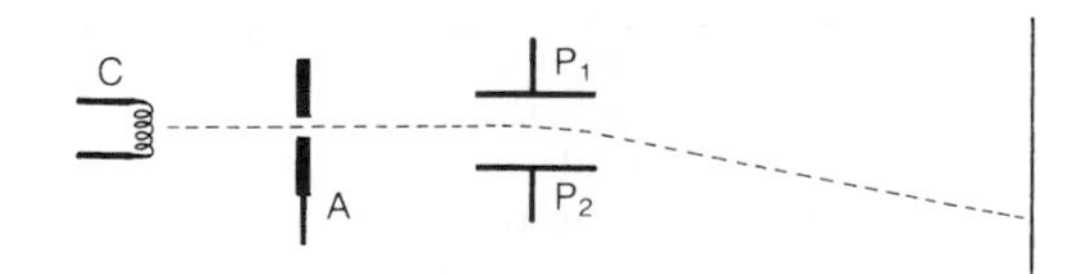

Figure 2

What change, if any, would you expect to observe in the brightness or the position of the spot on the screen if each of the following changes is made? (Assume in every case the original condition is afterwards restored.)
   a) The potential difference between P_1 and P_2 is increased.
   b) The potential difference between P_1 and P_2 is reversed.
   c) The potential difference between C and A is increased.
   d) The potential difference between C and A is reversed.
   e) The potential difference across the cathode heater is increased.
   f) The potential difference across the cathode heater is reversed. (O)

5 An evacuated cathode ray tube has a circular flat screen of area 0.04 m^2. Calculate the force due to air pressure on the screen if atmospheric pressure is 10^5 pascal. Give reasons why such a tube should be evacuated.
If the potential difference between anode and cathode is 8 kV and the electron beam current is 1.5 mA, calculate the energy supplied per second to this beam. Describe the changes in energy when the electrons hit the screen. What would you connect to the X and Y deflector plates to a) move the central spot vertically upwards, b) obtain a vertical line trace, c) obtain a sine wave trace? (SUJB)

6 A radioactive source emitting alpha (α-) particles is introduced into a cloud chamber. Sketch what you would expect to see when the chamber is in operation. (AEB)

7 Explain why α-particles produce visible tracks in a cloud chamber. Why do these tracks have a length of only a few centimetres?

8 Compare the properties of α-particles, β-particles and γ-radiation by copying and completing the following table:

	relative mass	sign and relative size of charge	approx. range in air
α-particle			
β-particle			
γ-radiation			

9 A radioactive source is placed in front of a Geiger-Muller tube, which can detect α-, β- and γ-radiations. Three absorbers (a sheet of paper, 3 mm thick aluminium, and 25 mm thick lead) are placed in turn

between the source and the detector and the following results are obtained:

Absorber	Counts/min
—	1000
paper	900
aluminium	900
lead	100

Does the source emit **A** α-radiation only? **B** β-radiation only? **C** α- and γ-radiation only? **D** β- and γ-radiation only? **E** α-, β- and γ-radiation?

10 A manufacturer wishes to check that the thickness of the steel sheets he produces is uniform. Describe how this could be done, using a radioactive source and a counter. What radiation would you expect the source to emit? (L)

11 Explain how it is possible for atoms of different elements to have the same mass numbers.

12 A particular atom of neon (Ne) has an atomic number 10 and a mass number 20.
   a) What does each of these numbers indicate about the atom?
   b) Write down the symbol used for this atom.
   c) Another atom of neon has a mass number of 22. Describe the composition of the nucleus of this atom.
   d) Explain what is meant when an atom is said to be ionised. (O)

13 What changes in the mass and charge of the nucleus of an atom take place if it emits i) an α-particle, ii) a β-particle, iii) a γ-ray? (O)

14 a) An α-particle is a helium nucleus that is emitted at high speed when a radioactive nucleus decays. Write down the atomic number Z and the mass number A of an α-particle.
   b) Each decay of a nucleus of a particular gas emits two α-particles in very quick succession. The original gas has $Z = 86$ and $A = 220$. Write down the values of Z and of A for the nucleus produced by this radioactive decay. (O)

15 a) The isotope $^{238}_{92}U$ decays by alpha-emission to an isotope of thorium (Th). Compare the $^{238}_{92}U$ and thorium nucei, explaining the changes which have occurred in the uranium nucleus.
   b) The thorium nucleus decays by beta-emission to an isotope of protactinium (Pa). Compare the thorium and Pa nuclei, accounting for the changes you describe. (JMB)

16 You are told that a radioactive source has a half-life of twenty-four hours. State what you understand by half-life.
Assuming that you had a counter and could use it over a period of three days, describe how you would check that the half-life was twenty-four hours. (L)

17 If the half-life of thoron is 52 s, how long will it take for the activity of a thoron sample to be reduced to $\frac{1}{32}$ of its initial value? (O)

18 A specimen of a radioactive isotope with a half-life of 12 minutes has an initial activity of 480 counts per second. Draw up a table and plot a graph accurately to show how the activity of this isotope in the specimen changes during the next hour. (O)

19 From the graph showing activity of a particular radioactive source against time (figure 3), determine the half-life of the source. (O&C)

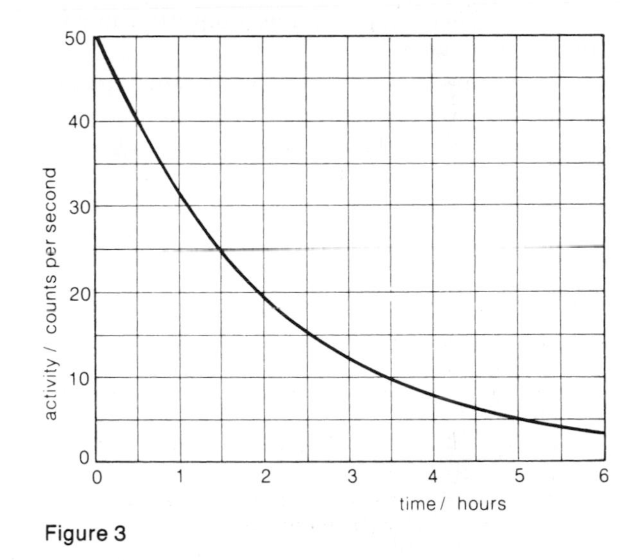

Figure 3

20 The following symbols represent five nuclides (nuclei):

$^{58}_{29}A$, $^{54}_{27}B$, $^{59}_{29}C$, $^{58}_{31}D$, $^{59}_{30}E$,

   a) Which nuclides are isotopes of each other?
   b) Which nuclide could be produced from which other by the emission of an α-particle?
   c) Which nuclide could be produced from which other by the emission of a β-particle?
   d) One nuclide emits γ-rays. Is it possible to determine which one from the above information? If so state which one; if not give a reason for your answer.
   e) Which nuclide possesses most neutrons? (L)

Revision check list

Different GCSE Examination Boards have different syllabuses, so not all of the topics below will be found in every one. Page numbers are given after each topic.
Ask your teacher which of the topics you should revise.

Section 4: Molecular motion and heat

Section 5: Waves: light and sound

Section 6: Electrical energy

Negative (−) and positive (+) charges (**page 230**)
Unlike charges attract, like charges repel (**page 230**)
Attraction between charged and uncharged objects (**page 232**)

Charging by rubbing (**page 231**)
1) Negative charge: more electrons than normal
2) Positive charge: less electrons than normal

Conductors and insulators (**page 231**)

Charge collects near sharp points (**page 234**)
Electric fields (**page 236**)

Coulomb, unit of charge (**page 237**)
Potential $= \frac{W}{Q}$ (**page 238**)

Capacitors (**page 240**)

Primary and secondary cells (**page 243**)
Cells in series and parallel (**page 260**)
Internal resistance of cell $= \frac{\text{'lost voltage'}}{\text{current}}$ (**page 260**)

Current, amperes (**page 245**)
P.D., volts (**page 247**)
Charge = current × time (**page 247**)

Ohm's law (**page 249**)

V,I,R equations

$R = \frac{V}{I}$ $\quad I = \frac{V}{R}$ $\quad V = IR$ (**page 250**)

Measuring resistance (**page 251**)
Effect of temperature on resistance (**page 252**)
Resistors in series $R = R_1 + R_2$ (**page 253**)

Resistors in parallel $\frac{1}{R} = \frac{1}{R_1} + \frac{1}{R_2}$ (**page 253**)

Circuit rules for current and p.d.:
1) p.d. across battery
   = sum of p.d.'s around circuit
2) when resistors are in series,
   p.d. is shared, current is the same
3) when resistors are in parallel,
   current is shared, p.d. is the same

Power $= VI = I^2R = \frac{V^2}{R}$ (**page 263**)

Electrical energy $= Pt = VIt = I^2Rt = \frac{V^2t}{R}$ (**page 266**)

A.C. mains electricity
1) Electrical energy in kilowatt hour (**page 270**)
2) Why fuse and switch must be in live line (**page 269**)
3) Need for an earth wire (**page 269**)
4) Three-pin plugs (**page 270**)
5) Fuse values (**page 270**)
6) Two-way switch (**page 272**)

Section 7: Magnets and currents

Unlike poles attract, like poles repel (**page 280**)
Induced magnetism (**page 281**)
Making magnets (**page 281**)
Hard and soft magnetic materials (**page 282**)
Demagnetization (**page 283**)
Magnetic fields (**page 284**)
Earth's magnetic field (**page 286**)
Field around current in a straight wire (**page 288**)
Maxwell's screw rule (**page 288**)
Field around a solenoid (**page 289**)
Electromagnets (**page 290**)
Magnetic relay (**page 290**)
Electric bell (**page 291**)
Fleming's left-hand rule (**page 292**)
Force between current-carrying conductors (**page 294**)

Coil in a magnetic field: how the turning effect depends on
1) current
2) number of turns
3) strength of field
4) area (**page 295**)
Use of the turning effect in a meter (**page 298**)

D.C. electric motor: purpose of
1) commutator
2) brushes (**page 295**)

Converting a milliammeter into
1) an ammeter (**page 300**)
2) a voltmeter (**page 300**)

Moving iron meters (**page 302**)

Induced e.m.f.: how it depends on
1) speed of movement
2) strength of field (**page 303**)

Direction of induced current (**page 304**)
Eddy currents (**page 305**)

A.C., root mean square current (**page 308**)
A.C. generators (alternators) (**page 306**)
D.C. generators (**page 308**)

Mutual induction (**page 310**)
Step-up and step-down transformers:
$\frac{V_1}{V_2} = \frac{n_1}{n_2}$ (**page 311**)

Power through transformer:
$V_1 I_1 = V_2 I_2$ (**page 312**)

Construction of a transformer (**page 312**)

Advantages of a.c. transmission (**page 316**)

Section 8: Electrons and atoms

Answers

Section 1

1.1 Units of measurement (page 11)
3. 1000, 1000, 10^6, 10^6, 1000
5. 10^3kg, 10^6m, 10^{-2}s, $3{\cdot}7 \times 10^{-1}$kg, $2{\cdot}5 \times 10^{-4}$m
6. 2700mm, 224mm, 0·330mm, $5{\cdot}6 \times 10^{-2}$mm
7. 5·000s, 4×10^4s
8. a) 5×10^{-3}kg b) 5000mg
9. 1000 000, 1000, 1000
10. a) 200l b) 2×10^5cm^3 c) 2×10^5ml
11. a) 96m^3 b) 31.4 m^3

1.2 Speed, velocity and acceleration (page 15)
1. 20m/s
5. 4m/s^2, 50m, 150m, 15m/s
6. 42m/s, 208m
7. 140m
8. 8m

1.3 Measuring with ticker tape(page 17)
2. 5mm, 50mm, 250mm, 250mm/s
3. 10mm, 50mm/s, 50mm, 250mm/s, 200mm/s, 250mm/s^2

1.4 Acceleration of free fall (page 20)
1. 3s, 36m
2. 3s, 45m, 30m/s
3. 2s, 20m/s
4. a) 75m b) 80m c) 75m d) 0m; 100m
5. 1·6m/s^2

1.5 Force, mass and acceleration (page 23)
2. 40N, 80N
3. 0·25m/s^2, 1m/s^2
4. 3m/s^2, 24N
5. 8m/s^2, 100m
6. 10N, 20N, 100N

1.6 Gravitational force (page 27)
3. 20N, 2kg
4. 10N; 20N, 250N, 500N, masses on Moon 2kg, 25kg, 50kg
5. 1·6N/kg, 32N
6. a) greater for A b) same c) greater for A

1.7 Balanced and unbalanced forces (page 30)
6. 500N
7. 5N, 0·5kg
8. 10 000N, 5000N, 5m/s^2, 40 000N

1.9 Changing momentum (page 36)
3. 48kg m/s, 72kg ms, 24kg m/s, 8kg m/s, 8N, $\frac{2}{3}$m/s^2
4. 100kg m/s, 500kg m/s, 500Ns, 25m/s

1.10 Action and reaction (page 38)
1. 500N

1.11 Conserving momentum (page 41)
1. 12kg m/s to left and 12kg m/s to right, 2m/s, 0, 24N
2. 80kg m/s to right, 20kg m/s to left, 60kg m/s to right, 3m/s to right
3. 6m/s

1.12 Rockets and jets (page 43)
3. 7500N, 7500N, 200m/s
4. 2500N, 2000N, 500N, 2·5m/s^2, 15m/s^2

Section 1 further questions A (pages 44–45)
1. a) 200m b) 11s c) 80m d) 15s e) 8 m/s f) same g) 0
2. a) 55s
   b) i) 8 m/s ii) 40s iii) 320m
   c) 380m
3. a) 5km
   b) i) 10m/s ii) 8min 20s
   c) 2m/s^2
4. B
5. D
6. C
7. D
8. a) 4s b) 3000N
9. b) 25m d) 4m/s^2
10. b) 1400kg
    c) i) 400N iii) 0.29m/s^2
11. c) 4s
12. D, A

Section 1 further questions B (pages 46–48)
1. a) 50s b) 10m/s c)0·5m/s^2 d) 350m
2. c) 16cm/s d) 20·7cm/s
3. b) i) 1·33m/s^2 ii) 225m
4. b) 3100m c) 20·7m/s
5. a) 5m b) 1·2m/s down slope
   c) back at starting point
   d) 0.4m/s^2 down slope
6. a) 9·6m/s^2
7. a) 0·63s b) 0.32m
8. 1s
9. b) i) 40m/s ii) 80m
10. a) 2s b) 5m
12. b) 500N c) 1·125m/s^2
15. 120m, 12s
16. 780N
17. a) weight 5N b) weight 5N, air resistance 5N
22. 24 000kg m/s, 8s
23. a) OA, 0·5m/s^2 b) 525m
    c) 10·5m/s d) 30 000kg m/s
25. 1500N
26. 5m/s
27. 1·2m/s
28. 45 000kg
29. 1·8m/s

Section 2

2.1 Combining forces (page 52)
1. a) 17N b) 7N c) 13N
2. a) 50N vertical, 87N horizontal
   b) 350N c) 250N

2.2 Turning effects of forces (page 57)
3. 12Nm; 12Nm; 4N and 10N; 24Nm, 30Nm, 6Nm; 30Nm; 30Nm
4. 700Nm and 800Nm, Y = 500N, X + Y = 1100N, X = 600N
   At tipping, Y = 0; 1·14m
5. 4Nm
6. 400N

2.3 Centre of gravity (page 60)
22. 1N; 5N
3. 475N, 225N

2.4 Stretching forces (page 63)
3. 50g, 100g, 500g
4. a) 50mm d) at 65mm extension approx. f) 2·3N, 4·5N g) 0·9N h) 4·5N

2.5 Work and energy (page 68)
1. 10^4J, 5×10^5J, 200J, 10^{12}J
5. 60J
6. 10J, 10J
7. 200j, 400N, 300J

2.6 Energy for the world (page 72)
4. a) 50MJ
   b) 16%
   c) 40MJ

2.7 Gravitational P.E. and K.E. (page 75)
1. a) 240J b) 480J; 6m
2. 50J
3. 75J, 300J
4. 20 m/s
5. a) 20J b) 20J c) 4m/s
6. a) 72J b) 36J c) 36J d) 6m/s e) 8·5m/s f) 72J, 8·5/s
7. a) 25J b) 25J c) 5m d) 5m

2.8 Engines (page 79)
1. c) 1.2kW d) 6.0kW
2. 60kJ, 200kJ, 10kW
3. 300W
4. 50kW
5. 4kW, 5kW

2.9 Machines (page 83)
2. 4
4. M.A. = 2·5, 8N, 40J
5. V.R. = 2, M.A. = 1·5, 80N, 80J, 20J

Section 2 further questions A (pages 84-85)
1. P = 600N, Q = 1039N
2. D
3. a) A b) B c) greater
4. A

7. a) 50s b) 700KJ c) 14kW
8. D
9. C
10. D
11. D

Section 2 further questions B (pages 86–88)

2. 300N
3. a) 25N b) 36·9° c) 6·25kg
   d) i) $24m/s^2$ ii) 72m
4. b) i) 3480N ii) 2000N
6. b) i) 24N ii) 48N
7. 800N at A, 1200N at B
8. b) i) 0·36Nm ii) 0·21m from A
9. 49 000N at A, 31 000N at B
10. 700N
12. a) 54mm b) 40N, $3{\cdot}33m/s^2$
13. 0·9kg
14. a) 0·25N b) 90mm c) 12N
15. 1920J
16. 3000J, 150N
17. 120J
18. a) 80m b) 40m/s c) 1600J
19. 100J
20. 8m
21. a) 24kg m/s b) 24kg m/s c) 6m/s
    d) 96J and 72J e) 168J
23. a) 3000N b) 180kJ c) 4kW
24. 4000kg
25. a) 36kJ b) 24kJ
26. a) 16N b) 2.86m/s
27. a) 1020MJ b) $1{\cdot}02 \times 10^6$kW
29. a) 6 b) 300J
30. 1·5
31. a) 0·4m b) 2 d) 1·5 e) 20N f) 200J
    g) 1kg

Section 3

3.1 Density (page 93)

1. $1g/cm^3$
2. $13\,600kg/m^3$, $13{\cdot}6g/cm^3$, 13·6
5. $3m^3$
6. 650kg
7. 22 800kg
8. 200g, 80g, 280g, $300cm^3$, $0{\cdot}93g/cm^3$
9. 0·8, $800kg/m^3$

3.2 Pressure (page 98)

3. 50Pa, 100Pa
4. 200N
5. 7500Pa, 500Pa
6. 16 000Pa, 192 000N
7. 200Pa at A and B, 100N thrust and load, 5

3.3 Pressure from the atmosphere (page 103)

1. a) 10^5Pa b) 760mm Hg c) 1atm
7. 12·5m
8. $800kg/m^3$

3.4 Pumps and gauges (page 105)

4. 25mmHg, 780mmHg

3.5 Archimedes' principle and flotation (page 111)

2. 10 000N, 10 000N, 10 000N
3. 3000N
4. a) 6 b) 0·75
5. 5400N, 3400N
6. 5cm

Section 3 further questions A (pages 112–113)

1. a) $100cm^3$ b) $13.6g/cm^3$ c) 50g
3. C
4. a) 200 000Pa c) 10 000kg d) 200J
   f) 1600kW
5. a) 10 000N c) 1000Pa d) 2000N
   e) 2000N
6. a) $0.25N/cm^2$ b) 20N
7. a) i) 12.5N ii) $20cm^2$
   iii) $0.625N/cm^2$ iv) $0.313N/cm^2$
   b) i) $160cm^3$ ii) $7.81g/cm^3$
8. A
9. a) 11atmos
10. a) i) 50N ii) 50N
    b) i) $2N/cm^2$ ii) $2N/cm^2$ iii) 4000N
    c) i) B

Section 3 further questions B (page 114)

1. b) i) $1300kg/m^3$ ii) $6{\cdot}4 \times 10^{-5}\ m^3$ ($64cm^3$)
2. a) $0{\cdot}25m^2$ b) $0{\cdot}5m^3$ c) 1500kg
   d) 15 000N e) 60 000Pa
3. 25 000N, 250 000Pa
4. a) 0·75m b) 9000Pa
7. 108kg, 600N
8. b) 2000N c) 1 47 m/s^2
9. a) 2000Pa, 5000Pa
   b) 1000N, 2500N, 1500N
   d) rise

Section 4

4.1 Moving molecules (page 119)

5. 5×10^{-7}m

4.3 Temperature (page 125)

6. 0K, 273K, 300K, 373K

4.4 Expansion of solids (page 129)

2. a) 0·000 019m b) 0·000 019cm
   c) 0·0152m d) 0·0152cm
3. 0·00002/K
4. 40°C
5. 217°C

4.7 The gas laws (page 139)

4. 1800mmHg
5. $12{\cdot}5cm^3$
6. $2{\cdot}73m^3$

4.10 Specific heat capacity (page 147)

1. 450kJ, 2100kJ
3. 5°C
4. 240s
5. 420J/(kg K)

4.11 Melting and freezing (page 151)

3. 1670kJ
6. 1508kJ
7. 0·05kg

4.12 Latent heat of vaporization (page 155)

1. 4520kJ, 22 600kJ
3. 7410kJ
6. 0·6kg
7. 63.8°C

Section 4 further questions A (pages 160–161)

1. B
2. D
3. C
4. A
5. C
6. d) 8020kJ/h e) 3
8. D
11. a) i) 100 800J/s ii) 0.003 36kg
    iii) 253 000s
12. B

Section 4 further questions B (pages 162–164)

2. b) $1{\cdot}27 \times 10^{-9}$m
3. 2×10^{-9}m
4. b) i) 0·96mm ii) 0·000 016/K
5. 100·011m
6. $3{\cdot}06 \times 10^{-3}$m (0·00306m)
7. 0·000 024/K
8. 27·3K
12. b) $2{\cdot}14 \times 10^5$Pa
13. 1400K
14. a) 1000mmHg b) 150mm
    c) 180mm d) 240mm
15. $288cm^3$
16. a) 870mmHg b) 960mmHg
    c) 200mm
17. b) 0·026kg
20. c) 900kJ
21. i) 5K/min ii) 283K; 2500J/(kg K)
22. 600s
23. 40°C
24. 4200J/(kg K)
25. 18kJ/kg
26. a) 6680J b) 1260J c) 331s
27. a) 210J b) 3340J c) 3550J
28. 30 350J
29. 336s
31. c) i) 88°C ii) 90mmHg

Section 5

5.1 Light and shade (page 168)

3. 500s, $3{\cdot}9 \times 10^8$m

5.2 Reflection of light (page 171)

2. 60°
5. 7·5m

5.3 Curved mirrors (page 176)

6. 20cm from mirror, 2cm high
   12cm from mirror, 6cm high

5.4 Refraction (page 180)

2. a) 17° b) 37°
3. a) 37° b) 64°
5. 30°
6. 20cm, 53°

5.5 Lenses (page 186)

6. 12cm from lens, 1·5cm high, real
   4·8cm from lens, 4·8cm high, virtual

5.6 The camera and the eye (page 188)

9. 5cm, 5·3cm

5.8 Colour (page 194)
8. white, green
11. green
12. a) red b) black

5.9 Wave motion (page 199)
4. 50mm/s, 2·5mm

5.10 Light waves (page 203)
4. 4×10^{-7}m

5.11 Electromagnetic waves (page 207)
4. a) 16W/m^2 b) 1W/m^2
6. a) 3m b) $3{\cdot}33 \times 10^{-4}$s

5.13 Sound waves (page 213)
8. 15m
9. 320m/s

5.14 Hearing sounds (page 217)
3. a) 100Hz b) 800Hz
6. 5 beats/s, 2 beats/s, no beats, 5 beats/s

5.15 Vibrating strings and air columns (page 223)
2. 300Hz, 1 octave
3. 400Hz, pitch increase; 2 octaves
4. 0·75m, 330Hz, 220Hz, 660Hz
5. 320m/s

Section 5 further questions A (pages 224–225)
3. i) 40° ii) 40° iii) 116.8 MJ
4. a) B, C b) B
5. a) 0.005s b) 1.6m c) 320m/s
7. b) 2m/s c) 4m d) 0.5Hz
8. a) C b) A

Section 5 further questions B (pages (226–228)
2. a) virtual, same size as object, 15cm from mirror
   b) ii) virtual, 30cm from mirror
   iii) virtual, 10cm from mirror
3. a) 9cm b) 2cm c) inverted d) real
4. 2×10^8m/s; angle of refraction i) 0° ii) 35·5°
5. b) 49°
8. b) i) 30cm, 90cm, 22·5cm
   ii) 90cm from lamp, $\frac{1}{3}$ ×
9. 1·5cm high, virtual
10. 7·14cm, 24·5mm
11. 45cm, 2
12. 40cm from object, 32cm
14. a) 5cm approx, 0·8cm
16. b) 256mm/s
17. b) ii) 0·013mm iii) $6{\cdot}5 \times 10^{-7}$m
19. 5×10^{14}Hz
23. c) 0·8m
24. a) 0·13s b) 0·33s
25. a) 360m/s
28. b) 1 beat/s, 306Hz
29. b) 1·2m d) 307m/s

Section 6

6.3 Charge, potential, and capacitance (page 241)
4. 3000V, − to +
6. +2000V, 0·2μC
8. conc. A<B; pot. A = B

6.5 Current and voltage in a simple circuit (page 248)
2. 750mA, 20mA
3. a) 4C b) 12J c) 5C d) 1·25s
4. a) 8V b) 2·5A
5. X:8V, Y:3A

6.6 Ohm's law and resistance (page 252)
2. 6·25
5. 6Ω, 12V
6. 0·5A, 18V
7. 800V

6.7 Series and parallel circuits (page 257)
1. X:2A, Y:8V, Z:4V
2. a) 20Ω (b) 3.75Ω c) 2·4Ω d) 9Ω
3. B
4. 2V
5. a) 2A b) 6V c) 0·5A
6. 1·5A, 2A

6.8 E.m.f. and internal resistance (page 261)
3. a) 0·25Ω b) 0·75Ω c) 0·5J
4. 1·5V, 20A
6. a) 8V b) 2V c) 6V d) 4V

6.9 Power in an electric circuit (page 265)
1. a) 200W b) 10A c) 2Ω d) 50W
2. 2·4Ω, 5A
3. a) 36W b) 162W
4. a) 3A b) 18W c) 36W d) 54W

6.10 Electrical energy calculations and costs (page 268)
1. a) 60J b) 216kJ
2. 2000J, 10A
3. 10°C
4. 5400s
5. 0·6kg
6. a) 18kWh b) $6{\cdot}48 \times 10^7$J
7. a) 180p b) 4·8p c) 28p

6.11 Mains electricity (page 272)
2. zero
7. a) 10A, 13A fuse b) 1·5A, 3A fuse
   c) 0·5A 3A fuse d) 0.25A, 3A fuse
8. 1A

Section 6 further questions A (pages 273–274)
2. C
3. B
4. D
5. a) 1.2V b) 0.30A c) 0.50A d) 2.9V
6. a) B b) B c) D
8. a) 500kWh b) £25
10. b) i) S_3 ii) S_3 and S_1
    c) i) 120Ω ii) 4.5A iii) 7A

Section 6 further questions B (pages 275–278)
5. 7·5V
7. c) 2×10^{-6}F d) 400μC
8. a) 6×10^{-9}F, $1{\cdot}5 \times 10^{-8}$F b) 500V, 200V
10. a) 6A b) 1080C
12. a) 4·0V b) 20Ω c) 0·13A
13. X: A,B,C,D,E; Y: B; Z: D, E
14. 4A
15. c) 0·67A, 0·67A, 1·33A d) 4V, 4V, 8V
16. 3Ω
17. a) 4Ω
18. a) 0·15A b) 1·2V c) 0·38A d) 0·19A
19. 1Ω
20. 1Ω
21. a) 0·75Ω b) 3·0A c) 2·25A
23. a) 0·5A b) 45W c) 2V d) 94V
24. a) 3·0V b) i) 0·25A ii) 2·5V
    iii) 0·75W iv) 0·63W
25. a) i) 3A ii) 4Ω
    b) i) 2·5A ii) 25W
26. a) 2Ω b) 8Ω c 1·5A, 6V d) 3V
    e) 1·0A, 0·5A f) 18W g) 9W, 3W, 1·5W h) 4·5W
27. a) 90J b) 12J c) 78J d) 2·6Ω
28. a) i) 8A ii) 31·3Ω iii) 120kJ b) 50p
29. 60p
30. 720J
31. 150J, 120J, 0·8
32. a) 7·5Ω b) i) 2·0A ii) 2400J
33. 10Ω, 0·5A, 5V, 3000J, 2·25A
34. a) 24Ω, 10A b) 2400J c) 210s
35. a) 62·5Ω b) 1000W; 10A
36. a) 0·2A, 1250Ω c) 13·89V d) 0.22A
37. 480Ω, 4

Section 7

7.7 Electrical meters (page 302)
5. a) $\frac{5}{9}$Ω b) 45Ω c) 10V d) 2A

7.8 Electromagnetic induction (page 305)
1. a) c) current increased b) current reversed
   d) e) current zero

7.9 Generators (page 309)
4. 10A, 20V, 98W

7.10 Mutual induction and transformers (page 314)
4. a) 6V b) 2A c) 0·05A d) 12W e) 12W
5. 200, 0·05A

7.11 Generating and transmitting power for the mains (page 317)
6. a) 50kW b) 0·05W
7. a) 3A b) 1A c) 3A d) 3W e) 39W f) 13V

Section 7 further questions A (pages (318–319)
2. D
3. B
7. A
9. D
10. a) 10C b) 240J c) 180J
    d) i) 0.6m/s ii) 1.08J
11. A

Section 7 further questions B (pages 320–322)
6. a) N
9. a) 0·02V
10. 0·404Ω
11. a) 0·051Ω b) 1995Ω
12. 4960Ω
14. Current B to A
16. N
20. d) 500 turns e) 2·5A

21. a) i) 4A ii) 40:1 iii) 0·1A
b) i) 1·5Ω ii) 234V iii) 58·5Ω
iv) 936W
23. i) 50A ii) 10V iii) 500W;
0·5A, 0·1V, 0·05W
24. c) 41·7V, 3·47kW d) 0·083V,
0·014W

Section 8

8.1 Electron beams (page 327)
4. down
5. down

8.2 The cathode ray oscilloscope (page 331)
4. 125mV, 25Hz

8.3 Electronic components (page 333)
1. **A** 1500Ω, 5%
B 49kΩ, 10%
C 4700Ω, 5%
D 3.9Ω, 20%

8.5 Transistors (page 340)
1. c) 0·67V, zero

8.6 Microchips (page 344)
1. a) B b) A, C
2. a) A b) 5 c) 20 f) 2V
3. a) −12V b) +12V

8.7 Logic gates (page 348)
a. a) NOT b) AND c) NAND, NOR
3. NAND gate:
a) OFF b) ON c) ON
NOR gate:
a) OFF b) ON c) OFF

8.10 The structure of the atom (page 360)
7. a) 17 b) 17 c) 18, $^{37}_{17}Cl$
8. a) $^{12}_{6}C$, 6 neutrons b) $^{14}_{6}C$,
8 neutrons
c) $^{27}_{13}Al$, 14 neutrons
9. iron–56

8.11 Radioactive decay (page 365)
3. A = 228, Z = 88, radium–228
4. lead–206
5. nitrogen–14
7. a) 25m b) 25m, 25m, 50m

8.12 Nuclear energy (page 369)
3. 3, Z = 35, bromine–85

Section 8 further questions A (pages 370–371)
1. D
5. b) i) open ii) closed
6. a) logic 0 b) 3V c) 300Ω
8. a) 100 b) 500kΩ
9. g) 30s h) 90m

Section 8 further questions B (pages 372–373)
3. $2{\cdot}4 \times 10^{-2}$W
5. 4000N, 12J/s
9. C
14. b) Z = 82, A = 212
17. 260s
19. 1·5h
20. a) A and C b) B from A c) E from
C d) no e) C

Index